Electronic Design Automation

Electronic Design Automation: Synthesis, Verification, and Test

Edited by

Laung-Terng Wang
Yao-Wen Chang
Kwang-Ting (Tim) Cheng

AMSTERDAM • BOSTON • HEIDELBERG • LONDON
NEW YORK • OXFORD • PARIS • SAN DIEGO
SAN FRANCISCO • SINGAPORE • SYDNEY • TOKYO

Morgan Kaufmann Publishers is an imprint of Elsevier

ELSEVIER

MORGAN KAUFMANN

Morgan Kaufmann Publishers is an imprint of Elsevier.
30 Corporate Drive, Suite 400, Burlington, MA 01803, USA

This book is printed on acid-free paper.

Library of Congress Cataloging-in-Publication Data

Electronic design automation : synthesis, verification, and test/edited by
Laung-Terng Wang, Yao-Wen Chang, Kwang-Ting (Tim) Cheng.
    p. cm.
    ISBN: 978-0-12-374364-0 (alk. paper)
  1. Electronic circuit design–Data processing. 2. Computer-aided design.
I. Wang, Laung-Terng, II. Chang, Yao-Wen. III. Cheng, Kwang-Ting, 1961–
    TK7867.E4227 2008
    621.39′5–dc22

                                                              2008041788

For information on all Morgan Kaufmann publications,
visit our Web site at www.mkp.com

Printed and bound in the United Kingdom
Transferred to Digital Printing, 2010

Electronic Design Automation

The Morgan Kaufmann Series in Systems on Silicon

Series Editor
Wayne Wolf
Georgia Institute of Technology

Contents

Jie-Hong (Roland) Jiang and Srinivas Devadas

Contents **xi**

*Hung-Pin (Charles) Wen, Li-C. Wang, and
Kwang-Ting (Tim) Cheng*

Preface

New applications enabled by advances in semiconductor manufacturing technology continue to grow at an amazing rate. A wide spectrum of novel products, ranging from high-performance processors to a broad array of low-power portable devices to micro sense/communicate/actuate chips, facilitates various new applications that have changed, and will continue to change, our daily lives. However, as the semiconductor industry moves to ever-smaller feature sizes and the number of transistors embedded within a *very-large-scale integration* (VLSI) circuit continues to grow, under the relentless pressure of time-to-market for high-quality, reliable products, the semiconductor industry is increasingly dependent on design technology for design closure and for meeting productivity goals. The design technology we refer to here covers all of the core knowledge, software tools, algorithms, methodologies, and infrastructure required to assist in the synthesis, verification, testing, and manufacturing of a functioning and reliable integrated circuit.

Electronic design automation (EDA) has driven advances in design technologies for the past 30 years and will continue to do so. Traditional EDA tools support the design process starting from the *register-transfer level* (RTL) through to layout. The tasks assisted by these tools can be coarsely classified into RTL/logic synthesis, physical design, design verification, and *design for testability* (DFT). Since the late 1990s, the landscape of EDA has rapidly expanded such that it now includes an even broader range of tasks. These new tasks cover the support of *electronic-system-level* (ESL) design that includes system specification, transaction-level modeling, and behavioral synthesis as well as tasks related to manufacturing and post-silicon activities such as *design for manufacturability and reliability* (DFM/DFR), post-layout manipulations for yield optimization, and post-silicon debug. At the same time, the traditional RTL-to-layout tasks are also refined, resulting in a synthesis process which involves many steps of design refinements and employs highly complex optimizations and analysis. The design environment has evolved from a set of point tools to a highly sophisticated and integrated system able to manipulate a huge amount of design data at several different levels of design abstraction.

The fast and continuing evolution of design technology and the enormous growth in the complexity and sophistication of an EDA system has made it such that very few people can master all fronts of this field. New problems, new algorithms, new methodologies and tools, and new start-ups offering new solutions, emerge every year. This trend will continue, perhaps at an even faster pace in the future! As a result, it is becoming difficult even for experts to follow and comprehend the progress on a continuing basis. Training students to prepare them for careers in academia or industry as the next generation of leaders in VLSI design and EDA is a challenging task!

While a comprehensive treatment of all EDA subjects is infeasible for either the undergraduate or entry-level graduate VLSI curriculum, integrating more EDA subjects into existing VLSI and logic design courses is essential for giving the students a balanced, and more accurate, view of modern *system-on-chip* (SOC) design. To facilitate that goal and to evolve the VLSI design curriculum, this textbook selects a set of core EDA topics which, in our opinion, provides an essential, fundamental understanding of the EDA tasks and the design process. These topics range from the basics of *complementary metal oxide semiconductor* (CMOS) design to key algorithms used in EDA. Also covered are various modeling and synthesis techniques at the system, register-transfer, and gate levels, as well as physical synthesis, including floorplanning, placement, routing, and synthesis of clock and power/ground networks. We have also chosen key topics on functional verification, including both simulation and formal techniques, and a range of testing topics, such as design for testability, test synthesis, fault simulation, and test generation. The intent is to allow the readers to understand fundamental EDA algorithms as well as VLSI test principles and DFT architectures, preparing them to tackle EDA and test problems caused by advances in semiconductor manufacturing technology and complex SOC designs in today's nanometer era.

Each chapter of this book follows a specific format. The subject matter of the chapter is first introduced. Related methods are explained in detail next. Then, industry practices, if applicable, are described before concluding remarks. Each chapter contains a variety of exercises to allow the use of this book as a textbook for an entry-level EDA course. Every chapter concludes with acknowledgment to contributors and reviewers and a list of references.

Chapter 1 provides an introduction to *electronic design automation (EDA)*. It begins with an overview of the EDA historic perspective. This is followed by a discussion of the importance of EDA — why EDA plays a central role in meeting time-to-market pressure and manufacturing quality of the nanometer design era. Typical design flows and examples are illustrated at different levels of abstraction — how a system-level design is automated through the modeling, synthesis, verification, and test stages.

Chapter 2 covers fundamental *complementary metal oxide semiconductor* (CMOS) design principles and techniques that are required knowledge for the understanding of *system-on-chip* (SOC) designs and EDA applications. While the topic is quite broad, we mainly focus on the widely used CMOS design and automation techniques and introduce them in an easy-to-grasp manner with extensive illustrations and examples. Emerging low-power design techniques that can be utilized to lengthen battery life or to reduce system failures due to overheat are also included in the chapter.

Chapter 3 covers fundamental *design-for-testability* (DFT) architectures to ensure high product quality and low test cost for VLSI or SOC designs. This chapter puts great emphasis on three basic DFT techniques that have been widely used in industry today for digital circuit testing: *scan design*, *logic built-in self-test* (BIST), and *test compression*. *Testability analysis* methods to assess the testability of a logic circuit are first described. The three DFT techniques are then explained in detail including schemes for *at-speed testing* and practiced in industry.

Chapter 4 introduces the fundamentals of algorithms that are essential to EDA tasks including synthesis, verification, and test. This chapter starts with an introduction to *computational complexity*, followed by various *graph* algorithms that are commonly used to model and solve EDA problems. It also covers several heuristic algorithms for practical use on real-life designs. The remainder of the chapter briefly surveys the *mathematical programming* techniques that can provide a theoretical background on the optimization problems.

Chapter 5 begins with *electronic-system-level* (ESL) *design modeling and high-level synthesis* – the first step of EDA after a design is specified for implementation. The role of high-level synthesis in the context of ESL design modeling is discussed. An example is given to describe the generic structure required to build a high-level synthesis tool and the tasks involved. This is followed by a detailed description of the key algorithms, including *scheduling* and *binding*. Advanced topics are discussed at the end of the chapter.

Chapter 6 jumps into *logic synthesis* – the essential step bridging high-level synthesis and physical design. Important data structures for Boolean function representation and reasoning are first introduced, followed by the classical issues of *logic optimization* (which includes two-level and multilevel logic minimization), *technology mapping*, *timing analysis*, and *timing optimization*. Advanced and emerging topics are outlined for further reading.

Chapter 7 discusses the *test synthesis* process that automatically inserts the DFT circuits, discussed in Chapter 3, into a design during or after logic synthesis. Design rules specific to scan design and logic BIST are given to comply with DFT requirements. Test synthesis flows and examples are then described to show how the test automation is performed. The remainder of the chapter is devoted to illustrating the automation of DFT circuit insertion at the *register-transfer level* (RTL).

Chapter 8 covers various *logic and circuit simulation* techniques that allow a designer to understand the dynamic behavior of a system at different stages of the design flow. The chapter begins with a discussion of logic simulation techniques that are fundamental to software simulators. Next, hardware-accelerated logic simulation, which is commonly referred to as *emulation*, is introduced. Both logic simulation and emulation of systems are typically performed at a higher level of design abstraction. The last part of the chapter deals with the simulation of the most basic components of a circuit, namely, devices and interconnects.

Chapter 9 is devoted to *functional verification*. This chapter first introduces the verification processes at various design stages. Common structural and functional coverage metrics which measure the verification quality are described. This chapter also discusses the key tasks involved in *simulation-based verification*, such as *stimulus generation*, *assertion-based verification*, and *random testing*. The mathematical backgrounds and examples for various formal approaches are also provided. Advanced verification techniques are presented as supplements at the end of the chapter.

Chapter 10 addresses *floorplanning* of the physical design process. The two most popular approaches to floorplanning, *simulated annealing* and *analytical formulations*, are covered. Based on simulated annealing, three popular floorplan representations, *normalized Polished expression*, *B*-tree*, and *sequence pair* are further discussed and compared. Some modern floorplanning issues related to soft modules, fixed-outline constraints, and large-scale designs are also addressed.

Chapter 11 covers *placement* of the physical design process. This chapter focuses on techniques to solve the *global placement* problem. Algorithms for the most common global placement approaches, namely *partitioning-based approach*, *simulated annealing approach* and *analytical approach*, are presented. The analytical approach is particularly emphasized as the best global placement algorithms are all based on the analytical approach. Techniques for *legalization* and *detailed placement* are also discussed.

Chapter 12 covers *signal routing*. This chapter classifies the routing algorithms into three major categories: *general-purpose routing*, *global routing*, and *detailed routing*. For general-purpose routing, maze routing, line-search routing, and *A*-search* routing are discussed. For global routing, both sequential and concurrent techniques are covered. Steiner tree construction is also addressed to handle the interconnection of multi-terminal nets. Some modern routing considerations in signal integrity, manufacturability, and reliability such as crosstalk *optical proximity correction* (OPC), *chemical-mechanical polishing* (CMP), antenna effect, and double-via insertion, are also briefly discussed.

Chapter 13 addresses the *synthesis of clock and power/ground networks*, with a stronger emphasis on clock network synthesis. Following a discussion of the key issues that affect the integrity of clock networks and power/ground networks, the chapter delves into the automated analysis, synthesis, and optimization of both types of large-scale interconnection networks.

Chapter 14 consists of two major VLSI testing topics – *fault simulation* and *automatic test pattern generation* (ATPG) – for producing high-quality test patterns to detect defective chips during manufacturing test. The chapter starts with fault collapsing, which helps speed up fault simulation and ATPG. Several fault simulation techniques, including serial, parallel, concurrent, and differential fault simulation, are introduced and compared. Next, basic ATPG techniques, including Boolean difference, PODEM, and FAN, are described. The chapter concludes with advanced test generation techniques to meet the needs of covering defects that arise in deep-submicron devices, including sequential ATPG, delay fault ATPG, and bridging fault ATPG.

In the Classroom

This book is designed to be used as an entry-level text for undergraduate seniors and first-year graduate students in computer engineering, computer science, and electrical engineering. Selected chapters can also be used to complement existing logic or system design courses. The book is also intended for use as a reference book for researchers and practitioners. It is self-contained with most topics covered extensively from fundamental concepts to current techniques used in research and industry. However, we assume that students have had basic courses in logic design, computer programming, and probability theory. Attempts are made to present algorithms, wherever possible, in an easy-to-understand manner.

To encourage self-learning, the instructor or reader is advised to check the Elsevier companion Web site (http://www.elsevierdirect.com/companions/9780123743640) to access up-to-date software and lecture slides. Instructors will have additional privileges to assess the Solutions directory for all exercises given in each chapter by visiting www.textbooks.elsevier.com and registering a username and password.

Laung-Terng (L.-T.) Wang
Yao-Wen Chang
Kwang-Ting (Tim) Chang

Acknowledgments

The editors would like to acknowledge many of their colleagues who helped create this book. First and foremost are the 25 chapter/section contributors listed in the next two pages. Without their strong commitments to contributing the chapters and sections of their specialty to the book in a timely manner, it would not have been possible to publish this book.

We also would like to thank the external reviewers in providing invaluable feedback to improve the contents of this book. We would like to thank Prof. Robert K. Brayton (University of California, Berkeley), Prof. Hung-Ming Chen (National Chiao Tung University), Prof. Jiang Hu (Texas A&M University), Professors Alan J. Hu and Andre Ivanov (University of British Columbia, Canada), Prof. Jing-Yang Jou (National Chiao Tung University), Prof. Shinji Kimura (Waseda University, Japan), Prof. Chong-Min Kyung (Korea Advanced Institute of Science and Technology, Korea), Prof. Yu-Min Lee (National Chiao Tung University), Prof. Eric MacDonald (University of Texas at El Paso), Prof. Subhasish Mitra (Stanford University), Prof. Preeti Ranjan Panda (India Institute of Technology at Delhi, India), Prof. Kewal K. Saluja (University of Wisconsin - Madison), Prof. Tsutomu Sasao (Kyushu Institute of Technology, Japan), Prof. Sheldon X.-D. Tan (University of California at Riverside), Prof. Ren-Song Tsay (National Tsing Hua University, Taiwan), Prof. Natarajan Viswanathan (Iowa State University), Prof. Ting-Chi Wang (National Tsing Hua University, Taiwan), Prof. Martin D. F. Wong, (University of Illinois at Urbana-Champagne), Prof. Hiroto Yasuura (Kyushu University, Japan), Prof. Evangeline F. Y. Young (Chinese University of Hong Kong, China), Prof. Tian-Li Yu (National Taiwan University), Khader S. Abdel-Hafez (Synopsys, Mountain View, CA), Dr. Aiqun Cao (Synopsys, Mountain View, CA), Wen-Chi Chao and Tzuo-Fan Chien (National Taiwan University), Dr. Tsung-Hao (Howard) Chen (Mentor Graphics, San Jose, CA), William Eklow (Cisco, San Jose, CA), Dr. Farzan Fallah (Fujitsu Laboratories of America, Sunnyvale, CA), Dr. Patrick Girard (LIRMM/CNRS, Montpellier, France), Dr. Sumit Gupta (Nvidia, San Jose, CA), Meng-Kai Hsu and Po-Sen Huang (National Taiwan University), Dr. Rohit Kapur (Synopsys, Mountain View, CA), Dr. Brion Keller (Cadence Design Systems, Endicott, NY), Benjamin Liang (University of California, Berkeley), T. M. Mak (Intel, Santa Clara, CA), Dr. Alan Mishchenko (University of California at Berkeley), Dr. Benoit Nadeau-Dostie (LogicVision, Ottawa, Canada), Linda Paulson (University of California, Santa Barbara), Chin-Khai Tang (National Taiwan University), Jensen Tsai (SpringSoft, Hsinchu, Taiwan), Dr. Chung-Wen Albert Tsao (Cadence Design Systems, San Jose, CA), Natarajan Viswanathan (Iowa State University), Dr. Bow-Yaw Wang

(Academia Sinica, Taipei, Taiwan), Dr. Ming-Yang Wang (SpringSoft, Fremont, CA), Ho-Chun Wu (Cadence Design Systems, Hsinchu, Taiwan), Dr. Jin Yang (Intel, Hillsboro, OR), and all chapter/section contributors for cross-reviewing the manuscript. Special thanks also go to Wan-Ping Lee and Guang-Wan Liao of National Taiwan University and many colleagues at SynTest Technologies, Inc., including Dr. Ravi Apte, Boryau Sheu, Dr. Zhigang Jiang, Jianping Yan, Jianghao Guo, Fangfang Li, Lizhen Yu, Ginger Qian, Jiayong Song, Sammer Liu, and Teresa Chang who helped draw symbolic layouts, review the manuscript, solve exercises, develop lecture slides, and draw figures and tables.

Finally, we would like to acknowledge the generosity of SynTest Technologies (Sunnyvale, CA) for allowing Elsevier to put an exclusive version of the company's most recent VLSI Testing and DFT software on the Elsevier companion Web site for readers to use in conjunction with the book to become acquainted with DFT practices.

Contributors

Stephen F. Cauley, Ph.D. Student (Chapters 8 and 13)
School of Electrical and Computer Engineering, Purdue University,
West Lafayette, Indiana

Huang-Yu Chen, Ph.D. Student (Chapter 12)
Graduate Institute of Electronics Engineering, National Taiwan
University, Taipei, Taiwan

Tung-Chieh Chen, Post-Doctoral Fellow (Chapter 10)
Graduate Institute of Electronics Engineering, National Taiwan
University, Taipei, Taiwan

Xinghao Chen, Ph.D. (Chapters 2 and 3)
CTC Technologies, Endwell, New York

Chris Chu, Associate Professor (Chapter 11)
Department of Electrical and Computer Engineering, Iowa State
University, Ames, Iowa

Srinivas Devadas, Professor and Associate Head,
EECS, IEEE Fellow (Chapter 6)
Department of Electrical Engineering and Computer Science,
Massachusetts Institute of Technology, Cambridge, Massachusetts

Nikil Dutt, Chancellor's Professor, IEEE Fellow (Chapter 5)
Department of Computer Science, University of California,
Irvine, California

Yinhe Han, Associate Professor (Chapter 3)
Institute of Computing Technology, Chinese Academy
of Sciences, Beijing, China

Michael S. Hsiao, Professor and Dean's Faculty Fellow (Chapter 14)
Bradley Department of Electrical and Computer Engineering, Virginia
Tech, Blacksburg, Virginia

Chung-Yang (Ric) Huang, Assistant Professor (Chapter 4)
Graduate Institute of Electronics Engineering, National Taiwan
University, Taipei, Taiwan

Jiun-Lang Huang, Associate Professor (Chapter 8)
*Graduate Institute of Electronics Engineering, National Taiwan
University, Taipei, Taiwan*

Jitesh Jain, Post-Doctoral Fellow (Chapters 8 and 13)
*School of Electrical and Computer Engineering, Purdue University,
West Lafayette, Indiana*

Jie-Hong (Roland) Jiang, Assistant Professor (Chapter 6)
*Graduate Institute of Electronics Engineering, National Taiwan
University, Taipei, Taiwan*

Cheng-Kok Koh, Associate Professor (Chapters 8 and 13)
*School of Electrical and Computer Engineering, Purdue University,
West Lafayette, Indiana*

Chao-Yue Lai, Research Assistant (Chapter 4)
*Graduate Institute of Electronics Engineering, National Taiwan
University, Taipei, Taiwan*

James C.-M. Li, Associate Professor (Chapter 14)
*Graduate Institute of Electronics Engineering, National Taiwan
University, Taipei, Taiwan*

Xiaowei Li, Professor (Chapter 3)
*Institute of Computing Technology, Chinese Academy of Sciences,
Beijing, China*

Charles E. Stroud, Professor, IEEE Fellow (Chapter 1)
*Department of Electrical and Computer Engineering,
Auburn University, Auburn, Alabama*

Nur A. Touba, Professor, IEEE Fellow (Chapters 2 and 3)
*Department of Electrical and Computer Engineering,
University of Texas, Austin, Texas*

Li-C. Wang, Associate Professor (Chapter 9)
*Department of Electrical and Computer Engineering,
University of California, Santa Barbara, California*

Ruilin Wang, Ph.D. Student (Chapter 13)
*School of Electrical and Computer Engineering, Purdue University,
West Lafayette, Indiana*

Hung-Pin (Charles) Wen, Assistant Professor (Chapter 9)
*Department of Communication Engineering, National Chiao Tung
University, Hsinchu, Taiwan*

Xiaoqing Wen, Professor (Chapters 3 and 7)
Graduate School of Computer Science and Systems Engineering,
Kyushu Institute of Technology, Fukuoka, Japan

Shianling Wu, Vice President of Engineering (Chapter 7)
SynTest Technologies, Inc., Princeton Junction, New Jersey

Jianwen Zhu, Associate Professor (Chapter 5)
Department of Electrical and Computer Engineering,
University of Toronto, Toronto, Ontario, Canada

About the Editors

Laung-Terng (L.-T.) Wang, Ph.D., is chairman and chief executive officer (CEO) of SynTest Technologies (Sunnyvale, CA). He received his BSEE and MSEE degrees from National Taiwan University in 1975 and 1977, respectively, and his MSEE and EE Ph.D. degrees under the Honors Cooperative Program (HCP) from Stanford University in 1982 and 1987, respectively. He worked at Intel (Santa Clara, CA) and Daisy Systems (Mountain View, CA) from 1980 to 1986 and was with the Department of Electrical Engineering of Stanford University as Research Associate and Lecturer from 1987 to 1991. Encouraged by his advisor, Professor Edward J. McCluskey, a member of the National Academy of Engineering, he founded SynTest Technologies in 1990. Under his leadership, the company has grown to more than 50 employees and 250 customers worldwide. The design for testability (DFT) technologies Dr. Wang has developed have been successfully implemented in thousands of ASIC designs worldwide. He currently holds 18 U.S. Patents and 12 European Patents in the areas of scan synthesis, test generation, at-speed scan testing, test compression, logic built-in self-test (BIST), and design for debug and diagnosis. Dr. Wang's work in at-speed scan testing, test compression, and logic BIST has proved crucial to ensuring the quality and testability of nanometer designs, and his inventions are gaining industry acceptance for use in designs manufactured at the 90-nanometer scale and below. He spearheaded efforts to raise endowed funds in memory of his NTU chair professor, Dr. Irving T. Ho, cofounder of the Hsinchu Science Park and vice chair of the National Science Council, Taiwan. Since 2003, he has helped establish a number of chair professorships, graduate fellowships, and undergraduate scholarships at Stanford University, National Taiwan University and National Tsing Hua University in Taiwan, as well as Xiamen University, Tsinghua University, and Shanghai Jiaotong University in China. Dr. Wang has co-authored and co-edited two internationally used DFT textbooks – *VLSI Test Principles and Architectures: Design for Testability* (2006) and *System-on-Chip Test Architectures: Nanometer Design for Testability* (2007). A member of Sigma Xi, he received a Meritorious Service Award from the IEEE Computer Society in 2007 and is a Fellow of the IEEE.

Yao-Wen Chang, Ph.D., is a Professor in the Department of Electrical Engineering and the Graduate Institute of Electronics Engineering at National Taiwan University. He is currently also a Visiting Professor at Waseda University, Japan. He received his B.S. degree from National Taiwan University in 1988, and his M.S. and Ph.D. degrees from the University of Texas at Austin in 1993 and

1996, respectively, all in Computer Science. He was with the IBM T.J. Watson Research Center, Yorktown Heights, NY, in the summer of 1994. From 1996 to 2001, he was on the faculty of National Chiao Tung University, Taiwan. His current research interests include VLSI physical design, design for manufacturability, design automation for biochips, and field programmable gate array (FPGA). He has been working closely with industry on projects in these areas. He co-authored one book on routing and has published over 200 technical papers in these areas, including a few highly cited publications on floorplanning, routing, and FPGA. Dr. Chang is a winner of the 2006 ACM Placement Contest and the 2008 Global Routing Contest at the International Symposium on Physical Design (ISPD), Best Paper Awards at the IEEE International Conference on Computer Design (ICCD) in 1995 and the VLSI Design/CAD Symposia in 2007 and 2008, and eleven Best Paper Award Nominations from the ACM/IEEE Design Automation Conference (DAC) (2000, 2005, 2007, 2008), the IEEE/ACM International Conference on Computer Aided Design (ICCAD) (2002, 2007), ISPD (two in 2007), the IEEE/ACM Asia and South Pacific Design Automation Conference (ASP-DAC; 2004), ICCD (2001), and ACM Transactions on Design Automation of Electronic Systems (2003). He has received many research awards, such as the 2007 Distinguished Research Award, the inaugural 2005 First-Class Principal Investigator Award, and the 2004 Dr. Wu Ta You Memorial Award from National Science Council of Taiwan. He held the 2004 MXIC Young Chair Professorship sponsored by the MXIC Corp. and received excellent teaching awards from National Taiwan University (2004, 2006, 2007, 2008) and National Chiao Tung University (2000). He is an associate editor of the *IEEE Transactions on Computer-Aided Design of Integrated Circuits and Systems (TCAD)* and an editor of *Journal of Information Science and Engineering (JISE)* and *Journal of Electrical and Computer Engineering (JECE)*. He currently serves on the ICCAD Executive Committee, the ASPDAC Steering Committee, the ACM/SIGDA Physical Design Technical Committee, and the ISPD and FPT Organizing Committees. He has also served on the technical program committees of ASP-DAC (topic chair), DAC, IEEE/ACM Design Automation and Test in Europe Conference (DATE), IEEE International Conference on Field Programmable Logic and Applications (FPL), IEEE Field-Programmable Technology (FPT; program co-chair), ACM Great Symposium on VLSI (GLSVLSI), ICCAD, ICCD, The Annual Conference of the IEEE Industrial Electronics Society (IECON; topic chair), ISPD, IEEE SOC Conference (SOCC; topic chair), IEEE TENCON, and IEEE-TSA VLSI Design Automation and Test Conference (VLSI-DAT; topic chair). He is currently an independent board director of Genesys Logic Inc., a technical consultant of RealTek Semiconductor Corp., a principal reviewer of the SBIR project of the Ministry of Economics Affairs of Taiwan, and a member of board of governors of Taiwan IC Design Society.

Kwang-Ting (Tim) Cheng, Ph.D., is a Professor and Chair of the Electrical and Computer Engineering Department at the University of California, Santa Barbara. He received the B.S. degree in Electrical Engineering from National Taiwan University in 1983 and the Ph.D. degree in Electrical Engineering and

Computer Science from the University of California, Berkeley in 1988. He worked at Bell Laboratories in Murray Hill, NJ, from 1988 to 1993. His current research interests include design verification, test, silicon debug, and multimedia computing. He has published over 300 technical papers, co-authored three books, and holds ten U.S. Patents in these areas. He has also been working closely with U.S. industry and government agencies for projects in these areas. He serves on the Executive Committee of the MARCO/DARPA Gigascale System Research Center (sponsored by the Semiconductor Industry Association, U.S. semiconductor equipment, materials, software and services industries, and the U.S. Dept. of Defense) and is Co-Director of the International Center of System-on-Chip (jointly sponsored by National Science Foundation, USA, Chinese National Science Foundation, China, and National Science Council, Taiwan) leading their test and verification research efforts. He served on both Design Working Group (DWG) and Test Working Group (TWG) for the International Technology Roadmap for Semiconductors (ITRS). A fellow of the IEEE, he received Best Paper Awards at the AT&T Conference on Electronic Testing in 1987, the ACM/IEEE Design Automation Conference in 1994 and 1999, the Journal of Information Science and Engineering in 2001, and the IEEE Design Automation and Test in Europe Conference in 2003. He currently serves as Editor-in-Chief for IEEE Design and Test of Computers, Editor for IEEE Transactions on Very Large Scale Integration (VLSI) Systems, Associate Editor for ACM Transactions on Design Automation of Electronic Systems, Associate Editor for Formal Methods in System Design, Editor for Journal of Electronic Testing: Theory and Applications, and Editor for Foundations and Trends in Electronic Design Automation. He has been General Chairs and Program Chairs for a number of international conferences on design, design automation, and test.

Introduction

Charles E. Stroud
Auburn University, Auburn, Alabama

Laung-Terng (L.-T.) Wang
SynTest Technologies, Inc., Sunnyvale, California

Yao-Wen Chang
National Taiwan University, Taipei, Taiwan

ABOUT THIS CHAPTER

Electronic design automation (EDA) is at the center of technology advances in improving human life and use every day. Given an electronic system modeled at the *electronic system level* (ESL), EDA automates the design and test processes of verifying the correctness of the ESL design against the specifications of the electronic system, taking the ESL design through various synthesis and verification steps, and finally testing the manufactured electronic system to ensure that it meets the specifications and quality requirements of the electronic system. The electronic system can also be a *printed circuit board* (PCB) or simply an *integrated circuit* (IC). The integrated circuit can be a *system-on-chip* (SOC), *application-specific integrated circuit* (ASIC), or a *field programmable gate array* (FPGA).

On one hand, EDA comprises a set of hardware and software co-design, synthesis, verification, and test tools that check the ESL design, translate the corrected ESL design to a *register-transfer level* (RTL), and then takes the RTL design through the synthesis and verification stages at the gate level and switch level to eventually produce a physical design described in *graphics data system II* (GDSII) format that is ready to signoff for fabrication and manufacturing test (commonly referred to as **RTL to GDSII** design flow). On the other hand, EDA can be viewed as a collection of design automation and test automation tools that automate the design and test tasks, respectively. The design automation tools deal with the correctness aspects of the electronic system across all levels, be it ESL, RTL, gate level, switch level, or physical level. The test automation tools manage the quality aspects of the electronic system, be it defect level, test cost, or ease of self-test and diagnosis.

This chapter gives a more detailed introduction to the various types and uses of EDA. We begin with an overview of EDA, including some historical perspectives, followed by a more detailed discussion of various aspects of logic design, synthesis, verification, and test. Next, we discuss the important and essential process of physical design automation. The intent is to orient the reader for the remaining chapters of this book, which cover related topics from ESL design modeling and synthesis (including high-level synthesis, logic synthesis, and physical synthesis) to verification and test.

1.1 OVERVIEW OF ELECTRONIC DESIGN AUTOMATION

EDA has had an extraordinary effect on everyday human life with the development of conveniences such as cell phones, *global positioning systems* (GPS), navigation systems, music players, and *personal data assistants* (PDAs). In fact, almost everything and every daily task have been influenced by, and in some cases are a direct result of, EDA. As engineers, perhaps the most noteworthy inventions have been the microprocessor and the *personal computer* (PC), their progression in terms of performance and features, and the subsequent development of smaller, portable implementations such as the notebook computer. As a result, the computer has become an essential tool and part of everyday life—to the extent that current automobiles, including safety features in particular, are controlled by multiple microprocessors. In this section, we give a brief overview of the history of EDA in its early years.

1.1.1 Historical perspective

The history of *electronic design automation* (EDA) began in the early 1960s after the introduction of *integrated circuits* (ICs) [Kilby 1958]. At this very early stage, **logic design** and **physical design** of these ICs were mainly created by hand in parallel. Logic design constructed out of wired circuit boards that mimic the physical design of the IC was built to simulate and verify whether the IC will function as intended before fabrication. The ACM and IEEE cosponsored the **Design Automation Conference** (DAC) debut in 1964 in a joint effort to automate and speed up the design process [DAC 2008]. However, it was not until the mid-1970s when mainframe computers and minicomputers were, respectively, introduced by IBM and Digital Equipment Corporation (DEC) that design automation became more feasible.

During this period, EDA research and development was typically internal to large corporations such as Bell Labs, Hewlett Packard, IBM, Intel, and Tektronix. The first critical milestones in EDA came in the form of programs for **circuit simulation** and **layout verification**. Various proprietary simulation languages and device models were proposed. The SPICE models were used in circuit simulation (*commonly referred* to as **SPICE simulation** now) to verify whether the then so-called logic design specified at the transistor level (called **transistor-level**

design) will behave the same as the functional specifications. This removes the need to build wired circuit boards. At the same time, layout verification tools that took SPICE models as inputs were developed to check whether the physical design would meet **layout design rules** and then tape out the physical design in the *graphics data system II* (GDSII) format introduced by Calma in the mid-1970s.

Although circuit simulation and layout verification ensure that the logic design and physical design will function correctly as expected, they are merely **verification tools**; **design automation tools** are needed to speed up the design process. This requires **logic simulation** tools for logic design at the gate level (rather than at the transistor level) and *place and route* (P&R) tools that operate at the physical level to automatically generate the physical design. The **Tegas logic simulator** that uses the *Tegas description language* (TDL) was the first logic simulator that came to widespread use until the mid-1990s, when industry began adopting the two IEEE developed *hardware description language* (HDL) standards: **Verilog** [IEEE 1463-2001] and **VHDL** [IEEE 1076-2002]. The first graphical software to assist in the physical design of the IC in the late 1960s and early 1970s was introduced by companies like Calma and Applicon. The first automatic *place and route* tools were subsequently introduced in the mid-1970s. Proprietary schematic capture and waveform display software to assist in the logic design of the IC was also spurring the marketplace.

Although much of the early EDA research and development was done in corporations in the 1960s and 1970s, top universities including Stanford, the University of California at Berkeley, Carnegie Mellon, and California Institute of Technology had quietly established large *computer-aided design* (CAD) groups to conduct research spreading from process/device simulation and modeling [Dutton 1993; Plummer 2000] to logic synthesis [Brayton 1984; De Micheli 1994; Devadas 1994] and *analog and mixed signal* (AMS) design and synthesis [Ochetta 1994] to silicon compilation and physical design automation [Mead 1980]. This also marks the timeframe in which EDA began as an industry with companies like Daisy Systems, Mentor Graphics [Mentor 2008], and Valid Logic Systems (acquired by Cadence Design Systems [Cadence 2008]) in the early 1980s. Another major milestone for academic-based EDA research and development was the formation of the *Metal Oxide Semiconductor Implementation Service* (MOSIS) in the early 1980s [MOSIS 2008].

Since those early years, EDA has continued to not only provide support and new capabilities for electronic system design but also solve many problems faced in both design and testing of electronic systems. For example, how does one test an IC with more than one billion transistors to ensure with a high probability that all transistors are fault-free? *Design for testability* (DFT) and *automatic test pattern generation* (ATPG) tools have provided EDA solutions. Another example is illustrated in Figure 1.1 for mask making during deep-submicron IC fabrication. In this example, the lithographic process is used to create rectangular patterns to form the various components of transistors and their interconnections. However, sub-wavelength components in lithography cause problems in that the intended shapes become irregular as shown in

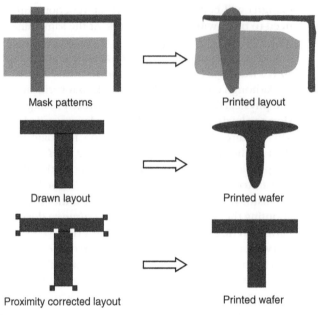

FIGURE 1.1

Sub-wavelength lithography problem and EDA solution.

Figure 1.1. This problem posed a serious obstacle to advances in technology in terms of reducing feature size, also referred to as shrinking design rules, which in turn increases the number of transistors that can be incorporated in an IC. However, EDA has provided the solution through *optical proximity correction* (OPC) of the layout to compensate for rounding off feature corners.

1.1.2 **VLSI design flow and typical EDA flow**

When we think of current EDA features and capabilities, we generally think of synthesis of *hardware description languages* (HDLs) to standard cell–based ASICs or to the configuration data to be downloaded into FPGAs. As part of the synthesis process, EDA also encompasses design audits, technology mapping, and physical design (including floorplanning, placement, routing, and design rule checking) in the intended implementation medium, be that ASIC, FPGA, PCB, or any other media used to implement electronic systems. In addition, EDA comprises logic and timing simulation and timing analysis programs for design verification of both pre-synthesis and post-synthesis designs. Finally, there is also a wealth of EDA software targeting manufacturing test, including testability analysis, *automatic test pattern generation* (ATPG), fault simulation, *design for testability* (DFT), logic/memory *built-in self-test* (BIST), and test compression.

In general, EDA algorithms, techniques, and software can be partitioned into three distinct but broad categories that include logic design automation, verification and test, and physical design automation. Although logic and physical design automation are somewhat disjointed in that logic design automation is performed before physical design automation, the various components and aspects of the verification and test category are dispersed within both logic and physical design automation processes. Furthermore, verification software is usually the first EDA tool used in the overall design method for simulation of the initial design developed for the intended circuit or system.

The two principal HDLs currently used include ***very high-speed integrated circuits*** (VHSIC) ***hardware description language*** (VHDL) [IEEE 1076-2002] and Verilog ***hardware description languages*** [IEEE 1463-2001]. VHDL originally targeted gate level through system-level design and verification. Verilog, on the other hand, originally targeted the design of ASICs down to the transistor level of design, but not the physical design. Since their introduction in the late 1980s, these two HDLs have expanded to cover a larger portion of the design hierarchy illustrated in Figure 1.2 to the extent that they both cover approximately the same range of the design hierarchy. These HDLs owe their success in current design methods to the introduction of synthesis approaches and software in the mid- to late 1980s. As a result, synthesis capabilities enabled VHDL and Verilog to become the design capture medium as opposed to "an additional step" in the design process.

There are many benefits of high-level HDLs such as VHDL and Verilog when used in conjunction with synthesis capabilities. They facilitate early design verification through high-level simulation, as well as the evaluation of alternate architectures for optimizing system cost and performance. These high-level simulations in turn provide baseline testing of lower level design representations such as gate-level implementations. With synthesis, top-down design methods are realized, with the high-level HDLs being the design capture medium independent of the implementation media (for example, ASIC *versus* FPGA).

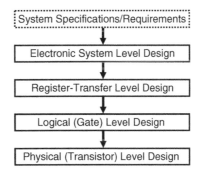

FIGURE 1.2

Design hierarchy.

This not only reduces design development time and cost but also reduces the risk to a project because of design errors. In addition, this provides the ability to manage and develop complex electronic systems and provides the basis for hardware/software co-design. As a result, *electronic system level* (ESL) design includes partitioning the system into hardware and software and the co-design and co-simulation of the hardware and software components. The ESL design also includes cost estimation and design-space exploration for the target system to make informed design decisions early in the design process.

The basic domains of most high-level HDLs include structural, behavioral, and RTL hierarchical descriptions of a circuit or system. The structural domain is a description of components and their interconnections and is often referred to as a **netlist**. The behavioral domain includes high-level algorithmic descriptions of the behavior of a circuit or system from the standpoint of output responses to a sequence of input stimuli. Behavioral descriptions are typically at such a high level that they cannot be directly synthesized by EDA software. The RTL domain, on the other hand, represents the clock cycle by clock cycle data flow of the circuit at the register level and can be synthesized by EDA software. Therefore, the design process often implies a manual translation step from behavioral to RTL with baseline testing to verify proper operation of the RTL design as illustrated in the example design flow for an *integrated circuit* (IC) in Figure 1.3. It should be noted that the behavioral domain is contained in the ESL design blocks in Figures 1.2 and 1.3. If the behavior of the ESL design is described in C, C++, **SystemC** [SystemC 2008], **SystemVerilog** [SystemVerilog 2008], or a mixture of these languages, modern verification

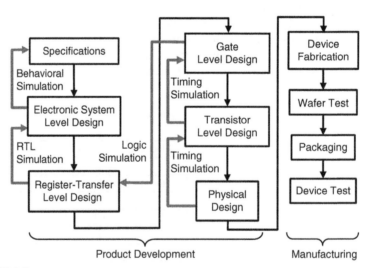

FIGURE 1.3

IC design and verification flow.

and simulation tools can either convert the language to VHDL or Verilog or directly accept the language constructs.

Although the simulation of behavioral and RTL descriptions is an essential EDA feature, most EDA encompasses the design and verification flow from the point of RTL design onward. This includes synthesis to a technology-independent gate-level implementation of the circuit followed by technology mapping to a specific implementation media such as a standard–cell–based ASIC design which in turn represents the transistor-level design in the IC design flow of Figure 1.3. Physical design then completes technology-specific partitioning, floorplanning, placement, and routing for the design. As the design flow progresses through various stages in the synthesis and physical design processes, regression testing is performed to ensure that the synthesized implementation performs the correct functionality of the intended design at the required system clock frequency. This requires additional simulation steps, as indicated in Figure 1.3, with each simulation step providing a more accurate representation of the manufactured implementation of the final circuit or system.

A number of points in the design process impact the testability and, ultimately, the manufacturing cost of an electronic component or system. These include consideration of DFT and BIST, as well as the development of test stimuli and expected good-circuit output responses used to test each manufactured product [Bushnell 2000; Stroud 2002; Jha 2003; Wang 2006, 2007]. For example, the actual manufacturing test steps are illustrated in the IC design flow of Figure 1.3 as wafer test and packaged-device test.

Physical design is one of the most important design steps because of its critical impact on area, power, and performance of the final electronic circuit or system. This is because layout (component placement) and routing are integral parts of any implementation media such as ICs, FPGAs, and PCBs. Therefore, physical design was one of the first areas of focus on EDA research and development. The result has been numerous approaches and algorithms for physical design automation [Preas 1988; Gerez 1998; Sait 1999; Sherwani 1999; Scheffer 2006a, 2006b]. The basic flow of the physical design process is illustrated in Figure 1.4.

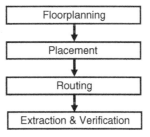

FIGURE 1.4

Typical physical design flow.

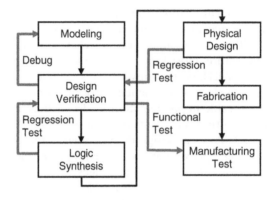

FIGURE 1.5

Typical EDA flow.

An alternate view of a typical EDA flow is illustrated in Figure 1.5, which begins with modeling and design verification. This implies a recursive process with debugging until the resultant models reach a level of detail that can be processed by logic synthesis. As a result of current EDA capabilities, the design process is highly automated from this point. This is particularly true for physical design but to a lesser extent for manufacture test and test development. Therefore, the functional stimuli developed for the device and output responses obtained from simulations during design verification typically form the basis for functional tests used during manufacturing test.

Many design space issues may be critical to a given project, and many of these issues can require tradeoffs. For example, area and performance are two of the most frequently addressed tradeoffs to be considered in the design space. Area considerations include chip area, how many ICs on a PCB, and how much board space will be required for a given implementation. Performance considerations, on the other hand, often require additional area to meet the speed requirements for the system. For example, the much faster carry-look-ahead adder requires significantly more area than the simple but slow ripple-carry adder. Therefore, EDA synthesis options include features and capabilities to select and control area and performance optimization for the final design. However, additional design space issues such as power consumption and power integrity must be considered. Inevitably, cost and anticipated volume of the final product are also key ingredients in making design decisions. Another is the design time to meet the market window and development cost goals. The potential risk to the project in obtaining a working, cost-effective product on schedule is an extremely important design issue that also hinges on reuse of resources (using the same core in different modes of operation, for example) and the target implementation media and its associated technology limits. Less frequently addressed, but equally important, design considerations include designer experience and EDA software availability and capabilities.

1.1.3 **Typical EDA implementation examples**

To better appreciate the current state of EDA in modern electronic system design, it is worth taking a brief look at the state and progression of EDA since the mid-1970s and some of the subsequent milestones. At that time, ASICs were typically hand-designed by creating a hand-edited netlist of standard cells and their interconnections. This netlist was usually debugged and verified using unit-delay logic simulation. Once functional verification was complete, the netlist was used as input to *computer-aided design* (CAD) tools for placement of the standard cells and routing of their interconnections. At that time, physical design was semiautomated with considerable intervention by physical design engineers to integrate *input/output* (I/O) buffers and networks for clocks, power, and ground connections. Timing simulation CAD tools were available for verification of the design for both pre-physical and post-physical design using estimated and extracted capacitance values, respectively. It is interesting to note that resistance was not considered in timing simulations until the mid-1980s, when design rules reached the point that sheet resistance became a dominant delay factor.

Graphical schematic entry CAD tools were available for PCBs for design capture, layout, and routing. However, schematic capture tools for ASIC design were generally not available until the early 1980s and did not significantly improve the design process other than providing a nicely drawn schematic of the design from which the standard-cell netlist was automatically generated. The actual digital logic continued to be hand-designed by use of state diagrams, state tables, Karnaugh maps, and a few simple CAD tools for logic minimization. This limited the complexity of ASICs in terms of the number of gates or transistors that could be correctly designed and verified by a typical designer. In the early 1980s, an ASIC with more than 100,000 transistors was considered to be near the upper limit for a single designer. By the late 1980s, the limit was significantly increased as a result of multi-designer teams working on a single IC and as a result of advances in EDA capabilities, particularly in the area of logic synthesis. Currently, the largest ICs exceed 1 billion transistors [Naffziger 2006; Stackhouse 2008].

One of the earliest approaches to EDA in terms of combinational logic synthesis was for implementing *programmable logic arrays* (PLAs) in *very large-scale integrated* (VLSI) circuits in the late 1970s [Mead 1980]. Any combinational logic function can be expressed as Boolean logic equations, *sum-of-products* (SOP) or *product-of-sums* (POS) expressions, and truth tables or Karnaugh maps. There are other representations, but these three are illustrated for the example circuit in Figure 1.6 and are important for understanding the implementation of PLAs and other programmable logic. We can program the truth table onto a *read-only memory* (ROM) with eight words and two bits/word and then use the ROM address lines as the three input signals (A, B, C) and the ROM outputs as the output signals (X, Y). Similarly, we can

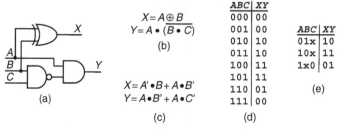

$$X = A \oplus B$$
$$Y = A \cdot (\overline{B \cdot C})$$

(b)

$$X = A' \cdot B + A \cdot B'$$
$$Y = A \cdot B' + A \cdot C'$$

(c)

ABC	XY
000	00
001	00
010	10
011	10
100	11
101	11
110	01
111	00

(d)

ABC	XY
01x	10
10x	11
1x0	01

(e)

FIGURE 1.6

Combinational logic implementation example: (a) Logic diagram. (b) Logic equations. (c) SOP expressions. (d) Truth table. (e) Connection array.

also write the truth table into a ***random-access memory*** (RAM) with eight words and two bits/word and then disable the write enable to the RAM and use the address lines as the inputs. Note that this is the same thing as the ROM, except that we can reprogram the logic function by rewriting the RAM; this also forms the basis for combinational logic implementations in FPGAs.

Another option for implementing a truth table is the PLA. In the connection array in Figure 1.6e, only three product terms produce logic 1 at the output signals. The PLA allows implementing only those three product terms and not the other five, which is much smaller than either the ROM or RAM implementation. Any SOP can be implemented as a 2-level AND-OR or NAND-NAND logic function. Any SOP can also be implemented as a 2-level NOR-NOR logic function if we invert the inputs and the output as illustrated in Figure 1.7a. Note that AB' is a shared product term and allows us to share an AND gate in the gate-level implementation. PLAs take advantage of this NOR-NOR implementation of logic equations and the large fan-in limit of ***N-channel metal oxide semiconductor*** (NMOS) NOR gates, as illustrated in Figure 1.7b, for PLA implementation of the example circuit. Note that there is a direct relationship between the crosspoints in the PLA and the AND-OR connection array in Figure 1.6e. A logic 1 (0) in the input columns of the connection array corresponds to a crosspoint between the bit (bit-bar) line and the AND line, also called the product term line. A logic 1 in the output columns corresponds to a crosspoint between the AND line and the OR line, also called the output line. Therefore, the physical design of the PLA is obtained directly from the connection array. It is also important to note that a connection array is derived from a minimized truth table but is not equivalent to a truth table as can be seen by considering the X output for the last two entries in the connection array.

PLAs are of historical importance because they not only led to the development of ***programmable logic devices*** (PLDs) including FPGAs, but they also led to the further development of CAD tools for logic minimization and automated physical design, because the physical design could be obtained directly

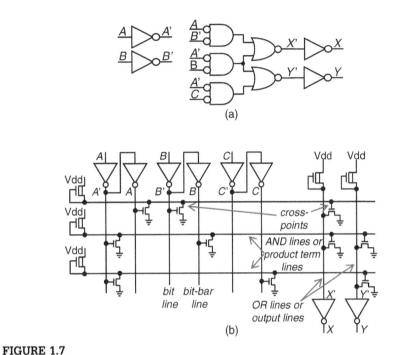

FIGURE 1.7

PLA implementation example: (a) 2-level NOR-NOR implementation. (b) PLA implementation.

from the connection array. For example, the outputs of early logic minimization tools, like **Espresso**, were usually in terms of a connection array for PLA implementations. These minimization CAD tools were the predecessors to high-level synthesis and many of the current physical design tools.

The quest for the ability to synthesize high-level descriptions of hardware began in earnest in the mid-1980s. One of the first successful synthesis tools, called **CONES**, was capable of synthesizing RTL models written in C to either standard–cell–based ASICs or to PLD-based PCBs and was used extensively internal to Bell Labs [Stroud 1986]. This timeframe also corresponds to the formation of EDA companies dedicated to synthesis, such as Synopsys [Synopsys 2008], as well as the introduction of VHDL and Verilog, which have been used extensively throughout industry and academia since that time.

The successful introduction of functional modeling into the VLSI design community was due, in part, to the development of logic synthesis tools and systems. Modeling a system at the functional level and simulating the resultant models had been previously used with simulation languages such as ADA [Ledgard 1983] to obtain a simulation environment that emulates the system to be designed. These simulation environments provided a platform from which the design of the various modules required for implementation of the system could proceed independently with the ability to regression test the detailed designs at various points in the design process. In addition, this model-based simulation environment could ensure a degree of coherence in the system long before hardware components were available for integration and testing in the final system. Despite the advantages of this approach, it did not receive widespread attention until logic synthesis tools and systems were developed to synthesize the detailed gate-level or transistor-level design from the functional description of the circuit. As a result, the design entry point for the designer became the functional model rather than gate-level or transistor-level descriptions of the VLSI device. Once removed from time-consuming and often error-prone gate-level or transistor-level design, designers had the ability to manage higher levels of design complexity. In addition, the speed at which the logic synthesis systems can implement the gate-level or transistor-level design significantly reduced the overall design interval.

1.1.4 **Problems and challenges**

With exponentially increasing transistor counts in ICs brought on by smaller feature sizes, there are also demands for increased bandwidth and functionality with lower cost and shorter time-to-market. The main challenges in EDA are well documented in the *International Technology Roadmap for Semiconductors* (ITRS) [SIA 2005, 2006]. One of the major challenges is that of design productivity in the face of large design teams and diversity in terms of heterogeneous components in system-level SOC integration. This includes design specification and verification at the system level and embedded system software co-design and *analog/mixed-signal* (AMS) circuitry in the hierarchy along with other system objectives such as fault or defect tolerance. To accurately verify the design before fabrication, the challenges include the ability to accurately extract physical design information to efficiently model and simulate full-chip interconnect delay, noise, and power consumption.

A primary trend in testing is an emphasis to provide information for *failure mode analysis* (FMA) to obtain yield enhancement. Another trend is **reliability screening** in which testing targets weak transistors and the location of non-uniformities in addition to hard defects; this includes detecting the symptoms and effects of line width variations, finite dopant distributions, and systemic process defects. Finally, there is a need to avoid potential yield losses as a result of tester inaccuracies, power droop, overly aggressive statistical postprocessing, defects occurring in test circuitry such as BIST, overtesting delay faults on non-functional paths, mechanical damages resulting from the testing process, and faulty repairs of repairable circuits, to name a few.

In the remaining sections of this chapter, we give a more detailed overview of the three fundamental components of EDA: logic design automation, testing, and physical design automation.

1.2 **LOGIC DESIGN AUTOMATION**

Logic design automation refers to all modeling, synthesis, and verification steps that model a design specification of an electronic system at an ***electronic system level*** (ESL), verify the ESL design, and then compile or translate the ESL representation of the design into an RTL or gate-level representation. The design hierarchy illustrated in Figure 1.2 has described all the levels of abstraction down to physical design, a step solely responsible for **physical design automation**. In the design hierarchy, a higher level description has fewer implementation details but more explicit functional information than a lower level description. The design process illustrated in Figure 1.3 and the EDA process flow illustrated in Figure 1.5 essentially represent the transforming of a higher level description of a design to a lower level description. The following subsections discuss the various steps associated with logic design automation, which include modeling, design verification, and logic synthesis.

1.2.1 **Modeling**

Starting from a design specification, a behavioral description of a system is developed in ESL languages, such as SystemC, SystemVerilog, VHDL, Verilog, and C/C++, and simulated to determine whether it meets the system requirements and specifications. The objective is to describe the behavior of the intended system in a number of behavioral models that can be simulated for design verification and then translated to RTL for logic synthesis. In addition, behavioral models representing existing or new hardware that interfaces with the system may be developed to create a simulation environment in which the behavioral models for the system to be designed can be verified in the presence of existing hardware. Alternately, such hardware can be directly embedded in an emulator for design verification [Scheffer 2006a, 2006b]. During design verification, a number of iterations of modeling and simulation steps are usually required to obtain a working behavioral description for the intended system to be implemented.

Once the requirements for the system to be designed have been defined, the designer faces the task of describing the functionality in models. The goal is to write models such that they can be simulated to verify the correct operation of the design and be synthesized to obtain the logic to implement the function. For a complex system, SOC, or VLSI device, this usually requires that the functionality be partitioned into multiple blocks that are more easily managed in terms of complexity. One or more functional models may represent each of these blocks. There are different ways to proceed to achieve the goal of functional models being synthesizable. One approach is to ignore the requirement that the models be synthesizable and to describe the function at as high a level as can be handled by the designer and by the simulation tools. These high-level descriptions can then be verified in the simulation environment to obtain correct functionality with respect to the system requirements and specifications. At that point, the high-level models can be partitioned and written as functional models of a form suitable for synthesis. In this case, the simulation environment is first used to verify the high-level models and later used as a baseline for **regression testing** of the synthesizable models to ensure that correct functionality has been maintained such that the synthesized design still meets the system requirements and specifications. At the other end of the spectrum, an alternate approach is to perform the partitioning and generation of the functional models at a level of detail compatible with the synthesis tools. Once sufficient design verification has been achieved, the design can move directly to the logic synthesis step.

Modeling the circuit to be simulated and synthesized is, in some respects, simply a matter of translating the system requirements and specifications to the ESL or HDL description. The requirements and specifications for the system or circuit to be modeled are sometimes quite specific. On the other hand, on the basis of inputs and outputs from other blocks necessary to construct the complete system, arbitrary values may be chosen by the designer.

1.2.2 **Design verification**

Design verification is the most important aspect of the product development process illustrated in Figures 1.3 and 1.5, consuming as much as 80% of the total product development time. The intent is to verify that the design meets the system requirements and specifications. Approaches to design verification consist of (1) **logic simulation/emulation and circuit simulation**, in which detailed functionality and timing of the design are checked by means of simulation or emulation; (2) **functional verification**, in which functional models describing the functionality of the design are developed to check against the behavioral specification of the design without detailed timing simulation; and (3) **formal verification**, in which the functionality is checked against a "golden" model. Formal verification further includes **property checking** (or **model checking**), in which the property of the design is checked against some presumed "properties" specified in the functional or behavioral model (*e.g.*, a finite-state machine should not enter a certain state), and **equivalence checking**, in which the functionality is checked against a "golden" model [Wile 2005]. Although equivalence checking can be used to verify the synthesis results in the lower levels of the EDA flow (denoted "regression test" in Figure 1.5), the original design capture requires property checking.

Simulation-based techniques are the most popular approach to verification, even though these are time-consuming and may be incomplete in finding design errors. Logic simulation is used throughout every stage of logic design automation, whereas circuit simulation is used after physical design. The most commonly used logic simulation techniques are compiled-code simulation and event-driven simulation [Wang 2006]. The former is most effective for cycle-based two-valued simulation; the latter is capable of handling various gate and wire delay models. Although versatile and low in cost, logic simulation is too slow for complex SOC designs or hardware/software co-simulation applications. For more accurate timing information and dynamic behavior analysis, device-level circuit simulation is used. However, limited by the computation complexity, circuit simulation is, in general, only applied to critical paths, cell library components, and memory analysis.

For simulation, usually, a number of different simulation techniques are used, including high-level simulation through a combination of behavioral modeling and testbenches. Testbenches are behavioral models that emulate the surrounding system environment to provide input stimuli to the design under test and process the output responses during simulation. RTL models of the detailed design are then developed and verified with the same testbenches that were used for verification of the architectural design, in addition to testbenches that target design errors in the RTL description of the design. With sufficient design verification at this point in the design process, functional vectors can be captured in the RTL simulation and then used for subsequent simulations (**regression testing**) of the more detailed levels of design, including synthesized gate-level design, transistor-level design, and physical design. These latter levels of design abstraction (gate, transistor, and physical design) provide the ability to perform additional design verification through logic, switch-level, and timing simulations. These three levels of design abstraction also provide the basis for fault models that can be used to evaluate the effectiveness of manufacturing tests.

The design verification step establishes the quality of the design and ensures the success of the project by uncovering potential errors in both the design and the architecture of the system. The objective of design verification is to simulate all functions as exhaustively as possible while carefully investigating any possibly erroneous behavior. From a designer's standpoint, this step deserves the most time and attention. One of the benefits of high-level HDLs and logic synthesis is to allow the designer to devote more time and concentration to design verification. Because much less effort is required to obtain models that can be simulated but not synthesized, design verification can begin earlier in the design process, which also allows more time for considering optimal solutions to problems found in the design or system. Furthermore, debugging a high-level model is much easier and faster than debugging a lower level description, such as a gate-level netlist.

An attractive attribute of the use of functional models for design verification (often called **functional verification**) is that HDL simulation of a collection of models is much faster than simulations of the gate-level descriptions that would correspond to those models. Although functional verification only verifies cycle accuracy (rather than timing accuracy), the time required to perform the design verification process is reduced with faster simulation. In addition, a more thorough verification of the design can be performed, which in turn improves the quality of the design and the probability of the success of the project as a whole. Furthermore, because these models are smaller and more functional than netlists describing the gate-level design, the detection, location, and correction of design errors are easier and faster. The reduced memory requirements and increased speed of simulation with functional models enable simulation of much larger circuits, making it practical to simulate and verify a complete hardware system to be constructed. As a result, the reduced probability of design changes resulting from errors found during system integration can be factored into the overall design schedule to meet shorter market windows. Therefore, design verification is economically significant, because it has a definite impact on time-to-market. Many tools are available to assist in the design verification process, including simulation tools, hardware emulation, and formal verification methods. It is interesting to note that many design verification techniques are borrowed from test technology, because verifying a design is similar to testing a physical product. Furthermore, the test stimuli developed for design verification of the RTL, logical, and physical levels of abstraction are often used, in conjunction with the associated output responses obtained from simulation, for functional tests during the manufacturing process.

Changes in system requirements or specifications late in the design cycle jeopardize the schedule and the quality of the design. Late changes to a design represent one of the two most significant risks to the overall project, the other being insufficient design verification. The quality of the design verification process depends on the ability of the testbenches, functional vectors, and the designers who analyze the simulated responses to detect design errors. Therefore, any inconsistency observed during the simulations at the various levels of design abstraction should be carefully studied to determine whether potential design errors to be corrected exist before design verification continues.

Emulation-based verification by use of FPGAs provides an attractive alternative to simulation-based verification as the gap between logic simulation capacity and design complexity continues growing. Before the introduction of FPGAs in the 1980s, ASICs were often verified by construction of a breadboard by use of *small-scale integration* (SSI) and *medium-scale integration* (MSI) devices on a wire-wrap board. This became impractical as the complexity and scale of ASICs moved into the VLSI realm. As a result, FPGAs became the primary hardware for emulation-based verification. Although these approaches are costly and may not be easy to use, they improve verification time by two to three orders of magnitude compared with software simulation. Alternately, a **reconfigurable emulation system** (or **reconfigurable emulator**) that automatically partitions and maps a design onto multiple FPGAs can be used to avoid building a prototype board and can be reused for various designs [Scheffer 2006a, 2006b].

Formal verification techniques are a relatively new paradigm for equivalence checking. Instead of input stimuli, these techniques perform exhaustive proof through rigorous logical reasoning. The primary approaches used for formal verification include *binary decision diagrams* (BDDs) and **Boolean satisfiability** (SAT) [Velev 2001]. These approaches, along with other algorithms specific to EDA applications, are extensively discussed in Chapter 4. The BDD approach successively applies **Shannon expansion** on all variables of a combinational logic function until either the constant function "0" or "1" is reached. This is applied to both the captured design and the synthesized implementation and compared to determine their equivalence. Although BDDs give a compact representation for Boolean functions in polynomial time for many Boolean operations, the size of BDD grows exponentially with input size, which is usually limited to 100 to 200 inputs. On the other hand, SAT techniques have been very successful in recent years in the verification area with the ability to handle million-gate designs and both combinational and sequential designs.

1.2.3 **Logic synthesis**

The principal goal of logic synthesis is to translate designs from the behavioral domain to the structural domain. This includes **high-level synthesis**, in which system behavior and/or algorithms are transformed into functional blocks such as processors, RAMs, *arithmetic logic units* (ALUs), etc. Another type of synthesis takes place at the *register-transfer level* (RTL), where Boolean expressions or RTL descriptions in VHDL or Verilog are transformed to logic gate networks.

Logic synthesis is initially technology independent where RTL descriptions are parsed for control/data flow analysis. Initial gate-level implementations are in terms of generic gate implementations (such as AND, OR, and NOT) with no relationship to any specific technology. As a result, the structure at this point is technology independent and can be ultimately implemented in any technology by means of **technology mapping** into specific libraries of cells as illustrated in Figure 1.8. Before technology mapping, however, a number of technology-independent optimizations can be made to the gate-level implementation by basic logic restructuring with techniques such as the **Quine-McCluskey method** for two-level logic optimization [McCluskey 1986] or methods for multilevel logic optimization that may be more appropriate for standard cell–based designs [Brayton 1984; De Michele 1994; Devadas 1994]. Once technology mapping has been performed, additional optimizations are performed such as for timing and power. This may be followed by insertion of logic to support *design for testability* (DFT) features and capabilities. However, it should be noted that once technology mapping is performed, most subsequent synthesis and optimizations fall into the domain of physical design automation.

Regression testing of the synthesized gate-level description ensures that there are no problems in the design that are not apparent from the functional model simulation, such as feedback loops that cannot be initialized.

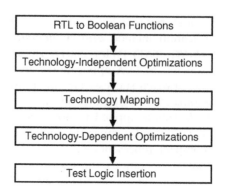

FIGURE 1.8

Logic synthesis flow.

This additional effort may seem to be avoidable with proper consideration given to undefined logic values in the function model. However, developing a functional model that initializes the same as a gate-level description requires considerable effort and knowledge of the gate-level structure of a circuit. Hence, the functional model may not behave exactly the same way as the synthesized circuit. Designers must be careful to avoid constructs in HDLs that allow the model to self-initialize but cannot be reproduced in the final circuit by the synthesis system. Therefore, regression testing is necessary and, fortunately, undefined logic values are relatively easy to trace to their source to determine the root cause. Good coding and reusability styles, as well as user-defined coding style rules, play an important role in avoiding many of the synthesis errors [Keating 1999].

1.3 TEST AUTOMATION

Advances in manufacturing process technology have also led to very complex designs. As a result, it has become a requirement that *design-for-testability* (DFT) features be incorporated in the ***register-transfer level*** (RTL) or gate-level design before physical design to ensure the quality of the fabricated devices. In fact, the traditional VLSI development process illustrated in Figure 1.3 involves some form of testing at each stage, including design verification. Once verified, the VLSI design then goes to fabrication and, at the same time, test engineers develop a test procedure based on the design specification and **fault models** associated with the implementation technology. Because the resulting product quality is in general unsatisfactory, modern VLSI test development planning tends to start when the RTL design is near completion. This test development plan defines what **test requirements** the product must meet, often in terms of **defect level** and **manufacturing yield**, test cost, and whether it is necessary to perform self-test and diagnosis. Because the test requirements mostly target manufacturing defects rather than **soft errors**, which would require **online fault detection and correction** [Wang 2007], one need is to decide what fault models should be considered.

The test development process now consists of (1) defining the targeted fault models for defect level and manufacturing yield considerations, (2) deciding what types of DFT features should be incorporated in the RTL design to meet the test requirements, (3) generating and fault-grading test patterns to calculate the final fault coverage, and (4) conducting manufacturing test to screen bad chips from shipping to customers and performing *failure mode analysis* (FMA) when the chips do not achieve desired defect level or yield requirements.

1.3.1 Fault models

A **defect** is a manufacturing flaw or physical imperfection that may lead to a **fault**, a fault can cause a circuit **error**, and a circuit error can result in a **failure** of the device or system. Because of the diversity of defects, it is difficult to generate tests for real defects. Fault models are necessary for generating and evaluating test patterns. Generally, a good fault model should satisfy two criteria: (1) it should accurately reflect the behavior of defects and (2) it should be computationally efficient in terms of time required for fault simulation and test generation. Many fault models have been proposed but, unfortunately, no single fault model accurately reflects the behavior of all possible defects that can occur. As a result, a combination of different fault models is often used in the generation and evaluation of test patterns. Some well-known and commonly used fault models for general sequential logic [Bushnell 2000; Wang 2006] include the following:

1. **Gate-level stuck-at fault model**: The stuck-at fault is a logical fault model that has been used successfully for decades. A stuck-at fault transforms the correct value on the faulty signal line to appear to be stuck-at a constant logic value, either logic 0 or 1, referred to as *stuck-at-0* (SA0) or *stuck-at-1* (SA1), respectively. This model is commonly referred to as the **line stuck-at fault model** where any line can be SA0 or SA1, and also referred to as the gate-level stuck-at fault model where any input or output of any gate can be SA0 or SA1.

2. **Transistor-level stuck fault model**: At the switch level, a transistor can be **stuck-off** or **stuck-on**, also referred to as **stuck-open** or **stuck-short**, respectively. The line stuck-at fault model cannot accurately reflect the behavior of stuck-off and stuck-on transistor faults in *complementary metal oxide semiconductor* (CMOS) logic circuits because of the multiple transistors used to construct CMOS logic gates. A stuck-open transistor fault in a CMOS combinational logic gate can cause the gate to behave like a level-sensitive latch. Thus, a stuck-open fault in a CMOS combinational circuit requires a sequence of two vectors for

detection instead of a single test vector for a stuck-at fault. Stuck-short faults, on the other hand, can produce a conducting path between power (V_{DD}) and ground (V_{SS}) and may be detected by monitoring the power supply current during steady state, referred to as I_{DDQ}. This technique of monitoring the steady state power supply current to detect transistor stuck-short faults is called $\mathbf{I_{DDQ}}$ **testing** [Bushnell 2000; Wang 2007].

3. **Bridging fault models**: Defects can also include opens and shorts in the wires that interconnect the transistors that form the circuit. Opens tend to behave like line stuck-at faults. However, a **resistive open** does not behave the same as a transistor or line stuck-at fault, but instead affects the propagation delay of the signal path. A short between two wires is commonly referred to as a **bridging fault**. The case of a wire being shorted to V_{DD} or V_{SS} is equivalent to the line stuck-at fault model. However, when two signal wires are shorted together, bridging fault models are needed; the three most commonly used bridging fault models are illustrated in Figure 1.9. The first bridging fault model proposed was the **wired-AND/wired-OR** bridging fault model, which was originally developed for bipolar technology and does not accurately reflect the behavior of bridging faults typically found in CMOS devices. Therefore, the **dominant bridging fault** model was proposed for CMOS where one driver is assumed to dominate the logic value on the two shorted nets. However, the dominant bridging fault model does not accurately reflect the behavior of a resistive short in some cases. The most recent bridging fault model, called the **4-way** bridging fault model and also known as the **dominant-AND/dominant-OR** bridging fault model, assumes that one driver dominates the logic value of the shorted nets for one logic value only [Stroud 2002].

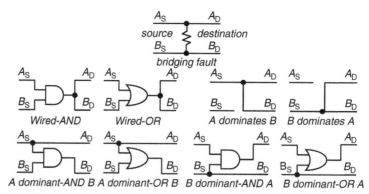

FIGURE 1.9

Bridging fault models.

4. **Delay fault models**: Resistive opens and shorts in wires and parameter variations in transistors can cause excessive delay such that the total propagation delay falls outside the specified limit. **Delay faults** have become more prevalent with decreasing feature sizes, and there are different delay fault models. In **gate-delay fault** and **transition fault** models, a delay fault occurs when the time interval taken for a transition through a single gate exceeds its specified range. The **path-delay fault** model, on the other hand, considers the cumulative propagation delay along any signal path through the circuit. The **small delay defect** model takes timing delay associated with the fault sites and propagation paths from the layout into consideration [Sato 2005; Wang 2007].

1.3.2 Design for testability

To test a given circuit, we need to control and observe logic values of internal nodes. Unfortunately, some nodes in sequential circuits can be difficult to control and observe. DFT techniques have been proposed to improve the controllability and observability of internal nodes and generally fall into one of the following three categories: (1) **ad-hoc DFT** methods, (2) **scan design**, and (3) *built-in self-test* (BIST). Ad-hoc methods were the first DFT technique introduced in the 1970s [Abramovici 1994]. The goal was to target only portions of the circuit that were difficult to test and to add circuitry (typically **test point** insertion) to improve the controllability and/or observability of internal nodes [Wang 2006].

Scan design was the most significant DFT technique proposed [Williams 1983]. This is because the scan design implementation process was easily automated and incorporated in the EDA flow. A scan design can be flip-flop based or latch based. The latch-based scan design is commonly referred to as *level-sensitive scan design* (LSSD) [Eichelberger 1978]. The basic idea to create a scan design is to reconfigure each flip-flop (*FF*) or latch in the sequential circuit to become a *scan flip-flop* (*SFF*) or **scan latch** (often called **scan cell**), respectively. These scan cells, as illustrated in Figure 1.10, are then connected in series to form a shift register, or **scan chain**, with direct access to a primary input (*Scan Data In*) and a primary output (*Scan Data Out*). During the shift operation (when *Scan Mode* is set to 1), the scan chain is used to shift in a test pattern from *Scan Data In* to be applied to the combinational logic. During one clock cycle of the normal system operation (when *Scan Mode* is set to 0), the test pattern is applied to the combinational logic and the output response is clocked back or captured into the scan cells. The scan chain is then used in scan mode to shift out the combinational logic output response while shifting in the next test pattern to be applied. As a result, scan design reduces the problem of testing sequential logic to that of testing combinational logic and, thereby, facilitates the use of *automatic test pattern generation* (ATPG) techniques and software developed for combinational logic.

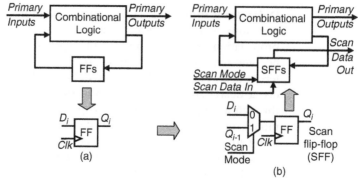

FIGURE 1.10

Transforming a sequential circuit to flip-flop-based scan design: (a) Example of a sequential circuit. (b) Example of a scan design.

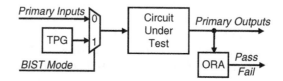

FIGURE 1.11

Simple BIST architecture.

BIST was proposed around 1980 to embed test circuitry in the device or system to perform self-test internally. As illustrated in Figure 1.11, a ***test pattern generator*** (TPG) is used to automatically supply the internally generated test patterns to the ***circuit under test*** (CUT), and an ***output response analyzer*** (ORA) is used to compact the output responses from the CUT [Stroud 2002]. Because the test circuitry resides with the CUT, BIST can be used at all levels of testing from wafer through system level testing. BIST is typically applied on the basis of the type of circuit under test. For example, scan-based BIST approaches are commonly used for general sequential logic (often called **logic BIST**); more algorithmic BIST approaches are used for regular structures such as memories (often called **memory BIST**). Because of the complexity of current VLSI devices that can include ***analog and mixed-signal*** (AMS) circuits, as well as hundreds of memories, BIST implementations are becoming an essential part of both system and test requirements [Wang 2006, 2007].

Test compression can be considered as a supplement to scan design and is commonly used to reduce the amount of test data (both input stimuli and output responses) that must be stored on the ***automatic test equipment*** (ATE) [Touba 2006]. Reduction in test data volume and test application time by 10× or more can be achieved. This is typically done by including a decompressor before the m scan chain inputs of the CUT to decompress the compressed input stimuli and a compactor after the *m* scan chain outputs of the CUT to compact output responses, as illustrated in Figure 1.12. The compressed input stimulus and compacted output response are each connected to *n* tester channels on the ATE, where *n* < *m* and *n* is typically at least 10× smaller than *m*. Modern test synthesis tools can now directly incorporate these test compression features into either an RTL design or a gate-level design as will be discussed in more detail in Chapter 3.

1.3.3 **Fault simulation and test generation**

The mechanics of testing for fault simulation, as illustrated in Figure 1.13, are similar at all levels of testing, including design verification. First, a set of target faults (fault list) based on the CUT is enumerated. Often, **fault collapsing** is applied to the enumerated fault set to produce a collapsed fault set to reduce fault simulation or fault grading time. Then, input stimuli are applied to the CUT, and the output responses are compared with the expected fault-free responses to determine whether the circuit is faulty. For fault simulation, the CUT is typically synthesized down to a gate-level design (or circuit netlist).

Ensuring that sufficient design verification has been obtained is a difficult step for the designer. Although the ultimate determination is whether or not the design works in the system, fault simulation, illustrated in Figure 1.13, can provide a rough quantitative measure of the level of design verification much earlier in the design process. Fault simulation also provides valuable information

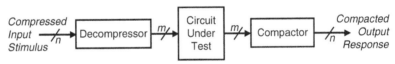

FIGURE 1.12

Test compression architecture.

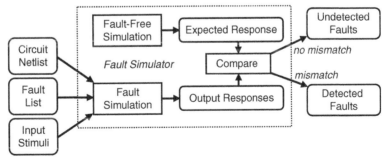

FIGURE 1.13

Fault simulation.

on portions of the design that need further design verification, because design verification vectors are often used as functional vectors (called **functional testing**) during manufacturing test.

Test development consists of selecting specific test patterns on the basis of circuit structural information and a set of **fault models**. This approach, called **structural testing**, saves test time and improves test efficiency, because the total number of test patterns is largely decreased since the test patterns target specific faults that would result from defects in the manufactured circuit. Structural testing cannot guarantee detection of all possible manufacturing defects, because the test patterns are generated on the basis of specific fault models. However, fault models provide a quantitative measure of the fault detection capabilities for a given set of test patterns for the targeted fault model; this measure is called **fault coverage** and is defined as:

$$fault\ coverage\ =\ \frac{number\ of\ detected\ faults}{total\ number\ of\ faults}$$

Any input pattern, or sequence of input patterns, that produces a different output response in a faulty circuit from that of the fault-free circuit is a test pattern, or sequence of test patterns, which will detect the fault. Therefore, the goal of **automatic test pattern generation** (ATPG) is to find a set of test patterns that detects all faults considered for that circuit. Because a given set of test patterns is usually capable of detecting many faults in a circuit, **fault simulation** is typically used to evaluate the fault coverage obtained by that set of test patterns. As a result, fault models are needed for fault simulation and for ATPG.

1.3.4 **Manufacturing test**

The tester, also referred to as the **automatic test equipment** (ATE), applies the functional test vectors and structural test patterns to the fabricated circuit and compares the output responses with the expected responses obtained from the design verification simulation environment for the fault-free (and hopefully, design error-free) circuit. A "faulty" circuit is now considered to be a circuit with manufacturing defects.

Some percentage of the manufactured devices, boards, and systems is expected to be faulty because of manufacturing defects. As a result, testing is required during the manufacturing process in an effort to find and eliminate those defective parts. The **yield** of a manufacturing process is defined as the percentage of acceptable parts among all parts that are fabricated:

$$yield\ =\ \frac{number\ of\ acceptable\ parts}{total\ number\ of\ parts\ fabricated}$$

A **fault** is a representation of a defect reflecting a physical condition that causes a circuit to fail to perform in a required manner. When devices or electronic systems are tested, the following two undesirable situations may occur: (1) a faulty circuit appears to be a good part passing the test, or (2) a good circuit fails the test and appears as faulty. These two outcomes are often due to a poorly designed test or the lack of DFT. As a result of the first case, even if all products pass the manufacturing test, some faulty devices will still be found in the manufactured electronic system. When these faulty circuits are returned to the manufacturer, they undergo *failure mode analysis* (FMA) or fault diagnosis for possible improvements to the manufacturing process [Wang 2006]. The ratio of field-rejected parts to all parts passing quality assurance testing is referred to as the **reject rate**, also called the **defect level**:

$$reject\ rate\ =\ \frac{number\ of\ faulty\ parts\ passing\ final\ test}{total\ number\ of\ parts\ passing\ final\ test}$$

Because of unavoidable statistical flaws in the materials and masks used to fabricate the devices, it is impossible for 100% of any particular kind of device to be defect free. Thus, the first testing performed during the manufacturing process is to test the devices fabricated on the wafer to determine which devices are defective. The chips that pass the wafer-level test are extracted and packaged. The packaged devices are retested to eliminate those devices that may have been damaged during the packaging process or put into defective packages. Additional testing is used to ensure the final quality before shipping to customers. This final testing includes measurement of parameters such as input/output timing specifications, voltage, and current. In addition, **burn-in** or **stress testing** is often performed when chips are subject to high temperature and supply voltage. The purpose of burn-in testing is to accelerate the effect of defects that could lead to failures in the early stages of operation of the device. FMA is typically used at all stages of the manufacturing test to identify improvements to processes that will result in an increase in the number of defect-free electronic devices and systems produced.

In the case of a VLSI device, the chip may be discarded or it may be investigated by FMA for yield enhancement. In the case of a PCB, FMA may be performed for yield enhancement or the board may undergo further testing for fault location and repair. A "good" circuit is assumed to be defect free, but this assumption is only as good as the quality of the tests being applied to the manufactured design. Once again, fault simulation provides a quantitative measure of the quality of a given set of tests.

1.4 PHYSICAL DESIGN AUTOMATION

Physical design refers to all synthesis steps that convert a circuit representation (in terms of gates and transistors) into a geometric representation (in terms of polygons and their shapes) [Sherwani 1999; Chang 2007]. An example is illustrated in

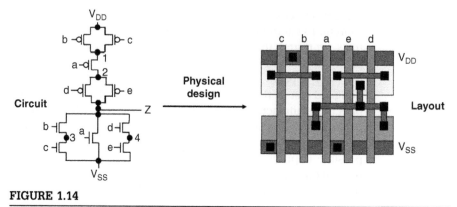

FIGURE 1.14

The function of physical design.

Figure 1.14. The geometric representation, also called **layout**, is used to design masks and then manufacture a chip. Because the design process is fairly complicated in nature, modern physical design typically is divided into three major steps: (1) floorplanning, (2) placement, and (3) routing. **Floorplanning** is an essential design step for a hierarchical, building block design method. It assembles circuit blocks into a rectangle (chip) to optimize a predefined cost metric such as area and wire length. The circuit blocks could be flexible or rigid in their shapes. **Placement** is the process of assigning the circuit components into a chip region. It can be considered as a restricted floorplanning problem for rigid blocks with some dimension similarity. After placement, the **routing** process defines the precise paths for conductors that carry electrical signals on the chip layout to interconnect all pins that are electrically equivalent. After routing, some **physical verification** processes (such as *design rule checking* [DRC]), **performance checking**, and **reliability checking**) are performed to verify whether all geometric patterns, circuit timing, and electrical effects satisfy the design rules and specifications.

As design and process technologies advance at a breathtaking speed, feature size and voltage levels associated with modern VLSI designs are decreasing drastically while at the same time die size, operating frequency, design complexity, and packing density keep increasing. Physical design for such a system must consider the integration of large-scale digital and *analog and mixed-signal* (AMS) circuit blocks, the design of system interconnections/buses, and the optimization of circuit performance, area, power consumption, and signal and power integrity. On one hand, designs with more than a billion transistors are already in production, and functional blocks are widely reused in nanometer circuit design, which all drive the need for a modern physical design tool to handle large-scale designs. On the other hand, the highly competitive IC market requires faster design convergence, faster incremental design turnaround, and better silicon area utilization. Efficient and effective design methods and tools capable of optimizing large-scale circuits are essential for modern VLSI physical designs.

1.4.1 **Floorplanning**

Floorplanning is typically considered the first stage of VLSI physical design. Given a set of **hard blocks** (whose shapes cannot be changed) and/or **soft blocks** (whose shapes can be adjusted) and a netlist, floorplanning determines the shapes of soft blocks and assembles the blocks into a rectangle (chip) so a predefined cost metric (such as the chip area, wire length, wire congestion) is optimized [Sait 1999; Chen 2006]. See Figure 1.15 for the floorplan of the Intel Pentium 4 microprocessor.

Floorplanning gives early feedback that suggests architectural modifications, estimates the chip area, and estimates delay and congestion caused by wiring [Gerez 1998]. As technology advances, designs with more than a billion transistors are already in production. To cope with the increasing design complexity, hierarchical design and functional blocks are widely used. This trend makes floorplanning much more critical to the quality of a VLSI design than ever. Therefore, efficient and effective floorplanning methods and tools are desirable for modern circuit designs.

1.4.2 **Placement**

Placement is the process of assigning the circuit components into a chip region. Given a set of fixed cells/macros, a netlist, and a chip outline, placement assigns the predesigned cells/macros to positions on the chip so that no two cells/macros overlap with each other (*i.e.*, legalization) and some cost functions (*e.g.*, wire length, congestion, and timing) are optimized [Nam 2007; Chen 2008].

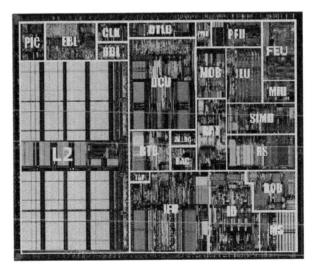

FIGURE 1.15

Floorplan of the Intel Pentium 4 microprocessor. (Courtesy of Intel Corporation.)

The traditional placement problem seeks to minimize wire length under the constraint that cells/macros do not overlap with each other. Two major challenges arise because of this high complexity for modern circuit design. First, the predesigned macro blocks (such as embedded memories, analog blocks, predesigned data paths) are often reused, and thus many designs contain hundreds of macro blocks and millions of cells. See Figure 1.16 for two example placements with large-scale cells and macros of very different sizes. Second, timing and routability (congestion) optimization become more challenging because of the design complexity and the scaling of devices and interconnects. As a result, modern design challenges have reshaped the placement problem. The modern placement problem becomes very hard, because we need to handle large-scale designs with millions of objects. Furthermore, the objects could be very different in their sizes. In addition to wire length, we also need to consider many placement constraints such as timing, routability (congestion), and thermal issues.

1.4.3 Routing

After placement, routing defines the precise paths for conductors that carry electrical signals on the chip layout to interconnect all pins that are electrically equivalent. See Figure 1.17 for a two-layer routing example [Chang 2004]. After routing, some physical verification processes (such as design rule checking, performance checking, and reliability checking) are performed to verify whether all geometric patterns, circuit timing, and electrical effects satisfy the design rules and specifications.

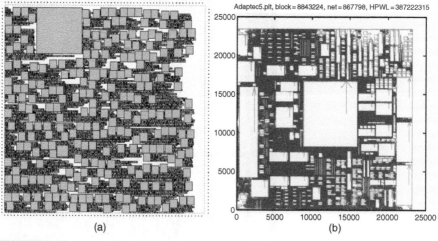

(a)   (b)

FIGURE 1.16

Two IBM placement examples: (a) The ibm01 circuit with 12,752 cells and 247 macros. (b) The adapetc5 circuit with 842 K cells, 646 macros, and 868 K nets.

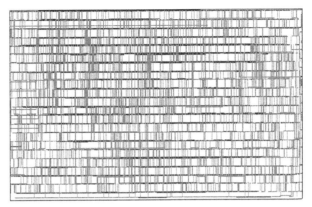

FIGURE 1.17

A two-layer routing example with 8109 nets. All horizontal wires are routed on one layer, and so are vertical ones.

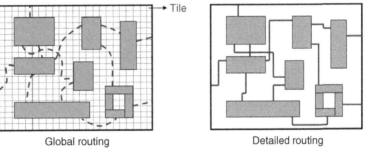

Global routing   Detailed routing

FIGURE 1.18

Global routing and detailed routing.

Typically, routing is a very complex problem. To make it manageable, a traditional routing system usually uses the two-stage technique of **global routing** followed by **detailed routing**. Global routing first partitions the entire routing region into tiles (or channels) and decides tile-to-tile paths for all nets while attempting to optimize some specified objective functions (*e.g.*, the total wire length and the critical timing constraints). Then, guided by the results of global routing, detailed routing determines actual tracks and routes for all nets according to the design rules. See Figure 1.18 for an illustration of the global and detailed routing [Ho 2007].

1.4.4 Synthesis of clock and power/ground networks

The specifications for clock and power/ground nets are significantly different from those for general signal nets. Generic routers cannot handle the requirements associated with clock and power/ground nets well. For example, we often need to synchronize the arrivals of the clock signals at all functional units for clock nets and minimize the IR (voltage) drops while satisfying the current density (electromigration) constraint for power/ground nets. As a result, it is desirable to develop specialized algorithms for routing such nets.

Two strategies are used to implement a digital system: synchronous and asynchronous systems. In a typical synchronous system, data transfer among circuit components is controlled by a highly precise clock signal. In contrast, an asynchronous system usually applies a data signal to achieve the communication for data transfer. The synchronous system dominates the on-chip circuit designs mainly because of its simplicity in chip implementation and easy debugging. Nevertheless, the realization and performance of the synchronous system highly rely on a network to transmit the clock signals to all circuit components that need to be synchronized for operations (*e.g.*, triggered with a rising edge of the clock signal). Ideally, the clock signals should arrive at all circuit components simultaneously so that the circuit components can operate and data can be transferred at the same time. In reality, however, the clock signals might not reach all circuit components at the same time. The maximum difference in the arrival times of the clock signals at the circuit components, referred to as **clock skew**, should be minimized to avoid the idleness of the component with an earlier clock signal arrival time. The smaller the clock skew, the faster the clock. Consequently, a **clock-net synthesis** problem arises from such a synchronous system: routing clock nets to minimize the clock skew (preferably zero) and delay [Tsay 1993]. More sophisticated synchronous systems might intentionally schedule nonzero clock skew to further reduce the clock period, called **useful clock skew**. More information can be found in Chapter 13. There are also some other important design issues for clock-net synthesis, for example, total wire length and power consumption optimization.

Example 1.1   Figure 1.19 shows two clock networks. The clock network in Figure 1.19a incurs a skew of 16 units and the maximum delay of 30 units, whereas the clock network in Figure 1.19b has zero clock skew and the same delay as that in Figure 1.19a.

For modern circuit design, the power and ground networks are usually laid out on metal layers to reduce the resistance of the networks. See Figure 1.20 for a popular two-layer meshlike power/ground network, in which parallel vertical power (V_{DD}) and ground (GND) lines run on the metal-4 layer, connected by horizontal power and ground lines on the metal-5 layer. All the blocks that need power supply or need to be connected to ground can thus connect to the appropriate power and ground lines.

The power and ground lines are typically much wider than signal nets because they need to carry much larger amounts of current. Therefore, we need to consider the wire widths of power/ground networks for the area requirement. As technology advances, the metal width decreases while the global wire length increases. This trend makes the resistance of the power line

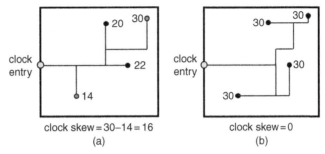

FIGURE 1.19

Two clock networks: (a) Clock network with a skew of 16 units and the maximum delay of 30 units. (b) Clock network with zero skew and 30-unit delay.

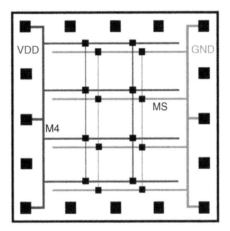

FIGURE 1.20

A typical power/ground network.

increase substantially. Furthermore, the threshold voltage scales nonlinearly, raising the ratio of the threshold voltage to the supply voltage and making the voltage drop in the power/ground network a serious challenge in modern circuit design. Because of the voltage drop, supply voltage in logic may not be an ideal reference. This effect may weaken the driving capability of logic gates, reduce circuit performance, slow down slew rate (and thus increase power consumption), and lower noise margin. As a result, power/ground network synthesis attempts to use the minimum amount of wiring area for a power/ground network under the power-integrity constraints such as voltage drops and electromigration. There are two major tasks for the synthesis: (1) power/ground network topology determination to plan the wiring topology of a power/ground network and (2) power/ground wire sizing to meet the current density and reliability constraints [Sait 1999; Sherwani 1999; Tan 2003].

Example 1.2 Figure 1.21a shows a chip floorplan of four modules and the power/ground network. As shown in the figure, we refer to a pad feeding supply voltage into the chip as a power pad, the power line enclosing the floorplan as a core ring, a power line branching from a core ring into modules inside as a power trunk, and a pin in a module that absorbs current (connects to a core ring or a power trunk) as a P/G pin. To ensure correct and reliable logic operation, we will minimize the voltage drops from the power pad to the P/G pins in a power/ground network. Figure 1.21a shows an instance of voltage drop in the power supply line, in which the voltage drops by almost 26% at the rightmost P/G pin. Figure 1.21b shows that by having a different chip floorplan, the worst-case voltage drop is reduced to approximately 5% [Liu 2007]. Recent research showed that a 5% voltage drop in supply voltage might slow down circuit performance by as much as 15% or more [Yim 1999]. Furthermore, it is typical to limit the voltage drop within 10% of the supply voltage to guarantee proper circuit operation. Therefore, voltage drop is a first-order effect and can no longer be ignored during the design process.

1.5 CONCLUDING REMARKS

The sophistication and complexity of current electronic systems, including **printed circuit boards** (PCBs) and **integrated circuits** (ICs), are a direct result of **electronic design automation** (EDA). Conversely, EDA is highly dependent on the power and performance of ICs, such as microprocessors and RAMs used to construct the computers on which the EDA software is executed. As a result, EDA is used to develop the next generation of ICs, which, in turn, are used to develop and execute the next generation of EDA, and so on in an ever-advancing progression of features and capabilities.

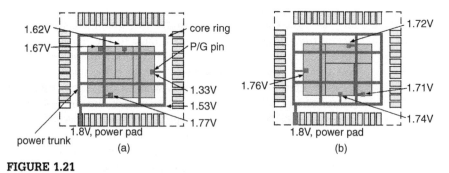

FIGURE 1.21

Two floorplans with associated power/ground network structures: (a) Worst-case voltage drop at the P/G pins approximately 26% of the supply voltage. (b) Worst-case voltage drop approximately only 5% [Liu 2007].

The current drivers for EDA include such factors as manufacturing volume, die size, integration heterogeneity, and increasing complexity [SIA 2005]. The primary influencing factors include **_system-on-chips_** (SOCs), microprocessors, **_analog/mixed-signal_** (AMS) circuits, and embedded memories, as well as continuing increases in both silicon and system complexity. Silicon complexity results from process scaling and introduction of new materials and device/interconnect structures. System complexity results from increasing transistor counts produced by the smaller feature sizes and demands for increased functionality, lower cost, and shorter time-to-market. Collectively, these factors and influences create major EDA challenges in the areas of design and verification productivity, power management and delivery, manufacturability, and manufacturing test, as well as product reliability [SIA 2005]. Other related challenges include higher levels of abstraction for ESL design, AMS codesign and automation, parametric yield at volume production, reuse and test of **_intellectual property_** (IP) cores in heterogeneous SOCs, cost-driven design optimization, embedded software design, and design process management. The purpose of this book is to describe more thoroughly the traditional and evolving techniques currently used to address these EDA challenges [SIA 2006, 2007].

The remaining chapters provide more detailed discussions of these topics. For example, general CMOS design techniques and issues are presented in Chapter 2 and fundamental design for testability techniques for producing quality CMOS designs are provided in Chapter 3. Most aspects of EDA in synthesis (including high-level synthesis, logic synthesis, test synthesis, and physical design), verification, and test, rely heavily on various algorithms related to the specific task at hand. These algorithms are described in Chapter 4. Modeling of a design at the _electronic system level_ (ESL) and synthesis of the ESL design to the high level are first presented in Chapter 5. The design then goes through logic synthesis (Chapter 6) and test synthesis (Chapter 7) to generate a testable design at the gate level for further verification before physical design is performed. Design verification that deals with logic and circuit simulation is presented in Chapter 8, and functional verification is discussed in Chapter 9. The various aspects of physical design are addressed in Chapter 10 (floorplanning), Chapter 11 (placement), Chapter 12 (routing), and Chapter 13 (synthesis of clock and power/ground networks). Finally, logic testing that includes the most important fault simulation and test generation techniques to guarantee high product quality is discussed in Chapter 14 in detail.

1.6 EXERCISES

1.1. (Design Language) What are the two most popular _hardware description languages_ (HDLs) practiced in the industry?

1.2. (Synthesis) Synthesis often implies high-level synthesis, logic synthesis, and physical synthesis. State their differences.

1.3. (Verification) Give three verification approaches that can be used to verify the correctness of a design. State the differences between model checking and equivalence checking.

1.4. (Fault Model) Assume a circuit has a total of n input and output nodes. How many single stuck-at faults, dominant bridging faults, 4-way bridging faults, and multiple stuck-at faults are present in the circuit?

1.5. (Design for Testability) Assume a sequential circuit contains n flip-flops and each state is accessible from an initial state in m clock cycles. If a sequential ATPG is used and p test patterns are required to detect all single stuck-at faults in the design, how many clock cycles would be required to load the sequential circuit with predetermined states? If all flip-flops have been converted to scan flip-flops and stitched together to form one scan chain, how many clock cycles would be required to load the combinational circuit with predetermined states?

1.6. (Testing) State the differences between fault simulation and test generation. Give three main reasons each why sequential test generation is difficult and why the industry widely adopts scan designs.

1.7. (Design Flow) As technology advances, interconnects dominate the circuit performance. When are the interconnect issues handled during the traditional VLSI design flow? How can we modify the design flow to better tackle the interconnect issues?

1.8. (Clock-net Synthesis) Give the clock entry point p_0 located at the coordinate $(3, 0)$ and four clock pins p_1, p_2, p_3, and p_4 located at $(1, 1)$, $(5, 1)$, $(1, 5)$, and $(5, 5)$, respectively. Assume that the delay is proportional to the path length and the wire can run only on the grid lines. Show how to interconnect the clock entry point p_0 to the other four clock pins p_i, $1 \leq i \leq 4$, such that the clock skew is zero and the clock delay is minimized. What is the resulting clock delay?

1.9. (Programmable Logic Array) A shorthand notation commonly used for *programmable logic arrays* (PLAs) and combinational logic in *programmable logic devices* (PLDs) is illustrated in Figure 1.22, which

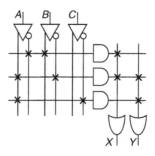

FIGURE 1.22

Shorthand notation for the connection array in Figure 1.6 and the PLA implementation in Figure 1.7.

corresponds to the connection array in Figure 1.6 and the PLA implementation in Figure 1.7. Give the connection array and draw the PLA shorthand diagram and PLA transistor-level implementation for the following set of Boolean equations, sharing product terms where possible:

$$O_2 = I_3 \bullet I_2$$
$$O_1 = I_3 \bullet I_2 \bullet I_1' \bullet I_0' + I_3' \bullet I_2' \bullet I_1 \bullet I_0$$
$$O_0 = I_3 \bullet I_2 + I_2 \bullet I_1 + I_1 \bullet I_0 + I_3 \bullet I_0$$

ACKNOWLEDGMENTS

We wish to thank Professor Ren-Song Tsay of National Tsing Hua University, Professor Jie-Hong (Roland) Jiang of National Taiwan University, and Professor Jianwen Zhu of University of Toronto for reviewing the Logic Design Automation section; Professor Wen-Ben Jone of University of Cincinnati for reviewing the Test Automation section; and Professor James C.-M. Li, Wen-Chi Chao, Po-Sen Huang, and Tzro-Fan Chien of National Taiwan University for reviewing the manuscript and providing very helpful comments.

REFERENCES
R1.0 Books

[Abramovici 1994] M. Abramovici, M. A. Breuer, and A. D. Friedman, *Digital Systems Testing and Testable Design*, IEEE Press, Revised Printing, Piscataway, NJ, 1994.

[Brayton 1984] R. Brayton, G. Hachtel, C. McMullen, and A. Sangiovanni-Vincentelli, *Logic Minimization Algorithms for VLSI Synthesis*, Kluwer Academic, Boston, 1984.

[Bushnell 2000] M. L. Bushnell and V. D. Agrawal, Essentials of Electronic Testing for Digital, Memory & Mixed-Signal VLSI Circuits, Springer, Boston, 2000.

[De Micheli 1994] G. De Micheli, *Synthesis and Optimization of Digital Circuits*, McGraw-Hill, New York, 1994.

[Devadas 1994] S. Devadas, A. Ghosh, and K. Keutzer, *Logic Synthesis*, McGraw-Hill, New York, 1994.

[Dutton 1993] R. Dutton and Z. Yu, *Technology CAD: Computer Simulation of IC Processes and Devices*, Kluwer Academic, Boston, 1993.

[Gerez 1998] S. Gerez, *Algorithms for VLSI Design Automation*, John Wiley & Sons, Chichester, England, 1998.

[Ho 2007] T.-Y. Ho, Y.-W. Chang, and S.-J. Chen, *Full-Chip Nanometer Routing Techniques*, Springer, New York, 2007.

[IEEE 1076-2002] *IEEE Standard VHDL Language Reference Manual*, IEEE, Std. 1076-2002, IEEE, New York, 2002.

[IEEE 1463-2001] *IEEE Standard Description Language Based on the Verilog Hardware Description Language*, IEEE, Std. 1463-2001, IEEE, New York, 2001.

[Jha 2003] N. Jha and S. Gupta, *Testing of Digital Systems*, Cambridge University Press, London, 2003.

[Keating 1999] M. Keating and P. Bricaud, *Reuse Methodology Manual for System-on-a-Chip Designs,* Springer, Boston, 1999.

[Ledgard 1983] H. Ledgard, *Reference Manual for the ADA Programming Language,* Springer, Boston, 1983.

[McCluskey 1986] E. J. McCluskey, *Logic Design Principles: With Emphasis on Testable Semiconductor Circuits,* Prentice-Hall, Englewood Cliffs, NJ, 1986.

[Mead 1980] C. Mead and L. Conway, *Physical Design Automation of VLSI Systems,* Addison Wesley, Reading, MA, 1980.

[Nam 2007] G.-J. Nam and J. Cong, editors, *Modern Circuit Placement: Best Practices and Results,* Springer, Boston, 2007.

[Plummer 2000] J. D. Plummer, M. Deal, and P. Griffin, *Silicon VLSI Technology–Fundamentals, Practice and Modeling,* Prentice-Hall, Englewood Cliffs, NJ, 2000.

[Preas 1988] B. Preas and M. Lorenzetti, *Physical Design Automation of VLSI Systems,* Benjamin/ Cummings, Menlo Park, CA, 1997.

[Sait 1999] S. Sait and H. Youssef, *VLSI Physical Design Automation: Theory and Practice,* World Scientific Publishing Company, 1999.

[Scheffer 2006a] L. Scheffer, L. Lavagno, and G. Martin, editors, *EDA for IC System Design, Verification, and Testing,* CRC Press, Boca Raton, FL, 2006.

[Scheffer 2006b] L. Scheffer, L. Lavagno, and G. Martin, editors, *EDA for IC Implementation, Circuit Design, and Process Technology,* CRC Press, Boca Raton, FL, 2006.

[Sherwani 1999] N. Sherwani, *Algorithms for VLSI Physical Design Automation,* 3rd Ed., Kluwer Academic, Boston, 1999.

[Stroud 2002] C. Stroud, *A Designer's Guide to Built-In Self-Test,* Springer, Boston, 2002.

[Wang 2006] L.-T. Wang, C.-W. Wu, and X. Wen, editors, *VLSI Test Principles and Architectures: Design for Testability,* Morgan Kaufmann, San Francisco, 2006.

[Wang 2007] L.-T. Wang, C. Stroud, and N. Touba, editors, *System-on-Chip Test Architectures: Nanometer Design for Testability,* Morgan Kaufmann, San Francisco, 2007.

[Wile 2005] B. Wile, J. Goss, and W. Roesner, *Comprehensive Functional Verification,* Morgan Kaufmann, San Francisco, 2005.

R1.1 Overview of Electronic Design Automation

[Cadence 2008] Cadence Design Systems, http://www.cadence.com, 2008.

[DAC 2008] Design Automation Conference, co-sponsored by Association for Computing Machinery (ACM) and Institute of Electronics and Electrical Engineers (IEEE), http://www.dac.com, 2008.

[Kilby 1958] J. Kilby, Integrated circuits invented by Jack Kilby, Texas Instruments, Dallas, TX, http://www.ti.com/corp/docs/company/history/timeline/semicon/1950/docs/58ic_kilby.htm, September 12, 1958.

[Mentor 2008] Mentor Graphics, http://www.mentor.com, 2008.

[MOSIS 2008] The MOSIS Service, http://www.mosis.com, 2008.

[Naffziger 2006] S. Naffziger, B. Stackhouse, T. Grutkowski, D. Josephson, J. Desai, E. Alon, and M. Horowitz, The implementation of a 2-core multi-threaded Itanium family processor, *IEEE J. of Solid-State Circuits Conf.,* 41(1), pp. 197–209, January 2006.

[Ochetta 1994] E. Ochetta, R. Rutenbar, and L. Carley, ASTRX/OBLX: Tools for rapid synthesis of high-performance analog circuits, in *Proc. ACM/IEEE Design Automation Conf.,* pp. 24–30, June 1994.

[SIA 2005] SIA, *The International Technology Roadmap for Semiconductors: 2005 Edition,* Semiconductor Industry Association, San Jose, CA, http://public.itrs.net, 2005.

[SIA 2006] SIA, *The International Technology Roadmap for Semiconductors: 2006 Update,* Semiconductor Industry Association, San Jose, CA, http://public.itrs.net, 2006.

[Stackhouse 2008] B. Stackhouse, B. Cherkauer, M. Gowan, P. Gronowski, and C. Lyles, A 65nm 2-billion-transistor quad-core Itanium processor, in *Digest of Papers, IEEE Int. Solid-State Circuits Conf.*, pp. 92, February 2008.

[Stroud 1986] C. Stroud, R. Munoz, and D. Pierce, CONES: A system for automated synthesis of VLSI and programmable logic from behavioral models, in *Proc. IEEE/ACM Int. Conf. on Computer-Aided Design*, pp. 428-431, November 1986.

[Synopsys 2008] Synopsys, http://www.synopsys.com, 2008.

[SystemC 2008] SystemC, http://www.systemc.org, 2008.

[SystemVerilog 2008] SystemVerilog, http://systemverilog.org, 2008.

R1.2 Logic Design Automation

[Velev 2001] M. N. Velev and R. Bryant, Effective use of Boolean satisfiability procedures in the formal verification of scalar and VLIW microprocessors, in *Proc. ACM/IEEE Design Automation Conf.*, pp. 226-231, June 2001.

R1.3 Test Automation

[Eichelberger 1978] E. Eichelberger and T. Williams, A logic design structure for LSI testability, *J. of Design Automation and Fault-Tolerant Computing*, 2(2), pp. 165-178, February 1978.

[Sato 2005] Y. Sato, S. Hamada, T. Maeda, A. Takatori, Y. Nozuyama, and S. Kajihara, Invisible delay quality–SDQM model lights up what could not be seen, in *Proc. IEEE Int. Test Conf.*, Paper 47.1, November 2005.

[Touba 2006] N. A. Touba, Survey of test vector compression techniques, *IEEE Design & Test of Computers*, 23(4), pp. 294-303, July-August 2006.

[Williams 1983] T. Williams and K. Parker, Design for testability—A survey, *Proceedings of the IEEE*, 71(1), pp. 98-112, January 1983.

R1.4 Physical Design Automation

[Chang 2004] Y.-W. Chang and S.-P. Lin, MR: A new framework for multilevel full-chip routing, *IEEE Trans. on Computer-Aided Design*, 23(5), pp. 793-800, May 2004.

[Chang 2007] Y.-W. Chang, T.-C. Chen, and H.-Y. Chen, Physical design for system-on-a-chip, in *Essential Issues in SOC Design*, Y.-L. Lin, editor, Springer, Boston, 2007.

[Chen 2006] T.-C. Chen and Y.-W. Chang, Modern floorplanning based on B*-trees and fast simulated annealing, *IEEE Trans. on Computer-Aided Design*, 25(4), pp. 637-650, April 2006.

[Chen 2008] T.-C. Chen, Z.-W. Jiang, T.-C. Hsu, H.-C. Chen, and Y.-W. Chang, NTUplace3: An analytical placer for large-scale mixed-size designs with preplaced blocks and density constraints, *IEEE Trans. on Computer-Aided Design*, 27(7), pp. 1228-1240, July 2008.

[Liu 2007] C.-W. Liu and Y.-W. Chang, Power/ground network and floorplan co-synthesis for fast design convergence, *IEEE Trans. on Computer-Aided Design*, 26(4), pp. 693-704, April 2007.

[Tan 2003] S. X-D. Tan and C.-J. R. Shi, Efficient very large scale integration power/ground network sizing based on equivalent circuit modeling, *IEEE Trans. on Computer-Aided Design*, 22(3), pp. 277-284, March 2003.

[Tsay 1993] R.-S. Tsay, An exact zero-skew clock routing algorithm, *IEEE Trans. on Computer-Aided Design*, 12(2), pp. 242-249, February 1993.

[Yim 1999] J. S. Yim, S. O. Bae, and C. M. Kyung, A Floorplan-based planning methodology for power and clock distribution in ASICs, in *Proc. ACM/IEEE Design Automation Conf.*, 766-771, June 1999.

R1.5 Concluding Remarks

[SIA 2005] SIA, *The International Technology Roadmap for Semiconductors: 2005 Edition*, Semiconductor Industry Association, San Jose, CA, http://public.itrs.net, 2005.

[SIA 2006] SIA, *The International Technology Roadmap for Semiconductors: 2006 Update*, Semiconductor Industry Association, San Jose, CA, http://public.itrs.net, 2006.

[SIA 2007] SIA, *The International Technology Roadmap for Semiconductors: 2007 Edition*, Semiconductor Industry Association, San Jose, CA, http://public.itrs.net, 2007.

Fundamentals of CMOS design

2

Xinghao Chen
CTC Technologies, Endwell, New York

Nur A. Touba
University of Texas, Austin, Texas

ABOUT THIS CHAPTER

The first ***integrated circuit*** (IC), called a ***phase shift oscillator*** composed of one transistor, one capacitor, and three resistors, was created by Jack Kilby of Texas Instruments on September 12, 1958. Today, a typical IC chip can easily contain several hundred millions of transistors and miles of interconnect wires. This ***very large-scale integration*** (VLSI) ability has been enabled by the modern use of the many ***electronic design automation*** (EDA) technologies and applications discussed in this book.

In this chapter, we discuss a few basic and very important concepts of ***complementary metal oxide semiconductor*** (CMOS) technology to aid in the learning process and facilitate greater understanding of the EDA subjects in the subsequent chapters. We first start with an overview of the fundamental integrated-circuit technology and CMOS logic design. Then, we discuss a few more advanced CMOS technologies that can be used to reduce transistor count, increase circuit speed, or reduce power consumption for modern VLSI designs. The physical design aspects, how to translate a CMOS logic design to a CMOS physical design for fabrication, is reviewed and included for completeness. For more in-depth study of specific CMOS technology areas, readers are referred to the various interesting topics thoroughly discussed in the references listed at the end of this chapter.

2.1 INTRODUCTION

The first ***integrated circuit*** (IC) was created by Jack Kilby of Texas Instruments on September 12, 1958. Called a **phase shift oscillator**, the integrated circuit consisted of only one transistor, one capacitor, and three resistors, as shown in Figure 2.1. Since then, IC technology has evolved from TTL (***transistor-transistor logic***) and nMOS to CMOS. Although CMOS was first introduced as an alternative to **39**

FIGURE 2.1

The first integrated circuit invented by Jack Kilby in 1950 (http://www.ti.com/corp/docs/company/history/timeline/semicon/1950/docs/58ic_kilby.htm, February 8, 2008. Courtesy of Texas Instruments.).

bipolar technologies (such as TTL and ECL), it soon overtook and became the dominant circuit implementation technology. This is because CMOS consumes much less power than TTL and nMOS, as well as the *very large-scale integration* (VLSI) capability it provides.

Now, with advanced CMOS process technologies, a chip can contain as many as 2 billion transistors (such as the Intel Quad-Core Itanium Processor, February 5, 2008). CMOS integrated circuits have been the primary digital system implementation technology for consumer electronics, personal, commercial, and enterprise computing systems, as well as electronic systems for scientific exploration.

However, the very large-scale integration ability of CMOS has also created problems that did not seem to be significant in the early days of CMOS technologies. We have seen more and more issues, such as power consumption, thermal effects, small delay defects, cost of test, and validation, dominating the agenda and schedule of a chip design project. Oftentimes, engineers have to make difficult tradeoffs to balance competing design parameters. Aside from providing the reader with fundamental CMOS design and layout principles, this chapter covers some advanced CMOS circuit technologies to assist the reader comprehend the learning process in designing modern VLSI circuits.

2.2 INTEGRATED CIRCUIT TECHNOLOGY

In this section, we first discuss the basic constructs and characteristics of a *metal oxide semiconductor* (MOS) transistor (*a.k.a.*, MOS device). Most transistors in digital circuits are switching devices that operate to perform desired Boolean functions. MOS transistors can also be configured as load devices that are used for circuit performance enhancements. Next, transistor equivalency is described, which is a widely used technique for analyzing large and complex circuits. We then discuss the wire and interconnects that connect the many transistors to form circuits and systems, followed by a discussion of the basic concepts related to noise margin, which is becoming ever more important in low-power applications.

2.2.1 **MOS transistor**

A MOS transistor is a 4-terminal device on a silicon substrate [Martin 2000]. Circuit schematic diagrams often show transistors in 3-terminal symbols, with the assumption that the fourth terminal (known as the substrate terminal) is either grounded or connected to power supply on the basis of the device type. Figure 2.2a shows the dimensions of a MOS transistor, where L is the n-channel length, W is the n-channel width, and t_{OX} is the thickness of the thin oxide layer under the gate. Figure 2.2b shows a cross-section view of a typical **n-channel transistor**. The three terminals of the devices are **Gate**, **Source**, and **Drain**. A fourth terminal connecting the **Substrate** is sometimes provided with devices as well. Common symbols used for n-channel and p-channel transistors are shown in Figure 2.3.

The switching characteristic of a MOS device is determined by its **threshold voltage**, denoted as V_{tn} for an n-channel transistor and V_{tp} for a p-channel transistor. When the effective gate-to-source voltage (V_{GS}) is greater than V_{tn}, a channel will form in a MOS transistor. For an n-channel device, this means $V_{eff} = V_{GS} - V_{tn} > 0$ and $V_{eff} = V_{SG} + V_{tp} > 0$ for a p-channel device, where

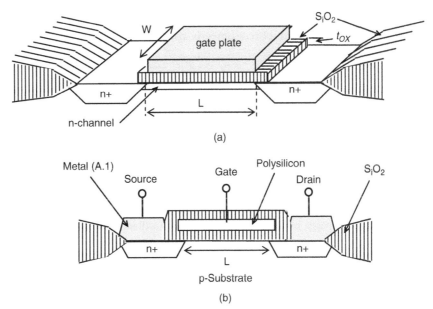

(a)

(b)

FIGURE 2.2

Illustrations of an n-channel transistor [Martin 2000]: (a) The dimensions of a MOS transistor. (b) A cross-section view of a MOS transistor.

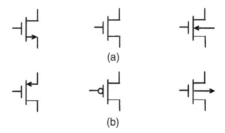

FIGURE 2.3

MOS transistor symbols: (a) For n-channel transistors. (b) For p-channel transistors.

typically $V_{tn} \approx 0.7V$ and $V_{tp} \approx -0.7V$. When the drain-to-source voltage (V_{DS}) is large, the channel current of an n-channel transistor is approximately

$$I_D = \mu_n \cdot C_{OX} \cdot \frac{W_n}{L_n} \cdot \left[(V_{GS} - V_{tn}) \cdot V_{DS} - \frac{V_{DS}^2}{2} \right] \tag{2.1}$$

(where $C_{OX} = \frac{\varepsilon_{OX}}{t_{OX}}$ is the gate-oxide capacitance) for $V_{DS} < V_{eff}$ and

$$I_D = \frac{\mu_n \cdot C_{OX}}{2} \cdot \frac{W_n}{L_n} \cdot (V_{GS} - V_{tn})^2 \tag{2.2}$$

for $V_{DS} > V_{eff}$. When V_{DS} is very small, the channel current is approximately

$$I_D = \mu_n \cdot C_{OX} \cdot \frac{W_n}{L_n} \cdot (V_{GS} - V_{tn}) \cdot V_{DS} \tag{2.3}$$

and the channel resistance is approximately

$$r_{ds} = \frac{V_{DS}}{I_D} \approx \frac{L_n}{\mu_n \cdot C_{OX} \cdot W_n \cdot (V_{GS} - V_{tn})} \tag{2.4}$$

Equations 2.1 and 2.2 are known as **large-signal equations**, whereas Equations 2.3 and 2.4 are known as **small-signal equations**. For p-channel devices, μ_n, W_n, L_n, V_{tn}, and V_{GS} in the preceding equations are replaced with μ_p, W_p, L_p, $V_{tp,}$ and V_{SG}, respectively. Note that the preceding equations assume the substrate to be zero-biased, where $V_{sb} = 0$. Considerations with body effect, channel-length modulation, and process variations, etc. can be found in the references with in-depth discussions.

With small V_{DS}, a MOS transistor's I_D is linearly related to V_{DS}. As V_{DS} increases beyond a certain value, I_D will start to tap off as illustrated in Figure 2.4. This means that a MOS transistor is essentially a nonlinear device.

Figure 2.5 illustrates the n-channel conditions with respect to V_{DS}. When voltage applied on the gate terminal is greater than V_{tn}, channel current I_D starts to flow between the drain and source terminals, as depicted in Figure 2.5a. When $V_{DG} >= -V_{tn}$, channel pinch-off takes place at the drain end, as depicted in Figure 2.5b.

There are several sources of capacitance within and in the periphery of a MOS transistor. Figure 2.6 illustrates their existences and notations. These capacitors are often known as **parasitic capacitors**, because their presence is due to the physical construction of the MOS device.

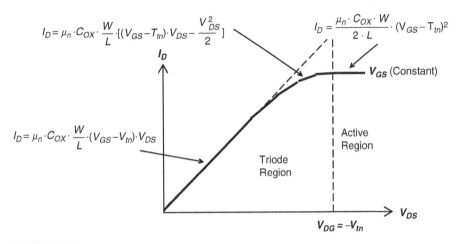

FIGURE 2.4

Nonlinear I_D *versus* V_{DS} relationship [Martin 2000].

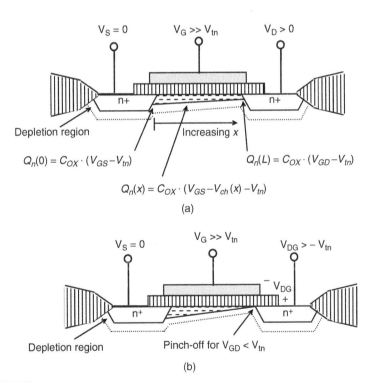

FIGURE 2.5

Illustration of n-channel conditions [Martin 2000]: (a) N-channel charge density. (b) N-channel pinch-off.

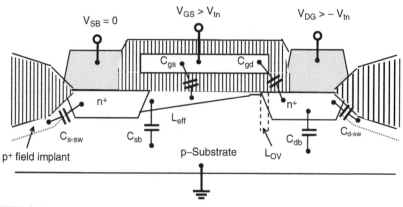

FIGURE 2.6

MOS device capacitance [Martin 2000].

It is worth noting that for IC engineering and manufacturing process control purposes, most transistors on the same chip are made with identical channel length. In addition, devices are often connected in parallel to form transistors having wider effective channels.

With nanometer technologies, process variations can affect the characteristics of individual transistors even on the same chips. We can no longer assume transistors on the same chip have the exact same threshold voltages. The ideal-case equations discussed in this section need to be adjusted to reflect process variation. We encourage readers to consult books on advanced CMOS modeling methods that take into account the effects of process variations.

2.2.2 **Transistor equivalency**

When a digital circuit uses many transistors, circuit analysis can get very complex and time-consuming. Transistor equivalency [Martin 2000] is a technique that simplifies larger circuits to smaller ones so that circuit analysis can be performed much more efficiently. The principles of transistor equivalency are illustrated in Figure 2.7. The first principle is **scaling**. When a MOS transistor's W and L are scaled by the same factor, as shown in Figure 2.7a, it has no effect on a first-order approximation. The second principle is called **parallel-connection equivalence**. When two MOS transistors T_1 and T_2 are connected in parallel, as shown in Figure 2.7b, the result is equivalent to a single transistor having the width equal to $W_1 + W_2$, with which

$$I_{D-eqv} = I_{D-1} + I_{D-2} = \frac{\mu \cdot C_{OX}}{2} \cdot \frac{W_1 + W_2}{L} \cdot (V_{GS} - V_t)^2 \qquad (2.5)$$

The third principle is called **serial-connection equivalence**, as depicted in Figure 2.7c, with which

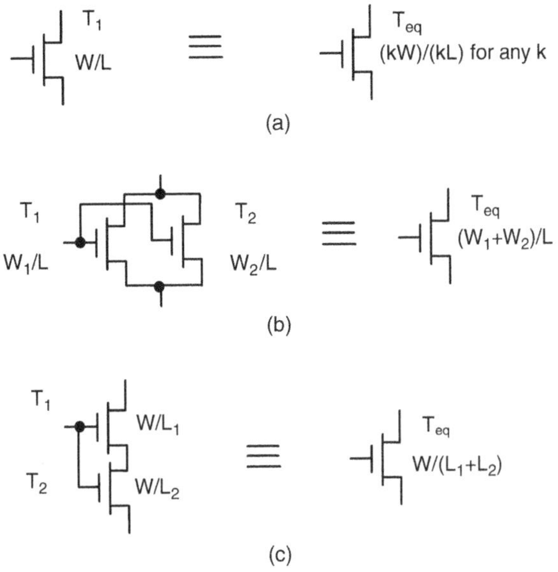

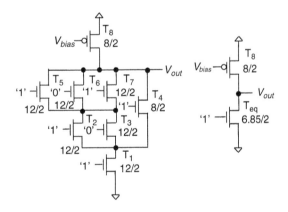

FIGURE 2.7

Illustration of transistor equivalency [Martin 2000]: (a) Scale equivalency. (b) Parallel-connection equivalency. (c) Serial-connection equivalency.

FIGURE 2.8

Application of transistor equivalency [Martin 2000].

$$I_{D-eqv} = I_{D-1} = I_{D-2} = \frac{\mu \cdot C_{OX}}{2} \cdot \frac{W}{L_1 + L_2} \cdot (V_{GS\_1} - V_t)^2 \quad (2.6)$$

Consider the circuit shown in Figure 2.8. It uses the classic **pseudo-nMOS** technology, with which a single p-channel transistor (set by a constant biasing voltage, V_{bias}) is used as the load, whereas the inputs determine the switching states of the n-channel transistors, which in turn determine the output of the circuit block. To apply transistor equivalency, the first step is to identify the n-channel transistors whose gate terminals are applied with "0" signals, because these transistors (T_3 and T_6, in this case) are set to the OFF state and can be ignored. Next, T_5 and T_7 are in parallel and are merged into a single one, T_5^*, with $W = 24/L = 2$. Because T_5^* and T_2 are in series, an equivalent transistor T_2^* can be determined by first scaling T_2 to $W = 24/L = 4$ and then computing T_2^* size as $W = 24/L = 6$. Repeat the same steps with T_4 followed by T_1. The resulting equivalent transistor, T_1^*, is to have the size $W = 6.857/L = 2$. The resulting equivalent circuit is much easier to analyze than the original circuit with the given inputs.

2.2.3 Wire and interconnect

With CMOS technologies scaling down to the nanometer arena, wires that connect transistors to each other are becoming a dominant factor in almost all aspects of IC manufacturing, ranging from complexity and timing to silicon area and yield. Advanced CMOS technologies today provide 9 to 11 metal layers in interconnect space. Many *application-specific integrated circuits* (ASICs) require at least 7 metal layers to connect transistors.

For a typical single wire, the *resistance-capacitance* (RC) effects are distributed along its length, as illustrated in Figure 2.9a. However, the lumped RC model, as illustrated in Figure 2.9b, is often used for circuit analysis. Figure 2.10 illustrates the RC tree network of a source driving a number of output branches (*a.k.a.* fanouts).

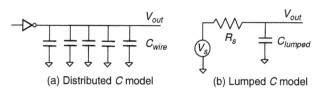

(a) Distributed C model   (b) Lumped C model

FIGURE 2.9

RC models for wire [Rabaey 2003].

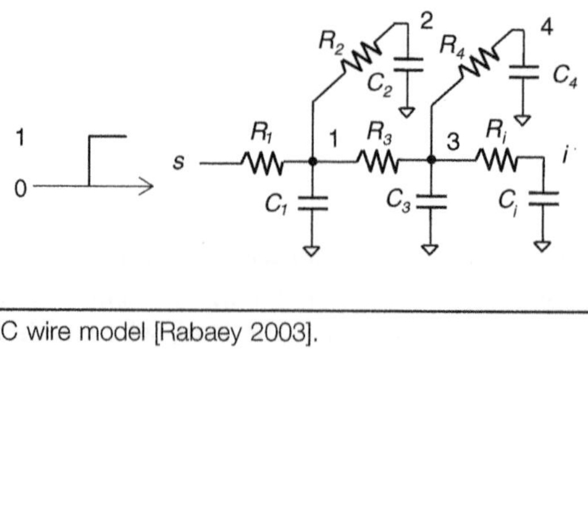

FIGURE 2.10

A tree-structured RC wire model [Rabaey 2003].

To calculate the RC effects between two nodes denoted as $\tau_{i,j}$ with i the source node and j the destination node, we have the following for the nodes in Figure 2.10:

$$\tau_{s,2} = C_1 \cdot R_1 + C_2 \cdot (R_1 + R_2) + (C_3 + C_4 + C_i) \cdot R_1$$
$$\tau_{s,4} = C_1 \cdot R_1 + C_2 \cdot R_1 + (C_3 + C_1) \cdot (R_1 + R_3) + C_4 \cdot (R_1 + R_3 + R_4)$$
$$\tau_{s,i} = C_1 \cdot R_1 + C_2 \cdot R_1 + (C_3 + C_4) \cdot (R_1 + R_3) + C_i \cdot (R_1 + R_3 + R_i)$$

As an exercise, readers are encouraged to figure out $\tau_{i,j}$ for other pairs of nodes.

In multilayer interconnect designs, wires placed in higher layers are usually wider and thicker than those in the lower layers, as illustrated in Figure 2.11, in which a six-metal layer hierarchy is depicted. This is to reduce resistance of long interconnects, because they are often placed in metal layers higher in the hierarchy. Lower metal layers are often reserved for shorter connections and for special purposes (such as distributing clocks). In addition, wires in higher layers are separated farther from each other to reduce coupling effects.

Coupling (inductive as well as capacitive) effects (*a.k.a.* **crosstalk**) between two or more parallel wires can affect signal integrity with unwanted circuit noise. **Coupling effects** also exist between wires on different layers. When long wires are placed in parallel next to each other, special care must be taken to reduce these effects.

Many of the IC routing technologies use two adjacent interconnect layers to complete one wiring. One layer would contain wires placed in North–South directions, and the other layer would contain wires placed in East–West directions. One advantage of this routing method is reduced interference between wires placed on adjacent layers. For this reason, wires on the two layers usually have the same width and thickness.

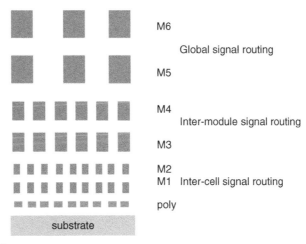

FIGURE 2.11

Multilayer interconnect hierarchy [Rabaey 2003].

2.2.4 **Noise margin**

Noise margin is a measure of design margins to ensure circuits functioning properly within specified conditions. Sources of noise include the operation environment, power supply, electric and magnetic fields, and radiation waves. On-chip transistor switching activity can also generate unwanted noise. To ensure that transistors switch properly under specified noisy conditions, circuits must be designed with specified **noise margins**.

Figure 2.12 illustrates noise margin and the terms, assuming that the signal generated by the driving device is wired to the input of the receiving device and that the wire is susceptible to noise. The minimum output voltage of the driving device for logic high, $V_{OH\ min}$, must be greater than the minimum input voltage, $V_{IH\ min}$, of the receiving device for logical high. Because of noise being induced on the wire, a logic high signal at the output of the driving device may arrive with lower voltage at the input of the receiving device. The noise margin, $NM_H = |V_{OH\ min} - V_{IH\ min}|$, for logical high is the range of tolerance for which a logical high signal can still be received correctly. The same can be said with noise margin, $NM_L = |V_{IL\ max} - V_{OL\ max}|$, for logical low, which specifies the range of tolerance for logical low signals on the wire. Smaller noise margins mean circuits are more sensitive to noise.

It is important to note that as CMOS technologies continue to advance, device feature size gets smaller, and channel length gets shorter. The miniaturization of transistors forces ever lower supply voltages, resulting in smaller noise margins. Table 2.1 shows the typical noise margin measurements with respect to technology advances.

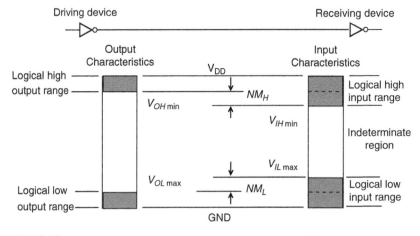

FIGURE 2.12

Noise margin and terms.

Table 2.1 Noise Margin Measures for Some Technologies [Wakerly 2001]

Technology	Noise-Margin Measures					
	V_{DD}	V_{OH}	V_{IH}	V_{TH}	V_{IL}	V_{OL}
5-V CMOS	5.0	4.44	3.5	2.5	1.5	0.5
5-V TTL	5.0	2.4	2.0	1.5	0.8	0.4
3.3-V LVTTL	3.3	2.4	2.0	1.5	0.8	0.4
2.5-V CMOS	2.5	2.0	1.7	1.2	0.7	0.4
1.8-V CMOS	1.8	1.45	1.2	0.9	0.65	0.45

2.3 CMOS LOGIC

In this section we highlight some CMOS circuit design principles. We first review the classic CMOS inverter, with which the major measurements are discussed. The principles are carried over to the design of elementary logic gates and complex circuit blocks. Next, we discuss the design of latches and flip-flops, followed by discussion of some simple circuit optimization techniques.

2.3.1 CMOS inverter and analysis

The CMOS inverter consists of a pair of p-channel and n-channel transistors, as shown in Figure 2.13. Unlike pseudo-nMOS circuits, the p-channel transistor in this CMOS inverter is also a switching device, always in a complement switching state of the n-channel transistor, as shown in the truth table in Figure 2.13. Timing characteristics of this CMOS inverter include three measurements: t_r as the **rise time** at the output, t_f as the **fall time**, and t_p as the **propagation time** (*a.k.a.* **delay**) between an input transition and the output response. Figure 2.14 illustrates these measurements in graphic form.

Note that t_r and t_f are measured graphically by the pair of 10% and 90% change points on the output transition curves. In practice, however, the two intersecting

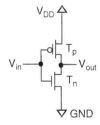

V_{in}	T_p	T_n	V_{out}
high	OFF	ON	low
low	ON	OFF	high

FIGURE 2.13

CMOS inverter and transistor state table.

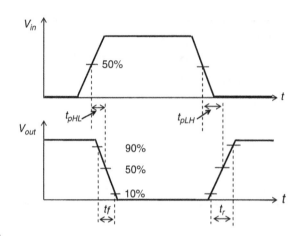

FIGURE 2.14

Illustrations of t_r, t_f, and t_p measurements [Rabaey 2003].

points on each transition curve by horizontally overlaying $V_{IH \min}$ and $V_{IL \max}$ are used. For $V_{DD} = 3.3V$, estimates of t_r and t_f can also be obtained as follows:

$$t_r = \frac{C_L}{I_{D-p}} \cdot \Delta V_{out} \approx \frac{2 \cdot C_L \cdot \Delta V_{out}}{\mu_p \cdot C_{OX} \cdot \frac{W_p}{L_p} \cdot \left(\frac{V_{DD}}{2} + V_{tp}\right)^2} \qquad (2.7)$$

and

$$t_f = \frac{C_L}{I_{D-n}} \cdot \Delta V_{out} \approx \frac{2 \cdot C_L \cdot \Delta V_{out}}{\mu_n \cdot C_{OX} \cdot \frac{W_n}{L_n} \cdot \left(\frac{V_{DD}}{2} - V_{tn}\right)^2} \qquad (2.8)$$

where C_L is the collective capacitance on the output of the CMOS inverter.

In practice, for process control and meeting engineering objectives (such as yield), both types of transistors are often manufactured with identical channel length. With this in mind and on the basis of Equations 2.7 and 2.8, making $t_r = t_f$ leads to

$$\left.\frac{W_p}{W_n}\right|_{t_r=t_f} = \frac{\mu_n \cdot (V_{DD} - V_{tn})}{\mu_p \cdot (V_{DD} + V_{tp})} \qquad (2.9)$$

With most CMOS technologies this W_p/W_n ratio (for $t_r = t_f$) is between 1.5 and 3. Readers are encouraged to substitute data for specific technologies and verify.

Instead of $t_r = t_f$ being used, sometimes the criteria can be to minimize the average rise and fall time, where

$$t_{avg-r-f} = \frac{t_r + t_f}{2} \qquad (2.10)$$

Substituting Equation 2.10 with Equations 2.7 and 2.8 and assuming $L_n = L_p = L$, we have

$$t_{avg\_r\_f} = C_L \cdot \Delta V_{out} \cdot \frac{L}{C_{OX}} \cdot \left(\frac{1}{\mu_p \cdot W_p \cdot \left(\frac{V_{DD}}{2} + V_{tp}\right)^2} + \frac{1}{\mu_n \cdot W_n \cdot \left(\frac{V_{DD}}{2} - V_{tn}\right)^2}\right) \qquad (2.11)$$

Assuming that $C_L \approx C_{OX} \cdot L \cdot (W_n + W_p)$ and $|V_{tn}| \simeq |V_{tp}|$, the optimal W_p/W_n ratio is obtained by first rearranging Equation 2.11 to:

$$
\begin{aligned}
t_{avg\_r\_f} &\approx \frac{\Delta V_{out} \cdot L^2}{\mu_n \cdot \left(\frac{V_{DD}}{2} - V_{tn}\right)^2} \cdot \left(1 + \frac{\mu_n \cdot W_n}{\mu_p \cdot W_p}\right) \cdot \left(1 + \frac{W_p}{W_n}\right) \\
&= \frac{\Delta V_{out} \cdot L^2}{\mu_p \cdot \left(\frac{V_{DD}}{2} + V_{tp}\right)^2} \cdot \left(1 + \frac{\mu_n \cdot W_n}{\mu_p \cdot W_p}\right) \cdot \left(1 + \frac{W_p}{W_n}\right)
\end{aligned}
\tag{2.12}
$$

and then differentiating Equation 2.12 with respect to W_p/W_n as:

$$
\begin{aligned}
\frac{\partial(t_{avg\_r\_f})}{\partial(W_p/W_n)} &= \frac{\Delta V_{out} \cdot L^2}{\mu_n \cdot \left(\frac{V_{DD}}{2} - V_{tn}\right)^2} \cdot \left[1 - \frac{\mu_n}{\mu_p} \cdot \left(\frac{W_p}{W_n}\right)^2\right] \\
&= \frac{\Delta V_{out} \cdot L^2}{\mu_p \cdot \left(\frac{V_{DD}}{2} + V_{tp}\right)^2} \cdot \left[1 - \frac{\mu_n}{\mu_p} \cdot \left(\frac{W_p}{W_n}\right)^2\right]
\end{aligned}
\tag{2.13}
$$

and finally setting Equation 2.13 to zero. Therefore, we have:

$$
\left.\frac{W_p}{W_n}\right|_{min-\_t_{avg\_r\_f}} = \sqrt{\frac{\mu_n}{\mu_p}}
\tag{2.14}
$$

For many CMOS technologies, this W_p/W_n ratio (minimizing $t_{avg\_r\_f}$) is approximately 2. In practice, Equations 2.9 and 2.14 are often applied in sizing transistors.

Compared with a pseudo-nMOS inverter, this CMOS inverter consumes much less energy, because there is no direct current path between V_{DD} and the ground. Power dissipation of the CMOS inverter has three types: static, dynamic, and short-circuit. The **static power dissipation** is proportional to the leakage current when the inverter is not switching; the **dynamic power dissipation** is proportional to the switching frequency; and the **short-circuit power dissipation** is proportional to t_r and t_f.

Ideally, when the CMOS inverter is in either output high (T_p is ON and T_n is OFF in Figure 2.13) or output low (T_p is OFF and T_n is ON) state, there should be no current passing through the two transistors. However, in either state, a small current (*a.k.a.* **leakage current**) passes through the OFF-state transistor, hence, causing static power dissipation. The channel leakage currents can be obtained by calculating the channel resistance in the OFF state. The average static power dissipation is then:

$$
P_{static\_avg} = V_{DD} \cdot \frac{I_{leak\_n} + I_{leak\_p}}{2}
\tag{2.15}
$$

Dynamic power dissipation is proportional to operating frequency, f_{clock}, which is the synchronization clock(s) in most digital circuits. Assuming V_{in} is a square wave signal running at f_{clock}, the average dynamic power dissipation is:

$$
P_{dyn\_avg} = C_L \cdot V_{DD}^2 \cdot f_{clock}
\tag{2.16}
$$

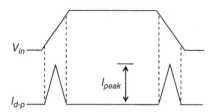

FIGURE 2.15

Illustration of direct-path current occurrences.

Short-circuit power dissipation is unique to CMOS circuits. It occurs while one of the two transistors is changing from the ON state to the OFF state and the other transistor from OFF to ON. During the transitions a direct-path current passes through both transistors. Figure 2.15 depicts the triangular $I_{d\text{-}p}$ waves.
The average short-circuit power dissipation is then:

$$P_{sc\_avg} = V_{DD} \cdot I_{peak} \cdot \frac{t_r + t_f}{2} \cdot f_{clock} \tag{2.17}$$

and

$$I_{peak} = \frac{\mu_n \cdot C_{ox}}{2} \cdot \frac{W_n}{L_n} \cdot (V_{tb} - V_{tn})^2 \tag{2.18}$$

where V_{tb} is the threshold voltage of the CMOS inverter and V_{tn} is the threshold voltage of the n-channel transistor. The total average dynamic power dissipation is then:

$$P_{total\_dyn\_avg} = P_{dyn\_avg} + P_{sc\_avg} \tag{2.19}$$

2.3.2 Design of CMOS logic gates and circuit blocks

An **elementary CMOS logic gate** consists of an N-block and a P-block, each containing the number of corresponding channel transistors equal to the number of inputs of the gate. For example, with the 1-input CMOS inverter, the N-block contains one n-channel transistor and the P-block contains one p-channel transistor. Furthermore, the gate terminal of each n-channel transistor in the N-block is always connected to a corresponding p-channel transistor in the P-block. In addition, if two (or more) inputs are connected to the gate terminals of two n-channel transistors whose drain and source terminals are connected in series in the N-block, the same inputs are also connected to the gates terminals of two (or more) p-channel transistors whose drain and source terminals are connected in parallel.

Consider a 2-input (*a* and *b*) 1-output (*c*) NAND gate whose Boolean function is defined as $c = \overline{a \cdot b}$. Its symbol and truth table are shown in Figure 2.16,

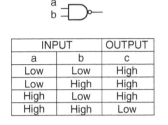

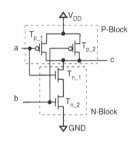

INPUT		OUTPUT
a	b	c
Low	Low	High
Low	High	High
High	Low	High
High	High	Low

FIGURE 2.16

A NAND gate, its truth table, and a CMOS circuit implementation.

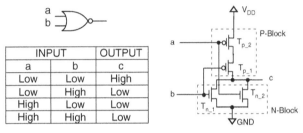

INPUT		OUTPUT
a	b	c
Low	Low	High
Low	High	Low
High	Low	Low
High	High	Low

FIGURE 2.17

A NOR gate, its truth table, and a CMOS circuit implementation.

along with a typical CMOS circuit implementation. The AND operator (shown as ·) indicates that the two n-channel transistors controlled by the inputs must be placed next to each other in series and the two p-channel transistors controlled by the same inputs must be placed next to each other in parallel. When inputs a and b are both set to high, transistors T_{n_1} and T_{n_2} are turned ON such that output c is pulled down by means of discharge through the N-block, while both transistors in the P-block are OFF. In other input conditions at least one of the two transistors in the N-block is OFF and at least one of the two transistors in the P-block is ON, such that output c is being charged to high through the P-block.

Estimation of t_f is straightforward by identifying W_{n_equ}, which comprises the width of both n-channel transistors. However, estimation of the rise time is somewhat complicated by the two p-channel transistors connected in parallel. Assuming that $W_{n\_1} = W_{n\_2}$ and $W_{p\_1} = W_{p\_2}$, which is often the case, then t_{r_min} is the rise time for both p-channel transistors to be turned ON and t_{r_max} is the rise time for only one of them to be turned ON, where $t_{r\_max} = 2\, t_{r\_min}$. It is often desired to make $t_f = t_{r\_max}$ in this and similar cases, for smaller W_{p_1} and W_{p_2}.

Figure 2.17 shows a typical CMOS implementation for a 2-input 1-output NOR gate whose Boolean function is defined as $c = \overline{a + b}$. When both inputs a and b are low, the output is driven to high by the P-block, because both p-channel transistors are turned to ON and both n-channel transistors are turned to OFF. In other input conditions, at least one of the n-channel transistors is ON, pulling the output c down to low.

Similar to the analysis of the NAND gate, estimation of t_r is straightforward by identifying W_{p_equ}, which comprises the width of both p-channel transistors. Because the two n-channel transistors are connected in parallel, the fall time comprises t_{f_min} (when both n-channel transistors are to be turned ON) and t_{f_max} (when only one of the two n-channel transistors is to be turned ON). Assuming that $W_{n\_1} = W_{n\_2}$, we have $t_{f\_max} = 2\,t_{f\_min}$. Oftentimes, it is desirable to also make $t_r = t_{f\_max}$ in this and similar cases.

To illustrate designing CMOS circuits implementing complex gates and random logic functions, as an example we use the carry bit circuit whose Boolean function is defined as $\overline{carry} = a \cdot b + (a + b) \cdot c$ and a typical CMOS implementation is shown in Figure 2.18. In the N-block, transistors T_{n_3} and T_{n_5} implement $a \cdot b$, T_{n_1} and T_{n_2} for $a + b$, which is ANDed with c (implemented by T_{n_4}). Note that to implement the two ORs, T_{n_3} and T_{n_5} are placed in parallel alongside the other three n-channel transistors (for the first OR); T_{n_1} and T_{n_2} are placed in parallel with each other (for the second OR); T_{n_3} is placed in series with T_{n_5} to implement the first AND; and T_{n_4} is placed in series with T_{n_1} and T_{n_2} to implement the second AND.

Configuring the p-channel transistors in the P-block is to complement the configurations of the n-channel transistors. Here, T_{p_3} and T_{p_5} are placed in parallel with each other to complement T_{n_3} and T_{n_5}; T_{p_1} and T_{p_2} are placed in series to complement T_{n_1} and T_{n_2}; and T_{p_4} complements T_{n_4} and is placed in parallel with T_{p_1} and T_{p_2}, which are then placed in series with T_{p_3} and T_{p_5}.

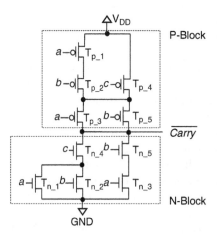

FIGURE 2.18

A CMOS implementation of a carry bit.

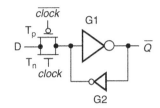

FIGURE 2.19

Implementation of a transmission-gate–based D latch.

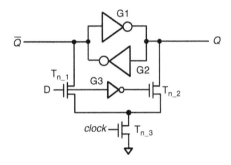

FIGURE 2.20

Implementation of an inverter-based D latch.

2.3.3 **Design of latches and flip-flops**

The simplest latch implementation uses two cross-coupled inverters and one **transmission gate**, as shown in Figure 2.19. The positive feedback allows the holding of a single bit of data at the output of G1 with its **collective load capacitance**. Transistors T_n and T_p are functioning together as a transmission gate. When the transmission gate is turned ON by the clock, the output bit $\bar{Q}$ is updated by the input D with $\bar{Q} = \bar{D}$. For this implementation to work reliably, the feedback inverter G2 must be significantly (approximately 10 times) smaller than the forward inverter G1. A smaller G2 will not interfere with input D to drive the G1 as desired.

Figure 2.20 shows an inverter-based D latch design with both Q and $\bar{Q}$ outputs. In this design, inverters G1 and G2 of identical sizes form the cross-coupled loop to hold a single bit of data. When the clock turns T_{n_3} to ON, input D will turn either T_{n_1} or T_{n_2} ON such that the outputs will be updated accordingly. When T_{n_3} is turned OFF, input D is disconnected from internal signals, and outputs Q and $\bar{Q}$ are driven by the **cross-coupled inverters** with the stored data. Note that G3 is a small inverter, because it only drives one transistor. By sizing the transistors properly, this inverter-based D latch can produce outputs Q and $\bar{Q}$ with similar timing characteristics. Figure 2.21 shows another inverter-based D latch implementation of two complementary outputs with the same timing measures—a characteristic important for **dual-rail processing**.

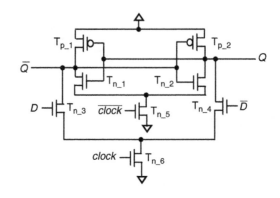

FIGURE 2.21

Implementation of a dual-rail inverter-based D latch.

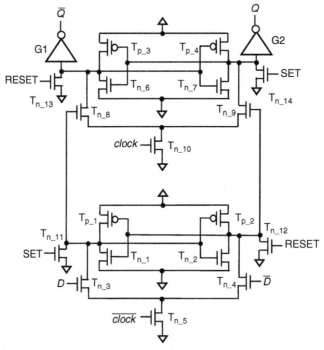

FIGURE 2.22

Implementation of a positive edge–triggered D flip-flop [Martin 2000].

A typical flip-flop contains two latches: one is called a master latch and the other is called a slave latch. The two latches work in complementary modes: when one latch is updating its content, the other is holding its outputs. Figure 2.22 shows a positive-edge-triggered dual-rail D flip-flop with asynchronous SET and RESET. Larger inverters G1 and G2 give greater driving capability. The SET and RESET functions are carried out in both the master and the slave latches.

2.3.4 **Optimization techniques for high performance**

In this section, we highlight several techniques for improving circuit performance. Other techniques that optimize circuits for low-power applications will be discussed in Section 2.6.

To improve circuit performance, it is often desirable to minimize the maximum number of transistors in series in the N-block and P-block. Consider the circuit shown in Figure 2.18. In the N-block, any path between the output and GND consists of two transistors. However, for the P-block there can be either two or three transistors between the output and V_{DD}. Carefully reviewing transistor configurations in the P-block, an equivalent implementation can be devised by rearranging the connections of the p-channel transistors as shown in Figure 2.23. This equivalent implementation has symmetric transistor configurations between the N-block and the P-block, hence improving performance.

Sometimes a small transistor is used to improve circuit performance. Figure 2.24 illustrates the concept of the use of a small **full-swing transistor** (*a.k.a.* keeper). As V_{out} goes low, T_p is turned ON, providing additional pulling of V_{in} to V_{DD}, which, in turn, speeds up V_{out} going low faster. When a CMOS logic block takes inputs from a pass-transistor logic block, the addition of this

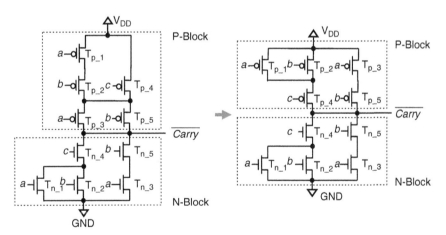

FIGURE 2.23

An optimized implementation of a carry bit.

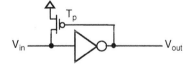

FIGURE 2.24

Application of a small full-swing transistor.

p-channel transistor eliminates the voltage drop because of the pass transistor. Note that the addition of T_p improves the t_f measure on V_{out}. Hence, it is a technique often used to balance circuit-timing measurements and optimize circuit implementations.

Because large digital systems often contain more than half a million latches in data path circuit structures and control logics, at times it becomes desirable to optimize their designs for a smaller area on silicon (*a.k.a.* footprint), as well as less power dissipation. Figure 2.25 shows a design known as an inverter-based three-state **dynamic latch**. T_{n_1} and T_{p_1} function as a traditional inverter. T_{n_2} and T_{p_2} control the periodical updating of the V_{out} node according to V_{in}. Capacitor C_{jp}, which is not explicitly included but rather is used to represent the junction and parasitic capacitance on the node, provides the single bit storage. This dynamic latch is approximately half the size of the transmission gate–based D latch shown in Figure 2.19 and approximately one fifth the size of the inverter-based D latch shown in Figure 2.20.

It should be pointed out that with the dynamic latch, as the data is stored on C_{jp}, the periodic updating (*a.k.a.* refresh) of V_{out} by *clock* must be performed before C_{jp} loses its charge through leakage to the substrate. Higher refresh rates mean higher power dissipation, which sometimes can be prohibitive. Meeting the clock frequency requirement with respect to C_{jp} and other design objectives can sometimes be challenging.

2.4 INTEGRATED CIRCUIT DESIGN TECHNIQUES

As modern digital systems demand more from circuit implementations, many new circuit technologies have emerged. These circuit technologies improve in one or more of the following areas: simplify implementation complexity, reduce silicon area, improve performance, and reduce power consumption. In this subsection, we highlight some of the techniques widely used in practice.

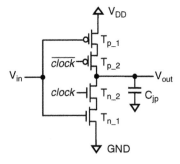

FIGURE 2.25

An inverter-based three-state dynamic latch.

2.4.1 **Transmission-gate/pass-transistor logic**

Transmission-gate/pass-transistor logic simplifies circuit implementations and yet does not require power supply to its circuit blocks. Consider a 2-to-1 multiplexer [Karim 2007]. Figure 2.26 compares a NAND gate implementation with a transmission-gate based implementation and a pass-transistor implementation.

The NAND-gate based implementation uses a total of 14 transistors, whereas the transmission-gate based and the pass-gate based implementations use 6 and 4 transistors, respectively. The NAND-gate based implementation incurs 2 gate delays between the data inputs and the output, whereas the transmission-gate based and the pass-transistor based implementations incur the channel resistance only.

One of the limiting factors with transmission-gate based and pass-transistor based implementations is the voltage drop when signals pass through them. Table 2.2 summarizes the transmission characteristics. Another is the higher internal capacitances in transmission-gate and pass-transistor configurations, because the junction capacitors are directly exposed to the signals passing through. Therefore, it is recommended that each transmission-gate based circuit block be followed with an active logic block, such as a CMOS inverter aided with a full-swing p-channel transistor (as shown in Figure 2.24).

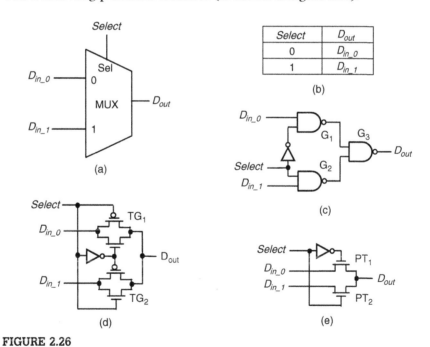

FIGURE 2.26

Comparison of 2-to-1 multiplexer implementations: (a) 2-to-1 MUX block symbol. (b) Truth table. (c) A NAND-gate-based implementation. (d) A transmission-gate-based implementation. (e) A pass-transistor-based implementaion.

Table 2.2 Measures of Transmission Characteristic [Wakerly 2001]

Device	Transmission Characteristic	
	High	Low
Transmission gate	Good	Good
N-channel pass transistor	Poor	Good
P-channel pass transistor	Good	Poor

a \ bc	00	01	11	10
0	0	0	1	0
1	1	1	1	0

(a)

a \ bc	00	01	11	10
0	a	a	c	c
0	a	a	c	c

(b)

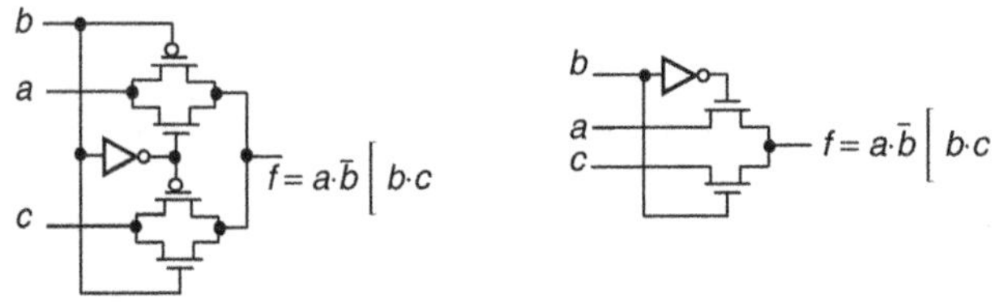

(c)

(d)

FIGURE 2.27

Comparison of 2-to-1 multiplexer implementations: (a) A normal Karnaugh map. (b) The modified Karnaugh map. (c) A transmission-gate-based design. (d) A pass-transistor-based design.

One of the key steps in the use of transmission gates and pass transistors for logic implementation is the identification of pass variable(s) to replace the 1's and 0's in normal Karnaugh maps. Instead of grouping 1's, as one would do in a normal Karnaugh map, variables are identified as pass variables or control variables and grouped accordingly. Pass variables are those to be connected to the data terminals of a multiplexer, whereas control variables are those to be connected to the select terminals. To illustrate this, consider a Boolean function $f(a, b, c) = a \cdot \bar{b} + b \cdot c$. Figure 2.27 shows the normal Karnaugh map (a) and its modified version (b) the use of pass variables, along with a transmission-gate based implementation (c) and a pass-transistor based implementation (d). After examining the normal Karnaugh map, one can conclude that when $b = 0$, the output f is determined by a; when $b = 1$, f is determined by c. This analysis results in the modified Karnaugh map, which indicates that b is the control variable, and a and c are the pass variables, resulting in the transmission-gate based and the pass-transistor based implementations shown in Figure 2.27. Readers are encouraged to try implementing other Boolean functions with this approach.

It should be noted that although transmission-gate based and pass-transistor based designs can reduce silicon area, placing a pass transistor on a normal signal path could lead to difficulty in testing, because a high-impedance state is introduced at the output of the pass transistor when the pass transistor is stuck at the OFF state.

2.4.2 Differential CMOS logic

Differential CMOS logic holds a unique place in dual-rail data processing circuits. This is because its two complementary outputs have identical timing characteristics. As illustrated in Figure 2.28, a differential CMOS circuit block consists of two symmetric left and right sub-blocks; each has one p-channel transistor in the P-block serving as the load device for the n-channel switching block below it. The two p-channel load devices are cross-coupled. The configurations of the n-channel transistors in the two sub–N-blocks follow the same AND-to-series OR-to-parallel constructions used with CMOS circuits. The symmetric circuit structures ensure identical timing characteristics at the two complementary outputs with respect to inputs.

Consider an XOR/XNOR combo block. Figure 2.29 compares three designs, an optimized CMOS NAND-based implementation (which is not for dual-rail), a differential CMOS logic implementation, and a hybrid of differential CMOS and pass-transistor implementation. With the CMOS NAND–based implementation shown in Figure 2.29b, the two complementary outputs have different delays. Hence, it is not suitable for dual-rail processing circuits. With the differential CMOS implementation shown in Figure 2.29c, the symmetric structures used by both output blocks ensure identical delay and, therefore, it is one of the desired circuit configurations for dual-rail processing. The implementation shown in Figure 2.29d simplifies the differential CMOS implementation by combining it with pass-transistor logic.

It should be noted that when complementary signals are not needed, the use of differential CMOS logic might result in a larger circuit footprint and more power consumption. Therefore, the circuit implementation must be chosen with respect to the requirements.

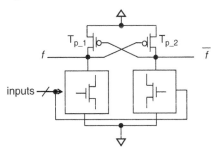

FIGURE 2.28

A generic diagram of a differential CMOS circuit block.

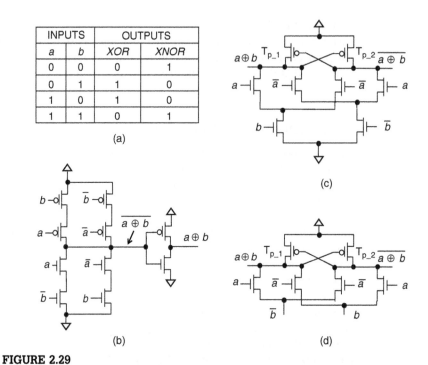

INPUTS		OUTPUTS	
a	b	XOR	XNOR
0	0	0	1
0	1	1	0
1	0	1	0
1	1	0	1

(a)

(b)

(c)

(d)

FIGURE 2.29

Comparison of implementations for XOR/XNOR: (a) Truth table for XOR/XNOR. (b) A differential CMOS implementation. (c) An optimized CMOS NAND-based implementation. (d) A hybrid implementation using differential CMOS and pass-transistor.

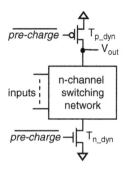

FIGURE 2.30

Generic structure of a dynamic pre-charge circuit block using n-channel switching transistors.

2.4.3 Dynamic pre-charge logic

Dynamic pre-charge logic has been widely used in high-performance micro-processors. Figure 2.30 illustrates the generic structure of a dynamic pre-charge circuit block, in which transistors T_{p_dyn} and T_{n_dyn} are **dynamic transistors**

and T_{p_dyn} is also known as the dynamic load. When the *pre-charge* signal is high, T_{p_dyn} is turned ON to charge the V_{out} node to high, while T_{n_dyn} is turned OFF to prevent currents going through the n-channel switching block to the ground. This period is called pre-charge phase, during which the output on V_{out} is ignored. This pre-charge phase is followed by an evaluation phase, during which T_{p_dyn} is turned OFF, T_{n_dyn} is turned ON, and V_{out} is determined by the n-channel switching network controlled by the inputs. If the inputs are evaluated for V_{out} to go low, the pre-charged voltage on V_{out} is discharged through the n-channel switching network, because it has at least one path connecting V_{out} to ground. Otherwise, V_{out} remains floating at the pre-charged high value.

Transistor configurations in the n-channel switching network follow the same design steps as those used for classic CMOS circuits. Figure 2.31 shows the NAND and NOR blocks using dynamic pre-charge logic.

Similarly, instead of using an n-channel switching network, dynamic pre-charge circuits can use p-channel switching transistors. A generic structure of dynamic pre-charge logic by use of a p-channel switching network is shown in Figure 2.32. During the pre-charge phases, T_{n_dyn} is turned ON and T_{p_dyn} is turned OFF, and V_{out} is discharged to low. During the evaluation phases, T_{n_dyn} is turned OFF and T_{p_dyn} is turned ON, and V_{out} is determined by the configurations of p-channel transistors in the p-channel switching network. If inputs are evaluated for V_{out} to go high, the output node gets charged from V_{DD} through at least one path in the p-channel switching network that connects V_{out} with V_{DD}. Otherwise, V_{out} remains low. Figure 2.33 shows the implementations for a 2-input NAND and 2-input NOR gate using p-channel switching transistors.

2.4.4 **Domino logic**

Cascading dynamic pre-charge logic blocks one after another may result in erroneous outputs because of a phenomenon known as **partial discharge**, as

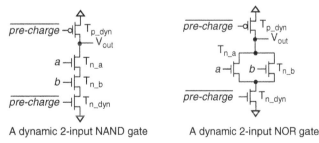

A dynamic 2-input NAND gate   A dynamic 2-input NOR gate

FIGURE 2.31

Dynamic 2-input NAND and NOR implementations using n-channel switching transistors gate.

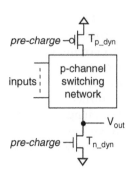

FIGURE 2.32

Generic structure of a dynamic pre-charge circuit block using p-channel switching transistors.

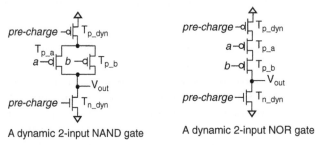

A dynamic 2-input NAND gate   A dynamic 2-input NOR gate

FIGURE 2.33

Dynamic 2-input NAND and NOR gate implementations using p-channel switching transistors.

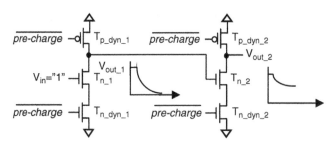

FIGURE 2.34

Partial discharge in cascaded dynamic pre-charge inverters.

illustrated in Figure 2.34 with respect to $V_{in} = 1$. First, both outputs of the two inverters will be pre-charged to high. Next, V_{out_1} is to be discharged to low. Ideally, V_{out_2} would remain high, because the input to the second inverter is going low. However, because T_{n_2} is initially in the ON state right after the evaluation phase begins, V_{out_2} may be partially discharged, potentially resulting in an erroneous output. (Readers are encouraged to analyze cascaded dynamic inverters by use of p-channel switching transistors.) To avoid this partial discharge problem in practice, a dynamic pre-charge block is often followed by a CMOS inverter, and the resulting circuit structure is known as **Domino CMOS logic** whose generic circuit structure is illustrated in Figure 2.35.

To demonstrate the applications of Domino logic, consider a 4-bit comparator. The truth table for a single-bit slice comparator is shown in Table 2.3, and the Boolean function is $f(C_{in}, A, B) = A \cdot \overline{B} + A \cdot C_{in} + \overline{B} \cdot C_{in} = A \cdot \overline{B} + (A + \overline{B}) \cdot C_{in}$. By use of Domino logic with n-channel switching transistors, the single-bit comparator circuit implementation is shown in Figure 2.36, along with the 4-bit block diagram.

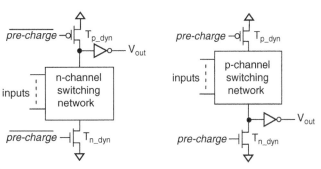

FIGURE 2.35

Generic structure of a Domino CMOS logic circuit block.

Table 2.3 Single-Bit Comparator

Inputs			Output
C_{in}	A	B	$A > B$
0	0	0	0
0	0	1	0
0	1	0	1
0	1	1	0
1	0	0	1
1	0	1	0
1	1	0	1
1	1	1	1

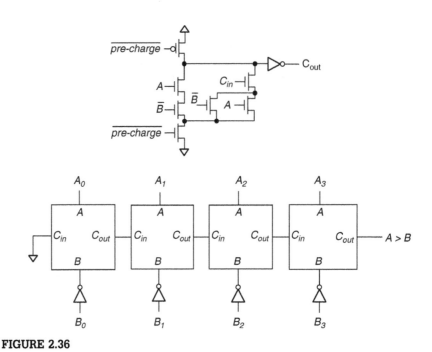

FIGURE 2.36

A 4-bit comparator implementation by use of Domino logic.

It should be pointed out that because transistor T_{p_dyn} acts as a dynamic load, the outputs of dynamic precharge logic and Domino logic will leak away over time and thus may not be valid in certain situations where clocking is halted. For example, when diagnosis of digital circuits is performed, it is often necessary for engineers to apply a certain number of clock cycles to a circuit, stop, and then probe selected signals to take necessary measurements. These and similar operations may not be possible with dynamic pre-charge and Domino logics, because they require constant pre-charge and evaluation cycles.

To overcome this shortcoming, a small (often of minimum size) static load p-channel transistor (*a.k.a.* **keeper**) is added alongside the dynamic load, as illustrated in Figure 2.37. This small keeper transistor provides just enough current to overcome the leakage current during probing, in the case with dynamic pre-charge logic, and it also improves the high-to-low transition at V_{out}.

For dynamic circuit blocks implementing complex logic functions, the n-channel switching network often contains many stacked transistors, which may cause erroneous outputs during the evaluation phases. The phenomenon is known as **charge sharing**, which is illustrated in Figure 2.38. During an evaluation phase, transistors A, B, and E are OFF and transistor D is ON, and the charge on C_1 is now shared with C_2, which is much bigger than C_1. This would cause the voltage at the input of the inverter to drop, which may lead to an erroneous V_{out}. To prevent this charge-sharing problem, selected internal nodes in

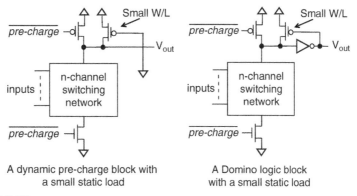

FIGURE 2.37

Illustration of dynamic circuit blocks with static load.

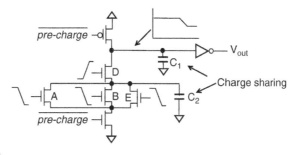

FIGURE 2.38

Charge sharing in a dynamic CMOS circuit.

the switching network can be pre-charged as well. This is illustrated in the implementation of a multi-output dynamic circuit block shown in Figure 2.39. No explicit dynamic transistor is placed at internal nodes where pre-charge is guaranteed. Readers are encouraged to identify these internal nodes as an exercise.

2.4.5 No-race logic

One of the limitations with Domino logic is the insertion of an inverter at each block's output. When Domino logic circuit blocks are cascaded, the added inverters can result in excessive delay. One way to reduce such delay is alternating between n-channel pre-charge blocks and p-channel pre-charge blocks, a technique known as **NORA** [Martin 2000] (for **no-race logic**), as illustrated in Figure 2.40, when dynamic circuit blocks are cascaded one after another.

A **dynamic latch** (*a.k.a.* clocked latch) has also been used in the place of the inverter in a Domino logic circuit block. During a pre-charge phase, the dynamic latch appears as high impedance. During an evaluation phase,

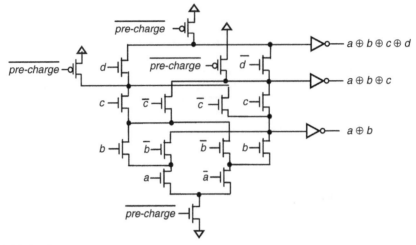

FIGURE 2.39

Precharge of selected internal nodes in a multi-output Domino logic circuit block.

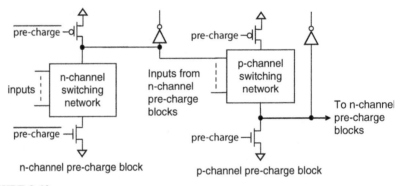

FIGURE 2.40

Altering n-channel pre-charge and p-channel pre-charge blocks.

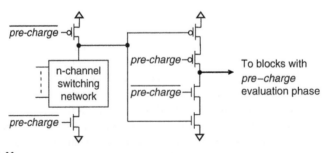

FIGURE 2.41

A dynamic circuit block with a dynamic latch output buffer.

the dynamic latch samples the output of the dynamic block and stores its output during the next pre-charge phase. The dynamic circuit block and the latch are pre-charged and evaluated in opposite phases, therefore, eliminating the partial discharge problem.

A circuit structure combining the preceding two approaches is known as *No-Race logic*, as illustrated in Figure 2.42 with two stages. The first is the $\overline{\textit{pre-charge}}$ evaluation stage because its circuit blocks are evaluated in that phase. This stage consists of an n-channel Domino block, which is followed by a p-channel Domino block, with the output being clocked by a dynamic latch. Outputs of the two Domino logic circuit blocks can feed other circuit blocks as indicated, without being latched. In the second stage, switching networks are evaluated in the *pre-charge* phase. Hence, this stage is called the

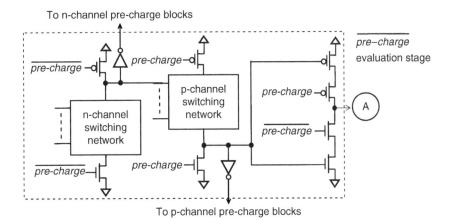

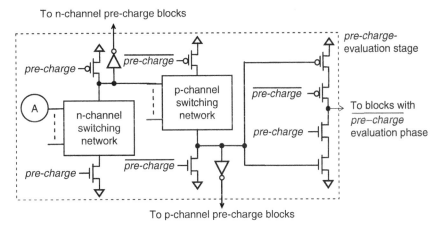

FIGURE 2.42

Circuit structure of No-Race logic.

pre-charge-evaluation stage. It consists of the same circuit components and structure as the first stage, except that the dynamic control signals are replaced with the complemented version. This two-stage section can be repeated several times to form highly efficient pipeline structures.

Note that the circuit blocks in the two-stage structure illustrated in Figure 2.42 use dynamic loads. When static loads are used, there are constraints on the number of inversions to guarantee race-free operation in the presence of clock skews. Techniques such as **reverse clock distribution** and **local clock generation** that use differential circuits are also used in practice to ensure race-free operation in high-performance CMOS circuits. For the analysis and design principles, readers are encouraged to explore further with the references listed at the end of this section.

2.4.6 **Single-phase logic**

As described and illustrated in the previous subsections on dynamic CMOS circuit implementations, both *pre-charge* and $\overline{pre\text{-}charge}$ phases are used. Techniques that use only one phase are known as **single-phase logic**, which simplifies dynamic implementations. Figure 2.43 illustrates the generic diagram of two basic single-phase logic components, with one that uses an n-channel switching network and the other that uses p-channel switching network. Note

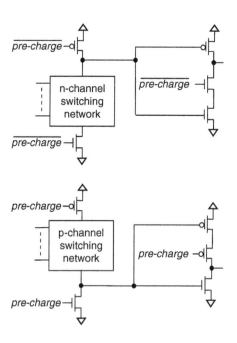

FIGURE 2.43

Generic diagram of single-phase logic blocks.

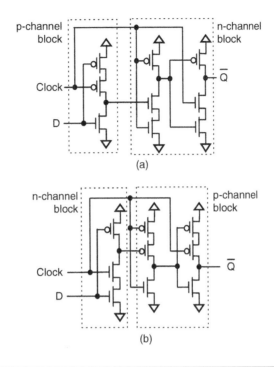

FIGURE 2.44

Single-phase edge-trigged dynamic D flip-flops: (a) Positive edge-triggered dynamic D flip-flop. (b) Negative edge-triggered dynamic D flip-flop.

that each dynamic circuit block uses one phase of the *pre-charge* signal. Figure 2.44 shows two single-phase edge-trigged dynamic D flip-flops. Readers are encouraged to analyze the way that these two dynamic flip-flops work. Single-phase logic can simplify the clock distribution that can be very complex in many large digital systems in which high-performance dynamic circuits are used.

2.5 CMOS PHYSICAL DESIGN

Once transistor schematics are ready, the next design step is to translate the circuit schematic designs into the device and wire placements on silicon. This design step is known as **physical design**, which produces silicon blueprints specifying the exact size and location of each transistor, wire, contact, and other components before manufacturing masks are generated.

Circuit simulation incorporating physical design specifics can more accurately mimic the real chip behavior than schematic-based circuit simulation. This is because at the circuit schematic level, oftentimes the exact length of each wire is not known yet. Therefore, circuit designs with small design margins are often simulated again with physical design data to further ensure that design metrics are satisfied.

In this section, we highlight some basic concepts and practices in physical design. For more in-depth study, readers are encouraged to explore the references further. To help with visualizing layout designs, the **Mead-Conway color-coordination** is often used to differentiate CMOS structures [Weste 1994]. Table 2.4 shows the color representation for the n-well CMOS process. When color display is not available, varying fill-in patterns and shades are used.

2.5.1 Layout design rules

Layout design rules specify geometric constraints with respect to physical constructs. These layout design rules are intended to ensure that designs can be properly manufactured through the manufacturing processes and satisfy all engineering metrics. Because layout design rules are technology and process specific, care must be taken to ensure that only certified layout design rules of the intended technology and processes are used.

Layout design rules are defined in terms of **feature sizes, separations,** and **overlaps**. Feature size defines the dimensions of constructs, such as the channel length and the width of wires. Separation defines the distance between two constructs on the same layer. Overlap defines the necessary overlap of two constructs on adjacent layers in a physical construction, such as a contact connecting a Poly wire with a Metal 1 wire, in which the Metal 1 wire must overlap with the Poly wire below. Table 2.5 lists two typical sets of CMOS layout design rules for an n-well–based process. One is called the **λ-Rule set** and the other is called the **μ-Rule set**. The λ-Rule set is scalable with λ (which is typically twice the channel feature size), therefore, giving designs much flexibility in choosing manufacturing facilities and stability in dealing with multiple manufacturing lines and vendors. The μ-Rule set specifies the exact feature

Table 2.4 N-Well CMOS Process Color-Layer Representation [Weste 1994]

Layer	Color	Symbolic
N-well	Brown	
Thin-oxide	Green	n-channel transistor
Poly	Red	Poly-silicon
p^+	Yellow	p-channel transistor
Contact-cut, via	Black	Contact
Metal 1	Blue	Metal 1
Metal 2	Tan	Metal 2
Metal 3	Gray	Metal 3
Metal 4	Purple	Metal 4

Table 2.5 CMOS Layout Design Rules [Weste 1994]

	λ-Rule	μ-Rule
A. N-well layer		
A.1 Minimum size	10λ	2μ
A.2 Minimum spacing (well at same potential)	6λ	2μ
A.3 Minimum spacing (well at different potential)	8λ	2μ
B. Active Area		
B.1 Minimum size	3λ	1μ
B.2 Minimum spacing	3λ	1μ
B.3 N-well overlap of p^+	5λ	1μ
B.4 N-well overlap of n^+	3λ	1μ
B.5 N-well space to n^+	5λ	5μ
B.6 N-well space to p^+	3λ	3μ
C. Poly		
C.1 Minimum size	2λ	1μ
C.2 Minimum spacing	2λ	1μ
C.3 Spacing to Active	1λ	0.5μ
C.4 Gate Extension	2λ	1μ
D. p^+/n^+		
D.1 Minimum overlap of Active	2λ	1μ
D.2 Minimum size	7λ	3μ
D.3 Minimum overlap of Active in substrate contact	1λ	2μ
D.4 Spacing of p^+/n^+ to n^+/p^+ gate	3λ	1.5μ
E. Contact		
E.1 Minimum size	2λ	0.75μ
E.2 Minimum space on Poly	2λ	1μ
E.3 Minimum space on Active	2λ	0.75μ
E.4 Minimum overlap of Active	2λ	0.5μ
E.5 Minimum overlap of Poly	2λ	0.5μ
E.6 Minimum overlap of Metal 1	1λ	0.5μ

continued

Table 2.5 CMOS Layout Design Rules [Weste 1994]—*cont.*

E.7 Minimum space to Gate	2λ	1μ
F. Metal 1		
F.1 Minimum size	3λ	1μ
F.2 Minimum spacing	3λ	1μ
G. Via		
G.1 Minimum size	2λ	0.75μ
G.2 Minimum spacing	3λ	1.5μ
G.3 Minimum Metal 1 overlap	1λ	0.5μ
G.4 Minimum Metal 2 overlap	1λ	0.5μ
H. Metal 2		
H.1 Minimum size	3λ	1μ
H.2 Minimum spacing	4λ	1μ
I. Via 2		
I.1 Minimum size	2λ	1μ
I.2 Minimum spacing	3λ	1.5μ
I.3 Minimum Metal 2 overlap	2λ	1μ
I.4 Minimum Metal 3 overlap	3λ	1.5μ
J. Metal 3		
J.1 Minimum size	8λ	4μ
J.2 Minimum spacing	5λ	2.5μ
J.3 Minimum Metal 2 overlap	2λ	1μ
J.4 Minimum Metal 3 overlap	2λ	1μ
K. Passivation		
K.1 Minimum opening		100μ
K.2 Minimum spacing		150μ

sizes, required separations, and overlaps for a targeted line of technology and processes. It is often used for high-volume designs.

Entries in Table 2.5 are mostly self-explanatory. For example, Rule A.1 specifies that, for the intended n-well technology, the dimensions of the n-well must be at least $10\lambda \times 10\lambda$ in a layout design following the λ-Rule set and $2\mu \times 2\mu$

following the μ-Rule set. Rule A.2 specifies that the minimum space between two separate n-wells of the same potential must be 6λ and 2μ, respectively. Rule C.1 specifies that a Poly section must be 2λ wide with λ-Rule and 1μ with μ-Rule. Rule C.2 specifies that there must be at least 2λ (or 1μ) separation between two neighboring Poly sections. As readers may observe in Table 2.5, layout designs following the λ-Rule set almost always end up occupying more silicon space than those following the μ-Rule set. This is because the λ-Rule set incorporates built-in scalability, whereas the μ-Rule does not have this flexibility (therefore, it can be optimized for minimum use of the silicon area). Figures 2.45 and 2.46 illustrate graphically the layout design rules in Color and Black/White, respectively.

2.5.2 Stick diagram

Stick diagrams are useful tools for planning custom physical layout designs of complex circuit blocks. In a stick diagram, transistors are represented by colored sticks, contacts are represented by black dots, and wires are represented by lines; all are placed on a square-grid background. Transistor representations in a stick diagram are the same regardless of their size. Figure 2.47 illustrates two stick diagrams of a CMOS inverter, illustrating that different transistor placement orientations result in layouts with different aspect ratios.

One of the applications of a stick diagram is to investigate the best placement of transistors, including their orientations and relative positions. This is an important step in designing layouts of complex circuit blocks, because transistor placements can affect wiring complexity and many circuit performance characteristics. The common objectives used in devising stick diagrams are minimizing the overall block area and the use of wires. Other objectives can be proper alignment of input and output signals, such that when a block is to be cascaded in series, the layout block can be repeated without much reconnection. Oftentimes, layout design engineers can find themselves in a position in which minimizing block area and the use of wires cannot be achieved at the same time, and hence a tradeoff must be made to proceed. The simplicity of stick diagrams gives layout design engineers a "quick-and-dirty" approach to investigate the potential impacts to aid in making layout design decisions.

Another application of stick diagrams is for estimating the block layout dimensions. In this case, the background grid X and Y dimensions are indexed. With a given layout stick diagram along with the set of layout design rules, sizes of constructs on the X and Y axis are added up to determine the total length on that index. For example: X(3) for the stick diagram in Figure 2.47a passes through the width of the GND wire and the source contact of the n-channel transistor, the n-channel length, the n-channel transistor drain terminal contact, the separation space of the terminal contacts, the p-channel drain terminal contact, the p-channel length, the p-channel source terminal contact, and the V_{DD} wire; X(8) for stick diagram in Figure 2.47b intersects with the GND wire,

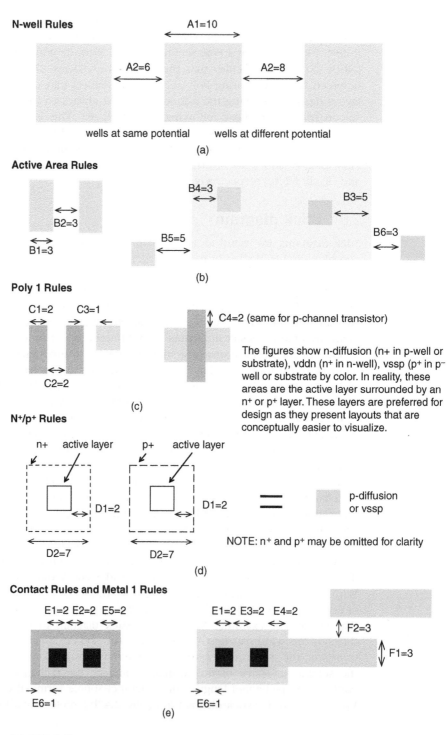

N-well Rules

A1=10

A2=6   A2=8

wells at same potential   wells at different potential

(a)

Active Area Rules

B2=3

B1=3

B4=3   B3=5

B5=5   B6=3

(b)

Poly 1 Rules

C1=2   C3=1

C2=2

C4=2 (same for p-channel transistor)

The figures show n-diffusion (n+ in p-well or substrate), vddn (n+ in n-well), vssp (p+ in p⁻ well or substrate by color. In reality, these areas are the active layer surrounded by an n+ or p+ layer. These layers are preferred for design as they present layouts that are conceptually easier to visualize.

(c)

N+/p+ Rules

n+   active layer   p+   active layer

D1=2   D1=2

D2=7   D2=7

p-diffusion or vssp

NOTE: n+ and p+ may be omitted for clarity

(d)

Contact Rules and Metal 1 Rules

E1=2 E2=2 E5=2   E1=2 E3=2 E4=2

F2=3

F1=3

E6=1   E6=1

(e)

FIGURE 2.45

Continued

Via Rules and F. Metal 2 Rules

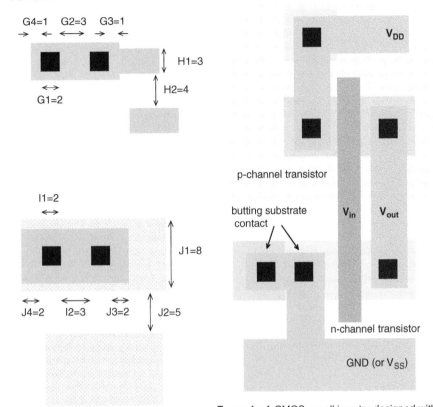

Example: A CMOS n-well inverter designed with Lambda Rules (with n+ and p+ layers omitted)

FIGURE 2.45

Illustration of layout rules and color designations [Weste 1994].

the source terminal contact of the n-channel transistor, spacing between M1 and the contact, the M1 wire, the source contact of the p-channel transistor, and the V_{DD} wire; X(9) goes through the GND wire, the n-channel gate extension, the width of the n-channel transistor, spacing between M1-Poly contact and the n-channel, the M1-Poly contact, spacing between p-channel and M1-Poly contact, the width of the p-channel transistor, the p-channel gate extension, the width of the V_{DD} wire. By use of the λ-Rule, Table 2.6 lists the estimates on the X and Y index for Figures 2.47a and 2.47b layouts, with the assumption that the transistors have an identical channel width of 2λ.

Because a custom physical layout design often requires several iterations of floorplanning, placement, and routing, estimates of block dimensions on the basis of stick diagrams can help to reduce the number of iterations, hence, improving the efficiency of design activities. Although in recent years, CAD

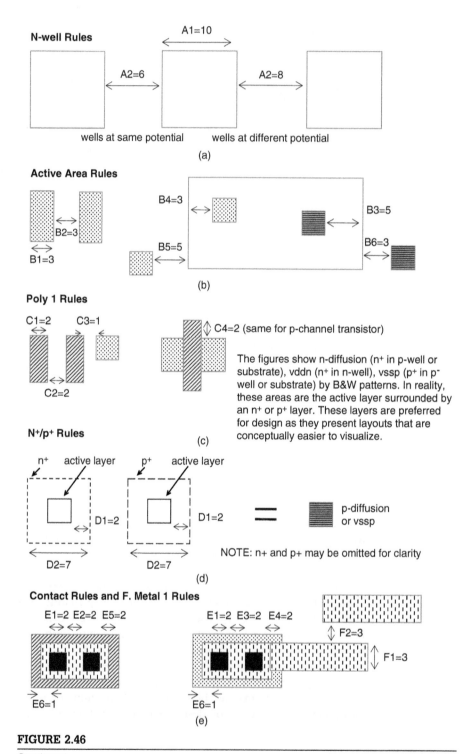

FIGURE 2.46

Continued

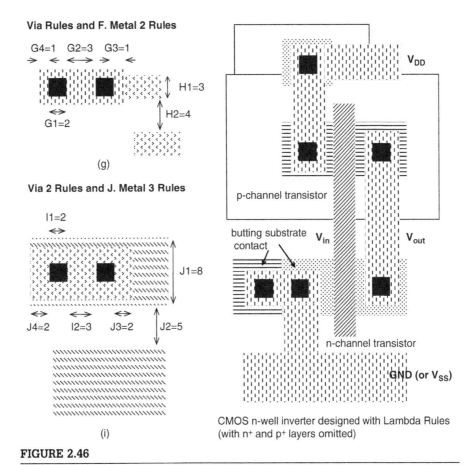

Via Rules and F. Metal 2 Rules

(g)

Via 2 Rules and J. Metal 3 Rules

(i)

CMOS n-well inverter designed with Lambda Rules
(with n+ and p+ layers omitted)

FIGURE 2.46

Illustration of layout rules with designated B&W patterns [Weste 1994].

tools have largely automated the floorplanning, placement, and routing tasks and processes, some designers still use stick diagrams in planning block layout designs and functional units.

2.5.3 **Layout design**

Although most of the chip-level physical layout design activities are done by running automated EDA tools, most physical layout design library cells (*a.k.a.* books) are still created and fine-tuned manually with the help of EDA tools such as a layout editor. In this subsection, we highlight a few physical layout design examples of small CMOS circuit blocks. The layer-overlapping color display seen on designers' computer screens is known as **symbolic layout**. A chip-level symbolic layout display is often called the **artwork**. Once a chip-level physical layout design is verified against engineering metrics (such as DRC, timing, yield)

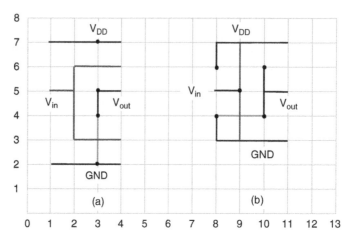

FIGURE 2.47

Stick diagrams for a CMOS inverter.

Table 2.6   Estimated Length on Stick Diagram X and Y Indexes

Index	Items	Length
For stick diagram of Figure 2.47a		
X(3)	$(4 + 1 + 2 + 1 + 4 + 2 + 4 + 1 + 2 + 1 + 4)\,\lambda$	26λ
Y(2)	$(2 + 4 + 2)\,\lambda$	8λ
Y(5)	$(2 + 2 + 2 + 4 + 2)\,\lambda$	12λ
Y(6)	$(2 + 2 + 2)\,\lambda$	6λ
	Estimated block layout dimensions: 26 λ by 12 λ	
For stick diagram of Figure 2.47b		
X(9)	$(4 + 2 + 2 + 2 + 4 + 2 + 2 + 2 + 4)\,\lambda$	24λ
X(11)	$(4 + 2 + 4 + 2 + 4 + 2 + 4)\,\lambda$	22λ
Y(4)	$(4 + 1 + 2 + 1 + 4)\,\lambda$	12λ
Y(5)	$(2 + 4 + 2 + 2 + 2)\,\lambda$	12λ
	Estimated block layout dimensions: 24λ by 12λ	

and approved, EDA tools are used to extract manufacturing mask data from the physical layout data for production masks.

Figure 2.48 shows a symbolic layout of a classic CMOS inverter that uses the n-well process. The layout design uses one metal layer. Typically, cells and blocks in a library have the same height so that wires for V_{DD} and GND can be aligned precisely throughout a chip. With this CMOS inverter, space is left between the n-channel transistor and the p-channel transistor so that this inverter cell maintains the same height as the other cells to be described in this subsection. Note that, whenever possible, n-well contacts (with V_{DD}) are placed along the V_{DD} supply line, and substrate contacts are placed along GND. These contacts are necessary to provide good grounding for the well and the substrate. Once a cell is created manually, it is important to check for any physical layout design rule violations. Typically, EDA tools provide such a function known as a ***design rule check*** (DRC). It is important to note that, when performing DRC with an EDA tool, a correct rule set must be specified. For example, to check this CMOS inverter layout design for any DRC violations, the n-well–based design rule set must be specified in the application. Inappropriate use of design rule set would result in either not discovering or wrongly identifying DRC violations.

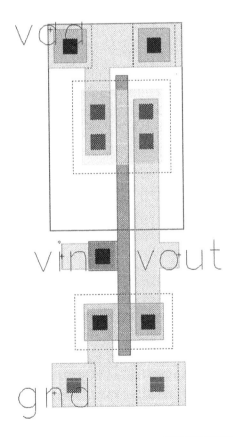

FIGURE 2.48

Symbolic layout of a CMOS inverter.

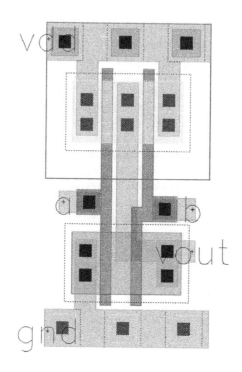

FIGURE 2.49

Symbolic layout of a 2-input 1-output CMOS NAND gate.

Figure 2.49 shows a symbolic layout for a 2-input NAND gate that uses one metal layer and the n-well process. Because of this limitation, its two inputs are accessed at different sides. Typically, library cells would have their inputs on one side and their outputs on the other side. This can effectively reduce the overall wire length when cells are used in functional blocks. When a second metal layer is available, input b in Figure 2.49 can easily be rerouted to the West along the side of input a.

Figure 2.50 shows a symbolic layout of a 3-input OR followed by a 2-input NAND block, which uses one metal layer and the n-well process. Because it also uses one metal layer, the inputs of the block are accessed from both sides, and the output goes out on the left side. When a second metal layer is available, one can reroute inputs to the West and the output to the East. As an alternative, the inputs can also be routed for access from the South by extending the Poly wires beyond GND.

Note that in Figure 2.50, the n-channel transistor controlled by input a is one third of the size of the p-channel transistors controlled by inputs b, c, and d. This is because the p-channel transistors of inputs b, c, and d are in series connection, and by the transistor equivalence theory, the equivalent transistor size

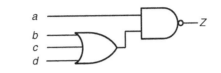

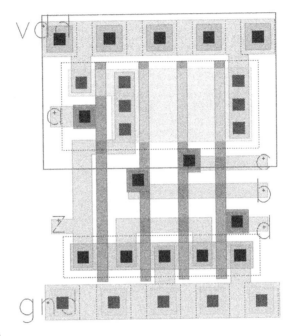

FIGURE 2.50

Symbolic layout of a 3-input-OR 2-input-NAND block.

of p-channel transistors controlled by inputs *b, c,* and *d* is the same as the size of p-channel transistor of input *a*.

Figure 2.51 shows a symbolic layout of grading-series transistors in an AND dynamic CMOS block [Weste 1994] with 4 inputs. The layout design uses transistors of varying sizes according to the position in the series structure to reduce delay. The n-channel transistor closest to the output is the smallest, with n-channel transistors increasing their size as they are placed nearer GND. The switching time is reduced, because there is less capacitance at the output. With older technologies, it provided 15% to 30% performance boost. However, with submicron technologies, this improvement is much less, at 2% to 4% in some cases. Nevertheless, the example demonstrates how layout designs of blocks can be optimized.

It is worth noting that often multiple techniques can be applied to a block. As an exercise, readers can attempt to improve the design of Figure 2.51 by first analyzing and identifying the problems associated with the design and then

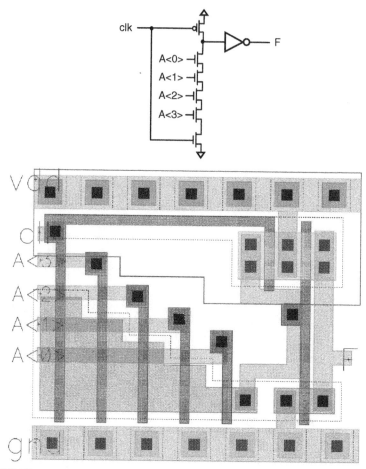

FIGURE 2.51

Symbolic layout of a 4-input AND gate by use of grading-series transistors. [Martin 2000].

modifying the circuit and layout designs that use the techniques discussed in this chapter to improve circuit speed, reduce transistor count, silicon area, and power consumption.

2.6 LOW-POWER CIRCUIT DESIGN TECHNIQUES

As mentioned earlier, there are three sources of power dissipation in CMOS circuits: dynamic power dissipation, short-circuit power dissipation, and static (leakage) power dissipation. Traditionally, dynamic power dissipation has been the dominant source of power dissipation. With continued scaling of CMOS technology, however, leakage power dissipation has become a significant source of power consumption as well. This subsection describes some commonly used circuit-level techniques for reducing power dissipation.

2.6.1 **Clock-gating**

One commonly used technique to reduce power dissipation is to use **clock-gating**. The idea is that clock lines to circuits that are not being used are ANDed with a gate-control signal that disables the clock line to avoid unnecessary charging and discharging of unused circuits. Not all circuits are used at all times. Individual circuit use varies widely across applications and time, so there are many opportunities to use clock-gating.

The clock tree distributes the clock to sequential elements like flip-flops and latches, as well as to dynamic logic gates. Portions of the clock tree can be pruned by gating them with an AND gate as illustrated in Figure 2.52. When the gate-control signal is set to 0, it holds the clock line at a constant 0. This avoids charging and discharging of the capacitive load on the clock line and also prevents latches from changing state, thereby avoiding additional switching activity in any combinational logic being driven by the latch. For dynamic logic circuits, holding the clock at a constant 0 prevents the evaluate phase from occurring, thereby preventing the output from switching values. In practice, transparent latches are often used to gate clocks and prevent potential glitches that can happen with logic AND.

Clock-gating is effective at reducing dynamic power dissipation in unused sequential circuits and dynamic logic gates. Some limitations of clock-gating are that it does not prevent switching in static logic gates that may occur because of changes in the primary input values, and it does not reduce leakage power consumption. These limitations can be addressed by the use of power-gating.

2.6.2 **Power-gating**

Another way to reduce power dissipation in unused circuits is to use **power-gating** [Mutoh 1993; Sakata 1993]. The idea in power-gating is to switch off the power supply to unused circuits, thereby putting them in a "sleep" mode. This is typically implemented by having a gating transistor that can be turned off when the circuit is to be idle for an extended period of time. The gating

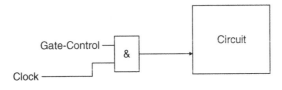

FIGURE 2.52

Clock-gating.

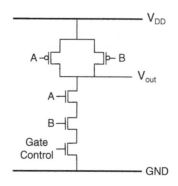

FIGURE 2.53

Power-gated 2-input NAND gate.

transistor can be either a header (p-channel transistor) or footer (n-channel transistor) transistor. Figure 2.53 illustrates a footer transistor. The gating transistor must be sized large enough to handle the amount of switching current at any given time so that there is no measurable amount of voltage drop across it. A footer transistor tends to require less area for a given switching current because of the higher mobility of electrons in an n-channel transistor compared with a p-channel header transistor. In a **multiple-V_T technology**, the gating transistor is typically implemented with a high V_T to minimize subthreshold leakage current through it. Power-gating can thus provide significant leakage power reduction, particularly when used in conjunction with circuits containing low V_T transistors.

Power-gating can be done at either a fine-grain or coarse-grain level. In **fine-grain power-gating**, the gating transistor is part of the standard cell logic. The advantage of this is that the burden of designing the gating transistor is left to the standard cell designer, and the cells can be easily handled by EDA tools. The drawback is that the gating transistor must be sized assuming worst-case conditions in which every cell is switching every clock cycle because nothing can be assumed about the module-level function. In **coarse-grain power-gating**, the gating transistor is part of the power distribution network rather than the standard cell and thus is shared among many gates. One advantage of this is that because only a fraction of the gates switch at any given time, the gating transistors can be sized smaller on aggregate compared with fine-grain power-gating. One issue for coarse-grain power-gating is that if too many gating transistors are switched simultaneously when going in and out of sleep mode, the current demand may overwhelm the power distribution network. Thus, some means for limiting the number of gating transistors that are simultaneously switched is needed.

Because the gating transistors are high V_{TH} devices, they can take several clock cycles to switch on and off and cause additional power dissipation. Thus, for power-gating to be efficient, the circuit must be idle for a sufficient number of clock cycles so that the power savings justifies the time and cost of switching in and out of sleep mode.

When power-gating is implemented in sequential circuits, a means for retaining the sequential state is needed when the circuit goes into sleep mode. One simple approach is to scan the values in the storage elements into a memory before going into sleep mode, and then scan them back from the memory when the circuit wakes up.

Whereas clock-gating can only reduce dynamic power dissipation, power-gating can reduce both dynamic and leakage power dissipation. Because leakage power dissipation has become a sizable portion of overall power dissipation, power-gating has become a very important power reduction method. A drawback of power-gating compared with clock-gating is that it takes several clock cycles to switch in and out of sleep mode, and hence it is only efficient if the circuit will be idle for a sufficiently long time.

2.6.3 **Substrate biasing**

Another way to reduce leakage current (hence, leakage power dissipation) when a circuit is not being used is through **substrate biasing** [Seta 1995], which is also known as **variable threshold CMOS**. The idea is to adjust the threshold voltage by changing the substrate bias voltage (V_{SB}). Increasing the substrate bias voltage induces a body effect on the transistor that increases its threshold voltage (V_T). By having a substrate bias control circuit as illustrated in Figure 2.54, the substrate bias can be adjusted for normal operation to minimize V_T and maximize performance, and then when the circuit is in standby mode, the substrate bias can be adjusted to increase V_T to reduce the subthreshold leakage current. For example, the voltage on V_{Bp} could be set to V_{DD} in normal mode and $2V_{DD}$ in standby mode. The voltage on V_{Bn} could be set to 0 in normal mode and $-V_{DD}$ in standby mode. This would significantly reduce the leakage power dissipation.

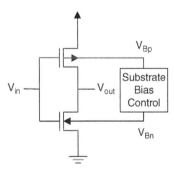

FIGURE 2.54

Substrate biasing.

One drawback of substrate biasing is that it requires a twin- or triple-well CMOS technology to apply different bias voltages to different parts of the chip. There is also a need to generate voltages outside of the normal 0 to V_{DD} power rail range that may require additional power pins on the chip.

2.6.4 Dynamic voltage and frequency scaling

The speed of a circuit depends linearly on the supply voltage. The idea in **dynamic voltage scaling** [Flautner 2001] is that during times when the circuit is not needing high performance, both its clock frequency and supply voltage can be scaled down. Because dynamic power dissipation depends on the square of the supply voltage and linearly on the frequency ($P = CV^2f$), if both the supply voltage and frequency are scaled down, there is a cubic reduction in power consumption.

Dynamic voltage scaling has been implemented in several commercial embedded microprocessors including the Transmeta Crusoe [Transmeta 2002], Intel Xscale [Intel 2003], and ARM IEM [ARM 2007]. When the processor is lightly loaded, the frequency and supply voltage are scaled down to save power, and when it is heavily executing, it is run at full frequency and voltage.

Figure 2.55 illustrates how a dynamic voltage–scaling scheme works. On the basis of the workload, the system requests a frequency change. First, the frequency is reduced, which takes on the order of hundreds of picoseconds, and then the voltage is ramped down, which takes on the order of hundreds of microseconds. Later, when switching back to high frequency, the voltage is first scaled back up to the normal voltage level, and then the frequency is raised back up.

Dynamic voltage scaling is a highly efficient way of reducing power consumption while still preserving functionality and meeting user expectations. It has been widely deployed.

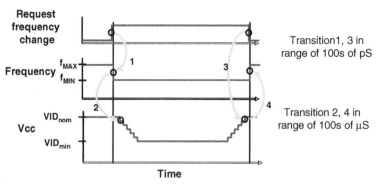

FIGURE 2.55

Dynamic voltage and frequency scaling.

2.6.5 **Low-power cache memory design**

Because microprocessor and ASIC chips contain cache memory often taking up more than half of the silicon space, power dissipation of these on-chip memory blocks can significantly contribute to the overall power consumption. In some cases, the static leakage power dissipation of cache memory contributes more than half of the chip's power consumption. Therefore, modern designs often use on-chip memory technologies with low-power features.

Power dissipation of on-chip memory blocks largely comes from the following functional units: the memory cells, the word and bit lines, and the peripheral circuits such as address decoders and sense amplifiers. In this subsection, we outline some of the low-power techniques applied with word and bit lines.

Figure 2.56 illustrates the memory cell of a typical on-chip cache SRAM memory block. A cell is being accessed (either READ or WRITE) by selected word and bit lines, which are connected to the outputs of address decoder circuits. The arrows indicate the leakage currents (because of bit lines being pre-charged to high) when the cell holds a 0 at the BL side and a 1 at the complementary side. For large on-chip memory, a word or bit line is a long interconnect that would connect to several thousands of cells. Longer word and bit lines not only require larger driving circuits at the outputs of address decoders but also cause concerns with respect to word/bit line delay and more power dissipation during word/bit line pre-charge.

To address these concerns, large on-chip memory is typically divided in many small sections so that each word or bit line drives a small number of cells. This technique is known as **banked cache design**. Both word and bit lines are also sectioned into a hierarchical structure such that each of the selected word

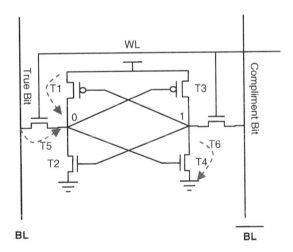

FIGURE 2.56

A typical SRAM cell.

and bit lines drives a few hundreds or fewer cells. A local sense amplifier bus is also used such that selected cache banks can connect to the nearest sense amplifiers, effectively reducing the length of active word and bit lines.

A technique known as **sub bit lines** [Karandikar 1998; Yong 2005] is illustrated in Figure 2.57. Each memory cell is connected to the main bit line by a sub bit line. A sub bit line is a short interconnect line that connects to a few cells. Only one selected sub bit line is connected to the main bit line at a time. Therefore, it significantly reduces the number of memory cells that load the main bit line at any time, which improves the bit-line response time. It also reduces leakage current, because inactive sub bit lines no longer need to be precharged. The disadvantage is that the addition of sub bit lines doubles the area used by bit line interconnects.

With multicore processor technologies becoming mainstream applications, more and more chips are making use of multi-port on-chip cache memory to maintain performance requirements. Classic hard-wired multi-port memory architecture usually uses dedicated word and bit lines to each memory cell for each port. Figure 2.58 illustrates a cell with 2 hard-wired ports. The addition of the second port not only increases the footprint of cache memory on silicon but also introduces additional leakage current (as indicated by arrows in Figure 2.58).

Figure 2.59 illustrates a new technique called **dynamic memory partitioning with isolation nodes** [Bajwa 2006, 2007; Chen 2007]. In theory, isolation nodes are placed on bit lines between neighboring memory cells. One port access is from the bottom of the bit line and the other port access is from the top of the bit line. When the two ports are accessing different cells, a selected isolation control line turns off the isolation nodes and divides the memory bank

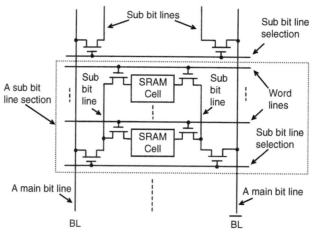

FIGURE 2.57

Illustration of sub bit lines.

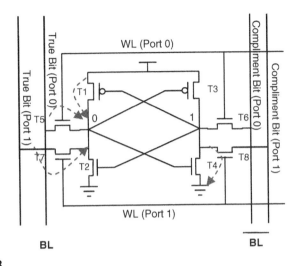

FIGURE 2.58

A typical hard-wired dual-port SRAM cell.

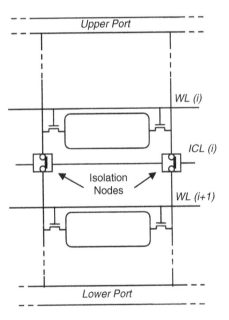

FIGURE 2.59

Illustration of energy-efficient and area-efficient dual-port SRAM.

into two virtually isolated sections to be accessed through the lower and upper ports. When the two ports are accessing the same memory location, all isolation nodes on the bit lines remain in the ON state.

One of the advantages of this dynamic memory partitioning technique that uses isolation nodes is the shared bit lines for the two ports. The length of active bit lines for both ports is shorter. Therefore, it reduces the silicon footprint of multi-port cache memory and improves bit-line response time. Another advantage is the low-power dissipation, because the shared bit line consumes no more power than the single-port configuration. In addition, leakage current remains the same as it is in a single-port configuration. This is because no dedicated bit lines and access transistors are used for the second port. By the use of local sense amplifiers and port multiplexing, this dynamic memory partitioning technique can be applied to on-chip cache memory with more than two ports. The same technique is applicable to DRAM. The disadvantage is that a port may need to pass through several isolation nodes to access a memory cell. The channel resistance of the pass transistors implementing the isolation nodes adds to the bit line response time. However, as the technology advances down to the 32-nanometer node and below, transistor channel resistance will become insignificant compared with wire resistance of the bit lines.

2.7 CONCLUDING REMARKS

CMOS technology has been the backbone of the many advances that have taken place in the past two decades, powering consumer appliances, automotives, personal and scientific computing, as well as many fascinating science and space explorations. Its advances have also made *electronic design automation* (EDA) tools possible and readily accessible to engineers. It is ironic that CMOS chips now power the computers on which engineers rely to design new chips. This chapter is intended to stimulate the reader's interest in the topic and provide background information for the reader to relate CMOS design to the EDA techniques to be discussed in the subsequent chapters.

New CMOS circuit technologies are still being developed. Currently, major improvements center on three fronts: transistors are used more efficiently to provide more computing and functionality, increasing circuit speed, and consuming less power. This chapter has provided some examples in all three of these improvements. For readers who wish to explore further on CMOS design, refer to more recent textbooks cited in the chapter and IEEE publications such as *IEEE Journal of Solid-State Circuits* (JSSC) and *IEEE International Solid-State Circuit Conference* (ISSCC).

2.8 EXERCISES

The following transistor parameters are used in **Exercises 2.1** to **2.13**:

For n-channel transistors:

$$\mu_n C_{ox} = 190 \ \mu A/V^2$$
$$C_{ox} = 3.4 \times 10^{-3} \ pF/(\mu m)^2$$
$$V_{tn} = 0.7 \ V$$
$$r_{ds}(\Omega) = 5000 \ L(\mu m)/I_D(mA) \ \ - - - \ \ \text{in active region}$$
$$C_j = 5 \times 10^{-4} \ pF/(\mu m)^2$$
$$C_{j-sw} = 2.0 \times 10^{-4} \ pF/\mu m$$
$$C_{gs(overlap)} = C_{gd(overlap)} = 2.0 \times 10^{-4} \ pF/\mu m$$

For p-channel transistors:

$$\mu_p C_{ox} = 50 \ \mu A/V^2$$
$$C_{ox} = 3.4 \times 10^{-3} \ pF/(\mu m)^2$$
$$V_{tn} = -0.8 \ V$$
$$r_{ds}(\Omega) = 6000 \ L(\mu m)/I_D(mA) \ \ - - - \ \ \text{in active region}$$
$$C_j = 6 \times 10^{-4} \ pF/(\mu m)^2$$
$$C_{j-sw} = 2.5 \times 10^{-4} \ pF/\mu m$$
$$C_{gs(overlap)} = C_{gd(overlap)} = 2.0 \times 10^{-4} \ pF/\mu m$$

2.1. (Integrated-Circuit Technology) An n-channel (or p-channel) transistor in the active region is measured to have $I_D = 20 \ \mu A$ when $V_{DS} = V_{eff}$. As V_{DS} increases by 0.5 V, I_D increases to 23 μA, estimate the output impedance r_{ds}.

2.2. (Integrated-Circuit Technology) Estimate the capacitances C_{gs}, C_{gd}, C_{db}, and C_{sb} for an n-channel transistor and a p-channel transistor with $W = 10 \ \mu m$ and $L = 1.2 \ \mu m$, assuming the junction areas A_s (at the source) and A_d (at the drain) are 40 $(\mu m)^2$ and the perimeter of each (P_s and P_d) is 12μm.

2.3. (Integrated-Circuit Technology) Consider the circuit below, when V_{in} is 1.2 V. Estimate V_{out} when the n-channel pass transistor ($W = 2.4 \ \mu m$ and $L = 1.2 \ \mu m$) is turned ON.

2.4. (Integrated-Circuit Technology) The effects of technology scaling are outlined in the following table. Now assume that all dimensions are scaled by S, but the voltage and doping levels are only scaled by $\sqrt{S}$, and estimate the scaling factor for other parameters listed in the Table 2.7.

2.5. (CMOS Logic) Design a CMOS circuit that implements $F = a \cdot b \cdot \bar{c} + \bar{a} \cdot c \cdot d$. Choose transistor sizes to give equal rise and fall times at the output.

Table 2.7 Effects of Scaling

Parameter	Scaling Factor
Device dimensions (t_{ox}, L, W, junction depth)	$1/S$
Doping concentration	S
Voltage	$1/S$
Current	$1/S$
Capacitance	$1/S$
Delay time	$1/S$
Power dissipation (per gate)	$1/S^2$
Power-delay product	$1/S^3$

2.6. **(CMOS Logic)** Design a circuit that converts 5.0 V TTL logic outputs to a CMOS logic block that uses a 3.3 V power supply.

2.7. **(CMOS Logic)** Design a circuit that interfaces the outputs of a 1.3 V CMOS logic block with the inputs of a 3.3 V CMOS block.

2.8. **(CMOS Logic)** Consider the circuit design in **Exercise 2.5** and analyze and estimate the static power dissipation. Also, assuming the circuit block switches at 5 MHz, estimate the dynamic power dissipation.

2.9. **(Advanced Integrated-Circuit Design)** Design a 2-input differential AND/NAND circuit block. Specify individual transistor sizes such that the rise and fall times at each output are roughly the same. Assume $V_{DD} = 3.3$ V and an external $C_L = 1$ pF is at each output.

2.10. **(CMOS Physical Design)** Construct a stick diagram of a transmission-gate and inverter-based D latch. Draw the transistor schematic first.

2.11. **(CMOS Physical Design)** Construct a stick diagram of a single-bit full-adder by first drawing its transistor schematic.

2.12. **(CMOS Physical Design)** Use a layout editor to design a physical layout for the D latch shown in Figure 2.21.

2.13. **(CMOS Physical Design)** Use a layout editor to design a physical layout for the single-bit carry circuit shown in Figure 2.23.

2.14. **(CMOS Physical Design)** Analyze the circuit block and layout design in Figure 2.51. Identify further improvements. Improve the circuit block by use of the techniques discussed in this chapter. Use an EDA layout editor to modify the original layout design by use of the same n-well process.

2.15. **(Low-Power Design)** List the advantages and disadvantages of power-gating versus clock-gating.

2.16. **(Low-Power Design)** Describe the advantages and disadvantages of substrate biasing.

ACKNOWLEDGMENTS

We thank Wan-Ping Lee, Guang-Wan Liao, and Professor Yao-Wen Chang of National Taiwan University for helping with generating the symbolic layouts, and Andrew Wu, Meng-Kai Hsu, and Professor James C.-M. Li for reviewing the manuscript. We also thank Professor Eric MacDonald of University of Texas at El Paso and Professor Martin Margala of University of Massachusetts at Lowell for their constructive comments and suggestions.

REFERENCES

R2.0 Books

[Karim 2007] M. Karim and X. Chen, *Digital Design: Basic Concepts and Principles*, CRC Press, New York, 2007.

[Martin 2000] K. Martin, *Digital Integrated Circuit Design*, Oxford University Press, New York, 2000.

[Rabaey 2003] J. M. Rabaey, A. Chandrakasan, and B. Nikolić, *Digital Integrated Circuits: A Design Perspective*, Second Edition, Prentice-Hall, Englewood Cliffs, NJ, 2003.

[Wakerly 2001] J. F. Wakerly, *Digital Design: Principles and Practices*, Third Edition, Prentice-Hall, Englewood Cliffs, NJ, 2001.

[Weste 1994] N. H. E. Weste and K. Eshraghian, *Principles of CMOS Design—A System Perspective*, Second Edition, Addison-Wesley, Reading, MA, 1994.

R2.6 Low-Power Design

[ARM 2007] ARM Ltd., 1176JZ(F)-S Documentation, http://www.arm.com/products/CPUs/ARM1176.html, 2007.

[Bajwa 2006] H. Bajwa and X. Chen, Area-efficient dual-port memory architecture for multi-core processors, in *Proc. Junior Scientists Conf.*, pp. 49–50, April 2006.

[Bajwa 2007] H. Bajwa and X. Chen, Low-power high-performance and dynamically reconfigured multi-port cache memory architecture, in *Proc. IEEE Int. Conf. on Electrical Engineering*, April, 2007.

[Chen 2007] X. Chen and H. Bajwa, Energy-efficient dual-port cache architecture with improved performances, Institution of Engineering and Technology. *J. of Electronics Letters*, 43(1), pp. 12–13, January, 2007.

[Flautner 2001] K. Flautner, S. Reinhardt, and T. Mudge, Automatic performance setting for dynamic voltage scaling, in *Proc. Int. Conf. on Mobile Computing and Networking*, pp. 260–271, May 2001.

[Intel 2003] Intel Corp., Intel Xscale Core Developer's Manual, http://developer.intel.com/design/intelxscale/, 2003.

[Karandikar 1998] A. Karandikar and K. K. Parhi, Low power SRAM design using hierarchical divided bitline approach, in *Proc. Int. Conf. Computer Design*, pp. 82–88, October 1998.

[Mutoh 1993] S. Mutoh, T. Douseki, Y. Matsuya, T. Aoki, and J. Yamada, 1V high-speed digital circuits technology with 0.5 μm multi-threshold CMOS, in *Proc. IEEE Int. ASIC Conf.*, pp. 186–189, September 1993.

[Sakata 1993] T. Sakata, M. Horiguchi, and K. Itoh, Subthreshold-current reduction circuits for multi-gigabit DRAM's, in *Proc. Symp. on VLSI Circuits*, pp. 45–46, May 1993.

[Seta 1995] K. Seta, H. Hara, T. Kuroda, M. Kakumu, and T. Sakurai, 50% active-power saving without speed degradation using standby power reduction (SPR) circuit, *Proc. Int. Solid-State Circuits Conf.*, pp. 318–319, February 1995.

[Transmeta 2002] Transmeta Corp., *Crusoe Processor Documentation*, http://www.transmeta.com, 2002.

[Yong 2005] B. D. Yong and L.-S. Kim, A low power SRAM using hierarchical bit line and local sense amplifier, *IEEE J. Solid-State Circuits*, 40(6), pp. 1366–1376, June 2005.

3

Design for testability

Laung-Terng (L.-T.) Wang
SynTest Technologies, Inc., Sunnyvale, California

ABOUT THIS CHAPTER

Design for testability (DFT) has become an essential part for designing *very-large-scale integration* (VLSI) circuits. The most popular DFT techniques in use today for testing the digital portion of the VLSI circuits include **scan** and **scan-based logic *built-in self-test*** (BIST). Both techniques have proved to be quite effective in producing testable VLSI designs. In addition, **test compression**, a supplemental DFT technique for scan, is growing in importance for further reduction in test data volume and test application time during manufacturing test.

To provide readers with an in-depth understanding of the most recent DFT advances in scan, logic BIST, and test compression, this chapter covers a number of fundamental DFT techniques to facilitate testing of modern digital circuits. These techniques are required to improve the product quality and reduce the defect level and test cost of a digital circuit, while at the same time simplifying the test, debug, and diagnosis tasks.

In this chapter, we first cover the basic DFT concepts and methods for performing testability analysis. Next, **scan design**, the most widely used structured DFT method, is discussed, including popular scan cell designs, scan architectures, and at-speed clocking schemes. After a brief introduction to the basic concept of logic BIST, we then discuss BIST pattern generation and output response analysis schemes along with a number of logic BIST architectures for in-circuit self-test. Finally, we present a number of test compression circuit structures for test stimuli compression and test response compaction. The chapter also includes a description of logic BIST and test compression architectures currently practiced in industry.

3.1 INTRODUCTION

With advances in semiconductor manufacturing technology, ***integrated circuits*** (ICs) can now contain tens to hundreds of millions of transistors running in the gigahertz range. The production and use of these integrated circuits has run into a variety of test challenges during wafer probe, wafer sort, preship screening, incoming test of chips and boards, test of assembled boards, system test, periodic maintenance, repair test, etc. During the early stages of IC production history, design and test were regarded as separate functions, performed by separate and unrelated groups of engineers. During these early years, a design engineer's job was to implement the required functionality on the basis of design specifications, without giving any thought to how the manufactured device was to be tested. Once the functionality was implemented, the design information was transferred to test engineers. A test engineer's job was to determine how to best test each manufactured device within a reasonable amount of time and to screen out the parts that may contain manufacturing defects while shipping all defect-free devices to customers. The final quality of the test was determined by keeping track of the number of defective parts shipped to the customers on the basis of customer returns. This product quality, measured in terms of ***defective parts per million*** (DPM) shipped, was a final test score for quantifying the effectiveness of the developed test.

Although this approach worked well for small-scale integrated circuits that mainly consisted of combinational logic or simple finite-state machines, it was unable to keep up with the circuit complexity as designs moved from ***small-scale integration*** (SSI) to ***very large-scale integration*** (VLSI). A common approach to testing these VLSI devices during the 1980s relied heavily on fault simulation to measure the fault coverage of the supplied functional patterns. Functional patterns were developed to navigate through the long sequential depths of a design, hoping to exercise all internal states and to detect all possible manufacturing defects. A **fault simulation** or **fault-grading** tool was used to quantify the effectiveness of the functional patterns. If the supplied functional patterns did not reach the target fault coverage goal, additional functional patterns were added. Unfortunately, this approach typically failed to improve the circuit's fault coverage beyond 80%, and the quality of the shipped products suffered.

Gradually, it became clear that designing devices without paying much attention to test resulted in increased test cost and decreased test quality. Some designs, which were otherwise best-in-class with regard to functionality and performance, failed commercially because of prohibitive test costs or poor product quality. These problems have since led to the development and deployment of DFT engineering in the industry.

The first challenge facing DFT engineers was to find simpler ways of exercising all internal states of a design and reaching the target fault coverage goal.

Various **testability measures** and **ad hoc testability enhancement** methods were proposed and used in the 1970s and 1980s to serve this purpose. These methods were mainly used to aid in the circuit's **testability** or to increase the circuit's **controllability** and **observability** [McCluskey 1986; Abramovici 1994]. Although attempts to use these methods have substantially improved the testability of a design and eased sequential *automatic test pattern generation* (ATPG), their end results at reaching the target fault coverage goal were far from satisfactory; it was still quite difficult to reach more than 90% fault coverage for large designs. This was mostly because even with these testability aids, deriving functional patterns by hand or generating test patterns for a sequential circuit is a much more difficult problem than generating test patterns for a combinational circuit [Fujiwara 1982; Bushnell 2000; Jha 2003].

Today, the semiconductor industry relies heavily on two techniques for testing digital circuits: *scan* and *logic built-in self-test* (BIST) [Abramovici 1994; McCluskey 1986]. **Scan** converts a digital sequential circuit into a scan design and then uses ATPG software [Bushnell 2000; Jha 2003; Wang 2006a] to detect faults that are caused by manufacturing defects (physical failures) and manifest themselves as errors, whereas logic BIST requires the use of a portion of the VLSI circuit to test itself on-chip, on-board, or in-system. To keep up with the design and test challenges [SIA 2005, 2006], more advanced *design-for-testability* (DFT) techniques, such as test compression, at-speed delay fault testing, and power-aware test generation, have been developed over the past few years to further address the test cost, delay fault, and test power issues [Gizopoulos 2006; Wang 2006a, 2007a].

Scan design is implemented by first replacing all selected storage elements of the digital circuit with **scan cells** and then connecting them into one or more shift registers, called **scan chains**, to provide them with external access. With external access, one can now control and observe the internal states of the digital circuit by simply shifting test stimuli into and test responses out of the shift registers during scan testing. The DFT technique has since proved to be quite effective in improving the product quality, testability, and diagnosability of scan designs [Crouch 1999; Bushnell 2000; Jha 2003; Gizopoulos 2006; Wang 2006a, 2007a]. Although scan has offered many benefits during manufacturing test, it is becoming inefficient to test deep submicron or nanometer VLSI designs. The reasons are mostly because (1) traditional test schemes that use ATPG software to target single faults have become quite expensive and (2) sufficiently high fault coverage for these deep submicron or nanometer VLSI designs is hard to sustain from the chip level to the board and system levels.

To alleviate these test problems, the scan approach is typically combined with **logic BIST** that incorporates BIST features into the scan design at the design stage [Bushnell 2000; Mourad 2000; Stroud 2002; Jha 2003]. With logic BIST, circuits that generate test patterns and analyze the output responses of the functional circuitry are embedded in the chip or elsewhere on the same board where the chip resides to test the digital logic circuit itself. Typically, pseudo-random patterns are applied to the ***circuit under test*** (CUT), while their test responses are compacted in a ***multiple-input signature register*** (MISR) [Bardell 1987; Rajski 1998; Nadeau-Dostie 2000; Stroud 2002; Jha 2003; Wang 2006a]. Logic BIST is crucial in many applications, in particular, for safety-critical and mission-critical applications. These applications, commonly found in the aerospace/defense, automotive, banking, computer, health-care, networking, and telecommunications industries, require on-chip, on-board, or in-system self-test to improve the reliability of the entire system, as well as the ability to perform in-field diagnosis.

Since the early 2000s, **test compression**, a supplemental DFT technique to scan, is gaining industry acceptance to further reduce test data volume and test application time [Touba 2006; Wang 2006a]. Test compression involves compressing the amount of test data (both test stimulus and test response) that must be stored on ***automatic test equipment*** (ATE) for testing with a deterministic ATPG-generated test set. This is done by use of **code-based schemes** or adding additional on-chip hardware before the scan chains to decompress the test stimulus coming from the ATE and after the scan chains to compress the test response going to the ATE. This differs from logic BIST in that the test stimuli that are applied to the CUT are a deterministic (ATPG-generated) test set rather than pseudo-random patterns. Typically, test compression can provide $10\times$ to $100\times$ or even more reduction in test application time and test data volume and hence can drastically save scan test cost.

3.2 TESTABILITY ANALYSIS

Testability is a relative measure of the effort or cost of testing a logic circuit. In general, it is based on the assumption that only primary inputs and primary outputs can be directly controlled and observed, respectively. Testability reflects the effort required to perform the main test operations of controlling internal signals from primary inputs and observing internal signals at primary outputs. **Testability analysis** refers to the process of assessing the testability of a logic circuit by calculating a set of numeric measures for each signal in the circuit.

One important application of testability analysis is to assist in the decision-making process during test generation. For example, if during test generation, it is determined that the output of a certain AND gate must be set to 0, testability analysis can help decide which AND gate input is the easiest to set to 0. The conventional application is to identify areas of poor testability to guide testability enhancement, such as test point insertion, for improving the testability of the design. For this purpose, testability analysis is performed at various design stages so that testability problems can be identified and fixed as early as possible.

Since the 1970s, many testability analysis techniques have been proposed [Rutman 1972; Stephenson 1976; Breuer 1978; Grason 1979]. The ***Sandia***

Controllability/Observability Analysis Program (SCOAP) [Goldstein 1979, 1980] was the first topology-based program that populated testability analysis applications. Enhancements based on SCOAP have also been developed and used to aid in test point selection [Wang 1984, 1985]. These methods perform testability analysis by calculating the **controllability** and **observability** of each signal line, where *controllability* reflects the difficulty of setting a signal line to a required logic value from primary inputs, and *observability* reflects the difficulty of propagating the logic value of the signal line to primary outputs.

Traditionally, gate-level topologic information of a circuit is used for testability analysis. Depending on a target application, deterministic and/or random testability measures are calculated. In general, **topology-based testability analysis**, such as SCOAP or probability-based testability analysis, is computationally efficient but can produce inaccurate results for circuits containing many reconvergent fanouts. **Simulation-based testability analysis**, on the other hand, can generate more accurate estimates by simulating the circuit behavior with deterministic, random, or pseudo-random test patterns, but may require a long simulation time.

In this section, we first describe the method for performing SCOAP testability analysis. Then, probability-based testability analysis and simulation-based testability analysis are discussed.

3.2.1 **SCOAP testability analysis**

The SCOAP testability analysis program [Goldstein 1979, 1980] calculates six numeric values for each signal s in a logic circuit:

- CC0(s): Combinational 0-controllability of s
- CC1(s): Combinational 1-controllability of s
- CO(s): Combinational observability of s
- SC0(s): Sequential 0-controllability of s
- SC1(s): Sequential 1-controllability of s
- SO(s): Sequential observability of s

Roughly speaking, the three combinational testability measures, CC0, CC1, and CO, are related to the number of signals that need to be manipulated to control or observe s from primary inputs or at primary outputs, whereas the three sequential testability measures, SC0, SC1, and SO, are related to the number of clock cycles required to control or observe s from primary inputs or at primary outputs [Bushnell 2000]. The values of controllability measures range between 1 and infinite, whereas the values of observability measures range between 0 and infinite. As a boundary condition, the CC0 and CC1 values of a primary input are set to 1, the SC0 and SC1 values of a primary input are set to 0, and the CO and SO values of a primary output are set to 0.

3.2.1.1 *Combinational controllability and observability calculation*

The first step in SCOAP is to calculate the combinational controllability measures of all signals. This calculation is performed from primary inputs toward primary outputs in a breadth-first manner. More specifically, the circuit is leveled from primary inputs to primary outputs to assign a *level order* for each gate. The output controllability for each gate is then scheduled in *level order* after the controllability measures of all of its inputs have been calculated. The rules for combinational controllability calculation are summarized in Table 3.1, where a 1 is added to each rule to indicate that a signal passes through one more level of logic gate. From this table, we can see that CC0 $(s) \geq 1$ and $CC1(s) \geq 1$ for any signal s. A larger $CC0(s)$ or $CC1(s)$ value implies that it is more difficult to control s to 0 or 1 from primary inputs.

Once the combinational controllability measures of all signals are calculated, the combinational observability of each signal can be calculated. This calculation is also performed in a breadth-first manner while moving from primary outputs toward primary inputs. The rules for combinational observability calculation are summarized in Table 3.2, where a 1 is added to each rule to indicate that a signal passes through one more level of logic. From this table, we can see that $CO(s) \geq 0$ for any signal s. A larger $CO(s)$ value implies that it is more difficult to observe s at any primary output.

Table 3.1   SCOAP Combinational Controllability Calculation Rules

	0-Controllability (Primary Input, Output, Branch)	**1-Controllability** (Primary Input, Output, Branch)
Primary Input	1	1
AND	*min* {input 0-controllabilities} + 1	Σ (input 1-controllabilities) + 1
OR	Σ (input 0-controllabilities) + 1	*min* {input 1-controllability} + 1
NOT	Input 1-controllability + 1	Input 0-controllability + 1
NAND	Σ (input 1-controllabilities) + 1	*min* {input 0-controllability} + 1
NOR	*min* {input 1-controllability} + 1	Σ (input 0-controllabilities) + 1
BUFFER	Input 0-controllability + 1	Input 1-controllability + 1
XOR	*min* {CC1(a) + CC1(b), CC0(a) + CC0(b)} + 1	*min* {CC1(a) + CC0(b), CC0(a) + CC1(b)} + 1
XNOR	*min* {CC1(a) + CC0(b), CC0(a) + CC1(b)} + 1	*min* {CC1(a) + CC1(b), CC0(a) + CC0(b)} + 1
Branch	Stem 0-controllability	Stem 1-controllability

a, b: inputs of an XOR or XNOR gate

Table 3.2 SCOAP Combinational Observability Calculation Rules

Observability (Primary Output, Input, Stem)	
Primary Output	0
AND/NAND	Σ (output observability, 1-controllabilities of other inputs) + 1
OR/NOR	Σ (output observability, 0-controllabilities of other inputs) + 1
NOT/BUFFER	Output observability + 1
XOR/XNOR	a: Σ (output observability, min {CC0(b), CC1(b)}) + 1 b: Σ (output observability, min {CC0(a), CC1(a)}) + 1
Stem	min {branch observabilities}

a, b: inputs of an XOR or XNOR gate

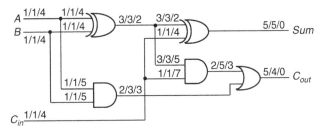

FIGURE 3.1

SCOAP full-adder example.

Figure 3.1 shows the combinational controllability and observability measures of a full-adder. The three-value tuple $v_1/v_2/v_3$ on each signal line represents the signal's 0-controllability (v_1), 1-controllability (v_2), and observability (v_3). The boundary condition is set by initializing the C0 and C1 values of the primary inputs A, B, and C_{in} to 1, and the CO values of the primary outputs *Sum* and C_{out} to 0. By applying the rules given in Tables 3.1 and 3.2 and starting with the given boundary condition, one can first calculate all combinational controllability measures forward and then calculate all combinational observability measures backward in *level order*.

3.2.1.2 *Sequential controllability and observability calculation*

Sequential controllability and observability measures are calculated in a similar manner as combinational measures, except that a 1 is not added as we move from one level of logic to another, but rather a 1 is added when a signal passes through a storage element. The difference is illustrated in the sequential circuit example shown in Figure 3.2, which consists of an AND gate and a positive

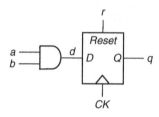

FIGURE 3.2

SCOAP sequential circuit example.

edge–triggered D flip-flop. The D flip-flop includes an active-high asynchronous reset pin r. SCOAP measures of a D flip-flop with a synchronous, as opposed to asynchronous, reset are shown in [Bushnell 2000].

First, we calculate the combinational and sequential controllability measures of all signals. To control signal d to 0, either input a or b must be set to 0. To control d to 1, both inputs a and b must be set to 1. Hence, the combinational and sequential controllability measures of signal d are:

$$CC0(d) = min \{CC0(a), CC0(b)\} + 1$$
$$SC0(d) = min \{SC0(a), SC0(b)\}$$
$$CC1(d) = CC1(a) + CC1(b) + 1$$
$$SC1(d) = SC1(a) + SC1(b)$$

To control the data output q of the D flip-flop to 0, the data input d and the reset signal r can be set to 0, while applying a rising clock edge (a 0-to-1 transition) to the clock CK. Alternately, this can be accomplished by setting r to 1 while holding CK at 0, without applying a clock pulse. Because a clock pulse is not applied to CK, a 1 is not added to the sequential controllability calculation in the second case. Therefore, the combinational and sequential 0-controllability measures of q are:

$$CC0(q) = min\{CC0(d) + CC0(CK) + CC1(CK) + CC0(r), CC1(r) + CC0(CK)\}$$
$$SC0(q) = min\{SC0(d) + SC0(CK) + SC1(CK) + SC0(r) + 1, SC1(r) + SC0(CK)\}$$

Here, $CC0(q)$ measures how many signals in the circuit must be set to control q to 0, whereas $SC0(q)$ measures how many flip-flops in the circuit must be clocked to set q to 0. To control the data output q of the D flip-flop to 1, the only way is to set the data input d to 1 and the reset signal r to 0, while applying a rising clock edge to the clock CK. Hence,

$$CC1(q) = CC1(d) + CC0(CK) + CC1(CK) + CC0(r)$$
$$SC1(q) = SC1(d) + SC0(CK) + SC1(CK) + SC0(r) + 1$$

Next, we calculate the combinational and sequential observability measures of all signals. The data input d can be observed at q by holding the reset signal r at 0 and applying a rising clock edge to CK. Hence,

$$CO(d) = CO(q) + CC0(CK) + CC1(CK) + CC0(r)$$
$$SO(d) = SO(q) + SC0(CK) + SC1(CK) + SC0(r) + 1$$

The asynchronous reset signal r can be observed by first setting q to 1, and then holding CK at the inactive state 0. Again, a 1 is not added to the sequential controllability calculation because a clock pulse is not applied to CK:

$$CO(r) = CO(q) + CC1(q) + CC0(CK)$$
$$SO(r) = SO(q) + SC1(q) + SC0(CK)$$

There are two ways to indirectly observe the clock signal CK at q: (1) set q to 1, r to 0, d to 0, and apply a rising clock edge at CK, or (2) set both q and r to 0, d to 1, and apply a rising clock edge at CK. Hence,

$$CO(CK) = CO(q) + CC0(CK) + CC1(CK) + CC0(r) +$$
$$min\{CC0(d) + CC1(q), CC1(d) + CC0(q)\}$$
$$SO(CK) = SO(q) + SC0(CK) + SC1(CK) + SC0(r) +$$
$$min\{SC0(d) + SC1(q), SC1(d) + SC0(q)\} + 1$$

To observe an input of the AND gate at d requires setting the other input to 1. Therefore, the combinational and sequential observability measures for both inputs a and b are:

$$CO(a) = CO(d) + CC1(b) + 1$$
$$SO(a) = SO(d) + SC1(b)$$
$$CO(b) = CO(d) + CC1(a) + 1$$
$$SO(b) = SO(d) + SC1(a)$$

It is important to note that controllability and observability measures calculated with SCOAP are heuristics, and only approximate the actual testability of a logic circuit. When scan design is used, testability analysis can assume that all scan cells are directly controllable and observable. It was also shown in [Agrawal 1982] that SCOAP may overestimate testability measures for circuits containing many reconvergent fanouts. However, with the capability of performing testability analysis in an $O(n)$ computational complexity for n signals in a circuit, SCOAP provides a quick estimate of the circuit's testability that can be used to guide testability enhancement and test generation.

3.2.2 Probability-based testability analysis

Topology-based testability analysis techniques, such as SCOAP, have been found to be extremely helpful in supporting test generation, which is a main topic of Chapter 14. These testability measures are able to analyze the **deterministic testability** of the logic circuit in advance and during the ATPG search process [Ivanov 1988]. On the other hand, in logic *built-in self-test* (BIST), which is the main topic of Section 3.4, random or pseudo-random test patterns are generated without specifically performing deterministic test pattern generation operations on any signal line. In this case, topology-based testability measures that use signal probability to analyze the **random testability** of the circuit can be used [Parker 1975; Savir 1984; Jain 1985; Seth 1985]. These measures are often referred to as **probability-based testability measures** or probability-based testability analysis techniques.

For example, given a random input pattern, one can calculate three measures for each signal s in a combinational circuit as follows:

- C0(s): Probability-based 0-controllability of s
- C1(s): Probability-based 1-controllability of s
- O(s): Probability-based observability of s

Here, C0(s) and C1(s) are the probability of controlling signal s to 0 and 1 from primary inputs, respectively. O(s) is the probability of observing signal s at primary outputs. These three probabilities range between 0 and 1. As a boundary condition, the C0 and C1 probabilities of a primary input are typically set to 0.5, and the O probability of a primary output is set to 1. For each signal s in the circuit, C0(s) + C1(s) = 1.

Many methods have been developed to calculate the probability-based testability measures. A simple method is given in the following, whose basic procedure is similar to the one used for calculating combinational testability measures in SCOAP, except that different calculation rules are used. The rules for probability-based controllability and observability calculation are summarized in Tables 3.3 and 3.4, respectively. In Table 3.3, p_0 is the initial 0-controllability chosen for a primary input, where $0 < p_0 < 1$.

Compared with SCOAP testability measures, where non-negative integers are used, probability-based testability measures range between 0 and 1. The smaller

Table 3.3 Probability-Based Controllability Calculation Rules

	0-Controllability (Primary Input, Output, Branch)	**1-Controllability** (Primary Input, Output, Branch)
Primary Input	p_0	$p_1 = 1 - p_0$
AND	1 − (output 1-controllability)	Π(input 1-controllabilities)
OR	Π(input 0-controllabilities)	1 − (output 0-controllability)
NOT	Input 1-controllability	Input 0-controllability
NAND	Π(input 1-controllabilities)	1 − (output 0-controllability)
NOR	1 − (output 1-controllability)	Π(input 0-controllabilities)
BUFFER	Input 0-controllability	Input 1-controllability
XOR	1 − 1-controllability	Σ (C1(a) × C0(b),C0(a) × C1(b))
XNOR	1 − 1-controllability	Σ (C0(a) × C0(b),C1(a) × C1(b))
Branch	Stem 0-controllability	Stem 1-controllability

a, b: inputs of an XOR or XNOR gate

Table 3.4 Probability-Based Observability Calculation Rules

Observability (Primary Output, Input, Stem)	
Primary output	1
AND/NAND	Π (output observability, 1-controllabilities of other inputs)
OR/NOR	Π (output observability, 0-controllabilities of other inputs)
NOT/BUFFER	Output observability
XOR/XNOR	a: Π (output observability, *max* {0-controllability of b, 1-controllability of b}) b: Π (output observability, *max* {0-controllability of a, 1-controllability of a})
Stem	*max* {branch observabilities}

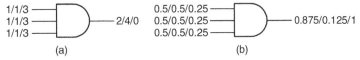

(a)   (b)

FIGURE 3.3

Comparison of SCOAP and probability-based testability measures: (a) SCOAP combinational measures. (b) Probability-based measures.

a probability-based testability measure of a signal, the more difficult it is to control or observe the signal. Figure 3.3 illustrates the difference between SCOAP testability measures and probability-based testability measures of a 3-input AND gate. The three-value tuple $v_1/v_2/v_3$ of each signal line represents the signal's 0-controllability (v_1), 1-controllability (v_2), and observability (v_3).

Signals with poor probability-based testability measures tend to be difficult to test with random or pseudo-random test patterns. The faults on these signal lines are often referred to as ***random pattern resistant*** (RP-resistant) [Savir 1984]. That is, either the probability of these signals randomly receiving a 0 or 1 from primary inputs, or the probability of observing these signals at primary outputs is low, assuming that all primary inputs have the equal probability of being set to 0 or 1.

The existence of such *RP-resistant faults* is the main reason why fault coverage that uses random or pseudo-random test patterns is low compared with the use of deterministic test patterns. In applications such as logic BIST, to solve this low fault coverage problem, test points are often inserted in the circuit to enhance the circuit's random testability. A few commonly used test point insertion techniques are discussed in [Wang 2006a].

3.2.3 **Simulation-based testability analysis**

In the calculation of SCOAP and probability-based testability measures as described previously, only the topologic information of a logic circuit is explicitly explored. These topology-based methods are static, in the sense that they do not use input test patterns for testability analysis. Their controllability and observability measures can be calculated in linear time, thus making them very attractive for applications that need fast testability analysis, such as test generation and logic BIST. However, the efficiency of these methods is achieved at the cost of reduced accuracy, especially for circuits that contain many reconvergent fanouts [Agrawal 1982].

As an alternative or supplement to static or topology-based testability analysis, dynamic or simulation-based methods that use input test patterns for testability analysis or testability enhancement can be performed through **statistical sampling**. Logic simulation and fault simulation techniques can be used [Bushnell 2000; Wang 2006a].

In statistical sampling, a sample set of input test patterns is selected, which is either generated randomly or derived from a given pattern set, and logic simulation is conducted to collect the responses of all or part of signal lines of interest. The commonly collected responses are the number of occurrences of 0's, 1's, 0-to-1 transitions, and 1-to-0 transitions, which are then used to profile statistically the testability of a logic circuit. These data are then analyzed to find locations of poor testability. If a signal line exhibits only a few transitions or no transitions for the sample input patterns, it might be an indication that the signal likely has poor controllability.

In addition to logic simulation, fault simulation has also been used to enhance the testability of a logic circuit with random or pseudo-random test patterns. For instance, a *random resistant fault analysis* (RRFA) method has been successfully applied to a high-performance microprocessor to improve the circuit's random testability in logic BIST [Rizzolo 2001]. This method is based on statistical data collected during fault simulation for a small number of random test patterns. Controllability and observability measures of each signal in the circuit are calculated by use of the probability models developed in the *statistical fault analysis* (STAFAN) algorithm [Jain 1985]. (STAFAN is the first method able to give reasonably accurate estimates of fault coverage in combinational circuits purely by use of input test patterns and without running fault simulation.) With these data, RRFA identifies signals that are difficult to control and/or observe, as well as signals that are statistically correlated. On the basis of the analysis results, RRFA then recommends test points to be added to the circuit to improve the circuit's random testability.

Because it can take a long simulation time to run through all input test patterns, these simulation-based methods are, in general, used to guide testability enhancement in test generation or logic BIST, when it is required to meet a very high fault coverage goal. This approach is crucial for life-critical and mission-critical applications, such as in the healthcare and defense/aerospace industries.

3.3 **SCAN DESIGN**

Scan design is currently the most widely used structured DFT approach. It is implemented by connecting selected storage elements of a design into one or more shift registers, called **scan chains**, to provide them with external access. Scan design accomplishes this task by replacing all selected storage elements with **scan cells**, each having one additional *scan input* (SI) port and one shared/additional *scan output* (SO) port. By connecting the SO port of one scan cell to the SI port of the next scan cell, one or more scan chains are created.

The scan-inserted design, called scan design, is now operated in three modes: **normal mode, shift mode**, and **capture mode**. Circuit operations with associated clock cycles conducted in these three modes are referred to as normal operation, shift operation, and capture operation, respectively.

In normal mode, all test signals are turned off, and the scan design operates in the original functional configuration. In both shift and capture modes, a **test mode** signal *TM* is often used to turn on all test-related fixes in compliance with scan design rules. A set of **scan design rules** that can be found in [Cheung 1997; Wang 2006a] are necessary to simplify the test, debug, and diagnose tasks, improve fault coverage, and guarantee the safe operation of the device under test. These circuit modes and operations are distinguished by use of additional test signals or test clocks. Fundamental scan architectures and at-speed clocking schemes are described in the following subsections.

3.3.1 **Scan architectures**

In this subsection, we first describe a few fundamental scan architectures. These fundamental scan architectures include (1) *muxed-D scan design*, in which storage elements are converted into muxed-D scan cells, (2) *clocked-scan design*, in which storage elements are converted into clocked-scan cells, and (3) *LSSD scan design*, in which storage elements are converted into *level-sensitive scan design* (LSSD) *shift register latches* (SRLs).

3.3.1.1 *Muxed-D scan design*

Figure 3.4 shows a sequential circuit example with three D flip-flops. The corresponding muxed-D full-scan circuit is shown in Figure 3.5. An edge-triggered **muxed-D scan cell** design is shown in Figure 3.5a. This scan cell is composed of a D flip-flop and a multiplexer. The multiplexer uses a *scan enable* (SE) input to select between the *data input* (DI) and the *scan input* (SI).

In normal/capture mode, *SE* is set to 0. The value present at the data input *DI* is captured into the internal D flip-flop when a rising clock edge is applied. In shift mode, *SE* is set to 1. The scan input *SI* is now used to shift in new data to the D flip-flop, while the content of the D flip-flop is being shifted out. Sample operation waveforms are shown in Figure 3.5b. The three D flip-flops,

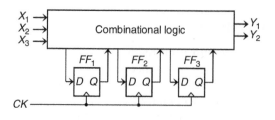

FIGURE 3.4

Sequential circuit example.

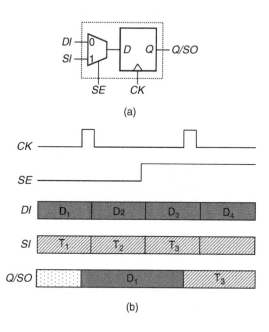

FIGURE 3.5

Edge-triggered muxed-D scan cell design and operation: (a) Muxed-D scan cell. (b) Sample waveforms.

FF_1, FF_2, and FF_3, shown in Figure 3.4, are replaced with three muxed-D scan cells, SFF_1, SFF_2, and SFF_3, respectively, shown in Figure 3.6.

In Figure 3.6, the data input DI of each scan cell is connected to the output of the combinational logic as in the original circuit. To form a scan chain, the scan inputs SI of SFF_2 and SFF_3 are connected to the outputs Q of the previous scan cells, SFF_1 and SFF_2, respectively. In addition, the scan input SI of the first scan cell SFF_1 is connected to the primary input SI, and the output Q of the last scan cell SFF_3 is connected to the primary output SO. Hence, in shift mode, SE is set to 1, and the scan cells operate as a single scan chain, which allows us to shift in any combination of logic values into the scan cells.

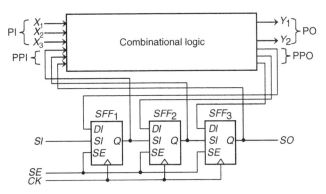

FIGURE 3.6

Muxed-D scan design.

In capture mode, *SE* is set to 0, and the scan cells are used to capture the test response from the combinational logic when a clock is applied.

In general, combinational logic in a full-scan circuit has two types of inputs: ***primary inputs*** (PIs) and ***pseudo primary inputs*** (PPIs). Primary inputs refer to the external inputs to the circuit, whereas pseudo primary inputs refer to the scan cell outputs. Both PIs and PPIs can be set to any required logic values. The only difference is that PIs are set directly in parallel from the external inputs, whereas PPIs are set serially through scan chain inputs. Similarly, the combinational logic in a full-scan circuit has two types of outputs: ***primary outputs*** (POs) and ***pseudo primary outputs*** (PPOs). Primary outputs refer to the external outputs of the circuit, whereas pseudo primary outputs refer to the scan cell inputs. Both POs and PPOs can be observed. The only difference is that POs are observed directly in parallel from the external outputs, whereas PPOs are observed serially through scan chain outputs.

3.3.1.2 *Clocked-scan design*

An edge-triggered ***clocked-scan cell*** can also be used to replace a D flip-flop in a scan design [McCluskey 1986]. Similar to a muxed-D scan cell, a clocked-scan cell also has a data input *DI* and a scan input *SI*; however, in the clocked-scan cell, input selection is conducted with two independent clocks, data clock *DCK* and shift clock *SCK*, as shown in Figure 3.7a.

In normal/capture mode, the data clock *DCK* is used to capture the contents present at the data input *DI* into the clocked-scan cell. In shift mode, the shift clock *SCK* is used to shift in new data from the scan input *SI* into the clocked-scan cell, while the content of the clocked-scan cell is being shifted out. Sample operation waveforms are shown in Figure 3.7b.

The major advantage of the use of a clocked-scan cell is that it results in no performance degradation on the data input. A major disadvantage, however, is that it requires additional shift clock routing.

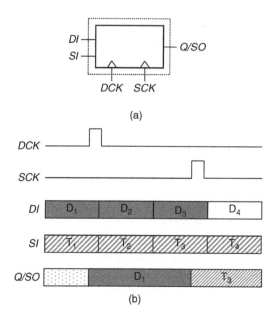

FIGURE 3.7

Clock-scan cell design and operation: (a) Clocked-scan cell. (b) Sample waveforms.

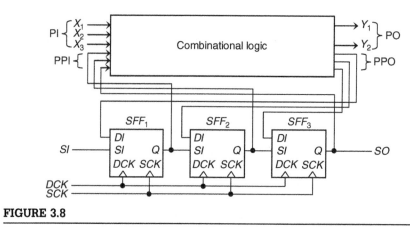

FIGURE 3.8

Clocked-scan design.

Figure 3.8 shows a clocked-scan design of the sequential circuit given in Figure 3.4. This clocked-scan design is tested with shift and capture operations, similar to a muxed-D scan design. The main difference is how these two operations are distinguished. In a muxed-D scan design, a scan enable signal SE is used, as shown in Figure 3.6. In the clocked scan shown in Figure 3.8, these two operations are distinguished by properly applying the two independent clocks *SCK* and *DCK* during shift mode and capture mode, respectively.

3.3.1.3 *LSSD scan design*

Figure 3.9a shows a polarity-hold *shift register latch* (SRL) design described in [Eichelberger 1977] that can be used as an LSSD scan cell. This scan cell contains two latches, a master two-port D latch L_1 and a slave D latch L_2. Clocks *C*, *A*, and *B* are used to select between the data input *D* and the scan input *I* to drive $+L_1$ and $+L_2$.

To guarantee race-free operation, clocks *A*, *B*, and *C* are applied in a nonoverlapping manner. In designs in which $+L_1$ is used to drive the combinational logic, the master latch L_1 uses the system clock *C* to latch system data from the data input *D* and to output this data onto $+L_1$. In designs in which $+L_2$ is

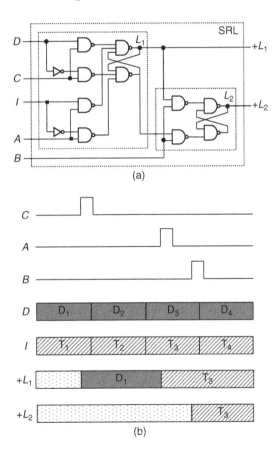

(a)

(b)

FIGURE 3.9

Polarity-hold SRL design and operation: (a) Polarity-hold SRL. (b) Sample waveforms.

used to drive the combinational logic, clock B is used after clock C to latch the system data from latch L_1 and to output these data onto $+L_2$. In both cases, capture mode uses both clocks C and B to output system data onto $+L_2$. Finally, in shift mode, clocks A and B are used to latch scan data from the scan input I and to output these data onto $+L_1$, and then latch the scan data from latch L_1 and to output these data onto $+L_2$, which is then used to drive the scan input of the next scan cell. Sample operation waveforms are shown in Figure 3.9b.

LSSD scan designs can be implemented with either a **single-latch design** or a **double-latch design**. In single-latch design [Eichelberger 1977], the output port $+L_1$ of the master latch L_1 is used to drive the combinational logic of the design. In this case, the slave latch L_2 is used only for scan testing. Because LSSD designs use latches instead of flip-flops, at least two system clocks C_1 and C_2 are required to prevent combinational feedback loops from occurring. In this case, combinational logic driven by the master latches of the first system clock C_1 are used to drive the master latches of the second system clock C_2, and vice versa. For this to work, the system clocks C_1 and C_2 should be applied in a nonoverlapping fashion. Figure 3.10a shows an LSSD single-latch design with the polarity-hold SRL shown in Figure 3.9.

Figure 3.10b shows an example of LSSD **double-latch design** [DasGupta 1982]. In normal mode, the C_1 and C_2 clocks are used in a nonoverlapping manner, where the C_2 clock is the same as the B clock. The testing of an LSSD scan design is conducted with shift and capture operations, similar to a muxed-D scan design. The main difference is how these two operations are distinguished. In a muxed-D scan design, a scan enable signal SE is used, as shown in Figure 3.6. In an LSSD scan design, these two operations are distinguished by properly applying nonoverlapping clock pulses to clocks C_1, C_2, A, and B. During the shift operation, clocks A and B are applied in a nonoverlapping manner, and the scan cells $SRL_1 \sim SRL_3$ form a single scan chain from SI to SO. During the capture operation, clocks C_1 and C_2 are applied in a nonoverlapping manner to load the test response from the combinational logic into the scan cells.

The major advantage of the use of an LSSD scan cell is that it allows us to insert scan into a latch-based design. In addition, designs that use LSSD are guaranteed to be race-free, which is not the case for muxed-D scan and clocked-scan designs. A major disadvantage, however, is that it requires routing for the additional clocks, which increases routing complexity.

The operation of a polarity-hold SRL is race-free if clocks C and B, as well as A and B, are nonoverlapping. This characteristic is used to implement LSSD circuits that are guaranteed to have race-free operation in normal mode and in test mode.

3.3.2 At-speed testing

Although scan design is commonly used in the industry for slow-speed stuck-at fault testing, its real value is in providing **at-speed testing** for high-speed and

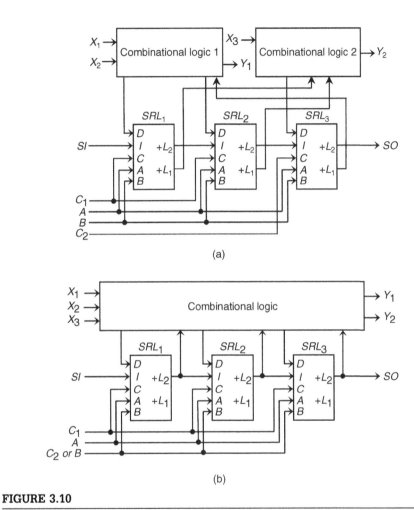

FIGURE 3.10

LSSD designs: (a) LSSD single-latch design. (b) LSSD double-latch design.

high-performance circuits. These circuits often contain multiple clock domains, each running at an operating frequency that is either synchronous or asynchronous to the other clock domains. Two clock domains are said to be **synchronous** if the active edges of both clocks controlling the two clock domains can be aligned precisely or triggered simultaneously. Two clock domains are said to be **asynchronous** if they are not synchronous.

There are two basic capture-clocking schemes for testing multiple clock domains at-speed: (1) *skewed-load* [Savir 1993] (also called *launch-on-shift* [LOS]) and (2) *double-capture* [Wang 2006a] (also called *launch-on-capture* [LOC] or *broad-side* [Savir 1994]). Both schemes can be used to test path-delay faults and transition faults within each clock domain (called **intra-clock-domain**

faults) or across clock domains (called **inter-clock-domain faults**). Skewed-load uses the last shift clock pulse followed immediately by a capture clock pulse to launch the transition and capture the output test response, respectively. Double-capture uses two consecutive capture clock pulses to launch the transition and capture the output test response, respectively. In both schemes, both launch and capture clock pulses must be running at the domain's operating speed or at-speed. The difference is that skewed-load requires the domain's scan enable signal *SE* to switch its value between the launch and capture clock pulses making *SE* act as a clock signal. Figure 3.11 shows sample waveforms that use the basic skewed-load and double-capture at-speed test schemes.

Scan designs typically include a few clock domains that will interact with one another. To guarantee the success of the capture operation, additional care must be taken in terms of the way the capture clocks are applied. This is mainly because the clock skew between different clock domains is typically large. To prevent this from happening, clocks can be applied sequentially (with the **staggered clocking** scheme [Wang 2005a, 2007b]), such that any clock skew that exists between the clock domains can be tolerated during the test generation process. It is also possible to apply only one clock during each capture operation by use of the **one-hot clocking** scheme. Most modern ATPG programs used currently can also automatically mask off unknown values (*X*'s) at the originating scan cells or receiving scan cells across clock domains. In this case, all clocks can also be applied simultaneously with the **simultaneous clocking** scheme [Wang 2007b]. During simultaneous clocking, if the launch clock pulses [Rajski 2003; Wang 2006a] or the capture clock pulses [Nadeau-Dostie 1994; Wang 2006a] can be aligned precisely, which applies only for synchronous clock domains, then the **aligned clocking** scheme can be used, and there is no need to mask off unknown values across these synchronous clock domains. These clocking schemes are illustrated in Figure 3.12.

In general, one-hot clocking produces the highest fault coverage at the expense of generating many more test patterns than other schemes. Simultaneous clocking can generate the smallest number of test patterns but may result

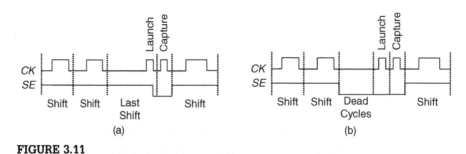

FIGURE 3.11

Basic at-speed test schemes: (a) Skewed-load. (b) Double-capture.

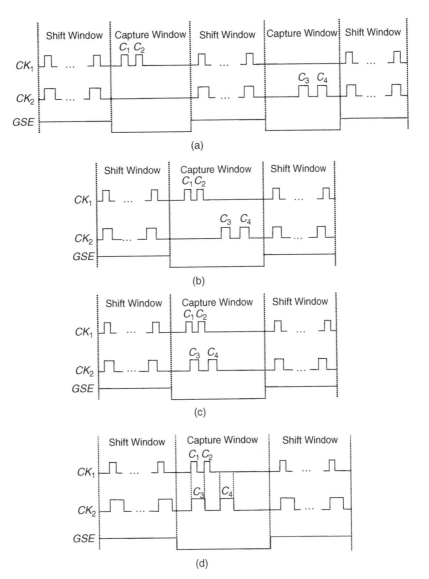

FIGURE 3.12

At-speed clocking schemes for testing two interacting clock domains: (a) One-hot clocking. (b) Staggered clocking. (c) Simultaneous clocking. (d) Aligned clocking.

in high fault coverage loss because of unknown (X) masking. The staggered clocking scheme is a happy medium because of its ability to generate test pattern count close to simultaneous clocking and fault coverage close to one-hot clocking. For large designs, it is no longer uncommon for transition fault ATPG to take more than 2 to 4 weeks to complete. To reduce test generation time while at the same time obtaining the highest fault coverage, modern ATPG programs tend to either (1) run simultaneous clocking followed by one-hot clocking or (2) use staggered clocking followed by one-hot clocking. As a result, modern **at-speed scan architectures** now start supporting a combination of at-speed clocking schemes for test circuits comprising multiple synchronous and asynchronous clock domains. Some programs can even generate test patterns by mixing skewed-load and double-capture schemes.

3.4 LOGIC BUILT-IN SELF-TEST

Logic built-in self-test (BIST) requires using a portion of the circuit to test itself on-chip, on-board, or in-system. A typical logic BIST system is illustrated in Figure 3.13. The *test pattern generator* (TPG) automatically generates test patterns for application to the inputs of the *circuit under test* (CUT). The *output response analyzer* (ORA) automatically compacts the output responses of the CUT into a *signature*. Specific BIST timing control signals, including scan enable signals and clocks, are generated by the **logic BIST controller** for coordinating the BIST operation among the TPG, CUT, and ORA. The logic BIST controller provides a pass/fail indication once the BIST operation is complete. It includes comparison logic to compare the *final signature* with an embedded *golden signature*, and often comprises **diagnostic logic** for fault diagnosis. Because compaction is commonly used for output response analysis, it is required that all storage elements in the TPG, CUT, and ORA be initialized to known states before self-test, and no unknown (*X*) values are allowed to propagate from the CUT to the ORA. In other words, the CUT must comply with more stringent **BIST-specific design rules** [Wang 2006a] in addition to those *scan design rules* required for scan design.

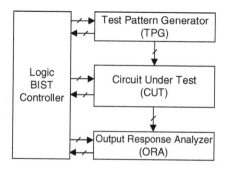

FIGURE 3.13

A typical logic BIST system.

3.4.1 **Test pattern generation**

For logic BIST applications, in-circuit TPGs constructed from ***linear feedback shift registers*** (**LFSRs**) are most commonly used to generate test patterns or test sequences for exhaustive testing, pseudo-random testing, and pseudo-exhaustive testing.

Exhaustive testing always guarantees 100% single-stuck and multiple-stuck fault coverage. This technique requires all possible 2^n test patterns to be applied to an n-input combinational CUT, which can take too long for combinational circuits where n is huge. Therefore, **pseudo-random testing** [Bardell 1987] is often used for generating a subset of the 2^n test patterns and uses fault simulation to calculate the exact fault coverage. In some cases, this might become quite time-consuming, if not infeasible. To eliminate the need for fault simulation while at the same time maintaining 100% single-stuck fault coverage, we can use **pseudo-exhaustive testing** [McCluskey 1986] to generate 2^w or $2^k - 1$ test patterns, where $w < k < n$, when each output of the n-input combinational CUT at most depends on w inputs. For testing delay faults, hazards must also be taken into consideration.

Standard LFSR

Figure 3.14 shows an n-stage **standard LFSR**. It consists of n D flip-flops and a selected number of *exclusive-OR* (XOR) gates. Because XOR gates are placed on the external feedback path, the standard LFSR is also referred to as an **external-XOR LFSR** [Golomb 1982].

Modular LFSR

Similarly, an n-stage **modular LFSR** with each XOR gate placed between two adjacent D flip-flops, as shown in Figure 3.15, is referred to as an **internal-XOR LFSR** [Golomb 1982]. The modular LFSR runs faster than its corresponding standard LFSR, because each stage introduces at most one XOR-gate delay.

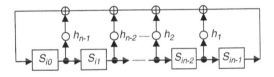

FIGURE 3.14

An n-stage (external-XOR) standard LFSR.

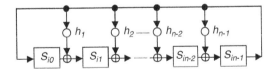

FIGURE 3.15

An n-stage (internal-XOR) modular LFSR.

LFSR Properties

The internal structure of the n-stage LFSR in each figure can be described by specifying a **characteristic polynomial** of degree n, $f(x)$, in which the symbol h_i is either 1 or 0, depending on the existence or absence of the feedback path, where

$$f(x) = 1 + h_1 x + h_2 x^2 + \ldots + h_{n-1} x^{n-1} + x^n$$

Let S_i represent the contents of the n-stage LFSR after ith shifts of the initial contents, S_0, of the LFSR, and $S_i(x)$ be the polynomial representation of S_i. Then, $S_i(x)$ is a polynomial of degree $n-1$, where

$$S_i(x) = S_{i0} + S_{i1} x + S_{i2} x^2 + \ldots + S_{in-2} x^{n-2} + S_{in-1} x^{n-1}$$

If T is the smallest positive integer such that $f(x)$ divides $1 + x^T$, then the integer T is called the **period** of the LFSR. If $T = 2^n - 1$, then the n-stage LFSR generating the **maximum-length sequence** is called a **maximum-length LFSR**.

For example, consider the four-stage standard and modular LFSRs shown in Figures 3.16a and 3.16b below. The characteristic polynomials, $f(x)$, used to construct both LFSRs are $1 + x^2 + x^4$ and $1 + x + x^4$, respectively.

The test sequences generated by each LFSR, when its initial contents, S_0, are set to {0001} or $S_0(x) = x^3$, are listed in Figures 3.16c and 3.16d, respectively.

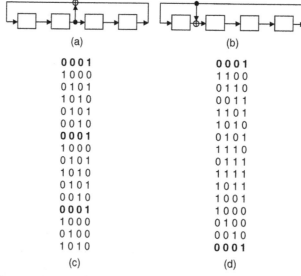

(a)	(b)

(c)	(d)
0 0 0 1	0 0 0 1
1 0 0 0	1 1 0 0
0 1 0 1	0 1 1 0
1 0 1 0	0 0 1 1
0 1 0 1	1 1 0 1
0 0 1 0	1 0 1 0
0 0 0 1	0 1 0 1
1 0 0 0	1 1 1 0
0 1 0 1	0 1 1 1
1 0 1 0	1 1 1 1
0 1 0 1	1 0 1 1
0 0 1 0	1 0 0 1
0 0 0 1	1 0 0 0
1 0 0 0	0 1 0 0
0 1 0 0	0 0 1 0
1 0 1 0	0 0 0 1

FIGURE 3.16

Example four-stage test pattern generators (TPGs): (a) Four-stage standard LFSR. (b) Four-stage modular LFSR. (c) Test sequence generated by (a). (d) Test sequence generated by (b).

Because the first test sequence repeats after 6 patterns and the second test sequence repeats after 15 patterns, the LFSRs have periods of 6 and 15, respectively. This further implies that $1 + x^6$ can be divided by $1 + x^2 + x^4$, and $1 + x^{15}$ can be divided by $1 + x + x^4$.

Define a **primitive polynomial** of degree n over **Galois field** $GF(2)$, $p(x)$, as a polynomial that divides $1 + x^T$, but not $1 + x^i$, for any integer $i < T$, where $T = 2^n - 1$ [Golomb 1982]. A primitive polynomial is **irreducible**. Because $T = 15 = 2^4 - 1$, the characteristic polynomial, $f(x) = 1 + x + x^4$, used to construct Figure 3.16b is a primitive polynomial, and thus the modular LFSR is a maximum-length LFSR. Let

$$r(x) = f(x)^{-1} = x^n f(x^{-1})$$

Then $r(x)$ is defined as a **reciprocal polynomial** of $f(x)$ [Peterson 1972]. A reciprocal polynomial of a primitive polynomial is also a primitive polynomial. Thus, the reciprocal polynomial of $f(x) = 1 + x + x^4$ is also a primitive polynomial, with $p(x) = r(x) = 1 + x^3 + x^4$.

Table 3.5 lists a set of primitive polynomials of degree n up to 100. It was taken from [Bardell 1987]. A different set was given in [Wang 1988a]. Each polynomial can be used to construct minimum-length LFSRs in standard or modular form. For primitive polynomials of degree up to 300, consult [Bardell 1987].

3.4.1.1 *Exhaustive testing*

Exhaustive testing requires applying 2^n exhaustive patterns to an n-input combinational CUT. Any **binary counter** can be used as an **exhaustive pattern generator** (EPG) for this purpose. Figure 3.17 shows an example of a 4-bit binary counter design for testing a 4-input combinational CUT.

Exhaustive testing guarantees that all detectable, combinational faults (those that do not change a combinational circuit into a sequential circuit) will be detected. This approach is especially useful for circuits in which the number of inputs, n, is a small number (*e.g.*, 20 or less). When n is larger than 20, the test time may be prohibitively long and is thus not recommended. The following techniques are aimed at reducing the number of test patterns. They are recommended when exhaustive testing is impractical.

3.4.1.2 *Pseudo-random testing*

One approach, which can reduce test length but sacrifices the circuit's fault coverage, uses a **pseudo-random pattern generator** (PRPG) for generating a pseudo-random sequence of test patterns [Bardell 1987; Rajski 1998; Bushnell 2000; Jha 2003]. **Pseudo-random testing** has the advantage of being applicable to both sequential and combinational circuits; however, there are difficulties in determining the required test length and fault coverage.

Table 3.5 Primitive Polynomials of Degree *n* up to 100

n	Exponents	*n*	Exponents	*n*	Exponents	*n*	Exponents
1	0	26	8 7 1 0	51	16 15 1 0	76	36 35 1 0
2	1 0	27	8 7 1 0	52	3 0	77	31 30 1 0
3	1 0	28	3 0	53	16 15 1 0	78	20 19 1 0
4	1 0	29	2 0	54	37 36 1 0	79	9 0
5	2 0	30	16 15 1 0	55	24 0	80	38 37 1 0
6	1 0	31	3 0	56	22 21 1 0	81	4 0
7	1 0	32	28 27 1 0	57	7 0	82	38 35 3 0
8	6 5 1 0	33	13 0	58	19 0	83	46 45 1 0
9	4 0	34	15 14 1 0	59	22 21 1 0	84	13 0
10	3 0	35	2 0	60	1 0	85	28 27 1 0
11	2 0	36	11 0	61	16 15 1 0	86	13 12 1 0
12	7 4 3 0	37	12 10 2 0	62	57 56 1 0	87	13 0
13	4 3 1 0	38	6 5 1 0	63	1 0	88	72 71 1 0
14	12 11 1 0	39	4 0	64	4 3 1 0	89	38 0
15	1 0	40	21 19 2 0	65	18 0	90	19 18 1 0
16	5 3 2 0	41	3 0	66	10 9 1 0	91	84 83 1 0
17	3 0	42	23 22 1 0	67	10 9 1 0	92	13 12 1 0
18	7 0	43	6 5 1 0	68	9 0	93	2 0
19	6 5 1 0	44	27 26 1 0	69	29 27 2 0	94	21 0
20	3 0	45	4 3 1 0	70	16 15 1 0	95	11 0
21	2 0	46	21 20 1 0	71	6 0	96	49 47 2 0
22	1 0	47	5 0	72	53 47 6 0	97	6 0
23	5 0	48	28 27 1 0	73	25 0	98	11 0
24	4 3 1 0	49	9 0	74	16 15 1 0	99	47 45 2 0
25	3 0	50	27 26 1 0	75	11 10 1 0	100	37 0

Note: "24 4 3 1 0" means $p(x) = x^{24} + x^4 + x^3 + x^1 + x^0 = x^{24} + x^4 + x^3 + x + 1.$

3.4.1.2.1 Maximum-length LFSR

Maximum-length LFSRs are commonly used for pseudo-random pattern generation. Each LFSR produces a sequence with 0.5 probability of generating 1's

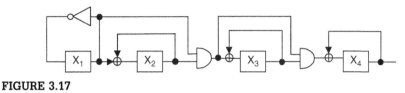

FIGURE 3.17

Example binary counter as EPG.

(or with probability distribution 0.5) at every output. The **LFSR pattern generation technique** that uses these LFSRs, in standard or modular form, to generate patterns for the entire design has the advantage of being very easy to implement. The major problem with this approach is that some circuits may be *random pattern resistant* (RP-resistant). For instance, consider a 5-input OR gate. The probability of applying an all-zero pattern to all inputs is 1/32. This makes it difficult to test the RP-resistant OR-gate output stuck-at-1.

3.4.1.2.2 Weighted LFSR

It is possible to increase fault coverage (and detect most RP-resistant faults) in RP-resistant designs. A **weighted pattern generation technique** that uses an LFSR and a combinational circuit was first described in [Schnurmann 1975]. The combinational circuit inserted between the output of the LFSR and the CUT is to increase the frequency of occurrence of one logic value while decreasing the other logic value. This approach may increase the probability of detecting those faults that are hard to detect with the typical LFSR pattern generation technique.

Implementation methods for realizing this scheme are further discussed in [Chin 1984]. The weighted pattern generation technique described in that paper modifies the maximum-length LFSR to produce an equally weighted distribution of 0's and 1's at the input of the CUT. It skews the LFSR probability distribution of 0.5 to either 0.25 or 0.75 to increase the chance of detecting those faults that are hard to detect with just a 0.5 distribution. Better fault coverage was also found in [Wunderlich 1987], where probability distributions in a multiple of 0.125 (rather than 0.25) are used. Figure 3.18 shows a four-stage weighted (maximum-length) LFSR with probability distribution 0.25 [Chin 1984].

3.4.1.2.3 Cellular automata

Cellular automata were first introduced in [Wolfram 1983]. They yielded better randomness property than LFSRs [Hortensius 1989]. The *cellular automaton based* (or CA-based) *pseudo-random pattern generator* (PRPG) is attractive for BIST applications [Khara 1987; Gloster 1988; Wang 1989; van Sas 1990] because it (1) provides patterns that look *more* random at the circuit inputs, (2) has higher opportunity to reach very high fault coverage in a circuit that is RP-resistant, and

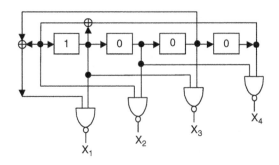

FIGURE 3.18

Example weighted LFSR as PRPG.

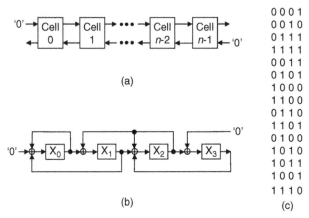

FIGURE 3.19

Example cellular automaton (CA) as PRPG: (a) General structure of an n-stage CA. (b) Four-stage CA. (c) Test sequence generated by (b).

(3) has implementation advantages because it only requires adjacent neighbor communication (no global feedback unlike the modular LFSR case).

A *cellular automaton* (CA) is a collection of **cells** with forward and backward connections. A general structure is shown in Figure 3.19a. Each cell can only connect to its local neighbors (adjacent left and right cells). The connections are expressed as **rules**; each rule determines the next state of a cell on the basis of the state of the cell and its neighbors. Assume cell i can only talk with its neighbors, $i - 1$ and $i + 1$. Define:

$$Rule\,90: x_i(t + 1) = x_{i-1}(t) + x_{i+1}(t)$$

and

$$Rule\ 150: x_i(t+1) = x_{i-1}(t) + x_i(t) + x_{i+1}(t)$$

Then the two rules, *rule 90* and *rule 150*, can be established on the basis of the following state transition table:

$x_{i-1}(t)x_i(t)x_{i+1}(t)$	111	110	101	100	011	010	001	000
Rule 90: $x_i(t+1)$	0	1	0	1	1	0	1	0

$$2^6 + 2^4 + 2^3 + 2^1 = 90$$

	111	110	101	100	011	010	001	000
Rule 150: $x_i(t+1)$	1	0	0	1	0	1	1	0

$$2^7 + 2^4 + 2^2 + 2^1 = 150$$

The terms *rule 90* and *rule 150* were derived from their decimal equivalents of the binary code for the next state of cell *i* [Hortensius 1989]. Figure 3.19b shows an example of a four-stage CA generated by alternating rules 150 (on even cells) and 90 (on odd cells). Similar to the four-stage modular LFSR given in Figure 3.16b, the four-stage CA generates a *maximum-length sequence* of 15 distinct states as listed in Figure 3.19c.

It has been shown in [Hortensius 1989] that by combining cellular automata rules 90 and 150, an *n*-stage CA can generate a maximum-length sequence of 2^n-1. The construction rules for $4 \leq n \leq 53$ can be found in [Hortensius 1989] and are listed in Table 3.6.

The CA-based PRPG can be programmed as a **universal CA** for generating different orders of test sequences. A **universal CA-cell** for generating patterns on the basis of *rule 90* or *rule 150* is given in Figure 3.20 [Wang 1989]. When the RULE150_SELECT signal is set to 1, the universal CA-cell will behave as a *rule 150* cell; otherwise, it will act as a *rule 90* cell. This *universal CA* structure is useful for BIST applications where it is required to obtain very high fault coverage for RP-resistant designs or detect additional classes of faults.

3.4.1.3 *Pseudo-exhaustive testing*

Another approach to reduce the test time to a practical value while retaining many of the advantages of exhaustive testing is the **pseudo-exhaustive test technique**. It applies fewer than 2^n test patterns to an *n*-input combinational CUT. The technique depends on whether any output is driven by all of its inputs. If none of the outputs depends on all inputs, a **verification test approach** proposed in [McCluskey 1984] can be used to test these circuits. In circuits in which there is one output that depends on all inputs or the test time that uses verification testing is still too long, a **segmentation test approach** must be used [McCluskey 1981]. Pseudo-exhaustive testing guarantees single-stuck fault coverage without any detailed circuit analysis.

Table 3.6 Construction Rules for Cellular Automat of Length *n* up to 53

n	Rule*	*n*	Rule*
4	05	29	2,512,712103
5	31	30	7,211,545,075
6	25	31	04,625,575,630
7	152	32	10,602,335,725
8	325	33	03,047,162,605
9	625	34	036,055,030,672
10	0,525	35	127,573,165,123
11	3,252	36	514,443,726,043
12	2,252	37	0,226,365,530,263
13	14,524	38	0,345,366,317,023
14	17,576	39	6,427,667,463,554
15	44,241	40	00,731,257,441,345
16	152,525	41	15,376,413,143,607
17	175,763	42	11,766,345,114,746
18	252,525	43	035,342,704,132,622
19	0,646,611	44	074,756,556,045,302
20	3,635,577	45	151,315,510,461,515
21	3,630,173	46	0,112,312,150,547,326
22	05,252,525	47	0,713,747,124,427,015
23	32,716,532	48	0,606,762,247,217,017
24	77,226,526	49	02,675,443,137,056,631
25	136,524,744	50	23,233,006,150,544,226
26	132,642,730	51	04,135,241,323,505,027
27	037,014,415	52	031,067,567,742,172,706
28	0,525,252,525	53	207,121,011,145,676,625

*Rule is given in octal format. For $n = 7$, Rule = $152 = 001,101,010 = 1,101,010$, where "0" denotes a *rule 90* cell and "1" denotes a *rule 150* cell, or vice versa.

RULE150_SELECT

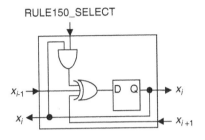

FIGURE 3.20

A universal CA-cell structure.

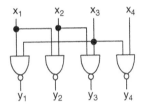

FIGURE 3.21

An $(n,w) = (4,2)$ CUT.

Verification testing [McCluskey 1984] divides the circuit under test into m cones, where m is the number of outputs. It is based on backtracing from each circuit output to determine the actual number of inputs that drive the output. Each cone will receive exhaustive test patterns, and all cones are tested concurrently.

Assume the combinational CUT has n inputs and m outputs. Let w be the maximum number of input variables on which any output of the CUT depends. Then, the n-input m-output combinational CUT is defined as an (n,w) CUT, where $w < n$. Figure 3.21 shows an $(n,w) = (4,2)$ CUT that will be used as an example for designing the *pseudo-exhaustive pattern generators* (PEPGs).

3.4.1.3.1 Syndrome driver counter

The first method for **pseudo-exhaustive pattern generation** was proposed in [Savir 1980]. ***Syndrome driver counters*** (SDCs) are used to generate test patterns [Barzilai 1981]. The SDC can be a binary counter, a maximum-length LFSR, or a complete LFSR. This method checks whether some circuit inputs can share the same test signal. If n-p inputs, $p < n$, can share the **test signals** with the other p inputs, then the circuit can be tested exhaustively with these p inputs. In this case, the test length becomes 2^p if $p = w$, or $2^p - 1$ if $p > w$. Figure 3.22 shows a three-stage SDC used to test the circuit given in Figure 3.21. Because both inputs x_1 and x_4 do

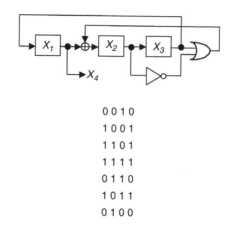

$$
\begin{array}{cccc}
0 & 0 & 1 & 0 \\
1 & 0 & 0 & 1 \\
1 & 1 & 0 & 1 \\
1 & 1 & 1 & 1 \\
0 & 1 & 1 & 0 \\
1 & 0 & 1 & 1 \\
0 & 1 & 0 & 0
\end{array}
$$

FIGURE 3.22

Example syndrome driver counter as PEPG.

not drive the same output, one test signal can be used to drive both inputs. In this case, p is 3, and the test length becomes $2^3 - 1 = 7$. Designs based on the SDC method for in-circuit test pattern generation are simple. The problem with this method is that when p is close to n, it may still take too long to test the circuit.

3.4.1.3.2 Condensed LFSR

The problem can be solved by use of the **condensed LFSR** approach proposed in [Wang 1986a]. Condensed LFSRs are constructed on the basis of **linear codes** [Peterson 1972]. An (n,k) *linear code over GF(2)* generates a code space C containing 2^k distinct code words (n-tuples) with the following property: if $c_1 \in C$ and $c_2 \in C$, then $c_1 + c_2 \in C$. Define an (n,k) *condensed LFSR* as an n-stage modular LFSR with period $2^k - 1$. A condensed LFSR for testing an (n,w) CUT is constructed by first computing the smallest integer k such that:

$$
w \le \lceil k/(n-k+1) \rceil + \lfloor k/(n-k+1) \rfloor
$$

where $\lceil x \rceil$ denotes the smallest integer equal to or greater than the real number x, and $\lfloor y \rfloor$ denotes the largest integer equal to or smaller than the real number y.

Then, by use of:

$$
f(x) = g(x)p(x) = (1 + x + x^2 + \ldots + x^{n-k})p(x)
$$

an (n,k) *condensed LFSR* can be realized, where $g(x)$ is a **generator polynomial** of degree n-k generating the (n,k) linear code, and $p(x)$ is a primitive polynomial of degree k.

Consider the $(n,k) = (4,3)$ condensed LFSR shown in Figure 3.23a used to test the $(n,w) = (4,2)$ CUT. Because $n = 4$ and $w = 2$, we obtain $k = 3$ and

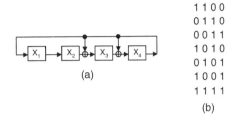

```
1 1 0 0
0 1 1 0
0 0 1 1
1 0 1 0
0 1 0 1
1 0 0 1
1 1 1 1
```
(b)

FIGURE 3.23

Example condensed LFSR as PEPG: (a) (4,3) condensed LFSR. (b) Test sequence generated by (a).

$(n - k) = 1$. Selecting $p(x) = 1 + x + x^3$, we have $f(x) = (1 + x)(1 + x + x^3) = 1 + x^2 + x^3 + x^4$. Figure 3.23b lists the generated period-7 test sequence. It is important to note that the seed polynomial $S_0(x)$ of the LFSR must be divisible by $g(x)$. In the example, we set $S_0(x) = g(x) = 1 + x$, or S_0 to {1100}.

For any given (n,w) CUT, this method uses at most two seeds and has shown to be effective when $w \geq n/2$. Designs based on this method are simple. However, this technique uses more patterns than the **combined LFSR/SR** approach, which uses a combination of an LFSR and a *shift register* (SR) [Barzilai 1983; Tang 1984; Chen 1987] and the **cyclic LFSR** approach [Wang 1987, 1988b] when $w < n/2$. For other verification test approaches, refer to [Abramovici 1994; Wang 2006a].

3.4.2 Output response analysis

For scan designs, our assumption was that output responses coming out of the *circuit under test* (CUT) are compared directly on a tester. For BIST operations, it is impossible to store all output responses on-chip, on-board, or in-system to perform bit-by-bit comparison. An *output response analysis* technique must be used such that output responses can be compacted into a **signature** and compared with a *golden signature* for the fault-free circuit either embedded on-chip or stored off-chip.

Compaction differs from *compression* in that compression is loss-less, whereas compaction is lossy. **Compaction** is a method for dramatically reducing the number of bits in the original circuit response during testing in which some information is lost. **Compression** is a method for reducing the number of bits in the original circuit response in which no information is lost, such that the original output sequence can be fully regenerated from the compressed sequence [Bushnell 2000]. Because all output response analysis schemes involve information loss, they are referred to as *output response compaction*. However, there is no general consensus in academia yet as to when the terms compaction or compression are to be used. However, for output response analysis, throughout the book, we will refer to the lossy compression as compaction.

In this section, we will present three different output response compaction techniques: (1) **ones count testing**, (2) **transition count testing**, and (3) **signature analysis**. We will also describe the architectures of the *output response analyzers* (ORAs) that are used. The signature analysis technique will be described in more detail, because it is the most popular compaction technique in use today.

When compaction is used, it is important to ensure that the faulty and fault-free signatures are different. If they are the same, the fault(s) can go undetected. This situation is referred to as **error masking**, and the erroneous output response is said to be an **alias** of the correct output response [Abramovici 1994]. It is also important to ensure that none of the output responses contains an unknown (*X*) value. If an unknown value is generated and propagated directly or indirectly to the ORA, then the ORA can no longer function reliably. Therefore, it is required that all unknown (*X*) propagation problems be fixed to ensure that the logic BIST system will operate correctly. Such **X-blocking** or **X-bounding** techniques have been extensively discussed in [Wang 2006a].

3.4.2.1 *Ones count testing*

Assume that the CUT has only one output and the output contains a stream of L bits. Let the fault-free output response, R_0, be $\{r_0 \, r_1 \, r_2 \dots r_{L-1}\}$. The **ones count test technique** will only need a counter to count the number of 1's in the bit stream. For instance, if $R_0 = \{0101100\}$, then the signature or ones count of R_0, $OC(R_0)$, is 3. If fault f_1 present in the CUT causes an erroneous response $R_1 = \{1100110\}$, then it will be detected because $OC(R_1) = 4$. However, fault f_2 causing $R_2 = \{0101010\}$ will not be detected because $OC(R_2) = OC(R_0) = 3$. Let the fault-free signature or ones count be m. There will be $C(L,m)$ possible ways having m 1's in an L-bit stream. Assuming all faulty sequences are equally likely to occur as the response of the CUT, the **aliasing probability** or **masking probability** of the use of ones count testing having m 1's [Savir 1985] can be expressed as

$$P_{OC}(m) = \Big(C(L, m) - 1\Big)/(2^L - 1)$$

In the previous example, where $m = OC(R_0) = 3$ and $L = 7$, $P_{OC}(m) = 34/127 = 0.27$. Figure 3.24 shows the ones count test circuit for testing the CUT with T patterns. The number of stages in the counter design must be equal to or greater than $\lceil \log_2(L + 1) \rceil$.

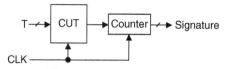

FIGURE 3.24

Ones counter as ORA.

3.4.2.2 *Transition count testing*

The theory behind transition count testing is similar to that for ones count testing, except the signature is defined as the number of 0-to-1 and 1-to-0 transitions. The **transition count test technique** [Hayes 1976] simply requires the use of a D flip-flop and an XOR gate connected to a ones counter (see Figure 3.25) to count the number of transitions in the output data stream. Consider the example given previously. Because $R_0 = \{0101100\}$, the signature or transition count of R_0, $TC(R_0)$, will be 4. Assume that the initial state of the D flip-flop, r_{-1}, is 0. Fault f_1 causing an erroneous response $R_1 = \{1100110\}$ will not be detected because $TC(R_1) = TC(R_0) = 4$, whereas fault f_2 causing $R_2 = \{0101010\}$ will be detected because $TC(R_2) = 6$.

Let the fault-free signature or transition count be m. Because a given L-bit sequence R_0 that starts with $r_0 = 0$ has $L - 1$ possible transitions, the number of sequences with m transitions can be given by $C(L - 1, m)$. Because R_0 can also start with $r_0 = 1$, there will be a total of $2C(L - 1, m)$ possible ways having m 0-to-1 and 1-to-0 transitions in an L-bit stream. Assuming all faulty sequences are equally likely to occur as the response of the CUT, the *aliasing probability* or *masking probability* of the use of transition count testing having m transitions [Savir 1985] is

$$P_{TC}(m) = \Big(2C(L - 1, m) - 1\Big)/(2^L - 1)$$

In the previous example, where $m = TC(R_0) = 4$ and $L = 7$, $P_{TC}(m) = 29/127 = 0.23$. Figure 3.25 shows the transition count test circuit. The number of stages in the counter design must be equal to or greater than $\lceil \log_2(L + 1) \rceil$.

3.4.2.3 *Signature analysis*

Signature analysis is the most popular response compaction technique used today. The compaction scheme, based on **cyclic redundancy checking** (CRC) [Peterson 1972], was first developed in [Benowitz 1975]. Hewlett-Packard commercialized the first logic analyzer, called HP 5004A Signature Analyzer, based on the scheme and referred to it as **signature analysis** [Frohwerk 1977].

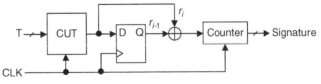

FIGURE 3.25

Transition counter as ORA.

In this subsection, we will discuss two signature analysis schemes: (1) **serial signature analysis** for compacting responses from a CUT having a single output and (2) **parallel signature analysis** for compacting responses from a CUT having multiple outputs.

3.4.2.3.1 Serial Signature Analysis

Consider the n-stage **single-input signature register** (SISR) shown in Figure 3.26. This SISR uses an additional XOR gate at the input for compacting an L-bit output sequence, M, into the *modular LFSR*. Let $M = \{m_0\, m_1\, m_2 \ldots m_{L-1}\}$, and define:

$$M(x) = m_0 + m_1 x + m_2 x^2 + \ldots + m_{L-1} x^{L-1}$$

After shifting the L-bit output sequence, M, into the *modular LFSR*, the contents (remainder) of the SISR, R, is given as $\{r_0\, r_1\, r_2 \ldots r_{n-1}\}$, or

$$r(x) = r_0 + r_1 x + r_2 x^2 + \ldots + r_{n-1} x^{n-1}$$

The SISR is basically a *CRC code generator* [Peterson 1972] or a *cyclic code checker* [Benowitz 1975]. Let the *characteristic polynomial* of the modular LFSR be $f(x)$. The authors in [Peterson 1972] have shown that the SISR performs polynomial division of $M(x)$ by $f(x)$, or

$$M(x) = q(x)f(x) + r(x)$$

The final state or **signature** in the SISR is the *polynomial remainder*, $r(x)$, of the division. Consider the four-stage SISR given in Figure 3.27 with $f(x) = 1 + x + x^4$. Assuming $M = \{10011011\}$, we can express $M(x) = 1 + x^3 + x^4 + x^6 + x^7$. By use of polynomial division, we obtain $q(x) = x^2 + x^3$ and $r(x) = 1 + x^2 + x^3$ or $R = \{1011\}$. The remainder $\{1011\}$ is equal to the *signature* derived from Figure 3.27a when the SISR is first initialized to a *starting pattern* (*seed*) of $\{0000\}$.

Now, assume fault f_1 produces an erroneous output stream $M' = \{11001011\}$ or $M'(x) = 1 + x + x^4 + x^6 + x^7$, as given in Figure 3.27b. By use of polynomial division, we obtain $q'(x) = x^2 + x^3$ and $r'(x) = 1 + x + x^2$ or $R' = \{1110\}$. Because the faulty signature R', $\{1110\}$, is different from the fault-free signature R, $\{1011\}$, fault f_1 is detected. For fault f_2 with $M'' = \{11001101\}$ or $M''(x) = 1 + x + x^4 + x^5 + x^7$ as given in Figure 3.27c, we have $q''(x) = x + x^3$ and $r''(x) = 1 + x^2 + x^3$ or $R'' = \{1011\}$. Because $R'' = R$, fault f_2 is not detected.

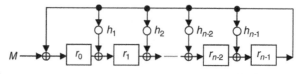

FIGURE 3.26

An n-stage single-input signature register (SISR).

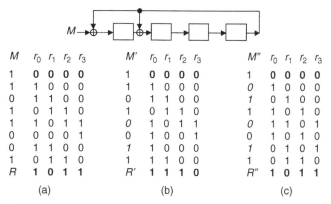

M	r_0 r_1 r_2 r_3	M'	r_0 r_1 r_2 r_3	M''	r_0 r_1 r_2 r_3
1	0 0 0 0	1	0 0 0 0	1	0 0 0 0
1	1 0 0 0	1	1 0 0 0	*0*	1 0 0 0
0	1 1 0 0	0	1 1 0 0	*1*	0 1 0 0
1	0 1 1 0	1	0 1 1 0	1	1 0 1 0
1	1 0 1 1	*0*	1 0 1 1	*0*	1 1 0 1
0	0 0 0 1	0	1 0 0 1	0	1 0 1 0
0	1 1 0 0	*1*	1 0 0 0	*1*	0 1 0 1
1	0 1 1 0	1	1 1 0 0	1	0 1 1 0
R	1 0 1 1	R'	1 1 1 0	R''	1 0 1 1
	(a)		(b)		(c)

FIGURE 3.27

A four-stage SISR: (a) Fault-free signature. (b) Signature for fault f_1. (c) Signature for fault f_2.

The *fault detection* or *aliasing* problem of an SISR can be better understood by looking at the *error sequence E* or *error polynomial $E(x)$* of the fault-free sequence M and a faulty sequence M'. Define $E = M + M'$, or:

$$E(x) = M(x) + M'(x)$$

If $E(x)$ is not divisible by $f(x)$, then all faults generating the faulty sequence M' will be detected. Otherwise, these faults are not detected. Consider fault f_1 again. We obtain $E = \{01010000\} = M + M' = \{10011011\} + \{11001011\}$ or $E(x) = x + x^3$. Because $E(x)$ is not divisible by $f(x) = 1 + x + x^4$, fault f_1 is detected. Consider fault f_2 again. We have $E = \{01010110\} = M + M'' = \{10011011\} + \{11001101\}$ or $E(x) = x + x^3 + x^5 + x^6$. Because $f(x)$ divides $E(x)$, i.e., $E(x) = (x + x^2)f(x)$, fault f_2 is not detected.

Assume the SISR consists of n stages. For a given L-bit sequence, $L > n$, there are $2^{(L-n)}$ possible ways of producing an n-bit signature of which one is the correct signature. Because there are a total of $2^L - 1$ erroneous sequences in an L-bit stream, the *aliasing probability* with an n-stage SISR for *serial signature analysis* (SSA) is:

$$P_{SSA}(n) = \left(2^{(L-n)} - 1\right)/(2^L - 1)$$

If $L >> n$, then $P_{SSA}(n) \approx 2^{-n}$. When $n = 20$, $P_{SSA}(n) < 2^{-20} = 0.0001\%$.

3.4.2.3.2 Parallel Signature Analysis

A common problem when using ones count testing, transition count testing, and serial signature analysis is the excessive hardware cost required to test an m-output CUT. It is possible to reduce the hardware cost by use of an m-to-1 multiplexer, but this increases the test time m times.

Consider the n-stage ***multiple-input signature register*** (MISR) shown in Figure 3.28. The MISR uses n extra XOR gates for compacting n L-bit output sequences, M_0 to M_{n-1}, into the *modular LFSR* simultaneously.

[Hassan 1984] has shown that the n-input MISR can be remodeled as a single-input SISR with *effective input sequence* $M(x)$ and *effective error polynomial* $E(x)$ expressed as:

$$M(x) = M_0(x) + xM_1(x) + \ldots + x^{n-2}M_{n-2}(x) + x^{n-1}M_{n-1}(x)$$

and

$$E(x) = E_0(x) + xE_1(x) + \ldots + x^{n-2}E_{n-2}(x) + x^{n-1}E_{n-1}(x)$$

Consider the four-stage MISR shown in Figure 3.29 that uses $f(x) = 1 + x + x^4$. Let $M_0 = \{10010\}$, $M_1 = \{01010\}$, $M_2 = \{11000\}$, and $M_3 = \{10011\}$. From this information, the signature R of the MISR can be calculated as $\{1011\}$. With $M(x) = M_0(x) + xM_1(x) + x^2M_2(x) + x^3M_3(x)$, we obtain $M(x) = 1 + x^3 + x^4 + x^6 + x^7$ or $M = \{10011011\}$ as shown in Figure 3.30. This is the same data stream we used in the SISR example in Figure 3.27a. Therefore, $R = \{1011\}$.

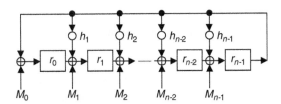

FIGURE 3.28

An n-stage multiple-input signature register (MISR).

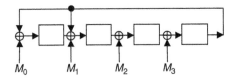

FIGURE 3.29

A four-stage MISR.

M_0	1 0 0 1 0			
M_1	0 1 0 1 0			
M_2	1 1 0 0 0			
M_3	1 0 0 1 1			
M	1 0 0 1 1 0 1 1			

FIGURE 3.30

An equivalent M sequence.

Assume there are m L-bit sequences to be compacted in an n-stage MISR, where $L > n \geq m \geq 2$. The *aliasing probability* for *parallel signature analysis* (PSA) now becomes:

$$P_{PSA}(n) = \left(2^{(mL-n)} - 1\right)/(2^{mL} - 1)$$

If $L >> n$, *then* $P_{PSA}(n) \approx 2^{-n}$. When $n = 20$, $P_{PSA}(n) < 2^{-20} = 0.0001\%$. The result suggests that $P_{PSA}(n)$ mainly depends on n, when $L >> n$. Hence, increasing the number of MISR stages or the use of the same MISR but with a different $f(x)$ can substantially reduce the *aliasing probability* [Hassan 1984; Williams 1987].

3.4.3 **Logic BIST architectures**

Several architectures for incorporating **offline BIST** techniques into a design have been proposed. These BIST architectures can be classified into two classes: (1) those that use the **test-per-scan BIST** scheme and (2) those that use the **test-per-clock BIST** scheme. The *test-per-scan BIST* scheme takes advantage of the already built-in scan chains of the scan design and applies a test pattern to the CUT after a shift operation is completed; hence, the hardware overhead is low. The *test-per-clock BIST* scheme, however, applies a test pattern to the CUT and captures its test response every system clock cycle; hence, the scheme can execute tests much faster than the test-per-scan BIST scheme but at an expense of more hardware overhead.

In this subsection, we only discuss three representative BIST architectures, the first two for pseudo-random testing and the last for pseudo-exhaustive testing. Although pseudo-random testing is commonly adopted in industry, the exhaustive and pseudo-exhaustive test techniques are applicable for designs that use the test-per-clock BIST scheme. For a more comprehensive survey of these BIST architectures, refer to [Abramovici 1994; Bardell 1987; McCluskey 1985; Wang 2006a]. Fault coverage enhancement with the pseudo-random test technique can also be found in [Tsai 1999; Wang 2006a; Lai 2007].

3.4.3.1 *Self-testing with MISR and parallel SRSG (STUMPS)*

A test-per-scan BIST design was presented in [Bardell 1982]. This design, shown in Figure 3.31, contains a PRPG (parallel *shift register sequence generator* [SRSG]) and a MISR. The scan chains are loaded in parallel from the PRPG. The system clocks are then triggered, and the test responses are shifted to the MISR for compaction. New test patterns are shifted in at the same time while test responses are being shifted out. This BIST architecture that uses the test-per-scan BIST scheme is referred to as *self-testing with MISR and parallel SRSG* (**STUMPS**) [Bardell 1982].

Because of the ease of integration with traditional scan architecture, the **STUMPS** architecture is the only BIST architecture widely used in industry to

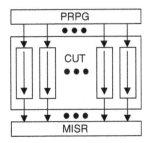

FIGURE 3.31

STUMPS.

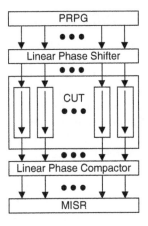

FIGURE 3.32

A STUMPS-based architecture.

date. To further reduce the lengths of the PRPG and MISR and improve the randomness of the PRPG, a STUMPS-based architecture that includes an optional linear phase shifter and an optional linear phase compactor is often used in industrial applications [Nadeau-Dostie 2000; Cheon 2005]. The linear phase shifter and linear phase compactor typically comprise a network of XOR gates. Figure 3.32 shows the STUMPS-based architecture.

3.4.3.2 *Built-in logic block observer (BILBO)*

The architecture described in [Könemann 1979, 1980] applies to circuits that can be partitioned into independent modules (logic blocks). Each module is assumed to have its own input and output registers (storage elements), or such registers are added to the circuit where necessary. The registers are redesigned so that for test purposes they act as PRPGs for test generation or MISRs for signature analysis. The redesigned register is called a ***built-in logic block observer*** (BILBO).

The BILBO is operated in four modes: normal mode, scan mode, test generation or signature analysis mode, and reset mode. A typical three-stage BILBO, which is reconfigurable into a TPG or a MISR during self-test is shown in Figure 3.33. It is controlled by two control inputs B_1 and B_2. When both control inputs B_1 and B_2 are equal to 1, the circuit functions in normal mode with the inputs Y_i gated directly into the D flip-flops. When both control inputs are equal to 0, the BILBO is configured as a shift register. Test data can be shifted in through the serial scan-in port or shifted out through the serial scan-out port. Setting $B_1 = 1$ and $B_2 = 0$ converts the BILBO into a MISR. It can then be used in this configuration as a TPG by holding every Y_i input to 1. The BILBO is reset after a system clock is triggered when $B_1 = 0$ and $B_2 = 1$.

This technique is most suitable for testing circuits, such as *random-access memories* (RAMs), *read-only memories* (ROMs), or bus-oriented circuits, where input and output registers of the partitioned modules can be reconfigured independently. For testing finite-state machines or pipeline-oriented circuits as shown in Figure 3.34, the signature data from the previous module must be

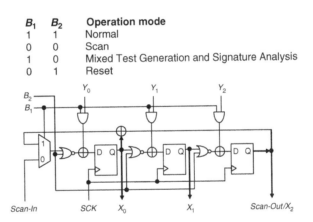

FIGURE 3.33

A three-stage built-in logic block observer (BILBO).

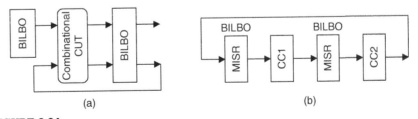

FIGURE 3.34

BILBO architectures: (a) For testing a finite-state machine. (b) For testing a pipeline-oriented circuit.

used as test patterns for the next module, because the test generation and signature analysis modes cannot be separated. In this case, a detailed fault simulation is required to achieve 100% single-stuck fault coverage.

3.4.3.3 *Concurrent built-in logic block observer (CBILBO)*

One technique to overcome the above BILBO fault coverage loss problem is to use the ***concurrent built-in logic block observer*** (CBILBO) approach [Wang 1986b]. Reconfigured from the BILBO design, the CBILBO is based on the test-per-clock BIST scheme and uses two registers to perform test generation and signature analysis simultaneously. A CBILBO design is illustrated in Figure 3.35, where only three modes of operation are considered: normal, scan, and test generation and signature analysis. When $B_1 = 0$ and $B_2 = 1$, the upper D flip-flops act as a MISR for signature analysis, whereas the lower two-port D flip-flops form a TPG for test generation. Because signature analysis is separated from test generation, an *exhaustive* or *pseudo-exhaustive pattern generator* (EPG/PEPG) can now be used for test generation; therefore, no fault simulation is required, and it is possible to achieve 100% single-stuck fault coverage with the CBILBO architectures for testing designs shown in Figure 3.36. However, the hardware cost associated with the use of the CBILBO approach is generally higher than for the STUMPS approach.

3.4.4 **Industry practices**

Logic BIST has a history of more than 30 years since its invention in the 1970s. Although it is only a few years behind the invention of scan, logic BIST has yet

B_1	B_2	Operation mode
–	0	Normal
1	1	Scan
0	1	Test Generation and Signature Analysis

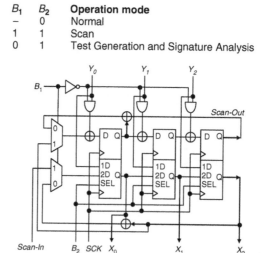

FIGURE 3.35

A three-stage concurrent BILBO (CBILBO).

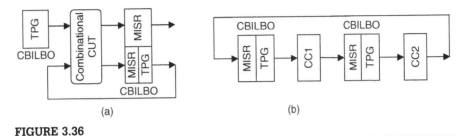

FIGURE 3.36

CBILBO architectures: (a) For testing a finite-state machine. (b) For testing a pipeline-oriented circuit.

to gain strong industry support. The worldwide market is estimated to be close to 10% of the scan market. The logic BIST products available in the marketplace now include **Encounter Test** from Cadence Design Systems [Cadence 2008], **ETLogic** from LogicVision [LogicVision 2008], **LBIST Architect** from Mentor Graphics [Mentor 2008], and **TurboBIST-Logic** from SynTest Technologies [SynTest 2008]. The logic BIST product offered in Encounter Test by Cadence currently includes support for test structure extraction, verification, logic simulation for signatures, and fault simulation for coverage. Unlike all other three BIST vendors that provide their own logic BIST structures in their respective products, Cadence offers a service to insert custom logic BIST structures or to use any customer-inserted logic BIST structures, including working with the customer to have custom on-chip clocking for logic BIST. A similar case exists in ETLogic from LogicVision when the double-capture clocking scheme is used.

All these commercially available logic BIST products support the STUMPS-based architectures. Cadence supports a weighted-random spreading network (XOR network) for STUMPS with multiple-weight selects [Foote 1997]. For at-speed delay fault testing, ETLogic [LogicVision 2008] uses a **skewed-load-based at-speed BIST architecture**; TurboBIST-Logic [Wang 2005b, 2006b; SynTest 2008] implements the **double-capture-based at-speed BIST architecture**; and LBIST Architect [Mentor 2008] adopts a **hybrid at-speed BIST architecture** that supports both skewed-load and double-capture. In addition, all products provide inter-clock-domain delay fault testing for synchronous clock domains. On-chip clock controllers for testing these inter-clock-domain faults at-speed can be found in [Rajski 2003; Furukawa 2006; Nadeau-Dostie 2006, 2007; Keller 2007], and Table 3.7 summarizes the capture-clocking schemes for at-speed logic BIST that is used by the EDA vendors.

3.5 TEST COMPRESSION

Test compression can provide 10× to 100× reduction or even more in the amount of test data (both test stimulus and test response) that must be stored on the *automatic test equipment* (ATE) [Touba 2006; Wang 2006a] for testing

Table 3.7 Summary of Industry Practices for At-Speed Logic BIST

Industry Practices	Skewed-Load	Double-Capture
Encounter test	Through service	Through service
ETLogic	√	Through service
LBIST Architect	√	√
TurboBIST-Logic		√

with a deterministic ATPG-generated test set. This greatly reduces ATE memory requirements and even more importantly reduces test time, because less data have to be transferred across the limited bandwidth between the ATE and the chip. Moreover, test compression methods are easy to adopt in industry because they are compatible with the conventional design rules and test generation flows used for scan testing.

Test compression is achieved by adding some additional on-chip hardware before the scan chains to decompress the test stimulus coming from the tester and after the scan chains to compact the response going to the tester. This is illustrated in Figure 3.37. This extra on-chip hardware allows the test data to be stored on the tester in a compressed form. Test data are inherently highly compressible because typically only 1% to 5% of the bits on a test pattern that is generated by an ATPG program have specified (*care*) values. Lossless compression techniques can thus be used to significantly reduce the amount of test stimulus data that must be stored on the tester. The on-chip **decompressor** expands the compressed test stimulus back into the original test patterns (matching in all the care bits) as they are shifted into the scan chains. The on-chip **compactor** converts long output response sequences into short signatures. Because the compaction is lossy, some fault coverage can be lost because

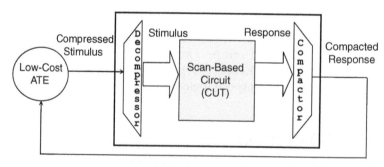

FIGURE 3.37

Architecture for test compression.

of unknown (X) values that might appear in the output sequence or aliasing where a faulty output response signature is identical to the fault-free output response signature. With proper design of the **circuit under test** (CUT) and the compaction circuitry, however, the fault coverage loss can be kept negligibly small.

3.5.1 Circuits for test stimulus compression

A **test cube** is defined as a deterministic test vector in which the bits that are not assigned values by the ATPG procedure are left as don't cares (X's). Normally, ATPG procedures perform *random fill* in which all the X's in the test cubes are filled randomly with 1's and 0's to create fully specified test vectors; however, for test stimulus compression, random fill is not performed during ATPG so the resulting test set consists of incompletely specified test cubes. The X's make the test cubes much easier to compress than fully specified test vectors.

As mentioned earlier, test stimulus compression should be an information lossless procedure with respect to the specified (care) bits to preserve the fault coverage of the original test cubes. After decompression, the resulting test patterns shifted into the scan chains should match the original test cubes in all the specified (care) bits.

Many schemes for compressing test cubes have been surveyed in [Touba 2006; Wang 2006a]. Two schemes based on linear decompression and broadcast scan are described here in greater detail mainly because the industry has favored both approaches over code-based schemes from area overhead and compression ratio points of view. These code-based schemes can be found in [Wang 2006a].

3.5.1.1 *Linear-decompression-based schemes*

A class of test stimulus compression schemes is based on the use of **linear decompressors** to expand the data coming from the tester to fill the scan chains. Any decompressor that consists of only XOR gates and flip-flops is a *linear decompressor* [Könemann 1991]. Linear decompressors have a very useful property: their *output space* (*i.e.*, the space of all possible test vectors that they can generate) is a linear subspace that is spanned by a Boolean matrix. In other words, for any linear decompressor that expands an m-bit compressed stimulus from the tester into an n-bit stimulus (test vector), there exists a Boolean matrix $A_{n \times m}$ such that the set of test vectors that can be generated by the linear decompressor is spanned by A. A test vector Z can be compressed by a particular linear decompressor if and only if there exists a solution to a system of linear equations, $AX = Z$, where A is the **characteristic matrix** of the linear decompressor and X is a set of **free variables** stored on the tester (every bit stored on the tester can be thought of as a "free variable" that can be assigned any value, 0 or 1).

The characteristic matrix for a linear decompressor can be obtained by symbolic simulation where each free variable coming from the tester is represented by a symbol. An example of this is shown in Figure 3.38, where a sequential linear decompressor containing an LFSR is used. The initial state of the LFSR is represented by free variables X_1 to X_4, and the free variables X_5 to X_{10} are shifted in from two channels as the scan chains are loaded. After symbolic simulation, the final values in the scan chains are represented by the equations for Z_1 to Z_{12}. The corresponding system of linear equations for this linear decompressor is shown in Figure 3.39.

The symbolic simulation goes as follows. Assume that the initial seed X_1 to X_4 has been already loaded into the flip-flops. In the first clock cycle, the top flip-flop is loaded with the XOR of X_2 and X_5; the second flip-flop is loaded with X_3; the third flip-flop is loaded with the XOR of X_1 and X_4; and the bottom flip-flop is loaded with the XOR of X_1 and X_6. Thus, we obtain $Z_1 = X_2 \oplus X_5$, $Z_2 = X_3$, $Z_3 = X_1 \oplus X_4$, and $Z_4 = X_1 \oplus X_6$. In the second clock cycle, the top flip-flop is loaded with the XOR of the contents of the second flip-flop (X_3) and X_7; the second flip-flop is loaded with the contents of the third flip-flop ($X_1 \oplus X_4$); the third flip-flop is loaded with the XOR of the contents of the first flip-flop ($X_2 \oplus X_5$) and the fourth flip-flop ($X_1 \oplus X_6$); and the bottom flip-flop is loaded with the XOR of the contents of the first flip-flop ($X_2 \oplus X_5$) and X_8. Thus, we obtain $Z_5 = X_3 \oplus X_7$, $Z_6 = X_1 \oplus X_4$, $Z_7 = X_1 \oplus X_2 \oplus X_5 \oplus X_6$, and $Z_8 = X_2 \oplus X_5 \oplus X_8$. In the third clock cycle, the top flip-flop is loaded with

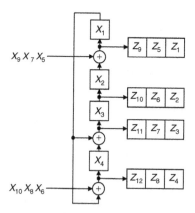

$Z_9 = X_1 \oplus X_4 \oplus X_9$	$Z_5 = X_3 \oplus X_7$	$Z_1 = X_2 \oplus X_5$
$Z_{10} = X_1 \oplus X_2 \oplus X_5 \oplus X_6$	$Z_6 = X_1 \oplus X_4$	$Z_2 = X_3$
$Z_{11} = X_2 \oplus X_3 \oplus X_5 \oplus X_7 \oplus X_8$	$Z_7 = X_1 \oplus X_2 \oplus X_5 \oplus X_6$	$Z_3 = X_1 \oplus X_4$
$Z_{12} = X_3 \oplus X_7 \oplus X_{10}$	$Z_8 = X_2 \oplus X_5 \oplus X_8$	$Z_4 = X_1 \oplus X_6$

FIGURE 3.38

Example of symbolic simulation for linear decompressor.

$$
\begin{pmatrix}
0 & 1 & 0 & 0 & 1 & 0 & 0 & 0 & 0 & 0 \\
0 & 0 & 1 & 0 & 0 & 0 & 0 & 0 & 0 & 0 \\
1 & 0 & 0 & 1 & 0 & 0 & 0 & 0 & 0 & 0 \\
1 & 0 & 0 & 0 & 0 & 1 & 0 & 0 & 0 & 0 \\
0 & 0 & 1 & 0 & 0 & 0 & 1 & 0 & 0 & 0 \\
1 & 0 & 0 & 1 & 0 & 0 & 0 & 0 & 0 & 0 \\
1 & 1 & 0 & 0 & 1 & 1 & 0 & 0 & 0 & 0 \\
0 & 1 & 0 & 0 & 1 & 0 & 0 & 1 & 0 & 0 \\
1 & 0 & 0 & 1 & 0 & 0 & 0 & 0 & 1 & 0 \\
1 & 1 & 0 & 0 & 1 & 1 & 0 & 0 & 0 & 0 \\
0 & 1 & 1 & 0 & 1 & 0 & 1 & 1 & 0 & 0 \\
0 & 0 & 1 & 0 & 0 & 0 & 1 & 0 & 0 & 1
\end{pmatrix}
\begin{pmatrix}
X_1 \\ X_2 \\ X_3 \\ X_4 \\ X_5 \\ X_6 \\ X_7 \\ X_8 \\ X_9 \\ X_{10}
\end{pmatrix}
=
\begin{pmatrix}
Z_1 \\ Z_2 \\ Z_3 \\ Z_4 \\ Z_5 \\ Z_6 \\ Z_7 \\ Z_8 \\ Z_9 \\ Z_{10} \\ Z_{11} \\ Z_{12}
\end{pmatrix}
$$

FIGURE 3.39

System of linear equations for the decompressor in Figure 3.38.

the XOR of the contents of the second flip-flop ($X_1 \oplus X_4$) and X_9; the second flip-flop is loaded with the contents of the third flip-flop ($X_1 \oplus X_2 \oplus X_5 \oplus X_6$); the third flip-flop is loaded with the XOR of the contents of the first flip-flop ($X_3 \oplus X_7$) and the fourth flip-flop ($X_2 \oplus X_5 \oplus X_8$); and the bottom flip-flop is loaded with the XOR of the contents of the first flip-flop ($X_3 \oplus X_7$) and X_{10}. Thus, we obtain $Z_9 = X_4 \oplus X_9$, $Z_{10} = X_1 \oplus X_6$, $Z_{11} = X_2 \oplus X_5 \oplus X_8$, and $Z_{12} = X_3 \oplus X_7 \oplus X_{10}$. At this point, the scan chains are fully loaded with a test cube, so the simulation is complete.

3.5.1.1.1 Combinational linear decompressors

The simplest linear decompressors use only combinational XOR networks. Each scan chain is fed by the XOR of some subset of the channels coming from the tester [Bayraktaroglu 2001, 2003; Könemann 2003; Mitra 2006; Han 2007; Wang 2004, 2008]. The advantage compared with sequential linear decompressors is simpler hardware and control. The drawback is that, to encode a test cube, each **scan slice** (the n-bits that are loaded into the n scan chains in each clock cycle) must be encoded with only the free variables that are shifted from the tester in a single clock cycle (which is equal to the number of channels). The worst-case most highly specified scan slices tend to limit the amount of compression that can be achieved, because the number of channels from the tester has to be sufficiently large to encode the most highly specified scan slices. Consequently, it is very difficult to obtain a high **encoding efficiency** (typically it will be less than 0.25); for the other less specified scan slices, a lot of the free variables end up getting wasted, because those scan slices could have been encoded with many fewer free variables.

One approach for improving the encoding efficiency of combinational linear decompressors that was proposed in [Krishna 2003] is to dynamically adjust the number of scan chains that are loaded in each clock cycle. So for a highly specified scan slice, four clock cycles could be used in which 25% of the scan chains are loaded in each cycle, whereas for a lightly specified scan slice, only one clock cycle can be used in which 100% of the scan slices are loaded. This allows a better matching of the number of free variables with the number of specified bits to achieve a higher encoding efficiency. Note that it requires that the scan clock be divided into multiple domains.

3.5.1.1.2 Sequential linear decompressors

Sequential linear decompressors are based on linear finite-state machines such as LFSRs, cellular automata, or ring generators [Mrugalski 2004]. The advantage of a sequential linear decompressor is that it allows free variables from earlier clock cycles to be used when encoding a scan slice in the current clock cycle. This provides much greater flexibility than combinational decompressors and helps avoid the problem of the worst-case most highly specified scan slices limiting the overall compression. The more flip-flops that are used in the sequential linear decompressor, the greater the flexibility that is provided. [Tobua 2006] classifies the sequential linear decompressors into two classes:

1. **Static reseeding** that computes a seed (an initial state) for each test cube [Touba 2006]. This seed, when loaded into an LFSR and run in autonomous mode, will produce the test cube in the scan chains [Könemann 1991]. This technique achieves compression by storing only the seeds instead of the full test cubes.
2. **Dynamic reseeding** calls for the injection of free variables coming from the tester into the LFSR as it loads the scan chains [Krishna 2001; Könemann 2001; Rajski 2004].

Figure 3.40 shows a generic example of a sequential linear decompressor that uses b channels from the tester to continuously inject free variables into the LFSR as it loads the scan chains through a combinational linear decompressor that typically is a combinational XOR network.

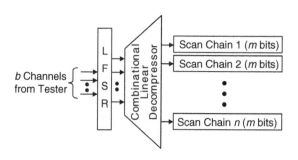

FIGURE 3.40

Typical sequential linear decompressor.

3.5.1.2 *Broadcast-scan-based schemes*

Another class of test stimulus compression schemes is based on broadcasting the same value to multiple scan chains. This was first proposed in [Lee 1998] and [Lee 1999]. Because of its simplicity and effectiveness, this method has been used as the basis of many test compression architectures, including some commercial *design for testability* (DFT) tools.

3.5.1.2.1 Broadcast scan

To illustrate the basic concept of **broadcast scan**, first consider two independent circuits C_1 and C_2. Assume that these two circuits have their own test sets $T_1 = <t_{11},t_{12},\dots,t_{1k}>$ and $T_2 = <t_{21},t_{22},\dots,t_{2l}>$, respectively. In general, a test set may consist of random patterns and deterministic patterns. In the beginning of the ATPG process, usually random patterns are initially used to detect the easy-to-detect faults. If the same random patterns are used when generating both T_1 and T_2, then we may have $t_{11} = t_{21}$, $t_{12} = t_{22}$, $\dots$, up to some ith pattern. After most faults have been detected by the random patterns, deterministic patterns are generated for the remaining difficult-to-detect faults. Generally, these patterns have many "don't care" bits. For example, when generating $t_{1(i + 1)}$, many "don't care" bits may still exist when no more faults in C_1 can be detected. By use of a test pattern with bits assigned so far for C_1, we can further assign specific values to the "don't care" bits in the pattern to detect faults in C_2. Thus, the final pattern would be effective in detecting faults in both C_1 and C_2.

The concept of pattern sharing can be extended to multiple circuits as illustrated in Figure 3.41. One major advantage of the use of *broadcast scan* for independent circuits is that all faults that are detectable in all original circuits will also be detectable with the broadcast structure. This is because if one test vector can detect a fault in a stand-alone circuit, then it will still be possible to apply this vector to detect the fault in the broadcast structure. Thus, the broadcast scan method will not affect the fault coverage if all circuits are independent. Note that broadcast scan can also be applied to multiple scan chains of a single circuit if all subcircuits driven by the scan chains are independent.

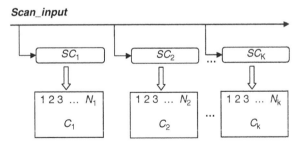

FIGURE 3.41

Broadcasting to scan chains driving independent circuits.

3.5.1.2.2 Illinois scan

If *broadcast scan* is used for multiple scan chains of a single circuit where the subcircuits driven by the scan chains are not independent, then the property of always being able to detect all faults is lost. The reason for this is that if two scan chains are sharing the same channel, then the ith scan cell in each of the two scan chains will always be loaded with identical values. If some fault requires two such scan cells to have opposite values to be detected, it will not be possible to detect this fault with broadcast scan.

To address the problem of some faults not being detected when broadcast scan is used for multiple scan chains of a single circuit, the **Illinois scan architecture** was proposed in [Hamzaoglu 1999] and [Hsu 2001]. This scan architecture consists of two modes of operations, namely a *broadcast mode* and a *serial scan mode,* which are illustrated in Figure 3.42. The *broadcast mode* is first used to detect most faults in the circuit. During this mode, a scan chain is divided into multiple subchains called *segments,* and the same vector can be shifted into all segments through a single shared scan-in input. The response data from all subchains are then compacted by a MISR or other space/time compactor. For the remaining faults that cannot be detected in broadcast mode, the *serial scan mode* is used where any possible test pattern can be applied. This ensures that complete fault coverage can be achieved. The extra logic required to implement the Illinois scan architecture consists of several multiplexers and some simple control logic to switch between the two modes. The area overhead of this logic is typically quite small compared with the overall chip area.

The main drawback of the Illinois scan architecture is that no test compression is achieved when it is run in *serial scan mode.* This can significantly degrade the overall **compression ratio** if many test patterns must be applied in serial scan mode. To reduce the number of patterns that need to be applied in serial scan mode, multiple-input broadcast scan or reconfigurable broadcast scan can be used. These techniques are described next.

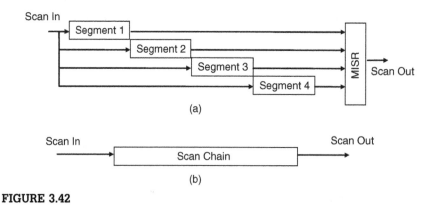

(a)

(b)

FIGURE 3.42

Two modes of Illinois scan architecture: (a) Broadcast mode. (b) Serial scan mode.

3.5.1.2.3 Multiple-input broadcast scan

Instead of the use of only one channel to drive all scan chains, a **multiple-input broadcast scan** could be used where there is more than one channel [Shah 2004]. Each channel can drive some subset of the scan chains. If two scan chains must be independently controlled to detect a fault, then they could be assigned to different channels. The more channels that are used and the shorter each scan chain is, the easier to detect more faults because fewer constraints are placed on the ATPG. Determining a configuration that requires the minimum number of channels to detect all detectable faults is thus highly desired with a multiple-input broadcast scan technique.

3.5.1.2.4 Reconfigurable broadcast scan

Multiple-input broadcast scan may require a large number of channels to achieve high fault coverage. To reduce the number of channels that are required, a **reconfigurable broadcast scan** method can be used. The idea is to provide the capability to reconfigure the set of scan chains that each channel drives. Two possible reconfiguration schemes have been proposed, namely **static reconfiguration** [Pandey 2002; Wang 2002; Samaranayake 2003; Chandra 2007], and **dynamic reconfiguration** [Li 2004; Sitchinava 2004; Wang 2004, 2008; Mitra 2006; Wohl 2007a]. In *static reconfiguration*, the reconfiguration can only be done when a new pattern is to be applied. For this method, the target fault set can be divided into several subsets, and each subset can be tested by a single configuration. After testing one subset of faults, the configuration can be changed to test another subset of faults. In *dynamic reconfiguration*, the configuration can be changed while scanning in a pattern. This provides more reconfiguration flexibility and hence can, in general, lead to better results with fewer channels. This is especially important for hard cores, when the test patterns provided by core vendor cannot be regenerated. The drawback of dynamic reconfiguration *versus* static reconfiguration is that more control information is needed for reconfiguring at the right time, whereas for static reconfiguration the control information is much less because the reconfiguration is done only a few times (only after all the test patterns that use a particular configuration have been applied).

Figure 3.43 shows an example *multiplexer* (MUX) network that can be used for dynamic configuration. When a value on the control line is selected, particular data at the four input pins are broadcasted to the eight scan chain inputs. For instance, when the control line is set to 0 (or 1), the scan chain 1 output will receive input data from Pin 4 (or Pin 1) directly.

3.5.1.2.5 Virtual scan

Rather than the use of MUX networks for test stimulus compression, combinational logic networks can also be used as decompressors. The combinational logic network can consist of any combination of simple combinational gates, such as buffers, inverters, AND/OR gates, MUXs, and XOR gates. This scheme, referred to as **virtual scan**, is different from *reconfigurable broadcast scan* and

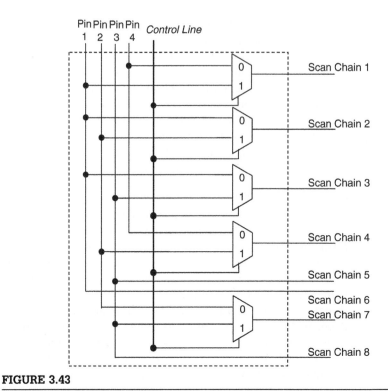

FIGURE 3.43

Example MUX network with control line(s) connected only to select pins of the multiplexers.

combinational linear decompression where pure MUX and XOR networks are allowed, respectively. The combinational logic network and the order of the scan chains can be specified as a set of constraints or just as an expanded circuit for ATPG. In either case, the test cubes that ATPG generates are the compressed stimuli for the decompressor itself. There is no need to solve a system of linear equations, and *dynamic compaction* can be effectively used during the ATPG process. Hence, only one-pass ATPG is required during test stimulus compression.

The *virtual scan* scheme was proposed in [Wang 2002, 2004, 2008]. In these papers, the decompressor was referred to as a **broadcaster**. The authors also proposed adding additional logic, when required, through *VirtualScan inputs* to reduce or remove the constraints imposed by the broadcaster on the circuit, thereby yielding very little or no fault coverage loss caused by test stimulus compression. For instance, a **scan connector** consisting of a set of multiplexers that places scan cells in the scan chains in a particular order can be connected to the outputs of the combinational logic network during each virtual scan test mode. Because the scan chains are reordered in each test mode, the imposed constraints of the combinational logic network on the circuit are reduced or removed.

In a broad sense, *virtual scan* is a generalized class of broadcast scan, Illinois scan, multiple-input broadcast scan, reconfigurable broadcast scan, and combinational linear decomposition. The advantage of the use of virtual scan is that it allows the ATPG to directly search for a test cube that can be applied by the decompressor and allows very effective dynamic compaction. Thus, virtual scan may produce shorter test sets than any test stimulus compression scheme based on solving linear equations; however, because this scheme may impose XOR or MUX constraints directly on the original circuit, it may take longer than those based on solving linear equations to generate test cubes or compressed stimuli. Two example virtual scan decompression circuits are shown in Figures 3.44a and 3.44b, respectively [Wang 2008]. Additional VirtualScan inputs are used to further reduce the XOR or MUX constraints imposed on the original circuit. An XOR network similar to the broadcaster shown in Figure 3.44a is sometimes referred to as a **space expander** or a **spreading network** in logic BIST applications.

3.5.2 **Circuits for test response compaction**

Test response compaction is performed at the outputs of the scan chains. The purpose is to reduce the amount of test response that needs to be transferred back to the tester. Although test stimulus compression must be lossless, test response compaction can be lossy. A large number of different test response compaction schemes and associated (response) compactors have been presented in the literature [Wang 2006a]. The effectiveness of each compaction scheme and the chosen compactor depends on its ability to avoid *aliasing* and tolerate *unknown test response bits* or X's. These schemes can be grouped into three categories: (1) **space compaction**, (2) **time compaction**, and (3) **mixed space and time compaction**.

A **space compactor** compacts an m-bit-wide output pattern to an n-bit-wide output pattern (where $n < m$). A **time compactor** compacts p output patterns to q output patterns (where $q < p$). A **mixed space and time compactor** has both space and time compaction performed concurrently. Typically, a space compactor is composed of XOR gates [Saluja 1983]; a time compactor includes a *multiple-input signature register* (MISR) [Frohwerk 1977]; and a mixed space and time compactor adds a space compactor at either the input or the output side of a time compactor [Saluja 1983; Wohl 2001]. Because test response compaction can be combinational-logic-based or sequential-logic-based, without loss of generality, we refer space compaction to as a **combinational compaction** scheme, and time compaction as well as mixed space and time compaction to as **sequential compaction** schemes.

There are three sources of aliasing according to [Wohl 2001]: (1) **combinational cancellation** occurs when two or more erroneous scan chain outputs (compactor inputs) are XORed in the compactor during the same cycle, which

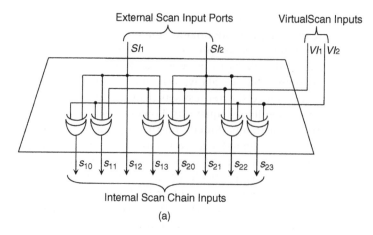

(a)

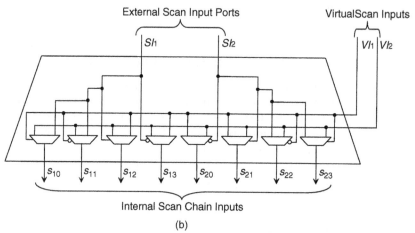

(b)

FIGURE 3.44

Example virtual scan decompression circuits: (a) Broadcaster that sees an example XOR network with additional VirtualScan inputs to reduce coverage loss. (b) Broadcaster that uses an example MUX network with additional VirtualScan inputs that can be also connected to data pins of the multiplexers.

cancel out the error effects in that cycle; (2) **shift cancellation** occurs when one or more erroneous scan chain output bits captured into the compactor are cancelled out by other erroneous scan chain output bits when the former are shifted down the shift path of the compactor; and (3) **feedback cancellation** occurs when one or more errors captured into the compactor during one cycle propagate through some feedback path of the compactor and cancel out with errors in later cycles. Combinational cancellation will exist in space compaction as well as mixed space and time compaction, because *non-aliasing* *space compactors* are impractical for real designs [Chakrabarty 1998; Pouya 1998]. On the other hand, shift cancellation and feedback cancellation are only present when either time compaction or mixed space and time compaction is used; however, shift cancellation is independent of the compactor feedback structure and its polynomial, whereas feedback cancellation depends on the compactor polynomial chosen.

Because unknown test response bits (*X*'s) can potentially reduce the fault coverage of the circuit under test when a combinational compactor is used and corrupt the final signature in a sequential compactor, one safe approach is to completely block these *X*'s before they reach the **response compactor** (combinational compactor or sequential compactor). During design, these potential **X-generators** (**X-sources**) can be identified with a scan design rule checker. When the *X* effects of an X-generator are likely to reach the response compactor, these *X*'s must be blocked before they reach the compactor [Gu 2001]. The process is often referred to as **X-blocking** or **X-bounding**.

In X-blocking, an X-source can be blocked either at the X-source or anywhere along its propagation paths before *X*'s reach the compactor. In case the X-source has been blocked at a nearby location during test and will not reach the compactor, there is no need to block the X-source; however, care must be taken to ensure that no observation points are added between the X-source and the location at which it is blocked to avoid capturing potential *X*'s into the compactor.

A simple example illustrating the X-blocking scheme for an X-source is shown in Figure 3.45. The output of the X-source is blocked and forced to 0 by setting the *select* signal of the multiplexer (MUX) to a fixed value (selecting the 0 input) in test mode. As a separate example, a non-scan flip-flop that is neither scanned nor initialized is a potential X-generator (X-source). If the flip-flop has two outputs (*Q* and *QB*), one can add two multiplexers forcing both outputs to opposite values in test mode. Alternately, if the flip-flip has an asynchronous set/reset pin, an AND/OR control point can be added to permanently force the flip-flip to 0 or 1 during test. Although an AND/OR control point can be added to force the non-scan flip-flop to a constant value, it is recommended that for

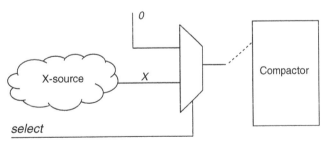

FIGURE 3.45

A simple illustration of the X-blocking scheme.

better fault coverage inserting a MUX control point driven by a nearby existing scan cell is preferred.

X-blocking can ensure that no X's will be propagated to the compactor; however, it also blocks the fault effects that can only propagate to an observable point through the now-blocked X-source (*e.g.*, the non-scan flip-flop). This can result in fault coverage loss. This problem can be addressed by use of a more flexible control on the *select* signal such that the X-source is blocked only during the cycles at which it may generate X's. Alternately, if the number of such faults for a given bounded X-generator justifies the cost, one or more observation points can be added before the X-source (*e.g.*, at the D input of the non-scan flip-flop) to provide an observable point to which those faults can propagate. These X-blocking or X-bounding methods have been extensively discussed in [Wang 2006a].

In this subsection, we only present some compactor designs that are widely used in industry along with some emerging compactors. For more information, refer to the key references cited in [Patel 2003; Mitra 2004b; Rajski 2004; Volkerink 2005; Wang 2006a; Touba 2007; Wohl 2007b].

3.5.2.1 *Combinational compaction*

A combinational compactor uses a combinational circuit to compact m outputs of the circuit under test into n test outputs, where $n < m$. If each output sequence contains only known (non-X) values (0's and 1's), then a combinational compactor that uses XOR gates with each internal scan chain output connected to only one XOR gate input is sufficient to guarantee no-fault coverage loss when the number of errors appearing at the m outputs is always odd [Saluja 1983]. A compactor that uses such XOR gates is referred to as a **conventional combinational compactor** or **simple space compactor**. An example is illustrated in Figure 3.46 [Wang 2008]. On the contrary, if any output sequence contains unknown values (X's), the combinational compaction scheme must have the capability to mask off or tolerate unknowns to prevent faults from going undetected. A compactor able to mask off or tolerate X's is referred to as an **X-tolerant combinational compactor** or **X-tolerant space compactor**. Two representative schemes currently practiced in industry are discussed in the following: (1) X-compact and (2) X-impact. Other schemes to further tolerate the amount of X's can be found in [Patel 2003; Rajski 2004; Wohl 2004, 2007b; Wang 2008].

3.5.2.1.1 X-compact

X-compact [Mitra 2004a] is an **X-tolerant space compaction** technique that connects each internal scan chain output to two or more external scan output ports through a network of XOR gates to tolerate unknowns. A response compaction circuit designed by use of the X-compact technique is called an **X-compactor**. Figure 3.47 shows an X-compactor with eight inputs and five outputs. It is composed of four 3-input XOR gates and eleven 2-input XOR gates.

Internal Scan Chain Outputs

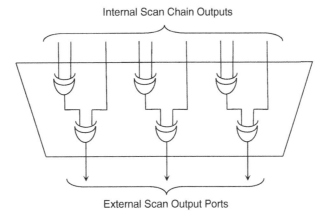

External Scan Output Ports

FIGURE 3.46

A conventional combinational compactor with nine inputs and three outputs.

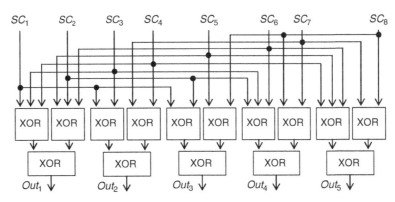

FIGURE 3.47

An X-compactor with eight inputs and five outputs.

Only one aliasing source, namely *combinational cancellation*, can exist in an X-compactor because of its combinational property. As an extreme example, if an X-compactor has only one output, it is, indeed, a **parity checker**, and any two error bits occurring simultaneously from the internal scan chain outputs will lead to aliasing.

Although aliasing may still exist when the X-compact technique is used, one can design an X-compactor that guarantees zero-aliasing in many practical cases. Consider Figure 3.47 again. If only one error bit occurs at the *SC* inputs, the error will be propagated to some output of the compactor and thus detected. One can also find that the compactor can detect any two or any odd number of errors that occur at the same cycle. In the following we use a binary matrix, called an **X-compact matrix**, to represent an X-compactor and to illustrate the fault detectability and X-tolerability of the compactor.

Suppose that the outputs of m scan chains are to be compacted into n bits for each scan cycle with an X-compactor. The associated *X-compact matrix* then contains n rows and k columns, in which each row corresponds to a scan chain output (*e.g.*, *SC* in Figure 3.47), and each column corresponds to an X-compactor output (*e.g.*, *Out* in Figure 3.47). The entry at row i and column j of the matrix is 1 if and only if the jth X-compactor output depends on the ith scan chain output; otherwise, the matrix entry is 0. Thus, the corresponding X-compact matrix M of the X-compactor shown in Figure 3.47 is:

$$M = \begin{bmatrix} 1 & 1 & 1 & 0 & 0 \\ 1 & 0 & 1 & 1 & 0 \\ 1 & 1 & 0 & 1 & 0 \\ 1 & 1 & 0 & 0 & 1 \\ 1 & 0 & 1 & 0 & 1 \\ 1 & 0 & 0 & 1 & 1 \\ 0 & 1 & 0 & 1 & 1 \\ 0 & 0 & 1 & 1 & 1 \end{bmatrix}$$

With the help of an X-compact matrix, it was shown in [Mitra 2004a] that errors from any one, two, or an odd number of scan chains at the same scan-out cycle are guaranteed to be detected by an X-compactor if every row of the corresponding X-compact matrix of the compactor is distinct and contains an odd number of 1's. This can be proved by the observation that (1) if all rows of the X-compact matrix are distinct, then a bitwise XOR of any two rows is nonzero, and (2) if each row further contains an odd number of 1's, then the bitwise XOR of any odd number of rows also contains an odd number of 1's.

The most distinctive feature of the X-compact technique is its X-tolerant capability (*i.e.*, detecting error bits even when the scan chain outputs have unknown bits). Refer to Figure 3.47 again. If one unknown bit occurs at SC_1, then the unknown value will be spread to Out_1, Out_2, and Out_3. Thus, after the XOR operation, the values at Out_1, Out_2, and Out_3 are masked (becoming unknown). However, if there is only one error bit in all other scan chain outputs, then the error bit will still be detected, because the error bit will be spread to at least one output that is not Out_1, Out_2, or Out_3. For example, an error bit occurring at SC_2 will be detected from Out_4. Thus, we have the following X-tolerant theorem:

Theorem 3.1:
An error from any scan chain with one unknown bit from any other scan chain at the same cycle is guaranteed to be observed at the outputs of an X-compactor if and only if:

1. No row of the X-compact matrix contains all 0's.
2. For any X-compact matrix row, the submatrix obtained by removing the row responding to the scan chain output with unknown bit and all columns having 1's in that row does not contain a row with all 0's.

The X-compact matrix of Figure 3.47 satisfies the preceding theorem. For example, if we remove row 1 and columns 1, 2, and 3, then each of the remaining rows in the submatrix contains at least a 1. Theorem 3.1 can be further extended to deal with errors from any k_1 or fewer scan chains with unknown bits from any k_2 or fewer scan chains ($k_1 + k_2 \leq n$) as follows:

Theorem 3.2:
Errors from any k_1 or fewer scan chains with unknown bits from any k_2 or fewer scan chains at the same cycle, where $k_1 + k_2 \leq n$ and n is the number of scan chains, are guaranteed to be observed at the outputs of an X-compactor if and only if:

1. No row of the X-compact matrix contains all 0's.
2. For any set S of k_1 X-compact matrix rows, any set of k_2 rows in the submatrix obtained by removing the rows in S and the X-compact matrix columns having 1's in the rows in S are linearly independent.

Designing an X-compact matrix to satisfy Theorem 3.2 is a complicated problem when an X-compactor is expected to tolerate three or more unknown bits. In some cycles, the number of actual knowns appearing at the scan chain outputs could exceed the number of unknowns designed to be tolerated by the X-compactor. Hence, the fault detectability and X-tolerability of an X-compactor highly depends on its actual implementation and the number of unknowns to be tolerated.

3.5.2.1.2 X-impact

Although X-blocking and X-compact each can achieve significant reduction in fault coverage loss caused by X's present at the inputs of a combinational compactor, the **X-impact** technique described in [Wang 2004] is helpful in that it can further reduce fault coverage loss simply by use of ATPG to algorithmically handle the impact of residual X's on the combinational compactor without adding any extra circuitry. The combinational compactor in use can be either a conventional combinational compactor or an X-tolerant combinational compactor.

Example 3.1   An example of algorithmically handling X-impact is shown in Figure 3.48. Here, SC_1 to SC_4 are scan cells connected to a conventional combinational compactor composed of XOR gates G_7 and G_8. Lines $a, b, \ldots, h$ are internal signals, and line f is assumed to be connected to an X-source (memory, non-scan storage element, etc.). Now consider the detection of the stuck-at-0 (SA0) fault f_1. Logic value 1 should be assigned to both lines d and e to activate f_1. The fault effect will be captured by scan cell SC_3. If the X on f propagates to SC_4, then the compactor output q will become X and f_1 cannot be detected. To avoid this, ATPG can try to assign either 1 to line g or 0 to line h to block the X from reaching SC_4. If it is impossible to achieve this assignment, ATPG can then try to assign 1 to line c, 0 to line b, and 0 to line a to propagate the fault effect to SC_2. As a result, fault f_1 can be detected. Thus, X-impact is avoided by algorithmic assignment without adding any extra circuitry.

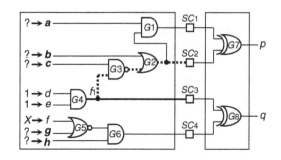

FIGURE 3.48

Handling of X-impact.

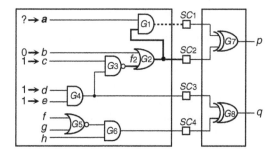

FIGURE 3.49

Handling of aliasing.

Example 3.2   It is also possible to use the X-impact approach to reduce combinational cancellation (an aliasing source). An example of algorithmically handling aliasing is shown in Figure 3.49. Here, SC_1 to SC_4 are scan cells connected to a conventional combinational compactor composed of XOR gates G_7 and G_8. Lines a, b, ..., h are internal signals. Now consider the detection of the stuck-at-1 fault f_2. Logic value 1 should be assigned to lines c, d, and e to activate f_2, and logic value 0 should be assigned to line b to propagate the fault effect to SC_2. If line a is set to 1, then the fault effect will also propagate to SC_1. In this case, aliasing will cause the compactor output p to have a fault-free value, resulting in an undetected f_2. To avoid this, ATPG can try to assign 0 to line a to block the fault effect from reaching SC_1. As a result, fault f_2 can be detected. Thus, aliasing can be avoided by algorithmic assignment without any extra circuitry.

3.5.2.2 *Sequential compaction*

In contrast to a combinational compactor that typically uses XOR gates to compact output responses, a sequential compactor uses sequential logic instead. The sequential compactor can be a **time-space compressor** or a **space-time compressor** as described in [Saluja 1983], although the authors only considered output bit streams of 0's and 1's. The type of sequential logic to be used for response compaction depends on whether the output responses contain unknown values (X's). A sequential compactor capable of masking off or tolerating these X's is often referred to as an **X-tolerant sequential compactor**.

3.5.2.2.1 Signature analysis

If X-bounding as described previously has been used such that each output response does not contain any unknown (X) values, then the *multiple-input signature register* (MISR) widely used for logic BIST applications can be simply used [Frohwerk 1977]. Referred to as a **conventional sequential compactor**, the MISR uses an XOR gate at each MISR stage input to compact the output sequences, M_0 to M_3, into the *linear feedback shift register* (LFSR) simultaneously. The final contents stored in the MISR after compaction is often called the (*final*) *signature* of the MISR. A conventional sequential compactor that uses a four-stage MISR is illustrated in Figure 3.50. For more information on signature analysis and the MISR design, the reader is referred to Section 3.4.2.3.

3.5.2.2.2 X-masking

On the contrary, if the output response contains unknown (X) values, then one must make sure when the sequential compactor is used that no X's from the circuit under test will reach the compactor. Although it may not result in fault coverage loss, the X-bounding scheme described previously does add area overhead and may impact delay because of the inserted logic. It is not surprising to find that, in complex designs, more than 25% of scan cycles could contain one or more X's in the test response. It is difficult to eliminate these residual X's by DFT; thus, an encoder with high X-tolerance is very attractive. Instead of blocking the X's where they are generated, the X's can also be masked off right before the sequential compactor. This scheme is referred to as **X-masking**. A typical X-masking circuit is shown in Figure 3.51. The mask controller applies a logic value 1 at the appropriate time to mask off any scan output that contains an X before the X reaches the compactor.

The **X-masking compactor** is one type of X-tolerant sequential compactors. Typically, it implies that sequential logic (comprising one or more MISRs or SISRs) is used in the compactor for response compaction. Almost all existing X-tolerant sequential compactors proposed in the literature use X-masking, including OPMISR+ [Barnhart 2002; Naruse 2003], ETCompression [Nadeau-Dostie 2004],

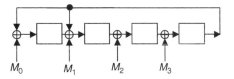

FIGURE 3.50

A conventional sequential compactor that uses a four-stage MISR.

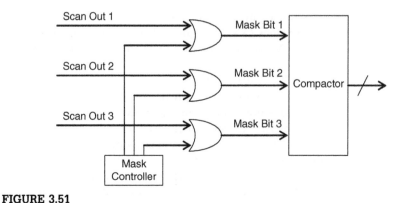

FIGURE 3.51

An example X-masking circuit in use with a compactor.

and convolutional compactors [Mitra 2004b; Rajski 2005, 2008]. In fact, combinational logic (such as XOR gates) can also be used in the compactor. Such an X-masking compactor that uses combinational logic is referred to as a **selective compactor** [Rajski 2004]. Mask data are needed to indicate when the masking should take place. These mask data can be stored in compressed format and can be decompressed with on-chip hardware. Possible compression techniques are *weighted pseudo-random LFSR reseeding* or *run-length encoding* [Volkerink 2005].

Another type of X-tolerant sequential compactor is an **X-canceling MISR** [Touba 2007, 2008] that does not mask the X's before they enter the MISR. It allows the X's to be compacted in a MISR and then selectively XORs together combinations of MISR signature bits that are linearly dependent in terms of the X's such that all the X's are canceled out.

3.5.2.2.3 *q*-compact

In case none of the X-bounding, X-masking, or X-canceling schemes is available to block, mask off, or cancel all X's, the sequential logic in use must not have a feedback path so these X's will only stay in the sequential compactor for a few clock cycles. Such an X-tolerant sequential compaction scheme is referred to as ***q*-compact**. A ***q*-compactor** that uses this X-tolerant compaction scheme is illustrated in [Han 2006].

Figure 3.52 shows an example of a *q*-compactor assuming the inputs are coming from internal scan chain outputs [Han 2006]. The spatial part of the *q*-compactor consists of single-output XOR networks (called *spread networks*) connected to the flip-flops by means of additional 2-input XOR gates interspersed between successive storage elements. As can be seen, every error in a scan cell can reach storage elements and then outputs in several possible ways. The spread network that determines this property is defined in terms of

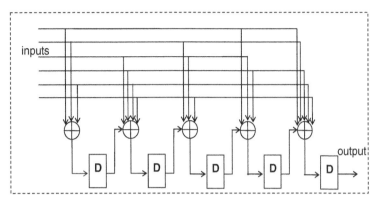

FIGURE 3.52

An example *q*-compactor with single output.

spread polynomials indicating how particular scan chains are connected to the register flip-flops.

Different from a conventional MISR, the *q*-compactor presented in Figure 3.52 does not have a feedback path; consequently, any error or X injected into the compactor is shifted out after at most five clock cycles. The shifted-out data will be compared with the expected data and then the error will be detected.

3.5.3 **Industry practices**

Several test compression products and solutions have been introduced by some of the major DFT vendors in the CAD industry. These products differ significantly with regard to technology, design overhead, design rules, and the ease of use and implementation. A few second-generation products have also been introduced by a few of the vendors [Kapur 2008]. This subsection summarizes a few of the products introduced by companies such as Cadence Design Systems [Cadence 2008], LogicVision [LogicVision 2008], Mentor Graphics [Mentor 2008], Synopsys [Synopsys 2008], and SynTest Technologies [SynTest 2008].

Current industry solutions can be grouped under two main categories for stimulus decompression. The first category uses *linear-decompression-based schemes*, whereas the second category uses *broadcast-scan-based schemes*. The main difference between the two categories is the manner in which the ATPG engine is used. The first category includes products, such as ETCompression [LogicVision 2008] from LogicVision, TestKompress [Rajski 2004] from Mentor Graphics, XOR Compression [Cadence 2008] from Cadence, and SOCBIST [Wohl 2003] from Synopsys. The second category includes products, such as OPMISR+ [Barnhart 2002; Cadence 2008] from Cadence, VirtualScan [Wang 2004, 2008] from SynTest, and DFT MAX [Sitchinava 2004; Wohl 2007a] from Synopsys.

For designs that use *linear-decompression–based schemes*, test compression is achieved in two distinct steps. During the first step, conventional ATPG is used to generate sparse ATPG patterns (called test cubes), in which *dynamic compaction* is performed in a nonaggressive manner, while leaving unspecified bit locations in each test cube as X. This is accomplished by not aggressively performing the *random fill* operation on the test cubes, which is used to increase coverage of individual patterns, and hence reduce the total pattern count. During the second step, a system of linear equations, describing the hardware mapping from the external scan input ports to the internal scan chain inputs, are solved to map each test cube into a compressed stimulus that can be applied externally. If a mapping is not found, a new attempt at generating a new test cube is required.

For designs that use *broadcast-scan–based schemes*, only a single step is required to perform test compression. This is achieved by embedding the constraints introduced by the decompressor as part of the ATPG tool, such that the tool operates with much more restricted constraints. Hence, whereas in conventional ATPG, each individual scan cell can be set to 0 or 1 independently, for *broadcast-scan–based schemes* the values to which related scan cells can be set are constrained. Thus, a limitation of this solution is that in some cases, the constraints among scan cells can preclude some faults from being tested. These faults are typically tested as part of a later top-up ATPG process if required, similar to the use of *linear-decompression–based schemes*.

On the response compaction side, industry solutions have used either combinational compactors such as XOR networks, or sequential compactors such as MISRs, to compact the test responses. At present, combinational compactors have a higher acceptance rate in the industry because they do not involve the process of guaranteeing that no unknown (X) values are generated in the circuit under test.

A summary of the different compression architectures used in the commercial products is shown in Table 3.8. Six products from five DFT companies are included. Since June 2006, Cadence has added XOR Compression as an alternative to the OPMISR+ product described in [Wang 2006a].

Table 3.8 Summary of Industry Practices for Test Compression

Industry Practices	Stimulus Decompressor	Response Compactor
XOR Compression or OPMISR+	Combinational XOR Network or Fanout Network	XOR Network with or without MISR
TestKompress	Ring Generator	XOR Network
VirtualScan	Combinational Logic Network	XOR Network
DFT MAX	Combinational MUX Network	XOR Network
ETCompression	(Reseeding) PRPG	MISR

Table 3.9 Summary of Industry Practices for At-Speed Delay Fault Testing

Industry Practices	Skewed-Load	Double-Capture
XOR Compression or OPMISR+	√	√
TestKompress	√	√
VirtualScan	√	√
DFT MAX	√	√
ETCompression	√	Through Service

It is evident that the solutions offered by the current EDA DFT vendors are quite diverse with regard to stimulus decompression and response compaction. For stimulus decompression, OPMISR+, VirtualScan, and DFT MAX are broadcast-scan–based, whereas TestKompress and ETCompression are linear-decompression–based. For response compaction, OPMISR+ and ETCompression can include MISRs, whereas four other solutions purely adopt (X-tolerant) XOR networks. What is common is that all six products provide their own diagnostic solutions.

Generally speaking, any modern ATPG compression program supports at-speed clocking schemes used in its corresponding at-speed scan architecture. For at-speed delay fault testing, ETCompression currently uses a **skewed-load–based at-speed test compression architecture** for ATPG. The product can also support the double-capture clocking scheme through service. All other ATPG compression products, including OPMISR+, TestKompress, VirtualScan, and DFT MAX, support the **hybrid at-speed test compression architecture** by use of both skewed-load (*a.k.a.* launch-on-shift) and double-capture (*a.k.a.* launch-on-capture). In addition, almost every product supports inter-clock-domain delay fault testing for synchronous clock domains. A few on-chip clock controllers for detecting these inter-clock-domain delay faults at-speed have been proposed in [Beck 2005; Nadeau-Dostie 2005, 2006; Furukawa 2006; Fan 2007; and Keller 2007].

The clocking schemes used in these commercial products are summarized in Table 3.9. It should be noted that compression schemes might be limited in effectiveness if there are a large number of unknown response values, which can be exacerbated during at-speed testing when many paths do not make the timing being used.

3.6 CONCLUDING REMARKS

Design for testability (DFT) has become vital for ensuring circuit testability and product quality. Scan design, which has proven to be the most powerful DFT technique ever invented, allowed the transformation of sequential circuit testing into combinational circuit testing and has since become an industry standard. Currently, a scan design can contain a billion transistors [Naffziger 2006; Stackhouse 2008]. To screen all possible physical failures (manufacturing defects) caused by manufacturing imperfection, test compression coupled to scan design has rapidly emerged, becoming a crucial DFT technique to address the explosive test data volume and long test application time problems. At the same time, scan-based logic *built-in self-test* (BIST) is of growing importance because of its inherent advantage of performing self-test on-chip, on-board, or in-system, which can substantially improve the reliability of the system and the ability of in-field diagnosis.

Whereas the STUMPS-based architecture [Bardell 1982] is the most popular logic BIST architecture practiced currently for scan-based designs, the efforts required to implement the BIST circuitry and the loss of the fault coverage for the use of pseudo-random patterns have prevented the BIST architecture from being widely used in industry. As the semiconductor manufacturing technology moves into the nanometer design era, it remains to be seen how the CBILBO-based architecture proposed in [Wang 1986b], which can always guarantee 100% single stuck-at fault coverage and has the ability of running 10 times more BIST patterns than the STUMPS-based architecture, will perform. Challenges lie ahead with regard to whether or not pseudo-exhaustive testing will become a preferred BIST pattern generation technique.

Because the primary objective of this chapter is to familiarize the reader with basic DFT techniques, many advanced DFT techniques, along with novel *design-for-reliability* (DFR), *design-for-manufacturability* (DFM), *design-for-yield* (DFY), *design-for-debug-and-diagnosis* (DFD), and low-power test techniques, are left out. For advanced reading, the reader is referred to [Gizopoulos 2006; Wang 2006a, 2007a]. These techniques are of growing importance to help us cope with the physical failures of the nanometer design era.

The DFT chapter is the first of a series of three chapters devoted to VLSI testing. These chapters are chosen to equip the reader with basic DFT skills to design quality digital circuits. Chapter 7 discusses the design rules and test synthesis steps required to implement testability logic into these digital circuits. Chapter 14 jumps into the important fault simulation and test generation techniques for generating quality test patterns to screen defective chips from manufacturing test.

3.7 EXERCISES

3.1. (**Testability Analysis**) Calculate the SCOAP controllability and observability measures for a 3-input XOR gate and for its NAND-NOR implementation.

3.2. (**Testability Analysis**) Use the rules given in Tables 3.3 and 3.4 to calculate the probability-based testability measures for a 3-input XNOR gate and for its NAND-NOR implementation. Assume that the probability-based controllability values at all primary inputs and the probability-based observability value at the primary output are 0.5 and 1, respectively.

3.3. (**Testability Analysis**) Repeat Exercise 3.2 for the full-adder circuit shown in Figure 3.1.

3.4. (**Muxed-D Scan Cell**) Show a possible CMOS implementation of the muxed-D scan cell shown in Figure 3.5a.

3.5. (**Low-Power Muxed-D Scan Cell**) Design a low-power version of the muxed-D scan cell given in Figure 3.5a by adding gated-clock logic that includes a lock-up latch to control the clock port.

3.6. (**At-Speed Scan**) Assume that a scan design contains three clock domains running at 100 MHz, 200 MHz, and 400 MHz, respectively. In addition, assume that the clock skew between any two clock domains is manageable. List all possible at-speed scan ATPG methods and compare their advantages and disadvantages in terms of fault coverage and test pattern count.

3.7. (**At-Speed Scan**) Describe two major capture-clocking schemes for at-speed scan testing and compare their advantages and disadvantages. Also discuss what will happen if three or more captures are used.

3.8. (**BIST Pattern Generation**) Implement a period-8 in-circuit *test pattern generator* (TPG) with a binary counter. Compare its advantages and disadvantages with a Johnson counter (twisted-ring counter).

3.9. (**BIST Pattern Generation**) Implement a period-31 in-circuit *test pattern generator* (TPG) with a modular *linear feedback shift register* (LFSR) with *characteristic polynomial* $f(x) = 1 + x^2 + x^5$. Convert the modular LFSR into a muxed-D scan design with minimum area overhead.

3.10. (**BIST Pattern Generation**) Implement a period-31 in-circuit *test pattern generator* (TPG) with a five-stage *cellular automaton* (CA) with *construction rule = 11001, where* "0" denotes a *rule 90* cell and "1" denotes a *rule 150* cell. Convert the CA into an LSSD design with minimum area overhead.

3.11. (**Cellular Automata**) Derive a construction rule for a cellular automaton of length 54, and then construction rules up to length 300 to match the list of primitive polynomials up to degree 300 reported in [Bardell 1987].

3.12. (**BIST Response Compaction**) Discuss in detail what errors can and cannot be detected by a MISR.

3.13. (**STUMPS** *versus* **CBILBO**) Compare the performance of a STUMPS design and a CBILBO design. Assume that both designs operate at 400 MHz and that the circuit under test has 100 scan chains each having 1000 scan cells. Compute the test time for each design when 100,000 test patterns are to be applied. In general, the shift (scan) speed is much slower than a circuit's operating speed. Assume that the scan shift frequency is 50 MHz, and compute the test time for the STUMPS design again. Explain further why the STUMPS-based architecture is gaining more popularity than the CBILBO-based architecture.

3.14. (Scan *versus* Logic BIST *versus* Test Compression) Compare the advantages and disadvantages of a scan design, a logic BIST design, and a test compression design in terms of fault coverage, test application time, test data volume, and area overhead.

3.15. (Test Stimulus Compression) Given a circuit with four scan chains, each having five scan cells, and with a set of test cubes listed:

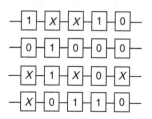

a. Design the multiple-input broadcast scan decompressor that fulfills the test cube requirements.

b. What is the compression ratio?

c. The assignment of X's will affect the compression performance dramatically. Give one X-assignment example that will unfortunately lead to no compression with this multiple-input broadcast scan decompressor.

3.16. (Test Stimulus Compression) Derive mathematical expressions for the following in terms of the number of tester channels, n, and the expansion ratio, k.

a. The probability of encoding a scan slice containing 2 specified bits with Illinois scan.

b. The probability of encoding a scan slice containing 3 specified bits, where each scan chain is driven by the XOR of a unique combination of 2 tester channels such that there are a total of $C_2^n = n(n-1)/2$ scan chains.

3.17. (Test Stimulus Compression) For the sequential linear decompressor shown in Figure 3.38 whose corresponding system of linear equations is shown in Figure 3.39, find the compressed stimulus, $X_1 - X_{10}$, necessary to encode the following test cube: $< Z_1, \ldots, Z_{12} > = <0 \cdots 1 \cdot 0 \cdots 010>$.

3.18. (Test Stimulus Compression) For the MUX network shown in Figure 3.43 and then the XOR network shown in Figure 3.44a, find the compressed stimulus at the network inputs necessary to encode the following test cube: $<1 \cdot 0 \cdots 01>$.

3.19. (Test Response Compaction) Explain further how many errors and how many unknowns (X's) can be detected or tolerated by the X-compactor and q-compactor as shown in Figures 3.47 and 3.52, respectively.

3.20. (Test Response Compaction) For the X-compact matrix of the X-compactor given below:

$$\begin{bmatrix} 0 & 1 & 1 & 1 & 0 \\ 0 & 1 & 0 & 1 & 1 \\ 1 & 1 & 0 & 0 & 1 \\ 1 & 1 & 0 & 1 & 0 \\ 1 & 0 & 1 & 0 & 1 \\ 1 & 0 & 0 & 1 & 1 \\ 1 & 0 & 1 & 1 & 0 \\ 0 & 0 & 1 & 1 & 1 \end{bmatrix}$$

a. What is the compaction ratio?

b. Which outputs after compaction are affected by the second scan chain output?

c. How many errors can be detected by the X-compactor?

ACKNOWLEDGMENTS

I wish to thank Dr. Xinghao Chen of CTC Technologies for contributing the Testability Analysis section; Professor Xiaowei Li and Professor Yinhe Han of Chinese Academy of Sciences, Professor Kuen-Jong Lee of National Cheng Kung University, Professor Nur A. Touba of the University of Texas at Austin for contributing a portion of the Circuits for Test Stimulus Compression and Circuits for Test Response Compaction sections. I also express my gratitude to Professor Xiaoqing Wen of Kyushu Institute of Technology, Professor Nur A. Touba of the University of Texas at Austin, Professor Kewal K. Saluja of the University of Wisconsin–Madison, Professor Subhasish Mitra of Stanford University, Dr. Rohit Kapur and Khader S. Abdel-Hafez of Synopsys, Dr. Brion Keller of Cadence Design Systems, and Dr. Benoit Nadeau-Dostie of LogicVision for reviewing the text and providing helpful comments, and Teresa Chang of SynTest Technologies for drawing most of the figures.

REFERENCES
R3.0 Books

[Abramovici 1994] M. Abramovici, M. A. Breuer, and A. D. Friedman, *Digital Systems Testing and Testable Design*, IEEE Press, Revised Printing, Piscataway, NJ, 1994.

[Bardell 1987] P. H. Bardell, W. H. McAnney, and J. Savir, *Built-In Test for VLSI: Pseudorandom Techniques,* John Wiley & Sons, Somerset, NJ, 1987.

[Bushnell 2000] M. L. Bushnell and V. D. Agrawal, *Essentials of Electronic Testing for Digital, Memory & Mixed-Signal VLSI Circuits*, Springer, Boston, 2000.

[Crouch 1999] A. Crouch, *Design for Test for Digital IC's and Embedded Core Systems,* Prentice-Hall, Englewood Cliffs, NJ, 1999.

[Gizopoulos 2006] D. Gizopoulos, editor, *Advances in Electronic Testing: Challenges and Methodologies,* Morgan Kaufmann, San Francisco, 2006.

[Golomb 1982] S. W. Golomb, *Shift Register Sequence,* Aegean Park Press, Laguna Hills, CA, 1982.

[Jha 2003] N. Jha and S. Gupta, *Testing of Digital Systems,* Cambridge University Press, London, 2003.

[McCluskey 1986] E. J. McCluskey, *Logic Design Principles: With Emphasis on Testable Semicustom Circuits,* Prentice-Hall, Englewood Cliffs, NJ, 1986.

[Mourad 2000] S. Mourad and Y. Zorian, *Principles of Testing Electronic Systems,* John Wiley & Sons, Somerset, NJ, 2000.

[Nadeau-Dostie 2000] B. Nadeau-Dostie, *Design for At-Speed Test, Diagnosis and Measurement,* Springer, Boston, 2000.

[Peterson 1972] W. W. Peterson and E. J. Weldon, Jr., *Error-Correcting Codes,* MIT Press, Cambridge, MA, 1972.

[Rajski 1998] J. Rajski and J. Tyszer, *Arithmetic Built-In Self-Test for Embedded Systems,* Prentice-Hall, Englewood Cliffs, NJ, 1998.

[Stroud 2002] C. E. Stroud, *A Designer's Guide to Built-In Self-Test,* Springer, Boston, 2002.

[Wang 2006a] L.-T. Wang, C.-W. Wu, and X. Wen, editors, *VLSI Test Principles and Architectures: Design for Testability,* Morgan Kaufmann, San Francisco, 2006.

[Wang 2007a] L.-T. Wang, C. E. Stroud, and N. A. Touba, editors, *System-on-Chip Test Architectures: Nanometer Design for Testability,* Morgan Kaufmann, San Francisco, 2007.

R3.1 Introduction

[Fujiwara 1982] H. Fujiwara and S. Toida, The complexity of fault detection problems for combinational circuits, *IEEE Trans. on Computers,* C-31(6), pp. 555–560, June 1982.

[SIA 2005] SIA, *The International Technology Roadmap for Semiconductors: 2005 Edition—Design,* Semiconductor Industry Association, San Jose, CA, http://public.itrs.net, 2005.

[SIA 2006] SIA, *The International Technology Roadmap for Semiconductors: 2006 Update,* Semiconductor Industry Association, San Jose, CA, http://public.itrs.net, 2006.

[Touba 2006] N. A. Touba, Survey of test vector compression techniques, *IEEE Design & Test of Computers,* 23(4), pp. 294–303, July–August 2006.

R3.2 Testability Analysis

[Agrawal 1982] V. D. Agrawal and M. R. Mercer, Testability measures—What do they tell us?, in *Proc. IEEE Int. Test Conf.,* pp. 391–396, November 1982.

[Breuer 1978] M. A. Breuer, New concepts in automated testing of digital circuits, in *Proc. EEC Symp. on CAD of Digital Electronic Circuits and Systems,* pp. 69–92, November 1978.

[Goldstein 1979] L. H. Goldstein, Controllability/Observability analysis of digital circuits, *IEEE Trans. on Circuits and Systems,* CAS-26(9), pp. 685–693, September 1979.

[Goldstein 1980] L. H. Goldstein and E. L. Thigpen, SCOAP: Sandia controllability/observability analysis program, in *Proc. ACM/IEEE Design Automation Conf.,* pp. 190–196, June 1980.

[Grason 1979] J. Grason, TMEAS—a testability measurement program, in *Proc. ACM/IEEE Design Automation Conf.,* pp. 156–161, June 1979.

[Ivanov 1988] A. Ivanov and V. K. Agarwal, Dynamic testability measures for ATPG, *IEEE Trans. on Computer-Aided Design,* 7(5), pp. 598–608, May 1988.

[Jain 1985] S. K. Jain and V. D. Agrawal, Statistical fault analysis, *IEEE Design & Test of Computers,* 2(2), pp. 38–44, February 1985.

[Parker 1975] K. P. Parker and E. J. McCluskey, Probability treatment of general combinational networks, *IEEE Trans. on Computers,* 24(6), pp. 668–670, June 1975.

[Rizzolo 2001] R. F. Rizzolo, B. F. Robbins, and D. G. Scott, A hierarchical approach to improving random pattern testability on IBM eServer z900 chips, in *Digest of Papers, IEEE North Atlantic Test Workshop*, pp. 84–89, May 2001.

[Rutman 1972] R. A. Rutman, Fault detection test generation for sequential logic heuristic tree search, *IEEE Computer Repository*, Paper R-72-187, September/October 1972.

[Savir 1984] J. Savir, G. S. Ditlow, and P. H. Bardell, random pattern testability, *IEEE Trans. on Computer*, C-33(1), pp. 79–90, January 1984.

[Seth 1985] S. C. Seth, L. Pan, and V. D. Agrawal, PREDICT—Probabilistic estimation of digital circuit testability, in *Proc. IEEE Fault-Tolerant Computing Symp.*, pp. 220–225, June 1985.

[Stephenson 1976] J. E. Stephenson and J. Garson, A testability measure for register transfer level digital circuits, in *Proc. IEEE Fault-Tolerant Computing Symp.*, pp. 101–107, June 1976.

[Wang 1984] L.-T. Wang and E. Law, Daisy testability analyzer (DTA), in *Proc. IEEE/ACM Int. Conf. on Computer-Aided Design*, pp. 143–145, November 1984.

[Wang 1985] L.-T. Wang and E. Law, An enhanced Daisy testability analyzer (DTA), in *Proc. Automatic Testing Conf.*, pp. 223–229, October 1985.

R3.3 Scan Design

[Cheung 1997] B. Cheung and L.-T. Wang, The seven deadly sins of scan-based designs, in *Integrated System Design*, www.eetimes.com/editorial/1997/test9708.html, August 1997.

[DasGupta 1982] S. DasGupta, P. Goel, R. G. Walther, and T. W. Williams, A variation of LSSD and its implications on design and test pattern generation in VLSI, in *Proc. IEEE Int. Test Conf.*, pp. 63–66, November 1982.

[Eichelberger 1977] E. B. Eichelberger and T. W. Williams, A logic design structure for LSI testability, in *Proc. ACM/IEEE Design Automation Conf.*, pp. 462–468, June 1977.

[Nadeau-Dostie 1994] B. Nadeau-Dostie, A. Hassan, D. Burek, and S. Sunter, Multiple Clock Rate Test Apparatus for Testing Digital Systems, U.S. Patent No. 5,349,587, September 20, 1994.

[Rajski 2003] J. Rajski, A. Hassan, R. Thompson, and N. Tamarapalli, Method and Apparatus for At-Speed Testing of Digital Circuits, U.S. Patent Application No. 20030097614, May 22, 2003.

[Savir 1993] J. Savir and S. Patil, Scan-based transition test, *IEEE Trans. on Computer-Aided Design*, 12(8), pp. 1232–1241, August 1993.

[Savir 1994] J. Savir and S. Patil, Broad-side delay test, *IEEE Trans. on Computer-Aided Design*, 13(8), pp. 1057–1064, August 1994.

[Wang 2005a] L.-T. Wang, M.-C. Lin, X. Wen, H.-P. Wang, C.-C. Hsu, S.-C. Kao, and F.-S. Hsu, Multiple-Capture DFT System for Scan-Based Integrated Circuits, U.S. Patent No. 6,954,887, October 11, 2005.

[Wang 2007b] L.-T. Wang, P.-C. Hsu, and X. Wen, Multiple-Capture DFT System for Detecting or Locating Crossing Clock-Domain Faults During Scan-Test, U.S. Patent No. 7,260,756, August 21, 2007.

R3.4 Logic Built-In Self-Test

[Bardell 1982] P. H. Bardell and W. H. McAnney, Self-testing of multiple logic modules, in *Proc. IEEE Int. Test Conf.*, pp. 200–204, November 1982.

[Barzilai 1981] Z. Barzilai, J. Savir, G. Markowsky, and M. G. Smith, The weighted syndrome sums approach to VLSI testing, *IEEE Trans. on Computers*, 30(12), pp. 996–1000, December 1981.

[Barzilai 1983] Z. Barzilai, D. Coppersmith, and A. Rosenberg, Exhaustive bit pattern generation in discontiguous positions with applications to VLSI testing, *IEEE Trans. on Computers*, 32(2), pp. 190–194, February 1983.

[Benowitz 1975] N. Benowitz, D. F. Calhoun, G. E. Alderson, J. E. Bauer, and C. T. Joeckel, An advanced fault isolation system for digital logic, *IEEE Trans. on Computers*, 24(5), pp. 489–497, May 1975.

[Cadence 2008] Cadence Design Systems, http://www.cadence.com, 2008.

[Chen 1987] C. L. Chen, Exhaustive test pattern generation with cyclic codes, *IEEE Trans. on Computers*, 37(3), pp. 329–338, March 1987.

[Cheon 2005] B. Cheon, E. Lee, L.-T. Wang, X. Wen, P. Hsu, J. Cho, J. Park, H. Chao, and S. Wu, At-speed logic BIST for IP cores, in *Proc. IEEE/ACM Design, Automation, and Test in Europe Conf.*, pp. 860–861, March 2005.

[Chin 1984] C. K. Chin and E. J. McCluskey, *Weighted Pattern Generation for Built-In Self-Test*, Center for Reliable Computing, Technical Report (CRC TR) No. 84-7, Stanford University, August 1984.

[Foote 1997] T. G. Foote, D. E. Hoffman, W. V. Huott, T. J. Koprowski, B. J. Robbins, and M. P. Kusko, Testing the 400 MHz IBM generation-4 CMOS chip, in *Proc. IEEE Int. Test Conf.*, pp. 106–114, November 1997.

[Frohwerk 1977] R. A. Frohwerk, Signature analysis: A new digital field service method, in *Hewlett-Packard J.*, 28, pp. 2–8, September 1977.

[Furukawa 2006] H. Furukawa, X. Wen, L.-T. Wang, B. Sheu, Z. Jiang, and S. Wu, A novel and practical control scheme for inter-clock at-speed testing, in *Proc. IEEE Int. Test Conf.*, Paper 17.2, October 2006.

[Gloster 1988] C. S. Gloster, Jr. and F. Brglez, Boundary scan with cellular built-in self-test, in *Proc. IEEE Int. Test Conf.*, pp. 138–145, September 1988.

[Hassan 1984] S. Z. Hassan and E. J. McCluskey, Increased fault coverage through multiple signatures, in *Proc. IEEE Fault-Tolerant Computing Symp.*, pp. 354–359, June 1984.

[Hayes 1976] J. P. Hayes, Transition count testing of combinational logic circuits, *IEEE Trans. on Computers*, C-25(6), pp. 613–620, June 1976.

[Hortensius 1989] P. D. Hortensius, R. D. McLeod, W. Pries, D. M. Miller, and H. C. Card, Cellular automata-based pseudorandom number generators for built-in self-test, *IEEE Trans. on Computer-Aided Design*, 8(8), pp. 842–859, August 1989.

[Keller 2007] B. Keller, A. Uzzaman, B. Li, and T. Snethen, Using programmable on-product clock generation (OPCG) for delay test, in *Proc. IEEE Asian Test Symp.*, pp. 69–72, October 2007.

[Khara 1987] M. Khara and A. Albicki, Cellular automata used for test pattern generation, in *Proc. IEEE Int. Conf. on Computer Design*, pp. 56–59, October 1987.

[Könemann 1979] B. Könemann, J. Mucha, and G. Zwiehoff, Built-in logic block observation techniques, in *Proc. IEEE Int. Test Conf.*, pp. 37–41, October 1979.

[Könemann 1980] B. Könemann, J. Mucha, and G. Zwiehoff, Built-in test for complex digital circuits, *IEEE J. of Solid-State Circuits*, 15(3), pp. 315–318, June 1980.

[Lai 2007] L. Lai, W.-T. Cheng, and T. Rinderknecht, Programmable scan-based logic built-in self test, in *Proc. IEEE Asian Test Symp.*, pp. 371–377, October 2007.

[LogicVision 2008] LogicVision, http://www.logicvision.com, 2008.

[McCluskey 1981] E. J. McCluskey and S. Bozorgui-Nesbat, Design for autonomous test, *IEEE Trans. on Computers*, 30(11), pp. 860–875, November 1981.

[McCluskey 1984] E. J. McCluskey, Verification testing—A pseudoexhaustive test technique, *IEEE Trans. on Computers*, 33(6), pp. 541–546, June 1984.

[McCluskey 1985] E. J. McCluskey, Built-in self-test structures, *IEEE Design & Test of Computers*, 2(2), pp. 29–36, April 1985.

[Mentor 2008] Mentor Graphics, http://www.mentor.com, 2008.

[Nadeau-Dostie 1994] B. Nadeau-Dostie, A. Hassan, D. Burek, and S. Sunter, Multiple Clock Rate Test Apparatus for Testing Digital Systems, U.S. Patent No. 5,349,587, September 20, 1994.

[Nadeau-Dostie 2006] B. Nadeau-Dostie and J.-F. Côté, Clock Controller for At-Speed Testing of Scan Circuits, U.S. Patent No. 7,155,651, December 26, 2006.

[Nadeau-Dostie 2007] B. Nadeau-Dostie, Method and Circuit for At-Speed Testing of Scan Circuits, U.S. Patent No. 7,194,669, March 20, 2007.

[Rajski 2003] J. Rajski, A. Hassan, R. Thompson, and N. Tamarapalli, Method and Apparatus for At-Speed Testing of Digital Circuits, U.S. Patent Application No. 20030097614, May 22, 2003.

[Savir 1980] J. Savir, Syndrome-testable design of combinational circuits, *IEEE Trans. on Computers*, 29(6), pp. 442–451, June 1980.

[Savir 1985] J. Savir and W. H. McAnney, On the masking probability with ones count and transition count, in *Proc. IEEE/ACM Int. Conf. on Computer-Aided Design*, pp. 111–113, November 1985.

[Schnurmann 1975] H. D. Schnurmann, E. Lindbloom, and R. G. Carpenter, The weighted random test-pattern generator, *IEEE Trans. on Computers*, 24(7), pp. 695–700, July 1975.

[SynTest 2008] SynTest Technologies, http://www.syntest.com, 2008.

[Tang 1984] D. T. Tang and C. L. Chen, Logic test pattern generation using linear codes, *IEEE Trans. on Computers*, 33(9), pp. 845–850, September 1984.

[Tsai 1999] H.-C. Tsai, K.-T. Cheng, and S. Bhawmik, Improving the test quality for scan-based BIST using a general test application scheme, in *Proc. ACM/IEEE Design Automation Conf.*, pp. 748–753, June 1999.

[van Sas 1990] J. van Sas, F. Catthoor, and H. D. Man, Cellular automata-based self-test for programmable data paths, in *Proc. IEEE Int. Test Conf.*, pp. 769–778, September 1990.

[Wang 1986a] L.-T. Wang and E. J. McCluskey, Condensed linear feedback shift register (LFSR) testing—A pseudoexhaustive test technique, *IEEE Trans. on Computers*, 35(4), pp. 367–370, April 1986.

[Wang 1986b] L.-T. Wang and E. J. McCluskey, Concurrent built-in logic block observer (CBILBO), in *Proc. IEEE Int. Symp. on Circuits and Systems*, 3(3), pp. 1054–1057, May 1986.

[Wang 1987] L.-T. Wang and E. J. McCluskey, Linear feedback shift register design using cyclic codes, *IEEE Trans. on Computers*, 37(10), pp. 1302–1306, October 1987.

[Wang 1988a] L.-T. Wang and E. J. McCluskey, Hybrid designs generating maximum-length sequences, *Special Issue on Testable and Maintainable Design, IEEE Trans. on Computer-Aided Design*, 7(1), pp. 91–99, January 1988.

[Wang 1988b] L.-T. Wang and E. J. McCluskey, Circuits for pseudo-exhaustive test pattern generation, *IEEE Trans. on Computer-Aided Design*, 7(10), pp. 1068–1080, October 1988.

[Wang 1989] L.-T. Wang, M. Marhoefer, and E. J. McCluskey, A self-test and self-diagnosis architecture for boards using boundary scan, in *Proc. IEEE European Test Conf.*, pp. 119–126, April 1989.

[Wang 2005b] L.-T. Wang, X. Wen, P.-C. Hsu, S. Wu, and J. Guo, At-speed logic BIST architecture for multi-clock designs, in *Proc. Int. Conf. on Computer Design*, pp. 475–478, October 2005.

[Wang 2006b] L.-T. Wang, P.-C. Hsu, S.-C. Kao, M.-C. Lin, H.-P. Wang, H.-J. Chao, and X. Wen, Multiple-Capture DFT System for Detecting or Locating Crossing Clock-Domain Faults During Self-Test or Scan-Test, U.S. Patent No. 7,007,213, February 28, 2006.

[Williams 1987] T. W. Williams, W. Daehn, M. Gruetzner, and C. W. Starke, Aliasing errors in signature analysis registers, *IEEE Design & Test of Computers*, 4(2), pp. 39–45, April 1987.

[Wolfram 1983] S. Wolfram, Statistical mechanics of cellular automata, in *Review of Modern Physics*, 55(3), pp. 601–644, July 1983.

[Wunderlich 1987] H.-J. Wunderlich, Self test using unequiprobable random patterns, in *Proc. IEEE Fault-Tolerant Computing Symp.*, pp. 258–263, July 1987.

R3.5 Test Compression

[Barnhart 2002] C. Barnhart, V. Brunkhorst, F. Distler, O. Farnsworth, A. Ferko, B. Keller, D. Scott, B. Koenemann, and T. Onodera, Extending OPMISR beyond 10x scan test efficiency, *IEEE Design & Test of Computers*, 19(5), pp. 65–73, May-June 2002.

[Bayraktaroglu 2001] I. Bayraktaroglu and A. Orailoglu, Test volume and application time reduction through scan chain concealment, in *Proc. ACM/IEEE Design Automation Conf.*, pp. 151–155, June 2001.

[Bayraktaroglu 2003] I. Bayraktaroglu and A. Orailoglu, Concurrent application of compaction and compression for test time and data volume reduction in scan designs, *IEEE Trans. on Computers*, 52(11), pp. 1480–1489, November 2003.

[Beck 2005] M. Beck, O. Barondeau, M. Kaibel, F. Poehl, X. Lin, and R. Press, Logic design for on-chip test clock generation—Implementation details and impact on delay test quality, in *Proc. IEEE/ACM Design, Automation, and Test in Europe Conf.*, pp. 56–61, March 2005.

[Cadence 2008] Cadence Design Systems, http://www.cadence.com, 2008.

[Chakrabarty 1998] K. Chakrabarty, B. T. Murray, and J. P. Hayes, Optimal zero-aliasing space compaction of test responses, *IEEE Trans. on Computers*, 47(11), pp. 1171-1187, November 1998.

[Chandra 2007] A. Chandra, H. Yan, and R. Kapur, Multimode Illinois scan architecture for test application time and test data volume reduction, in *Proc. IEEE VLSI Test Symp.*, pp. 84-92, May 2007.

[Fan 2007] X.-X. Fan, Y. Hu, and L.-T. Wang, An on-chip test clock control scheme for multi-clock at-speed testing, in *Proc. IEEE Asian Test Symp.*, pp. 341-348, October 2007.

[Frohwerk 1977] R. A. Frohwerk, Signature analysis: A new digital field service method, in *Hewlett-Packard J.*, 28, pp. 2-8, September 1977.

[Furukawa 2006] H. Furukawa, X. Wen, L.-T. Wang, B. Sheu, Z. Jiang, and S. Wu, A novel and practical control scheme for inter-clock at-speed testing, in *Proc. IEEE Int. Test Conf.*, Paper 17.2, October 2006.

[Gu 2001] X. Gu, S. S. Chung, F. Tsang, J. A. Tofte, and H. Rahmanian, An effort-minimized logic BIST implementation method, in *Proc. IEEE Int. Test Conf.*, pp. 1002-1010, October 2001.

[Hamzaoglu 1999] I. Hamzaoglu and J. H. Patel, Reducing test application time for full scan embedded cores, in *Proc. IEEE Fault-Tolerant Computing Symp.*, pp. 260-267, July 1999.

[Han 2006] Y. Han, X. Li, H. Li, and A. Chandra, Embedded test resource for SoC to reduce required tester channels based on advanced convolutional codes, *IEEE Trans. on Instrumentation and Measurement*, 55(2), pp. 389-399, April 2006.

[Han 2007] Y. Han, Y. Hu, X. Li, H. Li, and A. Chandra, Embedded test decompressor to reduce the required channels and vector memory of tester for complex processor circuit, *IEEE Trans. on Very Large Scale Integration Systems*, 15(5), pp. 531-540, May 2007.

[Hsu 2001] F. F. Hsu, K. M. Butler, and J. H. Patel, A case study on the implementation of Illinois scan architecture, in *Proc. IEEE Int. Test Conf.*, pp. 538-547, October 2001.

[Kapur 2008] R. Kapur, S. Mitra, and T. W. Williams, Historical perspective on scan compression, *IEEE Design & Test of Computers*, 25(2), pp. 114-120, March-April 2008.

[Keller 2007] B. Keller, A. Uzzaman, B. Li, and T. Snethen, Using programmable on-product clock generation (OPCG) for delay test, in *Proc. IEEE Asian Test Symp.*, pp. 69-72, October 2007.

[Könemann 1991] B. Koenemann, LFSR-coded test patterns for scan designs, in *Proc. IEEE European Test Conf.*, pp. 237-242, April 1991.

[Könemann 2001] B. Koenemann, C. Barnhart, B. Keller, T. Snethen, O. Farnsworth, and D. Wheater, A SmartBIST variant with guaranteed encoding, in *Proc. IEEE Asian Test Symp.*, pp. 325-330, November 2001.

[Könemann 2003] B. Koenemann, C. Barnhart, and B. Keller, Real-Time Decoder for Scan Test Patterns, U.S. Patent No. 6,611,933, August 26, 2003.

[Krishna 2001] C. V. Krishna, A. Jas, and N. A. Touba, Test vector encoding using partial LFSR reseeding, in *Proc. IEEE Int. Test Conf.*, pp. 885-893, October 2001.

[Krishna 2003] C. V. Krishna and N. A. Touba, Adjustable width linear combinational scan vector decompression, in *Proc. IEEE/ACM Int. Conf. on Computer-Aided Design*, pp. 863-866, September 2003.

[Lee 1998] K.-J. Lee, J. J. Chen, and C. H. Huang, Using a single input to support multiple scan chains, in *Proc. IEEE/ACM Int. Conf. on Computer-Aided Design*, pp. 74-78, November 1998.

[Lee 1999] K.-J. Lee, J. J. Chen, and C. H. Huang, Broadcasting test patterns to multiple circuits, *IEEE Trans. on Computer-Aided Design*, 18(12), pp. 1793-1802, December 1999.

[Li 2004] L. Li and K. Chakrabarty, Test set embedding for deterministic BIST using a reconfigurable interconnection network, *IEEE Trans. on Computer-Aided Design*, 23(9), pp. 1289-1305, September 2004.

[LogicVision 2008] LogicVision, http://www.logicvision.com, 2008.

[Mentor 2008] Mentor Graphics, http://www.mentor.com, 2008.

[Mitra 2004a] S. Mitra and K. S. Kim, X-Compact: An efficient response compaction technique, *IEEE Trans. on Computer-Aided Design*, 23(3), pp. 421-432, March 2004.

[Mitra 2004b] S. Mitra, S. S. Lumetta, and M. Mitzenmacher, X-tolerant signature analysis, in *Proc. IEEE Int. Test Conf.*, pp. 432-441, October 2004.

[Mitra 2006] S. Mitra and K. S. Kim, XPAND: An efficient test stimulus compression technique, *IEEE Trans. on Computers*, 55(2), pp. 163–173, February 2006.

[Mrugalski 2004] G. Mrugalski, J. Rajski, and J. Tyszer, Ring generators—new devices for embedded test applications, *IEEE Trans. on Computer-Aided Design*, 23(9), pp. 1306–1320, September 2004.

[Nadeau-Dostie 2004] B. Nadeau-Dostie, Method of Masking Corrupt Bits During Signature Analysis and Circuit for Use Therewith, U.S. Patent No. 6,745,359, June 1, 2004.

[Nadeau-Dostie 2005] B. Nadeau-Dostie, J.-F. Côté, and F. Maamari, Structural test with functional characteristics, in *Proc. IEEE Current and Defect-Based Testing Workshop*, pp. 57–60, May 2005.

[Nadeau-Dostie 2006] B. Nadeau-Dostie and J.-F. Côté, Clock Controller for At-Speed Testing of Scan Circuits U.S. Patent No. 7,155,651, December 26 2006.

[Naruse 2003] M. Naruse, I. Pomeranz, S. M. Reddy, and S. Kundu, On-chip compression of output responses with unknown values using LFSR reseeding, in *Proc. IEEE Int. Test Conf.*, pp. 1060–1068, October 2003.

[Pandey 2002] A. R. Pandey and J. H. Patel, Reconfiguration technique for reducing test time and test volume in Illinois scan architecture based designs, in *Proc. IEEE VLSI Test Symp.*, pp. 9–15, April 2002.

[Patel 2003] J. H. Patel, S. S. Lumetta, and S. M. Reddy, Application of Saluja-Karpovsky compactors to test responses with many unknowns, in *Proc. IEEE VLSI Test Symp.*, pp. 107–112, April 2003.

[Pouya 1998] B. Pouya and N. A. Touba, Synthesis of zero-aliasing space elementary-tree space compactors, in *Proc. IEEE VLSI Test Symp.*, pp. 70–77, April 1998.

[Rajski 2004] J. Rajski, J. Tyszer, M. Kassab, and N. Mukherjee, Embedded deterministic test, *IEEE Trans. on Computer-Aided Design*, 23(5), pp. 776–792, May 2004.

[Rajski 2005] J. Rajski, J. Tyszer, C. Wang, and S. M. Reddy, Finite memory test response compactors for embedded test applications, *IEEE Trans. on Computer-Aided Design*, 24(4), pp. 622–634, April 2005.

[Rajski 2008] J. Rajski, J. Tyszer, G. Mrugalski, W.-T. Cheng, N. Mukherjee, and M. Kassab, X-Press: Two-stage X-tolerant compactor with programmable selector, *IEEE Trans. on Computer-Aided Design*, 27(1), pp. 147–159, January 2008.

[Saluja 1983] K. K. Saluja and M. Karpovsky, Test compression hardware through data compression in space and time, in *Proc. IEEE Int. Test Conf.*, pp. 83–88, October 1983.

[Samaranayake 2003] S. Samaranayake, E. Gizdarski, N. Sitchinava, F. Neuveux, R. Kapur, and T. W. Williams, A reconfigurable shared scan-in architecture, in *Proc. IEEE VLSI Test Symp.*, pp. 9–14, April 2003.

[Shah 2004] M. A. Shah and J. H. Patel, Enhancement of the Illinois scan architecture for use with multiple scan inputs, in *Proc. IEEE Computer Society Annual Symp. on VLSI*, pp. 167–172, February 2004.

[Sitchinava 2004] N. Sitchinava, S. Samaranayake, R. Kapur, E. Gizdarski, F. Neuveux, and T. W. Williams, Changing the scan enable during shift, in *Proc. IEEE VLSI Test Symp.*, pp. 73–78, April 2004.

[Synopsys 2008] Synopsys, http://www.synopsys.com, 2008.

[SynTest 2008] SynTest Technologies, http://www.syntest.com, 2008.

[Touba 2006] N. A. Touba, Survey of test vector compression techniques, *IEEE Design & Test of Computers*, 23(4), pp. 294–303, July-August 2006.

[Touba 2007] N. A. Touba, X-canceling MISR—An X-tolerant methodology for compacting output responses with unknowns using a MISR, in *Proc. IEEE Int. Test Conf.*, Paper 6.2, October 2007.

[Touba 2008] N. A. Touba and L.-T. Wang, X-Canceling Multiple-Input Signature Register (MISR) for Compacting Output Responses with Unknowns, U.S. Patent Application No. 12,007,693, January 14, 2008.

[Volkerink 2005] E. H. Volkerink and S. Mitra, Response compaction with any number of unknowns using a new LFSR architecture, in *Proc. ACM/IEEE Design Automation Conf.*, pp. 117–122, June 2005.

[Wang 2002] L.-T. Wang, H.-P. Wang, X. Wen, M.-C. Lin, S.-H. Lin, D.-C. Yeh, S.-W. Tsai, K. S. Abdel-Hafez, Method and Apparatus for Broadcasting Scan Patterns in a Scan-Based Integrated Circuit, U.S. Patent Application No. 20030154433, January 16, 2002.

[Wang 2004] L.-T. Wang, X. Wen, H. Furukawa, F.-S. Hsu, S.-H. Lin, S.-W. Tsai, K. S. Abdel-Hafez, and S. Wu, VirtualScan: A new compressed scan technology for test cost reduction, in *Proc. IEEE Int. Test Conf.*, pp. 916–925, October 2004.

[Wang 2008] L.-T. Wang, X. Wen, S. Wu, Z. Wang, Z. Jiang, B. Sheu, and X. Gu, VirtualScan: Test compression technology using combinational logic and one-pass ATPG, *IEEE Design & Test of Computers*, 25(2), pp. 122–130, March-April 2008.

[Wohl 2001] P. Wohl, J. A. Waicukauski, and T. W. Williams, Design of compactors for signature-analyzers in built-in self-test, in *Proc. IEEE Int. Test Conf.*, pp. 54–63, October 2001.

[Wohl 2003] P. Wohl, J. A. Waicukauski, S. Patel, and M. B. Amin, Efficient compression and application of deterministic patterns in a logic BIST architecture, in *Proc. ACM/IEEE Design Automation Conf.*, pp. 566–569, June 2003.

[Wohl 2004] P. Wohl, J. A. Waicukauski, and S. Patel, Scalable selector architecture for X-tolerant deterministic BIST, in *Proc. ACM/IEEE Design Automation Conf.*, pp. 934–939, June 2004.

[Wohl 2007a] P. Wohl, J. A. Waicukauski, R. Kapur, S. Ramnath, E. Gizdarski, T. W. Williams, and P. Jaini, Minimizing the impact of scan compression, in *Proc. IEEE VLSI Test Symp.*, pp. 67–74, May 2007.

[Wohl 2007b] P. Wohl, J. A. Waicukauski, and S. Ramnath, Fully X-tolerant combinational scan compression, in *Proc. IEEE Int. Test Conf.*, Paper 6.1, October 2007.

R3.6 Concluding Remarks

[Bardell 1982] P. H. Bardell and W. H. McAnney, Self-testing of multiple logic modules, in *Proc. IEEE Int. Test Conf.*, pp. 200–204, November 1982.

[Naffziger 2006] S. Naffziger, B. Stackhouse, T. Grutkowski, D. Josephson, J. Desai, E. Alon, and M. Horowitz, The implementation of a 2-core multi-threaded Itanium family processor, *IEEE J. of Solid-State Circuits*, 41(1), pp. 197–209, January 2006.

[Stackhouse 2008] B. Stackhouse, B. Cherkauer, M. Gowan, P. Gronowski, and C. Lyles, A 65 nm 2-billion-transistor quad-core Itanium processor, *Digest of Papers, IEEE Int. Solid-State Circuits Conf.*, pp. 92, February 2008.

[Wang 1986b] L.-T. Wang and E. J. McCluskey, Concurrent built-in logic block observer (CBILBO), in *Proc. IEEE Int. Symp. on Circuits and Systems*, 3(3), pp. 1054–1057, May 1986.

Fundamentals of algorithms

4

Chung-Yang (Ric) Huang
National Taiwan University, Taipei, Taiwan

Chao-Yue Lai
National Taiwan University, Taipei, Taiwan

Kwang-Ting (Tim) Cheng
University of California, Santa Barbara, California

ABOUT THIS CHAPTER

In this chapter, we will go through the fundamentals of algorithms that are essential for the readers to appreciate the beauty of various EDA technologies covered in the rest of the book. For example, many of the EDA problems can be either represented in graph data structures or transformed into graph problems. We will go through the most representative ones in which the efficient algorithms have been well studied.

The readers should be able to use these graph algorithms in solving many of their research problems. Nevertheless, there are still a lot of the EDA problems that are naturally difficult to solve. That is to say, it is computationally infeasible to seek for the optimal solutions for these kinds of problems. Therefore, heuristic algorithms that yield suboptimal, yet reasonably good, results are usually adopted as practical approaches. We will also cover several selected heuristic algorithms in this chapter. At the end, we will talk about the mathematical programming algorithms, which provide the theoretical analysis for the problem optimality. We will especially focus on the mathematical programming problems that are most common in the EDA applications.

4.1 INTRODUCTION

An algorithm is a sequence of well-defined instructions for completing a task or solving a problem. It can be described in a natural language, pseudocode, a flowchart, or even a programming language. For example, suppose we are interested in knowing whether a specific number is contained in a given sequence of numbers. By traversing the entire number sequence from a certain beginning number

173

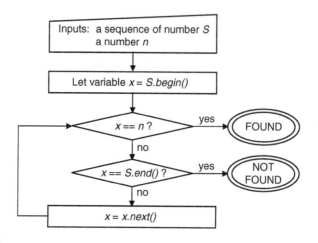

FIGURE 4.1

Flowchart of the "Linear Search" algorithm.

to a certain ending number, we use a search algorithm to find this specific number. Figure 4.1 illustrates this intuitive algorithm known as *linear search*.

Such kinds of algorithms can be implemented in a computer program and then used in real-life applications [Knuth 1968; Horowitz 1978]. However, the questions that must be asked before implementation are: "Is the algorithm efficient?" "Can the algorithm complete the task within an acceptable amount of time for a specific set of data derived from a practical application?" As we will see in the next section, there are methods for quantifying the efficiency of an algorithm. For a given problem, different algorithms can be applied, and each of them has a different degree of efficiency. Such metrics for measuring an algorithm's efficiency can help answer the preceding questions and aid in the selection of the best possible algorithm for the task.

Devising an efficient algorithm for a given EDA problem could be challenging. Because a rich collection of efficient algorithms already exists for a set of standard problems where data are represented in the form of graphs, one possible approach is to model the given problem as a graph problem and then apply a known, efficient algorithm to solve the modeled graph problem. In Section 4.3, we introduce several graph algorithms that are commonly used for a wide range of EDA problems.

Many EDA problems are intrinsically difficult, because finding an optimal solution within a reasonable runtime is not always possible. For such problems, certain *heuristic algorithms* can be applied to find an acceptable solution first. If time or computer resources permit, such algorithms can further improve the result incrementally.

In addition to modeling EDA problems in graphs, it is sometimes possible to transform them into certain mathematical models, such as linear inequalities or nonlinear equations. The primary advantage of modeling an EDA problem with a mathematical formula is that there are many powerful tools that can automatically handle these sorts of mathematical problems. They may yield better results than the customized heuristic algorithms. We will briefly introduce some of these useful mathematical programming techniques near the end of this chapter.

4.2 COMPUTATIONAL COMPLEXITY

A major criterion for a good algorithm is its **efficiency**—that is, how much **time** and **memory** are required to solve a particular problem. Intuitively, time and memory can be measured in real units such as seconds and megabytes. However, these measurements are not subjective for comparisons between algorithms, because they depend on the computing power of the specific machine and on the specific data set. To standardize the measurement of algorithm efficiency, the **computational complexity theory** was developed [Ullman 1984; Papadimitriou 1993, 1998; Wilf 2002]. This allows an algorithm's efficiency to be estimated and expressed conceptually as a mathematical function of its *input size*.

Generally speaking, the *input size* of an algorithm refers to the number of items in the input data set. For example, when sorting n words, the input size is n. Notice that the conventional symbol for input size is n. It is also possible for an algorithm to have an input size with multiple parameters. Graph algorithms, which will be introduced in Section 4.3, often have input sizes with two parameters: the number of vertices $|V|$ and the number of edges $|E|$ in the graph.

Computational complexity can be further divided into **time complexity** and **space complexity**, which estimate the time and memory requirements of an algorithm, respectively. In general, time complexity is considered much more important than space complexity, in part because the memory requirement of most algorithms is lower than the capacity of current machines. In the rest of the section, all calculations and comparisons of algorithm efficiency refer to time complexity as **complexity** unless otherwise specified. Also, time complexity and running time can be used interchangeably in most cases.

The time complexity of an algorithm is calculated on the basis of the number of required elementary computational steps that are interpreted as a function of the input size. Most of the time, because of the presence of conditional constructs (*e.g.*, if-else statements) in an algorithm, the number of necessary steps differs from input to input. Thus, *average-case complexity* should be a more meaningful characterization of the algorithm. However, its calculations are often difficult and complicated, which necessitates the use of a *worst-case complexity* metric. An algorithm's worst-case complexity is its complexity with respect to the worst possible inputs, which gives an upper bound on the average-case complexity. As we shall see, the worst-case complexity may sometimes provide a decent approximation of the average-case complexity.

The calculation of computational complexity is illustrated with two simple examples in Algorithm 4.1 and 4.2. Each of these entails the process of looking up a word in a dictionary. The input size n refers to the total number of words in the dictionary, because every word is a possible target. The first algorithm—linear search—is presented in Algorithm 4.1. It starts looking for the target word t from the first word in the dictionary *(Dic[0])* to the last word *(Dic[n-1])*. The conclusion "not found" is made only after every word is checked. On the other hand, the second algorithm—binary search—takes advantage of the alphabetic ordering of the words in a dictionary. It first compares the word in the middle of the dictionary *(Dic[mid])* with the target t. If t is alphabetically "smaller" than *Dic[mid]*, t must rest in the front part of the dictionary, and the algorithm will then focus on the front part of the word list in the next iteration (line 5 of `Binary_Search`), and *vice versa*. In every iteration, the middle of the search region is compared with the target, and one half of the current region will be discarded in the next iteration. Binary search continues until the target word t is matched or not found at all.

Algorithm 4.1   Linear Search Algorithm

Linear_Search(Array_of_words *Dic[n]*, Target *t*)

1. **for** counter *ctr* from 0 to *n*-1
2.   **if** (*Dic[ctr]* is *t*) **return** *Dic[ctr]*;
3. **return** NOT_FOUND;

Algorithm 4.2   Binary Search Algorithms

Binary_Search(Array_of_words *Dic[n]*, Target *t*)

1. Position *low* = 0, *high* = *n*-1;
2. **while** (*low* <= *high*) **do**
3.   Position *mid* = (*low* + *high*)/2;
4.   **if** (*Dic[mid]* < t) *low* = *mid*;
5.   **else if** (*Dic[mid]* > t) *high* = *mid*;
6.   **else** // *Dic[mid]* is *t*
7.   **return** *Dic[mid]*;
8.   **end if**
9. **end while**
10. **return** NOT_FOUND;

In linear search, the worst-case complexity is obviously n, because every word must be checked if the dictionary does not contain the target word at all. Different target words require different numbers of executions of lines 1-2 in `Linear_Search`, yet on average, $n/2$ times of checks are required.

Thus, the average-case complexity is roughly $n/2$. Binary search is apparently quicker than linear search. Because in every iteration of the *while* loop in `Binary_Search` one-half of the current search area is discarded, at most $\log_2 n$ (simplified as lg n in the computer science community) of lookups are required—the worst-case complexity. n is clearly larger than lg n, which proves that binary search is a more efficient algorithm. Its average-case complexity can be calculated as in Equation (4.1) by adding up all the possible numbers of executions and dividing the result by n.

$$average-case-complexity = \left(1{\cdot}1 + 2{\cdot}2 + 4{\cdot}3 + 8{\cdot}4 + \ldots + \frac{n}{2}{\cdot}\lg n\right)/n$$
$$= \lg n - 1 + \frac{3}{n}$$

(4.1)

4.2.1 **Asymptotic notations**

In computational complexity theory, not all parts of an algorithm's running time are essential. In fact, only the **rate of growth** or the **order of growth** of the running time is typically of most concern in comparing the complexities of different algorithms. For example, consider two algorithms A and B, where A has longer running time for smaller input sizes, and B has a higher rate of growth of running time as the input size increases. Obviously, the running time of B will outnumber that of A for input sizes greater than a certain number. As in real applications, the input size of a problem is typically very large, algorithm B will always run more slowly, and thus we will consider it as the one with higher computational complexity.

Similarly, it is also sufficient to describe the complexity of an algorithm considering only the factor that has highest rate of growth of running time. That is, if the computational complexity of an algorithm is formulated as an equation, we can then focus only on its dominating term, because other lower-order terms are relatively insignificant for a large n. For example, the average-case complexity of `Binary_Search`, which was shown in Equation (4.1), can be simplified to only lg n, leaving out the terms -1 and $3/n$. Furthermore, we can also ignore the dominating term's constant coefficient, because it contributes little information for evaluating an algorithm's efficiency. In the example of `Linear_Search` in Algorithm 4.1, its worst-case complexity and average-case complexity—n and $n/2$, respectively—are virtually equal under this criterion. In other words, they are said to have asymptotically equal complexity for larger n and are usually represented with the following *asymptotic notations*.

Asymptotic notations are symbols used in computational complexity theory to express the efficiency of algorithms with a focus on their orders of growth. The three most used notations are O-notation, Ω-notation, and Θ-notation.

	Also called	$n = 100$	$n = 10,000$	$n = 1,000,000$
$O(1)$	Constant time	0.000001 sec.	0.000001 sec.	0.000001 sec.
$O(\lg n)$	Logarithmic time	0.000007 sec.	0.000013 sec.	0.00002 sec.
$O(n)$	Linear time	0.0001 sec.	0.01 sec.	1 sec.
$O(n\lg n)$		0.00066 sec.	0.13 sec.	20 sec.
$O(n^2)$	Quadratic time	0.01 sec.	100 sec.	278 hours
$O(n^3)$	Cubic time	1 sec.	278 hours	317 centuries
$O(2^n)$	Exponential time	10^{14} centuries	10^{2995} centuries	10^{30087} centuries
$O(n!)$	Factorial time	10^{143} centuries	10^{35645} centuries	N/A

FIGURE 4.2

Frequently used orders of functions and their aliases, along with their actual running time on a million-instructions-per-second machine with three input sizes: $n = 100$, 10,000, and 1,000,000.

4.2.1.1 *O-notation*

O-notation is the dominant method used to express the complexity of algorithms. It denotes the ***asymptotic upper bounds*** of the complexity functions. For a given function $g(n)$, the expression $O(g(n))$ (read as "big-oh of g of n") represents the set of functions

$$O(g(n)) = \{f(n)\text{: positive constants } c \text{ and } n_0 \text{ exist such that}$$
$$0 \le f(n) \le cg(n) \text{ for all } n \ge n_0\}$$

A non-negative function $f(n)$ belongs to the set of functions $O(g(n))$ if there is a positive constant c that makes $f(n) \le cg(n)$ for a sufficiently large n. We can write $f(n) \in O(g(n))$ because $O(g(n))$ is a set, but it is conventionally written as $f(n) = O(g(n))$. Readers have to be careful to note that the equality sign denotes set memberships in all kinds of asymptotic notations.

The definition of *O*-notation explains why lower-order terms and constant coefficients of leading terms can be ignored in complexity theory. The following are examples of legal expressions in computational theory:

$$n^2 = O(n^2)$$
$$n^3 + 1000n^2 + n = O(n^3)$$
$$1000n = O(n)$$
$$20n^3 = O(0.5n^3 + n^2)$$

Figure 4.2 shows the most frequently used *O*-notations, their names, and the comparisons of actual running times with different values of n. The first order of functions, $O(1)$, or constant time complexity, signifies that the algorithm's running time is independent of the input size and is the most efficient. The other *O*-notations are listed in their rank order of efficiency. An algorithm can be considered feasible with quadratic time complexity $O(n^2)$ for a relatively small n, but when n = 1,000,000, a quadratic-time algorithm takes dozens of days to complete the task. An algorithm with a cubic time complexity may handle a problem with small-sized inputs, whereas an algorithm with exponential or factorial time complexity is virtually infeasible. If an algorithm's time complexity can be expressed with or is asymptotically bounded by a polynomial function, it has **polynomial time complexity**. Otherwise, it has **exponential time complexity**. These will be further discussed in Subsection 4.2.2.

4.2.1.2 Ω-*notation and Θ-notation*

Ω-notation is the inverse of O-notation. It is used to express the **asymptotic lower bounds** of complexity functions. For a given function $g(n)$, the expression $\Omega(g(n))$ (read as "big-omega of g of n") denotes the set of functions:

$$\Omega(g(n)) = \{f(n): \text{positive constants } c \text{ and } n_0 \text{ exist such that}$$
$$0 \leq cg(n) \leq f(n) \text{ for all } n \geq n_0\}$$

From the definitions of O- and Ω-notation, the following mutual relationship holds:

$$f(n) = O(g(n)) \text{ if and only if } g(n) = \Omega(f(n))$$

Ω-notation receives much less attention than O-notation, because we are usually concerned about how much time *at most* would be spent executing an algorithm instead of the *least* amount of time spent.

Θ-notation expresses the **asymptotically tight bounds** of complexity functions. Given a function $g(n)$, the expression $\Theta(g(n))$ (read as "big-theta of g of n") denotes the set of functions

$$\Theta(g(n)) = \{f(n): \text{positive constants } c_1, c_2, \text{ and } n_0 \text{ exist such that}$$
$$0 \leq c_1 g(n) \leq f(n) \leq c_2 g(n) \text{ for all } n \geq n_0\}$$

A function $f(n)$ can be written as $f(n) = \Theta(g(n))$ if there are positive coefficients c_1 and c_2 such that $f(n)$ can be squeezed between $c_1 g(n)$ and $c_2 g(n)$ for a sufficiently large n. Comparing the definitions of all three asymptotic notations, the following relationship holds:

$$f(n) = \Theta(g(n)) \text{ if and only if } f(n) = O(g(n)) \text{ and } f(n) = \Omega(g(n))$$

In effect, this powerful relationship is often exploited for verifying the asymptotically tight bounds of functions [Knuth 1976].

Although Θ-notation is more precise when characterizing algorithm complexity, O-notation is favored over Θ-notation for the following two reasons: (1) upper bounds are considered sufficient for characterizing algorithm complexity, and (2) it is often much more difficult to prove a tight bound than it is to prove an upper bound. In the remainder of the text, we will stick with the convention and use O-notation to express algorithm complexity.

4.2.2 Complexity classes

In the previous subsection, complexity was shown to characterize the efficiency of algorithms. In fact, complexity can also be used to characterize the problems themselves. A problem's complexity is equivalent to the time complexity of the most efficient possible algorithm. For instance, the dictionary lookup problem mentioned in the introduction of Section 4.2 has a complexity of $O(\lg n)$, the complexity of `Binary_Search` in Algorithm 4.2.

To facilitate the exploration and discussion of the complexities of various problems, those problems that share the same degree of complexity are grouped, forming complexity classes. Many complexity classes have been established in the history of computer science [Baase 1978], but in this subsection we will only discuss those that pertain to problems in the EDA applications. We will make the distinction between optimization and decision problems first, because these are key concepts within the area of complexity classes. Then, four fundamental and important complexity classes will be presented to help readers better understand the difficult problems encountered in the EDA applications.

4.2.2.1 *Decision problems versus optimization problems*

Problems can be categorized into two groups according to the forms of their answers: decision problems and optimization problems. Decision problems ask for a "yes" or "no" answer. The dictionary lookup problem, for example, is a decision problem, because the answer could only be whether the target is found or not. On the other hand, **an optimization problem** seeks for an optimized value of a target variable. For example, in a combinational circuit, a critical path is a path from an input to an output in which the sum of the gate and wire delays along the path is the largest. Finding a critical path in a circuit is an optimization problem. In this example, optimization means the *maximization* of the target variable. However, optimization can also be *minimization* in other types of optimization problems.

An example of a simple decision problem is the HAMILTONIAN CYCLE problem. The names of decision problems are conventionally given in all capital letters [Cormen 2001]. Given a set of nodes and a set of lines such that each line connects two nodes, a HAMILTONIAN CYCLE is a loop that goes through all the nodes without visiting any node twice. The HAMILTONIAN CYCLE problem asks whether such a cycle exists for a given graph that consists of a set of nodes and lines. Figure 4.3 gives an example in which a Hamiltonian cycle exists.

A famous optimization problem is the traveling salesman problem (TSP). As its name suggests, TSP aims at finding the shortest route for a salesman who needs to visit a certain number of cities in a round tour. Figure 4.4 gives a simple example of a TSP. There is also a version of the TSP as a decision problem: TRAVELING SALESMAN asks whether a route with length under a constant k exists. The optimization version of TSP is more difficult to solve than its

FIGURE 4.3

A graph with one HAMILTONIAN CYCLE marked with thickened lines.

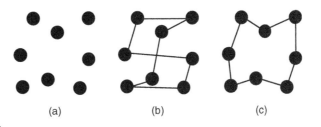

     (a)   (b)   (c)

FIGURE 4.4

(a) An example of the traveling salesman problem, with dots representing cities. (b) A non-optimal solution. (c) An optimal solution.

decision version, because if the former is solved, the latter can be immediately answered for any constant k. In fact, an optimization problem usually can be decomposed into a series of decision problems by use of a different constant as the target for each decision subproblem to search for the optimal solution. Consequently, the optimization version of a problem always has a complexity equal to or greater than that of its decision version.

4.2.2.2 *The complexity classes P versus NP*

The complexity class P, which stands for polynomial, consists of problems that can be solved with known polynomial-time algorithms. In other words, for any problem in the class P, an algorithm of time complexity $O(n^k)$ exists, where k is a constant. The dictionary lookup problem mentioned in Section 4.2 lies in P, because `Linear_Search` in Algorithm 4.1 has a complexity of $O(n)$.

The nondeterministic polynomial or NP complexity class involves the concept of a *nondeterministic computer*, so we will explain this idea first. A nondeterministic computer is not a device that can be created from physical components but is a conceptual tool that only exists in complexity theory. A deterministic computer, or an ordinary computer, solves problems with deterministic algorithms. The characterization of determinism as applied to an algorithm means that at any point in the process of computation the next step is always determined or uniquely defined by the algorithm and the inputs. In other words, given certain inputs and a deterministic computer, the result is always the same no matter how many times the computer executes the algorithm. By contrast, in a nondeterministic computer multiple possibilities for the next step are available at each point in the computation, and the computer will make a *nondeterministic* choice from these possibilities, which will somehow magically lead to the desired answer. Another way to understand the idea of a nondeterministic computer is that it can execute all possible options in parallel at a certain point in the process of computation, compare them, and then choose the optimal one before continuing.

Problems in the NP complexity class have three properties:

1. They are decision problems.
2. They can be solved in polynomial time on a nondeterministic computer.
3. Their solution can be verified for correctness in polynomial time on a deterministic computer.

The TRAVELING SALESMAN decision problem satisfies the first two of these properties. It also satisfies the third property, because the length of the solution route can be calculated to verify whether it is under the target constant k in linear time with respect to the number of cities. TRAVELING SALESMAN is, therefore, an NP class problem. Following the same reasoning process, HAMIL-TONIAN CYCLE is also in this class.

A problem that can be solved in polynomial time by use of a deterministic computer can also definitely be solved in polynomial time on a nondeterministic computer. Thus, $P \subseteq NP$. However, the question of whether $NP = P$ remains unresolved—no one has yet been able to prove or disprove it. To facilitate this proof (or disproof), the most difficult problems in the class NP are grouped together as another complexity class, NP-complete; proving $P = NP$ is equivalent to proving $P = NP$-complete.

4.2.2.3 *The complexity class NP-complete*

Informally speaking, the complexity class NP-complete (or NPC) consists of the most difficult problems in the NP class. Formally speaking, for an arbitrary problem P_a in NP and any problem P_b in the class NPC, a *polynomial transformation* that is able to transform an example of P_a into an example of P_b exists.

A polynomial transformation can be defined as follows: given two problems P_a and P_b, a *transformation* (or *reduction*) from P_a to P_b can express any example of P_a as an example of P_b. Then, the transformed example of P_b can be solved by an algorithm for P_b, and its answer can then be mapped back to an answer to the problem of P_a. A *polynomial* transformation is a transformation with a polynomial time complexity. If a polynomial transformation from P_a to P_b exists, we say that P_a is *polynomially reducible* to P_b. Now we illustrate this idea by showing that the decision problem HAMILTONIAN CYCLE is polynomially reducible to another decision problem—TRAVELING SALESMAN.

Given a graph consisting of n nodes and m lines, with each line connecting two nodes among the n nodes, a HAMILTONIAN CYCLE consists of n lines that traverse all n nodes, as in the example of Figure 4.3. This HAMILTONIAN CYCLE problem can be transformed into a TRAVELING SALESMAN problem by assigning a distance to each pair of nodes. We assign a distance of 1 to each pair of nodes with a line connecting them. For the rest of node pairs, we assign a distance greater than 1, say, 2. With such assignments, the TRAVELING SALESMAN problem of finding whether a round tour of a total distance not greater than n exists is equal to finding a HAMILTONIAN CYCLE in the original graph. If such a tour exists, the total length of the route must be exactly n, and all the distances between the neighboring cities on the route must be 1, which corresponds to existing lines in the original graph; thus, a HAMILTONIAN CYCLE is found. This transformation from HAMILTONIAN CYCLE to TRAVELING SALESMAN is merely based on the assignments of distances, which are of polynomial time complexity—or, more precisely, quadratic time complexity—with respect to the number of nodes. Therefore the transformation is a polynomial transformation.

Now that we understand the concept of a polynomial transformation, we can continue discussing NP-completeness in further detail. Any problem in NPC should be polynomially reducible from any NP problem. Do we need to examine *all* NP problems if a polynomial transformation exists? In fact, a property of the NPC class can greatly simplify the proof of the NP-completeness of a problem: all problems in the class NPC are polynomially reducible to one another. Consequently, to prove that a problem P_t is indeed NPC, only two properties have to be checked:

1. The problem P_t is an NP problem, that is, P_t can be solved in polynomial time on a nondeterministic computer. This is also equivalent to showing that the solution checking of P_t can be done in polynomial time on a deterministic computer.
2. A problem already known to be NP-complete is polynomially reducible to the target problem P_t.

For example, we know that HAMILTONIAN CYCLE is polynomially reducible to TRAVELING SALESMAN. Because the former problem is an NPC problem, and TRAVELING SALESMAN is an NP problem, TRAVELING SALESMAN is, therefore, proven to be contained in the class of NPC.

Use of transformations to prove a problem to be in the NPC class relies on the assumption that there are already problems known to be NP-complete. Hence, this kind of proof is justified only if there is one problem proven to be NP-complete in another way. Such a problem is the SATISFIABILITY problem. The input of this problem is a Boolean expression in the product of sums form such as the following example: $(x_1 + x_2 + x_3)(x_2 + \overline{x_4})(\overline{x_1} + \overline{x_3})(\overline{x_2} + x_3 + x_4)$. The problem aims at assigning a Boolean value to each of the input variables x_t so that the overall product becomes true. If a solution exists, the expression is said to be *satisfiable*. Because the answer to the problem can only be true or false, SATISFIABILITY, or SAT, is a decision problem.

The NP-completeness of the SAT problem is proved with *Cook's theorem* [Cormen 2001] by showing that all NP problems can be polynomially reduced to the SAT problem. The formal proof is beyond the scope of this book [Garey

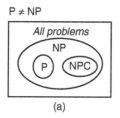 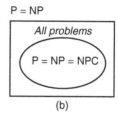

FIGURE 4.5

Relationship of complexity classes if (a) P $\neq$ NP or (b) P $=$ NP.

1979], so we will only informally demonstrate its concept. We have mentioned that all NP problems can be solved in polynomial time on a nondeterministic computer. For an arbitrary NP problem, if we record all the steps taken on a nondeterministic computer to solve the problem in a series of statements, Cook's theorem proves that the series of statements can be polynomially transformed into a product of sums, which is in the form of an SAT problem. As a result, all NP problems can be polynomially reduced to the SAT problem; consequently, the SAT problem is NP-complete.

An open question in computer science is whether a problem that lies in both the P and the NPC classes exists. No one has been able to find a deterministic algorithm with a polynomial time complexity that solves any of the NP-complete problems. If such an algorithm can be found, all of the problems in NPC can be solved by that algorithm in polynomial time, because they are polynomially reducible to one another. According to the definition of NP-completeness, such an algorithm can also solve all problems in NP, making P $=$ NP, as shown in Figure 4.5b. Likewise, no one has been able to prove that for any of the problems in NPC no polynomial time algorithm exists. As a result, although the common belief is that P $\neq$ NP, as shown in Figure 4.5a, and decades of endeavors to tackle NP-complete problems suggest this is true, no hard evidence is available to support this point of view.

4.2.2.4 *The complexity class NP-hard*

Although NP-complete problems are realistically very difficult to solve, there are other problems that are even more difficult: *NP-hard* problems. The NP-hard complexity class consists of those problems at least as difficult to solve as NP-complete problems. A specific way to define an NP-hard problem is that the solution checking for an NP-hard problem cannot be completed in polynomial time. In practice, many optimization versions of the decision problems in NPC are NP-hard. For example, consider the NP-complete TRAVELING SALESMAN problem. Its optimization version, TSP, searches for a round tour going through all cities with a minimum total length. Because its solution checking requires computation of the lengths of all possible routes, which is a $O(n \cdot n!)$ procedure, with n being the number of cities, the solution definitely cannot be found in

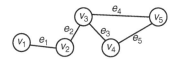

FIGURE 4.6

A combinational circuit and its graph representation.

FIGURE 4.7

An exemplar graph.

polynomial time. Therefore, TSP, an optimization problem, belongs to the NP-hard class.

4.3 GRAPH ALGORITHMS

A *graph* is a mathematical structure that models pairwise relationships among items of a certain form. The abstraction of graphs often greatly simplifies the formulation, analysis, and solution of a problem. Graph representations are frequently used in the field of Electronic Design Automation. For example, a combinational circuit can be efficiently modeled as a *directed graph* to facilitate structure analysis, as shown in Figure 4.6.

Graph algorithms are algorithms that exploit specific properties in various types of graphs [Even 1979; Gibbons 1985]. Given that many problems in the EDA field can be modeled as graphs, efficient graph algorithms can be directly applied or slightly modified to address them. In this section, the terminology and data structures of graphs will first be introduced. Then, some of the most frequently used graph algorithms will be presented.

4.3.1 Terminology

A graph G is defined by two sets: a vertex set V and an edge set E. Customarily, a graph is denoted with $G(V, E)$. Vertices can also be called *nodes*, and edges can be called *arcs* or *branches*. In this chapter, we use the terms *vertices* and *edges*.

Figure 4.7 presents a graph G with $V = \{v_1, v_2, v_3, v_4, v_5\}$ and $E = \{e_1, e_2, e_3, e_4, e_5\}$. The two vertices connected by an edge are called the edge's *endpoints*. An edge can also be characterized by its two endpoints, u and v, and denoted as (u, v). In the example of Figure 4.7, $e_1 = (v_1, v_2)$, $e_2 = (v_2, v_3)$, etc. If there is an edge e connecting u and v, the two vertices u and v are *adjacent* and edge e is

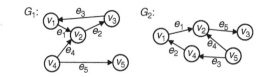

FIGURE 4.8

Two examples of directed graphs.

incident with u (and also with *v*). The *degree* of a vertex is equal to the number of edges incident with it.

A *loop* is an edge that starts and ends at the same vertex. If plural edges are incident with the same two vertices, they are called *parallel edges*. A graph without loops and parallel edges is called a *simple graph*. In most discussions of graphs, only simple graphs are considered, and, thus, a graph implicitly means a simple graph. A graph without loops but with parallel edges is known as a *multigraph*.

The number of vertices in a graph is referred to as the *order* of the graph, or simply $|V|$. Similarly, the *size* of a graph, denoted as $|E|$, refers to its number of edges. It is worth noting that inside asymptotic notations, such as O and Θ, and *only* inside them, $|V|$ and $|E|$ can be simplified as V and E. For example, $O(|V| + |E|)$ can be expressed as $O(V + E)$.

A *path* in a graph is a sequence of alternating vertices and edges such that for each vertex and its next vertex in the sequence, the edge between these vertices connects them. The *length* of a path is defined as the number of edges in a path. For example, in Figure 4.7, $<v_5, e_4, v_3, e_3, v_4>$ is a path with a length of two. A path in which the first and the last vertices are the same is called a *cycle*. $<v_5, e_4, v_3, e_3, v_4, e_5, v_5>$ is a cycle in Figure 4.7. A path, in which every vertex appears once in the sequence is called a *simple* path. The word "simple" is often omitted when this term is used, because we are only interested in simple paths most of the time.

The terms defined so far are for *undirected* graphs. In the following, we introduce the terminology for *directed* graphs. In a *directed* graph, every edge has a direction. We typically use arrows to represent directed edges as shown in the examples in Figure 4.8. For an edge $e = (u, v)$ in a directed graph, u and v cannot be freely exchanged. The edge e is directed from u to v, or equivalently, *incident from u* and *incident to v*. The vertex u is the *tail* of the edge e; v is the *head* of the edge e. The degree of a vertex in a directed graph is divided into the *in-degree* and the *out-degree*. The *in-degree* of a vertex is the number of edges incident *to* it, whereas the *out-degree* of a vertex is the number of edges incident *from* it. For the example of G_2 in Figure 4.8, the in-degree of v_2 is 2 and its out-degree is 1.

The definitions of paths and cycles need to be revised as well for a directed graph: every edge in a path or a cycle must be preceded by its tail and followed by its head. For example, $<v_4, e_4, v_2, e_2, v_3>$ in G_1 of Figure 4.8 is a path and $<v_1, e_1, v_2, e_2, v_3, e_3, v_1>$ is a cycle, but $<v_4, e_4, v_2, e_1, v_1>$ is not a path.

If a vertex u appears before another vertex v in a path, u is v's *predecessor* on that path and v is u's *successor*. Notice that there is no cycle in G_2. Such directed graphs without cycles are called *directed acyclic graphs* or *DAGs*. DAGs are powerful tools used to model combinational circuits, and we will dig deeper into their properties in the following subsections.

In some applications, we can assign values to the edges so that a graph can convey more information related to the edges other than their connections. The values assigned to edges are called their *weights*. A graph with weights assigned to edges is called a *weighted graph*. For example, in a DAG modeling of a combinational circuit, we can use weights to represent the time delay to propagate a signal from the input to the output of a logic gate. By doing so, critical paths can be conveniently determined by standard graph algorithms.

4.3.2 Data structures for representations of graphs

Several data structures are available to represent a graph in a computer, but none of them is categorically better than the others [Aho 1983; Tarjan 1987]. They all have their own advantages and disadvantages. The choice of the data structure depends on the algorithm [Hopcroft 1973].

The simplest data structure for a graph is an *adjacency matrix*. For a graph $G = (V, E)$, a $|V| \times |V|$ matrix A is needed. $A_{ij} = 1$ if $(v_i, v_j) \in E$, and $A_{ij} = 0$ if $(v_i, v_j) \notin E$. For an undirected graph, the adjacency matrix is symmetrical, because the edges have no directions. Figure 4.9 shows the adjacency matrices for the graph in Figure 4.7 and G_2 in Figure 4.8.

One of the strengths of the use of an adjacency matrix is that it can easily represent a weighted graph by changing the ones in the matrix to the edges' respective weights. However, the weight cannot be a zero in this representation (otherwise we cannot differentiate zero-weight edge from "no connection" between two vertices). Also, an adjacency matrix requires exactly $\Theta(V^2)$ space. For a *dense* graph for which $|E|$ is close to $|V|^2$, this could be a memory-efficient representation. However, if the graph is *sparse*, that is, $|E|$ is much smaller than $|V|^2$, most of the entries in the adjacency matrix would be zeros, resulting in a waste of memory.

A sparse graph is better represented with an *adjacency list*, which consists of an array of size $|V|$, with the ith element corresponding to the vertex v_i. The ith element points to a *linked list* that stores those vertices adjacent to v_i

$$
\begin{bmatrix} 0 & 1 & 0 & 0 & 0 \\ 1 & 0 & 1 & 0 & 0 \\ 0 & 1 & 0 & 1 & 1 \\ 0 & 0 & 1 & 0 & 1 \\ 0 & 0 & 1 & 1 & 0 \end{bmatrix}
\qquad
\begin{bmatrix} 0 & 1 & 0 & 0 & 0 \\ 0 & 0 & 1 & 0 & 0 \\ 0 & 0 & 0 & 0 & 0 \\ 1 & 0 & 0 & 0 & 0 \\ 0 & 1 & 0 & 1 & 0 \end{bmatrix}
$$

$\qquad\qquad\qquad$ (a) $\qquad\qquad\qquad\qquad$ (b)

FIGURE 4.9

The adjacency matrices: (a) for Figure 4.7. (b) for G_2 in Figure 4.8.

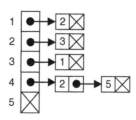

FIGURE 4.10

The adjacency list for G_1 of Figure 4.8.

in an undirected graph. For a directed graph, any vertex v_j in the linked list of the ith element satisfies the condition $(v_i, v_j) \in E$. The adjacency list for G_1 in Figure 4.8 is shown in Figure 4.10.

4.3.3 Breadth-first search and depth-first search

Many graph algorithms rely on efficient and systematic traversals of vertices and edges in the graph. The two simplest and most commonly used traversal methods are *breadth-first search* and *depth-first search*, which form the basis for many graph algorithms. We will examine their generic structures and point out some important applications.

4.3.3.1 *Breadth-first search*

Breadth-first search (BFS) is a systematic means of visiting vertices and edges in a graph. Given a graph G and a specific ***source*** vertex s, the BFS searches through those vertices adjacent to s, then searches the vertices adjacent to those vertices, and so on. The routine stops when BFS has visited all vertices that are reachable from s. The phenomenon that the vertices closest to the source s are visited earlier in the search process gives this search its name. Several procedures can be executed when visiting a vertex. The function BFS in Algorithm 4.3 adopts two of the most frequently used procedures: building a *breadth-first* tree and calculating the distance, which is the minimum length of a path, from the source s to each reachable vertex.

Algorithm 4.3   Breadth-first Search Algorithm

BFS (Graph G, Vertex s)

1. FIFO_Queue Q = {s};
2. **for** (each v ∈ V) do
3.   v.visited = false;   // visited by BFS
4.   v.distance = ∞;   // distance from source s
5.   v.predecessor = NIL; // predecessor of v
6. **end for**
7. s.visited = true;

8. *s*.distance = 0;

9. **while** $(Q \neq \emptyset)$ **do**

10.   Vertex *u* = Dequeue(*Q*);

11.   **for** (each $(u, w) \in E$) **do**

12.   **if** (!(*w*.visited))

13.   *w*.visited = true;

14.   *w*.distance = *u*.distance + 1;

15.   *w*.predecessor = *u*;

16.   Enqueue(*Q*, *w*);

17.   **end if**

18.   **end for**

19. **end while**

The function BFS implements breadth-first search with a queue Q. The queue Q stores the indices of, or the links to, the visited vertices whose adjacent vertices have not yet been examined. The first-in first-out (FIFO) property of a queue guarantees that BFS visits every reachable vertex once, and all of its adjacent vertices are explored in a breadth-first fashion. Because each vertex and edge is visited at most once, the time complexity of a generic BFS algorithm is $O(V + E)$, assuming the graph is represented by an adjacency list.

Figure 4.11 shows a graph produced by the BFS in Algorithm 4.3 that also indicates a breadth-first tree rooted at v_1 and the distances of each vertex to v_1. The distances of v_7 and v_8 are infinity, which indicates that they are *disconnected* from v_1. In contrast, subsets of a graph in which the vertices are connected to one another and to which no additional vertices are connected, such as the set from v_1 to v_6 in Figure 4.11, are called *connected components* of the graph. One of the applications of BFS is to find the connected components of a graph. The attributes `distance` and `predecessors` indicate the lengths and the routes of the shortest paths from each vertex to the vertex v_1. A BFS algorithm

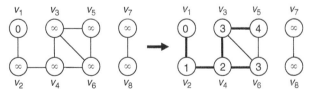

FIGURE 4.11

Applying BFS on an undirected graph with source v_1. The left is the graph after line 8 and the right shows the graph after the completion of the BFS. Numbers in the vertices are their distances to the source v_1. Thick edges are breadth-first tree edges.

can also compute the shortest paths and their lengths from a source vertex to all other vertices in an *unweighted* graph. The calculation of the shortest paths in a weighted graph will be discussed in Subsection 4.3.6.

4.3.3.2 *Depth-first search*

While BFS traverses a graph in a breadth-first fashion, *depth-first search* (DFS) explores the graph in an opposite manner. From a predetermined source vertex *s*, DFS traverses the vertex as deep as possible along a path before backtracking, just as the name implies. The recursive function `DFSPrototype`, shown in Algorithm 4.4, is the basic structure for a DFS algorithm.

Algorithm 4.4   A Prototype of the Depth-first Search Algorithm

---

DFSPrototype(Vertex *v*)

1. // Pre-order process on *v*;
2. mark *v* as visited;
3. **for** (each unvisited vertex *u* adjacent to *v*)
4.   DFSPrototype(*u*);
5.   // In-order process on *v*;
6. **end for**
7. // Post-order process on *v*

The terms *pre-order*, *in-order*, and *post-order* processes on the lines 1, 5, and 7 in Algorithm 4.4 refer to the traversal patterns on a conceptual tree formed by all the vertices in the graph. DFS performs a pre-order process on all the vertices in the exact same order as a pre-order tree traversal in the resulting "*depth-first forest*." This is also the case for in-order and post-order processes. The functionality of these processes, which will be tailor-designed to an application, is the basis of DFS algorithms. The function `DFS` in Algorithm 4.5 provides an example of a post-order process.

Algorithm 4.5   A Complete Depth-first Search Algorithm

DFS(Graph *G*)

1. **for** (each vertex *v* ∈ *V*) **do**
2.   *v*.visited = false;
3.   *v*.predecessor = NIL;
4. **end for**
5. *time* = 0;
6. **for** (each vertex *v* ∈ *V*)
7.   **if** (!(*v*.visited))
8.   DFSVisit(*v*);

9. **end if**

10. **end for**

DFSVisit(Vertex *v*)

1. *v*.visited = true;
2. **for** (each (*v*, *u*) ∈ *E*)
3.   **if** (!(*u*.visited)) **do**
4.   *u*.predecessor = *v*;
5.   DFSVisit(*u*);
6.   **end if**
7. **end for**
8. *time* = *time* + 1;
9. *v*.PostOrderTime = *time*;

Notice that it is guaranteed that every vertex will be visited by lines 6 and 7 in DFS. This is another difference between DFS and BFS. For most applications of DFS, it is preferred that all vertices in the graph be visited. As a result, a *depth-first forest* is formed instead of a tree. Moreover, because each vertex and edge is explored exactly once, the time complexity of a generic DFS algorithm is $O(V + E)$ assuming the use of an adjacency list.

Figure 4.12 demonstrates a directed graph on which DFS(G_1) is executed. The PostOrderTimes of all vertices and the tree edges of a depth-first forest, which is constructed from the predecessor of each vertex, are produced as the output. PostOrderTimes have several useful properties. For example, the vertices with a lower post-order time are never predecessors of those with a higher post-order time on any path. The next subsection uses this property for sorting the vertices of a DAG. In Subsection 4.3.5, we will introduce some important applications of the depth-first forest.

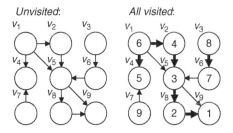

FIGURE 4.12

Applying DFS on a directed graph G_1. The numbers in the vertices are their PostOrderTimes. Thickened edges show how a depth-first forest is built.

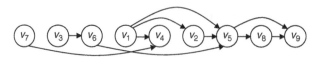

FIGURE 4.13

A topological sort of the graph in Figure 4.12.

4.3.4 **Topological sort**

A *topological sort* is a linear ordering of vertices in a **directed acyclic graph** (DAG). Given a DAG $G = (V, E)$, a topological sort algorithm returns a sequence of vertices in which the vertices never come before their predecessors on any paths. In other words, if $(u, v) \in E$, v never appears before u in the sequence. A topological sort of a graph can be represented as a horizontal line of ordered vertices, such that all edges point only to the right (Figure 4.13).

DAGs are used in various applications to show precedence among events. In the EDA industry, DAGs are especially useful because they are capable of modeling the input-output relationships of combinational circuits, as shown in Figure 4.6. To effectively simulate a combinational circuit with EDA tools, inputs of a gate should usually be examined before the output is analyzed. A topological sort of a DAG provides an appropriate ordering of gates for simulations.

The simple algorithm in Algorithm 4.6 topologically sorts a DAG by use of the depth-first search. Note that line 2 in Algorithm 4.6 should be embedded into line 9 of the function `DFSVisit` in Algorithm 4.5 so that the complexity of the function `TopologicalSortByDFS` remains $O(V + E)$. The result of running `TopologicalSortByDFS` on the graph in Figure 4.12 is shown in Figure 4.13. The vertices are indeed topologically sorted.

Algorithm 4.6   A Simple DFS-based Topological Sort Algorithm

---

TopologicalSortByDFS(Graph *G*)

1. call DFS(*G*) in Algorithm 4.5;

2. as PostOrderTime of each vertex *v* is computed, insert *v* onto the front of a linked list *ll*;

3. **return** *ll*;

Another intuitive algorithm, shown in Algorithm 4.7, can sort a DAG topologically without the overhead of recursive functions typically found in DFS. With careful programming, it has a linear time complexity $O(V + E)$. This version of a topological sort is also superior because it can detect cycles in a directed graph. One application of this feature is efficiently finding feedback loops in a circuit, which should not exist in a combinational circuit.

Algorithm 4.7   A Topological Sort Algorithm that can Detect Cycles

---

TopologicalSort(Graph G)

   1. FIFO_Queue Q = {vertices with in-degree 0};

   2. LinkedList ll = Ø;

   3. **while** (Q is not empty) **do**

   4.   Vertex v = Dequeue(Q);

   5.   insert v into ll;

   6.   **for** (each vertex u such that $(v, u) \in E$) **do**

   7.   remove (v, u) from E;

   8.   **if** (in-degree of u is 0) Enqueue(Q, u);

   9.   **end for**

 10.   **end while**

 11. **if** ($E \neq$ Ø) **return** "G has cycles";

 12. **else return** ll;

4.3.5 **Strongly connected component**

A connected component in an undirected graph has been defined in Subsection 4.3.3.1. For a directed graph, connectivity is further classified into *"strong connectivity"* and *"weak connectivity."* A directed graph is *weakly* connected if all vertices are connected provided all directed edges are replaced as undirected edges. For a *strongly connected* directed graph, every vertex must be reachable from every other vertex. More precisely, for any two vertices u and v in a strongly connected graph, there exists a path from u to v, as well as a path from v to u. A *strongly connected component* (SCC) in a directed graph is a subset of the graph that is strongly connected and is maximal in the sense that no additional vertices can be included in this subset while still maintaining the property of strong connectivity. Figure 4.14a shows a weakly connected graph with four strongly connected components. As an SCC consisting of more than one vertex must contain cycles, it follows naturally that a directed *acyclic* graph has no SCCs that consist of more than one vertex.

The algorithm used to extract SCCs, SCC in Algorithm 4.8, requires the knowledge of the *transpose* of a directed graph (line 2). A transpose of a directed graph G, G^{T}, contains the same vertices of G, but the directed edges are reversed. Formally speaking, for $G = (V, E)$, $G^{\mathrm{T}} = (V, E^{\mathrm{T}})$ with $E^{\mathrm{T}} = \{(u, v): (v, u) \in E\}$. Transposing a graph incurs a linear time complexity $O(V + E)$, which preserves the efficiency of the algorithm for finding SCCs.

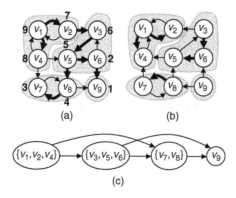

(a)   (b)

(c)

FIGURE 4.14

(a) A directed graph G after running DFS with depth-first tree edges thickened. Post-order times are labeled beside each vertex and SCC regions are shaded. (b) The graph G^T, the transpose of G, after running scc in Algorithm 4.8 (c) Finding SCCs in G as individual vertices result in a DAG.

Algorithm 4.8 An Algorithm to Extract SCCs from a Directed Graph

---

SCC(Graph G)

1. call DFS(G) in Algorithm 4.5 for PostOrderTime;
2. G^T = transpose(G);
3. call DFS(G^T), replacing line 6 of DFS with a procedure examining vertices in order of decreasing PostOrderTime;
4. **return** different trees in depth-first forest built in DFS(G^T) as separate SCCs;

SCC is simple: a DFS, then a transpose, then another DFS. It is also efficient because DFS and transpose incur only a linear time complexity, resulting in a time complexity of $O(V + E)$. Figure 4.14 gives an example of running SCC on a graph G. The four SCCs are correctly identified by the four depth-first trees in G^T. Moreover, if we view an SCC as a single vertex, the resultant graph, shown in Figure 4.14, is a DAG. We also observe that examining vertices in a descending order of the post-order times in DFS is equivalent to visiting the resultant SCCs in a topologically sorted order.

If we model a sequential circuit as a directed graph where vertices represent registers and edges represent combinational signal flows between registers, extracting SCCs from the graph identifies clusters of registers, each of which includes a set of registers with strong functional dependencies among themselves. Extracting SCCs also enables us to model each SCC as a single element, which greatly facilitates circuit analysis because the resultant graph is a DAG.

4.3.6 **Shortest and longest path algorithms**

Given a combinational circuit in which each gate has its own delay value, suppose we want to find the critical path—that is, the path with the longest delay—from an input to an output. A trivial solution is to explicitly evaluate all paths from the input to the output. However, the number of paths can grow exponentially with respect to the number of gates. A more efficient solution exists: we can model the circuit as a directed graph whose edge weights are the delays of the gates. The *longest path algorithm* can then give us the answer more efficiently.

In this subsection, we present various shortest and longest path algorithms. Not only can they calculate the delays of critical paths, but they also can be applied to other EDA problems, such as finding an optimal sequence of state transitions from the starting state to the target state in a state transition diagram. In the *shortest-path problem* or the *longest-path problem*, we are given a **weighted**, **directed** graph. The weight of a path is defined as the sum of the weights of its constituent edges. The goal of the shortest-/longest-path problem is to find the path from a source vertex s to a destination vertex d with minimum/maximum weight. Three algorithms are capable of finding the shortest paths from a source to all other vertices, each of which works on the graph with different constraints. First, we will present a simple algorithm used to solve the shortest-path problem on DAGs. *Dijkstra's algorithm* [Dijkstra 1959], which functions on graphs with non-negative weights, will then be presented. Finally, we will introduce a more general algorithm that can be applied to all types of directed graphs—the *Bellman-Ford algorithm* [Bellman 1958]. On the basis of these algorithms' concepts, we will demonstrate how to modify them to apply to longest-path problems.

4.3.6.1 *Initialization and relaxation*

Before explaining these algorithms, we first introduce two basic techniques used by all the algorithms in this subsection: initialization and relaxation.

Before running a shortest-path algorithm on a directed graph $G = (V, E)$, we must be given a source vertex s and the weight of each edge $e \in E$, $w(e)$. Also, two attributes must be stored for each vertex $v \in V$: the **predecessor** $pre(v)$ and the **shortest-path estimate** $est(v)$. The predecessor $pre(v)$ records the predecessor of v on the shortest path, and $est(v)$ is the current estimation of the weight of the shortest path from s to v. The procedure in Algorithm 4.9, known as *initialization*, initializes $pre(v)$ and $est(v)$ for all vertices.

Algorithm 4.9 Initialization Procedure for Shortest-path Algorithms

---

Initialize(graph G, Vertex s)

  1. **for** (each vertex $v \in V$) **do**
  2. $pre(v)$ = NIL; // predecessor
  3. $est(v) = \infty$; // shortest-path estimate
  4. **end for**
  5. $est(s) = 0$;

The other common procedure, **relaxation**, is the kernel of all the algorithms presented in this subsection. The *relaxation* of an edge (u, v) is the process of determining whether the shortest path to v found so far can be shortened or relaxed by taking a path through u. If the shortest path is, indeed, improved by use of this procedure, *pre(v)* and *est(v)* will be updated. Algorithm 4.10 shows this important procedure.

Algorithm 4.10   Relaxation Procedure for Shortest-path Algorithms

---

Relax(Vertex u, Vertex v)

1. **if** $(est(v) > est(u) + w(u, v))$ **do**
2. $est(v) = est(u) + w(u, v)$;
3. $pre(v) = u$;
4. **end if**

4.3.6.2 *Shortest path algorithms on directed acyclic graphs*

DAGs are always easier to manipulate than the general directed graphs, because they have no cycles. By use of a topological sorting procedure, as shown in Algorithm 4.11, this $\Theta(V + E)$ algorithm calculates the shortest paths on a DAG with respect to a given source vertex s.

The function DAGShortestPaths, used in Algorithm 4.11, sorts the vertices topologically first; in line 4, each vertex is visited in the topologically sorted order. As each vertex is visited, the function relaxes all edges incident from it. The shortest paths and their weights are then available in *pre(v)* and *est(v)* of each vertex v. Figure 4.15 gives an example of running DAGShortestPaths on a DAG. Notice that the presence of negative weights in a graph does not affect the correctness of this algorithm.

Algorithm 4.11   A Shortest-path Algorithm for DAGs

DAGShortestPaths(Graph G, vertex s)

1. topologically sort the vertices of G;
2. Initialize(G, s);
3. **for** (each vertex u in topological sorted order)
4.   **for** (each vertex v such that $(u, v) \in E$)
5.   Relax(u, v);
6.   **end for**
7. **end for**

4.3.6.3 *Dijkstra's algorithm*

Dijkstra's algorithm solves the shortest-path problem for any weighted, directed graph with **non-negative** weights. It can handle graphs consisting of cycles,

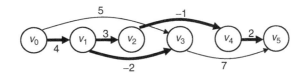

visited	Predecessors					Shortest-Path Estimates				
vertex	v_1	v_2	v_3	v_4	v_5	v_1	v_2	v_3	v_4	v_5
non	NIL	NIL	NIL	NIL	NIL	∞	∞	∞	∞	∞
v_0	v_0	NIL	v_0	NIL	NIL	4	∞	5	∞	∞
v_1	v_0	v_1	v_1	NIL	NIL	4	7	2	∞	∞
v_2	v_0	v_1	v_1	v_2	NIL	4	7	2	6	∞
v_3	v_0	v_1	v_1	v_2	v_3	4	7	2	6	9
v_4	v_0	v_1	v_1	v_2	v_4	4	7	2	6	8
v_5	v_0	v_1	v_1	v_2	v_4	4	7	2	6	8

FIGURE 4.15

The upper part is a DAG with its shortest paths shown in thickened edges, and the lower part is the changes of predecessors and shortest-path estimates when different vertices are visited in line 3 of the function DAGShortestPaths.

but negative weights will cause this algorithm to produce incorrect results. Consequently, we assume that $w(e) \geq 0$ for all $e \in E$ here.

The pseudocode in Algorithm 4.12 shows Dijkstra's algorithm. The algorithm maintains a priority queue minQ that is used to store the unprocessed vertices with their shortest-path estimates $est(v)$ as key values. It then repeatedly extracts the vertex u which has the minimum $est(u)$ from minQ and relaxes all edges incident from u to any vertex in minQ. After one vertex is extracted from minQ and all relaxations through it are completed, the algorithm will treat this vertex as processed and will not touch it again. Dijkstra's algorithm stops either when minQ is empty or when every vertex is examined exactly once.

Algorithm 4.12 Dijkstra's shortest-path algorithm

Dijkstra(Graph G, Vertex s)

1. Initialize(G, s);

2. Priority_Queue minQ = {all vertices in V};

3. **while** (minQ $\neq \emptyset$) **do**

4.   Vertex u = ExtractMin(minQ); // minimum $est(u)$

5.   **for** (each $v \in$ minQ such that $(u, v) \in E$)

6.   Relax(u, v);

7.   **end for**

8. **end while**

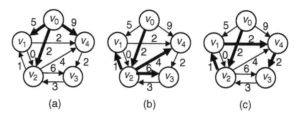

(a)   (b)   (c)

vertex	Predecessors					Shortest-Path Estimates				
	v_0	v_1	v_2	v_3	v_4	v_0	v_1	v_2	v_3	v_4
non	NIL	NIL	NIL	NIL	NIL	0	∞	∞	∞	∞
v_0	NIL	v_0	v_0	NIL	v_0	0	5	2	∞	9
v_2	NIL	v_2	v_0	v_2	v_2	0	3	2	8	6
v_1	NIL	v_2	v_0	v_2	v_1	0	3	2	8	5
v_4	NIL	v_2	v_0	v_4	v_1	0	3	2	7	5
v_3	NIL	v_2	v_0	v_4	v_1	0	3	2	7	5

FIGURE 4.16

An example of Dijkstra's algorithm: (a), (b), and (c) respectively show the edges belonging to the shortest paths when v_0, v_2, and v_3 are visited. The table exhibits the detailed data when each vertex is visited.

	Predecessors			Shortest-Path Estimates		
	v_0	v_1	v_2	v_0	v_1	v_2
Dijkstra's	NIL	v_0	v_0	0	2	3
Correct path	NIL	v_2	v_0	0	1	3

FIGURE 4.17

Running Dijkstra's algorithm on a graph with negative weights causes incorrect results on v_1.

Dijkstra's algorithm works correctly, because all edge weights are non-negative, and the vertex with the least shortest-path estimate is always chosen. In the first iteration of the `while` loop in lines 3 through 7, the source s is chosen and its adjacent vertices have their $est(v)$ set to $w((s, v))$. In the second iteration, the vertex u with minimal $w((s, u))$ will be selected; then those edges incident from u will be relaxed. Clearly, there exists no shorter path from s to u than the single edge (s, u), because all weights are not negative, and any path traced that uses an intermediate vertex is longer. Continuing this reasoning brings us to the conclusion that the algorithm, indeed, computes the shortest paths.

Figure 4.16 illustrates the execution of Dijkstra's algorithm on a directed graph with non-negative weights and containing cycles. However, a small example in Figure 4.17 shows that Dijkstra's algorithm fails to find the shortest paths when negative weights exist.

Dijkstra's algorithm necessitates the use of a priority queue that supports the operations of extracting a minimum element and decreasing keys. A linear array can be used, but its complexity will be as much as $O(V^2 + E) = O(V^2)$. If a more efficient data structure, such as a *binary* or *Fibonacci heap* [Moore 1959], is used to implement the priority queue, the complexity can be reduced.

4.3.6.4 *The Bellman-Ford algorithm*

Cycles should never appear in a shortest path. However, if there exist negative-weight cycles, a shortest path can have a weight of $-\infty$ by circling around negative-weight cycles infinitely many times. Therefore, negative-weight cycles should be avoided before finding the shortest paths. In general, we can categorize cycles into three types according to their weights: negative-weight, zero–weight, and positive-weight cycles. Positive-weight cycles would not appear in any shortest paths and thus will never be threats. Zero-weight cycles are unwelcome in most applications, because we generally want a shortest path to have not only a minimum weight, but also a minimum number of edges.

Because a shortest path should not contain cycles, it should traverse every vertex at most once. It follows that in a directed graph $G = (V, E)$, the maximum number of edges a shortest path can have is $|V| - 1$, with all the vertices visited once. The *Bellman-Ford algorithm* takes advantage of this observation and relaxes all the edges $(|V| - 1)$ times. Although this strategy is time-consuming, with a runtime of $O((|V| - 1) \times |E|) = O(VE)$, it helps the algorithm handle more general cases, such as graphs with negative weights. It also enables the discovery of negative-weight cycles.

The pseudocode of the Bellman-Ford algorithm is shown in Algorithm 4.13. The negative-weight cycles are detected in lines 5 through 7. They are identified on the basis of the fact that if any edge can still be relaxed after $(|V| - 1)$ times of relaxations (line 6), then a shortest path with more than $(|V| - 1)$ edges exists; therefore, the graph contains negative-weight cycles.

Algorithm 4.13 Bellman-Ford algorithm

Bellman-Ford(Graph G, Vertex s)

1. Initialize(G, s);
2. **for** (*counter* = 1 to |V| - 1)
3.   **for** (each edge $(u, v) \in E$)
4.   Relax(u, v);
5.   **end for**
6. **end for**
7. **for** (each edge $(u, v) \in E$)
8.   **if** (est(v) > est(u) + w(u, v))
9.   report "negative-weight cycles exist";
10.   **end if**
11. **end for**

4.3.6.5 *The longest-path problem*

The longest-path problem can be solved by use of a modified version of the shortest-path algorithm. We can multiply the weights of the edges by -1 and feed the graph into either the shortest-path algorithm for DAGs or the Bellman-Ford algorithm. We cannot use Dijkstra's algorithm, which cannot handle graphs with negative-weight edges. Rather than finding the shortest path, these algorithms discover the longest path. If we do not want to alter any attributes in the graph, we can alter the algorithm by initializing the value of $est(v)$ to $-\infty$ instead of ∞, as shown in the `Initialize` procedure of Algorithm 4.9, and changing a line in the `Relaxation` procedure of Algorithm 4.10 from:

1. if $(est(v) > est(u) + \mathrm{w}(\mathrm{u}, v))\{$

to

1. if $(est(v) < est(u) + \mathrm{w}(\mathrm{u}, v))\{$

Again, this modification cannot be applied to Dijkstra's algorithm, because positive-weight cycles should be avoided in the longest paths, but avoiding them is difficult, because all or most weights are positive in most applications. As a result, the longest-path version of the Bellman-Ford algorithm, which can detect positive-weight cycles, is typically favored for use. If we want to find the longest *simple* paths in those graphs where positive cycles exist, then no efficient algorithm yet exists, because this problem is NP-complete.

4.3.7 Minimum spanning tree

Spanning trees are defined on **connected**, **undirected** graphs. Given a graph $G = (V, E)$, a spanning tree connects all of the vertices in V by use of some edges in E without producing cycles. A spanning tree has exactly $(|V| - 1)$ edges. For example, the thickened edges shown in Figure 4.18 form a spanning tree. The *tree weight* of a spanning tree is defined as the sum of the weights of the tree edges. There would be many spanning trees in a connected, weighted graph with different tree weights. The *minimum spanning tree (MST) problem* searches for a spanning tree whose tree weight is minimized. The MST problem can model the construction of a power network with a minimum wire length in an integrated circuit. It can also model the clock network, which connects the clock source to each terminal with the least number of clock delays. In this subsection, we present an algorithm for the MST problem, *Prim's algorithm* [Prim 1957].

Prim's algorithm builds an MST by maintaining a set of vertices and edges. This set initially includes a starting vertex. The algorithm then adds edges (along with vertices) one by one to the set. Each time the edge closest to the set—with the least edge weight to any of the vertices in the set—is added. After the set contains all the vertices, the edges in the set form a minimum spanning tree.

The pseudocode of Prim's algorithm is given in Algorithm 4.14. The function `PrimMST` uses a priority queue `minQ` to store those vertices not yet included in

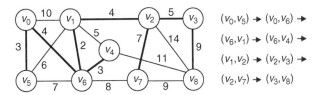

FIGURE 4.18

An example of an MST returned by Prim's algorithm. The MST consists of the thickened edges. The order of choices is shown on the right.

the partial MST. Every vertex in `minQ` is keyed with its minimum edge weight to the partial MST. In line 7, the vertex with the minimum key is extracted from `minQ`, and the keys of its adjacent vertices are updated accordingly, as shown in lines 8 through 11. The parameter `predecessor` refers to MST edges.

Algorithm 4.14 Prim's MST algorithm

PrimMST(Graph G)

1. Priority_Queue minQ = {all vertices in V};

2. for(each vertex $u \in$ minQ) u.key = ∞;

3. randomly select a vertex r in V as root;

4. r.key = 0;

5. r.predecessor = NIL;

6. **while** (minQ $\neq \emptyset$) **do**

7.   Vertex u = ExtractMin(minQ);

8.   **for** (each vertex v such that $(u, v) \in$ E) **do**

9.   **if** ($v \in$ minQ and w$(u, v) < v$.key) **do**

10. v.predecessor = u;

11. v.key = w(u, v);

12.   **end if**

13.   **end for**

14. **end while**

Like Dijkstra's algorithm, the data structure of `minQ` determines the runtime of Prim's algorithm. `PrimMST` has a time complexity of $O(V^2 + E)$ if `minQ` is implemented with a linear array. However, less time complexity can be achieved by use of a more sophisticated data structure.

Figure 4.18 shows an example in which Prim's MST algorithm selects the vertex v_0 as the starting vertex. In fact, an MST can be built from any starting vertex. Moreover, an MST is not necessarily unique. For example, if the edge (v_7, v_8) replaces the edge (v_3, v_8), as shown in Figure 4.18, the new set of edges still forms an MST.

The strategy used by Prim's algorithm is actually very similar to that of Dijkstra's shortest-path algorithm. Dijkstra's algorithm implicitly keeps a set of processed vertices and chooses an unprocessed vertex that has a minimum shortest-path estimate at the moment to be the next target of relaxation. This strategy follows the principle of a *greedy algorithm*. This concept will be explained in Subsection 4.4.1.

4.3.8 Maximum flow and minimum cut

4.3.8.1 *Flow networks and the maximum-flow problem*

A *flow network* is a variant of *connected*, *directed* graphs that can be used to model physical flows in a network of terminals, such as water coursing through interconnecting pipes or electrical currents flow through a circuit. In a flow network $G = (V, E)$, every edge $(u, v) \in E$ has a non-negative *capacity* $c(u, v)$ that indicates the quantity of flow this edge can hold. If $(u, v) \notin E$, $c(u, v) = 0$. There are two special vertices in a flow network, the *source s* and the *sink t*. Every flow must start at the source s and end at the sink t. Hence, there is no edge incident to s and neither an edge leaving t. For convenience, we assume that every vertex lies on some path from the source to the sink. Every edge (u, v) in a flow network has another attribute, *flow f(u, v)*, which is a real number that satisfies the following three properties:

Capacity constraint: For every edge $(u, v) \in E$, $f(u, v) \leq c(u, v)$.
Skew symmetry: For every flow $f(u, v)$, $f(u, v) = -f(v, u)$.
Flow conservation: For all vertices in V, the flows entering it are equal to the flows exiting it, making the *net flow* of every vertex zero. There are two exceptions to this rule: the source s, which generates the flow, and the sink t, which absorbs the flow. Therefore, for all vertices $u \in V - \{s, t\}$, the following equality holds:

$$\sum_{v \in V} f(u, v) = 0$$

Notice that the flow conservation property corresponds to Kirchhoff's Current Law, which describes the principle of conservation in electric circuits. Therefore, the flow networks can naturally model electric currents.

The *value* of a flow f is defined as:

$$|f| = \sum_{v \in V} f(s, v)$$

which is the total flow out of the source. In a ***maximum-flow problem***, the goal is to find a flow with the maximal value in a flow network. Figure 4.19 is an example of a flow network G with a flow f. The values shown on every edge (u, v) are $f(u, v)/c(u, v)$. In this example, $|f| = 19$, but it is not a maximum flow, because we can push more flow into the path $s \rightarrow v_2 \rightarrow v_3 \rightarrow t$.

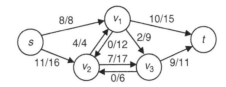

FIGURE 4.19

A flow network G with a flow $f = 19$. The flow and the capacity of each edge are denoted as $f(u, v)/c(u, v)$.

4.3.8.2 *Augmenting paths and residual networks*

The path $s \to v_2 \to v_3 \to t$ in Figure 4.19 can accommodate more flow and, thus, it can enlarge the value of the total flow. Such paths from the source to the sink are called ***augmenting paths***. An intuitive maximum-flow algorithm operates by iteratively finding augmenting paths and then augmenting a corresponding flow until there is no more such path. However, finding these augmenting paths on flow networks is neither easy nor effective. ***Residual networks*** are hence created to simplify the process of finding augmenting paths.

In the flow network $G = (V, E)$ with a flow f, for every edge $(u, v) \in E$ we define its *residual capacity* $c_f(u, v)$ as the amount of additional flow allowed without exceeding $c(u, v)$, given by

$$c_f(u, v) = c(u, v) - f(u, v) \tag{4.2}$$

Given a flow network $G = (V, E)$, its corresponding residual network $G_f = (V, E_f)$ with respect to a flow f consists of the same vertices in V but has a different set of edges, E_f. The edges in the residual network, called the *residual edges*, are weighted edges, whose weights are the residual capacities of the corresponding edges in E. The weights of residual edges should always be positive. For every pair of vertices in E, there exist up to two residual edges connecting them with opposite directions in G_f. Figure 4.20 shows the residual network G_f of the flow network G in Figure 4.19. Notice that, for the vertex pair v_1 and v_3 in G, there are two residual edges in G_f, (v_1, v_3) and (v_3, v_1). We see that $c_f(v_3, v_1) = 2$, because we can push a flow with a value of two in G to cancel out its original flow. On the other hand, there should be three residual edges between v_2 and v_3 in G_f, one from v_2 to v_3 and two from v_3 to v_2. However, the residual edges of the same direction will be merged as one edge only. Therefore, $c_f(v_3, v_2) = 7 + 6 = 13$.

We can easily find augmenting paths in the residual network, because they are just simple paths from the source to the sink. The amount of additional flow that can be pushed into an augmenting path p is determined by the residual capacity of p, $c_f(p)$, which is defined as the minimum residual capacity of all edges on the path. For example, $s \to v_2 \to v_3 \to t$ is an augmenting path p in Figure 4.20. Its residual capacity $c_f(p) = 2$ is determined by the residual edge (v_3, t). Therefore, we can push extra flow with a value of two through p in the original flow network. By repeatedly finding augmenting paths in the

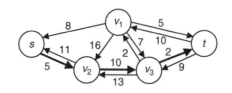

FIGURE 4.20

The residual network G_f of the flow network G in Figure 4.19 in which the augmenting path is shown by the thickened lines.

residual network and updating the residual network, a maximum-flow problem can be solved. The next Subsection shows two algorithms implementing this idea.

4.3.8.3 *The Ford-Fulkerson method and the Edmonds-Karp algorithm*

The *Ford-Fulkerson method* is a classical means of finding maximum flows [Ford 1962]. It simply finds augmenting paths on the residual network until no more paths exist. The pseudocode is presented in Algorithm 4.15.

Algorithm 4.15 Ford-Fulkerson method

Ford-Fulkerson(Graph G, Source s, Sink t)

1. **for** (each $(u, v) \in$ E) $f[u, v] = f[v, u] = 0$;
2. Build a residual network G_f based on flow f;
3. **while** (there is an augmenting path p in G_f) **do**
4. $c_f(p) = \min(c_f(u, v) : (u, v) \in p)$;
5.   **for** (each edge $(u, v) \in p$) **do**
6. $f[u, v] = f[u, v] + c_f(p)$;
7. $f[v, u] = -f[u, v]$;
8.   **end for**
9.   Rebuild G_f based on new flow f;
10. **end while**

We can apply the Ford-Fulkerson method to the flow network G in Figure 4.19. Figure 4.21a shows the result of adding the augmenting path to G in Figure 4.20. The function `Ford-Fulkerson` gives us the result in Figure 4.21c. The maximum flow, denoted as f^*, has a value of 23.

We call this the Ford-Fulkerson *method* rather than *algorithm*, because the approach to finding augmenting paths in a residual graph is not fully specified. This ambiguity costs precious runtime. The Ford-Fulkerson method has a time complexity of $O(E \cdot |f^*|)$. It takes $O(E)$ time to construct a residual network and each augmenting path increases the flow by at least 1. Therefore, we build the residual

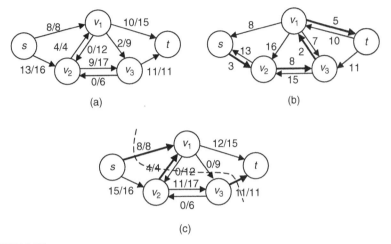

FIGURE 4.21

(a) Adding the augmenting path found in Figure 4.20 to G of Figure 4.19. (b) The resultant residual network of (a) with an augmenting path p. (c) Adding p to (a) results in a maximum flow of value 23. The dashed line is the minimum cut with a value of 23.

networks at most $|f^*|$ times. $|f^*|$ is not an input parameter for the maximum-flow problem, so the Ford-Fulkerson method does not have a polynomial-time complexity. It will be a serious problem if $|f^*|$ is as great as, say, 1,000,000,000.

The ambiguity present in the Ford-Fulkerson method is fixed by the *Edmonds-Karp algorithm* [Edmonds 1972]. Instead of blindly searching for any augmenting paths, the Edmonds-Karp algorithm uses *breadth-first search* to find the augmenting path with a minimum number of edges in the residual network. For an edge in the residual work, there can be many augmenting paths passing through it in different iterations. It can be proven that for every edge in the residual network, the lengths of the augmenting paths passing through it will only increase with the advancement of iterations [Ahuja 1993; Cormen 2001]. Because the upper limit of the length of an augmenting path is $|V| - 1$, there exist $O(V)$ different augmenting paths passing through a specific edge. Therefore, there exist $O(VE)$ different augmenting paths and thus $O(VE)$ constructions of residual networks, resulting in a time complexity of $O(E \cdot VE) = O(VE^2)$.

4.3.8.4 *Cuts and the max-flow min-cut theorem*

Until now we have not proven the correctness of finding the maximum flow by use of residual networks. In this subsection, we introduce an important concept in the flow network—*cuts*. The *max-flow min-cut theorem* is used to prove the correctness of the Ford-Fulkerson method and the Edmonds-Karp algorithm.

A *cut* (S, T) of the flow network $G = (V, E)$ is a partition of V that divides V into two subsets, S and $T = V - S$, such that the source $s \in S$ and the sink $t \in T$. The *net flow* across the cut (S, T) is denoted as $f(S, T)$:

$$f(S,T) = \sum_{u \in S, v \in T} f(u,v) \tag{4.3}$$

The *capacity* of the cut (S, T), $c(S, T)$, is defined as

$$c(S,T) = \sum_{u \in S, v \in T} c(u,v) \tag{4.4}$$

Notice that only those edges incident from S to T are counted according to (4.4). Take Figure 4.21a as an example. For the cut $(\{s, v_2, v_3\}, \{v_1, t\})$, its net flow is:

$$f(s,v_1) + f(v_2,v_1) + f(v_3,v_1) + f(v_3,t) = 8 + 4 + (-2) + 11 = 21$$

and its capacity is:

$$c(s,v_1) + c(v_2,v_1) + c(v_3,t) = 8 + 4 + 11 = 23$$

We can observe that for any cut (S, T), the property $f(S, T) \leq c(S, T)$ always holds. The number of possible cuts in a flow network grows exponentially with the number of vertices. We are particularly interested in finding a **minimum cut**, which is the cut with a minimum capacity among all possible cuts in a network.

With the knowledge of cuts in a flow network, we can explain the max-flow min-cut theorem. For a flow f in a flow network $G = (V, E)$, the **max-flow min-cut theorem** states that the following three conditions are equivalent:

(1) f is a maximum flow in G.
(2) The residual network G_f has no augmenting paths.
(3) $|f| = c(S, T)$ for some cut of G.

We first prove (1)⇒(2). If f is a maximum flow in G and there is still an augmenting path p in G_f, then the sum of flow $|f| + c_f(p) > |f|$, which is a contradiction. Secondly, we prove (2)⇒(3). Suppose G_f has no augmenting path or, equivalently, there is no path in G_f from s to t. We define $S = \{v \in V$ such that v is reachable from s in $G_f\}$ and $T = V - S$. The partition (S, T) is a cut. For any edge (u, v) across the cut, we have $f(u, v) = c(u, v)$ because $(u, v) \notin G_f$, so $f(S, T) = c(S, T)$. It can be reasoned that $|f| = f(S, T)$ as follows:

$$|f| = f(s,V) = f(s,V) + f(S - s, V) = f(S, V) = f(S, V) - f(S, S) = f(S, T)$$

with $f(S - s, V) = 0$, because the source s is excluded. As a result, we can see that $|f| = f(S, T) = c(S, T)$. Finally, we prove (3)⇒(1) by use of the property $|f| \leq c(S, T)$ of any cut (S, T). Because $f(u, v) \leq c(u, v)$ for any edge across the cut (S, T), $|f| = f(S, T) \leq c(S, T)$. And if a flow f^* has $|f^*| = c(S^*, T^*) \geq |f|$ for a specific cut (S^*, T^*), then the flow f^* must be a maximum flow and the cut (S^*, T^*) must be a minimum cut.

The *max-flow min-cut theorem* not only proves that finding augmenting paths in a residual network is a correct way to solve the maximum-flow problem, it also proves that finding a maximum flow is equivalent to finding a minimum cut. In Figure 4.21c, we see that the maximum flow found indeed has the same value as the cut ($\{s, v_2, v_3\}, \{v_1, t\}$).

Finding a minimum cut has many EDA applications, such as dividing a module into two parts with a minimum interconnecting wire length. We can thus solve this kind of problem with a maximum-flow algorithm.

4.3.8.5 *Multiple sources and sinks and maximum bipartite matching*

In some applications of the maximum-flow problem, there can be more than one source and more than one sink in the flow network. For example, if we want to count the number of paths from a set of inputs to a set of outputs in an electrical circuit, there would be multiple sources and multiple sinks. However, we can still model those flow networks as a single-source, single-sink network by use of a *supersource* and a *supersink*. Given a flow network with sources s_i, $1 \leq i \leq m$ and sinks t_j, $1 \leq j \leq n$, a supersource s connects the sources with edges (s, s_i) and capacities $c(s, s_i) = \infty$. Similarly, a supersink t is created with edges (t_j, t) and capacities $c(t_j, t) = \infty$. With this simple transformation, a flow network with multiple sources and sinks can be solved with common maximum-flow algorithms.

Maximum bipartite matching is an important application of the multiple-source, multiple-sink maximum flow problem. A *bipartite graph* $G = (V, E)$ is an undirected graph whose vertices are partitioned into two sets, L and R. For each edge $(u, v) \in E$; if $u \in L$, then $v \in R$, and vice versa. Figure 4.22a gives an example of a bipartite graph. A *matching* on an undirected graph $G = (V, E)$ is a subset of edges $M \subseteq E$ such that for all $v \in V$, at most one edge of M is incident on V. *Maximum matching* is a matching that contains a maximum number of edges. The *maximum bipartite matching* problem is the problem of finding a maximum matching on a bipartite graph. Figure 4.22a shows such a maximum matching with three edges on a bipartite graph.

The maximum bipartite graph problem itself has many useful applications in the field of EDA. For example, technology mapping can be modeled as a

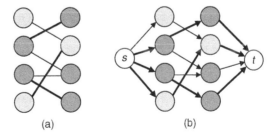

(a)　　　　(b)

FIGURE 4.22

(a) A bipartite graph with its maximum matching indicated by thickened lines. (b) The corresponding flow network provides the solution to the maximum bipartite matching problem. Every edge has unit capacity.

bipartite graph. The functional modules to be mapped are modeled as vertices on one side, and all cell libraries of the target technology are vertices on the other side. We can solve the maximum bipartite graph problem by solving the corresponding multiple-source, multiple-sink maximum graph problem as shown in Figure 4.22b. The Ford-Fulkerson method can solve this problem with a time complexity of $O(VE)$ because $|f^*| \leq |V|/2$.

4.4 HEURISTIC ALGORITHMS

Heuristic algorithms are algorithms that apply heuristics, or rules of thumb, to find a good, but not necessarily optimal, solution for the target problem. The heuristics in such algorithms function as guidelines for selecting good solutions from possible ones. Notice that good solutions, rather than optimal solutions, are found in heuristic algorithms, which is the biggest difference between heuristics and other types of algorithms. To compensate for this disadvantage, heuristic algorithms generally have much lower time complexity. For problems that are either large in size or computationally difficult (NP-complete or NP-hard, or both) other types of algorithms may find the best solutions but would require hours, days, or even years to identify such a solution. Heuristic algorithms are the preferred method for these types of problems because they sacrifice some solution quality while saving a huge amount of computational time.

NP-complete and NP-hard problems are currently prevalent in the EDA applications. For example, the Traveling Salesman Problem (TSP, see Section 4.2) has many EDA applications such as routing, but TSP optimization is an NP-hard problem. In a TSP problem with n cities (nodes), a brute-force search for the shortest route results in an overwhelmingly high time complexity of $O(n!)$. For these sorts of problems, heuristic algorithms are often a better and necessary choice.

Heuristic algorithms empirically yield good, and sometimes optimal, solutions. The solution quality, however, cannot be guaranteed. For example, there is a greedy algorithm (see Subsection 4.4.1 for more details) called the Nearest Neighbor (NN) algorithm that can be used to solve the TSP problem. NN lets the salesman start from any one city and then travel to the nearest unvisited city at each step. NN quickly generates a short route with a $O(n^2)$ time complexity, given n as the number of cities. Nevertheless, there are some examples showing that this intuitive algorithm yields inefficient routes. In Figure 4.23, applying NN and starting from city C results in the route C→B→D→A→E→C whose total length is $1 + 3 + 7 + 15 + 10 = 36$; however, traversing the cities in the loop C→D→E→A→B→C is a shorter route: $2 + 8 + 15 + 4 + 1 = 31$. This example shows that we have to be cautious when we use heuristic algorithms, because they can sometimes yield poor solutions.

In this section, we discuss several frequently used heuristic algorithms. *Greedy algorithms*, *dynamic programming*, and *branch-and-bound* algorithms are heuristic algorithms that direct the search toward a solution space

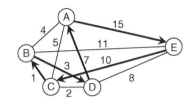

FIGURE 4.23

An inefficient route yielded by the Nearest Neighbor algorithm.

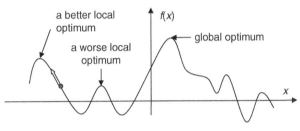

FIGURE 4.24

Local versus global optima for a one-dimensional function. From a current solution (gray dot), greedy algorithms try to make a greedy choice that bring it toward a local optimum, which may be different from a global optimal one.

that promises a better solution quality. *Simulated annealing* and *genetic algorithms* exert a series of perturbations on current solutions, trying to ameliorate them through the process. These heuristic algorithms have extensive EDA applications [Reeves 1993].

4.4.1 Greedy algorithm

Algorithms targeting an optimization problem typically consist of a series of stages with choices made at each of these stages. A *greedy algorithm*, which aims to solve an optimization problem, makes choices at every stage toward a local optimum and with the hope of eventually reaching a globally optimal solution. Greedy algorithms get their name from the fact that these algorithms always make a choice that looks like the best possible solution at the moment without thoroughly considering the underlying conditions and consequences that may result from that choice, acting much like a greedy person. Figure 4.24 illustrates the difference between local and global optima for a one-dimensional function.

In fact, we often exploit the concept of greedy algorithms in our daily lives without knowing it. For instance, making change in sequence by use of the minimum number of coins is a typical situation illustrating this concept. Suppose we want to give change of 36 cents in U.S. currency. The coins that can be used consist of the 25-cent quarter, the 10-cent dime, the 5-cent nickel, and the

1-cent penny. Then, we apply a rule of thumb: pick the coin of the greatest value that is less than the change amount first. The change will consequently be made in this sequence: a quarter (25 cents), a dime (10 cents), and a penny (1 cent)—a total of three coins. This rule of thumb leads to the minimum number of coins, three, because it perfectly embodies the essence of greedy algorithms: making greedy choices at each moment. In this particular problem, a greedy algorithm yields the optimal solution.

However, greedy algorithms do not always produce optimal solutions. Let us revisit the making change example. If a coin with a value of 20 cents exists, the rule of thumb just mentioned would not lead to the minimum number of coins if the amount of change needed was 40 cents. By applying the rule of picking the coin of highest value first, we would be giving change of a quarter (25 cents), a dime (10 cents) and a nickel (5 cents), a total of three coins, but, in fact, two, 20-cent coins would be the optimal solution for this example. The greedy algorithm fails to reach the optimal solution for this case.

Actually, the example given previously is not ideal for illustrating the concept of greedy algorithms, because it violates the *optimal substructure* property. In general, problems suitable for greedy algorithms must exhibit two characteristics: the *greedy-choice* property and the *optimal substructure* property. If we can demonstrate that a problem has these two properties, then a greedy algorithm would be a good choice.

4.4.1.1 *Greedy-choice property*

The ***greedy-choice property*** states that a globally optimal solution can always be achieved by making locally optimal, or greedy, choices. By locally optimal choices we mean making choices that look best for solving the current problem without considering the results from other subproblems or the effect(s) that this choice might have on future choices.

In Section 4.4, we introduced the Nearest Neighbor (NN) algorithm for solving—more precisely, for approximating—an optimal solution to TSP. NN is a greedy algorithm that picks the nearest city at each step. NN violates the greedy-choice property and thus results in suboptimal solutions, as indicated in the example of Figure 4.23. In Figure 4.23, the choice of B→D is a greedy one, because the other remaining cities are further from B. In a globally optimal solution, the route of either D→C→B or B→C→D is a necessity, and the choice of B→D is suboptimal. Hence, NN is not an optimal greedy algorithm, because TSP does not satisfy the greedy-choice property.

Making change with a minimum number of coins is an interesting example. On the basis of the current U.S. coins, this problem satisfies the greedy-choice property. But when a 20-cent coin comes into existence, the property is violated—when making change for 40 cents, the greedy choice of picking a quarter affects the solution quality of the rest of the problem.

How do we tell if a particular problem has the greedy-choice property? In a greedy algorithm designed for a particular problem, if any greedy choice can be proven better than all of the other available choices at the moment in terms of solution quality, we can say that the problem exhibits the greedy-choice property.

4.4.1.2 *Optimal substructure*

A problem shows ***optimal substructure*** if a globally optimal solution to it consists of optimal solutions to its subproblems. If a globally optimal solution can be partitioned into a set of subsolutions, optimal substructure requires that those subsolutions must be optimal with respect to their corresponding subproblems. Consider the previous example of making change of 36 cents with a minimum number of coins. The optimal solution of a quarter, a dime, and a penny can be divided into two parts: (1) a quarter and a penny and (2) a dime. The first part is, indeed, optimal in making change of 26 cents, as is the second part for making change of 10 cents.

The NN algorithm for TSP lacks both greedy-choice and optimal substructure properties. Its global solutions cannot be divided into solutions for its subproblems, let alone optimal solutions.

To determine whether a particular problem has an optimal substructure, two aspects have to be examined: *substructure* and *optimality*. A problem has substructures if it is divisible into subproblems. *Optimality* is the property that the combination of optimal solutions to subproblems is a globally optimal solution.

Greedy algorithms are highly efficient for problems satisfying these two properties. On top of that, greedy algorithms are often intuitively simple and easy to implement. Therefore, greedy algorithms are very popular for solving optimization problems. Many graph algorithms, mentioned in Section 4.3, are actually applications of greedy algorithms—such as Prim's algorithm used for finding minimum spanning trees. Greedy algorithms often help find a lower bound of the solution quality for many challenging real-world problems.

4.4.2 **Dynamic programming**

Dynamic programming (DP) is an algorithmic method of solving optimization problems. *Programming* in this context refers to mathematical programming, which is a synonym for optimization.

DP solves a problem by combining the solutions to its subproblems. The famous divide-and-conquer method also solves a problem in a similar manner. The divide-and-conquer method divides a problem into independent subproblems, whereas in DP, either the subproblems depend on the solution sets of other subproblems or the subproblems appear repeatedly. DP uses the dependency of the subproblems and attempts to solve a subproblem only once; it then stores its solution in a table for future lookups. This strategy spares the time spent on recalculating solutions to old subproblems, resulting in an efficient algorithm.

To illustrate the superiority of DP, we show how to efficiently multiply a chain of matrices by use of DP. When multiplying a chain of matrices, the order of the multiplications dramatically affects the number of scalar multiplications. For example, consider multiplying three matrices A, B, and C whose dimensions are 30×100, 100×2, and 2×50, respectively. There are two ways to start the multiplication: either $A \cdot B$ or $B \cdot C$ first. The numbers of necessary scalar multiplications are:

$$(A \cdot B) \cdot C : 30 \times 100 \times 2 + 30 \times 2 \times 50 = 6000 + 3000 = 9000,$$
$$A \cdot (B \cdot C) : 100 \times 2 \times 50 + 30 \times 100 \times 50 = 10,000 + 150,000 = 160,000$$

$(A \cdot B) \cdot C$ is clearly more computationally efficient.

The **matrix-chain multiplication problem** can be formulated as follows: given a chain of n matrices, $<M_1, M_2, \ldots, M_n>$, where M_i is a $v_{i-1} \times v_i$ matrix for $i = 1$ to n, we want to find an order of multiplication that minimizes the number of scalar multiplications.

To solve this problem, one option is to exhaustively try all possible multiplication orders and then select the best one. However, the number of possible multiplication orders grows exponentially with respect to the number of matrices n. There are only two possibilities for three matrices, but it increases to 1,767,263,190 possibilities for 20 matrices. A brute-force search might cost more time finding the best order of multiplications than actually performing the multiplication.

Here, we define $m[i, j]$ as the minimum number of scalar multiplications needed to calculate the matrix chain $M_i M_{i+1} \ldots M_j$, for $1 \leq i \leq j \leq n$. The target problem then becomes finding $m[1, n]$. Because a matrix chain can be divided into two smaller matrix chains, each of which can be multiplied into a single matrix first, the following recurrent relationship holds:

$$m[i,j] = \begin{cases} 0 & \text{if } i = j \\ \min_{i \leq k < j} \left\{ m[i,k] + m[k+1,j] + v_{i-1} v_k v_j \right\} & \text{if } i < j \end{cases} \qquad (4.5)$$

A simple recursive algorithm on the basis of recurrence (4.5) can provide the answer to $m[1, n]$; however, such an algorithm will be extremely inefficient because, in the process of computing $m[1, n]$, many entries of $m[i, j]$ are computed multiple times. For example, if we wish to compute $m[1, 6]$, the value of $m[3, 4]$ will be repeatedly computed in the process of calculating $m[1, 4]$, $m[2, 5]$, and $m[3, 6]$. However, we could store the values in a table, which leads to the dynamic programming algorithm `BottomUpMatrixChain` shown in Algorithm 4.16.

Algorithm 4.16 A dynamic programming algorithm for solving the matrix-chain multiplication problem

---

BottomUpMatrixChain(Vector *v*)

   1. *n* = *v*.size − 1;

   2. **for** (*i* = 1 to *n*) *m*[i, i] = 0;

   3. **for** (*p* = 2 to *n*) **do** // p is the chain length

4.   **for** $(i = 1$ to $n - p + 1)$ **do**

5. $j = i + p - 1$;

6. $m[i, j] = \infty$;

7.   **for** $(k = i$ to $j - 1)$ **do**

8. $temp = m[i, k] + m[k + 1, j] + v_{i-1}v_kv_j$;

9.   **if** $(temp < m[i, j])$ **do**

10. $m[i, j] = temp$;

11. $d[I, j] = k$;

12.   **end if**

13.   **end for**

14.   **end for**

15. **return** m and d;

The `BottonUpMatrixChain` perfectly embodies the property of recurrence (4.5). A triangular table $m[i, j]$, where $1 \leq i \leq j \leq n$, records the minimum numbers of scalar multiplications for its respective matrix chains, whereas another triangular table $d[i, j]$, where $1 \leq i < j \leq n$, tracks where the separations of matrix chains should be. We can see in line 3 that the m table is filled in the ascending order of the length of the matrix chains, so that in line 8, the items to be added are already in place. Finally, the fully filled m and d tables are returned as answers in line 15.

`BottomUpMatrixChain` handles recurrence (4.5) by making use of the repetitive nature of the subproblems. The three loops in lines 3, 4, and 7 indicate that this algorithm has a time complexity of $O(n^3)$. Compared with the exponential time needed to search through all possible multiplication orders, `BottomUpMatrixChain` is highly efficient.

`BottomUpMatrixChain` is a typical example of dynamic programming. It solves the matrix-chain multiplication problem by systematically combining solutions to multiplication of smaller matrix chains. In fact, the matrix-chain multiplication problem contains two key ingredients that make `BottomUpMatrixChain` a successful function: *overlapping subproblems* and *optimal substructure*. These two properties are indispensable for any DP algorithm to work.

4.4.2.1 *Overlapping subproblems*

We say that a problem has **overlapping subproblems** when it can be decomposed into subproblems that are not independent of one another. Often several subproblems share the same smaller subproblems. For example, running a recursive algorithm often requires solving the same subproblem multiple times. DP solves each subproblem only once and stores the answer in a table, so that recurrences of the same subproblems take only constant time to get the answer (by means of a table lookup).

The matrix-chain multiplication problem is an instance of this property. Repeated multiplications of smaller matrix chains cause a high complexity for a simple recursive algorithm. In contrast, the DP algorithm `BottomUpMatrixChain` creates the m table for the overlapping subproblems to achieve high efficiency.

4.4.2.2 *Optimal substructure*

A problem exhibits an ***optimal substructure*** if its globally optimal solution consists of optimal solutions to the subproblems within it. Recall that in Subsection 4.4.1, having an optimal substructure ensures that greedy algorithms yield optimal solutions. It fact, if a problem has an optimal substructure, both greedy algorithms and DP could yield optimal solutions. One key consideration in choosing the type of algorithm is determining whether the problem has the greedy-choice property, the overlapping subproblems, or neither. If the problem shows overlapping subproblems but not the greedy-choice property, DP is a better way to solve it. On the other hand, if the problem exhibits the greedy-choice property instead of overlapping subproblems, then a greedy algorithm fits better. A problem rarely has both of the properties because they contradict each other. The matrix-chain multiplication problem has an optimal substructure, reflected in recurrence (4.4), but it does not have the greedy-choice property. It consists of overlapping subproblems. Therefore, DP is a suitable approach to address this problem.

4.4.2.3 *Memoization*

`BottomUpMatrixChain`, as its name suggests, solves the problem iteratively by constructing a table in a bottom-up fashion. A top-down approach, on the other hand, seems infeasible, from this simple recursive algorithm. In fact, the unnecessary recomputations that prevent the recursive algorithm from being efficient can be avoided by recording all the computed solutions along the way. This idea of constructing a table in a top-down recursive fashion is called ***memoization***. The pseudocode of a *memoized* DP algorithm to solve the matrix-chain multiplication problem is shown in Algorithm 4.17.

Algorithm 4.17   Solving matrix-chain multiplication problems with memoization

---

TopDownMatrixChain(Vector v)

   1. $n = v.\text{size} - 1$;

   2. **for** ($i = 1$ to n)

   3. **for** ($j = i$ to n) $m[i, j] = \infty$;

   4.   **return** Memoize(v, 1, n);

Memoize(Vector v, Index i, Index j)

   1. **if** ($m[i, j] < \infty$) **return** $m[i, j]$;

2. **if** $(i = j)$ $m[i, j] = 0$;

3. **else**

4.   **for** $(k = i$ to $j - 1)$ **do**

5. $temp$ = Memoize(v, i, k) + Memoize$(v, k + 1, j)$ + $v_{i-1}v_kv_j$;

6.   if$(temp < m[i, j])$ $m[i, j] = temp$;

7.   **end for**

8. **end if**

9. **return** $m[i, j]$;

The time complexity of the `TopDownMatrixChain` shown in Algorithm 4.17 is still $O(n^3)$, because it maintains the m table. The actual runtime of the `TopDownMatrixChain` will be slightly longer than the `BottomUpMatrixChain` because of the overhead introduced by recursion. In general, memorization can outperform a bottom-up approach only if some subproblems need not be visited. If every subproblem has to be solved at least once, the bottom-up approach should be slightly better.

4.4.3 Branch-and-bound

Branch-and-bound is a general technique for improving the searching process by systematically enumerating all candidate solutions and disposing of obviously impossible solutions.

Branch-and-bound usually applies to those problems that have finite solutions, in which the solutions can be represented as a sequence of options. The first part of branch-and-bound, **branching**, requires several choices to be made so that the choices *branch out* into the solution space. In these methods, the solution space is organized as a treelike structure. Figure 4.25 shows an instance of TSP and a solution tree, which is constructed by making choices on the next cities to visit.

Branching out to all possible choices guarantees that no potential solutions will be left uncovered. But because the target problem is usually NP-complete or even NP-hard, the solution space is often too vast to traverse. The branch-and-bound algorithm handles this problem by **bounding** and **pruning**. Bounding refers to setting a bound on the solution quality (*e.g.*, the route length for TSP), and pruning means trimming off branches in the solution tree whose solution quality is estimated to be poor. Bounding and pruning are the essential concepts of the branch-and-bound technique, because they are used to effectively reduce the search space. We demonstrate in Figure 4.25 how branch-and-bound works for the TSP problem.

The number under a leaf node of the solution tree represents the length of the corresponding route. For incomplete branches, an expression in the form of $a + b$ is shown. In this notation, a is the length of the traversed edges, and

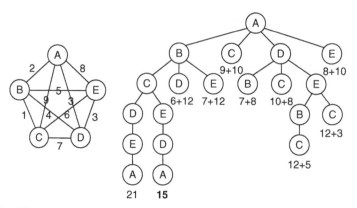

FIGURE 4.25

A TSP and its solution tree after applying branch-and-bound.

b is a lower bound for the length of the remaining route that has not been explored. The lower bound is derived by use of a *minimum spanning tree* that consists of the unvisited vertices, as well as the root and leaf vertices of the partial route. For example, for the unfinished route A→B→E, a minimum spanning tree is built for nodes A, C, D, and E, and its value is 12. This lower bound is a true underestimate for the length of the remaining route. The sum of these two numbers provides the basis for bounding.

The solution tree is traversed depth-first, with the length of the current shortest route as the upper bound for future solutions. For example, after A→B→C→D→E→A is examined, the upper bound is 21, and after the next route is explored, the bound drops to 15. Every time a partial route is extended by a vertex, a lower bound for the length of the rest of the route is computed. If the sum $a + b$ is over or equal to the current upper bound, the solutions on that branch guarantees to be worse than the current best solution, and the branch can be pruned. Most branches are pruned in Figure 4.25.

An exhaustive search will build a search tree with 89 nodes,[1] but the solution tree with branch-and-bound has only 20 nodes. Branch-and-bound accelerates the search process by reducing the solution space *en masse*. Although branch-and-bound algorithms generally do not possess proven time complexity, their efficiency has made them the first choice for many problems, especially for NP-complete problems.

Branch-and-bound mainly addresses optimization problems, because bounding is often based on numerical comparisons. TSP that uses the route length as the bound is a classical application; however, it can also be applied to some decision problems. In these cases, the bounding criteria are often restrictions or

---

[1]Let n be the number of cities and $f(n)$ be the number of nodes in the exhausted search tree. Then $f(2) = 3$, $f(3) = 7$, and $f(n) = (n-1)f(n-1) + 1$.

additional descriptions of possible solutions. The **Davis-Putnam-Logemann-Loveland (DPLL)** search scheme for the Boolean Satisfiability problem is a typical and important application for this kind of branch-and-bound algorithm.

4.4.4 Simulated annealing

Simulated annealing (SA) is a general probabilistic algorithm for optimization problems [Wong 1988]. It uses a process searching for a global optimal solution in the solution space analogous to the physical process of *annealing*. In the process of annealing, which refines a piece of material by heating and controlled cooling, the molecules of the material at first absorb a huge amount of energy from heating, which allows them to wander freely. Then, the slow cooling process gradually deprives them of their energy, but grants them the opportunity to reach a crystalline configuration that is more stable than the material's original form. The idea to use simulated annealing on optimization problems was first proposed by S. Kirkpatrick, C. D. Gelatt, and M. P. Vecchi in [Kirkpatrick 1983] for the placement and global routing problems.

Simulated annealing (SA) is analogous to annealing in three ways:

1. The energy in annealing corresponds to the **cost function** in SA. The cost function evaluates every solution, and the cost of the best-known solution generally decreases during the SA process. The goal of an optimization problem is to find a solution with a minimum cost.

2. The movements of molecules correspond to small **perturbations** in the current solution, such as switching the order of two consecutive vertices in a solution to TSP. SA repeatedly perturbs the current solution so that different regions in the solution space are explored.

3. The temperature corresponds to a control parameter **temperature** T in SA. T controls the probability of accepting a new solution that is worse than the current solution. If T is high, the acceptance probability is also high, and *vice versa*. T starts at the peak temperature, making the current solution changes almost randomly at first. T then gradually decreases, so that more and more suboptimal perturbations are rejected. The algorithm normally terminates when T reaches a user-specified value.

An SA algorithm typically contains two loops, an outer one and an inner one. In the outer loop, T dwindles every time, and the outer loop terminates when T reaches some user-specified value. In the inner loop, the solution is perturbed, and the cost function of the perturbed solution is evaluated. If the new solution has a lower cost, it directly replaces the current solution. Otherwise, to accept or reject the new, higher-cost solution is based on a probability function that is positively related to T and negatively related to the cost difference between the current and new solutions. The inner loop continues until a *thermal equilibrium* is reached, which means that T also controls the number of iterations of the inner loop. After both loops terminate, the best solution visited in the process is returned as the result.

The pseudocode in Algorithm 4.18 outlines the SA algorithm. There are a few details worth discussion: in line 2 of the function `Accept`, the number $e^{\frac{-\Delta c}{T}}$ ensures that a higher cost solution has a greater likelihood of acceptance if T is high or the cost difference (Δc) is small. Although there is no strong theoretical justification for the need of strictly following this exact formula, this formula has been popular among SA users.

Algorithm 4.18   Simulated annealing algorithm

Accept(temperature T, cost Δc)

1. Choose a random number *rand* between 0 and 1;

2. **return** ($e^{-\Delta c/T} > rand$);

SimulatedAnnealing()

1. solution *sNow*, *sNext*, *sBest*;

2. temperature T, *endingT*;

3. Initialize *sNow*, T and *endingT*;

4. **while** ($T > endingT$) **do**

5.   **while** (!ThermalEquilibrium(T))**do**

6.   *sNext* = Perturb(*sNow*);

7.   **if** (cost(*sNext*) < cost(*sNow*))

8.   *sNow* = *sNext*;

9.   **if** (cost(*sNow*) < cost(*sBest*))

10.   *sBest* = *sNow*;

11.   **else if** (Accept(T, cost(*sNext*)-cost(*sNow*)))

12.   *sNow* = *sNext*;

13.   **end if**

14.   **end while**

15.   Decrease(T);

16. **end while**

17. **return** *sBest*;

The combination of the functions `ThermalEquilibrium`, `Decrease`, and the parameter `endingT` in Algorithm 4.18 characterize an SA algorithm. In combination, they determine the *cooling schedule* or the *annealing schedule*. The cooling schedule can be tuned in many ways, such as making T drop faster at first and slower afterwards in the function `Decrease` or allowing more perturbations when T is small in the function `ThermalEquilibrium`. Every adjustment in the cooling schedule affects the solution quality and the time taken to find a solution. In practice, empirical principles and a trial-and-error strategy are commonly used to find a good cooling schedule [Hajek 1988].

SA has many advantages over other optimization algorithms. First, because there is a non-zero probability of accepting higher cost solutions in the search process, SA avoids becoming stuck at some local minima, unlike some greedy approaches. Also, the runtime of SA is controllable through the cooling schedule. One can even abruptly terminate this algorithm by changing the parameter *endingT* in line 4 of `SimulatedAnnealing`. Finally, there is always a best-known solution available no matter how little time has elapsed in the search process. With SA, the user can always get a solution. In general, a longer runtime would result in a better-quality solution. This flexibility explains SA's wide popularity. SA is considered the top choice for several EDA problems, such as placement and Binary Decision Diagram (BDD) variable reordering.

4.4.5 Genetic algorithms

Just like simulated annealing, *genetic algorithms* are another general randomized algorithm catering to optimization problems [Goldberg, 1989; Davis 1991]. They also perform a series of computations to search for a global optimal solution in the solution space. As the name suggests, genetic algorithms use techniques inspired by operations found in evolutionary biology such as selection, crossover, and mutation.

Genetic algorithms are different from other global search heuristics in many ways. First of all, other global search algorithms, such as simulated annealing, perform a series of perturbations on a single solution to approach a global optimum. Genetic algorithms simultaneously operate on a set of feasible solutions or a *population*. Moreover, the solutions in a genetic algorithm are always encoded into strings of mathematical symbols, which facilitate future manipulations on them. Many types of coding symbols can be used, such as bits, integers, or even permutations. In the simplest versions of genetic algorithms, fixed-length bit strings are used to represent solutions. A bit string that specifies a feasible solution is called a *chromosome*. Each bit in a chromosome is called a *gene*.

Genetic algorithms have many variations [Holland 1992]. Here we will focus on the *simple genetic algorithm* (SGA) to get a taste of the mechanics of genetic algorithms. SGA can be separated into six phases: initialization, evaluation, selection, crossover, mutation, and replacement. After the initial population is generated in the **initialization** phase, the other five actions take place in turns until termination. Figure 4.26 shows the flow of SGA.

In the **evaluation** phase, chromosomes in the population are evaluated with a *fitness function*, which indicates how good the corresponding solutions are. Their *fitness values* are the criteria of selection in the next phase. Advanced

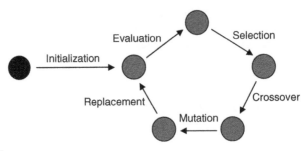

FIGURE 4.26

The flow of a simple genetic algorithm.

genetic algorithms can even handle multi-purposed optimization problems with plural fitness functions.

The **selection** phase aims at finding the best parents or a group of solutions to generate the next population. Many schemes can be implemented to exercise selection in SGA. The simplest scheme is *truncation selection*, in which the s chromosomes with the highest fitness values are chosen, and l/s copies are duplicated for each of them, in which l is the population size. Notice that the population size will not change after selection. Another simple selection scheme, Roulette-Wheel selection, chooses a chromosome with the probability of the ratio of its fitness value to the sum of all fitness values of the population.

In the **crossover** phase, children chromosomes are produced by inheriting genes from pairs of parent chromosomes. As always, there are many methods to implement the crossover, each with its pros and cons. *Uniform crossover* states that every gene of a child chromosome comes from a dad with a probability of p (usually 0.5) and from a mom with a probability of $(1 - p)$. Conventionally, two parents give birth to two children so that the population size remains unchanged.

Mutation means changing a tiny fraction of the genes in the chromosomes. Although in biology mutations rarely happen, they do prevent genetic algorithms from getting stuck in local minima. After the processes of evaluation, selection, crossover, and mutation are complete, the new population **replaces** the old one and the next iteration begins.

Figure 4.27 shows a tiny example of an SGA, with a population size of four and chromosome length of six. The fitness function simply counts "1" genes. *Truncation selection* and *uniform crossover* with a probability of 0.5 are used in this example. Notice that the average and highest fitness values increase after one generation.

In this example, the best solution seems very easy to achieve, so an SGA seems unnecessary; however, in real-life applications of SGA, a population size can be as large as 100,000 and a chromosome can contains up to 10,000 genes. The fitness function will be much more complex as well.

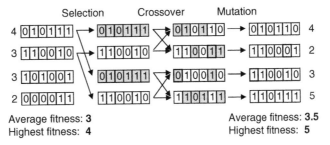

FIGURE 4.27

One-generation simulation of a simple genetic algorithm.

SGA is just a small part of the broad subject of genetic algorithms. Genetic algorithms remain an active research topic for various applications. In addition to EDA, they have applications in a variety of fields, including designing shapes for high-speed trains and human face recognition.

4.5 MATHEMATICAL PROGRAMMING

Mathematical programming, or mathematical optimization, is a systematic approach used for optimizing (minimizing or maximizing) the value of an objective function with respect to a set of constraints. The problem in general can be expressed as:

$$\text{Minimize (or maximize) } f(x);$$
$$\text{Subject to } X = \{X | g_i(x) \leq b_i, i = 1 \ldots m\}$$

where

$x = (x_1, \ldots, x_n)$ are optimization (or decision) variables,

$f : R^n \to R$ is the objective function, and

$g_i : R^n \to R$ and $b_i \in R$ form the constraints for the valid values of x

4.5.1 Categories of mathematical programming problems

According to the natures of f and X, mathematical programming problems can be classified into several different categories:

1. If $X = R^n$, the problem is unconstrained;
2. If f and all the constraints are linear, the problem is called a *linear programming* (LP) problem. The linear constraints can then be represented in the matrix form:

$$Ax \leq b$$

where A is an $m \times n$ matrix corresponding to the coefficients in $g_i(x)$.

3. If the problem is linear, and all the variables are constrained to integers, the problem is called an *integer linear programming* (ILP) problem. If only some of the variables are integers, it is called a *mixed integer linear programming* (MILP or MIP) problem.

4. If the constraints are linear, but the objective function f contains some quadratic terms, the problem is called a *quadratic programming* (QP) problem.

5. If f or any of $g_i(x)$ is not linear, it is called a *nonlinear programming* (NLP) problem.

6. If all the constraints have the following convexity property:

$$g_i(\alpha x_a + \beta x_b) \leq \alpha g_i(x_a) + \beta g_i(x_b)$$

where $\alpha \geq 0$, $\beta \geq 0$, and $\alpha + \beta = 1$, then the problem is called a *convex programming* or *convex optimization* problem.

7. If the set of feasible solutions defined by f and X are discrete, the problem is called a *discrete* or *combinatorial optimization* problem.

Intuitively speaking, different categories of mathematical programming problems should involve different solving techniques, and, thus, they may have different computational complexities. In fact, most of the mathematical optimization problems are generally intractable—algorithms to solve the preceding optimization problems such as the *Newton method, steepest gradient, branch-and-bound*, etc., often require an exponential runtime or an excessive amount of memory to find the *global* optimal solutions. As an alternative, people turn to heuristic techniques such as *hill climbing, simulated annealing, genetic algorithms*, and *tabu search* for a reasonably good local optimal solution.

Nevertheless, some categories of mathematical optimization problems, such as linear programming and convex optimization, can be solved efficiently and reliably. Therefore, it is feasible to examine whether the original optimization problem can be modeled or approximated as one of these problems. Once the modeling is completed, the rest should be easy—there are numerous shareware or commercial tools available to solve these standard problems.

In the following, we will briefly describe the problem definitions and solving techniques of the linear programming and convex optimization problems. For more theoretical details, please refer to other textbooks or lecture notes on this subject.

4.5.2 Linear programming (LP) problem

Many optimization problems can be modeled or approximated by linear forms. Intuitively, solving LP problems should be simpler than solving the general mathematical optimization problems, because they only deal with linear constraint and objective functions; however, it took people several decades to develop a polynomial time algorithm for LP problems, and several related theoretical problems still remain open [Smale 2000].

The *simplex algorithm*, developed by George Dantzig in 1947, is the first practical procedure used to solve the LP problem. Given a set of n-variable linear constraints, the simplex algorithm first finds a basic feasible solution that satisfies all the constraints. This basic solution is conceptually a *vertex* (*i.e.,* an *extreme point*) of the convex polyhedron expanded by the linear constraints in R^n hyperspace. The algorithm then moves along the *edges* of the polyhedron in the direction toward finding a better value of the objective function. It is guaranteed that the procedure will eventually terminate at the optimal solution.

Although the simplex algorithm can be efficiently used in most practical applications, its worst-case complexity is still exponential. Whether a polynomial time algorithm for LP problems exists remained unknown until the late 1970s, when Leonid Khachiyan applied the ellipsoid method to this problem and proved that it can be solved in $O(n^4w)$ time. Here n and w are the number and width of variables, respectively.

Khachiyan's method had theoretical importance, because it was the first polynomial-time algorithm that could be applied to LP problems; however, it did not perform any better than the simplex algorithm for most practical cases. Many researchers who followed Khachiyan focused on improving the average case performance, as well as the computational worst-case complexity. The most noteworthy improvements included Narendra Karmarkar's interior point method and many other revised simplex algorithms [Karmarkar 1984].

4.5.3 Integer linear programming (ILP) problem

Many of the linear programming applications are concerned with variables only in the integral domain. For example, signal values in a digital circuit are under a modular number system. Therefore, it is very likely that optimization problems defined with respect to signals in a circuit can be modeled as ILP problems. On the other hand, problems that need to enumerate the possible cases, or are related to scheduling of certain events, are also often described as ILP.

The ILP problem is in general much more difficult than is LP. It can be shown that ILP is actually one of the NP-hard problems. Although the formal proof of the computational complexity of the ILP problem is beyond the scope of this book, we will use the following example to illustrate the procedure and explain the difficulty in solving the ILP problem.

The ILP problem in Figure 4.28 is to maximize an objective function f, with respect to four linear constraints $\{g_1, g_2, g_3, g_4\}$. Because the problem consists of only two variables, x and y, it can be illustrated on a two-dimensional plane, where each constraint is a straight line, the four constraints form a closed region C, and the feasible solutions are the lattice or integral points within this region. The objective function f, represented as a stright line to the right of region C, moves in parallel with respect to different values of k. Intuitively, to obtain

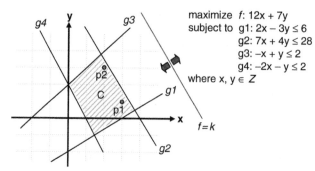

maximize f: $12x + 7y$
subject to $g1$: $2x - 3y \leq 6$
  $g2$: $7x + 4y \leq 28$
  $g3$: $-x + y \leq 2$
  $g4$: $-2x - y \leq 2$
where $x, y \in Z$

FIGURE 4.28

An ILP example.

the maximum value of f, we can move the line $f = k$ from where it is located in the figure until it intersects the region C on a lattice point for the first time.

From the figure, it is clear that the maximum value must occur on either point p_1 (3, 1) or p_2 (2, 3). For p_1, $f = 12 \times 3 + 7 \times 1 = 43$, and for p_2, $f = 12 \times 2 + 7 \times 3 = 45$. Therefore, the maximum value of f is 45, which occurs at $(x, y) = (2, 3)$.

This solving procedure is not applicable for ILP problems with more variables—it will be impossible to visualize the constraints and to identify the candidate integral points for the optimum solutions. In fact, to find a feasible assignment that satisfies all the constraints of an ILP problem is already an NP-complete problem. Finding an optimal solution is even more difficult.

4.5.3.1 *Linear programming relaxation and branch-and-bound procedure*

Because it is very difficult to directly find a feasible solution that satisfies all the constraints of the ILP problem, one popular approach is to relax the integral constraints on the variables and use a polynomial-time linear programming solver to find an approximated nonintegral solution first. Then, on the basis of the approximated solution, we can apply a branch-and-bound algorithm to further narrow the search [Wolsey 1998].

In the previous example, the LP relaxation tells us that the optimal solution occurs at $(x, y) = (108/29, 14/29)$. Because x is an integer, we can branch on variable x into two conditions: $x \leq 3$ and $x \geq 4$. For $x \geq 4$, the LP solver will report infeasibility because the union of the constraints is an empty set. On the other hand, for the $x \leq 3$ case we will have the optimal solution at $(x, y) =$ (3, 7/4). Because y is not yet an integer, we further branch on y—$y \leq 1$ and $y \geq 2$. For $y \leq 1$, we obtain an integral solution $(x, y) = (3, 1)$ and $f = 43$. For $y \geq 2$, the LP optimal solution will be $(x, y) = (20/7, 2)$. Repeating the above process, we will eventually acquire the integral optimal solution

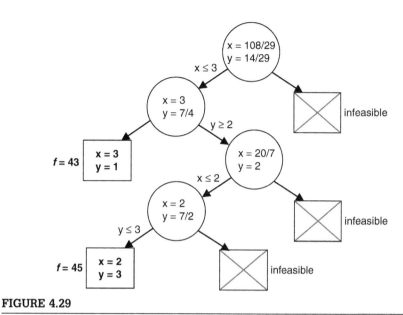

FIGURE 4.29

Decision tree of the LP-based branch-and-bound.

$(x, y) = (2, 3)$ and $f = 45$. The decision graph of the branch-and-bound process is shown in Figure 4.29.

4.5.3.2 *Cutting plane algorithm*

Another useful approach for solving ILP problems is the cutting plane algorithm. This algorithm iteratively adds valid inequalities to the original problem to narrow the search area enclosed by the constraints while retaining the feasible points. Figure 4.30 illustrates an example of such valid inequalities.

In Figure 4.30, the cuts c_1 and c_2 are said to be *valid inequalities,* because all the feasible points (*i.e.,* the integral points within the dash region C) are still valid after adding the new constraints. On the other hand, cut c_3 is not a valid inequality because one feasible point p_1 becomes invalid afterward.

It is clear to see that the addition of the valid inequality c_2 will not help the search for the optimal solution because it does not narrow the search region. On the contrary, cut c_1 is said to be a *strong* valid inequality because it makes the formulation "stronger." The goal of the cutting plane algorithm is to add such strong valid inequalities in the hope that the optimal solution will eventually become an extreme point of the polyhedron so that it can be found by the polynomial-time LP algorithm.

There are many procedures to generate valid inequalities such as *Chvátal-Gomory* [Gomory 1960], *0-1 Knapsack* [Wolsey 1999], and *lift-and-project* [Balas 1993] *cuts*. However, sheer use of these valid inequality generation procedures in the cutting plane algorithm will not go too far in solving difficult ILP

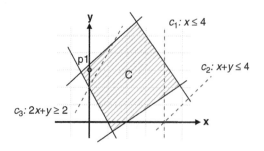

FIGURE 4.30

Valid and invalid inequalities.

problems—it may take an exponential number of steps to approach an integral extreme point. A better approach would be combining the cutting plane algorithm with the branch-and-bound process. This combined technique is called the branch-and-cut algorithm.

4.5.4 **Convex optimization problem**

As mentioned in Subsection 4.5.1, the constraints in the convex optimization problem are convex functions with the following convexity property (Figure 4.31):

$$g_i(\alpha x_a + \beta x_b) \le \alpha g_i(x_a) + \beta g_i(x_b)$$

where $\alpha \ge 0$, $\beta \ge 0$, and $\alpha + \beta = 1$. Conceptually, the convexity property can be illustrated as follows:

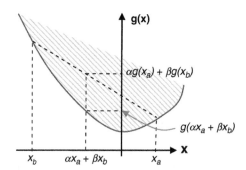

FIGURE 4.31

The convexity property.

In other words, given two points x_a and x_b from the set of points defined by a convex function, all the points on the line segment between x_a and x_b will also belong to the set (*i.e.*, the dash region), which is called a *convex set*. Moreover, it can be shown that for a convex function, a local optimal solution is also a global optimal solution. In addition, the intersection of multiple convex sets is also convex [Boyd 2004].

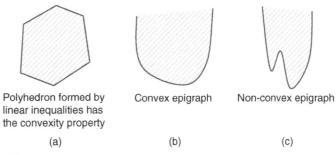

Polyhedron formed by
linear inequalities has
the convexity property

Convex epigraph

Non-convex epigraph

(a)

(b)

(c)

FIGURE 4.32

Examples of convex functions.

More examples of convex functions can be found in Figure 4.32. The LP problem, where its constraints form a polyhedron in the n-dimensional hyperspace, is a special case of the convex optimization problem.

4.5.4.1 *Interior-point method*

Similar to linear programming, there is, in general, no analytical formula for the solution of a convex optimization problem. However, there are many effective methods that can solve the problems in polynomial time within a reasonably small number of iterations. The *interior-point* method is one of the most successful approaches.

Although detailed comprehension of the interior-point method requires the introduction of many mathematical terms and theorems, we can get a high-level view of the method by comparing it with the simplex method as shown in Figure 4.33. In the simplex method, we first obtain an initial feasible solution and then refine it along the edge of the polyhedron until the optimal solution is reached. In the interior-point method, the initial feasible solution is approximated as an interior point. Then, the method iterates along a path, called a *central path*, as the approximation improves toward the optimal solution.

One popular way to bring the interior-point solution to the optimal one is by the use of a *barrier function*. The basic idea is to rewrite the original problem into an *equality formula* so that Newton's method can be applied to find the optimal solution.[2]

Let's first define an *indicator function $I(u)$* such that $I(u) = 0$ if $u \leq 0$, and $I(u) = \infty$ otherwise (Figure 4.34). We can then combine the convex objective function *min f(x)*, and the constraints $g_i(x) \leq 0 \mid_i = 1 \sim m$ as:

$$\min\left(f(x) + \sum_{1}^{m} I(g_i(x))\right)$$

---

[2]To apply the Newton's method, the formula needs to be an equality and twice differentiable.

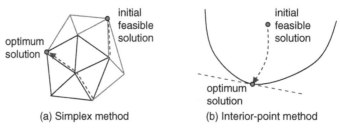

(a) Simplex method   (b) Interior-point method

FIGURE 4.33

Comparison of simplex and interior-point methods.

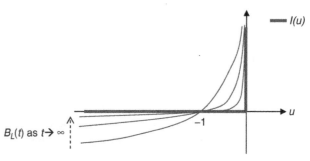

FIGURE 4.34

Indicator $I(u)$ and logarithmic functions B_L.

This formula describes the same problem as the original convex optimization problem and after the rewrite, there are no more inequalities. However, this formula is not twice differentiable (*i.e.*, not *smooth*) near $u = 0$, so Newton's method cannot work. One solution is to use the following *logarithmic barrier function* to approximate the indicator function:

$$B_L(u, t) = -(1/t)\log(-u)$$

where $t > 0$ is a parameter to control the approximation. As t approaches infinity, the logarithmic barrier function $B_L(u)$ gets closer to the indicator function $I(u)$.

By use of the logarithmic barrier function, the objective function then becomes:

$$\min\left(f(x) + \sum_{1}^{m} -(1/t)\log(-g_i(x))\right)$$

Please note that now the optimization formula is convex and twice differentiable (we assume that both $f(x)$ and $g_i(x)$ are twice differentiable here). Therefore, we can apply Newton's method iteratively and eventually reach an optimal

InteriorMethod (objFunction f, Constraints g)

1. Let $(x, t) = \min\left(f(x) + \sum_{1}^{m} - (1/t)\log(-g_i(x))\right)$
2. Given initial t, tolerance e
3. Find an interior feasible point x_p s.t. $\forall i. g_i(x_p) < 0$
4. Starting from x_p, apply Newton's method to find the optimal solution x_{opt}
5. If $(\frac{1}{t} < e)$ return optimality as $\{x_{opt}, (x_{opt}, t)\}$;
6. Let $x_p = x_{opt}$, $t = k \cdot t$ for $k > 1$, repeat 4

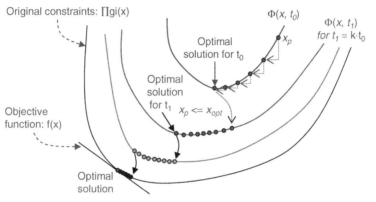

FIGURE 4.35

Interior-point algorithm and an illustration of its concept.

solution. However, please remember that this will be just an approximate solution because of the introduction of the logarithmic barrier function.

The questions then arise: How close is this solution to the solution of the original problem? What is the effect of t? Intuitively, if t gets larger, the final solution will be closer to the solution of the original convex optimization problem. However, with a larger t, it will take a longer time for Newton's method to converge. On the other hand, the use of a smaller t will lead to a faster solution at the cost of accuracy.

The pseudocode in Figure 4.35 is an interior-point algorithm that gives a solution balancing runtime and accuracy. We first start with a smaller t so that Newton's method converges faster. Once the optimal solution for this t value is obtained, we then increase t so that the optimal solution gradually approaches the real optimization of the original problem. This process terminates when the inverse of the variable t becomes less than the specified tolerance e.

4.6 CONCLUDING REMARKS

In this chapter, we present various fundamental algorithms to the EDA research and development—from the classic graphic theories, the practical heuristic approaches, and then to the theoretical mathematical programming techniques. The readers are advised to get acquainted with these algorithms to completely appreciate the spirit of the research in different areas of the later chapters.

In addition, please note that a good EDA algorithm is usually *hybrid*. In other words, it should act as a *strategy*, or say *problem-solving tactic* that is able to apply different algorithms in different situations. It should be working efficiently for the most common cases, taking advantage of the easy ones, and at the same time, handling the worst-case scenarios gracefully. In summary, do not just take the algorithms in this chapter as ready solutions; instead, thoroughly understand the problems first, consider the trade-offs between runtime and memory, and then treat the algorithms as different utilities, or weapons, for the different challenges in the EDA problem solving process.

4.7 EXERCISES

4.1. (Computational Complexity) Rank the following functions by order of growth by use of asymptotic notations. One function is neither $O(f_i)$ nor $\Omega(f_i)$ for any other functions f_i. Which is that?
   a. $4^{\lg n}$
   b. $n \cdot 2^n$
   c. $n^{n \cdot \cos n}$
   d. $n \cdot \lg n$
   e. $(n + 1)!$ f. $\lg^{999} n$ g. $n \cdot \lg n$ h. $n^{1/\lg n}$

4.2. (Computational Complexity) A *Hamiltonian path* in a graph is a simple path that visits every vertex exactly once. The decision problem HAMILTONIAN PATH for a graph G and vertices u and v asks whether a Hamiltonian path exists from u to v in G.
   a. Prove that HAMILTONIAN PATH is NP.
   b. Given that HAMILTONIAN PATH is NP-complete, prove that HAMILTONIAN CYCLE is also NP-complete.

4.3. (Graph Algorithms) Figure 4.36 shows a directed graph of 10 vertices. How many strongly connected components does this graph have and which are they?

4.4. (Graph Algorithms) Given an undirected, weighted graph $G = (V, E)$ and two vertices u and v in V. Find an efficient path from u to v such that the biggest edge weight on the path is minimized.

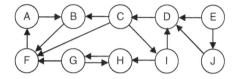

FIGURE 4.36

A directed graph to find strongly connected components.

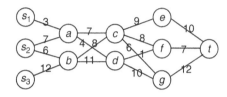

FIGURE 4.37

A model for a combinational circuit in which vertices represent gates and edge weights stand for the number of connecting wires, where gates s_1, s_2, and s_3 have to be in one module and t in the other, and meanwhile minimizing the number of wires crossing two modules. What is the minimal number of crossing wires and where should the cut of two modules be?

4.5. (Graph Algorithms) The weighted, undirected graph illustrated in Figure 4.37 models a combinational circuit. We want to divide these gates (vertices) into two modules.

4.6. (Heuristic Algorithms) In a dance class, n male students and n female students should be paired. If we want to minimize the sum of height differences of the n pairs,

   a. Design a *greedy algorithm* to efficiently solve this problem.
   b. Prove that the algorithm works because the problem exhibits both the greedy-choice and optimal substructure properties.

4.7. (Heuristic Algorithms) Solve the matrix-chain multiplication problem if the dimensions of the matrices are 5×10, 10×3, 3×12, 12×5, 5×50, and 50×6. What are the minimum number of scalar multiplications needed and the order of the multiplications?

4.8. (Heuristic Algorithms) Use the *branch-and-bound* technique to solve the TSP problem in Figure 4.38. What is the length of the shortest route? If only the branching technique is used to form the search tree, what is the number of tree nodes?

4.9. (Linear Programming) Given an $n \times m$ rectangular, which is composed of equal-length (length = 1) matches as shown in Figure 4.39.

   In this problem, we will try to remove as few as possible matches so that all the squares (including 1×1, 2×2, 3×3, ...) in the

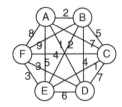

FIGURE 4.38

A TSP instance.

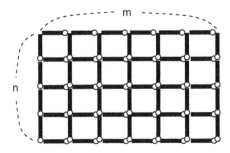

FIGURE 4.39

A square-breaking problem.

rectangular are broken. Please model this problem as an integer linear programming (ILP) problem.

4.10. **(Convex Optimization)** Prove that a local optimum of a convex function must be a global optimum.

ACKNOWLEDGMENTS

We thank Dr. Bow-Yaw Wang of Academia Sinica, Taiwan, and Mr. Benjamin Liang of University of California, Berkeley, for their thorough reviews of the entire chapter, and Professor Tian-Li Yu of National Taiwan University for providing valuable comments on the "Genetic Algorithm" subsection.

REFERENCES
R4.1 Books

[Aho 1983] A. V. Aho, J. E. Hopcroft, and J. D. Ullman, *Data Structures and Algorithms*, Addison-Wesley, Reading, MA, January 1983.

[Ahuja 1993] R. K. Ahuja, T. L. Magnanti, and J. B. Orlin, *Network Flows: Theory, Algorithms and Applications*, Prentice Hall, Englewood Cliffs, NJ, February 1993.

[Baase 1978] S. Baase, *Computer Algorithms: Introduction to Design and Analysis*, Addison-Wesley, Reading, MA, December 1978.

[Boyd 2004] S. Boyd and L. Vandenberghe, *Convex Optimization*, Cambridge University Press, Cambridge, UK, March 2004.

[Cormen 2001] T. H. Cormen, C. E. Leiserson, R. L. Rivest, and C. Stein, *Introduction to Algorithms*, Second Edition, MIT Press, Cambridge, MA, September 2001.

[Davis 1991] L. Davis, *Handbook of Genetic Algorithms*, Van Nostrand Reinhold, New York, NY, January 1991.

[Even 1979] S. Even, *Graph Algorithms*, Computer Science Press, Rockville, MD, June 1979.

[Ford 1962] R. W. Ford and D. R. Fulkerson, *Flows in Networks*, Princeton University Press, Princeton, NJ, June 1962.

[Garey 1979] M. R. Garey and D. S. Johnson, *Computers and Intractability: A Guide to the Theory of NP-Completeness*, W.H. Freeman and Company, San Francisco, CA, January 1979.

[Gibbons 1985] A. Gibbons, *Algorithmic Graph Theory*, Cambridge University Press, Cambridge, UK, July 1985.

[Goldberg 1989] D. E. Goldberg, *Genetic Algorithms in Search, Optimization and Machine Learning*, Reading, MA, Addison-Wesley, January 1989.

[Holland 1992] J. H. Holland, *Adaptation in Natural and Artificial Systems*, MIT Press, Cambridge, CA, April 1992.

[Horowitz 1978] E. Horowitz and S. Sahni, *Fundamentals of Computer Algorithms*, Computer Science Press, Rockville, MD, January 1978.

[Knuth 1968] D. E. Knuth, *Fundamental Algorithms: The Art of Computer Programming, Volume 1*, Reading, MA, Addison-Wesley, February 1968.

[Papadimitriou 1993] C. H. Papadimitriou, *Computational Complexity*, Reading, MA, Addison-Wesley, December 1993.

[Papadimitriou 1998] C. H. Papadimitriou and K. Steiglitz, *Combinatorial Optimization, Algorithms and Complexity*, Englewood Cliffs, NJ, Prentice Hall, January 1998.

[Reeves 1993] C. R. Reeves, *Modern Heuristic Techniques for Combinatorial Problems*, McGraw-Hill, London, UK, February 1993.

[Tarjan 1987] R. E. Tarjan, *Data Structures and Network Algorithms*, Society for Industrial and Applied Mathematics, Philadelphia, PA, January 1987.

[Ullman 1984] J. D. Ullman, *Computational Aspects of VLSI*, Computer Science Press, Rockville, MD, August 1984.

[Wilf 2002] H. S. Wilf, *Algorithms and Complexity*, Second Edition, A. K. Peters, Ltd., Wellesley, MA, December 2002.

[Wong 1988] D. F. Wong, H. W. Leong, and C. L. Liu, *Simulated Annealing for VLSI Design*, Kluwer Academic, Boston, MA, March 1988.

[Wolsey 1998] L. A. Wolsey, *Integer Programming*, Wiley-Interscience, Hoboken, NJ, September 1998.

[Wolsey 1999] L. A. Wolsey and G. L. Nemhauser, *Integer and Combinatorial Optimization*, Wiley-Interscience, Hoboken, NJ, November 1999.

R4.2 Computational Complexity

[Knuth 1976] D. E. Knuth, Big Omicron and big Omega and big Theta, in *ACM SIGACT News*, 8(2), pp. 18-24, April-June 1976.

R4.3 Graph Algorithms

[Bellman 1958] R. Bellman, On a routing problem, in *Quarterly of Applied Mathematics*, 16(1), pp. 87-90, December 1958.

[Dijkstra 1959] E. W. Dijkstra, A note on two problems in connection with graphs, in *Numerische Mathematik*, 1, pp. 269-271, June 1959.

[Edmonds 1972] J. Edmonds and R. M. Karp, Theoretical improvements in the algorithmic efficiency for network flow problems, in *J. of the ACM*, 19, pp. 248-264, April 1972.

[Hopcroft 1973] J. E. Hopcroft and R. E. Tarjan, Efficient algorithms for graph manipulation, in *Communications of the ACM*, 16(6), pp. 372-378, March 1973.

[Moore 1959] E. F. Moore, The shortest path through a maze, in *Proc. Int. Symp. on the Theory of Switching*, pp. 285-292, November 1959.

[Prim 1957] R. C. Prim, Shortest connection networks and some generalizations, in *Bell System Technical J.*, 36, pp. 1389-1401, November 1957.

R4.4 Heuristic Algorithms

[Hajek 1988] B. Hajek, Cooling schedules for optimal annealing, in *Mathematics of Operational Research*, 13(2), pp. 311-329, May 1988.

[Kirkpatrick 1983] S. Kirkpatrick, C. D. Gelatt, and M. P. Vecchi, Optimization by simulation annealing, in *Science*, 220(4598), pp. 671-690, May 1983.

R4.5 Mathematical Programming

[Balas 1993] S. Balas, S. Ceria, and G. Cornuéjols, A lift-and-project cutting plane algorithm for mixed 0-1 programs, in *Mathematical Programming*, 58(1-3), pp. 295-324, January 1993.

[Gomory 1960] R. E. Gomory, *An algorithm for the mixed integer problem*, Research Memorandum, RM-2597, The Rand Corp., 1960.

[Karmarkar 1984] N. Karmarkar, A new polynomial-time algorithm for linear programming, in *Proc. ACM Symposium on Theory of Computing*, pp. 302-311, April 1984.

[Smale 2000] S. Smale, Mathematical Problems for the Next Century, in *Mathematics: Frontiers and Perspectives*, pp. 271-294, 2000.

Electronic system-level design and high-level synthesis

Jianwen Zhu
University of Toronto, Toronto, Canada

Nikil Dutt
University of California, Irvine, California

ABOUT THIS CHAPTER

System designers conceptualize designs at an abstract functional level where outputs are typically described as (algorithmic or transfer) functions of the system inputs. This design abstraction level, called ***electronic system level*** (ESL), enables ease of design capture and early design space exploration of multiple design implementation alternatives. ESL designs can be refined into lower levels of abstraction through a number of steps that gradually map abstract functions into ***register-transfer level*** (RTL) components. An enabling technology for ESL design is ***high-level synthesis*** (HLS), also known as behavioral synthesis. A high-level synthesis tool bridges the gap between an algorithmic description of a design and its structural implementation at the register transfer level and is the next natural step in design automation, succeeding logic synthesis. The need for high-level synthesis becomes more pressing with the proliferation of billion-transistor designs. High-level synthesis has been researched actively since the 1980s, and has yielded several promising results. However, it also faces a number of challenges that has prevented its wide adoption in practice.

In this chapter, we first introduce the notion of ESL design and an ESL design method. Next, we describe high-level synthesis in the context of an ESL design method. We then describe the generic structure of the high-level synthesis process and the basic tasks accomplished by high-level synthesis. This is followed by a detailed description of the key high-level synthesis algorithms and exercises designed to reinforce understanding. The reader will have been exposed to the basic principles of high-level synthesis and its applicability in an ESL design flow by the end of the chapter.

5.1 INTRODUCTION

A key goal of *electronic design automation* (EDA) is to shrink the rapidly growing "designer productivity gap" that exists between how many transistors we can manufacture per chip, and how many person-years we need to complete a design with that many transistors. Collectively, EDA provides to chip designers a *design method*, which can be considered as a set of complementary *design tools* built on a *design abstraction* (*i.e.*, a mechanism to conceptualize the chip design), as well as a set of processes and guidelines that indicate the flow of design, ordering of tool application, strategies for incorporating late engineering changes, etc. The software design tools include *design entry* tools, which capture design specification; *design synthesis* tools, which target different parts of the design specification and bring them down to low level implementation; and *design verification* tools, which either simulate/verify the specification or compare a specification against its implementation.

As discussed in Chapter 1, the EDA design method has traditionally progressed by raising the abstraction at which designs are conceptualized and specified. As a step in this progression, the basic components used by designers grow in complexity, which results in fewer, but more complex, components; therefore, designer productivity improves because designers need to manipulate fewer components and can reason about the design at abstractions that are closer to how systems are specified and conceptualized. Historically, the basic design component has evolved from polygons, to transistors, to gates, and then to register transfer level blocks. To cope with the challenges of designing emerging *billion transistor system-on-chips* (BTSOCs), it is widely believed that the chips have to be designed at an abstraction level well above RTL. Indeed, system designers typically reason about and conceptualize designs at an abstract functional level where system outputs are described as algorithms or transfer functions of the system inputs. This design abstraction, called *electronic system level* (ESL), enables ease of design capture and early design space exploration of multiple design implementation alternatives. ESL designs can be refined into lower levels of abstraction through a number of steps that gradually map abstract functions into *register-transfer level* (RTL) components, which is the next level of design abstraction. In this section, we discuss the main drivers and the basic elements of the emerging ESL design method.

5.1.1 ESL design methodology

Moore's law, which states that chip complexity doubles every 18 months, has been the key driver behind the paradigm shift in EDA methodology. Although RTL design methods are dominant currently, rapid growth in chip complexity coupled with shrinking time-to-market windows result in RTL design methods not being able to scale with the complexity of emerging designs. This trend necessitates fundamental changes that force the move toward higher levels of abstraction. Let us examine two such fundamental changes.

The first fundamental change is that the cost of a new chip design by use of an RTL design method is no longer economically viable.

Example 5.1 Consider a startup company designing a new chip at 65-nm technology. The average design cost with the RTL design method is $30 million. Assuming a fivefold return on investment, the company has to make at least $150 million in sales. Assuming a 10% market share (a respectable goal for a startup), the chip design has to target a $1.5 billion market. The reality is that few such markets exist, and if they do, they would be very crowded.

To dramatically reduce the design cost, the basic building blocks have to be one order of magnitude larger than what is used in RTL. It has been suggested that the basic component becomes a design block with 10,000 to 50,000 gates. The 50,000 gate limit is set so that contemporary RTL-to-GDSII tools can comfortably handle each block without running into issues relating to design complexity explosion that lead to excessive memory requirements and long run times. Note that the complexity of these new building blocks coincides with the complexity of an embedded processor. Indeed, many view processors to be the basic building blocks (*i.e.*, the "gates") of an ESL method.

Example 5.2 Figure 5.1 shows a prediction by the **international technology roadmap for semiconductors** (ITRS) as an implication of Moore's law [SIA 2007]. Assume a constant

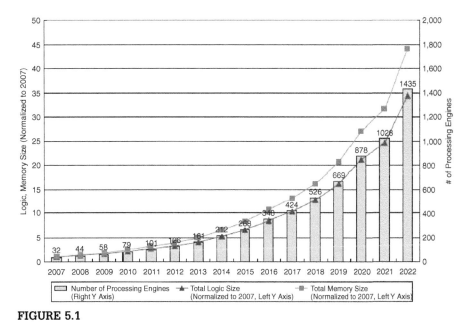

FIGURE 5.1

ITRS 2007 prediction of processing engine count.

die size of 66 mm^2 and that the sizes of the main processor and peripherals remain unchanged. The number of **processing engines** (PEs)—*i.e.*, processors that perform fixed functions—will grow from 32 in 2007 to 79 in 2010 and 268 in 2015. Conceivably, chip designs with below a thousand components are easier to conceptualize and implement.

The second fundamental change is the rapid growth in chip complexity that enables integration of previously separate components on a printed circuit board (*e.g.*, CPU, Ethernet controller, or memory) into a single chip. This integration of heterogeneous "content" comprises not only hardware but also software. Indeed, a chip is more often designed as a programmable computer system, requiring not only hardware but also firmware, operating systems, and application software, with applications often downloaded after the end product is deployed to consumers.

Example 5.3 Figure 5.2 shows a block diagram of Texas Instrument's OMAP platform designed for the cellular phone market. On this chip, one can find a mixture of heterogeneous components, including an ARM programmable processor, an image signal processor, numerous hardware processing engines for acceleration, and peripherals that interface with the outside world.

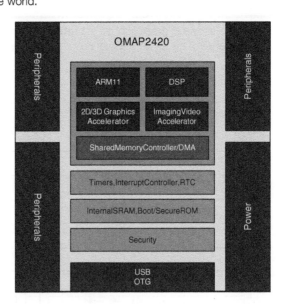

FIGURE 5.2

Texas Instrument's OMAP platform.

To better understand emerging ESL methods, we examine three necessary elements of a design abstraction that allow a chip design to be conceptualized as a connected set of components:

- *Computation*: How to specify what each component does?
- *Composition*: How to assemble different components and coordinate their computation to form a larger system?
- *Communication*: How do the components exchange information?

It is instructive to first examine the RTL design abstraction against these three elements, and consider the design method we build around the RTL abstraction.

Example 5.4 A component in an RTL abstraction can be captured with objects in a **hardware description language** (HDL) (*e.g.*, a module in Verilog, or an entity in VHDL). The computation of a component is captured by its per-cycle behavior, specifying the transformation of register values at each clock cycle. The components can be composed together into a larger design by connecting the ports of components with wires (*e.g.*, with the port map construct in Verilog or VHDL). The communication between components is effected by the transfer of values through wires. With the three design abstraction elements clearly defined, an RTL method can be constructed: RTL synthesis tools convert each module into a gate level design according to its per-cycle behavior, and all modules are stitched together with wires, after which gate level optimizations are performed; RTL simulation tools convert each module into a concurrent software process triggered by events such as the rising edge of a clock.

The ESL abstraction typically conceptualizes the computation of a component at a more abstract level by use of *untimed* behavior. In other words, chip designers do not make the decision as to how computation is mapped to a particular clock step. Currently C/C++ and Matlab are among the most commonly used languages for capturing system behavior. However, the mechanisms for composition and communication differ widely on the basis of the semantics of each specification language.

Next, we examine two ESL methods that are becoming increasingly accepted. For each method, we examine the design abstraction used for composition and communication. We also examine how we map each component to RTL (called *component synthesis*), how we map the full system into RTL (called *system synthesis*), and how we verify the full system.

5.1.2 **Function-based ESL methodology**

The function-based ESL design method uses a *computational model* [Lee 1996] to compose different functional components into a complete system. The computational model determines how the components execute in parallel and how they communicate with each other. Note that the manner in which components exchange information is also more abstract than RTL.

Example 5.5 A process network [Kahn 1974] is a design abstraction in which functional components, called processes, communicate through data items called tokens. On each execution of a process, input tokens are consumed, and output tokens are produced. A process is executed whenever its input tokens are available. Compared with RTL, all three elements of computation, composition, and communication are more abstract: the computation of each process is an untimed algorithm; the execution of processes follow a partial order; and the communication between processes are through unbounded **first-in first-outs** (FIFOs).

A plethora of computational models has been developed in the past. Some are developed as special cases of a general model. For example, synchronous data flow [Lee 1987] is a special case of process networks that imposes the constraint in which the number of input and output tokens consumed and produced by each process is statically determined to be constant. The synchronous data flow model is useful to capture multi-rate signal processing systems. Other computational models are designed for particular application domains. For example, synchronous models [Halbwachs 1991; Berry 1992], which capture a system as atomic actions in response to external events, are very expressive in capturing control-dominated applications.

In practice, the most widely used computational models arise from models used for the simulation of dynamic physical systems. This is not surprising: HDLs at the RTL abstraction (*e.g.*, Verilog/VHDL) were in fact languages originally used for the simulation of digital systems. One of the most widely used ESL computation models in the industry is Simulink developed by MathWorks. Like a process network, Simulink graphically captures a system as a connected network of components, called a *block diagram*. Here each block captures the instantaneous behavior of a component, in other words, how the component output changes given component input and state variable. The connections between components, or *signals*, serve as constraints, subject to which the entire dynamic system can be solved to find the relationship between signals and state variables over a time period. As a result, Simulink can be used to simulate both continuous time systems (often analog subsystems) and discrete time systems (often digital subsystems). A certain execution order of the blocks is imposed in each iteration of system solving. For example, if block A's output drives the input of another block B, it is required, according to the execution semantics of Simulink block diagram, that A is executed before B. This execution order coincides with process network. Therefore, Simulink can serve as an executable modeling environment for process network and other computational models.

Verification in a function-based method is achieved through simulation. A simulator respecting the underlying computational model is typically used.

In a function-based design method, component synthesis can be achieved through a number of paths:

1. Direct translation to RTL
2. Direct mapping to predesigned intellectual property component

3. High-level synthesis to RTL
4. Compilation to software programs

The most widely used forms of system component synthesis are to directly map each component into separate hardware or map all components into software running on a single processor. In the former case, point-to-point communication is implemented by hardware FIFOs or ping-pong memories.[1] In the latter case, communication is trivially implemented by shared memory.

Example 5.6 Consider the **digital signal processor** (DSP) builder product from Altera. It uses Simulink as the design specification and simulation environment. In addition, it supplies a large library of predesigned, configurable intellectual property (IP) components, each with both a Simulink simulation model, and an RTL implementation. When a DSP designer uses Simulink to design a system with the predesigned components, DSP builder can automatically generate a full-system RTL by mapping each component to its corresponding IP component, as well as the top level interconnection. Xilinx' SystemGenerator product and Berkeley's Chip-In-A-Day, share a similar methodology.

5.1.3 Architecture-based ESL methodology

The architecture-based ESL method follows closely the traditional discipline of computer organization. Here a design is conceptualized as a set of components, including processors, peripherals, and memories. Because these components are often available as reused designs, either from previous projects or acquired from third parties, they are referred to in the industry as *intellectual property* components, or simply IPs. The components are connected through buses, switch fabric, or point-to-point connections. Components connected to buses and switches typically contain a set of master and/or slave ports. Each slave port is assigned an address range so that targeted communication can be identified. Several bus-based communication protocols are commonly used for this purpose, including AMBA from ARM and CoreConnect from IBM. Furthermore, the industry-wide SPIRIT consortium has developed IP-XACT, a standard to help specify the composition of architecture-based systems with IP blocks.

A key difference in the architecture-based method is that a processor component is not required to implement a fixed function. Instead, the computation of the processor can be "programmed." Such programming can happen in the traditional sense in which the component is a programmable processor and is programmed post-silicon. Or it can happen in the design exploration phase in which an accelerator is assigned a certain function to be synthesized into hardware. In this context, it is important for a processor to present a *programming*

---

[1]Ping-pong memories are two SRAMs alternately accessed by a producer and a consumer component. By switching the access of the SRAMs, data can be exchanged between the producer and the consumer without the need for copying data.

model to facilitate software development. For example, a RISC processor defines an instruction set into which application software can be compiled. We use the term *program* broadly to refer to system software, application software, as well as the accelerator functions.

Communication in the architecture-based method is abstracted as a set of *transactions*. Typically, a transaction is either a bus transaction, which represents a (burst) *read* or *write* operation to peripherals or memories, or a point-to-point data transfer. **Transaction level modeling** is a popular modeling style in which a construct called *channel* is used to encapsulate or abstract away the concrete protocols for such transactions [Zhu 1997]. The use of abstract channels allows faster high-level design exploration, because unnecessary details are abstracted away. In addition, it significantly eases the task of verification because testbenches can be written at a high abstraction level. The abstract channel can be replaced easily by a *transactor*, which encapsulates the timed protocol, whenever a detailed simulation or implementation is required.

In the architecture-based method, the system architecture and its program are simulated together. Two popular approaches for simulation exist currently. The *SystemC-based approach* [Grötker 2002] models the architecture and program together in a single environment. Here the program for a processor is directly modeled as a C++ class by extending a predefined SystemC class. The communication ports (an architectural element) of the processor are directly accessible to the program. On the other hand, the *virtual prototyping approach* has a clear separation between the architecture and the program it runs. Here a system is first constructed by connecting a set of virtual components predesigned for emulating hardware devices. Such virtual components are referred to in the industry as verification IPs. Together, they present a *programmer's view* or a programming model. The program is typically captured as a binary executable and can be loaded for simulation. The SystemC approach is suitable for the design phase when the system architecture is not yet well defined and requires extensive exploration. Virtual prototyping is more suitable for use when the system architecture is relatively well defined, enabling concurrent development of the software and the hardware.

In the architecture-based design method, component (IP) synthesis can be achieved by the following approaches:

1. Direct instantiation of a predesigned IP component.
2. Extension of the processor design with processor configuration and instruction set extension.
3. Synthesis of the processor from an ***architecture description language*** (ADL) specification [Mishra 2008].
4. High-level synthesis to RTL.

Approach 1 requires the least involvement with users. However, mechanisms have to be established to ensure that the simulation model and the instantiated RTL model are consistent with each other. Approaches 2 and 3 are used for application-specific instruction processors, with varying degrees of user freedom in defining the processor instruction set and microarchitecture. Typically, suppliers of such components create the compiler and simulator tool chain according to the instruction set definition or extension. Approach 4 is used when a hardwired accelerator has to be created to meet exacting performance, power/energy, or cost constraints, when an existing IP cannot satisfy such constraints.

In the architecture-based design method, system synthesis mainly involves the synthesis of system-level interconnect and the generation of the top level design. Such a method is often referred to as a *communication-centric design method*. Many **computer-aided design** (CAD) vendors offer *IP assembly* tools to help assemble these architectural components into a computer system. For instance, given a system description in the form of IP-XACT, tools such as Synopsys *coreAssembler* and Mentor Graphics *Platform Express* can be used to generate the top level netlist, as well as the interconnect fabric. Xilinx's *EDK*, and Altera's *SOPC Builder*, fill the same role for **field programmable gate arrays** (FPGAs). The generated fabric typically respects an on-chip bus protocol standard. One of the most widely used bus standards is ARM's AMBA/AXI bus. Commercial products are available to generate specialized circuits for on-chip buses conforming to the AMBA bus standards.

5.1.4 Function architecture codesign methodology

A more ambitious form of ESL design methodology is function architecture codesign. As shown in Figure 5.3, this method follows a top-down, stepwise refinement approach in which designers start with design requirements followed by the development of a functional model. As in the case of the function-based ESL method, here the functional model consists of a network of functional components under a specific computational model, thereby capturing the system function as a relation between system output and input. This functional model is gradually refined into an architectural model, which as in the case of architecture-based ESL method, consists of a network of architectural components that communicate at the transaction level. The architectural model can be further refined into RTL by a step of high-level synthesis as described in the next section. Finally, the RTL model can be implemented in silicon by means of *"RTL-handoff"* to the ensuing steps of logic synthesis and physical design.

An important aspect of this method is verifiability. In an ideal function architecture codesign method, a common executable modeling language is used for both functional and architectural modeling. This ensures that at every refinement step, an executable model is created that can be simulated to confirm correctness or to collect performance metrics. Thus, this method is particularly useful for system architects to perform architecture exploration. A pioneering modeling language to support such methodology is SpecC developed at University of California at Irvine [Gajski 2000]. A pioneering commercial environment

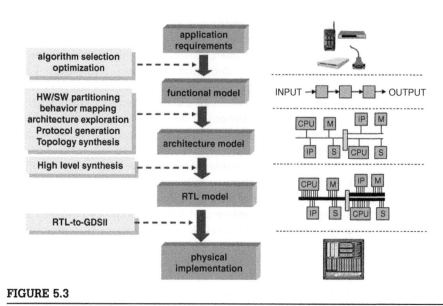

FIGURE 5.3

Function architecture codesign method.

is the VCC tool developed in Cadence [Krolikoski 1999]. Today, SystemC-based environments are widely used for this purpose.

Although top-down function architecture codesign is one of the earliest ESL methods advocated by several researchers, its deployment as a commercially viable method has not yet been realized. The reasons are twofold. First, many *system-on-chips* (SOCs) are typically designed as a platform that can be extended with many derivatives to serve many applications. The idealized top-down approach, which ties the architecture to a particular function or application, makes it difficult to extend the platform for derivatives. In this case, the architecture-based method is more practical. Second, when a system is captured in a domain-specific computational model, it is often the case that the best architecture template is already known, and the architecture instance is best generated by automation rather than through manual refinement. In this case, the functional-based method may be more practical. However, in general, a more practical meet-in-the-middle method (that combines the top-down and bottom-up approaches) may be best suited for a number of applications.

5.1.5 High-level synthesis within an ESL design methodology

We now examine the role of high-level synthesis within an ESL design method. High-level synthesis is an automated method of creating RTL designs from algorithmic descriptions. Within an ESL design method flow, we consider the following usage models of high-level synthesis:

1. Functional component synthesis
2. Co-processor synthesis
3. Application processor synthesis

Functional component synthesis is used in a function-based ESL method. Because the communication semantics are well defined under a computation model, high-level synthesis can be used to create the internal design of each component, while respecting the communication constraints at the inputs and outputs of each component. This enables the construction of larger systems. For example, in a process network–based ESL method, each component has a set of FIFO ports for communication. Therefore, each component can be synthesized to run asynchronously with respect to other components.

Coprocessor synthesis is used in an architecture-based ESL method when part of an application is implemented as software running on a programmable processor and part of the application is implemented as a hardware accelerator. In this case, *hardware/software partitioning*, be it manual or automated, has to be performed to decide on the division of responsibility. Two criteria are typically used to make partitioning decisions: performance/power and flexibility. It is preferable to implement performance/power critical portions of the application in hardware, and it is preferable to implement those that require post silicon changes into software. Once the partitioning is performed, high-level synthesis is used to create the accelerator, and interface synthesis is used to create the software/hardware communications.

Example 5.7　Consider the implementation of an MPEG video encoder. The algorithm divides a video stream into frames, and each frame is divided into slices, where each slice contains a sequence of macro blocks, which are 8×8 pixels. Although most of the algorithm deals with management and configuration, the profiling result shows that much of the program run time is spent on processing each macro block: including stages such as motion estimation (ME), discrete cosine transform, Huffman encoding, and run length encoding. Consequently, a typical hardware/software partitioner would synthesize the processing pipeline (with high-level synthesis) into hardware accelerators, whereas the rest of the design is implemented as software.

In many occasions where performance or power is more important than post silicon programmability, application processor synthesis is used in an architecture-based method to synthesize the entire, stand-alone program into custom hardware. Because in an architecture-based method, component processors communicate through transaction-level ports (such as streaming or bus ports), the mapping from a C programming model to interface hardware is well defined. RTL synthesized by high-level synthesis can be considered as "drop-in" replacement of programmable processors. Large systems can thus be constructed with the well-established IP assembly method.

High-level synthesis has seen intensive research in academia since the 1980s, and many believed that it would succeed RTL synthesis as the dominant design method. Unfortunately, this transition did not happen for a number of reasons. One factor limiting the commercial success of high-level synthesis is its lack of scalability: although large RTL designs can be constructed by composing smaller RTL designs, it is not as easy to compose designs created by high-level synthesis together with legacy RTL designs. Without a well-defined system-level design method, there is no standard way of defining how algorithmic components communicate with each other. Given today's complex heterogeneous systems-on-chip, high-level synthesis in practice has to become a component synthesis tool in the context of an ESL method as opposed to a full chip synthesis method originally envisioned in the early days of HLS research.

Furthermore, both functional and architecture-based ESL methods have matured sufficiently that the standard design flow is converging in the industry. Because the design of new hardware logic is often where it costs chip companies labor and where they achieve product differentiation, it has become increasingly important to add high-level synthesis to the emerging design flow.

5.2 FUNDAMENTALS OF HIGH-LEVEL SYNTHESIS

In a nutshell, **_high-level synthesis_** (HLS) takes as input an algorithmic description (*e.g.,* in C/C++) and generates as output a hardware implementation of a microarchitecture (*e.g.,* in VHDL/Verilog). Algorithmic languages such as C/C++ capture what we refer to as the *behavioral-level* (or high-level) description of the design; whereas hardware description languages such as VHDL/Verilog capture the **_register-transfer level_** (RTL) description of the design.

Figure 5.4a shows a typical example of an input behavioral level description. Here a design is described as a sequence of statements and expressions operating on program variables. Such description captures the function of the design without any hardware implementation detail.

Figure 5.4b shows a typical output of HLS as an RTL description in a form known as **_finite state machine with datapath_** (FSMD) [Gajski 1992]. The FSM controller sequences the design through states of the machine by following the flow of control in the algorithmic behavior, whereas the datapath performs computations on the abstract data types specified in the behavior. The datapath contains a set of registers, functional units, and multiplexers connecting the output of registers to the inputs of functional units, and *vice versa*. The controller takes as input a set of status signals from the datapath and outputs a set of control signals to the datapath. The control signals include those that control the datapath multiplexers, the loading of datapath registers, and the opcode used to select different functions in a functional unit. The state, that is, the control

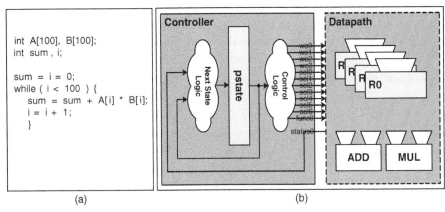

```
int A[100], B[100];
int sum, i;

sum = i = 0;
while ( i < 100 ) {
    sum = sum + A[i] * B[i];
    i = i + 1;
}
```

(a)   (b)

FIGURE 5.4

High-level synthesis input/output.

step a circuit is currently in, is remembered in the controller with a set of flip-flops. Thus, the datapath performs *register transfers*, or computations, that transform values retrieved from some registers and stores the results to other registers; whereas the controller determines "when" certain register transfers are executed by specific valuation of control signals.

HLS effectively transforms an untimed behavioral specification into a clocked RTL design, resulting in a substantial semantic gap between the RTL description and the behavioral description. To bridge this gap, HLS involves a complex sequence of design steps that gradually refines the design behavior into an FSMD design. As shown in Figure 5.5, the structure of an HLS tool naturally resembles a software compiler: it includes a set of *compilation components*. Each component transforms one *program form* to another more detailed program form. We distinguish between two types of program forms: we use *code* to refer to the textual human-readable form, and *representation* to refer to the in-memory machine-readable form (*e.g.*, data structures such as graphs).

The *front-end* component performs the lexical and syntactical analysis of the behavioral program code to build an ***intermediate representation*** (IR) in memory. The IR can be considered as a sequence of operations transforming values retrieved from memory or generated as the result of other operations. The operations typically correspond to arithmetic (addition, subtraction, etc.) or logic computations (AND, OR, etc.).

The *optimizer* component performs program analysis on the IR to extract useful information about the program and transforms the IR to a semantically equivalent but improved IR. Virtually all code optimization algorithms found in software compilers can be applied here, although their effects may differ [Gupta 2003]. For example, common subexpression elimination and dead code elimination can be used to remove redundant code. Tree height reduction, strength reduction, and algebraic

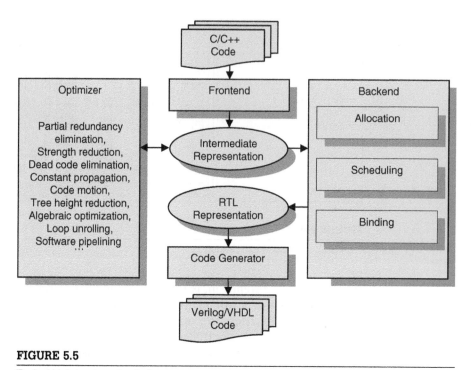

FIGURE 5.5

Typical high-level synthesis flow.

transformations can be used to simplify or speed up the evaluation of expressions. Loop unrolling and software pipelining can significantly improve code in loops.

The core of high-level synthesis resides in the backend, which includes several critical steps to transform the IR into an *RTL representation*. To facilitate the generation of an FSMD in the form of Verilog/VHDL, which performs computation one clock cycle at a time, the RTL representation has to capture more detailed implementation information than initially available in the IR. These critical steps include allocation, scheduling, and binding.

The *allocation* step determines the hardware resources required to implement the operations within the IR to satisfy certain performance requirements. The resources required include storage resources such as registers and functional units such as adders and multipliers. Often, a library of such resources, called modules, is created and characterized in advance for a specific standard cell library. Then allocation involves the choice of which module to use, among potentially many alternatives with differing area and timing, and how many of them.

The *scheduling* step maps each operation in the IR into a particular clock cycle, called a control step. Note that multiple operations can be mapped to the same control step, in which case they run in parallel. In fact, the key objective of scheduling is to find and exploit the parallelism in the sequential code represented by the IR. The extent to which operations can run in parallel is limited by the dependency among them. For example, if operation A uses the result of operation B, then A can only be executed in later steps than B. It is not always easy to exactly determine the dependency relationship, especially when indirect memory references are involved. In addition, the availability of resources (determined in the allocation step) also constrains what can be scheduled in parallel.

The *binding* step maps each operation to a functional unit (functional unit binding) and the value it computes into a storage resource (storage binding). The purpose of binding is to recognize the fact that not all operations and values require the use of hardware resources at the same time. The nonoverlapping use can be exploited for hardware sharing, thereby minimizing the hardware circuitry demand.

It is important to note that each of the allocation, scheduling, and binding steps involves the solution of a complex optimization problem. These problems can all be elegantly formulated as mathematical optimization problems solving for some decision variables to minimize an objective subject to certain constraints. There are many challenges in solving these problems. First, abstracting each problem mathematically inevitably introduces approximations. For example, the objective to optimize area through the minimization of functional units is often an indirect measure of the true objective designers' concern about reducing silicon area, because this includes not only functional units but also the area of interconnects. Second, it has been shown that all allocation, scheduling and binding problems are NP hard problems. Therefore, one often devises heuristics-based algorithms for practical solution of large-scale problems. Third, the allocation, scheduling, and binding tasks are tightly interdependent. In fact, the preceding three-phase approach is not necessarily the best. Often, one faces the *phase-ordering* problem, where the optimal solution of one phase leads to the suboptimal solution of the other phase. It is thus often necessary to have an iterative improvement solution where these tasks are applied multiple times.

Finally, the *code generator* generates the Verilog/VHDL code that is ready for RTL and logic synthesis. The output code is in FSMD form. The datapath can be derived directly from the allocation and binding decisions. The controller FSM can be derived from scheduling and binding.

Without HLS, designers have to manually generate the RTL design, a painstaking process that requires the manual tasks of scheduling, allocation, and binding. In contrast, HLS automates the task of RTL design, allowing designers to more productively focus their design activities at the behavioral level. Often, there is an order of magnitude difference between the size of the RTL description and the corresponding behavioral description. In addition, the automatically generated RTL can be guaranteed to be correct by construction. This design automation process results in a significant reduction of RTL development and verification effort, yielding a large gain in design productivity.

In the sequel, we use an example to illustrate the process of transforming a behavioral code into Verilog code. For now, we treat compilation components as black boxes and leave the discussion until Section 5.3, while focusing on the series of program forms that are generated at each stage. To enhance understanding and allow for a hands-on treatment of HLS that will permit construction of simple HLS tasks/components, we present simple but completed program forms, called TinyC, TinyIR, and TinyRTL; these can be used as vehicles to construct software implementations of high-level synthesis tools. We begin by defining the three exemplar program forms.

5.2.1 TinyC as an example for behavioral descriptions

The behavioral description of a design is typically captured by a program in a procedural programming language, also known as an imperative programming language. An imperative program captures the computation as a sequence of actions on program state. The program state is defined by the set of all program *variables*. The actions are *statements* that change program states by updating the values of one or many variables. Because this style of programming prescribes the steps to solve a problem, an imperative program is often called an *algorithmic description*.

We use an instruction language, TinyC, throughout this text. TinyC resembles the commonly used C language, but a number of simplifying assumptions are made to facilitate the discussion, as follows:

- All statements are captured in a single procedure, and there are no procedural calls.
- Only 32-bit signed integer (int) and Boolean (bool) primitive types are supported.
- Only one-dimensional arrays are supported for aggregate types.
- No pointers are supported, as a result, the address variables cannot be taken, and no memory is allocated dynamically.

Although TinyC is a fully functional language that captures the essence of imperative languages, it is important to recognize that the excluded features present in the real-world languages make the job of high-level synthesis substantially more difficult. We will deal with this subject in Section 5.4.

Example 5.8 Figure 5.4a shows the dot product algorithm in TinyC. Dot product is widely used in engineering. The dot product of two vectors, A and B, is defined to be the sum of products of their respective elements. In TinyC, the two vectors, A and B, are declared as arrays. Two scalar variables, sum and i are declared. A while-loop is used to calculate the result, in which each loop iteration is used to calculate the partial sum up until the ith iteration.

The language definition of TinyC can be defined with a set of production rules in ***Backus-Naur Form*** (BNF). Each production rule defines a language construct, called a non-terminal, by the composition of other non-terminals, or terminals. Like a word in English, terminals are atomic units of a language. The non-terminals can be recursively defined.

Definition 5.1 The syntax of TinyC is defined as follows, which includes the following constructs:

- **Declarations**, *which define scalars or array variables*
- **Statements**, *which are either assignment statements, leading to a change of program state, or control flow statements*
- **Expressions**, *which are either transformations of scalar values using predefined operators, or defined primitive values, such as integer or Boolean literals, or program variable values.*

```
program:                                expression:
    declaration* statement*                 '-' expression , '!' expression ,
                                            expression '+' expression ,
statement:                                  expression '-' expression ,
    variable '=' expression ';',            expression '*' expression ,
    'if' '(' expression ')' statement       expression '/' expression ,
            ( 'else' statement )* ,         expression '^' expression ,
    'while' '(' expression ')' statement ,  expression '>>' expression ,
    'break' ';',                            expression '<<' expression ,
    '{' declaration* statement* '}'         expression '&' expression ,
                                            expression '|' expression ,
declaration:                                expression '=' expression ,
    type  identifier ['=' expression]';',   expression '!=' expression ,
    type  identifier '[' expression ']' ';' expression '<' expression ,
                                            expression '<=' expression ,
type:                                       expression '>' expression ,
    'int' , 'bool'                          expression '>=' expression ,
                                            '(' expression ')' ,
                                            integer , identifier , 'TRUE' , 'FALSE' ,
                                            identifier '[' expression ']'
```

The language can be used to construct a parser, which parses the textual program into a data structure suitable for machine manipulation. In the case of HLS, the parser generates an intermediate representation (as represented by the front-end component in Figure 5.5).

5.2.2 Intermediate representation in TinyIR

The purpose of the intermediate representation (IR) is to separate the optimization algorithms from input languages and target architectures.

Before we discuss the IR in detail, we need a notation to capture an IR. In practice, an IR is implemented in software as a complex data structure. However, presenting the IR in this form introduces unnecessary implementation details. To be concise and precise, we elect to use a more abstract mathematical notation to describe the IR. In the following, we outline how we can replace data structures with mathematical objects. The readers should do the opposite when they translate the algorithms presented in this chapter into software implementation.

- A data type T corresponds to a **set** T; in particular the integer type Z corresponds to set Z.
- A linked list or arrays whose elements are of type T corresponds to the **power set** of T, or the set of all subsets of T, denoted as $T[]$.
- A record with fields a of type A, and b of type B corresponds to a set of **named tuples**, denoted as $\langle a : A, b : B \rangle$.
- A graph R whose nodes are of type A corresponds to a **relation** $R : A \times A$.
- A hash table or dictionary F that maps a value of type A to a value of type B corresponds to a **function** $F : A \mapsto B$.

This mathematical notation is used later in Section 5.3 to present HLS algorithms, because we can use it to represent complex operations on data structures. For example,

- Use function application $F(a)$ to represent a dictionary lookup.
- Use $a \in A$ to perform a set membership test, and use $\forall a \in A$ to enumerate all members in A; and use $A[i]$ to retrieve the ith element in A.
- Use $a\ R\ b$ to check whether b is a predecessor of a in graph (relation) R.

We are now ready to discuss TinyIR, a simple IR designed to sufficiently capture programs in TinyC.

Definition 5.2 A TinyIR is a tuple $\langle O, S, V, B \rangle$ with the following elements:

- *A set $O = \{lds, sts, lda, sta, ba, br, cnst, +, -, *, /, <<, >>, \ldots \}$ of operation codes, which corresponds to the set of all virtual instruction types.*
- *A set S of **symbols**, which corresponds to the scalar and array variables.*
- *A set V: $\langle opcode$: O, src1: V, src2: V, symb: $S \cup B \cup Z \rangle$ of **virtual instructions**, which corresponds to the expressions and control transfers in the program.*
- *A set B: $V[]$ of **basic blocks**, each containing a sequence of virtual instructions.*

Constructs in TinyC (*e.g.*, declarations, statements, and expressions) have equivalent representation in TinyIR: declarations correspond to symbols, and statements and expressions correspond to virtual instructions. In particular, the control transfer statements are converted into *ba* (branch always) or *br* (branch if true) instructions. Variable accesses are converted to *lds* (load scalar) or *lda* (load array) instructions, and assignments are converted into *sts* (store scalar) or *sta* (store array) instructions. Expressions in TinyC have a one-to-one correspondence to virtual instructions in TinyIR.

Each instruction is a tuple with an *opcode* field; zero, one, or two source operands; and optionally a *symb* field. For branch instructions, the *symb* field defines its branching target, that is, which basic block it branches to. For load/store instructions, the *symb* field defines the symbol corresponding to the scalar or array variable it accesses. For constants, it defines the Boolean or integer constant value.

Although the sequence of virtual instructions completely defines the behavior of the program, in TinyIR the virtual instructions are grouped within different basic blocks. Only the last instruction of a basic block could be a branch instruction. In the other words, the instructions in a basic block are what we usually referred to as "straight line code."

Example 5.9 In the following we show the dot product program in Example 5.8 in TinyIR form. There are two basic blocks, B1 and B2. Each virtual instruction is uniquely numbered. When the result of an instruction is used as an operand for another instruction, its number is used. Note that B2 is a loop, indicated by the bt instruction (15), which is a branch branching to the beginning of B2.

scalar sum;	(4) lda (3), A
scalar i;	(5) lda (3), B
array A[100];	(6) * (4) (5)
array B[100];	(7) lds sum
	(8) + (6) (7)
B1:	(9) sts (8), sum
(0) cnst 0	(10) cnst 1
(1) sts (0), sum	(11) + (3) (10)
(2) sts (0), i	(12) sts (11), i
	(13) cnst 100
B2:	(14) < (11) (13)
(3) lds i	(15) bt (14), B2

5.2.3 **RTL representation in TinyRTL**

An RTL representation is to HLS what assembly is to a software compiler. Like assembly, the RTL representation captures key microarchitectural information. In the case of a software compiler, the processor architecture is predetermined; therefore, the assembly only exposes these microarchitecture features, for example, the architectural registers available, the instruction set, etc. In the case of HLS, the microarchitecture is synthesized; therefore, the RTL representation has to convey what architecture resources are allocated and how they are used.

One key difference between an RTL representation and an IR is that microarchitecture resources are introduced. There are two types of microarchitectural resources: **computational resources**, which are functional units that perform logical, relational, and/or arithmetic functions; and **storage resources**, which include memories and registers. A memory typically corresponds to a on-chip static RAM used to store scalar or array variables, whereas a register is an array of flip-flops used to store scalar or temporary values.

An instruction in RTL representation is referred to as a *register transfer*. A key difference between a register transfer and a virtual instruction in an IR is that the former is annotated with microarchitecture resource use. For example, most register transfers designate a destination register. Likewise, the source operands of a register transfer are registers. In addition, each register transfer designates the functional unit executing the instruction.

The register transfer representation is **cycle accurate** in the sense that the clock cycle (control step) at which a register transfer is executed is fully specified. This level of detail makes it possible to generate a sequential circuit implementation of the program.

Definition 5.3 A TinyRTL is a tuple $\langle M, R, U, I, C \rangle$ with the following elements:

- A set M of **memories** used to store scalar and array variables.
- A set R of **registers** used to store scalar variables or temporary instruction results.
- A set U of **functional units**, such as adders, subtractors, multipliers, shifters, etc.
- A set $I : \langle unit : U, opcode : O, dest : R, src1 : R \cup S \cup Z \cup C, src2 : R \rangle$ of **register transfers**, each of which uses a functional unit to transform values, which are either constants or retrieved from registers, and stores the result back to a register.
- A set $C: I[]$ of **control steps**, each of which contains a set of register transfers.

Example 5.10 The dot product algorithm of Example 5.8 is shown in TinyRTL form. Here C0-C4 corresponds to the set of control steps. Each control step contains one or more register transfers. It is instructive to compare the TinyIR form and TinyRTL form. Note that most virtual instructions in TinyIR are translated into register transfers. Some virtual instructions (e.g., constants) degenerate into direct operands, because it takes nothing to compute them. Also note that almost all register transfers are annotated with the computational resources they use, with the exception of scalar store (in C0) and branch instructions (in C4), because they involve simply copying values to registers but not computation.

```
register R0, R1, R2, R3;
memory M;
unit U0, U1;

C0: sts 0, R0; sts 0, R1;
C1: M.lda R2, R1, A;
C2: M.lda R3, R1, B; U0. + R1, R1, 1;
C3: U1. * R2, R2, R3; U0. < R3, R1, 100;
C4: U0. + R0, R0, R2; bt C1, R3;
```

5.2.4 Structured hardware description in FSMD

Given an RTL representation in memory, we are ready to produce Verilog/VHDL code to drive the downstream RTL-to-GDSII design flow. The key task of the code generator is thus to convert the RTL representation, still in a functional form, to a structural form, that is, as a connected network of components. We use FSMD as a template and generate the controller and datapath separately.

Example 5.11 The FSM diagram and Verilog code for the controller are shown in the following. Note the correspondence between the control steps in the RTL representation and the states in the FSM. Also note the correspondence between the register transfers in each control step and the control signal valuations.

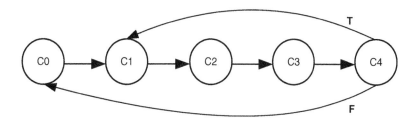

```
`define C0  3'b000
`define C1  3'b001
`define C2  3'b010
`define C3  3'b011
`define C4  3'b100

module ctrl(
  clk, rst, status0,
  we0, we1, we2, we3,
  sel0, sel1, sel2, sel3, sel4, sel5, sel6,
  func0
  );

  input      clk, rst;
  input      status0;
  output     we0, we1, we2, we3;
  output     sel0, sel1, sel2, sel3, sel4, sel5;
  output [1:0] sel6;
  output     func0;

  reg  [2:0] pstate, nstate;
  reg        we0, we1, we2, we3;
  reg        sel0, sel1, sel2, sel3, sel4, sel5;
  reg  [1:0] sel6;
  reg        func0;

  // present state register
  always@( posedge clk or negedge rst )
    if( !rst )
      pstate <= `C0;
    else
      pstate <= nstate;

  // next state logic
  always@( pstate or status0 )
    case( pstate )
      `C0: nstate = `C1;
      `C1: nstate = `C2;
      `C2: nstate = `C3;
      `C3: nstate = `C4;
      `C4: if( status0 )
              nstate = `C1;
           else
              nstate = `C0;
      default: nstate = `C0;
    endcase
```

```
// control signals
always@( pstate ) begin
  we0 = 1'b0; we1 = 1'b0;
  we2 = 1'b0; we3 = 1'b0;
  sel0 = 1'bx; sel1 = 1'bx;
  sel2 = 1'bx; sel3 = 1'bx;
  sel4 = 1'bx; sel5 = 1'bx;
  sel6 = 2'bxx; func0 = 1'bx;
  case( pstate )
    `C0:begin
        we0 = 1'b1; we1 = 1'b1; sel0 = 1'b0;
        sel1 = 1'b0;
      end
    `C1:begin
        we2 = 1'b1; sel2 = 1'b0; sel4 = 1'b0;
      end
    `C2:begin
        we1 = 1'b1; we3 = 1'b1; sel1 = 1'b1;
        sel3 = 1'b0; sel4 = 1'b1; sel5 = 1'b0;
        sel6 = 2'b00; func0 = 1'b0;
      end
    `C3:begin
        we2 = 1'b1; we3 = 1'b1; sel2 = 1'b1;
        sel3 = 1'b1; sel5 = 1'b0; sel6 = 2'b01;
        func0 = 1'b1;
      end
    `C4:begin
        we0 = 1'b1; sel0 = 1'b1; sel5 = 1'b1;
        sel6 = 2'b10; func0=1'b0;
      end
  endcase
end

endmodule
```

Example 5.12 The datapath of the dot product example is shown in the following as a schematic diagram and its corresponding Verilog code. Note that each r ∈ R *and* u ∈ U is mapped directly to a hardware resource. Multiplexers are inserted at the input of each register and inputs of each functional unit. A multiplexer degenerates into a wire if it has only one input. For example, both inputs of the multiplier connect only to R2 and R3, respectively; therefore, there is no need for multiplexers.

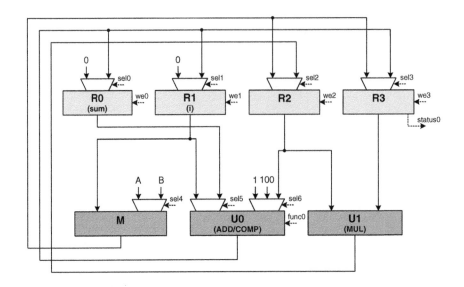

```
`define A 8'd0
`define B 8'd100

module datapath(
clk, rst,
we0, we1, we2, we3,
sel0, sel1, sel2, sel3, sel4, sel5, sel6,
func0, rdata,
status0, base, offs
);

    input       clk, rst;
    input       we0, we1, we2, we3;
    input       sel0, sel1, sel2, sel3, sel4, sel5;
    input [1:0] sel6;
    input       func0;
    input [7:0] rdata;
    output      status0;
    output [7:0] base, offs;

    reg  [7:0] R0, R1, R2, R3;
    reg  [7:0] nR0, nR1, nR2, nR3;
    wire [7:0] U0, U1;
    reg  [7:0] baseIn, offsIn;
    reg  [7:0] U1In0, U1In1;
    reg  [7:0] U0In0, U0In1;

    // registers
    always@( posedge clk or negedge rst )
      if( !rst ) begin
        R0 <= 8'b0;
        R1 <= 8'b0;
        R2 <= 8'b0;
        R3 <= 8'b0;
      end
      else begin
        if( we0 ) R0 <= nR0;
        if( we1 ) R1 <= nR1;
        if( we2 ) R2 <= nR2;
        if( we3 ) R3 <= nR3;
      end

    // registers' input multiplexers
    always@( sel0 or sel1 or sel2 or sel3 or
             U0 or U1 or rdata ) begin
      case( sel0 )
        1'b0: nR0 = 8'b0;
        1'b1: nR0 = U0;
      endcase
```

```
      case( sel1 )
        1'b0: nR1 = 8'b0;
        1'b1: nR1 = U0;
      endcase
      case( sel2 )
        1'b0: nR2 = rdata;
        1'b1: nR2 = U1;
      endcase
      case( sel3 )
        1'b0: nR3 = rdata;
        1'b1: nR3 = U0;
      endcase
    end

    // functional units
    assign U1 = U1In0 * U1In1;
    assign U0 = !func0 ? U0In0 + U0In1 :
                         U0In0 < U0In1;

    // functional units'/memory input multiplexers
    always@( sel4 or sel5 or sel6 or
             R0 or R1 or R2 or R3 ) begin
      baseIn = R1;
      case( sel4 )
        1'b0: offsIn = `A;
        1'b1: offsIn = `B;
      endcase
      case( sel5 )
        1'b0: U0In0 = R1;
        1'b1: U0In0 = R0;
      endcase
      case( sel6 )
        2'b00: U0In0 = 8'd1;
        2'b01: U0In0 = 8'd100;
        2'b10: U0In0 = R2;
      endcase
      U1In0 = R2;
      U1In1 = R3;
    end

    // outputs
    assign status0 = R3[0];
    assign base = baseIn;
    assign offs = offsIn;

endmodule
```

5.2.5 **Quality metrics**

When using high-level synthesis to create RTL hardware for a given application, it is important to, first, satisfy the performance requirements of the application and, second, choose among the best implementation alternatives. To quantify both requirements and quality, we need to establish certain metrics.

Because most applications targeted by high-level synthesis consume and produce large volumes of data, the performance requirement is often dictated by input/output *bandwidth*, defined to be the amount of data consumed/produced per second. The unit of bandwidth varies, depending on the domain of application. For example, triangles per second for 3-D graphics; packets per second for networking; and pixels per second for imaging or video.

The bandwidth requirement can be translated into a performance requirement for the datapath. Typically, the *work*, or the amount of computation an algorithm applies per unit of data, can be measured by the number of instructions required for the computation. Although the precise meaning of an instruction varies depending on the architecture used for implementation, we can use elementary operations, such as those defined by the virtual instruction set of TinyIR, as a rough but implementation-independent measure. Combining bandwidth and work, we can obtain the performance requirement for the datapath in ***millions of instructions per second*** (MIPS).

$$perf(TinyIR) = bandwidth \times work$$

Example 5.13 Consider an Ethernet application in which the wire speed is 4 gigabits per second (raw bandwidth). Assuming a minimum packet size of 64 bytes and a 20-byte preamble between packets, the application needs to process 5.95 million packets per second (bandwidth). Assuming the work is 1000 instructions per packet, then the performance requirement is 5.95 * 1000 = 5950 MIPS.

We now turn to quality metrics of the RTL implementation. Although it is possible to evaluate the metrics accurately after the Verilog/VHDL is generated and synthesized, these metrics are often too late to be useful for HLS at these downstream stages. We, therefore, need an early, fast, yet reasonably accurate estimation of important metrics. In this section, we develop a "back-of-the-envelope" method that can quickly assess quality metrics directly from the RTL representation.

To calculate the performance of the RTL, we need to estimate the "wall clock" time used for completion of the synthesized algorithm. Assuming the algorithm processes one unit of data, we then have:

$$perf(TinyRTL) = \frac{work}{CycleCount(TinyRTL) * CycleTime(TinyRTL)}$$

Here *CycleCount* is the number of clock cycles it takes to complete the algorithm, whereas *CycleTime* is the shortest clock period for correct operation of the synthesized circuit.

To estimate *CycleCount(TinyRTL)*, we need to know the number of times each control step is executed. This can be obtained by statically examining the RTL. Alternately, functional simulation can be performed to collect execution count statistics, a process known as *profiling*.

Example 5.14 Consider the dot product algorithm in TinyRTL shown in Example 5.12. It can be shown that C0 will be executed once, whereas C1-C4 will be executed 100 times. Therefore CycleCount(TinyRTL) = 1 + 4 * 100 = 401. Note that this is significantly less than the total number of virtual instructions, which can be calculated from TinyIR as 3 + 13 * 100 = 1303. This type of speedup is achieved by executing multiple instructions in parallel in RTL.

CycleTime(RTL) is more difficult to estimate. Recall that the cycle time of a sequential circuit equals the worst-case delay along all register-to-register paths. To calculate such a delay, we first have to establish the delay of individual components along a path.

The delay of a component (in nanoseconds or picoseconds)—if not already available—can be obtained by precharacterizing commonly used components. However, such a method depends on the cell library, as well as the fabrication process. We choose to use fanout four delay (FO4) as the delay unit. FO4 is the delay of a minimal sized inverter driving four identical copies of itself. It has been shown that FO4 delay scales linearly with the feature size (the drawn gate length) of the fabrication process and can be estimated with the following:

$$FO4 = 0.36ns/um * L_{drawn}$$

With FO4 delay, we can express our delay estimation in a process-independent manner and use the preceding formula when we need to find out the absolute delay value.

Example 5.15 The FO4 delay of commonly used 32-bit components is shown in Table 5.1. Under 90nm fabrication technology, FO4 = 0.36ns/um*0.09um = 32.4ps. Therefore, Delay (adder) = 10*32.4ps = 0.324ns. Delay(multiplier) = 35*32.4ps = 1.13ns.

Table 5.1 Component delays

register	FO4	functional unit	FO4	multiplexer	FO4	SRAM	FO4
setup	2.0	adder	10.0	2-input	2.4	1KB	10.5
hold	0.0	comparator	6.0	4-input	3.2	4KB	12.0
clock skew/jitter	4.0	multiplier	35.0	8-input	4.8	8KB	12.8
clock to Q	4.0	inverter	1.0	16-input	8.0	64KB	15.0

Without examining the structural representation, we attempt to estimate the cycle time from TinyRTL representation. We can find all *true* register-to-register paths by examining each register transfer in TinyRTL.

$$CycleTime(TinyRTL) = MAX_{rt \in T} Delay(rt)$$

We now consider the delay of a register transfer $rt = \langle u, op, dest, src1, src2 \rangle$. Let us denote the state register of controller to be *pstate* and the multiplexers for the two operands of the functional unit and the destination register to be *mux1*, *mux2*, and *muxd*, respectively. Recall that in an FSMD model, the register-to-register delays involve paths within the datapath, to the FSM controller, and between the datapath and the FSM controller. Accordingly, we have the following paths to consider:

- Path 1: src1→mux1→u→muxd→dest
- Path 2: src2→mux2→u→muxd→dest
- Path 3: pstate→c1→mux1→u→muxd→dest
- Path 4: pstate→c2→mux2→u→muxd→dest
- Path 5: pstate→cu→u→muxd→dest
- Path 6: pstate→cd→muxd→dest

Paths 1 and 2 are from source operand to destination register in the datapath. Paths 3 and 4 are from state register (FSM) to destination register (datapath) through the source operand multiplexer. Here *c1* and *c2* correspond to control logic for the select signals of the respective multiplexers. Path 5 is from state register (FSM) to destination register (datapath) through the control logic *cu* used for selecting the function used in the functional unit. Path 6 is from state register (FSM) to destination register (datapath) through control logic *cd* used to select the destination operand in multiplexer. Note that not all paths are realizable. For example, *mux1* might not exist because there is no need for a multiplexer before the unit's only input. This happens in the case that only one register is ever used as its first operand. In this case, the delay of *mux1* should be taken as 0 in Path 1, and Path 3 should be ignored. As another example, if a unit performs a single function, the controller does not have to generate a control signal for selecting the function to perform; therefore, Path 5 can be ignored.

For every path, time has to be reserved for the correct functioning of the registers. Their sum is called the *sequential overhead*.

$$SeqOverhead = T_{setup} + T_{CQ} + T_{skew}$$

We can, therefore, calculate the worst case delay of all paths in a register transfer as follows:

$$Delay(rt) = SeqOverhead +$$
$$Max \left[Max \begin{bmatrix} Delay(c1) + Delay(mux1), \\ Delay(c2) + Delay(mux2), \\ Delay(cu) \end{bmatrix} + Delay(u), \\ Delay(cd) \\ Delay(muxd) \right] +$$

Although we can determine the delay of the functional unit by looking up Table 5.1, delays of the multiplexers and the control logic are not readily available. There are two complications. First, as discussed earlier, there are degenerate cases where *mux1*, *mux2*, *muxd*, *c1*, *c2*, *cd*, and *cu* are not needed. In these cases, they assume a zero delay value. Second, when the multiplexers do exist, their delays depend on the number of inputs they have. This requires us to find, for each functional unit, the total number of different operands for each input and the total number of different opcodes by scanning the entire RTL representation.

Assuming the delay of all control signals assume a value of T_c, we then have:

$$Delay(cu) = \begin{cases} 0 & |Opcodes(u)| = 1 \\ T_c & |Opcodes(u)| > 1 \end{cases}$$

where

$$Opcodes(u) = \{rt.opcode|\ \forall rt \in T.[rt.u = u]\}$$

Similarly, we have

$$Delay(c1) = \begin{cases} 0 & |Src1(u)| = 1 \\ T_c & |Src1(u)| > 1 \end{cases}$$

$$Delay(c2) = \begin{cases} 0 & |Src2(u)| = 1 \\ T_c & |Src2(u)| > 1 \end{cases}$$

$$Delay(cd) = \begin{cases} 0 & |Srcd(dest)| = 1 \\ T_c & |Srcd(dest)| > 1 \end{cases}$$

where

$$Src1(u) = \{rt.src1|\forall rt \in T.[rt.u = u]\}$$
$$Src2(u) = \{rt.src2|\forall rt \in T.[rt.u = u]\}$$
$$Srcd(dest) = \{rt.u|\forall rt \in T.[rt.dest = dest]\}.$$

Because the numbers of inputs of *mux1*, *mux2*, and *muxd*, are given by $|Src1(u)|$, $|Src2(u)|$, and $|Srcd(dest)|$ respectively, we can determine their delays by looking up the multiplexer delays, which are listed according to the input count in Table 5.1.

Example 5.16 To estimate the cycle time of the dot product example, we assume $T_c = 5$ FO4. With the method developed previously, we can find the delay of each register transfer and determine that the cycle time of the RTL is 47.4 FO4. In 90nm technology, this is equivalent to 32.4ps*47.4 = 1.54ns. In other words, the maximum speed at which the circuit can run is 649Mhz. Recall that the cycle count is 401, and the work (number of virtual instructions) is 1303, we can conclude that

$$Perf(TinyRTL) = \frac{1303}{401 * 1.54} = 2108 MIPS$$

Register transfer	Seq overhead	u	muxd	cd	cu	c1	c2	mux1	mux2	Delay
sts 0, R0	10.0	0.0	2.4	5.0	0.0	0.0	0.0	0.0	0.0	17.4
sts 0, R1	10.0	0.0	2.4	5.0	0.0	0.0	0.0	0.0	0.0	17.4
M.lda R2, R1, A	10.0	10.5	2.4	5.0	0.0	0.0	5.0	0.0	2.4	30.3
M.lda R3, R1, B	10.0	10.5	2.4	5.0	0.0	0.0	5.0	0.0	2.4	30.3
U0.+ R1, R1, 1	10.0	10.0	2.4	5.0	5.0	5.0	5.0	2.4	3.2	33.2
U1.* R2, R2, R3	10.0	35.0	2.4	5.0	0.0	0.0	0.0	0.0	0.0	47.4
U0.< R3, R1, 100	10.0	10.0	2.4	5.0	5.0	5.0	5.0	2.4	3.2	33.2
U0.+ R0, R0, R2	10.0	10.0	2.4	5.0	5.0	5.0	5.0	2.4	3.2	33.2

Of course, other quality metrics such as silicon area and power consumption are also very important. Similar procedures can be developed to estimate these metrics directly from RTL representation and currently are active areas of research.

Walking through the dot product example reveals that to transform an application in behavioral (C/C++) form to RTL (Verilog) form, several key decisions have to be made to convert the application into an IR form and eventually into RTL form. Performance analysis gives us further insight that these decisions impact the performance the RTL implementation can achieve, both in terms of cycle count, and cycle time, in a non-trivial manner. In the next section, we develop CAD algorithms that allow these key decisions to be made to optimize the various quality metrics.

5.3 HIGH-LEVEL SYNTHESIS ALGORITHM OVERVIEW

In the previous section, we outlined the steps required to transform a behavioral description of the dot product algorithm in TinyC into its RTL implementation in Verilog. Although we defined the intermediate program forms, we did not describe how the input form is transformed into the output form. For the frontend and optimizer components, the techniques used are largely no different from those used by software compilers. We refer the readers to [Aho 2006] for a detailed treatment. For the code generator component, it is relatively simple to output the HDL code from the RTL representation. Thus in this section we focus on the backend component containing the core synthesis algorithms that take as input an IR in terms of virtual instructions, and generate as output an RTL representation in terms of register transfers.

As discussed in Section 5.2.5, we are not only interested in generating the correct RTL that is functionally equivalent to the IR, but also with the best *quality of result* (QoR). Therefore, the synthesis problem can be formulated as an optimization problem targeting multiple objectives such as performance, area, and power. For example, the backend synthesis problem can be formulated as the following.

PROBLEM 5.1

Given: $TinyIR = \langle O, S, V, B \rangle$
Find: $TinyRTL = \langle M, R, U, I, C \rangle$
Maximize: $Perf(TinyRTL)$
Minimize: $Area(TinyRTL)$

Recall that because this is a complex, phase-coupled optimization, it is not obvious how this problem can be solved. As discussed in Section 5.2, a divide-and-conquer strategy is usually followed, and high-level synthesis is further decomposed into allocation, scheduling, and binding problems, and solved in sequence. To simplify the presentation, in this section we make further simplifications:

- Assumption 1: The functional unit allocation (*i.e.*, the hardware resources used in the implementation), is specified by the user as a constraint. The rationale behind this assumption is that the user could try out different hardware resource allocations, and let the automated synthesis tool generate solutions for comparison, a process known as *design exploration*. We further assume that each allocated unit can implement all virtual instructions.
- Assumption 2: Storage allocation and binding is performed partially by assuming all array variables are mapped to a single memory, and all scalar variables are mapped to separate registers. Therefore, only the temporary values produced by virtual instructions need to be mapped to registers.
- Assumption 3: Other than the constant instruction, each virtual instruction is implemented by one and only one register transfer. This does not have to be the case in a production synthesis tool, because a subset of virtual instructions can be grouped together and implemented by a single register transfer, a process known as *chaining*.
- Assumption 4: Each register transfer can be completed in a single clock cycle without degrading cycle time. In practice, this is not true because a virtual instruction could be implemented by a functional unit with longer delay than the desired cycle time, a process known as *multicycling*.

We can then refine Problem 5.1 as follows.

PROBLEM 5.2

Given: $\text{TinyIR} = \langle O, S, V, B \rangle$

Find:
(1) Schedule $Sched: B \mapsto (V \mapsto Z)$
(2) Register binding $B^R: V \mapsto Z$
(3) Functional unit binding $B^U: V \mapsto U$

Minimize:
Objective (1) $\forall b \in B, |range\ Sched(b)|$
Objective (2) $|range\ B^R|$
Objective (3) $\Sigma_{u \in U}|Src1(u)| + \Sigma_{u \in U}|Src2(u)| + \Sigma_{r \in R}|Srcd(r)|$

Subject to: Constraint $\forall b \in B, \forall s \in Z, |Sched(b)^{-1}(s)| \leq |U|$

Here we attempt to make three key decisions. For each basic block, the schedule *Sched* maps each instruction contained in the basic block to a control step. The register binding B^R maps the value computed by each instruction to a register. The functional unit binding B^U maps each instruction to a functional unit. The decisions have to satisfy the resource constraint; in other words, the number of instructions scheduled at each control step cannot exceed the number of available functional units. Combined with the simplifying assumptions, it is trivial to find the TinyIR representation $\langle M,R,U,I,C \rangle$ from *Sched*, B^R, and B^U.

We now relate the objectives in Problem 5.1 to the objectives in Problem 5.2. To maximize performance, Problem 5.2 states that it is equivalent to minimizing the number of control steps in each basic block (objective 1). Recall performance is the product of cycle count and maximum clock frequency. Although cycle count is not the same as control step count (because a control step could be executed many times), minimizing the latter does minimize the former. Here we assumed that maximum clock frequency is independent of scheduling and binding, which does not hold in general. To minimize area, Problem 5.2 states that it is equivalent to minimizing the register count (objective 2), and minimizing the total number of multiplexer inputs (objective 3). This makes sense because the functional units, memory, and registers for scalar variables are pre-allocated and therefore fixed. Here we have assumed that the area of the synthesized circuit is dominated by the datapath (controller area is therefore ignored), and areas of multiplexers are proportional to the input count.

The next two sections describe the scheduling and register allocation steps.

5.4 **SCHEDULING**

5.4.1 **Dependency test**

Because the objective of scheduling is to minimize the total number of control steps (*i.e.*, maximize performance), we wish to schedule as many instructions in the same step as possible, thereby executing all of them in parallel to maximize design performance. However, this is not always possible. For an RTL implementation to preserve the semantics of the original algorithm, *data dependencies have* to be respected. We illustrate the notion of data dependency below.

Consider the following scenarios in a basic block, where virtual instruction *A* precedes virtual instruction *B*:

1. Instruction *A* is the operand of instruction *B*, in other words, instruction *A* produces a value consumed by instruction *B*;
2. Instruction *A* stores a value to symbol *x*, whereas instruction *B* loads a value from symbol *y*;
3. Instruction *A* loads a value to symbol *x*, whereas instruction *B* stores a value to symbol *y*;
4. Instruction *A* stores a value to symbol *x*, whereas instruction *B* stores a value to symbol *y*.

For scenario 1, there is a *definite data dependency* between *A* and *B*: *B* has to be scheduled at least one step later than *A*, because the value of *A* has to be produced first before it can be consumed. This relation is explicitly represented in, and easily extracted from, the IR.

For each of the scenarios 2, 3, and 4, there is a *potential data dependency* between A and B: as soon as one can determine that symbols x and y are the same (*i.e.*, they are *aliased* to each other), then B has to be scheduled at least one step later than A. This relation is implicitly induced by the runtime value of memory addresses, and thus not easy to extract from the IR.

A dependency tester is an algorithm that statically determines whether two instructions depend on each other. In TinyRTL, all scalar symbols are explicitly named, in other words, symbols x and y either have the same name, or they are not aliased to each other. It is therefore straightforward to compute the data dependency induced by scalar variables by comparing symbol names. Dependency through indexed accesses to arrays is more difficult to detect. Given array access $x[i]$, and array access $x[j]$, the dependency test amounts to determining whether values i and j can be equal to each other at runtime. The supercomputing research community has developed comprehensive methods for array-based dependency tests [Aho 2006]. A simple dependency tester for TinyRTL can be constructed by simply comparing symbol names, even for array accesses (in other words, conservatively assume that all indices are potentially equal).

In a real-world language (*e.g.*, C/C++), anonymous symbols exist through the use of pointer dereferences. Pointers can be created either by taking the address of a named symbol, or by dynamic memory allocation. Because pointers can be copied, manipulated, and stored as any other value, two pointers in a program can assume the same value at runtime, in which case the corresponding pointer dereferences become aliases. Computing runtime pointer values statically, known as *pointer analysis*, or statically detecting if two pointer dereferences alias, known as *alias analysis*, are both undecidable. Many pointer/alias analysis algorithms have been developed, with varying precision and scalability. The simplest pointer analysis algorithm collects the set of all symbols in the program whose addresses have been taken, as well as the set of all dynamic memory allocation sites, and assumes all pointers in the program can carry one of those values.

To facilitate scheduling, a precedence graph is first constructed to capture the dependency relation among the instructions in a basic block. The precedence graph is named so because it captures the partial order of instructions.

Definition 5.4 *A precedence graph $\langle E,s,t \rangle$ is a polar graph where $E \subseteq V \times V$ is the set of edges, and $s,t \in V$ is the source and sink node, respectively.*

The precedence graph is sometimes also referred to as the *dataflow graph* in the literature. A minor difference is that the dataflow graph captures the data dependency of all instructions in a procedure, whereas the precedence graph is its subgraph for instructions within a basic block. In particular, the source and sink nodes are introduced to lump all instructions defined outside the basic block under consideration. All instructions outside the basic block that are depended on by the instructions in the basic block are lumped into the source node. All instructions outside the basic block that depend on the instructions in the basic block are lumped into the sink node.

Example 5.17 Consider the TinyC code fragment in Figure 5.6a. The corresponding TinyIR is shown in Figure 5.6b. A simple dependency test algorithm can establish the dependency relation by examining the chain of operations in each instruction. For example, consider instruction (28) in basic block B4, whose chain of operands is shown in Figure 5.7a. It can be inferred that it depends on instruction (26), which in turn depends on instructions (24) and (25), and so on. This process can be repeated, which yields the precedence graph for basic block B4, as graphically depicted in Figure 5.7b. Note that instructions (4) and (6) are defined outside B4, and they are lumped together as source node s. Likewise instructions (27) and (28) are also defined outside B4, and they are lumped together as sink node t. Thus, we have the following edges defined $E = \{\langle s,15\rangle, \langle s,16\rangle, \langle s,24\rangle, \langle s,25\rangle, \langle 15,17\rangle, \langle 16,17\rangle, \langle 24,26\rangle, \langle 25,26\rangle, \langle 17,18\rangle, \langle 18,20\rangle, \langle 20,t\rangle, \langle 26,t\rangle\}$.

```
int a, b, c, d;                        scalar c;
        .                              scalar d;
        .
        .                              B3:
c = ...;                                   ...
d = ...;                                   (4)  lds a
                                           (6)  lds b
if( ... ) {                                ...
    c = (((a + b) * (a - b)) * 13) + 16;   (13) bt ..., B5
    d = (a + 12) * (a * 12);           B4:
}                                          (14) cnst 13
                                           (15) + (4) (6)
... = c + d;                               (16) - (4) (6)
                                           (17) * (15) (16)
        .                                  (18) * (14) (17)
        .                                  (19) cnst 16
        .                                  (20) + (18) (19)
                                           (23) cnst 12
                                           (24) + (4) (23)
                                           (25) * (23) (4)
                                           (26) * (24) (25)
                                           (27) sts (20), c
                                           (28) sts (26), d
                                       B5:
                                           (30) lds c
                                           (31) lds d
                                           (32) + (30) (31)
                                           ...
            (a)                                    (b)
```

FIGURE 5.6

(a) TinyC code. (b) TinyIR representation.

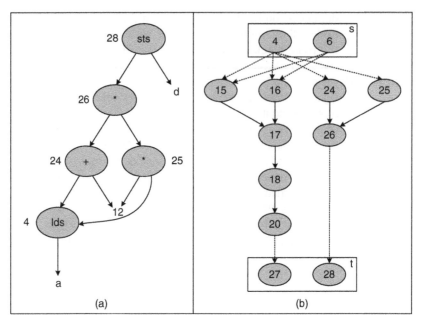

FIGURE 5.7

(a) Chain of instructions. (b) Precedence graph of basic block B4.

We are now ready to examine some commonly used scheduling formulations.

5.4.2 Unconstrained scheduling

We first consider the simple case where the allocation constraints are ignored. In other words, we assume an unlimited number of functional units are available. The unconstrained scheduling problem for a basic block can then be formulated as follows.

PROBLEM 5.3

Given: *Precedence graph $\langle E, s, t \rangle$*
Find: $S: V \mapsto Z$
Minimize: $S(t) - S(s)$
Subject to: $\forall \langle u, v \rangle \in E, S(v) - S(u) > 0$

Assuming the control steps are sequentially numbered, then the total number of steps is defined as $S(t) - S(s)$, which becomes the objective to minimize. To respect data dependency, for every edge $\langle u, v \rangle$, the schedule of the sink, $S(v)$, has to be "later" than the schedule of the source, $S(u)$.

To solve Problem 5.3, one can use an iterative approach. In each iteration, a set of nodes (instructions) in the precedence graph is scheduled to a control step.

The set of nodes that can be scheduled, or are "ready," should be those whose predecessors are all scheduled; otherwise, the dependency constraint would be violated. In addition, we will schedule all nodes as soon as they are ready. This strategy is referred to as **as-soon-as-possible** (ASAP) scheduling.

To implement ASAP scheduling, one can maintain a set *Ready*, representing the set of nodes ready to be scheduled in the current control step. Initially, *Ready* contains a single element, *s*. In each iteration, one chooses a node *v* from *Ready*, and assigns its schedule as the current control step. In addition, it needs to examine each successor *w* of *v*, and add it to *NextReady*, if it becomes ready because *v* is scheduled. To judge if *w* becomes ready, one could check if all its predecessors have been scheduled. This results in an algorithm with a complexity of $O(|V| + |E|^2)$.

A better approach is shown in Algorithm 5.1. The key insight is that one only needs to maintain the number of unscheduled predecessors for each node, called *counter* in line 4 of Algorithm 5.1. It is initialized to be the number of incoming edges for each node in line 7-8. When node *v* is scheduled, this number is decremented for each of its successors *w*. When this number becomes 0, node *w* becomes ready (lines 14-16). Algorithm 5.1, therefore, has a complexity of $O(|V| + |E|)$.

Algorithm 5.1 ASAP Scheduling

```
algorithm asapSched(E : (V × V)[ ], s : V, t : V) returns V ↦ Z
1. var      S : V ↦ Z;
2. var      Ready, NextReady : V[ ];
3. var      step : Z;
4. var      counter : V ↦ Z;
5. step = −1;
6. Ready = {s};
7. foreach (v ∈ V)
8.    counter(v) = |{u|⟨u,v⟩ ∈ E}|;
9. while (Ready ≠ ∅) do
10.   NextReady = ∅;
11.   foreach (v ∈ Ready) begin
12.      S(v) = step;
13.      foreach (⟨v,w⟩ ∈ E) begin
14.         counter(w) = counter(w) −1;
15.         If (counter(w) == 0)
16.            NextReady = NextReady ∪ {w};
17.      end foreach
18.   end foreach
19.   step = step + 1;
20.   Ready = NextReady;
21. end while
22. return S;
```

Example 5.18 Consider applying Algorithm 5.1 to the precedence graph shown in Figure 5.7b. The source node s is first scheduled (step -1). Because all its successors {15, 16, 24, 25} have only one predecessor s, they become ready next. Scheduling node 15 to step 0 decrements counter for node 17 from 2 to 1. Scheduling node 16 to step 0 further decrements counter for node 17 from 1 to 0, which triggers node 17 to become ready in the next step. The same would apply to node 24 and node 25 to node 26. So, {17, 26} are scheduled at step 1. In the subsequent iterations, node 18, with node 17 as the only predecessor, is scheduled at step 2; and its dependent, node 20, is scheduled at step 3. The complete ASAP schedule is shown in Figure 5.8a; it shows that it takes 4 steps at minimum to schedule all instructions, excluding s and t.

Note that ASAP scheduling is not the only possible solution to the unconstrained scheduling problem. An equally viable solution is the *as-late-as-possible* (ALAP) algorithm. Opposite to the ASAP, the ALAP starts scheduling in reverse time order.

Example 5.19 Consider applying ALAP scheduling to the precedence graph in Figure 5.7b. In the first iteration of the algorithm, all immediate predecessors of the sink will be scheduled at the last step of the schedule. So, {20, 26} are scheduled at step 3. In the subsequent iterations, an instruction is scheduled as soon as all of its successors in the precedence graph have been scheduled at some later step. For example, in the second iteration, node 18 is scheduled at step 2, because it has the already scheduled node 20 as its only successor. Figure 5.8b is the complete ALAP schedule.

5.4.3 Resource-constrained scheduling

We now turn to solving the scheduling problem under resource constraints. In particular, a preallocation of functional units U is available. To simplify discussion,

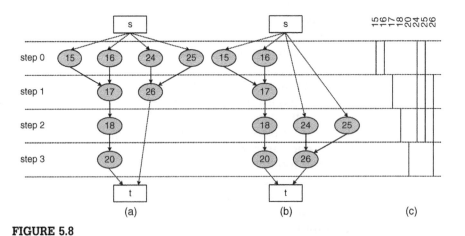

FIGURE 5.8

(a) ASAP schedule. (b) ALAP schedule. (c) Mobility.

we assume each unit $u \in U$ can implement a subset of virtual instructions. The resource-constrained scheduling problem can then be formulated as follows.

PROBLEM 5.4

Given: Precedence graph $\langle E, s, t \rangle$
Find: Schedule $S: V \mapsto Z$
Minimize: $S(t)-S(s)$
Subject to: Constraint (a) $\forall \langle u,v \rangle \in E : S(v)-S(u) > 0$
             Constraint (b) $\forall i \in Z, |S^{-1}(i)| \leq |U|$

Note that compared with Problem 5.3, a new constraint is added such that the number of instructions scheduled at any control step cannot exceed the number of functional units available. This seemingly simple constraint dramatically changes the combinatorial structure of the problem. Although Problem 5.3 can be optimally solved in linear time, Problem 5.4 is shown to be an NP complete problem and requires heuristics for practical implementations.

Example 5.20 Consider again the scheduling problem that we solved in Example 5.18 and Example 5.19. The precedence graph is redrawn in Figure 5.9, where each node is annotated with its corresponding opcode. Assuming that we only have 2 add/sub units and 1 multiplier, both of our ASAP and ALAP schedules will become infeasible. Referring to the ASAP schedule in Figure 5.8a, our resource constraint is violated by the schedule at step 0 and that at step 1, because 3 add/sub units and 2 multipliers will be needed,

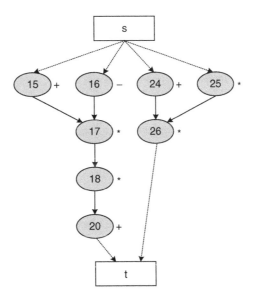

FIGURE 5.9

Precedence graph annotated with opcodes.

respectively. Referring to Figure 5.8b, the schedule at step 2 requires at least 2 multipliers, so it also violates the constraint. To satisfy the resource constraint, we will use a list-scheduling algorithm to schedule these operations again.

List scheduling, shown in Algorithm 5.2, is a modified version of the ASAP scheduling algorithm. Like ASAP, a list of nodes ready for scheduling is maintained, hence its name. The difference is that in each iteration, only a subset of nodes can be scheduled depending on the availability of resources at the current control step. The availability information is maintained with a "reservation table" at line 5 of Algorithm 5.2. We define a Boolean vector *restab*, in which each entry indicates the availability of a unit. At each control step, *restab* is initialized to be all false. Whenever a resource u is available, one of the instructions is selected for assignment to the current step. The case in which a unit can only implement a subset of instructions can be trivially handled by an additional test *impl*. Meanwhile, *restab(u)* is assigned to true, indicating it is "occupied." This ensures constraint (b) is always satisfied.

Algorithm 5.2 List Scheduling

---

algorithm *listSched* $(E : (V \times V)[\ ], s : V, t : V)$ **returns** $V \mapsto \mathbf{Z}$
1. **var** $S : V \mapsto \mathbf{Z}$;
2. **var** $Ready, NextReady : V[\ ]$;
3. **var** $step : \mathbf{Z}$;
4. **var** $counter : V \mapsto \mathbf{Z}$;
5. **var** $restab : U \mapsto \{true, false\}$;
6. $step = 0$;
7. $Ready = \{s\}$;
8. **foreach** $(v \in V)$
9. $counter(v) = |\{u | \langle u,v \rangle \in E\}|$;
10. **while** $(Ready \neq \varnothing)$ **do**
11. $NextReady = \varnothing$;
12.   **foreach** $(u \in U)$
13. $restab(u) = false$;
14.   **while** $(\exists u \in U, \exists y \in Ready \mid !restab(u) \wedge impl(u, y))$ **do**
15. $v = choose(Ready, u)$;
16. $restab(u) = true$;
17. $S(v) = step$;
18. $Ready = Ready - \{v\}$;
19.   **foreach** $(\langle v,w \rangle \in E)$ **begin**
20. $counter(w) = counter(w) - 1$;
21.   **If** $(counter(w) == 0)$
22. $NextReady = NextReady \cup \{w\}$;
23.   **end foreach**
24.   **end while**
25. $step = step + 1$;

26.   *Ready = Ready ∪ NextReady*;
27. **end while**
28. **return** *S*;

Note that at each scheduling step, the number of ready nodes is often more than the resources available to implement them. Therefore, we need to decide which subset of nodes should be chosen for scheduling, in other words, how we implement the *choose* function in line 15 of Algorithm 5.2. It is this key step that impacts the quality of the solution. If a node is chosen too late, then it can potentially lengthen the total schedule. If a node is chosen too early, then potentially more clock steps are needed to keep its value in a register before it is consumed by all its successors. This is referred to as "register pressure," which can lead to an excessive use of registers.

A common way to solve this problem is to assign a priority for each node indicating the desirability of scheduling the node early. The priority can be assigned according to several heuristics or as a weighted sum of them.

One heuristic is to exploit the flexibility of the nodes: in general, there can be potentially many different clock steps a node can be scheduled to. We have already seen that ASAP and ALAP give different solutions, both satisfy the dependency constraint. However, the degree of flexibility can differ. The less-flexible-first heuristics says that we should assign high priority to those nodes that are less flexible. On the other hand, we can afford to schedule highly flexible nodes later because there are more options. To quantify the schedule flexibility, one can use *mobility range*, defined to be the difference between the ALAP schedule and the ASAP schedule for each node.

Example 5.21 The mobility range of all nodes in Figure 5.7b is shown in Figure 5.8c. It can be observed that nodes {15, 16, 17, 18, 20} have zero mobility, whereas the others have a mobility of 2.

We now consider the use of mobility range as the priority function for list scheduling.

Example 5.22 Consider the application of Algorithm 5.2 on the precedence graph in Figure 5.7b, modified in Figure 5.9 with each node annotated with the opcode they require. At step 0, each of {15, 16, 24} in the ready list is requesting an add/sub unit, and {15, 16} are chosen to fill the two available resource slots because both of them have lower mobility than node 24. At the same step, operation 25 is the only candidate requesting a multiplier in the ready list, so it is selected without competition. Such competition happens again at step 2: the multiplier is requested by both ready operations. With lower mobility than operation 26, operation 18 wins. As shown in the following, the schedule generated by this algorithm takes 4 steps, which yields the same minimum latency for ASAP and ALAP scheduling without resource constraints.

Ready List	STEP	ADD/SUB 1	ADD/SUB 2	MULT
{15, 16, 24, 25}	0	15	16	25
{24, 17}	1	24	-	17
{18, 26}	2	-	-	18
{26, 20}	3	20	-	26

We can also deploy a priority function that selects randomly; it is instructive to compare list scheduling with a mobility-based priority function and with a random priority function.

Example 5.23 Assume the priority function in list scheduling gives random selections. This could lead to two possible schedules. In the first case, nodes {15, 24} are selected to fill the two add/sub slots instead of {15, 16}.

Ready List	STEP	ADD/SUB 1	ADD/SUB 2	MULT
{15, 16, 24, 25}	0	15	24	25
{16, 26}	1	16	-	26
{17}	2	-	-	17
{18}	3	-	-	18
{20}	4	20	-	-

In Figure 5.10b, we can see that the result of delaying operation 16 is taking 1 more cycle than optimal to complete the computation. In the second case, we choose to delay operation 18 at step 3 as shown in the following table, and it also leads to an increase in latency by 1 as shown in Figure 5.10c.

Ready List	STEP	ADD/SUB 1	ADD/SUB 2	MULT
{15, 16, 24, 25}	0	15	24	25
{24, 17}	1	24	-	17
{18, 26}	2	-	-	26
{18}	3	-	-	18
{20}	4	20	-	-

Because 16→17→18→20 is a critical path in the precedence graph as shown in Figure 5.9, it is impossible to generate a schedule that takes less than 5 cycles if we delay any node on this path.

Many other heuristics can be developed and are used in practice as well. For example, the distance of a node v to the sink node; in other words, the difference of unconstrained schedules of t and v, wither ASAP or ALAP, can be used as penalty (the inverse of priority). The rationale being that the closer a node is to the sink, the more likely it is to extend the overall schedule if not scheduled early. As another example, the out degree of a node can be used as a priority

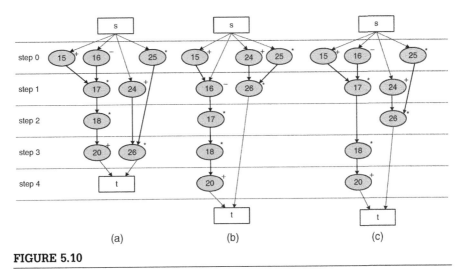

FIGURE 5.10

List scheduling with priority functions.

function: the more number of successors a node has, the more likely additional nodes would become ready after the scheduling of such a node.

5.5 REGISTER BINDING

In an IR, a virtual instruction may compute a *value*, which needs to be kept in a register and later used by other instructions as operands.[2] In a simplistic implementation, one could allocate a distinct register to hold each value. This leads to excessive use of storage resources. In contrast, one could exploit the fact that the values do not need to be held all the time, because they have limited *lifetimes*. The values that have nonoverlapping lifetimes can then share the same register to save silicon area. The task of mapping values to registers to maximize sharing is called *register binding*.

5.5.1 Liveness analysis

To enable register binding, we have to establish the condition under which variables can share a common register.

Definition 5.5 *A value (instruction) v is **live** at a control step s1 if there exists another control step s2 reachable from s1, such that v is used as an*

---

[2]Not all virtual instructions compute a value, for example, the memory store instructions. For most instructions that do compute a value, we do not distinguish the virtual instruction and the value it computes.

*operand by one of the instructions scheduled at s2. A **live set** at control step s1 is the set of all values alive at s1.*

Clearly, a value v scheduled at control step s cannot share a common register with a value w live at s, otherwise, the new value v would corrupt the value w, that is used later.

We use a liveness analysis algorithm to compute the live set. We first consider a single basic block.

The strategy we take is to start at the end of the schedule and scan each control step backwards (in reverse order). At each control step s, we define,

Live(s):	Set of live values at the beginning of step s
Def(s):	Set of values defined at step s
Use(s):	Set of values used at step s.

The relationship between them can be established as follows, assuming all control steps are sequentially numbered.

$$Live(s) \;=\; Use(s) \cup [Live(s + 1) - Def(s)]$$

Note that, *Def(s)* is the set of all instruction scheduled at step s; and *Use(s)* is the set of all operands used by instructions scheduled at step s.

A liveness analysis algorithm for a basic block can then be developed, as shown in Algorithm 5.3. It takes the schedule of the basic block as input. An additional input to the algorithm is the set of live values leaving the basic block, called *liveOut*.

Algorithm 5.3 Basic Block Liveness Analysis

---

algorithm *liveBB* $(S : V \mapsto \mathbf{Z},\ liveOut : V[\,])$ **returns** $\mathbf{Z} \mapsto V[\,]$
1. **var**   *Live*: $\mathbf{Z} \mapsto V[\,]$;
2. **var** $I : \mathbf{Z}$;
3. $I = |\ range\ S\ |$; $Live(I) = liveOut$;
4. **foreach** $(s \in [I-1 \ldots 0])$ **begin**
5. $Live(s) = Live(s + 1) - S^{-1}(s)$;
6.   **foreach** $(v \in S^{-1}(s))$
7. $Live(s) = Live(s) \cup \{v.src1\} \cup \{v.src2\}$;
8. **end foreach**
9. **return** *Live*;

Example 5.24 Consider the scheduled basic block in Example 5.22, re-created in Figure 5.11a. The live range of each value can be visualized by an interval, starting from just after the control step when it is defined and ending at the control step when it is last used as shown in Figure 5.11b. The live set of each control step can then be visualized as a horizontal cut line through the live ranges, in other words, the set of all values crossing the control

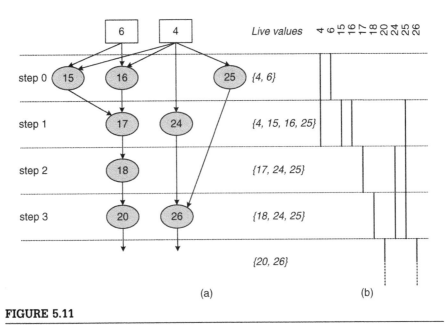

FIGURE 5.11

(a) Results of liveness analysis. (b) Live ranges of values.

step. We now consider how Algorithm 5.3 can compute the correct live set. In the beginning, the liveOut value is {20, 26}, because they are used by other basic blocks. As we scan step 3, we remove {20, 26} from the live set (Line 5, where $S^{-1}(s)$ applies the inverse function of schedule S on step s, or returning the set of a value scheduled at step s), because they are defined in step 3, and add {18, 24, 25}, because they are used by {20, 26}. This leaves {18, 24, 25} as the live set for step 3. This process repeats until we reach step 0, where the live set is {4, 6}, defined in other basic blocks.

STEP	Def	Use	Live
0	{15, 16, 25}	{4, 6}	{4, 6}
1	{17, 24}	{4, 15, 16}	{4, 15, 16, 25}
2	{18}	{17}	{17, 24, 25}
3	{20, 26}	{18, 24, 25}	{18, 24, 25}
4			{20, 26}

We now extend liveness analysis to the whole program. As shown in Algorithm 5.4, we are now given the schedule for all basic blocks and attempt to find the live set for each control step in all basic blocks. We use a standard data-flow analysis framework deployed for software compiler analysis. In this framework, a ***control flow graph*** (CFG) is constructed so that there is one edge from basic block A to basic block B if there is an instruction in block A that branches or jumps to B. The framework traverses all basic blocks following the CFG order and derives the information of interest by processing each basic block. In this case, we use the post depth first order to make sure all successors of a basic block are processed before a given basic block. As each basic block is processed by calling *liveBB* (in Line 12), the *liveOut* of the basic block is computed by combining the set of live values flowing out of all its successors (Lines 8–13). This process repeats and is terminated when reaching a fixed point; in other words, the computed live set values no longer change.

Algorithm 5.4 Liveness Analysis

---

algorithm *live(Sched*: $B \mapsto (V \mapsto \mathbf{Z})$) **returns** $B \mapsto (\mathbf{Z} \mapsto V[\ ])$
1.   **var**   *Live*: $B \mapsto (\mathbf{Z} \mapsto V[\ ])$;
2.   **var**   *LiveOut*: $B \mapsto V[\ ]$;
3.   **var**   *New*: $V[\ ]$;
4.   **var**   *changed*: {*true, false*} = *true*;
5.   **while** (*changed*) **do**
6.   *changed* = *false*;
7.   **foreach** ($b \in B$ in postorder) **begin**
8. $New = \bigcup\limits_{s \in succ(b)} Live\ (s, 0)$;
9.   **if** ($New \neq LiveOut(b)$) **begin**
10. $LiveOut(b) = New$;
11. $changed = true$;
12. $Live(b) = liveBB(Sched(b), liveOut(b))$;
13.   **end if**
14.   **end foreach**
15. **end while**
16. **return** *Live*;

With liveness information, we can then capture the relation between values with a graph, called an *interference graph*. In an interference graph, a node represents a value, and an edge between two nodes indicates that they cannot share a common register. The interference graph can be derived from liveness information with Algorithm 5.5.

Algorithm 5.5 Interference Graph Construction

---

algorithm *intf(Sched*: $B \mapsto (V \mapsto \mathbf{Z})$, *Live*: $B \mapsto (\mathbf{Z} \mapsto V[\ ])$) **returns** $(V \times V)[\ ]$
1. **var** $E_{intf} : (V \times V)[\ ]$;
2. **foreach** ($b \in B$) **foreach** ($v \in b$) **begin**
3. $s = Sched(b, v)$;
4. $E_{intf} = E_{intf} \cup \{\langle u, v \rangle, \langle v, u \rangle | u \in Live(b, s + 1) \wedge u \neq v\}$;
5. **end foreach**
6. **return** E_{intf};

Example 5.25 Figure 5.12 shows the interference graph constructed from the liveness information in Example 5.24. For example, for instruction 20, it is scheduled at control step 3. The live set for step 4 is {20, 26}. Therefore an interference graph edge between 20 and 26 is created. This process is repeated for every instruction.

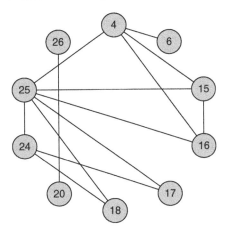

FIGURE 5.12

Interference graph.

5.5.2 **Register binding by coloring**

Given an interference graph, the register-binding problem reduces to assigning each node to a register number, while ensuring that two nodes connected by an edge are assigned different register numbers. This is equivalent to the classic *graph coloring* problem, if each register number is treated as a color.

PROBLEM 5.5

Given: Interference graph E_{intf}
Find: Register binding $B^R : V \mapsto Z$
Minimize: $| \text{range } B^R |$
Subject to: $\forall \langle u,v \rangle \in E_{intf}, B^R(u) \neq B^R(v)$

Minimizing the number of registers is then equivalent to minimizing the *chromatic number*, or the minimum number of colors used to color the interference graph.

The coloring problem is an NP-complete problem and thus requires heuristic solutions. A typical heuristic algorithm colors one node at a time by choosing the minimum color not used by its neighbors. Of course, one can only choose a color different from the neighbors that have already been colored. Therefore, it is sufficient to consider only the *remainder graph* (*i.e.*, the subgraph of the interference graph where all uncolored nodes and their incident edges are removed).

Algorithm 5.6 uses this strategy by first finding the so-called *vertex elimination order* σ (Line 6), which can be considered as the order in which the sequence of the remainder graph is generated by removing one node at a time starting from the full interference graph. The inverse of the vertex elimination order is, therefore, the order in which the nodes are colored. In fact, in each iteration of the loop in Line 7, a node v is selected according to σ and added to the remainder graph (Lines 9–10) and colored (Line 11).

Algorithm 5.6 Register Binding by Coloring

---

algorithm *color*(E_{intf} : $(V \times V)[\ ]$) **returns** $V \mapsto \mathbf{Z}$
1. **var** $C : V \mapsto \mathbf{Z}$;
2. **var** $\sigma : V \mapsto \mathbf{Z}$;
3. **var** $V' : V[\ ]$;
4. **var** $E' : E_{intf}[\ ]$;
5. **var** $v : V$;
6. $\sigma = vertexElim(E_{intf})$;
7. **foreach** ($i \in [1 \ldots |V|]$) **begin**
8. $v = \sigma^{-1}(i)$;
9. $V' = V' \cup \{v\}$;
10. $E' = E' \cup \{\langle u,v \rangle, \langle v,u \rangle \mid u \in V' \wedge \langle u,v \rangle \in E_{intf}\}$;
11. $C(v) = \min(\{c \in \mathbf{Z} \mid \forall \langle u,v \rangle \in E', C(u) \neq c\})$;
12. **end foreach**
13. **return** C;

Example 5.26 Consider the interference graph in Example 5.25. Assume the vertex elimination order is {6, 20, 26, 17, 18, 24, 4, 15, 16, 25}. Then the coloring order is {25, 16, 15, 4, 24, 18, 17, 26, 20, 6}. We start with an empty remainder graph. The first node chosen is node 25. Because it has no neighbors, color 1 is chosen. In the next iteration, node 16 is chosen, and the remainder graph is expanded with the node, as well as the incident edges, shown in Figure 5.13a. Color 2 is chosen as the minimum number besides 1. In the next iteration, node 15 is added to the remainder graph, shown in Figure 5.13b, and assigned color 3, the minimum color different from its neighbors 25 and 16. This process is repeated in Figure 5.13b–e. In the end, all 10 nodes in the interference graph are colored with 4 colors. In other words, 10 values can share 4 registers. The mapping is as follows.

> R0: {6, 25, 26}
> R1: {16, 20, 24}
> R2: {15, 17, 18}
> R3: {4}

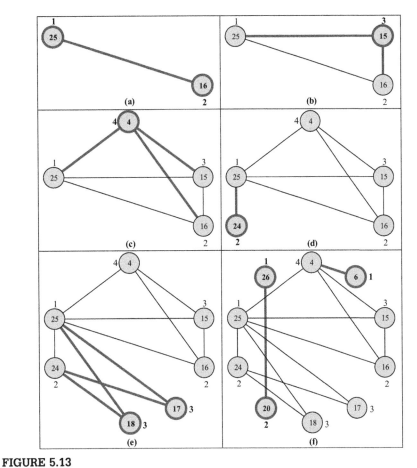

FIGURE 5.13

Coloring with σ = {6, 20, 26, 17, 18, 24, 4, 15, 16, 25}.

Although the coloring process itself is rather straightforward, the key step requires computing the appropriate vertex elimination order, or the coloring order.

We first consider the special case in which the interference graph is derived from a basic block. As shown in Example 5.24, each value is associated with a live range characterized by a control step when the value is defined, and a control step when it is last used. This live range can be considered as an interval on the integer set. An interference graph constructed by creating one edge for each pair of overlapping intervals is called an *interval graph*. For an interval graph, one could pick a coloring order by sorting all intervals according to the left edge of the interval. This *left-edge algorithm* is optimal for an interval graph.

We now consider the general case in which the interference graph may not have the structural property of an interval graph. This may partly be due to the presence of complex control structures, such as branches and loops. This may also be due to the special requirements that certain values have to be mapped to the same register.

A typical heuristic we can use is called less-flexible-first, which we have used in list scheduling. Here, the more neighbors a node has, the less choices it may have to assign a color, and, therefore, the earlier we should color it. We can, therefore, pick a vertex elimination order according to the degree of a node. This strategy is used in Algorithm 5.7.

Algorithm 5.7 Vertex Elimination

---

algorithm *vertexElim* $(E_{intf} : (V \times V)[\ ])$ **returns** $V \mapsto \mathbf{Z}$
1. **var** $\sigma : V \mapsto \mathbf{Z}$;
2. **var** $V' : V[\ ]$;
3. **var** $E' : (V \times V)[\ ]$;
4. **var** $v : V$;
5. $V' = V$;
6. $E' = E_{intf}$;
7. **foreach** $(i \in [1..|V|])$ **begin**
8. $v = \text{argmin}_{v \in V'} \ | \ \{\langle u,v \rangle \in E'\}|$;
9. $\sigma(v) = i$;
10. $V' = V' - \{v\}$;
11. $E' = E' - \{\langle u,v \rangle, \langle v,u \rangle \in E'\}$
12. **end foreach**
13. **return** σ;

Example 5.27 The vertex elimination process is illustrated in Figure 5.14. Here we use a stack to keep track of the vertex elimination order, which facilitates the use of its inverse as coloring order. With one adjacent node, 6, 20, and 26 are pushed to the stack first, as shown in Figure 5.14a. These nodes, as well as their incident edges, are removed, which results in a remainder graph shown in Figure 5.14b. With the degree of 2, vertices 17 and 18 score highest to be the next two nodes to be eliminated. After removing both 17 and 18, vertex 24 has its degree reduced to 1 and becomes the next candidate to be removed, as shown in Figure 5.14c. As we can see in Figure 5.14d, the remaining nodes have the same degree, and vertex 4 is selected arbitrarily. The same happens in the subsequent iterations, and we choose to push vertices 15, 16, and 25 to stack in order, as shown in Figure 5.14e–f. With the vertex elimination order stored in the stack, coloring can be performed by popping one node at a time from the stack and reconstructing the remainder graph, as illustrated in Example 5.25.

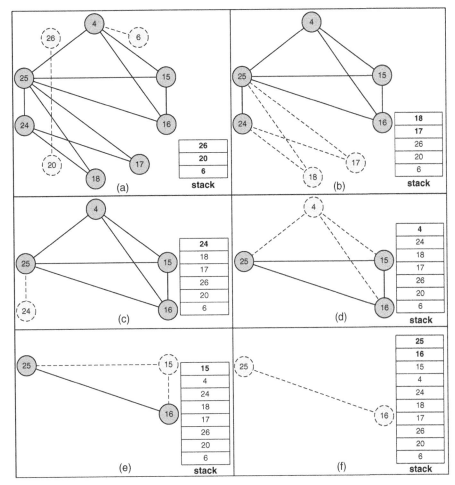

FIGURE 5.14

Vertex elimination.

5.6 **FUNCTIONAL UNIT BINDING**

Although the coloring algorithm can minimize the register count, the register sharing decision may affect the size of multiplexers. The silicon area of multiplexers can be quite significant. For example, in an ***application-specific integrated circuit*** (ASIC), the area of a six-input multiplexer is close to the area of an adder. For today's FPGAs, the area of multiplexers, typically implemented as lookup tables, is often larger than adders, typically available as hardwired logic.

Example 5.28 To see how register binding may affect multiplexer area, consider the code fragment in Figure 5.15a. Assume both instructions will be mapped to the same adder. If all possible input operands, which are values {t0, t1, t3, t4}, are mapped to different registers, as shown in Figure 5.15b, two 2-to-1 multiplexers will be needed. However, as shown in Figure 5.15c, the multiplexers can be eliminated by mapping values {t0, t3} to the same register and values {t1, t4} to another one. In general, even in the cases that only a subset of possible input operands share the same register, there is still a benefit from a smaller multiplexer. Sharing registers among the possible output values of a functional unit can also simplify the interconnection between the functional unit and the registers. This is as illustrated by the mapping of values {t2, t5} to register R3 in Figure 5.15c.

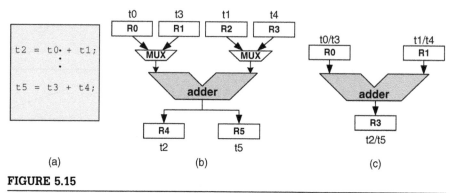

(a)   (b)   (c)

FIGURE 5.15

Impact of register sharing on multiplexers.

Likewise, although the silicon area of functional units is determined by allocation, functional unit binding affects the area of multiplexers. Ideally, if the source operand of one instruction A maps to the same register of the source operand of another instruction B, then it is desirable to map A and B to the same functional unit, because no extra multiplexer input needs to be introduced at the functional unit input. This scenario is called *common source*. Likewise, if the destination operand of instruction A maps to the same register as the destination operand of another instruction B, then it is desirable to map A and B to the same functional unit, because no extra multiplexer input needs to be introduced at the destination register input. This scenario is called *common destination*.

Example 5.29 Consider the design in TinyRTL in Figure 5.16a. Here two addition operations scheduled at C0 are bound to U0 and U1, respectively; whereas the one scheduled at C1 is bound to U0. To share unit U0, as shown in Figure 5.16b, three 2-input multiplexers are needed in the corresponding datapath. However, because the operation scheduled at C1 shares common sources and common destinations with the other scheduled operation at C0, the multiplexers can be eliminated by binding both instructions to the same unit, as shown in Figure 5.16c.

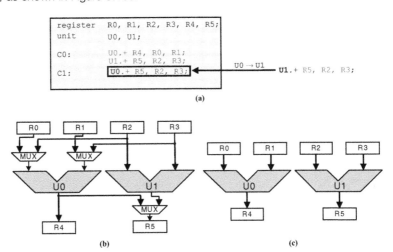

(a)

(b)   (c)

FIGURE 5.16

Impact of functional unit binding on multiplexers.

Mathematically, the register binding and functional unit binding problems for multiplexer area minimization are equivalent, although they depend on each other (called the phase ordering problem). Recall in Section 5.5 that the register-binding problem can be formulated as the coloring of an interference graph. Each edge indicates that the connected nodes cannot be assigned the same color, or they cannot share the same register. We now take a different perspective. Two instructions are said to be *compatible* if they can be mapped to the same resource. Like the interference graph, we can establish a *compatibility graph* whose nodes are instructions, and edges between nodes indicate that they are compatible. The binding problem can then be formulated as a clique partitioning problem, or an integer labeling of nodes in the compatibility graph, such that all nodes with the same label are fully connected to each other. In other words, all nodes with the same label form a *clique*, or a complete subgraph of the compatibility graph.

Interestingly, the interference graph and compatibility graph are dual of each other: one can find the compatibility graph as the inverse of interference graph and vice versa, and clique partitioning of a compatibility graph is equivalent to the coloring of the corresponding interference graph. Thus, an alternate way of performing register binding is to inverse the interference graph to obtain the compatibility graph, and then solve the clique partitioning problem. Likewise, we can solve the functional unit binding problem by clique partition. The compatibility graph for functional unit binding can be established directly: an edge is created for every pair of instructions that satisfy the following:

- Their opcodes are of the same class.[3]
- They are not scheduled at the same control step.

Example 5.30 Consider the example in Figure 5.6. Consider only the instructions that perform additions and subtractions, which belong to the same class. This gives the set of instructions {15, 16, 20, 24}. Given the schedule in Example 5.22, shown in Figure 5.10a, we conclude that only 15 and 16 interfere with each other, because they are scheduled at the same clock step and, as a result, cannot be bound to the same functional unit. This is reflected in the interference graph in Figure 5.17a. The compatibility graph is shown in Figure 5.17b, essentially by complementing the edges in the interference graph.

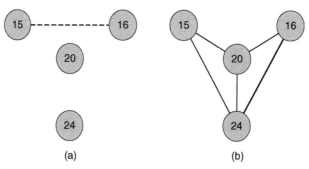

(a)   (b)

FIGURE 5.17

Interference graph and compatibility graph.

We can then refine functional unit binding problem as follows.

PROBLEM 5.6

Given: Compatibility graph E_{comp}
Find: Unit binding $B^U: V \mapsto U$
Minimize: $\sum_{u \in U}|Src1(u)| + \sum_{u \in U}|Src2(u)| + \sum_{u \in U}|Dest(u)|$
Subject to: $\forall u,v \in V, B^U(u) = B^U(v) \Rightarrow \langle u,v \rangle \in E_{comp}$

---

[3]Two instruction opcodes are of the same class if there is significant chance of logic sharing if they are bound to the same functional unit. Addition and subtraction are of the same class, because it requires only a few extract logic gates to convert an adder into an adder/subtractor.

Note that the third term $\Sigma_{u \in U}|Dest(u)|$ of the objective in Problem 5.6, where $Dest(u)$ is defined as the different registers a functional unit u connects to, is derived from $Srcd(r)$ of each register r from the third term of objective (3) in Problem 5.2, where $Srcd(r)$ is the different functional units outputs to register r. In fact, they are just different ways of estimating the total number of multiplexer inputs to the registers.

Like graph coloring, clique partitioning that minimizies the number of cliques is also an NP-hard problem. Problem 5.6 is harder, because its objective is the total number of multiplexers, which nontrivially depends both on functional unit binding and register binding.

To solve the problem heuristically, we again take the iterative approach, which makes one decision at a time. More specifically, as we show in Algorithm 5.8, we start by assuming each node in the compatibility graph forms its own clique. In each iteration, we select and *contract* one edge $\langle u,v \rangle$ in the graph (Lines 11–17); in other words, we merge the pair of nodes incident to the edge into a larger clique. With some bookkeeping (Line 16), the larger clique is represented by one of the pair, say u; therefore, edges incident to the other node v are removed (Line 14). To ensure further merging leads to cliques, any edges incident to u that do not share a common neighbor with v should also be removed (Line 13). This process repeats until all edges are removed, and what is left is a set of nodes, each representing a clique. After this, all virtual instructions in a clique v are assigned a common unit u (line 18–22).

The key step of the algorithm is the criterion used to select the edge in each iteration, so that it positively improves, if doesn't optimizes, our objective. Algorithm 5.8 uses the *partial binding result* to approximate the objective. It assumes that each clique corresponds to a functional unit and maintains its *Src1*, *Src2*, and *Dest*, calculated as the set of registers for the corresponding operands of all nodes in the clique they are mapped to, according to register binding B^R. These sets are called the *operand sets*. In each iteration, when two nodes (cliques) are merged, their corresponding operand sets are merged as well by unions (Line 15). With operand sets defined for each node, we can, in turn, define the *edge weight* as the total number of common operands in respective operand sets. The edge that leads to the least changes, that is, having the most number of common sources and destinations, is greedily selected (Line 11).

Algorithm 5.8 Weighted Clique Partitioning

---

algorithm *CliquePartition*(E_{cmpat} : $(V \times V)[\,]$, B^R : $V \mapsto \mathbf{Z}$) **returns** $V \mapsto \mathbf{Z}$
  1. **var** V' : $V[\,] = V$;
  2. **var** E' : $(V \times V)[\,] = E$;
  3. **var**   *Clique* : $V \mapsto V[\,]$;
  4. **var** B^U : $V \mapsto \mathbf{Z}$;
  5. **var**   *Src1,Src2,Dest* : $V \mapsto \mathbf{Z}[\,]$;
  6. **foreach** ($v \in V'$) **begin**

7. $Src1(v) = \{B^R(v.src1)\}$; $Src2(v) = \{B^R(v.src2)\}$; $Dest(v) = \{B^R(v.dest)\}$;
8. $Clique(v) = \{v\}$;
9. **end foreach**
10. **while** $(E' \neq \emptyset)$ **do**
11. $\langle u,v \rangle = \text{argmax}_{\langle u,v \rangle \in E'} |Src1(u) \cap Src1(v)| + |Src2(u) \cap Src2(v)|$
       $+ |Dest(u) \cap Dest(v)|$;
12. $V' = V' - \{v\}$;
13. $E' = E' - \{\langle u,w \rangle, \langle w,u \rangle | \langle w,v \rangle \notin E'\}$;
14. $E' = E' - \{\langle v,w \rangle, \langle w,v \rangle \in E'\}$;
15. $Src1(u) = Src1(u) \cup Src1(v)$; $Src2(u) = Src2(u) \cup Src2(v)$;
       $Dest(u) = Dest(u) \cup Dest(v)$;
16. $Clique(u) = Clique(u) \cup Clique(v)$;
17. **end while**
18. **foreach** $(v \in V')$ **begin**
19. $u = next(U)$;
20.   **foreach** $(w \in Clique(v))$
21. $B^U(w) = u$;
22. **end foreach**
23. **return** B^U;

We now illustrate the application of Algorithm 5.8 for Example 5.30. We start by computing the initial operand sets.

Example 5.31 From the TinyIR representation in Figure 5.6b, we can find the source operands of the instructions as follows,

INSTRUCTION	SRC 1	SRC 2
15	4	6
16	4	6
20	18	⟨16⟩
24	4	⟨12⟩

Note: constant inputs are bracketed by ⟨ ⟩

From the register allocation result of the previous section (Example 5.26), we have

R0: {6, 25, 26}
R1: {16, 20, 24}
R2: {15, 17, 18}
R3: {4}

We can, therefore, establish the operand sets in Figure 5.18a. In addition, we mark each edge with a weight, valued as the number of common elements in the respective operand sets.

We can now start the iterative clique partitioning process.

Iteration 1

INSTRUCTION	SRC 1	SRC 2	DEST
15	R3	R0	R2
16	R3	R0	R1
20	R2		R1
24	R3		R1

(a)

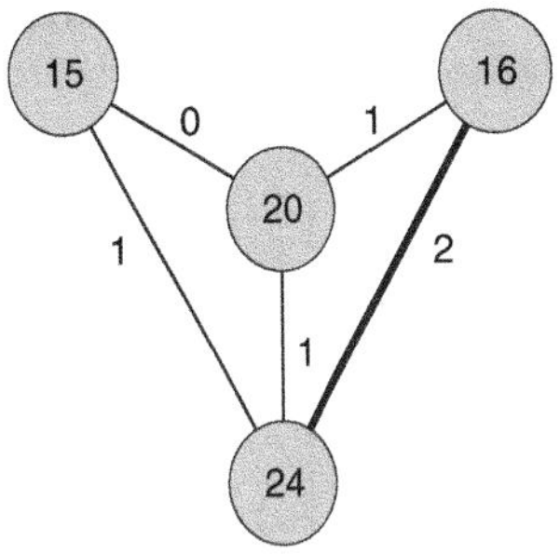

(b)

Iteration 2

INSTRUCTION	SRC 1	SRC 2	DEST
15	R3	R0	R2
16, 24	R3	R0	R1
20	R2		R1

(c)

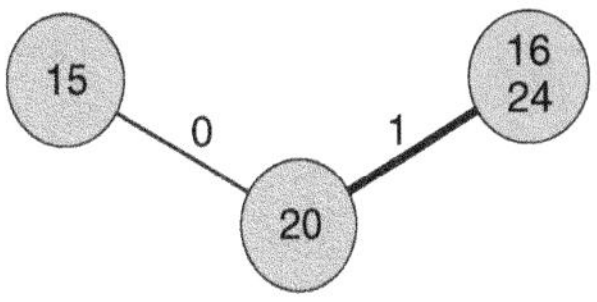

(d)

After iteration 2

INSTRUCTION	SRC 1	SRC 2	DEST
15	R3	R0	R2
16, 24, 20	R3, R2	R0	R1

(e)

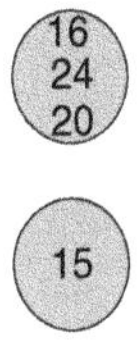

(f)

FIGURE 5.18

Operand sets and iterative clique partitioning.

Example 5.32 From Figure 5.18a, the edge ⟨16,24⟩ is the only edge with the maximum weight of 2. So in the first iteration it is selected first for contraction. In doing so, nodes 16 and 24 are merged into one node. Note in particular that the original edge ⟨15,24⟩ is removed, because 15 and 16 are not compatible. Also note that the operand sets are updated in Figure 5.18c, as well as the edge weights in Figure 5.18d. In the second iteration, the edge between 20 and the merged node in the first iteration has the maximum weight of 1 and is selected next for contraction. The result of this step is shown in Figure 5.18e-f. Examining the remainder of the compatibility graph, there are no more edges left, and the iterative process terminates.

We now have enough information to generate the full datapath.

Example 5.33 Combining scheduling, register binding, and functional unit binding, we can summarize the decisions made in high-level synthesis by the following tables.

REGISTER	VALUES
R0	6, 25, 26
R1	16, 20, 24
R2	15, 17, 18
R3	4

UNIT	INSTRUCTIONS
ADD/SUB 0	16, 20, 24
ADD/SUB 1	15
MULT	17, 18, 25, 26

With the resource binding result, we can complete the datapath of the design by adding multiplexers before functional unit and register input ports. To accomplish this, we need to identify the set of all possible unit-to-register and register-to-unit transfers, given the virtual instructions, as well as the binding result. According to the resource binding result in the above tables, we can identify the sources of registers as in the following table. For example, register R0 takes values {25, 26} from the MULT unit and takes value 6 from an external input, so it needs a 2-input multiplexer to take values from the both sources.

REGISTER	INPUTS	VALUES
R0	external input	6
	MULT	25, 26
R1	ADD/SUB 0	16, 20, 24
R2	ADD/SUB 1	15
	MULT	17, 18
R3	external input	4

To identify the sources of functional units, we first identify the source registers of each instruction, as follows,

INSTRUCTION	SRC 1	SRC$_{REG}$ 1	SRC 2	SRC$_{REG}$ 2
15	4	R3	6	R0
16	4	R3	6	R0
17	15	R2	16	R1
18	⟨13⟩		17	R2
20	18	R2	⟨16⟩	
24	4	R3	⟨12⟩	
25	⟨12⟩		4	R3
26	24	R1	25	R0

Note: constant inputs are bracketed by ⟨ ⟩

Then, the sources of each functional unit are the union of the sources of all instructions bound to the unit. For example, the sources of input port 1 of the add/sub 0 unit is the union of {R3}, {R2}, and {R3}, which are the corresponding sources of instructions 16,

20, and 24, respectively. So, a 2-input multiplexer is needed at the port. The sources of the rest of the functional unit ports are summarized as follows,

UNIT	INPUT PORT 1	INPUT PORT 2
ADD/SUB 0	R2, R3	R0, ⟨12⟩, ⟨16⟩
ADD/SUB 1	R3	R0
MULT	R1, R2, ⟨12⟩, ⟨13⟩	R0, R1, R2, R3

The complete synthesized datapath is as illustrated in Figure 5.20a.

It is constructive to examine whether the heuristic in Algorithm 5.8 is effective. To see this, we apply the same clique partitioning process, except that in each iteration, an edge is randomly selected for contraction. Figure 5.19 shows a possible result, and the corresponding datapath is as illustrated in Figure 5.20b. It shows that the random unit binding–based datapath takes 20 multiplexer inputs compared with 17 multiplexer inputs from the clique-partitioning unit binding. In terms of wiring complexity at the output port, it has a total of 16 destinations from all sources, whereas the clique partitioning unit binding synthesizes a datapath with only 15, which shows that in this instance the clique partitioning heuristic yielded a superior design.

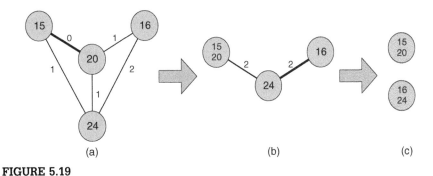

FIGURE 5.19

Random unit binding.

5.7 CONCLUDING REMARKS

In this chapter, we described a complete, although simplified high-level synthesis system. The presented algorithms are distilled from a rich body of research since the 1980s. The representative early academic efforts include CMU [McFarland 1978; Gyrczyc 1984; Thomas 1988], IMEC [De Man 1986], USC [Parker 1986], and Illinois [Pangrle 1987; Brewer 1988]. The representative early industry efforts include IBM [Camposano 1991] and Bell Lab [Bhasker 1990]. Readers are referred

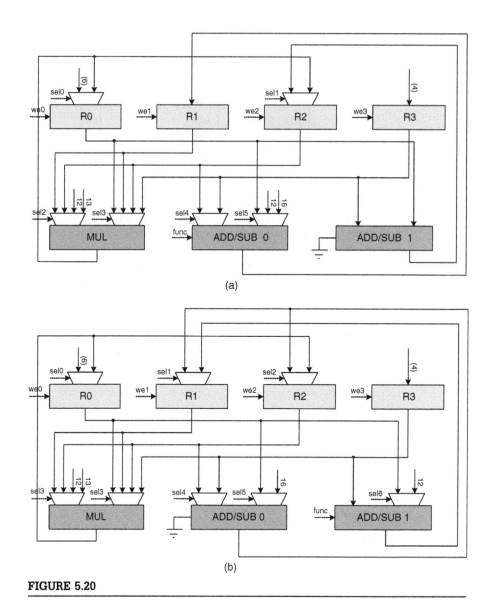

FIGURE 5.20

Synthesized datapaths.

to [Gajski 1992] and [De Micheli 1994] for a comprehensive treatment of development in this period. In particular, list scheduling is due to [Landskov 1980]. Left edge algorithm for register binding is due to [Kurdahi 1987]. Clique partitioning based binding is due to [Tseng 1986].

Although in this chapter we chose to present a resource-constrained–based formulation of high-level synthesis, a large body of literature was devoted to performance-constrained (or time-constrained) formulations of scheduling. A representative work in this area is force-directed scheduling [Paulin 1989], which attempts to minimize resource count under a cycle count constraint of a basic block.

Despite the intensive research efforts, high-level synthesis was not embraced by the design community as much as it was intended. As discussed in Section 5.1, one reason behind the reluctance of acceptance is methodologic: because the chip content has become increasingly heterogeneous, high-level synthesis has to be integrated into an electronic system–level design method before it can replace register transfer level synthesis as the dominant synthesis technology. Toward this end, *hardware/software codesign* [Gupta 1992; Gajski 1994] emerged in the mid-1990s as a research field to address the issue of how one can partition an application into software and hardware components, select the processors, and generate the interface between them. Attempts to fully automate the task, called *hardware/software cosynthesis* [Ernst 1996; Yen 1996], largely failed in practice because the partitioning decisions are often dictated by non-technical factors, such as the availability of legacy *intellectual property* (IP) components that populate the ecosystem of processor IPs. Although the cosynthesis approach will remain effective for domain-specific subsystems, the full system-level design method has given way to the architecture-based method described in Section 5.2, which gives more discretion to system architects, enables concurrent development of hardware and software, and allows for derivative implementations to amortize the design cost to multiple products.

A restricted form of hardware/software codesign that has gained success in practice is *application-specific instruction set processor* (ASIP) design. Here a programmable processor is designed; however, the instruction set is adapted to a family of applications. The main attractiveness of ASIPs is that it allows for post-silicon programmability (which is not possible for traditional high-level synthesis that generates custom hardware); furthermore, ASIPs can achieve a certain level of application acceleration through the use of custom instructions. The ASIP based design methodologies pioneered in [Marwedel 1986], [Fauth 1995], [Halambi 1999], and [Hoffmann 2001] are summarized in [Mishra 2008], where different forms of *architecture description languages* (ADLs) are advocated to specify processor architectures. Often, both the processor RTL and a compiler/simulator tool chain can be generated automatically. Recent efforts attempt to automatically extract custom instructions from application, given a tight budget of available instruction slots. Because the main mechanism of application customization is through instruction extension, the amount of acceleration an ASIP can achieve is limited. Therefore, ASIP designs cannot completely replace what can be offered by high-level synthesis. Furthermore, they have to compete against the existing design ecosystem (including legacy software) made available by dominant embedded processor vendors who continually enhance their instruction sets for each processor generation.

For high-level synthesis to be successful as a component technology within an ESL design method, many advanced issues beyond the scope of this chapter have to be solved. The key drivers for these issues are that ***design productivity*** and ***quality of result*** (QoR) have to be substantially better than RTL to justify the departure from a mature, well-tested design flow.

On the design productivity front, classical high-level synthesis only tackles behavioral description with loop kernel level complexity and accepts only a subset of software program constructs. As a result, the design flow requires significant manual partitioning and rewriting effort from designers, diminishing the productivity gain promised by high-level synthesis. To succeed, high-level synthesis has to scale analysis and optimization algorithms to handle applications in their entirety and target architectures beyond FSMD. For example, performance-constrained–based algorithms work only on a basic block, and it is impractical to ask users to specify the cycle count constraint of every basic block in a complex program. Therefore, either resource-constrained algorithms driven by a design exploration environment have to be applied or new performance-budgeting algorithms capable of distributing end-to-end performance constraints to individual blocks have to be developed. As another example, classical high-level synthesis does not permit the use of pointers in the behavioral description, yet pointers are pervasively used in C/C++ programs. Pointer analysis [Hind 2000] was demonstrated to relax such limitations [Panda 2001a; Semeria 2001; Zhu 2002]. As another example, ***multi-processor system-on-chip*** (MPSOC) architectures [Dutt 2001; Helmig 2002; Intel 2002; Artieri 2003] were explored to enable coarse-grained parallelism, and both bus-based communication schemes [Pasricha 2008] and ***network-on-chips*** (NOC) [Dally 2001] were proposed to provide on-chip communication support among the processing elements.

Although it is debatable whether the new acronyms above truly advance the state-of-the-art in high-level synthesis, it is the QoR a synthesis tool can achieve that would finally earn acceptance by designers. As a discipline, high-level synthesis sits at the intersection of multiple domains, including parallelizing compilers, computer architecture, and circuit optimization. Therefore, it has to exploit and adapt existing techniques in these domains, and innovations likely result by crossing boundaries of these domains. For example, *presynthesis transformations* were shown to have significant impact on QoR [Bhasker 1990; Nicolau 1991; Gupta 2003], yet a different strategy needs to be taken from those in optimizing compilers. As another example, like in general purpose computing, memory accesses are often the performance bottleneck. The freedom in creating a customized memory system in high-level synthesis led to many innovative *memory optimization algorithms* exploiting memory access patterns at the application side and available bandwidth at the architecture and circuit level [Panda 2001b; Wolf 2003]. Another problem that led to the poor performance of classical high-level synthesis is the lack of a link to downstream logic synthesis and place-and-route tools. The classical methods abstract away the effects of downstream tools by use of area and timing estimation models that are often too crude to be useful. This leads to the well-known timing closure problem. A promising direction is the so-called *C-to-gate methodology* in which behavioral and logic synthesis are integrated in an effective fashion. Finally, there is an urgent need to tightly couple physical design with high-level synthesis to allow for better predictability of design results at the later stages of chip design [Xu 1998; Um 2003].

5.8 EXERCISES

5.1. (Frontend/IR Design) Use lex/yacc to build a frontend for TinyC, which is given in Section 5.2.1.

5.2. (Resource-Constrained Scheduler) Implement the list scheduler with the frontend built in Problem 5.1. The resource constraint is passed as command line in the following format:

- behsyn -R constraint_spec foo.c
- constraint_spec: = component [; component]*
- component: = opcode [, opcode]* : num

For example, *behsyn -R "OP_ADD: 2; OP_MUL: 1" foo.c* specifies that two adder components (which implements OP_ADD) and one multiplier component (which implements OP_MUL) are allocated. Your program should optimize the expected cycle count under the specified resource constraint.

5.3. (Register Binder) Implement a register binder with one of the following algorithms:

- Left edge algorithm
- Coloring algorithm
- Weighted clique partitioning algorithm

Your program should optimize the number of registers used while respecting the result of scheduling in Exercise 5.2.

5.4. (Functional Unit Binder) The goal of this exercise is to implement a functional unit binder. The result of binding should respect the result of scheduling (Problem 5.2) and register allocation (Problem 5.3) in the previous exercises, while minimizing the cost of multiplexers.

5.5. (HDL Generation) The goal of this exercise is to export the result of behavioral synthesis to a form that is acceptable to commercial logic synthesis and backend tools:

- A component generator that can output VHDL/Verilog code that uses the Synopsys DesignWare components to implement the necessary RTL component in your synthesis result.
- A controller/datapath generator that outputs the VHDL/Verilog code for the controller, datapath, and the top-level design, respectively.

5.6. (Multicycling and Functional Unit Pipelining) Modify Algorithm 5.1 to incorporate realistic timing:

- Functional unit latency is larger than one.
- Functional unit latency is larger than one, but can process data every cycle.

5.7. (Register Binding) It has been shown in [Golumbic 1980] that the coloring problem can be optimally solved in linear time, if the interference graph is chordal.

- Show that the interference graph for values in a basic block is chordal.
- Show that the interference graph for values in a TinyC program is chordal.

5.8. (Phase Ordering) Create an example to demonstrate that scheduling can significantly impact the register allocation result. Devise a strategy to mitigate this so-called phase ordering problem.

ACKNOWLEDGMENTS

We thank Rami Beidas and Wai Sum Mong of University of Toronto for their help in preparing the examples used in the text. We also thank Professor Jie-Hong (Roland) Jiang of National Taiwan University, Professor Preeti Ranjan Panda of Indian Institute of Technology, Delhi, and Dr. Sumit Gupta of Nvidia, for their valuable feedback on this chapter.

REFERENCES

R5.0 Books

[Aho 2006] A. Aho, R. Sethi, J. Ullman, and M. Lam, *Compilers: Principles and Techniques and Tools,* Second, Addison-Wesley, Reading, MA, 2006.

[De Micheli 1994] G. De Micheli, *Synthesis and Optimization of Digital Circuits,* McGraw-Hill, Hightstown, NJ, 1994.

[Gajski 1992] D. D. Gajski, N. D. Dutt, A. C.-H. Wu, and S. Y.-L. Lin, *High-level Synthesis: Introduction to Chip and System Design,* Kluwer Academic, Norwell, MA, 1992.

[Gajski 1994] D. D. Gajski, F. Vahid, Narayan, and J. Gong, *Specification and Design of Embedded Systems,* Prentice-Hall, Englewood Cliffs, NJ, 1994.

[Gajski 2000] D. D. Gajski, J. Zhu, R. Domer, A. Gerstlauer, and S. Zhao, *SpecC: Specification Language and Methodology,* Kluwer Academic, Norwell, MA, 2000.

[Golumbic 1980] M. C. Golumbic, *Algorithmic Graph Theory and Perfect Graphs,* Academic Press, 1980.

[Grötker 2002] T. Grötker, S. Liao, G. Martin, and S. Swan, *System Design with SystemC,* Kluwer Academic, Norwell, MA, 2004.

[Mishra 2008] P. Mishra and N. Dutt, *Processor Description Languages,* Morgan Kauffman, San Francisco, 2008.

[Pasricha 2008] S. Pasricha and N. Dutt, *On-Chip Communication Architectures: System on Chip Interconnect,* Morgan Kauffman, San Francisco, 2008.

[Yen 1996] T.-Y. Yen and W. Wolf, *Hardware-Software Co-synthesis of Distributed Embedded Systems,* Kluwer Academic, Norwell, MA, 1996.

R5.1 Introduction

[Berry 1992] G. Berry and G. Gonthier, The Esterel Synchronous Programming Language: Design, Semantics, Implementation, *Science of Computer Programming*, 19(2), pp. 87–152, November 1992.

[Halbwachs 1991] N. Halbwachs, P. Caspi, P. Raymond, and D. Pilaud, The synchronous data flow programming language LUSTRE, *Proceedings of the IEEE*, 79(9), pp. 1305–1320, September 1991.

[Kahn 1974] G. Kahn, The semantics of a simple language for parallel programming, in *Information Processing*, pp. 471–475, August 1974.

[Krolikoski 1999] S. J. Krolikoski, F Schirrmeister, B. Salefski, J. Rowson, and G. Martin, Methodology and technology for virtual component driven hardware/software co-design on the system-level, in *Proc. IEEE Int. Symp. on Circuits and Systems*, 6, pp. 456–459, July 1999.

[Lee 1996] E. Lee and A. Sangiovanni-Vincentelli, Comparing models of computation, in *Proc. IEEE/ACM Int. Conf. on Computer-Aided Design*, pp. 234–241, November 1996.

[Lee 1987] E. A. Lee and D. G. Messerschmitt, Static scheduling of synchronous data flow programs for digital signal processing, *IEEE Trans. on Computers*, 36(1), pp. 24–35, January 1987.

[SIA 2007] Semiconductor Industry Association, *The International Technology Roadmap for Semiconductors*: 2007 Edition, http://public.itrs.net, 2007.

[Zhu 1997] J. Zhu, R. Doemer, and D. Gajski, Syntax and semantics of SpecC+ language, in *Proc. Seventh Workshop on Synthesis and System Integration of Mixed Technologies*, pp. 75–82, December 1997.

R5.7 Concluding Remarks

[Artieri 2003] A. Artieri, V. D'Alto, R. Chesson, M. Hopkins, and M. C. Rossi, Nomadik™ open multimedia platform for next-generation mobile devices, *STMicroelectronics Technical Article TA305*, http://www.si.com, 2003.

[Bhasker 1990] J. Bhasker and H.-C. Lee, An optimizer for hardware synthesis, *IEEE Design & Test of Computers*, 7(5), pp. 20–36, September–October 1990.

[Brewer 1988] F. D. Brewer, Constraint driven behavioral synthesis, Ph.D. thesis, Dept. of Computer Science, University of Illinois, May 1988.

[Camposano 1991] R. Camposano, Path-based scheduling for synthesis, *IEEE Trans. on Computer-Aided Design*, 10(1), pp. 85–93, January 1991.

[Dally 2001] W. J. Dally and B. Towles, Route packets, not wires: On chip interconnection networks, in *Proc. ACM/IEEE Design Automation Conf.*, pp. 684–689, June 2001.

[De Man 1986] H. De Man, J. Rabaey, P. Six, and L. Claesen, Cathedral-II: A silicon compiler for digital signal processing, *IEEE Design & Test of Computers*, 3(6), pp. 73–85, November-December 1986.

[Dutt 2001] S. Dutt, R. Jensen, and A. Rieckmann, Viper: A multiprocessor SOC for advanced set-top box and digital TV systems, *IEEE Design & Test of Computers*, 18(5), pp. 21–31, September-October 2001.

[Ernst 1996] R. Ernst, J. Henkel, T. Benner, W. Ye, U. Holtmann, D. Herrmann, and M. Trawny, The COSYMA environment for hardware/software cosynthesis of small embedded systems, *J. Microprocessors and Microsystems*, 20(3), pp. 159–166, May 1996.

[Fauth 1995] A. Fauth, J. Van Praet, and M. Freericks, Describing instruction set processors with nML, in *Proc. IEEE/ACM Design, Automation and Test in Europe Conf.*, pp. 503–507, March 1995.

[Gomez 2004] J. I. Gomez, P. Marchal, S. Verdoorlaege, L. Pinuel, and F. Catthoor, Optimizing the memory bandwidth with loop morphing, in *Proc. IEEE Int. Conf. on Application-Specific Systems, Architectures and Processors*, pp. 213–223, September 2004.

[Grun 2001] P. Grun, N. Dutt, and A. Nicolau, APEX: Access pattern based memory architecture exploration, in *Proc. Int. Symp. on System Synthesis*, pp. 25–32, September 2001.

[Gupta 1992] R. K. Gupta, C. N. Coelho, and G. De Micheli, Synthesis and simulation of digital systems containing interacting hardware and software components, in *Proc. ACM/IEEE Design Automation Conf.*, pp. 225–230, June 1992.

[Gupta 2003] S. Gupta, N. D. Dutt, R. K. Gupta, and A. Nicolau, SPARK: A high-level synthesis framework for applying parallelizing compiler transformations, in *Proc. IEEE Int. Conf. on VLSI Design*, pp. 461–466, January 2003.

[Gyrczyc 1984] E. Gyrczyc, Automatic generation of micro-sequenced data paths to realize ADA circuit descriptions, Ph.D. thesis, Carleton University, 1984.

[Halambi 1999] A. Halambi, P. Grun, V. Ganesh, A. Khare, N. D. Dutt, and A. Nicolau, EXPRESSION: A language for architectural exploration through compiler/simulator retargetability, in *Proc. IEEE/ACM Design, Automation and Test in Europe Conf.*, pp. 485–490, March 1999.

[Helmig 2002] J. Helmig, Developing core software technologies for TI's OMAP[TM] platform, *Texas Instruments*, http://www.ti.com, 2002.

[Hind 2000] M. Hind and A. Pioli, Which pointer analysis should I use?, in *Proc. ACM SIGSOFT Int. Symp. on Software Testing and Analysis*, pp. 113–123, August 2000.

[Hoffmann 2001] A. Hoffmann, O. Schliebusch, A. Nohl, G. Braun, O. Wahlen, and H. Meyr, A methodology for the design of application specific instruction set processors (ASIP) with the machine description language LISA, in *Proc. IEEE/ACM Int. Conf. on Computer-Aided Design*, pp. 625–630, November 2001.

[Intel 2002] Intel, *Product Brief: Intel IXP2850 Network Processor*, http://www.intel.com, 2002.

[Kurdahi 1987] F. J. Kurdahi and A. C. Parker, REAL: A program for register allocation, in *Proc. ACM/IEEE Design Automation Conf.*, pp. 210–215, June 1987.

[Landskov 1980] D. Landskov, S. Davidson, B. Shriver, and P. W. Mallett, Local microcode compaction techniques, *ACM Computing Surveys*, 12(3), pp. 261–294, September 1980.

[Marwedel 1986] P. Marwedel, A new synthesis for the MIMOLA software system, in *Proc. ACM/IEEE Design Automation Conf.*, pp. 271–277, June 1986.

[McFarland 1978] M. C. McFarland, The Value Trace: A database for automated digital design, Technical Report DRC-01-4-80, Design Centre, Carnegie-Mellon University, December 1978.

[Nicolau 1991] A. Nicolau and R. Potasman, Incremental tree height reduction for high-level synthesis, in *Proc. ACM/IEEE Design Automation Conf.*, pp. 770–774, June 1991.

[Panda 2001a] P. R. Panda, L. Semeria, and G. De Micheli, Cache-efficient memory layout of aggregate data structures, in *Proc. Int. Symp. on System Synthesis*, pp. 101–106, September 2001.

[Panda 2001b] P. R. Panda, F. Catthoor, N. D. Dutt, K. Danckaert, E. Brockmeyer, C. Kulkarani, A. Vandercappelle, and P. G. Kjeldsberg, Data and memory optimization techniques for embedded systems, *ACM Trans. on Design Automation of Electronic Systems*, 6(2), pp. 149–206, February 2001.

[Pangrle 1987] B. M. Pangrle and D. D. Gajski, Slicer: A state synthesizer for intelligent silicon compilation, in *Proc. IEEE/ACM Int. Conf. on Computer-Aided Design*, pp. 42–45, November 1987.

[Parker 1986] A. C. Parker, J. Pizarro, and M. Mlinar, MAHA: a program for datapath synthesis, in *Proc. ACM/IEEE Design Automation Conf.*, pp. 461–466, June 1986.

[Paulin 1989] P. Paulin and J. Knight, Force-directed scheduling for the behavioral synthesis of ASIC's, *IEEE Trans. on Computer-Aided Design*, 8(6), pp. 661–679, June 1989.

[Semeria 2001] L. Semeria and G. De Micheli, Resolution, optimization, and encoding of pointer variables for the behavioral synthesis from C, *IEEE Trans. on Computer-Aided Design*, 20(2), pp. 213–233, February 2001.

[Thomas 1988] D. E. Thomas, E. M. Dirkes, R. A. Walker, J. V. Rajan, J. A. Nestor, and R. L. Blackburn, The system architect's workbench, in *Proc. ACM/IEEE Design Automation Conf.*, pp. 337–343, June 1988.

[Tseng 1986] C.-J. Tseng and D. P. Siewiorek, Automated synthesis of data paths in digital systems, *IEEE Trans. on Computer-Aided Design*, 5(3), pp. 379–395, March 1986.

[Um 2003] J. Um and T. Kim, Synthesis of arithmetic circuits considering layout effects, *IEEE Trans. on Computer-Aided Design*, 22(11), pp. 1487–1503, November 2003.

[Wolf 2003] W. Wolf and M. Kandemir, Memory system optimization of embedded software, *Proceedings of The IEEE*, 91(1), pp. 165–182, January 2003.

[Wuytack 1999] S. Wuytack, F. Catthoor, G. D. Jong, and H. J. De Man, Minimizing the required memory bandwidth in VLSI system realizations, *IEEE Trans. on Very Large Scale Integration Systems*, 7(4), pp. 433–441, April 1999.

[Xu 1998] M. Xu and F. J. Kurdahi, Layout-driven high level synthesis for FPGA based architectures, in *Proc. IEEE/ACM Design, Automation and Test in Europe*, pp. 446–450, February 1998.

[Zhu 2002] J. Zhu, Symbolic pointer analysis, in *Proc. IEEE/ACM Int. Conf. on Computer-Aided Design*, pp. 150–157, November 2002.

Logic synthesis in a nutshell

6

Jie-Hong (Roland) Jiang
National Taiwan University, Taipei, Taiwan

Srinivas Devadas
*Massachusetts Institute of Technology, Cambridge
Massachusetts*

ABOUT THIS CHAPTER

What is **logic synthesis?** As the name itself suggests, logic synthesis is the process of automatic production of logic components, in particular digital circuits. It is a subject about how to abstract and represent logic circuits, how to manipulate and transform them, and how to analyze and optimize them. *Why* does logic synthesis matter? Not only does it play a crucial role in the electronic design automation flow, its techniques also find broader and broader applications in formal verification, software synthesis, and other fields. *How* is logic synthesis done? Read on!

This chapter covers classic elements of logic synthesis for combinational circuits. After introducing basic data structures for Boolean function representation and reasoning, we will study technology-independent logic minimization, technology-dependent circuit optimization, timing analysis, and timing optimization. Some advanced subjects and important trends are presented as well for further exploration.

6.1 INTRODUCTION

Since Jack Kilby's invention of the first *integrated circuit* (IC) in 1958, there have been unprecedented technological advances. Intel co-founder Gordon E. Moore in 1965 predicted an important miniaturization trend for the semiconductor industry, known as Moore's Law, which says that the number of available transistors being economically packed into a single IC grows exponentially, doubling approximately every two years. This trend has continued for more than four decades, and perhaps will continue for another decade or even longer. At this time of 2008, the number of transistors in a single IC can be as many as several billion. This continual increase in design complexity under stringent time-to-market constraints is the primary driving force for changes in design tools and methodologies. To manage the ever-increasing complexity, people seek to maximally automate the design process and deploy techniques such as abstraction and hierarchy. Divide-and-conquer approaches are typical in the **electronic design automation** (EDA) flow and lead to different abstraction levels, such as the behavior level, **register-transfer level** (RTL), gate level, transistor level, and layout level from abstract to concrete.

Logic synthesis is the process that takes place in the transition from the register-transfer level to the transistor level. It is a highly automated procedure bridging the gap between high-level synthesis and physical design automation. Given a digital design at the register-transfer level, logic synthesis transforms it into a gate-level or transistor-level implementation. The highly engineered process explores different ways of implementing a logic function optimal with respect to some desired design constraints. The physical positions and interconnections of the gate layouts are then further determined at the time of physical design.

The main mathematical foundation of logic synthesis is the intersection of logic and algebra. The "algebra of logic" created by George Boole in 1847, a.k.a. *Boolean algebra*, is at the core of logic synthesis. (In our discussion we focus on two-element Boolean algebra [Brown 2003].) One of the most influential works connecting Boolean algebra and circuit design is Claude E. Shannon's M.S. thesis, *A Symbolic Analysis of Relay and Switching Circuits*, completed at the Massachusetts Institute of Technology in 1937. He showed that the design and analysis of switching circuits can be formalized using Boolean algebra, and that switching circuits can be used to solve Boolean algebra problems. Modern electronic systems based on digital (in contrast to analog) and two-valued (in contrast to multi-valued) principles can be more or less attributed to Shannon. The minimization theory of Boolean formulas in the two-level **sum-of-products** (SOP) form was established by Willard V. Quine in the 1950s. The minimization of SOP formulas found its wide application in IC design in the 1970s when **programmable logic arrays** (PLAs) were a popular design style for control logic implementation. It was the earliest stage of logic design minimization. When multilevel logic implementation became viable in the 1980s, the minimization theory and practice were broadened to the multi-level case.

Switching circuits in their original telephony application were strictly *combinational*, containing no memory elements. Purely combinational circuits however are not of great utility. For pervasive use in computation a combinational circuit needs to be augmented by memory elements that retain some of the state of a circuit. Such a circuit is *sequential* and implements a **finite state machine** (FSM). FSMs are closely related to finite automata, introduced in the theory of computation. Finite automata and finite state machines as well as their state minimization were extensively studied in the 1950s. Even though FSMs have limited computation power, any realistic electronic system as a whole can be seen as a large FSM because, after all, no system can have infinite memory resources. FSM state encoding for the two-level and multilevel logic implementations was studied extensively in the 1980s.

In addition to two-level and multilevel logic minimization, important algorithmic developments in logic synthesis in the 1980s include retiming of synchronous sequential circuits, algorithmic technology mapping, reduced ordered binary decision diagrams, and symbolic sequential equivalence checking using characteristic functions, just to name a few. Major logic synthesis tools of this period include, for example, ESPRESSO [Rudell 1987] and later MIS [Brayton 1987], developed at the University of California at Berkeley. They soon turned out to be the core engines of commercial logic synthesis tools.

In the 1990s, the subject of logic synthesis was much diversified in response to various IC design issues: power consumption, interconnect delay, testability, new implementation styles such as *field programmable gate array* (FPGA), etc. Important algorithmic breakthroughs over this period include, for instance, sequential circuit synthesis with retiming and resynthesis, don't care computation, image computation, timing analysis, Boolean reasoning techniques, and so on. Major academic software developed in this period include, *e.g.,* SIS [Sentovich 1992], the descendant of MIS.

In the 2000s, the directions of logic synthesis are driven by design challenges such as scalability, verifiability, design closure issues between logic synthesis and physical design, manufacture process variations, etc. Important developments include, for instance, effective satisfiability solving procedures, scalable logic synthesis and verification algorithms, statistical static timing analysis, statistical optimization techniques, and so on. Major academic software developed in this period include, *e.g.,* MVSIS [Gao 2002] and the ABC package [ABC 2005], with first release in 2005.

The advances of logic synthesis have in turn led to blossoming of EDA companies and the growth of the EDA industry. One of the first applications of logic optimization in a commercial use was to remap a netlist to a different standard cell library (in the first product, *remapper*, developed by Synopsys, an EDA company founded in 1986). It allowed an IC designer migrate a design from one library to another. Logic optimization could be used to optimize a gate-level netlist and map it into a target library. While logic optimization was finding its first commercial use for remapping, designers at major corporations, such as IBM, had already been demonstrating the viability of a top-down design methodology based on logic synthesis. At these corporations, internal simulation languages were coupled with synthesis systems that translated the simulation model into a gate-level netlist. Designers at IBM had demonstrated the utility of this synthesis-based design methodology on thousands of real industrial ICs. Entering a simulation model expressed using a *hardware description language* (HDL) makes logic synthesis and optimization move from a minor tool in a gate-level schematic based design methodology to the cornerstone of a highly productive IC design methodology. Commercial logic synthesis tools evolve and continue to incorporate developments addressing new design challenges.

The scope of logic synthesis can be identified as follows. An IC may consist of digital and analog components; logic synthesis is concerned with the digital part. For a digital system with sequential behavior, its state transition can be implemented in a synchronous or an asynchronous way depending on the existence of synchronizing clock signals. (Note that even a combinational circuit can be considered as a single-state sequential system.) Most logic synthesis algorithms focus on the synchronous implementation, and a few on the asynchronous one.

A digital system can often be divided into two portions: datapath and control logic. The former is concerned with data computation and storage, and often consists of arithmetic logic units, buses, registers/register files, etc.; the latter is concerned with the control of these data processing units. Unlike control logic, datapath circuits are often composed of regular structures. They are typically laid out manually by IC designers with full custom design to ensure that design constraints are satisfied, especially for high performance applications. Hence datapath design involves less logic synthesis efforts. In contrast, control logic is typically designed using logic synthesis. As the strengths of logic synthesis are its capabilities in logic minimization, it simplifies control logic. Consequently logic synthesis is particularly good for control-dominating applications, such as protocol processing, but not for arithmetic-intensive applications, such as signal processing.

Aside from the design issues related to circuit components, market-oriented decisions influence the design style chosen in implementing a product. The amount of design automation and logic synthesis efforts depends heavily on such decisions. Design styles based on full custom design, standard cells, and FPGAs represent typical trade-offs. In full custom design, logic synthesis is of limited use, mainly only in synthesizing performance non-critical controllers. For standard cell and FPGA based designs, a great portion of a design may be processed through logic synthesis. It is not surprising that logic synthesis is widely applied in ***application specific ICs*** (ASICs) and FPGA-based designs.

6.2 DATA STRUCTURES FOR BOOLEAN REPRESENTATION AND REASONING

The basic mathematical objects to be dealt with in this chapter are Boolean functions. How to compactly represent Boolean functions (the subject of logic minimization) and how to efficiently solve Boolean constraints (the subject of

Boolean reasoning) are closely related questions that play central roles in logic synthesis. There are several data structures for Boolean function representation and manipulation. For *Boolean representation*, we introduce some of the most commonly used ones, in particular, ***sum-of-products*** (SOP), ***product-of-sums*** (POS), ***binary decision diagrams*** (BDDs), ***and-inverter graphs*** (AIGs), and **Boolean networks**, among many others. For Boolean reasoning, we discuss how BDD, SAT, and AIG packages can serve as the core engines for Boolean function manipulation and for automatic reasoning of Boolean function properties. The efficiency of a data structure is mainly determined by its succinctness in representing Boolean functions and its capability of supporting Boolean manipulation. Each data structure has its own strengths and weaknesses; there is not a single data structure that is universally good for all applications. Therefore, conversion among different data types is a necessity in logic synthesis, where various circuit transformation and verification techniques are applied.

6.2.1 Quantifier-free and quantified Boolean formulas

We introduce (quantifier-free) Boolean formulas for Boolean function representation and ***quantified Boolean formulas*** (QBFs) for Boolean reasoning.

A **Boolean variable** is a variable that takes on binary values $\mathbb{B} = \{false, true\}$, or $\{0, 1\}$, under a truth assignment; a **literal** is a Boolean variable or its complement. In the n-**dimensional Boolean space** or **Boolean n-space** $\mathbb{B}^n$, an atomic element (or vertex) $a \in \mathbb{B}^n$ is called a **minterm**, which corresponds to a truth assignment on a vector of n Boolean variables.

An n-ary **completely specified Boolean function** $f : \mathbb{B}^n \to \mathbb{B}$ maps every possible truth assignment on the n input variables to either true or false. Let symbol "-", "X", or "2" denote the don't care value. We augment $\mathbb{B}$ to $\mathbb{B}_+ = \mathbb{B} \cup \{-\}$ and define an **incompletely specified Boolean function** $f : \mathbb{B}^n \to \mathbb{B}_+$, which maps every possible truth assignment on the n input variables to true, false, or don't care. For some $a \in \mathbb{B}^n$, $f(a) = -$ means the function value of f under the truth assignment a does not matter. That is, a is a don't care condition for f. Unless otherwise stated, we shall assume that a Boolean function is completely specified.

The mapping induced by a set of Boolean functions can be described by a **functional vector** or a **multiple-output function $\boldsymbol{f}$**, which combines $m > 1$ Boolean functions into a mapping $\boldsymbol{f} : \mathbb{B}^n \to \mathbb{B}^m$ if $\boldsymbol{f}$ is completely specified, or a mapping $\boldsymbol{f} : \mathbb{B}^n \to \mathbb{B}_+^m$ if $\boldsymbol{f}$ is incompletely specified.

For a completely specified function f, we define its **onset** $f^{\text{on}} = \{a \in \mathbb{B}^n \mid f(a) = 1\}$ and **offset** $f^{\text{off}} = \{a \in \mathbb{B}^n \mid f(a) = 0\}$. For an incompletely specified function f, in addition to the onset and offset, we have the **dcset** $f^{\text{dc}} = \{a \in \mathbb{B}^n \mid f(a) = -\}$. Although the onset, offset, and dcset are named sets rather than functions, we will see later that sets and functions can be unified through the use of the so-called **characteristic functions**.

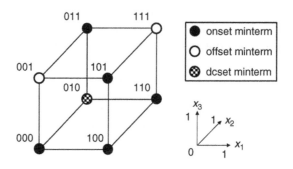

FIGURE 6.1

Boolean 3-space and a 3-ary Boolean function.

Example 6.1 The Boolean 3-space spanned by the variable vector (x_1, x_2, x_3) can be viewed as a combinatorial cube as shown in Figure 6.1, where the labeled vertices represent the minterms and two minterms are connected by an edge if their Hamming distance is one (that is, their binary codes differ in one position). The onset $f^{\text{on}} = \{000, 011, 100, 101, 110\}$, offset $f^{\text{off}} = \{001, 111\}$, and dcset $f^{\text{dc}} = \{010\}$ of some function f are embedded in the combinatorial cube.

A completely specified Boolean function f is a **tautology**, written as $f \equiv 1$ or $f \Leftrightarrow 1$, if its onset equals the universal set, *i.e.*, the entire Boolean space. In other words, the output of f equals 1 under every truth assignment on the input variables.

Any Boolean function can be expressed in a **Boolean formula**. Table 6.1 shows the building elements (excluding the last two symbols, $\exists$ and $\forall$) of a Boolean formula. Symbols $\neg, \wedge, \vee, \Rightarrow, \Leftrightarrow$ are Boolean connectives. A Boolean formula φ can be built recursively through the following formation rules:

$$\varphi ::= 0 \,|\, 1 \,|\, A \,|\, \neg\varphi_1 \,|\, \varphi_1 \wedge \varphi_2 \,|\, \varphi_1 \vee \varphi_2 \,|\, \varphi_1 \Rightarrow \varphi_2 \,|\, \varphi_1 \Leftrightarrow \varphi_2 \qquad (6.1)$$

where the symbol "::=" is read as "can be" and symbol "|" as "or". That is, a Boolean formula φ can be a constant 0, a constant 1, an atomic Boolean variable from a variable set A, $\neg\varphi_1$, $\varphi_1 \wedge \varphi_2$, $\varphi_1 \vee \varphi_2$, $\varphi_1 \Rightarrow \varphi_2$, or $\varphi_1 \Leftrightarrow \varphi_2$, built from Boolean formulas φ_1 and φ_2. To save on parentheses and enhance readability, we assume the precedence of the Boolean connectives $\Leftrightarrow, \Rightarrow, \vee, \wedge, \neg$ is in an ascending order. Also we often omit expressing the conjunction symbol $\wedge$ in a formula.

Example 6.2 The Boolean formula

$$((x_1 \vee (\neg x_2)) \vee ((\neg x_1) \wedge x_3)) \wedge (x_1 \wedge (\neg x_2))$$

can be shortened to

$$((x_1 \vee \neg x_2) \vee \neg x_1 x_3)(x_1 \neg x_2)$$

Table 6.1 Symbolic Notation and Meaning

Symbol	Symbol Name	English Meaning
(	left parenthesis	for punctuation
)	right parenthesis	for punctuation
$\neg$, $'$	complement symbol	logical "not"
$\wedge$, .	conjunction symbol	logical "and"
$\vee$, $+$	disjunction symbol	logical "(inclusive) or"
$\Rightarrow$	implication symbol	logical "if . . . , then . . . "
$\Leftrightarrow$, $\equiv$	bi-implication symbol	logical ". . . if and only if . . . "
$\exists$	existential quantifier	"there exists . . . "
$\forall$	universal quantifier	"for all . . . "

Using the associativity of disjunction and conjunction, we can further shorten the formula to

$$(x_1 \vee \neg x_2 \vee \neg x_1 x_3) x_1 \neg x_2$$

but we can no longer trace a unique sequence of rules used to derive this formula.

A set of Boolean operators is called **functionally complete** if they are sufficient to generate any Boolean function. Note that not all of the above Boolean connectives are necessary to form a set of functionally complete operators. For example, the sets $\{\neg, \wedge\}$ and $\{\neg, \Rightarrow\}$ are functionally complete, whereas $\{\wedge, \Rightarrow\}$ is not.

We may consider a Boolean function as the semantics of some Boolean formulas. There are different (syntactical) Boolean formulas representing the same (semantical) Boolean functions. It is this flexibility that makes logic synthesis an art.

Boolean operations over Boolean functions can be defined in terms of set operations, such as union $\cup$, intersection $\cap$, and complement over sets. Boolean function $b = f \wedge g$ has onset $b^{on} = f^{on} \cap g^{on}$ and offset $b^{off} = f^{off} \cup g^{off}$; Boolean function $b = f \vee g$ has onset $b^{on} = f^{on} \cup g^{on}$ and offset $b^{off} = f^{off} \cap g^{off}$; Boolean function $b = \neg f$ (also denoted as $\bar{f}$ or f') has onset $b^{on} = f^{off}$ and offset $b^{off} = f^{on}$. The dcset of function b can be derived using the fact that the union of the onset, offset, and dcset is equal to the universal set.

Quantified Boolean formulas (QBFs) generalize (quantifier-free) Boolean formulas with the additional universal and existential quantifiers: $\forall$ and $\exists$, respectively. In writing a QBF, we assume that the precedences of the quantifiers are lower than those of the Boolean connectives. In a QBF, variables being quantified are called **bound variables**, whereas those not quantified are called **free variables**.

Example 6.3 Consider the QBF $\forall x_1, \exists x_2.f(x_1, x_2, x_3)$, where f is a Boolean formula. It is read as *"For every* (truth assignment of) x_1, *there exists some* (truth assignment of) x_2, $f(x_1, x_2, x_3)$." In this case, x_1 and x_2 are bound variables, and x_3 is a free variable.

Any QBF can be rewritten as a quantifier-free Boolean formula through **quantifier elimination** by formula expansion (among other methods), *e.g.,*

$$\forall x.f(x,y) = f(0,y) \wedge f(1,y)$$

and

$$\exists x.f(x,y) = f(0,y) \vee f(1,y)$$

where f is a Boolean formula. Consequently, for any QBF φ, there exists an equivalent quantifier-free Boolean formula that *refers only to the free variables* of φ. For a QBF of size n with k bound variables, its quantifier-free Boolean formula derived by formula expansion can be of size $O\,(2^n \cdot k)$. QBFs are thus of the same expressive power as quantifier-free Boolean formulas, but can be exponentially more succinct.

Example 6.4 The QBF $\forall x_1, \exists x_2.f(x_1, x_2, x_3)$ can be rewritten as

$$
\begin{aligned}
&\forall x_1.(\,f(x_1,0,x_3) \vee f(x_1,1,x_3)) \\
=\ &(\exists x_2.f(0,x_2,x_3)) \wedge (\exists x_2.f(1,x_2,x_3)) \\
=\ &(f(0,0,x_3) \vee f(0,1,x_3)) \wedge (f(1,0,x_3) \vee f(1,1,x_3))
\end{aligned}
$$

Note that $\forall x_1, \exists x_2.f(x_1, x_2, x_3)$ differs from and is, in fact, weaker than $\exists x_2, \forall x_1.f(x_1, x_2, x_3)$. That is, $(\exists x_2, \forall x_1.f(x_1, x_2, x_3)) \Rightarrow (\forall x_1, \exists x_2.f(x_1, x_2, x_3))$. In contrast, $\forall x_1, \forall x_2.f(x_1, x_2, x_3)$ is equivalent to $\forall x_2, \forall x_1.f(x_1, x_2, x_3)$, and similarly $\exists x_1, \exists x_2.f(x_1, x_2, x_3)$ is equivalent to $\exists x_2, \exists x_1.f(x_1, x_2, x_3)$.

Moreover, it can be verified that the universal quantification $\forall$ commutes with the conjunction $\wedge$, whereas the existential quantification $\exists$ commutes with the disjunction $\vee$. That is, for any QBFs φ_1 and φ_2, we have

$$\forall x.(\varphi_1 \wedge \varphi_2) = \forall x.\varphi_1 \wedge \forall x.\varphi_2$$

whereas

$$\exists x.(\varphi_1 \vee \varphi_2) = \exists x.\varphi_1 \vee \exists x.\varphi_2$$

Nonetheless in general $\forall$ does not commute with $\vee$, whereas $\exists$ does not commute with $\wedge$. That is, in general

$$\forall x.(\varphi_1 \vee \varphi_2) \neq \forall x.\varphi_1 \vee \forall x.\varphi_2$$

and

$$\exists x.(\varphi_1 \wedge \varphi_2) \neq \exists x.\varphi_1 \wedge \exists x.\varphi_2$$

On the other hand, for any QBF φ, we have

$$\neg \forall x.\varphi = \exists x.\neg \varphi \qquad (6.2)$$

and
$$\neg \exists x.\varphi = \forall x.\neg\varphi \tag{6.3}$$

Because $\forall$ and $\exists$ can be converted to each other through negation, either quantifier solely is suffcient to represent QBFs.

An important fact about QBFs is that they are equivalent under renaming of bound variables. For example, $\forall x.f(x, y) = \forall z.f(z, y)$ and $\exists x.f(x, y) = \exists z.f(z, y)$. Renaming bound variables is often necessary if we want to rewrite a QBF in a different way. Being able to identify the scope of a quantifier is crucial for such renaming.

Example 6.5 In the QBF

$$Q_1 x, Q_2 y.(f_1(x,y,z) \vee f_2(y,z) \wedge Q_3 x.f_3(x,y,z))$$

with $Q_i \in \{\forall, \exists\}$, quantifier Q_1 is applied only to the variable x of f_1, quantifier Q_2 is applied to the y variables of all the functions, and quantifier Q_3 is applied only to the variable x of f_3. The QBF can be renamed as

$$Q_1 a, Q_2 b.(f_1(a,b,z) \vee \neg f_2(b,z) \wedge Q_3 x.f_3(x,b,z))$$

In studying QBFs, it is convenient to introduce a uniform representation, the so-called **prenex normal form**, where the quantifiers of a QBF are moved to the left leaving a quantifier-free Boolean formula on the right. That is,
$$Q_1 x_1, Q_2 x_2, \ldots, Q_n x_n f(x_1, x_2, \ldots, x_n)$$
where $Q_i \in \{\forall, \exists\}$ and f is a quantifier-free Boolean formula. Such movement is always possible by Equations (6.2) and (6.3) as well as the following equalities: For QBFs φ_1 and φ_2,
$$(\varphi_1 \Diamond Qx.\varphi_2) = Qx.(\varphi_1 \Diamond \varphi_2) \text{ if } x \text{ is not a free variable in } \varphi_1 \tag{6.4}$$

where $Q \in \{\forall, \exists\}$ and $\Diamond \in \{\wedge, \vee\}$,
$$(\varphi_1 \Rightarrow \forall x.\varphi_2) = \forall x.(\varphi_1 \Rightarrow \varphi_2) \text{ if } x \text{ is not a free variable in } \varphi_1 \tag{6.5}$$

$$(\varphi_1 \Rightarrow \exists x.\varphi_2) = \exists x.(\varphi_1 \Rightarrow \varphi_2) \text{ if } x \text{ is not a free variable in } \varphi_1 \tag{6.6}$$

$$((\forall x.\varphi_1) \Rightarrow \varphi_2) = \exists x.(\varphi_1 \Rightarrow \varphi_2) \text{ if } x \text{ is not a free variable in } \varphi_2. \quad \text{and} \tag{6.7}$$

$$((\exists x.\varphi_1) \Rightarrow \varphi_2) = \forall x.(\varphi_1 \Rightarrow \varphi_2) \text{ if } x \text{ is not a free variable in } \varphi_2 \tag{6.8}$$

With the renaming of bound variables, we know that the above conditions, x not a free variable in φ_i, can always be satisfied. Thereby any QBF can be converted into an equivalent formula in prenex normal form.

Prenex normal form is particularly suitable for the study of **computational complexity**. The number of alternations between existential and universal quantifiers in a QBF in prenex normal form directly reflects the difficulty in solving the formula. (In solving a QBF φ, we shall assume that all variables of φ are quantified, *i.e.*, no free variables in φ.) For instance, there are three alternations of quantifiers in the QBF $\forall x_1, \forall x_2, \exists x_3, \forall x_4, \exists x_5.f(x_1, \ldots, x_5)$. The more alternations of quantifiers are in a QBF in prenex normal form, the higher the computational complexity is in solving it. The levels of difficulties induce the **polynomial hierarchy**, a hierarchy of complexity classes, in complexity theory (see, *e.g.*, [Papadimitriou 1993] for comprehensive introduction). The problem of solving QBFs is known as *quantified satisfiability* (QSAT); in particular, the problem is known as QSAT$_i$ for QBFs in prenex normal form with *i* alternations of quantifiers. The entire polynomial hierarchy is contained by the PSPACE complexity class; the problem QSAT (without an *a priori* alternation bound *i*) is among the hardest in PSAPCE, *i.e.*, PSPACE-complete. A particularly interesting special case is QSAT$_0$ with all variables quantified existentially. It is known as the *Boolean satisfiability* (SAT) problem, which is NP-complete [Garey 1979]. Solving QBFs is much harder than solving the satisfiability of Boolean formulas.

In the above discussion of QBF solving, we assumed all variables are not free. For a QBF φ with free variables, we say that it is **satisfiable** (respectively **valid**) if it is true under *some* (respectively *every*) truth assignment on the set of free variables. Hence asking about the *satisfiability* of a Boolean formula $f(\boldsymbol{x})$ is the same as asking about the *validity/satisfiability* of the QBF $\exists \boldsymbol{x}.f(\boldsymbol{x})$; asking about the *validity* of a Boolean formula $f(\boldsymbol{x})$ is the same as asking about the *validity/satisfiability* of the QBF $\forall \boldsymbol{x}.f(\boldsymbol{x})$. Note that the validity and satisfiability of a formula are the same if there are no free variables.

Although QBFs are not directly useful for circuit representation, many computational problems in logic synthesis and verification (such as image computation, don't care computation, Boolean resubstitution, combinational equivalence checking, etc.) can be posed as QBF solving. Once a computational task is written in a QBF, its detailed algorithmic solution is almost apparent and can be derived using Boolean reasoning engines.

6.2.2 Boolean function manipulation

In addition to Boolean AND, OR, NOT operations, **cofactor** is an elementary Boolean operation. For a function $f(x_1, \ldots, x_i, \ldots, x_n)$, the **positive cofactor** and **negative cofactor** of f with respect to x_i are $f(x_1, \ldots, 1, \ldots, x_n)$, denoted as f_{x_i} or $f|_{x_i = 1}$, and $f(x_1, \ldots, 0, \ldots, x_n)$, denoted as $f_{\neg x_i}$ or $f|_{x_i = 0}$, respectively. We can also cofactor a Boolean function with respect to a **cube**, namely the conjunction of a set of literals, by iteratively cofactoring the function with each literal in the cube.

Example 6.6 Cofactoring the Boolean function $f = x_1 x_2 \neg x_3 \vee x_4 \neg x_5 x_6$ with respect to the cube $c = x_1 x_2 \neg x_5$ yields function $f_c = \neg x_3 \vee x_4 x_6$.

Universal and existential quantifications can be expressed in terms of cofactor, with

$$\forall x_i.f = f_{x_i} \wedge f_{\neg x_i} \tag{6.9}$$

and

$$\exists x_i f = f_{x_i} \vee f_{\neg x_i} \tag{6.10}$$

Moreover, the **Boolean difference** $\frac{\partial f}{\partial x_i}$ of f with respect to variable x_i is defined as

$$\frac{\partial f}{\partial x_i} = \neg(f_{x_i} \equiv f_{\neg x_i}) = f_{x_i} \oplus f_{\neg x_i} \tag{6.11}$$

where $\oplus$ denotes an exclusive-or (XOR) operator. Using the Boolean difference operation, we can tell whether a Boolean function functionally depends on a variable. If $\frac{\partial f}{\partial x_i}$ equals constant 0, then the valuation of f does not depend on x_i, that is, x_i is a redundant variable for f. We call that x_i is a **functional support variable** of f if x_i is not a redundant variable.

By **Shannon expansion**, every Boolean function f can be decomposed with respect to some variable x_i as

$$f = x_i f_{x_i} \vee \neg x_i f_{\neg x_i} \tag{6.12}$$

Note that the variable x_i needs not be a functional support variable of f.

6.2.3 **Boolean function representation**

Below we discuss different ways of representing Boolean functions.

6.2.3.1 *Truth table*

The mapping of a Boolean function can be exhaustively enumerated with a **truth table**, where every truth assignment has a corresponding function value listed.

Example 6.7 Figure 6.2 shows the truth table of the majority function $f(x_1, x_2, x_3)$, which valuates to true if and only if at least two of the variables $\{x_1, x_2, x_3\}$ valuate to true.

Truth tables are **canonical** representations of Boolean functions. That is, two Boolean functions are equivalent if and only if they have the same truth table.

x_1	x_2	x_3	f
0	0	0	0
0	0	1	0
0	1	0	0
0	1	1	1
1	0	0	0
1	0	1	1
1	1	0	1
1	1	1	1

FIGURE 6.2

Truth table of the 3-ary majority function.

Canonicity is an important property that may be useful in many applications of logic synthesis and verification.

For practical implementation, a truth table is effective in representing functions with a few input variables (often no more than 5 or 6 variables for modern computers having a word size 32 or 64 bits). By storing a truth table as a computer word, basic Boolean operations over two small functions can be done in constant time by parallel bitwise operation over their truth tables. Truth tables however are impractical to represent functions with many input variables.

6.2.3.2 *SOP*

Sum-of-products (SOP), or *disjunctive normal form* (DNF) as it is called in computer science, is a special form of Boolean formulas consisting of disjunctions (sums) of conjunctions of literals (product terms or cubes). It is a flat structure corresponding to a two-level circuit representation (the first level of AND gates and the second level of an OR gate). In two-level logic minimization, the set of product terms (*i.e.*, cubes) of an SOP representation of a Boolean function is called a **cover** of the Boolean function. A Boolean function may have many different covers, and a cover uniquely determines a Boolean function.

Example 6.8 The expression $f = ab\neg c + a\neg bc + \neg abc + \neg a\neg b\neg c$ is in SOP form. The set $\{ab\neg c,$ $a\neg bc, \neg abc, \neg a\neg b\neg c\}$ of cubes forms a cover of function f.

In our discussion, we often do not distinguish a cover and its represented function.

Every Boolean formula can be rewritten in an SOP representation. Unlike the truth table representation, the SOP representation is not canonical. In fact, how to express a Boolean function in the most concise SOP-form is intractable (in fact, NP-complete), and is termed **two-level logic minimization**.

Given SOP as the underlying Boolean representation, we study its usefulness for Boolean manipulation. Consider the conjunction of two cubes. It is computable in time linear in the number of literals because, having defined cubes as sets of literals, we compute the conjunction of cubes c and d, denoted $q = c \cap d$, by actually taking the union of the literal sets in c and d. However if $q = c \cap d$ computed in this fashion contains both a literal l and its complement $\neg l$, then the intersection is empty. Similarly the conjunction of two covers can be obtained by taking the conjunction of each pair of the cubes in the covers. Therefore, the AND operation of two SOP formulas is of quadratic time complexity. On the other hand, the OR operation is of constant time complexity since the disjunction of two SOP formulas is readily in SOP form. The complement operation is of exponential time complexity in the worst case.

Example 6.9 Complementing the function

$$f = x_1 \cdot y_1 + x_2 \cdot y_2 + \ldots + x_n \cdot y_n$$

will result in 2^n product terms in the SOP representation.

In addition to the above basic Boolean operations, SAT and TAUTOLOGY checkings play a central role in Boolean reasoning. Checking whether an SOP formula is satisfiable is of constant time complexity since any (irredundant) SOP formula other than constant 0 must be satisfiable. In contrast, checking whether an SOP formula is tautological is intractable, in fact, coNP-complete. When compared with other data structures to be introduced, SOP is not commonly used as the underlying representation in Boolean reasoning engines, but mainly used in two-level and multilevel logic minimization.

For the purposes of minimizing two-level logic functions, efficient procedures for performing Boolean operations on SOP representations or covers are desirable. A package for performing various Boolean operations such as conjunction, disjunction, and complementation is available as part of the ESPRESSO program [Rudell 1987].

6.2.3.3 *POS*

Product-of-sums (POS), or ***conjunctive normal form*** (CNF) as it is called in computer science, is a special form of Boolean formulas consisting of conjunctions (products) of disjunctions of literals (clauses). It is a flat structure corresponding to a two-level circuit representation (the first level of OR gates and the second level of an AND gate).

Example 6.10 The formula $(a + b + \neg c)(a + \neg b + c)(\neg a + b + c)(\neg a + \neg b + \neg c)$ is in POS form.

Every Boolean formula has an equivalent formula in POS form. Even though POS seems just the dual of SOP, it is not as commonly used in circuit design as SOP partly due to the characteristics of CMOS circuits, where NMOS is preferable to PMOS. Nevertheless it is widely used in Boolean reasoning. Satisfiability (SAT) solving over CNF formulas is one of the most important problems in computer science. In fact, every NP-complete problem can be reformulated in polynomial time as a SAT problem.

Given POS as the underlying data structure, we study its usefulness for Boolean function manipulation. For the AND operation, it is of constant time complexity since the conjunction of two POS formulas is readily in POS. For the OR operation, it is of quadratic time complexity since in the worst case the disjunction of two POS formulas must be converted to a POS formula by the distributive law.

Example 6.11 Given POS formulas $\varphi_1 = (a) \cdot (b)$ and $\varphi_2 = (c) \cdot (d)$, their disjunction $\varphi_1 + \varphi_2$ equals $(a + c) \cdot (a + d) \cdot (b + c) \cdot (b + d)$.

On the other hand, the complement operation is of exponential time complexity since in the worst case a POS formula may need to be complemented with De Morgan's Law followed by the distributive law.

Example 6.12 Complementing the $2n$-input Achilles heel function

$$f = (x_1 + y_1) \cdot (x_2 + y_2) \cdots (x_n + y_n)$$

will result in 2^n clauses in the POS representation.

As for the SAT and TAUTOLOGY checkings of POS formulas, the former is NP-complete, and the latter is of constant time complexity because any (irredundant) POS formula other than constant 1 cannot be a tautology. The POS representation is commonly used as the underlying representation in Boolean reasoning engines, called **SAT solvers**.

6.2.3.4 *BDD*

Binary decision diagrams (BDDs) were first proposed by Lee [Lee 1959] and further developed by Akers [Akers 1978]. In their original form, BDDs are not canonical in representing Boolean functions. To canonicalize the representation, Bryant [Bryant 1986, 1992] introduced restrictions on BDD variable ordering and proposed several reduction rules, leading to the well-known ***reduced ordered BDDs*** (ROBDDs). Among various types of decision diagrams, ROBDDs are the most widely used, and will be our focus.

Consider using an n-level binary tree to represent an arbitrary n-input Boolean function $f(x_1, \ldots, x_n)$. The binary tree, called a BDD, contains two types of nodes. A **terminal** node, or leaf, υ has as an attribute a value $value(\upsilon) \in \{0, 1\}$. A **non-terminal** node υ has as attributes an argument level-index $index(\upsilon) \in \{1, \cdots, n\}$ and two children: the 0-child, denoted $else(\upsilon) \in V$, and the 1-child, denoted *then* $(\upsilon) \in V$. If $index(\upsilon) = i$, then x_i is called the **decision variable** for node υ. Every node υ in a BDD corresponds to a Boolean function $f[\upsilon]$ defined recursively as follows.

1. For a terminal node υ,
   (a) If $value(\upsilon) = 1$, then $f[\upsilon] = 1$.
   (b) If $value(\upsilon) = 0$, then $f[\upsilon] = 0$.

2. For a non-terminal node υ with $index(\upsilon) = i$,
   $$f[\upsilon](x_1, \ldots, x_n) = \neg x_i \cdot f[else(v)](x_1, \ldots x_n) + x_i \cdot f[then(v)](x_1, \ldots, x_n)$$

Recall that, in Shannon expansion, a Boolean function f can be written as $f = x_i f_{x_i} + \neg x_i f_{\neg x_i}$. Suppose a BDD node representing some function f is controlled by variable x_i. Then its 0-child and 1-child represent functions $f_{\neg x_i}$ and f_{x_i}, respectively. Accordingly a BDD in effect represents a recursive Shannon expansion. For a complete binary tree, it is easily seen that we can always find some value assignment to the leaves of a BDD to implement any n-input function $f(x_1, \ldots, x_n)$ because every truth assignment of variables $x_1, \ldots, x_n$ activates exactly one path from the root node to a unique leaf with the right function value. Note that a BDD represents the offset and the onset of a function as disjoint covers, where each cube in the cover corresponds to a path from the root node to some terminal node.

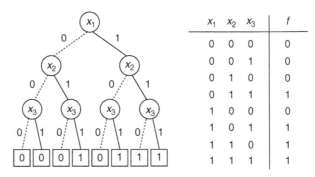

x_1	x_2	x_3	f
0	0	0	0
0	0	1	0
0	1	0	0
0	1	1	1
1	0	0	0
1	0	1	1
1	1	0	1
1	1	1	1

FIGURE 6.3

Binary tree representation of the majority function.

Example 6.13 The binary tree representation of the majority function is shown in Figure 6.3, where a circle (square) represents a non-terminal (terminal) node and a dotted (solid) edge indicates the pointed 0-child (1-child) of its parent node.

Definition 6.1. *A BDD is* **ordered** *(i.e., an OBDD) if the nodes on every path from the root node to a terminal node of the BDD follow the same variable ordering.*

Definition 6.2. *Two OBDDs D_1 and D_2 are* isomorphic *if there exists a one-to-one function σ from the nodes of D_1 onto the nodes of D_2 such that for any node υ if $\sigma(\upsilon) = w$, then either both υ and w are terminal nodes with value(υ) = value(w), or both υ and w are non-terminal nodes with index(υ) = index(w), $\sigma(else(\upsilon)) = else(w)$ and $\sigma(then(\upsilon)) = then(w)$.*

Since an OBDD only contains one root and the children of any non-terminal node are distinguished, the isomorphic mapping σ between OBDDs D_1 and D_2 is constrained and easily checked for. The root in D_1 must map to the root in D_2, the root's 0-child in D_1 must map to the root's 0-child in D_2, and so on all the way to the terminal nodes. Testing two OBDDs for isomorphism is thus a simple linear-time check.

Definition 6.3. ([Bryant 1986]). *An OBDD D is* **reduced** *if it contains no node υ with else(υ) = then(υ) nor does it contain distinct nodes υ and w such that the subgraphs rooted in υ and w are isomorphic.*

An **reduced OBDD** (ROBDD) can be constructed from an OBDD with the following three reduction rules:

1. Two terminal nodes with the same value attribute are merged.
2. Two non-terminal nodes u and υ with the same decision variable, the same 0-child, *i.e.*, $else(u) = else(\upsilon)$, and the same 1-child, $then(u) = then(\upsilon)$ are merged.
3. A non-terminal node υ with $else(\upsilon) = then(\upsilon)$ is removed, and its incident edges are redirected to its child node.

Iterating the reduction steps bottom-up on an OBDD until no further modification can be made, we obtain its unique corresponding ROBDD. These rules ensure that no two nodes of the ROBDD are structurally (also functionally) isomorphic, and that the derived ROBDD has fewest nodes under a given variable ordering. It can be shown that no two nodes of an ROBDD represent the same Boolean function, and thus two ROBDD of the same Boolean function must be isomorphic. That is, ROBDDs are a canonical representation of Boolean functions. Every function has a unique ROBDD for a given variable ordering.

Theorem 6.1 (ROBDD Canonicity [Bryant 1986]). *For any Boolean function f, there is a unique (up to isomorphism) ROBDD denoting f, and any other OBDD denoting f contains more nodes.*

Proof. A sketch of the proof is given using induction on the number of inputs.

Base case: If f has zero inputs, it can be either the unique 0 or 1 ROBDD.

Induction hypothesis: Any function g with a number of inputs $< k$ has a unique ROBDD.

Choose a function f with k inputs. Let D and D' be two ROBDDs for f under the same ordering. Let x_i be the input with the lowest index in the ROBDDs D and D'. Define the functions f_0 and f_1 as f_{x_i} and $f_{\neg x_i}$, respectively. Both f_0 and f_1 have less than k inputs, and by the induction hypothesis these are represented by unique ROBDDs D_0 and D_1.

We can have nodes in common between D_0 and D_1 or have no nodes in common between D_0 and D_1. If there are no nodes in common between D_0 and D_1 in D, and no nodes in common between D_0 and D_1 in D', then clearly D and D' are isomorphic.

Consider the case where there is a node u that is shared by D_0 and D_1 in D. There is a node u' in the D_0 of D' that corresponds to u. If u' is also in D_1 of D', then we have a correspondence between u in D and u' in D'. However, there could be another node u'' in the D_1 of u'' that also corresponds to u. While the existence of this node implies that D and D' are not isomorphic, the existence of u' and u'' in D' is a contradiction to the statement of the theorem, since the two nodes root isomorphic subgraphs corresponding to u. (This would imply that D' is not reduced.) Therefore, u'' cannot exist, and D and D' are isomorphic. □

Example 6.14 Figure 6.4, from 6.4a to 6.4c, shows the derivation of the ROBDD from the binary tree of the majority function.

Example 6.15 Consider the OBDD of Figure 6.5a. By the first reduction rule, we can merge all the terminal nodes with value 0 and all the terminal nodes with value 1. The functions rooted in the two nodes with control variable x_3 are identical, namely x_3. By the second reduction rule, we can delete one of the identical nodes and make the nodes that were pointing to the deleted node (those nodes whose 0- or 1-child correspond to the deleted node) point instead to the other node. This does not change the Boolean function corresponding to the OBDD. The simplified OBDD is shown in Figure 6.5b. In Figure 6.5b there is a node with control variable x_2 whose 0-child and 1-child both point to the same node. This node is redundant because the function f rooted in the node corresponds to function

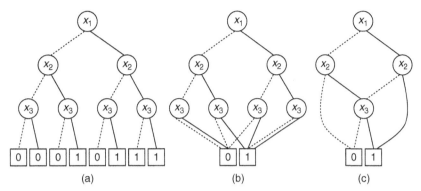

FIGURE 6.4

From binary tree to ROBDD.

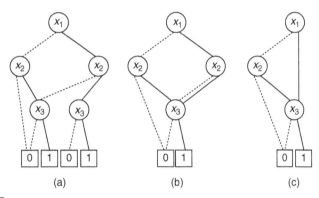

FIGURE 6.5

OBDD and simplified OBDDs.

$$f = x_2 \cdot x_3 + \neg x_2 \cdot x_3 = x_3$$

Thus, by the third reduction rule, all the nodes that point to f can be made to point to its 0- or 1-child without changing the Boolean function corresponding to the OBDD as illustrated in Figure 6.5c.

Example 6.16 Figure 6.6 shows a reduction example using a labeling technique for the ROBDD taken from [Bryant 1986]. We first assign the 0 and 1 terminal nodes a and b labels, respectively, in Figure 6.6a. Next, the right node with control variable x_3 is assigned label c. Upon encountering the other node with node with control variable x_3, we find that the second reduction rule is satisfied and assign this node the label c as well. Proceeding upward we assign the label c to the right node with control variable x_2 since the third reduction rule is satisfied for this node. (The 0-child and the 1-child of this node have the same label.) The left node with control variable x_2 is assigned label d, and the root node is assigned the label e. Note that the nodes are labeled in such a way that each label indicates a unique (sub-)ROBDD. Sorting and deleting redundant nodes results in the ROBDD of Figure 6.6b.

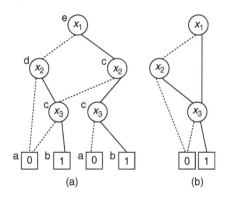

FIGURE 6.6

Reduction example.

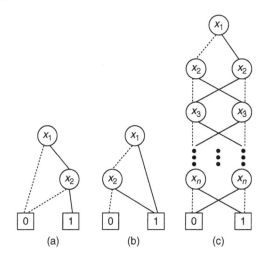

FIGURE 6.7

ROBDD examples: (a) ROBDD of function $f = x_1 \wedge x_2$. (b) ROBDD of function $f = x_1 \vee x_2$. (c) ROBDD of the n-ary odd parity function.

Example 6.17 To see that ROBDDs represent the offset and the onset of a function as disjoint covers, consider the examples of Figure 6.7. The ROBDD in (a) represents the function $f = x_1 \wedge x_2$. There are exactly two paths leading to the 0 terminal node. If x_1 is a 0, then the function represented by the ROBDD evaluates to a 0 since the 0-child of the node with index x_1 is the 0 terminal node. If x_1 is a 1 and x_2 is a 0, the function evaluates to a 0. Thus, the offset is represented as $\{\neg x_1, x_1 \neg x_2\}$. The two cubes in the cover are disjoint. If x_1 and x_2 are both 1, the function evaluates to a 1. The onset is the singleton $\{x_1 x_2\}$. Note that a cube of these covers corresponds to a single path from the root node to some terminal node. Similar analysis can be applied for the ROBDDs in (b) and (c).

In representing a Boolean function, different variable orderings may result in ROBDDs with very different sizes (in terms of the number of nodes).

Example 6.18 If the variables in the function $f = ab + cd$ are ordered as $index(a) < index(b) < index(c) < index(d)$ (a on top and d at bottom), the resulting ROBDD has only 4 non-terminal nodes. However, if the order $index(a) < index(c) < index(b) < index(d)$ is chosen, there are 7 non-terminal nodes.

Due to the sensitivity of ROBDD sizes to the chosen variable ordering, finding a suitable ordering becomes an important problem to obtain a reasonably sized ROBDD representing a given logic function. Finding the best variable ordering that minimizes the ROBDD size is coNP-complete [Bryant 1986]. However, there are good heuristics. For example, practical experience suggests that symmetric and/or correlated variables should be ordered close to each other. Other heuristics attempt to generate an ordering such that the structure of the ROBDD under this ordering mimics the given circuit structure.

It is not surprising that there exists a family of Boolean functions whose BDD sizes are exponential in their formula sizes under all BDD variable orderings. For instance, it has been shown that ROBDDs of certain functions, such as integer multipliers, have exponential sizes irrespective of the ordering of variables [Bryant 1991]. Fortunately for many practical Boolean functions, there are variable orderings resulting in compact BDDs. This phenomenon can be explained intuitively by the fact that a BDD with n nodes may contain up to 2^n paths, which correspond to all possible truth assignments. ROBDD representations can be considerably more compact than SOP and POS representations.

Example 6.19 The odd parity function of Figure 6.7c is an example of function which requires $2n - 1$ nodes in an ROBDD representation but 2^{n-1} product terms in a minimum SOP representation.

We examine how well ROBDDs support Boolean reasoning. Complementing the function of an ROBDD can be done in constant time by simply interchanging the 0 and 1 terminal nodes.

In cofactoring an ROBDD with respect to a literal x_i (respectively $\neg x_i$), the variable x_i is effectively set to 1 (respectively 0) in the ROBDD. This is accomplished by determining all the nodes whose 0- or 1-child corresponds to any node υ with $index(\upsilon) = i$, and replacing their 0- or 1-child by $then(\upsilon)$ (respectively $else(\upsilon)$).

Example 6.20 Figure 6.8 illustrates a cofactor example, where the given ROBDD of (a) has been cofactored with respect to x_3 yielding the ROBDD of (b). Similarly, an ROBDD can be cofactored with respect to $\neg x_i$ by using $else(\upsilon)$ to replace all nodes υ with $index(\upsilon) = i$.

Binary Boolean operations, such as AND, OR, XOR, and so on, over two ROBDDs (under the same variable ordering) can be realized using the recursive BddApply operation. In the generic BddApply operation, ROBDDs D_1 and D_2 are combined as $D_1 \langle op \rangle D_2$ where $\langle op \rangle$ is a Boolean function of two arguments.

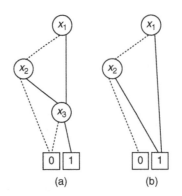

(a)   (b)

FIGURE 6.8

Cofactor example.

The result of the BDDAPPLY operation is another ROBDD. The operation can be customized by replacing $\langle op \rangle$ with a specific operator, *e.g.*, AND, OR, XOR, etc.

The algorithm proceeds from the roots of the two argument graphs downward, creating nodes in the resultant graph. It is based on the following recursion

$$f \langle op \rangle \, g = x_i \cdot (f_{x_i} \langle op \rangle \, g_{x_i}) + \neg x_i \cdot (f_{\neg x_i} \langle op \rangle \, g_{\neg x_i})$$

From an ROBDD perspective we have

$$f[v] \langle op \rangle g[w] = x_i \cdot (f[then(v)] \langle op \rangle g[then(w)]) + \neg x_i \cdot (f[else(v)] \langle op \rangle g[else(w)]) \quad (6.13)$$

where $f[v]$ and $g[w]$ are the Boolean functions rooted in the nodes v and w. There are several cases to consider.

1. If v and w are terminal nodes, we simply generate a terminal node u with *value(u) = value(v) $\langle op \rangle$ value(w)*.
2. Else, if *index(v) = index(w) = i*, we follow Equation (6.13). Create node u with *index(u) = i*, and apply the algorithm recursively on *else(v)* and *else(w)* to generate *else(u)* and on *then(v)* and *then(w)* to generate *then(u)*.
3. If *index(v) = i* but *index(w) > i*, we create a node u having index i, and apply the algorithm recursively on *else(v)* and w to generate *else(u)* and on *then(v)* and w to generate *then(u)*.
4. If *index(v) > i* and *index(w) = i* we create a node u having index i and apply the algorithm recursively on v and *else(w)* to generate *else(u)* and on v and *then(w)* to generate *then(u)*.

Implementing the above algorithm directly results in an algorithm of exponential complexity in the number of input variables, since every call in which one of the arguments is a non-terminal node generates two recursive calls. Two refinements can be applied to reduce this complexity. Firstly, if the algorithm is applied to two nodes where one is a terminal node, then we can return the result based on some Boolean identities. For example, we have $f \vee 1 = 1$ and

$f \vee 0 = f$ for $\langle op \rangle$ = OR, $f \wedge 0 = 0$ and $f \wedge 1 = f$ for $\langle op \rangle$ = AND and $f \oplus 0 = f$ and $f \oplus 1 = \neg f$ for $\langle op \rangle$ = XOR. Secondly, more importantly the algorithm need not evaluate a given pair of nodes more than once. We can maintain a hash table containing entries of the form (υ, w, u) indicating that the result of applying the algorithm to subgraphs with roots v and w was u. Before applying the algorithm to a pair of nodes we first check whether the table contains an entry for these two nodes. If so, we can immediately return the result. Otherwise we make the two recursive calls, and upon returning, add a new entry to the table. This refinement drops the time complexity to $O(|D_1| \cdot |D_2|)$, where $|D_1|$ and $|D_2|$ are the number of nodes in the two given graphs.

Example 6.21 We illustrate the BDDAPPLY algorithm with an example taken from [Bryant 1986]. The two ROBDDs to be operated on by an OR operator are shown in Figure 6.9a and 6.9b. Each node in the two ROBDDs has been assigned a unique label. This label could correspond to the labels generated during ROBDD reduction. The labels are required to maintain the table entries described immediately above.

The OBDD resulting from the OR of the two ROBDDs is shown in Figure 6.9c. First, we choose the pair of root nodes labeled a1 and b1. We create a node with control variable x_1 and recursively apply the algorithm to the node pairs a3, b1 and a2, b1. Since a3 corresponds to the 1 terminal node, we can immediately return the 1 terminal node as the result of the OR. We must still compute the OR of the a2, b1 node pair. This involves the computation of the OR of a2, b3 and a2, b2, and so on. Note that a3, b3 will appear as a node pair twice during the course of the algorithm.

Reducing the OBDD of Figure 6.9c results in the ROBDD of Figure 6.9d.

On the other hand, SAT and TAUTOLOGY checkings using BDDs are of constant time complexity due to the canonicity of BDDs. More specifically, SAT (respectively TAUTOLOGY) checking corresponds to checking if the BDD is not equal to the 0-terminal (respectively 1-terminal) node. Another application of BDDs is

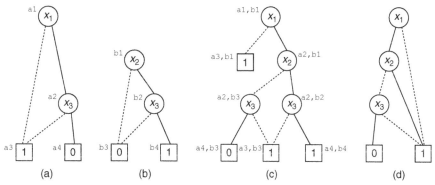

(a) (b) (c) (d)

FIGURE 6.9

ROBDD examples for the BDDAPPLY operation: (a) ROBDD of function $f_1 = \neg x_1 \vee \neg x_3$. (b) ROBDD of function $f_2 = x_2 \wedge x_3$. (c) Intermediate OBDD after the BDDAPPLY operation for $f_1 \vee f_2$. (d) Final ROBDD of $f_1 \vee f_2$.

checking if two functions f_1 and f_2 are equivalent. The problem is of constant time complexity given that f_1 and f_2 are already represented in BDDs under the same variable ordering. Two BDDs (under the same variable ordering) represent the same function if and only if they have the same root node.

As all the above Boolean manipulations are efficiently solvable (*i.e.*, in polynomial time), BDDs are a powerful tool in logic synthesis and verification. We are by no means saying that Boolean reasoning is easy because the BDD size of a function can be exponential in the number of variables. Building the BDD itself risks exponential memory blow-up. Consequently BDD shifts the difficulty from Boolean reasoning to Boolean representation. Nevertheless once BDDs are built, Boolean manipulations can be done efficiently. In contrast, CNF-based SAT solving is memory efficient but risks exponential runtime penalty. Depending on problem instances and applications, the capability and capacity of state-of-the-art BDD packages vary. Just to give a rough idea, BDDs with hundreds of Boolean variables are still manageable in memory but not with thousands of variables. In contrast, state-of-the-art SAT solvers typically may solve in reasonable time the satisfiability problem of CNF formulas with up to tens of thousands of variables.

For the implementation of effective BDD packages, there are several important techniques. Firstly, complemented edges can be used to compactly represent a function as well as its complement [Madre 1988]. A complemented edge indicates that the function rooted in the node that the edge points to has be complemented. Introducing complemented edges does not destroy the canonicity of the ROBDD if the edges to be complemented are selected properly.

Example 6.22 The ROBDDs for a function with and without complemented edges are shown in Figure 6.10. Complemented edges are indicated by dots on them.

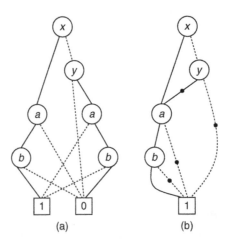

(a)   (b)

FIGURE 6.10

ROBDDs (a) without and (b) with complemented edges.

Secondly, a global unique table can be maintained wherein every node representing a unique function is given a unique label. Before creating a new node the table is checked to see if the function corresponding to this new node exists in the table. If not, the node is created, given a new label, and added to the unique table. If the function already exists, the node in the table corresponding to this function is returned.

Thirdly, dynamic variable ordering [Rudell 1993] can effectively reduce BDD sizes. A BDD variable ordering good for some functions may be bad for other functions. In the manipulation of ROBDD, new functions can be created. As a result, originally good variable ordering may become inadequate. Dynamic variable ordering provides a way of adjusting variable ordering to keep BDD sizes small. The description of an efficient implementation of an ROBDD package can be found in [Rudell 1990].

6.2.3.5 *AIG*

An *and-inverter graph* (AIG) is a *directed acyclic graph* (DAG) $G = (V, E)$ consisting of vertices V representing AND2 (two-input AND) gates and directed edges $E \subseteq V \times V$ connecting gates. Inverters are denoted by markers on edges. Since operators $\{\wedge, \neg\}$ are functionally complete, any Boolean function can be represented in an AIG. Most Boolean functions can be represented compactly using AIGs.

The simple AIG data structure allows quick and cheap **structural hashing** among AIG nodes. Two AIG nodes with the same inputs under the same complementation conditions are merged (similar to the second reduction rule of ROBDD). Unlike ROBDD, however, the AIG representation is not canonical even when structural hashing is applied.

Example 6.23 Figure 6.11 shows the AIGs of function $f = a\neg cd + \neg b\neg cd$ without and with structural hashing in 6.11a and 6.11b, respectively.

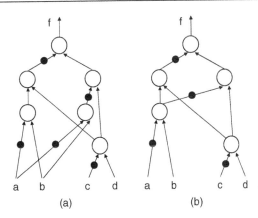

(a)   (b)

FIGURE 6.11

AIGs (a) without and (b) with structure hashing.

From the practical point of view, what make AIGs distinct from circuit netlists composed of AND2 gates and inverters are threefold:

1. Structural hashing — Structural hashing is applied during AIG construction; it propagates constants and ensures that each node is structurally unique. Accordingly AIGs are stored in a compact form.

2. Complemented edges — AIGs represent inverters as attributes on edges and thus do not require extra memory. Such complemented edges facilitate fast manipulation of AIGs and, in particular, lead to efficient structural hashing.

3. Regularity — As a result of regularity, memory management of an AIG package can be done by a simple customized memory manager which uses *fixed* amount of memory for each node (thanks to the fixed number of inputs of each node). By allocating memory for nodes in a topological order, we can optimize AIG *traversal*, which is repeatedly performed in many logic synthesis algorithms, in the same order. Experience suggests that many AIG-based applications have reduced **memory footprint** (namely, the amount of main memory used or referenced during a program's execution).

These features make a modern AIG package particularly efficient for Boolean function representation and reasoning.

We analyze the usefulness of AIGs for Boolean manipulation. The AND operation has a constant time complexity since the conjunction of two given AIGs can be done by adding an AIG node. The OR operation is essentially the same as the AND operation except for the markings on the input and output edges of the added AIG node, and thus is of constant time complexity. The complementation corresponds to marking an edge and is therefore of constant time complexity, too.

SAT and TAUTOLOGY checkings using AIGs are NP-complete and coNP-complete, respectively. When used as a Boolean reasoning engine, an AIG package can be viewed as a solver performing satisfiability checking over circuits rather than over CNF formulas, and is similar to *automatic test pattern generation* (ATPG).

AIGs can also be used in verification applications, such as equivalence checking and even model checking. For instance, checking if two given AIGs under comparison are functionally equivalent can be reduced to TAUTOLOGY (SAT) checking by adding an XNOR (XOR) gate, which can be expressed in terms of AND2 and INV gates, with its two-inputs fed in by the outputs of the two AIGs. The two AIGs are equivalent if and only if the output of the XNOR (XOR) gate is tautological (unsatisfiable). Hence the equivalence checking problem is coNP-complete. When it comes to synthesis, AIGs are used in multilevel logic minimization and technology mapping. In the academic system ABC [ABC 2005], AIGs are used as a unifying data structure for both logic synthesis and verification.

A new binary format called AIGER [Biere 2007] was recently proposed to enable compact representation of AIGs in files and memory. With memory requirements of about three bytes per AIG node, AIGER has become a standard representation for circuit-based problems in *SAT Competitions* and *Hardware Model Checking Competitions*, organized annually as satellite events of *International Conference on Theory and Applications of Satisfiability Testing* and *International Conference on Computer Aided Verification*, respectively.

6.2.3.6 *Boolean network*

A (combinational) logic circuit can be represented with a **Boolean network**, a directed graph $G = (V, E)$ with nodes V and directed edges E. Every node $i \in V$ is associated with a logic function f_i and a Boolean variable x_i, called the **output variable** of node i, representing the output of function f_i. Hence the relation between variable x_i and function f_i obeys $(x_i \equiv f_i)$. Every edge $(i, j) \in E$ connecting from node i to node j signifies that variable x_i is an input to function f_j, and we call that node i (j) is a **fanin (fanout)** of node j (i). That is, variable x_i syntactically appears in the Boolean expression of f_j as x_i or $\neg x_i$. We say x_i is a **(structural) support variable** of f_j. If, in addition, the Boolean difference $\frac{\partial f}{\partial x_i}$ is satisfiable, then x_i is a **functional support variable** of f_j, as defined previously.

A node i without any fanin is a **primary input** and its associated logic function is x_i, i.e., identical to its output variable. Moreover, a subset of V is specified as **primary outputs**. Among the variables of node outputs, we say those of the primary inputs are the **primary input variables**, those of the primary outputs are the **primary output variables**, and others are **local** (or **intermediate**) **variables**.

The sets of fanins and fanouts of node i are denoted as $FI(i)$ and $FO(i)$, respectively. The **transitive fanins** $TFI(i)$ and **transitive fanouts** $TFO(i)$ of a node i are defined recursively as

$$TFI(i) = \{k \in V \mid k = i, \text{or } k \in FI(j) \text{ for } j \in TFI(i)\}$$

and

$$TFO(i) = \{k \in V \mid k = i, \text{or } k \in FO(j) \text{ for } j \in TFO(i)\}$$

respectively.

A (combinational) Boolean network can be acyclic or cyclic. Any acyclic circuit must behave combinationally because no internal states can be maintained and the output only depends on the current input assignment, rather than on the prior input assignments; a cyclic circuit, in contrast, may possibly exhibit combinational behavior as well [Kautz 1970]. Because the existence of cyclic structures substantially complicates the analysis and optimization of logic design, most logic synthesis systems assume that combinational circuits are acyclic. In the sequel we shall assume that a Boolean network is acyclic. Therefore, $TFI(i) \cap TFO(i) = \{i\}$.

A node function f_i is a **local function**, in the sense that it is in terms of the output variables of the immediate fanins of node i. The function of node i can be alternatively expressed purely in terms of the primary input variables. In this case, it is called the **global function** g_i of node i. Function g_i can be derived from f_i by recursively substituting f_j for y_j, for $j \in TFI(i)$, until no further substitution is possible. This substitution process is guaranteed to terminate because of the assumption of *acyclic* combinational Boolean networks.

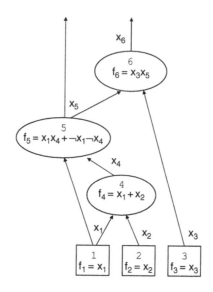

FIGURE 6.12

Boolean network example.

Example 6.24 Figure 6.12 shows a Boolean network example, where nodes 1, 2 and 3 are the primary inputs, and nodes 5 and 6 are the primary outputs. A local function f_i is shown in the corresponding node i. The global function of node i can be obtained by either recursive composition or quantification. For instance, the global function

$$g_5 = x_1(x_1 + x_2) + \neg x_1 \neg (x_1 + x_2)$$

by recursive composition, or equivalently

$$g_5 = \exists x_4.(x_1 x_4 + \neg x_1 \neg x_4)(x_4 \equiv (x_1 + x_2))$$

by quantification.

As for the implementation issue, how to represent the logic function f_i of a node i in a Boolean network is a matter of choice. Our previously mentioned data structures, such as the truth table, SOP, BDD, AIG, and Boolean network representations, can be adopted. Compared with AIGs, generic Boolean networks may lack special structures to be exploited for effective Boolean reasoning. They however are suitable for generic circuit representation.

6.2.4 **Boolean representation conversion**

6.2.4.1 *CNF vs. DNF*

SOP-to-POS and POS-to-SOP conversions can be achieved by applying double complements. By applying De Morgan's Law, an SOP (a POS) formula φ becomes a POS (an SOP) one φ' after the first complement. We can then convert the POS (SOP) formula φ' to an SOP (a POS) one φ'' by the distributive law. Finally, applying De Morgan's Law again for the second complement, we convert the SOP (POS) formula φ'' to a POS (an SOP) one φ'''. Note that the conversions may suffer from an exponential blow-up in formula sizes due to the intermediate step of applying the distributive law.

Example 6.25 The $2n$-input Achilles heel function $(x_1 + y_1)(x_2 + y_2) \cdots (x_n + y_n)$ has 2^n product terms in an SOP representation but has a linear-sized POS representation.

There exist Boolean functions whose SOP- and POS-formula sizes are inevitably exponential in the number of input variables. For example, the n-input odd parity function $(x_1 \oplus x_2 \cdots \oplus x_n)$ has 2^{n-1} product terms in an SOP representation and is equally large in a POS representation. As another example, integer multiplication over n-bit operands, comparison of two n-bit operands, and addition and subtraction of n-bit operands all have SOP and POS realizations that grow exponentially with n.

An interesting application of Boolean representation conversion is on Boolean reasoning. Recall that SAT (respectively TAUTOLOGY) checking is trivial for DNF (respectively CNF) formulas. If we are interested in knowing the satisfiability of a CNF formula, we may covert it into DNF and then check the satisfiability of the DNF formula, which is a constant time checking. Similarly we may check the tautology of a DNF formula by converting it into CNF. The hardness of Boolean reasoning, of course, is shifted to the representation conversion process. Another application of Boolean representation conversion is on quantifier elimination for QBFs. Observe that the universal (respectively existential) quantification is easy for CNF (respectively DNF) formulas. The QBF $\forall x_i.\varphi(\boldsymbol{x})$ with $\varphi(\boldsymbol{x})$ in CNF equals the induced quantifier-free Boolean formula of removing every appearance of literals x_i and $\neg x_i$ in $\varphi(\boldsymbol{x})$; similarly the QBF $\exists x_i.\varphi(\boldsymbol{x})$ with $\varphi(\boldsymbol{x})$ in DNF equals the induced quantifier-free Boolean formula of removing every appearance of literals x_i and $\neg x_i$ in $\varphi(\boldsymbol{x})$. It is thus of linear time complexity. Therefore given a QBF, we can convert the formula back and forth between CNF and DNF to eliminate quantifiers. As a consequence, any SOP-POS converter can be used as a Boolean reasoning engine and QBF solver.

Example 6.26 The QBF

$$\forall a.(a + b + \neg c)(a + \neg b + c)(\neg a + b + c)$$

equals the quantifier-free Boolean formula

$$(b + \neg c)(\neg b + c)(b + c)$$

The QBF

$$\exists a.(ab\neg c + a\neg bc + \neg abc)$$

equals the quantifier-free Boolean formula

$$b\neg c + \neg bc + bc$$

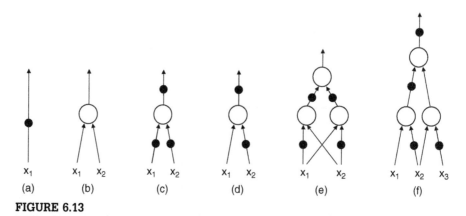

FIGURE 6.13

(a) AIG of $\neg x_1$. (b) AIG of $(x_1 \wedge x_2)$. (c) AIG of $(x_1 \vee x_2)$. (d) AIG of $(x_1 \Rightarrow x_2)$. (e) AIG of $(x_1 \Leftrightarrow x_2)$. (f) AIG of $(x_1 \wedge \neg x_2) \vee (x_2 \Rightarrow x_3)$.

6.2.4.2 *Boolean formula vs. circuit*

A Boolean formula φ can be translated into a circuit, *e.g.*, an AIG, in linear time. The translation can be done by following the inductive construction of φ with the rules of Equation (6.1).

Example 6.27 Figure 6.13a-e show the AIGs of $\neg x_1$, $x_1 \wedge x_2$, $x_1 \vee x_2$, $x_1 \Rightarrow x_2$, and $x_1 \Leftrightarrow x_2$. They form the templates of the basic formation rules of Equation (6.1). Given an arbitrary Boolean formula, its AIG can be built from these templates, *e.g.*, the AIG of $(x_1 \wedge \neg x_2) \vee (x_2 \Rightarrow x_3)$ is shown in Figure 6.13f.

Any (combinational) circuit, on the other hand, represents some Boolean function $f : \mathbb{B}^n \to \mathbb{B}$, which can be specified with a Boolean formula. Recall Example 6.24, which shows how an output function of a circuit can be obtained.

6.2.4.3 *BDD vs. Boolean network*

A two-input multiplexor is a switch with two data inputs i_0, i_1, one control input c, and one output o, with $o = i_0$ if $c = 0$ and $o = i_1$ if $c = 1$. Because a non-terminal node in a BDD can be seen as a two-input multiplexor and BDDs are universal for functional representation, any Boolean function can be implemented using a circuit whose only constituent gates are two-input multiplexors. Translating a BDD to a multiplexor-based Boolean network is a straightforward process by substituting every BDD node with a multiplexor, and can be accomplished in time linear in the size of the BDD.

Given a Boolean network, the ROBDD of a primary output function in terms of the primary input variables can be constructed. A naïve approach is to build an OBDD representing the global function of the Boolean network and then reduce it. Rather, a more effective way is to traverse the circuit from primary inputs to primary outputs using a series of Boolean manipulations over ROBDDs based on node

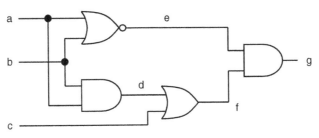

FIGURE 6.14

Multilevel circuit.

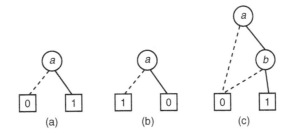

FIGURE 6.15

(a) ROBDD for primary input a. (b) ROBDD for $\neg a$. (c) ROBDDs for $a \wedge b$.

functions. For a primary input, its ROBDD is a graph with a single non-terminal node and two terminal nodes. For a functional node, its ROBDD can be constructed using a series of complement and/or BddApply operations.

Example 6.28 Consider the circuit of Figure 6.14. The ROBDD for primary input a is shown in Figure 6.15a. Similarly, the ROBDD for primary input b will have one node with control variable b with a 0-child (1-child) corresponding to the 0 (1) terminal node. The ROBDD for $\neg a$ is shown in Figure 6.15b. We can create the ROBDD for signal d by performing an AND operation on the ROBDDs for the primary inputs a and b. This ROBDD is shown in Figure 6.15c. We can create the ROBDD for signal f by performing an OR operation on the ROBDD for signal d and the ROBDD for the primary input c.

As an application, ROBDD-based circuit equivalence checking can be achieved by the conversion from Boolean networks to ROBDDs. Since ROBDDs are a canonical representation of Boolean functions, in order to check two circuits C_1 and C_2 for equivalence, we can use the following method.

1. Choose an ordering for the primary inputs of the circuits.
2. Create ROBDDs for the primary outputs of the two circuits.
3. Check if the ROBDDs are isomorphic. If so, the circuits are equivalent. If not, the circuits are not equivalent.

In order to check two ROBDDs for equivalence, we can use the canonicity property of ROBDDs and perform a linear-time graph isomorphism check as per Definition 6.2. Notice that any ordering will suffce, as long as the same ordering is chosen for both circuits. However, the size of the ROBDDs created is strongly dependent on the ordering chosen.

6.2.5 Isomorphism between sets and characteristic functions

A very profound application of Boolean functions is the concept of characteristic functions in representing sets. It is a very important idea leading to a leap in capacity of many logic synthesis and verification algorithms. A **characteristic function** is a (total) function $\chi_A : U \to \mathbb{B}$, where U is a finite set often in the form of $\mathbb{B}^n$ for some n, such that $\chi_A(e) = 1$ if and only if $e \in A$, that is, the onset of χ_A equals A. It serves as a predicate indicating the membership property. In other words, the function χ_A answers a query, whether an element $e \in U$ is in $A \subseteq U$. Essentially, any finite set $A \subseteq U$ can be represented with a characteristic function χ_A. Thereby set operations (*e.g.*, intersection $\cap$, union $\cup$, and complement) over sets are in effect Boolean operations (*e.g.*, conjunction $\wedge$, disjunction $\vee$, and negation $\neg$, respectively) over characteristic functions. Note that constant functions 0 and 1 are characteristic functions of the empty set $\emptyset$ and universal set U, respectively. Some applications of characteristic functions are given below.

Incompletely Specified Function as Characteristic Function. To represent an incompletely specified Boolean function $I : \mathbb{B}^n \to \{0, 1, -\}$, three characteristic functions r, f, d can be used to represent its onset, offset and dcset, respectively. That is, for a minterm $m \in \mathbb{B}^n$,

$$
\begin{aligned}
r(m) &= 1 \quad \text{if and only if} \quad I(m) &= 0 \\
f(m) &= 1 \quad \text{if and only if} \quad I(m) &= 1, \text{ and} \\
d(m) &= 1 \quad \text{if and only if} \quad I(m) &= -
\end{aligned}
$$

As the three sets form a **partition** on $\mathbb{B}^n$, *i.e.*, the three sets are pairwise disjoint and union to $\mathbb{B}^n$, two characteristic functions are suffcient in representing an incompletely specified function. However, even so three characteristic functions are often used for the sake of convenience in Boolean manipulation.

Boolean Relation as Characteristic Function. A relation is more general than a function as it allows one-to-many mappings, which are prohibited in a function. A Boolean relation can be treated as a set of input-output mapping pairs, and thus can be represented by a characteristic function.

Example 6.29 Given a set of Boolean functions $f_1(\boldsymbol{x}), \ldots, f_m(\boldsymbol{x})$, they can be converted into a Boolean relation

$$
R(\mathbf{x}, \mathbf{y}) = \bigwedge_{i=1}^{m} (y_i \equiv f_i(\mathbf{x}))
$$

by introducing a vector of output variables $\boldsymbol{y} = (y_1, \ldots, y_m)$. For truth assignments $a \in \mathbb{B}^n$ and $b \in \mathbb{B}^m$ on variables $\boldsymbol{x}$ and $\boldsymbol{y}$, respectively, relation $R(a, b)$ valuates to true if and only if the ith bit of b equals the value of $f_i(a)$ for $i = 1, \ldots, m$. In other words, $R(a, b) = 1$ if and only if a and b are consistent assignments under the mapping of functions $f_1, \ldots, f_m$.

Circuit Consistency Condition as Characteristic Function. The consistency condition imposed by a circuit can be converted into a Boolean formula, in particular, a CNF formula by Tseitin's procedure [Tseitin 1970], where every gate of a circuit translates into a set of clauses of fixed sizes and, further, the CNF formula of a circuit is the conjunction of the clauses of all gates. Therefore the conversion is done in time linear to the circuit size.

Example 6.30 The CNF formula of the consistency condition imposed by an AND2 gate with inputs a, b and output c is

$$
\begin{aligned}
& (a \wedge b) \Leftrightarrow c \\
= \; & ((a \wedge b) \Rightarrow c)(c \Rightarrow (a \wedge b)) \\
= \; & (\neg a \vee \neg b \vee c)(\neg c \vee (a \wedge b)) \\
= \; & (\neg a \vee \neg b \vee c)(\neg c \vee a)(\neg c \vee b)
\end{aligned}
$$

Using the above three clauses for an AND2 gate, we can obtain the CNF formula

$$
\begin{aligned}
& (\neg x_1 \vee x_2 \vee x_4)(\neg x_4 \vee x_1)(\neg x_4 \vee x_2)\wedge \\
& (\neg x_2 \vee x_3 \vee x_5)(\neg x_5 \vee x_2)(\neg x_5 \vee \neg x_3)\wedge \\
& (x_4 \vee \neg x_5 \vee x_6)(\neg x_6 \vee \neg x_4)(\neg x_6 \vee x_5)\wedge \\
& (x_6 \vee x_7)(\neg x_6 \vee \neg x_7)
\end{aligned}
$$

for the consistency condition imposed by the AIG of Figure 6.16. Note that the first three clauses correspond to the AIG node of x_4, the second three clauses correspond to the AIG node of x_5, the third three clauses correspond to the AIG node of x_6, and the last two clauses correspond to the inversion of x_6 for x_7. Hence for given an AIG, the so-constructed CNF formula is of size linear in the number of nodes.

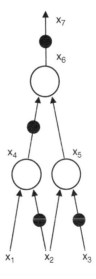

FIGURE 6.16

AIG example for CNF conversion.

Note that the function represented by the so-constructed CNF formula is not the same as the primary output functions of a given circuit. A circuit and its CNF formula are equivalent only in the sense that the CNF formula is true under a truth assignment if and only if the truth assignment is consistent in the circuit. A circuit implements some Boolean functions whereas such a CNF formula represents a Boolean relation.

At first glance, Tseitin's linear-time translation from circuits to CNF formulas seems contradictory to the exponential cost of the SOP-to-POS conversion because we may covert in linear time any SOP formula to an AIG and then further convert the AIG to a CNF formula by Tseitin's procedure. This paradox can be clarified by observing that in Tseitin's conversion new extra variables are present in the resultant POS/CNF formula. It differs from the previous SOP-to-POS conversion where no new variables are created. In fact, a Boolean relation derived from the new conversion reduces to a Boolean function as derived from the old one when the intermediate variables (those other than the primary input and output variables) are existentially quantified out and further a positive co-factor is performed on the Boolean relation with respect to the primary output variable. The existential quantification and conversion back to a POS formula, however, may result in exponential blow-up in formula sizes.

Example 6.31 Figure 6.17 shows the AIG of function $f = x_1x_2 + x_3x_4 + \ldots + x_{2n-1}x_{2n}$. By Tseitin's conversion, the CNF formula is of size linear to n due to the allowance of intermediate variables. Without intermediate variables, the POS representation of f must have 2^n clauses.

Set Manipulation as Boolean Manipulation. By dealing with characteristic functions, we are able to manipulate sets of elements simultaneously rather than manipulate individual elements separately. For instance, the intersection of

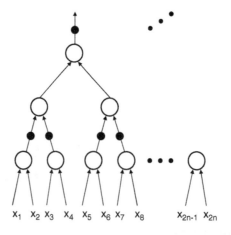

FIGURE 6.17

AIG of function $f = x_1x_2 + x_3x_4 + \ldots + x_{2n-1}x_{2n}$.

two sets A and B can be done by performing $\chi_A \wedge \chi_B$ instead of examining, for every element $e \in A$, whether e is in B as well. It leads to substantial improvements to many logic synthesis and verification algorithms. Such approaches that manipulate sets of objects simultaneously are known as **(implicit) symbolic algorithms**, in contrast to the traditional **(explicit) enumerative algorithms** (which enumerate individual objects separately).

Example 6.32 Let set U be the universe $\{0, 1, 2, 3, 4, 5, 6, 7\}$, set $A \subseteq U$ be $\{0, 1, 2, 4\}$, and set $B \subseteq U$ be $\{2, 3, 4, 6\}$. Consider the binary encoding with Boolean variables x_1, x_2, and x_3 such that element 0 is encoded as $\neg x_1 \neg x_2 \neg x_3$, 1 as $\neg x_1 \neg x_2 x_3$, 2 as $\neg x_1 x_2 \neg x_3$, 3 as $\neg x_1 x_2 x_3$, 4 as $x_1 \neg x_2 \neg x_3$, 5 as $x_1 \neg x_2 x_3$, 6 as $x_1 x_2 \neg x_3$, and 7 as $x_1 x_2 x_3$. Then the characteristic functions of these sets with respect to the binary encoding are

$$
\begin{aligned}
\chi_U &= 1 \\
\chi_A &= \neg x_1 \neg x_2 + \neg x_1 \neg x_3 + \neg x_2 \neg x_3, \text{ and} \\
\chi_B &= \neg x_1 x_2 + x_1 \neg x_3
\end{aligned}
$$

It can be checked that formula $\neg \chi_A$ corresponds to the characteristic function of the set $U \setminus A$, formula $\chi_A \wedge \chi_B$ corresponds to that of $A \cap B$, and formula $\chi_A \vee \chi_B$ corresponds to that of $A \cup B$.

Example 6.33 Image and pre-image computations are key operations in logic synthesis and formal verification. The **image** of $A \subseteq \mathbb{B}^n$ under the functional vector $f = (f_1, \ldots, f_m)$ is the set $\{q \in \mathbb{B}^m \mid q = \boldsymbol{f}(p), p \in A\}$. The characteristic function of the image is

$$
Img_f(A) = \exists \boldsymbol{x}. \bigwedge_{i=1}^{m} (y_i \equiv f_i(\boldsymbol{x})) \wedge \chi_A(\boldsymbol{x})
$$

which refers to the newly introduced $\boldsymbol{y}$ variables taking on the function values. In contrast, the **pre-image** of $B \subseteq \mathbb{B}^m$ under the functional vector $\boldsymbol{f} = \{f_1, \ldots, f_m\}$ is the set $\{p \in \mathbb{B}^n \mid q = \boldsymbol{f}(p), q \in B\}$. The characteristic function of the pre-image is

$$
PreImg_f(B) = \exists \boldsymbol{y}. \bigwedge_{i=1}^{m} (y_i \equiv f_i(\boldsymbol{x})) \wedge \chi_B(\boldsymbol{y})
$$

which refers to the $\boldsymbol{x}$ variables only.

6.2.6 Boolean reasoning engines

Among the introduced data structures, BDD packages and SAT solvers are the most widely used Boolean reasoning engines. They are extensively used in various *symbolic*, or called *implicit*, algorithms, such as image computation, don't care computation, state reachability analysis, and so on. Any Boolean reasoning engine can be more or less used in developing symbolic algorithms. In the sequel when a computational task is expressed in terms of a QBF, we should be aware that its computation is already achievable by Boolean manipulation using a BDD package.

Although BDD-based algorithms and symbolic algorithms were once almost synonymous in the 1990s, recently other data structures were developed as alternatives to BDDs. Due to the capacity limit of BDDs, more and more symbolic algorithms are based on other data structures. Notably, Boolean reasoning engines using SAT and AIGs, for instance, are gaining in popularity in hardware synthesis and verification. Moreover, hybrid Boolean reasoning engines combining complementary data structures may become important tools. In fact, combinational equivalence checking of multi-million gate designs has been demonstrated in an industrial setting through such hybrid solvers combining BDD and AIG [Kuehlmann 1997].

6.3 COMBINATIONAL LOGIC MINIMIZATION

Logic synthesis is typically divided into two phases: **technology independent optimization** and **technology dependent optimization**. The former aims at simplifying Boolean expressions and logic netlist structures regardless of the target technology node for manufacturing, whereas the latter aims at optimizing circuits under the target implementation technology. This divide-and-conquer separation is often beneficial in orthogonalizing various design concerns. Simplified Boolean expressions are often good for optimization with respect to the target implementation technology. Also it allows a designer to migrate a design from one technology node to another without substantial re-optimization. Our study will begin with the first phase, and then proceed to the second one in Section 6.4.

In technology independent optimization, combinational logic minimization consists of two-level and multilevel logic minimization. Two-level logic minimization is a relatively simple and well-studied subject in both theory and practice. As a multilevel logic netlist can be seen as a network of two-level logic components, the results of two-level minimization are in part applicable to multilevel minimization. Not only optimized two-level SOP representations can be used as a starting point for multilevel synthesis, but two-level minimization techniques can also be used in minimizing multilevel netlists. Hence we delve into two-level logic minimization before considering the multilevel counterpart.

6.3.1 Two-level logic minimization

There are a variety of two-level logic implementations. The most common one is the SOP implementation, where the first level of logic corresponds to AND gates and the second level to OR gates. NOR-NOR structures, NAND-NAND structures, AND-XOR structures, and OR-AND structures are also possible.

Example 6.34 The function of Figure 6.18a can be reexpressed in POS form and implemented as the circuit shown in Figure 6.18c. An SOP implementation can be directly converted into an equivalent NAND-NAND implementation by replacing all the AND gates and OR gates by NAND gates. A NAND-NAND implementation of the function of Figure 6.18a is shown

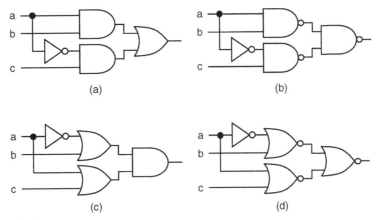

FIGURE 6.18

Two-level logic implementations.

in Figure 6.18b. Similarly, a POS implementation can be directly converted into a NOR-NOR implementation as shown in Figure 6.18d.

Two-level logic is typically implemented as a ***programmable logic array*** (PLA) [Fleisher 1975] in a NOR-NOR form followed by inverters at the outputs. PLAs have the advantage of being very structured and are therefore amenable to automated logic and layout synthesis. Even though PLAs are no longer a popular IC implementation style, they can be an important ingredient in modern system designs because their regular structures [Mo 2004] provide a solution to alleviate the infamous process variation problem of IC manufacturing in the nanometer regime.

6.3.1.1 *PLA implementation vs. SOP minimization*

Despite the fact that many regular functions have a minimum two-level logic representation whose size grows exponentially with the number of inputs to the function (*e.g.*, parity functions and adders), two-level logic circuits can efficiently implement control logic.

The hardware cost of a PLA implementing some SOP formula is directly reflected in the formula. The number of literals (respectively product terms) of the formula corresponds to the number of transistors (respectively product lines) of the PLA. Therefore, minimizing an SOP expression not only reduces PLA area cost, but also improves circuit performance due to the reduction in capacitive loads.

Example 6.35 An NMOS PLA is shown in Figure 6.19a, whose output marked f implements the logic function of Figure 6.18. Note that while the input plane and output plane are both NOR-planes, we have inverters at the outputs. An SOP representation can be directly mapped to a NOR-NOR PLA with output inverters by complementing each literal in the input plane. The function $f = a \cdot b + \neg a \cdot c$ has been implemented as

$$\neg(\neg(\neg(\neg a + \neg b) + \neg(a + \neg c)))$$

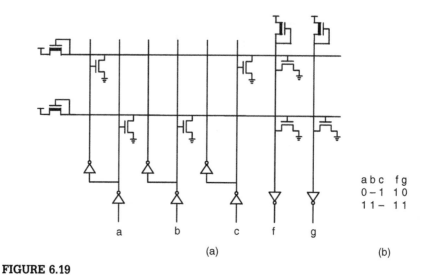

$$
\begin{array}{cc}
a\,b\,c & f\,g \\
0-1 & 1\,0 \\
1\,1- & 1\,1
\end{array}
$$

(a) (b)

FIGURE 6.19

(a) Programmable logic array. (b) Multiple-output cover.

PLAs can implement multiple-output functions that share product terms across outputs as shown in Figure 6.19a. The multiple-output cover is represented as shown in Figure 6.19b. The two outputs share the cube $a \cdot b$ in their onsets. Therefore, in the PLA of Figure 6.19a the first row from the bottom feeds transistors in both columns in the output plane. The number of columns in a PLA equals two times the number of inputs plus the number of outputs, the number of rows equals the number of product terms in the cover, the number of transistors in the input plane equals the number of "1" or "0" literals in the input part of the multiple-output cover, and the number of transistors in the output plane equals the number of 1's in the output part of the multiple-output cover.

6.3.1.2 *Terminology*

We define terminology and notation used for two-level logic minimization.

As a notational convention, we write a cube (*i.e.*, a product term) c in a bit-vector form $c = [c_1 \ldots c_n]$, where c_i is "0" if the ith variable x_i appears complemented in c, c_i is "1" if variable x_i appears uncomplemented in c, and c_i is "–" if variable x_i does not appear in c.

Example 6.36 A cube $c = x_1 \neg x_2$ in the Boolean space spanned by variables x_1, x_2, x_3 can be represented as $[10–]$.

For multi-output functions, the notion of cubes is slightly generalized. A cube of a Boolean function f with n inputs and m outputs is written as $c = [c_1 \cdots c_n | c_{n+1} \cdots c_{n+m}]$, which consists of the *input part* with c_i's for $1 \leq i \leq n$ and *output part* with c_i's for $n+1 \leq i \leq n+m$. In the input part, c_i is defined the same as before; in the output part, c_i is "0," "1," and "–" if the input part of c belongs to the offset, onset, and dcset, respectively, of the $(i - n)^{th}$ output of f. For single-output functions, we may not write the $(n+1)^{st}$ bit of the cube if the function is fully specified.

A **minterm**, defined in Section 6.2, corresponds to a cube in which every variable of a Boolean space appears. Minterms and cubes may be used to represent the values of a set of input variables, *e.g.*, $x\neg yz$ is shorthand for $x = 1$, $y = 0$, and $z = 1$. Therefore, there is a natural correspondence between an input assignment and a vertex in the Boolean n-space. This correspondence may be extended to cubes where absent variables are assumed to be unassigned. Thus, if a circuit C has inputs v, w, x, y, and z then applying the cube $x\neg yz$ to C is shorthand for applying $v = X, w = X, x = 1, y = 0$, and $z = 1$, where "X" is used to denote an unknown value.

A cube q *contains* another cube r if the literals in the input part of cube q are a subset of the literals in the input part of cube r and the outputs in the output part of q are a superset of the outputs in the output part of cube r. In bit-vector notation, the cube $[0-|1]$ of a two-input, single-output function contains the cube $[00|1]$. Similarly, the cube $[0-|11]$ of a two-input, two-output function contains the cube $[0-|10]$. A cube is said to be contained by a cover if every minterm contained by the cube is contained by some cube in the cover. For example, the cover $\{00--, -1-1\}$ contains the cube $[0--1]$.

If a cube q contains only onset and dcset vertices of a Boolean function f, then q is called an **implicant** of f. A **prime implicant** (or **prime**) of f is an implicant which is not contained by any other implicant of f and which is not entirely contained in the dcset of f. An alternate operational definition, which is crucial in ESPRESSO, of a prime implicant is as follows. An implicant is prime if no 0- or 1-literal can be "raised" (to include more minterms) to a "$-$" without resulting in the implicant intersecting the offset of any component of the multiple-output function. For instance, a cube $[111]$ of a three-input, single-output function would be a prime cube if each of $[11-]$, $[1-1]$ and $[-11]$ intersected the offset. A literal in a cube is said to be prime if raising that particular literal to a "$-$" results in a cube that intersects the offset. Thus, $[110]$ may not be a prime cube of a function f because $[11-]$ is an implicant of f, but the first two literals may be prime in the implicant $[110]$ because $[-10]$ and $[1-0]$ intersect the offset of f. All the literals contained in a cube have to be prime in order for the cube to be prime.

An **essential prime implicant** (or **essential prime**) is a prime implicant which includes one or more onset vertices which are not included in any other prime implicant. These vertices are termed **essential vertices**. An **optional prime implicant** is a prime implicant for which all vertices are included in other prime implicants.

A minimal cover for a function f is generated by selecting all of the essential prime implicants and a minimal set of optional prime implicants such that all vertices in the onset of f are included in the cover.

Example 6.37 For the example in Figure 6.1b, there are three essential prime implicants and no optional prime implicants. The minimal cover would be $f = \neg x_3 + \neg x_1 x_2 + x_1 \neg x_2$.

A **relatively essential vertex** of a cube q in a cover C is a vertex in the onset that is contained by q and is not contained in any other cube in C.

Example 6.38 In Figure 6.1b, $x_1 \neg x_2 x_3$ is a relatively essential vertex of the cube $x_1 \neg x_2$, while the other vertex in this cube, $x_1 \neg x_2 \neg x_3$, is not a relatively essential vertex since it is also contained in the cube $\neg x_3$.

A two-input, two-output function can also be represented as a multiple-output cover, with cubes that have input as well as output parts.

Example 6.39 The two-output function $F = \{11|01, 00|10, 10|11\}$ has two cubes in each of its components F_1 and F_2. If the inputs are a and b, then F_1 can be represented as $\neg a \neg b + a \neg b$, and F_2 is $ab + a \neg b$. The cube $a \neg b$ is shared by F_1 and F_2, because its output part indicates that it belongs to both their onsets.

In order to keep cover sizes small, it is desirable to ensure some form of minimality for the cover. An easily satisfiable property is that no cube c of a cover contains another cube d of the cover. Such a cover is minimal with respect to **single cube containment**.

An implicant in a cover is **irredundant** if it contains an essential or a relatively essential vertex. Thus, removing the implicant changes the functionality of the cover. Else it is **redundant** and can be safely removed from the cover. A cover is prime if each of the implicants in the cover is prime. A cover is irreundant if each of the implicants is irredundant. The definitions apply to both completely specified and incompletely specified functions.

6.3.2 **SOP minimization**

Two-level Boolean minimization is used to find an SOP representation for a Boolean function that is optimum according to a given cost function. The typical cost functions used are the number of product terms, the number of literals, or a combination of both.

With any of these cost functions, the problem of two-level minimization contains the subproblem of finding the solution of a minimum covering problem which has been shown to be NP-complete [Garey 1979]. Nevertheless, sophisticated exact minimizers (*e.g.*, [Dagenais 1986; Rudell 1987]) have been developed whose average-case behavior for most commonly encountered functions is acceptable. Furthermore, heuristic minimization methods exist (*e.g.*, [Hong 1974; Brayton 1984]) which have been shown to produce results that are close to the minimum within reasonable amounts of time, even for large Boolean functions.

Two-level Boolean minimization for a given function consists of two steps:

1. generating the set of prime implicants, and
2. selecting a minimum set of prime implicants to cover all onset minterms.

6.3.2.1 *The Quine-McCluskey method*

The first algorithmic method proposed for two-level minimization is the Quine-McCluskey method [McCluskey 1956], which follows the two steps outlined above.

Prime Implicant Generation. The set of prime implicants can be generated by iteratively merging two cubes which differ in exactly one position, where one is of literal x and the other is of literal $\neg x$ assuming variable x is the corresponding variable in the position. For instance, two cubes $c_1 = [00{-}1]$ and $c_2 = [01{-}1]$ can be merged as $[0{-}{-}1]$. This merging process continues until no more merging is possible. Initially all onset and dcset minterms are the cubes to start with. Upon termination, a maximal cube (not contained by every other cube) is a prime implicant provided that it is not entirely contained by the dcset.

Example 6.40 Consider the completely specified Boolean function shown in Figure 6.20a. It has been represented as a list of minterms. Each minterm has an associated decimal value obtained by converting the binary number represented by the minterm into a decimal number; for instance the value of 0000 is 0 and that of 1100 is 12. The cubes generated by merging the pairs of cubes are shown in Figure 6.20b and 6.20c. We have five prime implicants, marked as **A, B, C, D**, and **E**, for the function in this example.

Prime Implicant Table. A **prime implicant table** is a table with rows indexed by onset minterms and columns indexed by prime implicants. An entry at position (i, j) in the table is marked "X" if prime implicant j contains onset minterm i.

Example 6.41 Figure 6.21 shows the prime implicant table of the previous example.

Since we want a minimum set of prime implicants that covers all the onset minterms, we have to select a minimum set of columns in a prime implicant table such that there is at least one X in every row. This is the classical *minimum unate covering problem* which has been shown to be NP-complete [Garey 1979]. Nevertheless there are several reduction techniques that help simplify solving the unate covering problem:

Simplification by Essential Prime Implicants. A row with a single X represents a (relatively) essential vertex, and the corresponding column represents a (relatively) essential prime implicant. The column must be selected in the final cover because any prime cover for the function will have to contain

```
                              0, 8      -000 → E
         0    0000            5, 7      01-1 → D
         5    0101            7, 15     -111 → C
         7    0111            8, 9      100-
         8    1000            8, 10     10-0
         9    1001            9, 11     10-1
        10    1010           10, 11     101-
        11    1011           10, 14     1-10
        14    1110           11, 15     1-11      8,  9,  10, 11    10--  → B
        15    1111           14, 15     111-     10, 11, 14, 15    1-1-  → A
             (a)                       (b)                (c)
```

FIGURE 6.20

Prime implicant generation.

	A	B	C	D	E
0000					X
0101				X	
0111			X	X	
1000		X			X
1001		X			
1010	X	X			
1011	X	X			
1110	X				
1111	X		X		

FIGURE 6.21

Prime implicant table.

the prime that contains the onset minterm corresponding to this row. Therefore we can simplify the prime implicant table by removing the columns corresponding to (relatively) essential prime implicants and removing the rows covered by these removed columns.

Example 6.42 In the prime implicant table of Figure 6.21 **A, B, D**, and **E** are essential prime implicants. We select the essential prime implicants since they have to be contained in any prime cover. This results in a cover for the function, since selecting columns **A, B, D**, and **E** results in the presence of X in every row.

Some functions may not have essential prime implicants. Consider the hypothetical prime implicant table of Figure 6.22a. There is no row with a single X. It is necessary to make an arbitrary selection of a prime to begin with. Assume that prime **A** is selected. We obtain the reduced table of Figure 6.22b after deleting column **A** and the first two rows contained by **A** from the table of Figure 6.22a.

Simplification by Column Dominance. A column **U** of a prime implicant table is said to dominate another column **V** if **U** contains every row contained by **V**. We can delete the dominated columns, since selecting the dominating column will result in covering more uncontained minterms than the dominated column. Note that the dominating column might not exist in a minimum solution. Further if minimizing the literal count was our objective, then we can only delete dominated columns that correspond to primes with equal or more literals than the dominating prime.

(a)

	A	B	C	D	E	F	G	H
0000	x							x
0001	x	x						
0101		x	x					
0111			x	x				
1000							x	x
1010						x	x	
1110					x	x		
1111				x	x			

(b)

	B	C	D	E	F	G	H
0101	x	x					
0111		x	x				
1000						x	x
1010					x	x	
1110				x	x		
1111			x	x			

(c)

	C	D	E	F	G
0101	x				
0111	x	x			
1000					x
1010				x	x
1110			x	x	
1111		x	x		

FIGURE 6.22

Cyclic prime implicant table.

Example 6.43 In the reduced table of Figure 6.22b column **B** is dominated by column **C** and column **H** is dominated by column **G**. Reducing the table of Figure 6.22b yields the table of Figure 6.22c. In this table **C** and **G** are relatively essential prime implicants. Choosing **C** and **G** results in the selection of **E**, which completes the cover $f = \{$**A, C, E, G**$\}$. We are not guaranteed that this cover is minimum; we have to backtrack to our arbitrary choice of selecting prime **A** and delete prime **A** from the table, *i.e.*, explore the possibility of constructing a cover that does not have **A** in it. This results in $f = \{$**B, D, F, H**$\}$.

Simplification by Row Dominance. A row i of a prime implicant table is said to dominate another row j if i has a 1 in every column in which j has a 1. Any minimum expression derived from a table which contains both rows i and j can be derived from a table which only contains the dominated row.

Example 6.44 In Figure 6.22c, row 0111 dominates row 0101 and can be deleted; row 1010 dominates row 1000 and can be deleted as well.

A Branch-and-Bound Covering Strategy. The covering procedure of the Quine-McCluskey method is summarized below. The input to the procedure is the prime implicant table T.

1. Delete the dominated primes (columns) and the dominating minterms (rows) in T. Detect essential primes[1] in T by checking to see if any minterm is contained by a single prime implicant. Add these essential prime implicants to the selected set. Repeat until no new essential primes are detected.
2. If the size of the selected set of prime implicants equals or exceeds the best solution thus far, return from this level of recursion. If there are no elements left to be contained, declare the selected set as the best solution recorded thus far.

[1]These primes may not be essential primes of the original function or table.

3. Heuristically select a prime implicant.
4. Add this prime implicant to the selected set and recur for the sub-table resulting from deleting the prime implicant and all minterms that are contained by this prime implicant. Then, recur for the sub-table resulting from deleting this prime implicant without adding it to the selected set.

6.3.2.2 *Other methods*

State-of-the-art exact two-level logic minimization algorithms, such as ESPRESSO [Rudell 1987] and Scherzo [Coudert 1995], are all based on the Quine-McCluskey method, but are able to outperform the Quine-McCluskey method significantly due to superior prime generation, implicant table generation, and covering techniques. In particular, with decision diagram based data structures, Scherzo [Coudert 1995] was able to outperform ESPRESSO by two orders of magnitude in terms of speed. Introductions to ESPRESSO and decision diagram based two-level logic minimization can be found in [Devadas 1994] and [Minato 1996], respectively. A good overview on two-level logic minimization can be found in [Coudert 1994].

6.3.3 **Multilevel logic minimization**

Two-level logic is limited because not all Boolean functions can be efficiently represented in the SOP form. Multilevel logic implementation of a function is often faster and smaller than two-level logic. Therefore multilevel realizations are the preferred means of implementing combinational logic in *very large scale integrated* (VLSI) systems. Because of the increased potential for reusing sub-circuits, there are more degrees of freedom in implementing a Boolean function than in the two-level case. This increased freedom, however, largely expands the search space in identifying an optimal solution.

The area of multilevel logic synthesis has blossomed since the mid-1980s. Many of the methods developed have been successfully used in commercially available computer-aided design packages. There are two types of basic approaches, rule-based local transformations and algorithmic transformations. Rule-based local transformations were developed at IBM in the late 1970s, known as the LSS system [Darringer 1981]. A rule transforms a pattern for a local set of gates and interconnections into another equivalent one when certain patterns are recognized in logic netlists. The transformations have somewhat limited optimization capability since they are local in nature and do not have a global perspective of the design.

Algorithmic transformations began to evolve in about 1981, in parallel with activity in two-level logic synthesis and influenced by it. The algorithmic counterpart uses two phases: a technology-independent step based on algorithms for manipulating general Boolean functions [Brayton 1982] and a technology mapping step (the subject of Section 6.4) where the design described in terms of generic Boolean functions is mapped into a set of gates that can be implemented in the design method of choice (gate arrays, standard cells, or macrocells). Both rule-based methods (*e.g.*, [Darringer 1984; Bartlett 1986]) and algorithmic methods (*e.g.,* [Brayton 1987; Bostick 1987]) have been successful. Algorithmic methods for logic synthesis are our main focus.

We describe the various logic transformations used in algorithmic logic synthesis systems, most of which use algebraic [Brayton 1982, 1984] and Boolean [Bostick 1987; Devadas 1989] operations in technology-independent optimization, and use graph covering methods [Keutzer 1987] in technology mapping. We first introduce technology-independent optimization and focus primarily on area minimization. Implementation details of the algorithms can be found in [Brayton 1987, 1990].

6.3.3.1 *Logic transformations*

The goal of multilevel logic optimization is to obtain multilevel representation of a Boolean function optimal with respect to some design constraints. In order to restructure a logic function, a collection of different operations is helpful. The operations described below are commonly used and can be composed in a script file for orchestrated optimization.

Decomposition. Decomposition of a Boolean function is the process of reexpressing a single function as a composition of new functions.

Example 6.45 The process of translating the expression

$$F = a \cdot b \cdot c + a \cdot b \cdot d + \neg a \cdot \neg c \cdot \neg d + \neg b \cdot \neg c \cdot \neg d$$

to the set of expressions

$$
\begin{aligned}
F &= X \cdot Y + \neg X \cdot \neg Y \\
X &= a \cdot b, \text{ and} \\
Y &= c + d
\end{aligned}
$$

is decomposition.

Extraction. Extraction, related to decomposition, is applied to multiple functions. It is the process of identifying and creating new intermediate functions and their corresponding output variables, and reexpressing the original functions in terms of the original as well as the new variables.

Extraction creates nodes which feed multiple outputs. The operation identifies common subexpressions among different logic functions forming a network. New nodes corresponding to the common subfunctions are created and each of the logic functions in the original network is simplified with respect to these new nodes. The optimization problem of extraction is to find a set of intermediate functions such that the resulting network has minimum area, delay, or power.

Example 6.46 Extraction applied to the following three functions

$$
\begin{aligned}
F &= (a + b) \cdot c \cdot d + e \\
G &= (a + b) \cdot \neg e, \text{ and} \\
H &= c \cdot d \cdot e
\end{aligned}
$$

may yield

$$
\begin{aligned}
F &= X \cdot Y + e \\
G &= X \cdot \neg e \\
H &= Y \cdot e \\
X &= a + b, \text{ and} \\
Y &= c \cdot d
\end{aligned}
$$

Factoring. A factored form is a parenthesized representation of a tree network where each internal node is an AND or an OR gate and each leaf is a literal. Like SOP, factored forms are a way of representing Boolean functions and are perhaps a more natural way for multilevel circuits than the SOP representation.

A factored-form Boolean expression can be implemented using a complex CMOS gate. The number of transistors of the logic gate is closely related to the number of literals of the factored form as can be seen from the following example.

Example 6.47 Figure 6.23 shows a complex CMOS gate implementing the factored form $f = a + (b + c)d$. In general, excluding the possible output buffer, $2n$ transistors are needed to implement a factored form with n literals.

Consequently the literal count of a factored form can be used as a good estimate of hardware cost. The optimization problem associated with factoring is to find a factored form with a minimum number of literals.

Factoring is the process of deriving a factored form from an SOP representation of a function.

Example 6.48 The expression

$$
F = a \cdot c + a \cdot d + b \cdot c + b \cdot d + e
$$

can be factored into

$$
F = (a + b) \cdot (c + d) + e
$$

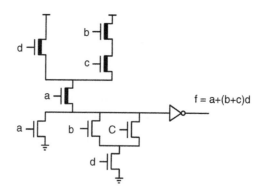

FIGURE 6.23

Factored form *vs.* complex CMOS gate implementation.

Substitution. Substitution, also called **resubstitution**, of a function G into F is the process of reexpressing F as a function of its original inputs and G.

Example 6.49 Substituting

$$G = a + b$$

into

$$F = a + b \cdot c$$

produces

$$F = G \cdot (a + c)$$

This operation creates an arc in the Boolean network connecting the node of the substituting function, namely G, to the node of the function being substituted into, namely F.

Elimination. Elimination, collapsing, or **flattening** is the the inverse operation of substitution. If G is a fanin node of F, collapsing G into F reexpresses F without G. It undoes the operation of substituting G into F.

Example 6.50 If

$$
\begin{aligned}
F &= G \cdot a + \neg G \cdot b \text{ and} \\
G &= c + d
\end{aligned}
$$

then collapsing G into F results in

$$
\begin{aligned}
F &= a \cdot c + a \cdot d + b \cdot \neg c \cdot \neg d \text{ and} \\
G &= c + d
\end{aligned}
$$

If the node G is not a primary output and does not fan out to other nodes, then it may be removed from the Boolean network, resulting in a network with one less node.

Flattening a logic function into the SOP form could result in an exponential growth in representation.

Example 6.51 Consider the flattening of the nodes g_1 through g_k into F with

$$
\begin{aligned}
F &= g_1 \cdot g_2 \cdots g_k \\
g_1 &= a_1 + b_1 \\
g_2 &= a_2 + b_2 \\
&\vdots \\
g_k &= a_k + b_k
\end{aligned}
$$

After flattening, the SOP representation for F will have 2^k product terms.

Given a Boolean network, we may compute the **value** of a node, which represents the saved literal count due to the existence of this node rather than collapsing this node into its fanout nodes. For nodes with little or negative values, we may eliminate them from the Boolean network by collapsing them into their fanouts. It should be noted that eliminating a node may change other nodes' values.

6.3.3.2 *Division and common divisors*

To realize the above logic transformations, it is important to define operations which, when given functions f and p, find functions q and r such that $f = p \cdot q + r$, if such q and r exist. This operation is called the **division** of f by p generating **quotient** q and **remainder** r. The function p is called a **divisor** of f if r is not null and a **factor** if r is null.

The conditions for p being a Boolean factor or a Boolean divisor are stated in the following propositions.

Proposition 6.1. *A logic function p is a Boolean factor of a logic function f if and only if $f \cdot \neg p = 0$ (that is, the onset of f is contained in the onset of p).*

Proposition 6.2. *If $f \cdot p \neq 0$, then p is a Boolean divisor of f.*

For a given division operation, the resulting q and r may depend upon the particular representation of f and p. Moreover for any logic function, there are many Boolean factors and divisors. This fact poses a problem in choosing a good factor and divisor. If the domain is restricted to a particular subset of expressions, then the division operation is unique and much easier to carry out. A restricted version of such division is called **algebraic division**.

6.3.3.3 *Algebraic division*

We begin the description of algebraic division with some definitions. The **support** of a Boolean expression f denoted as $sup(f)$ is the set of all variables v that syntactically occur in f as v or $\neg v$. For example, if $f = a + \neg a + b \cdot c$, then $sup(f) = \{a, b, c\}$. We say that f is **orthogonal** to g, written as $f \perp g$, if $sup(f) \cap sup(g) = \emptyset$. For example, $f = a + b$ and $g = c + d$ are orthogonal.

The function g is an **algebraic divisor** of f if there exist h and r such that $f = g \cdot h + r$, where $h \neq 0$, $g \perp h$, and the remainder r is minimal, *i.e.*, has as *few cubes as possible*. Under this condition on the remainder, the **quotient** h, denoted as f/g, is in fact unique. We say the function g divides f *evenly* if $f = g \cdot h$, where $h \neq 0$, $g \perp h$, and $r = 0$.

We consider two main problems of algebraic optimization, namely computing quotients f/g given f and g, and determining divisors g of a given function f.

Computing the Quotient. Given two covers (*i.e.*, *sets* of cubes) $f = \{b_1, b_2, \ldots, b_{|f|}\}$ and $g = \{a_1, a_2, \ldots, a_{|g|}\}$, we define $b_i = \{c_j \mid a_i \cdot c_j \in f\}$ for all $i = 1, 2, \ldots, |g|$, *i.e.*, b_i corresponds to all the multipliers of the cube a_i in g that produce elements of f. It is easy to see that

$$f/g = \bigcap_{i=1}^{|g|} b_i = b_1 \cap b_2 \ldots \cap b_{|g|}$$

Example 6.52 Consider two covers

$$f = a \cdot b \cdot c + a \cdot b \cdot d + d \cdot e, \text{ and}$$
$$g = a \cdot b + e.$$

We have $|g| = 2$ and $|f| = 3$. With $3 \times 2 = 6$ comparisons, we obtain

$$h_1 = \{c, d\}, \text{ and}$$
$$h_2 = \{d\}$$

Hence $h_1 \cap h_2 = d$, and

$$f = (a \cdot b + e) \cdot d + a \cdot b \cdot c$$

The above algorithm requires $O(|f| \cdot |g|)$ operations. Encoding and sorting the cubes of f and g can reduce the complexity to $O((|f| + |g|) \log(|f| + |g|))$ [McGeer 1987].

Kernels and Algebraic Divisors. Given an efficient method for algebraic division, optimization can be carried out if good algebraic divisors can be found. The set of algebraic divisors is defined as $D(f) = \{g \mid f/g \neq 0\}$. The **primary divisors** of f are defined as $P(f) = \{f/c \mid c \text{ is a cube}\}$.

Example 6.53 If

$$f = a \cdot b \cdot c + a \cdot b \cdot d \cdot e$$

then

$$f/a = b \cdot c + b \cdot d \cdot e$$

is a primary divisor.

Proposition 6.3. *Every divisor of f is contained in a primary divisor, i.e., if g divides f, then $g \subseteq p \in P(f)$.*

Proof. Let $c \in f/g$ be a cube. Then $g \subseteq f/(f/g)$ and $f/(f/g) \subseteq f/c \in P(f)$. ▫

A function g is termed **cube-free** if the only cube that divides g evenly is 1. The **kernels** of f are defined as $K(f) = \{k \mid k \in P(f), k \text{ is cube-free}\}$. For a kernel $k \in K(f)$, its **cokernel** is the cube c with $f/c = k$.

Example 6.54 If

$$f = a \cdot b \cdot c + a \cdot b \cdot d \cdot e$$

then

$$f/a = b \cdot c + b \cdot d \cdot e$$

is a primary divisor but not cube-free since b is a factor of $f/a = b \cdot (c + d \cdot e)$. However, $f/(a \cdot b) = c + d \cdot e$ is a kernel, and $a \cdot b$ is a cokernel.

The following theorem (originally proven in [Brayton 1982]) is the basis of algebraic optimization methods.

Theorem 6.2. *Two expressions f and g have a non-cube common divisor d if and only if there exist kernels $k_f \in K(f)$ and $k_g \in K(g)$ such that $k_f \cap k_g$ has two or more terms (i.e., $k_f \cap k_g$ is not a cube).*

Proof. For the "if" part, $k_f \cap k_g$ is clearly a common divisor of f and g. It remains to prove the "only if" part. □

Assume d divides both f and g, and d has two or more terms. Then there is a cube-free SOP expression e such that e divides d. Also e divides f and g as well. By Proposition 6.3, $e \subseteq k_f \in P(f)$ and $e \subseteq k_g \in P(g)$ for some k_f and k_g. Since e is cube-free, k_f and k_g are cube-free as well. Hence, $k_f \in K(f)$ and $k_g \in K(g)$. Finally, since $e \subseteq k_f \cap k_g$, $k_f \cap k_g$ must have two or more terms. □

We can therefore use the kernels of f and g to locate common divisors. Note that these are not the only common divisors of f and g, but they are good common divisors to consider during logic optimization. We compute the set of kernels for each logic expression, then form intersections among kernels from the different logic expressions. If this intersection set contains no non-cube elements, then by Theorem 6.2, we need only look for divisors consisting of single cubes. Otherwise, we have found an algebraic divisor common to two or more expressions.

Computing the Kernels. The kernels of a function f can be computed using the algorithm of Figure 6.24. The kernel generation algorithm first makes f cube-free by finding its largest cube factor. It then selects the literals of f in a lexicographical order and divides them into f; the resulting quotient is a kernel if it is cube-free. (Note that this kernel might contain other kernels, too.) If it is not cube-free, then it is made cube-free by selecting its largest cube factor. Note that in this context the largest cube is the cube with the most number of

```
KERNELS(f){
    c_f = largest cube (with maximum number of literals) factor of f ;
    K = KERNEL1(0, f/c_f ) ;
    if (f is cube-free)
        return(f ∪ K) ;
    return(K) ;
}


KERNEL1(j, g){
    R = g ;
    N = Maximum index of variables in g ;
    for(i = j + 1; i ″ N; i = i + 1) {
        if (l_i in 1 or no cubes of g) continue ;
        c = largest cube dividing g/l_i evenly ;
        if (for all k ″ i, l_k ∉ c) /* Pruning Condition */
            R = R ∪ KERNEL1 (i, g/(l_i ∩ c)) ;
    }
    return(R) ;
}
```

FIGURE 6.24

Procedure to determine all the kernels of a single-output logic function.

literals. The procedure is repeated on the resulting functions until functions with no kernels (called the level-0 kernels of f) are found. A major efficiency is obtained by noting that if the largest cube factor extracted contains an already selected literal, then the current branch can be terminated, since all the kernels that can be found by continuing have already been generated. This leads to an algorithm in which no cokernel is duplicated.

Example 6.55 Consider

$$f = a \cdot b \cdot c \cdot d + a \cdot b \cdot c \cdot e + a \cdot b \cdot e \cdot f$$

In the routine **KERNELS** $c_f = a \cdot b$. Therefore,

$$f/c_f = c \cdot d + c \cdot e + e \cdot f$$

In the next step we call **KERNEL1**$(0, c \cdot d + c \cdot e + e \cdot f)$.

In **KERNEL1** we set $R = \{c \cdot d + c \cdot e + e \cdot f\}$. Since the ordering is lexicographic, we have $l_1 = a, l_2 = b$, etc. Note that $N = 6$. The literals l_1 and l_2 are in none of the terms of R, and we move to $l_3 = c$. The largest cube dividing $(c \cdot d + c \cdot e + e \cdot f)/c$, which is $d + e$, is 1.

We therefore make a recursive call to **KERNEL1**$(3, (c \cdot d + c \cdot e + e \cdot f) = (c \cap 1))$. This call returns with $\{d + e\}$. In the parent **KERNEL1** R is set to $\{c \cdot d + c \cdot e + e \cdot f, d + e\}$. We skip $l_4 = d$ and move to $l_5 = e$. The largest cube evenly dividing $(c \cdot d + c \cdot e + e \cdot f)/e$, which is $c + f$, is 1. We next call **KERNEL1**$(5, (c \cdot d + c \cdot e + e \cdot f)/(e \cap 1))$. This returns with $c + f$. We end with $K = R = \{c \cdot d + c \cdot e + e \cdot f, d + e, c + f\}$.

If the largest cube factor extracted contains an already selected literal, then the current branch can be terminated, since all kernels that can be found by continuing have already been generated. We illustrate the pruning condition with the following example.

Example 6.56 Consider

$$f = a \cdot b \cdot c \cdot (d + e) \cdot (k + l) + a \cdot f \cdot g + h$$

In the first call to **KERNEL1**, we will generate the kernels corresponding to

$$f/a = b \cdot c \cdot (d + e) \cdot (k + l) + f \cdot g$$

KERNEL1 calls itself recursively to compute

$$f/(a \cdot b) = c \cdot (d + e) \cdot (k + l)$$

Since $f/(a \cdot b)$ is not cube-free, the next recursive call to **KERNEL1** will use $(d + e) \cdot (k + l)$. All the kernels of this expression will be generated.

We move up one level in the recursion and compute

$$f/(a \cdot c) = b \cdot (d + e) \cdot (k + l)$$

At this stage, we note that $f/(a \cdot c)$ is not cube-free, and the largest cube dividing this expression evenly is b. However, b is an already selected literal implying that we have already generated the kernels for the cube-free expression $(d+e)\cdot(k+l)$. We do not have to recursively call **KERNEL1** for this branch and can go ahead to $f/(a \cdot d)$.

It is possible to modify the **KERNEL1** procedure to generate only the level-0 kernels which do not contain other kernels. This modification is based on the observation that if no kernels of g are found in the **for** loop, then g is a level-0 kernel.

Factoring Algorithm. A function can be algebraically factored using the generic factoring algorithm shown in Figure 6.25.

The procedure **DIVIDE** performs algebraic division and reexpresses f as $g \cdot b + r$. The procedure **CHOOSE_DIVISOR** is critical to obtaining a good factorization. One alternative is to select an arbitrary level-0 kernel as a divisor. This may not produce the best final result. Another alternative is to select a kernel which when substituted into the original function maximally reduces the total number of literals.

Example 6.57 Given

$$X = a \cdot c + a \cdot d + a \cdot e + a \cdot g + b \cdot c + b \cdot d + b \cdot e + b \cdot f + c \cdot e + c \cdot f + d \cdot f + d \cdot g$$

if, in the procedure **CHOOSE_DIVISOR**, we choose literals in lexicographical order, we obtain

$$X = a \cdot (c + d + e + g) + b \cdot (c + d + e + f) + c \cdot (e + f) + d \cdot (f + g)$$

However, if we choose kernels, we obtain a better factorization

$$X = (c + d + e) \cdot (a + b) + f \cdot (b + c + d) + g \cdot (a + d) + c \cdot e$$

which has fewer literals.

Extraction and Resubstitution Algorithm. To identify cube-free expressions that occur in multiple functions $\{f_i\}$, we do the following.

1. Generate kernels for each f_i.
2. Select a pair of kernels $k_1 \in K(f_i)$ and $k_2 \in K(f_j)$ for $i \neq j$ such that $k_1 \cap k_2$ is not a cube. If no such pair exists, stop.

```
GFACTOR(f){
    if (number of terms in f is 1 )
        return(f) ;
    g = CHOOSE_DIVISOR(f) ;
    (h, r) = DIVIDE(f, g) ;
    f = GFACTOR(g) · GFACTOR(h) + GFACTOR(r) ;
    return(f) ;
}
```

FIGURE 6.25

Procedure to algebraically factor a function.

3. Set a new variable v equal $k_1 \cap k_2$.

4. Update the associated functions to

$$f_i = v \cdot (f_i/(k_1 \cap k_2)) + r_i$$

where r_i is the remainder of the division $f_i/(k_1 \cap k_2)$.

Common cubes are extracted as follows.

1. Select a pair of cubes $c_1 \in f_i$, $c_2 \in f_j$ for $i \neq j$ such that $c_1 \cap c_2$ consists of two or more literals. If no such pair exists, stop.

2. Set a new variable u equal $c_1 \cap c_2$.

3. Update each function f_i with the new variable u wherever possible in the network.

Example 6.58 Consider the factored functions

$$\begin{aligned} X &= a \cdot b \cdot (c \cdot (d + e) + f + g) + h, \quad \text{and} \\ Y &= a \cdot i \cdot (c \cdot (d + e) + f + j) + k \end{aligned}$$

We have $d + e$ being a level-0 kernel of both functions. Extraction results in

$$\begin{aligned} L &= d + e, \\ X &= a \cdot b \cdot (c \cdot L + f + g) + h, \quad \text{and} \\ Y &= a \cdot i \cdot (c \cdot L + f + j) + k \end{aligned}$$

Now, we select $c \cdot L + f + g$ as a level-0 kernel of the reexpressed X and $c \cdot L + f + j$ as a level-0 kernel of reexpressed Y. We obtain

$$\begin{aligned} M &= c \cdot L + f \\ L &= d + e \\ X &= a \cdot b \cdot (M + g) + h, \quad \text{and} \\ Y &= a \cdot i \cdot (M + j) + k \end{aligned}$$

Now X and Y have no kernel intersections that are not cubes. We now extract common cubes. The cubes $a \cdot b \cdot M$ in X and $a \cdot i \cdot M$ in Y have two literals in common. Extraction produces

$$\begin{aligned} N &= a \cdot M \\ M &= c \cdot L + f \\ L &= d + e \\ X &= b \cdot (N + a \cdot g) + h, \quad \text{and} \\ Y &= i \cdot (N + a \cdot j) + k \end{aligned}$$

Because we are continually recomputing level-0 kernels on the reexpressed functions, it is possible to obtain decompositions corresponding to level-k kernels for $k > 0$. If we collapse L into M into N above, we obtain

$$\begin{aligned} N &= a \cdot (c \cdot (d + e) + f) \\ X &= b \cdot (N + a \cdot g) + h, \quad \text{and} \\ Y &= i \cdot (N + a \cdot j) + k \end{aligned}$$

where N contains a level-1 kernel of the original X and Y, since it contains the level-1 kernel M which contains the level-0 kernel $d + e$.

Algebraic Resubstitution with Complement. Algebraic factorization and resubstitution can be performed with the complement of a given divisor.

Example 6.59 Consider

$$f = a \cdot b + a \cdot c + \neg b \cdot \neg c \cdot d$$

where we choose $b + c$ as a level-0 kernel of f and decompose f as

$$\begin{aligned} f &= a \cdot X + \neg b \cdot \neg c \cdot d, \quad \text{and} \\ X &= b + c \end{aligned}$$

In many cases it is useful to check if the complement of the new variable is an algebraic divisor for the function. In this case we can obtain

$$\begin{aligned} f &= a \cdot X + \neg X \cdot d, \quad \text{and} \\ X &= b + c. \end{aligned}$$

6.3.3.4 *Common divisors*

One of the key problems in algebraic optimization is the identification of good (common) divisors. We have described the use of kernels for determining a good set of divisors for algebraic factoring, decomposition, and extraction. The problem of finding a kernel and finding a single-cube or multiple-cube divisor can be reduced to the combinatorial optimization problem of rectangle covering [Rudell 1989]. This formulation of the problem is not only elegant, but it also favors the development of fast and effective algorithms.

Before introducing the method, we give some definitions.

A (**combinatorial**) **rectangle** (R, C) of a matrix B, with entries $B_{ij} \in \{0,1,*\}$, is a subset of rows R and subset of columns C such that $B_{ij} \in \{1, *\}$ for all $i \in R$ and $j \in C$. Note that the rows and columns forming the rectangle do not have to be contiguous.

A rectangle (R_1, C_1) is said to *strictly contain* rectangle (R_2, C_2) if $R_2 \subseteq R_1$ and $C_2 \subset C_1$, or $R_2 \subset R_1$ and $C_2 \subseteq C_1$.

A rectangle (R, C) of B is said to be a **prime rectangle** if it is not strictly contained in any other rectangle of B.

The **corectangle** of a rectangle (R, C) is the pair (R, C') where C' is the set of columns not in C.

A set of rectangles $\{(R^k, C^k)\}$ forms a **rectangle cover** of matrix B if $B_{ij} = 1$ implies that $i \in R^k$ and $j \in C^k$ for some k. Thus, each 1-entry in B must be covered by at least one rectangle from the cover. A covering need not be disjoint, and therefore a 1-entry in B can be covered by more than one rectangle. The *-entries of B are not required to be covered by any rectangle in the cover and therefore represent don't-care points in the matrix.

Example 6.60 In the following matrix

	1 2 3 4 5
1	1 1 1 0 0
2	1 * 1 0 *
3	0 1 1 0 1
4	1 0 1 1 1

The tuple ({1,2}, {2,3}) is a rectangle, but it is not prime as it is contained by the prime rectangle ({1,2}, {1,2,3}). The tuple ({2,4}, {1,3,5}) is another prime rectangle while ({2,3}, {1,2}) is not a rectangle.

Each rectangle (R^k, C^k) has an associated weight or cost defined by a weight function $w(R^k, C^k)$. The weight of a rectangle cover is then defined as the sum

$$\sum_k w\left(R^k, C^k\right)$$

The *minimum-weighted rectangle covering problem* is that of finding a rectangle cover of a matrix with minimum total weight.

Rectangles and Kernels. Rectangles provide an alternate way of looking at the kernels of a function. By representing a Boolean expression as a **cube-literal matrix**, where each row corresponds to a cube in the expression and the columns correspond to all the distinct literals, each prime rectangle is a cokernel while each corectangle of a prime rectangle is a kernel of the expression.

Example 6.61 Consider the expression $g = a \cdot b \cdot e + a \cdot c \cdot d + b \cdot c \cdot d$. It can be represented using a cube-literal matrix shown below.

	a b c d e
a · b · e	1 1 0 0 1
a · c · d	1 0 1 1 0
b · c · d	0 1 1 1 0

Consider the prime rectangle $(R, C) = (\{2,3\}, \{3,4\})$ and its corectangle $(R, C') = (\{2,3\}, \{1,2,5\})$. The rectangle obviously corresponds to a cube $c \cdot d$ that is common to all the product terms corresponding to rows in R. Since the rectangle is prime, it is the largest cube common to all the product terms in R. If this cube is extracted from these product terms, the resulting expression is cube-free and is also a divisor of the original function g. In other words, the resulting expression is a kernel of g. The expression resulting from the extraction of the cube corresponds to the corectangle $(R, C') = (\{2,3\}, \{1,2,5\})$, which is $a + b$.

From the rectangle interpretation of kernels, it is also possible to understand more clearly the notion of the level of a kernel. A level-0 kernel is the corectangle of a prime rectangle which has no other rectangle containing its column set, *i.e.,* a rectangle of maximal width. The corectangle of a prime rectangle of maximal height, *i.e.,* one whose row set is not contained in any other rectangle, corresponds to a kernel of maximal level.

Common-Cube Extraction. Common-cube extraction is the process of finding cubes common to two or more expressions and extracting the common cube to simplify each of the expressions. To optimize the network it is necessary to find the particular cubes to introduce that provide an optimal decomposition. The optimal decomposition can be defined as minimizing the total number of literals summed over all expressions or minimizing the total number of literals given a bound on the number of levels of logic in the final circuit.

Common cubes can be easily identified using the cube-literal matrix described above.

Example 6.62 Consider the equations

$$F = a \cdot b \cdot c + a \cdot b \cdot d + e \cdot g$$
$$G = a \cdot b \cdot f \cdot g, \text{ and}$$
$$H = b \cdot d + e \cdot f$$

The cube-literal matrix for these expressions is

	a b c d e f g
$F_1 : a \cdot b \cdot c$	1 1 1 0 0 0 0
$F_2 : a \cdot b \cdot d$	1 1 0 1 0 0 0
$F_3 : e \cdot g$	0 0 0 0 1 0 1
$G_1 : a \cdot b \cdot f \cdot g$	1 1 0 0 0 1 1
$H_1 : b \cdot d$	0 1 0 1 0 0 0
$H_2 : e \cdot f$	0 0 0 0 1 1 0

The rectangle ({1,2,4}, {1,2}) corresponds to the common cube $a \cdot b$ which is present in functions F and G. If this common cube is extracted as a new function X, the equations can be rewritten as

$$F = X \cdot c + X \cdot d + e \cdot g$$
$$G = X \cdot f \cdot g$$
$$H = b \cdot d + e \cdot f, \text{ and}$$
$$X = a \cdot b$$

The process of extracting a cube modifies a Boolean network. A new node is added to the Boolean network with a logic function which is the common-cube divisor. All functions which the cube divides are replaced with the algebraic division of the function by the single cube. In order to extract cubes efficiently in an iterative algorithm, it is necessary to modify the cube-literal matrix incrementally to reflect the extraction of the cube. The advantage is that the cube-literal matrix does not have to be recreated as each cube is extracted.

The modifications required to form the new cube-literal matrix are the following. A new row is added to reflect the new single cube expression added to the network. The entries covered by the rectangle are marked with a * to reflect that the position has been covered. However, the * allows other rectangles to cover the same position.

The choice of the weight function for a rectangle measures the optimization goal for cube extraction. To minimize the total number of literals in the network, the weight of a rectangle is chosen so that the weight of a rectangle cover of the cube-literal matrix equals the total number of literals in the network after the new single-cube functions are added to the network. Hence, a minimum weighted rectangle cover corresponds to the optimal simultaneous extraction of a collection of cubes. The weight of a rectangle is defined as:

$$w(R,C) = \begin{cases} |C| & \text{if } |R| = 1 \\ |C| + |R| & \text{if } |R| > 1 \end{cases}$$

If there is a single row in the rectangle, then it corresponds to leaving the cube unchanged in the network. Hence, the weight of the rectangle counts the number of literals in the cube, which equals the number of columns. When the number of rows is greater than one, this corresponds to creating a new single cube function with $|C|$ literals and substituting this new function into $|R|$ other cubes at a cost of $|R|$ literals.

Note that the above weight does not reflect the savings obtained in terms of the number of literals by extracting a common cube. Therefore, when searching for a cube to extract it is useful to define a second function called the **value** of the rectangle. For cube extraction, the value of the rectangle should indicate the savings obtained from extracting the corresponding cube. Since the number of literals before cube extraction is the number of 1-entries in the rectangle and the number of literals after cube extraction is the weight of the rectangle, the value $v(R, C)$ of a rectangle is defined as

$$v(R, C) = \left|\{(i,j)\,|\,B_{ij} = 1, i \in R, j \in C\}\right| - w(R, C)$$

Example 6.63 For the rectangle ({1,2,4}, {1,2}) in the cube-literal matrix of the previous example, the weight is the number of rows plus the number of columns, which equals 5. There are 6 positions in this rectangle and each of them has a 1. Therefore, the value of the rectangle is $6 - 5 = 1$. Therefore only one literal can be saved by extracting this rectangle, as illustrated in the previous example.

Kernel Intersection. As described previously, intersections among the kernels of a collection of expressions are useful for finding common multiple-cube divisors between two or more expressions. If two functions share a common multiple-cube divisor, then the common divisor can be found as the intersection of a kernel from each of the functions.

The Boolean matrix associated with the optimal kernel intersection problem is called the **cokernel-cube matrix**. A row in this matrix corresponds to a cokernel (and its associated kernel) and each column corresponds to a cube present in some kernel, called a **kernel-cube**. The entry B_{ij} is set to 1 if the kernel associated with row i contains the cube associated with column j. Then a rectangle of the cokernel-cube matrix identifies an intersection of kernels. The columns of the rectangle identify the cubes in the subexpression, and the rows in the rectangle identify the particular functions the subexpression divides.

Example 6.64 Consider the functions

$$
\begin{aligned}
F &= a \cdot f + b \cdot f + a \cdot g + c \cdot g + a \cdot d \cdot e + b \cdot d \cdot e + c \cdot d \cdot e \\
G &= a \cdot f + b \cdot f + a \cdot c \cdot e + b \cdot c \cdot e, \text{ and} \\
H &= a \cdot d \cdot e + c \cdot d \cdot e
\end{aligned}
$$

The kernels and cokernels of each of the functions are shown below.

Function	Cokernel	Kernel
F	a	$d \cdot e + f + g$
F	b	$d \cdot e + f$
F	$d \cdot e$	$a + b + c$
F	f	$a + b$
F	c	$d \cdot e + g$
F	g	$a + c$
G	a	$c \cdot e + f$
G	b	$c \cdot e + f$
G	f	$a + b$
G	$c \cdot e$	$a + b$
H	$d \cdot e$	$a + c$

Note that functions F and G are themselves kernels but have not been shown above for ease of presentation. Let us number the cubes in the original function from 1 to 13, with $a \cdot f$ being 1, $b \cdot f$ being 2, and so on. The cokernel-cube matrix for this set of kernels is shown below. Note that instead of 1's in the matrix, we have numbers. These numbers indicate a cube of the original functions formed by multiplying the cokernel corresponding to a row and the cube corresponding to a column. For example, in the third row under column a we have the number 5 corresponding to the the the fifth cube $a \cdot d \cdot e$.

	a	b	c	ce	de	f	g
F : a	0	0	0	0	5	1	3
F : b	0	0	0	0	6	2	0
F : d·e	5	6	7	0	0	0	0
F : f	1	2	0	0	0	0	0
F : c	0	0	0	0	7	0	4
F : g	3	0	4	0	0	0	0
G : a	0	0	0	10	0	8	0
G : b	0	0	0	11	0	9	0
G : f	8	9	0	0	0	0	0
G : c·e	10	11	0	0	0	0	0
H : d·e	12	0	13	0	0	0	0

Rectangle ({3,4,9,10}, {1,2}) identifies the subexpression $a + b$. This corresponds to the factorization of the equations into the form

$$
\begin{aligned}
F &= d \cdot e \cdot X + f \cdot X + a \cdot g + c \cdot g + c \cdot d \cdot e \\
G &= c \cdot e \cdot X + f \cdot X \\
H &= a \cdot d \cdot e + c \cdot d \cdot e \text{ and} \\
X &= a + b
\end{aligned}
$$

Whenever a new subexpression is identified, it is inserted into the Boolean network. This insertion consists of adding a new node to the network and dividing the node into each of the expressions which this node divides. A new cokernel-cube matrix is then created for the modified Boolean network.

To reduce the complexity of extracting each factor from the network it is desirable to modify the cokernel-cube matrix incrementally as each subexpression is identified. To do this, new rows are added to the cokernel-cube matrix for each kernel of the new subexpression. The cubes which are formed by the insertion of this new factor into the network are then marked as covered. This includes the points directly contained in the rectangle and other points which are labeled with the same number. These points are marked * so that other rectangles can cover them.

The weight of a rectangle of the cokernel-cube matrix is chosen to reflect the number of literals in the network if the corresponding common subexpression is inserted into the network. A minimum weighted rectangle cover of the cokernel-cube matrix then corresponds to a simultaneous selection of a set of subexpressions to add to the network in order to minimize the total number of literals.

Let w_j^c be the number of literals in the kernel-cube for column j. w_j^c is also called the column weight of column j. If a rectangle (R, C) is used to identify a subexpression, then a new function is formed from the columns of C. This new function has $\sum_{j \in C} w_j^c$ literals. Let w_i^r be 1 plus the number of literals the cokernel corresponding to row i. w_i^r is also called the row weight of row r. The chosen subexpression divides the expressions indicated by the rows R of the rectangle. After algebraic division by the subexpression, each of these expressions consists of a sum of the corresponding cokernel cubes multiplying the literal for the new expression. The number of literals in the affected functions after the extraction of the subexpression corresponding to the rectangle is $\sum_{i \in R} w_i^r$. Therefore, the weight of a rectangle (R, C) in the cokernel-cube matrix is defined as:

$$
w(R, C) = \sum_{i \in R} w_i^r + \sum_{j \in C} w_j^c
$$

The value of a rectangle measures the difference in the number of literals in the network if the particular rectangle is selected. The number of literals after the rectangle is selected is the weight of the rectangle as defined above. Let V_{ij} be the number of literals in the cube which is covered by position (i, j) of the cokernel-cube matrix. Then the number of literals before extraction of the rectangle is simply $\sum_{i \in R, j \in C} V_{ij}$. As elements of the cokernel-cube matrix are covered, their values V_{ij} are set to 0. This includes the elements V_{ij} covered by the matrix and all other elements which represent the same cube in the network. The value of a rectangle (R, C) of the cokernel-cube matrix is thus defined as

$$v(R, C) = \sum_{i \in R, j \in C} V_{ij} - w(R, C)$$

Example 6.65 For the rectangle ({3,4,9,10}, {1,2}) of the cokernel-cube matrix in the previous example, $\sum_{i \in R, j \in C} V_{ij} = 3 + 3 + 2 + 2 + 2 + 2 + 3 + 3 = 20$, $\sum_{i \in R} w_i^r = 3 + 2 + 2 + 3 = 10$, $\sum_{j \in C} w_j^c = 1 + 1 = 2$. Therefore, the value of the rectangle is $20 - 10 - 2 = 8$. Eight literals can be saved by extracting the expression corresponding to the rectangle, as can be verified in the example above.

Rectangle Covering. Since minimum-weighted rectangle covering corresponds to optimum algebraic extraction, it offers a unified approach to the extraction, factorization, and decomposition of Boolean expressions. However, the minimum-weighted rectangle covering problem is NP-complete [Rudell 1989] and thus heuristic algorithms are resorted.

There are two types of algorithms for rectangle covering. The first type of algorithm is greedy and selects one rectangle at a time and modifies the matrix to reflect the extraction of the rectangle. The advantage of this technique is that it immediately takes into account common factors between the newly extracted function and the rest of the logic network. The disadvantage of this approach is that it selects only one rectangle at a time and does not easily account for the simultaneous extraction of multiple rectangles. The second type of algorithm finds the best collection of factors to extract at each step by solving the minimum-weighted rectangle covering problem heuristically. First, all the prime rectangles are generated, and a collection of rectangles are then extracted. Second, the matrix is updated, and the entire process is repeated to find factors between the new expressions and the remainder of the logic network. A detailed exposition of this approach can be found in [Rudell 1989].

6.3.3.5 *Boolean division*

So far we have primarily described algebraic optimization methods. Apparently the optimality of algebraic division is limited. For example, the Boolean expression $f = a \neg b + ad + \neg ab + bd + \neg ac + \neg bc + cd$ can not be factored into $f = (a + b + c)(\neg a + \neg b + d)$ through algebraic division. It motivates the development of Boolean division.

To do so, in **Boolean resubstitution** we would like to reexpress a given Boolean function $f(x)$ in terms of a given divisor $g(x)$. The computation can be done by first building the function

$$h(x, y) = f(x) \wedge (y \equiv g(x))$$

where y is a newly introduced Boolean variable representing the output signal of function g. We then minimize function h with respect to the don't care set $y \not\equiv g(\boldsymbol{x})$ while insisting y to be a support variable of h. If h after minimization is "simpler" than function f, then the resubstitution is successful.

Boolean resubstitution can be formalized more generally as **functional dependency** [Jiang 2004]. We say that a function $f(\boldsymbol{x})$ **functionally depends** on a set of functions $g_1(\boldsymbol{x}), \ldots, g_m(\boldsymbol{x})$ if there exists some function h such that

$$f(\boldsymbol{x}) = h(g_1(\boldsymbol{x}), \ldots, g_m(\boldsymbol{x}))$$

The necessary and sufficient condition, informally speaking, is that the set $\{g_1, \ldots, g_m\}$ of functions must be more distinguishing than f on the domain elements. That is, for every $a, b \in \mathbb{B}^n$ with $f(a) \neq f(b)$ there must exist some g_i such that $g_i(a) \neq g_i(b)$. ROBDD and SAT based computation of functional dependency can be found in [Jiang 2004] and [Lee 2007; Mishchenko 2007a], respectively.

To see that Boolean resubstitution is a special case of functional dependency, for $\boldsymbol{x} = (x_1, \ldots, x_n)$ we set $g_i(\boldsymbol{x}) = x_i$ for $i = 1, \ldots, n$ and $g_{n+1}(\boldsymbol{x}) = g(\boldsymbol{x})$. Thus functional dependency reduces to Boolean resubstitution $f(\boldsymbol{x}) = h(\boldsymbol{x}, g(\boldsymbol{x}))$. In fact, we can minimize the support variables of h by setting as many $g_i(\boldsymbol{x}) = 0$ (or 1) as possible to remove x_i from the support set of h.

6.3.4 **Combinational complete flexibility**

The aforementioned multilevel logic minimization approaches, such as decomposition, extraction, factoring, substitution, and elimination, change the structure of a Boolean network. In contrast, in this section we study how to perform logic minimization without changing a multilevel network structure. More specifically, given a structurally optimized multilevel network, we may further minimize it by simplifying the logic expression within every node.

To minimize the logic function of a node u in a Boolean network, we would like to characterize the don't care conditions of the node u, such that we may choose the best among the set of valid functions, called *permissible functions*, that can implement u without changing the functionality of the entire Boolean network. Notice that node u imposes a topological constraint on a permissible function whose inputs are restricted to the fanins of node u in the Boolean network.

In fact, don't cares exist pervasively in a multilevel logic netlist because the Boolean space is largely expanded due to the existence of many intermediate variables. Let X be the set of primary input variables and Y the set of all other variables of a Boolean network. In the $\mathbb{B}^{|X|+|Y|}$ Boolean space, only $2^{|X|}$ valuations are consistent because the valid valuations are determined by the assignments on the primary input variables. Consequently a lot of invalid valuations may not appear in the Boolean network and can be exploited for logic minimization. Moreover the effect of one signal may be conditionally blocked by other signals and cannot affect the valuations of primary outputs. Based on these reasons, flexibility may exist to some extent in a multilevel logic network.

The don't-care conditions arising in multilevel logic can either be specified by the user or can be an artifact of the network structure. Essentially there are three types of don't cares: satisfiability don't cares (SDC), observability don't cares (ODC), and external don't cares (XDC). *Internal* don't-cares arise in multilevel logic because of the structure of a Boolean network. They are divided into satisfiability and observability don't-cares. User specified don't-cares or don't-cares derived from considerations other than the network structure are called *external* don't-cares.

In the following discussion, for a Boolean network, let X be the set of primary input variables, Y the set of all other variables, and $Z \subseteq Y$ the set of primary output variables. For a node i in a Boolean network, its output variable is denoted as y_i and its local or intermediate input variables, other than primary input variables, are denoted as Y_i; its local function is denoted as $f_i(X, Y_i)$ and its global function, in terms of only primary input variables, is denoted as $g_i(X)$. Of course, since we consider only acyclic Boolean networks, f_i depends only on a subset of the Y variables that are not in the transitive fanout cone TFO_i of node i.

External Don't-Cares. External don't-cares are specified for every primary output, which indicate under what valuations on the primary input variables X the value of the output is immaterial.

Satisfiability Don't-Cares. Satisfiability don't-cares are a result of the existence of the additional intermediate variables introduced at the intermediate nodes of a Boolean network. A node with output variable y_i and immediate function $f_i(X, Y_i)$ of a Boolean network imposes the relation

$$y_i \equiv f_i(X, Y_i) \qquad (6.14)$$

which characterizes the set of valuations on variables X and Y that are consistent under the constraint imposed by node i. Therefore the set of satisfiability don't cares of the entire Boolean network is given by

$$SDC(X, Y) = \bigvee_i (y_i \neq f_i(X, Y_i)) \qquad (6.15)$$

which gives all the valuations on variables X and Y that will never occur due to the network structure and is so called because each of the relations $y_i \equiv f_i(X, Y_i)$ must be satisfied during the correct operation of the network. In order to optimize a given node i we are typically interested in the satisfiability don't cares imposed by the transitive fanin cone TFI_i of node i.

Example 6.66 Consider the network

$$
\begin{aligned}
y_1 &= x_1 \wedge x_2 \\
y_2 &= x_2 \vee x_3, \text{ and} \\
y_3 &= y_1 \oplus y_2 = \neg y_1 y_2 \vee y_1 \neg y_2
\end{aligned}
$$

It implements function $g_3 = (x_1 \wedge x_2) \oplus (x_2 \vee x_3)$. We have the option of eliminating y_1 and y_2 or expanding the Boolean space to include these variables. If we do the latter there are assignments of variables which will never occur. For example, the assignment $y_1 = 1$ and $y_2 = 0$ will never happen. The assignments that will never occur are expressed by

$$SDC = (y_1 \not\equiv (x_1 \wedge x_2)) \vee (y_2 \not\equiv (x_2 \vee x_3)) \vee (y_3 \not\equiv (y_1 \oplus y_2))$$

To optimize f_3, the satisfiability don't care set

$$SDC_3 = (y_1 \not\equiv (x_1 \wedge x_2)) \vee (y_2 \not\equiv (x_2 \vee x_3))$$

imposed by the fanin nodes 1 and 2 of node 3 is of particular interest. Furthermore, SDC_3 in terms of the local input variables of node 3 can be computed by

$$\forall x_1, x_2, x_3.(y_1 \not\equiv (x_1 \wedge x_2)) \vee (y_2 \not\equiv (x_2 \vee x_3)) = y_1 \neg y_2$$

which ensures that the computed SDC in term of variables y_1 and y_2 is valid under any valuation on the X variables. Accordingly, we may optimize f_3 using the impossible condition $y_1 \neg y_2$. So $f_3 = \neg y_1 y_2$ is another permissible function for node 3.

Observability Don't-Cares. Observability don't-cares occur in a network because at each node there is a network structure that limits the observability of the value of the node as seen at primary outputs.

To compute the observability don't cares ODC_i of a node i in a Boolean network N. We construct a new Boolean network N' from N by treating y_i as a (pseudo) primary input and removing node i and other induced nodes without fanouts from N. The condition that node i is observable at primary output j is given by

$$\frac{\partial g_j'}{\partial y_i} = \left[g_j(X, y_i = 0) \not\equiv g_j'(X, y_i = 1) \right] \tag{6.16}$$

where g_j' is the global function of j in network N'. That is, Formula (6.16) gives the input conditions under which the g_j' produces different values under different y_i values, *i.e.*, the conditions under which output j is sensitive to y_i. Therefore the conditions under which the value of y_i cannot be observed at any output are characterized by

$$\begin{aligned} ODC_i(X) &= \bigwedge_{y_j \in Z} \left(g_j'(X, y_i = 0) \equiv g_j'(X, y_i = 1) \right) \\ &= \bigwedge_{y_j \in Z} \neg \left(\frac{\partial g_j'}{\partial y_i} \right) \end{aligned}$$

The above computation assumes the external don't-care set is empty. For non-empty XDC, the observability of a node at some primary output should be conditioned on the external don't care set of the primary output.

Local Don't-Cares and Node Minimization. Note that SDC is in terms of X and Y variables; XDC and ODC are in terms of X variables. To minimize a node i, they are not directly useful unless they are expressed in terms of the local input variables of node i. Don't cares in terms of the local input variables are called *local don't cares*. Let

$$DC_i(X) = \bigwedge_{y_k \in Z} XDC_k(X) \vee ODC_i(X)$$

Let D_i be the local don't cares of node i. Then it can be computed by

$$D_i(Y_i) = \neg\left(\exists X. \bigwedge_{y_j \in Y_i} \left(y_j \equiv g_j(X)\right) \wedge \neg DC_i(X)\right) \tag{6.17}$$

It should be noted that we cannot simply project $DC_i(X)$ to the local space spanned by Y_i using image computation. Rather we should project the care set into the local input space and then take the complement. It is because the former may mistakenly include some care minterm in the local space if there exists some care minterm and don't care minterm in the global space mapping to the same image. On the other hand, notice that, even though SDC is absent from Formula (6.17), it has been implicitly computed in the image computation.

With the local don't cares $D_i(Y_i)$ of node i, we can minimize the SOP expression of node i using two-level logic minimization methods. The don't-care generation and logic minimization procedure can be summarized as follows.

1. Select a node i in the Boolean network.
2. Compute its local don't care set D_i.
3. Minimize the cover of node i with respect to D_i.

Therefore by treating a multilevel netlist as a network of PLAs, two-level minimization methods can be applied as a baseline tool for multilevel logic minimization.

The above computation assumes that the rest of the Boolean network is not changed. One generalization is to consider *compatible don't cares* among multiple nodes simultaneously. Since the don't care conditions of different nodes may be conflicting with each other, they must be made compatible. The high computational complexity however restricts the application of compatible don't cares. Often a network is iteratively optimized one node at a time with respect to its local don't cares.

Complete Flexibility. The characterization of don't cares, including SDC, ODC, and XDC, can be unified through the concept of *complete flexibility* [Mishchenko 2002]. The *complete flexibility* (CF) of a node in a Boolean network is a Boolean relation that characterizes the set of all possible input-output behaviors of the node assuming that the rest of the network is not changed. The complete flexibility subsumes all the above don't cares. In addition, it is more powerful in capturing non-determinism, and can be generalized for a non-deterministic Boolean network [Mishchenko 2006a] where each node represents some relation allowing one-to-many mappings, not possible for functions.

Consider computing the complete flexibility of node i in a Boolean network N. Let $S(X, Z)$, given from specification, be the *specification relation* specifying all the allowed input-output behavior of the Boolean network. Hence $S(X, Z)$ subsumes XDC. Let

$$E_i(X, Y_i) \;=\; \bigwedge_{y_j \in Y_i} \left(y_j \equiv g_j{}'(X) \right)$$

be the *environment relation* characterizing the set of consistent assignments on variables X and Y_i. Hence $\neg E_i(X, Y_i)$ subsumes the SDC of node i. Let

$$I_i(X, y_i, Z) = \bigwedge_{y_j \in Z} \left(y_j \equiv g_j{}'(X, y_i) \right)$$

be the *influence relation* characterizing the allowed valuations on y_i consistent with those on X and Z, where $g_j{}'$ is a primary output function of network N', same as that obtained in the ODC computation. Hence

$$R_i(X, y_i) = \forall Z. [I_i(X, y_i, Z) \Rightarrow S(X, Z)]$$

subsumes the ODC of node i. The complete flexibility CF_i of node i in terms of the local input variables Y_i can be obtained by

$$
\begin{aligned}
CF_i(Y_i, y_i) \;&=\; \forall X. [E_i(X, Y_i) \Rightarrow R_i(X, y_i)] \\
&=\; \forall X. [E_i(X, Y_i) \Rightarrow \forall Z. [I_i(X, y_i, Z) \Rightarrow S(X, Z)]] \\
&=\; \forall X, Z. [\neg E_i(X, Y_i) \vee \neg I_i(X, y_i, Z) \vee S(X, Z)] \\
&=\; \forall X, Z. \neg [E_i(X, Y_i) \wedge I_i(X, y_i, Z) \wedge \neg S(X, Z)]
\end{aligned}
\tag{6.18}
$$

ROBDD Implementation. Notice that all of the above computations can be realized using ROBDDs as operations over Boolean functions.

6.3.5 Advanced subjects

AIG-based Multilevel Logic Minimization. In addition to the division-based transformations, we may approach the multilevel logic minimization problem with a new view using the AIG representation.

Any Boolean expression can be converted into an AIG in polynomial time while structural hashing can be applied during the AIG construction. The obtained AIG can then be further simplified through *rewriting* [Bjesse 2004; Mishchenko 2006b]. This simplification is in terms of AIG nodes and/or levels, rather than the conventional literal or cube counts.

By grouping the nodes of the AIG into clusters (such that each cluster consists of a set of connected nodes rooted at some node producing its output, and the fanins of a cluster are outputs of some other clusters), each cluster can be seen as a complex logic node in a Boolean network. Therefore an AIG can be considered as a data structure that encompasses a set of multilevel logic netlists subject to different interpretations of cluster boundaries, called **cuts**. Given an AIG, the problem of multilevel logic minimization now boils down to the enumeration of good cuts, see, *e.g.*, [Ling 2007; Mishchenko 2007b]. This approach to logic minimization is taken by the ABC package [ABC 2005].

Sequential Logic Minimization. The aforementioned combinational logic minimization methods can be applied to simplify sequential circuits. For a given sequential circuit, treating the register outputs as primary inputs and register inputs as primary outputs results in the combinational methods being applicable to sequential circuit optimization. The optimization, of course, does not take full advantage of sequential flexibilities.

We can in fact pursue more progressive logic transformations. *State minimization* [Kohavi 1978], *state encoding* [Villa 1997], and logic minimization using *unreachable states* or *state equivalence* [Kohavi 1978] as don't cares, for example, are valid transformation methods because they do not change the input-output behavior of a sequential circuit. Furthermore, it is possible to characterize complete flexibility in the sequential domain [Yevtushenko 2001; Mishchenko 2005], similar to the combinational counterpart. In the computation, however, we have to manipulate finite automata, rather than Boolean formulas.

The above approaches are *state-based* in the sense that we have to know some state information for a given sequential circuit. The expensive derivation of state information limits their applicability to large designs. In contrast, there are *structure-based* transformations, which are carried out according to circuit structures and do not rely on state information. *Retiming* [Leiserson 1983, 1991] and *resynthesis* [Malik 1991], for example, are practical transformation methods for sequential logic minimization.

Although most designs are sequential and practical sequential optimization techniques are available, logic synthesis flows for the industrial design typically consist of only combinational optimization methods. This phenomenon can be attributed to the hardness of sequential circuit equivalence verification [Jiang 2006]. From the complexity viewpoint, sequential equivalence checking is PSPACE-complete, which is considered much harder than the coNP-complete combinational equivalence checking problem. In industrial practice, combinational equivalence checking is considered "solvable." (In fact, equivalence checking of industrial circuits with multi-million gates has been demonstrated [Kuehlmann 1997]. Of course there are special cases of combinational circuits that are hard to verify, *e.g.*, multipliers with different circuit structures.) On the contrary, for sequential equivalence checking, there are almost no good approaches that are general enough and work for the majority of practical test-cases. Making sequential circuit optimization scalable and verifiable is an important research subject.

6.4 TECHNOLOGY MAPPING

The logic optimization algorithms described thus far operate on Boolean networks. The optimization aims at simplifying logic expressions and is independent of the target implementation technology. To finish the logic synthesis steps, we need to implement logic gates with physical layouts. One solution to it is to perform

technology mapping, which is one of the most important tasks in technology dependent optimization. It takes on a technology-independently optimized logic netlist, and expresses the netlist using a set of pre-designed and pre-characterized gate layouts from a technology library. Typically, the goal is to make optimal use of all of the gates in the library to produce a circuit with minimum area subject to the delay constraint for critical-path delay no greater than a target value.

Technology mapping algorithms are constrained by the structure of the logic netlists produced by technology-independent optimization. It is not the role of technology mapping to change the structure of the circuit radically, for example, by finding common sub-expressions between two or more parts of the circuit. Likewise, it is not the main role of technology mapping to reduce the number of levels of logic along the critical path. The role of technology mapping is to make the actual gate choice to implement the logic netlist, for example, choosing the fastest gates along the critical path and using the most area-efficient combination of gates off the critical path.

A technology mapping algorithm should ideally achieve several goals. It should be able to adapt to a variety of different libraries because an algorithm which depends on characteristics of a particular library is of limited use, and an algorithm which is geared to a subset of the gates in a library is limited in its optimization potential. To practically achieve this goal of adaptability, a user must be able to provide new gates to the technology mapper without understanding its detailed operation, and these gates should be used effectively.

6.4.1 **Technology libraries**

The introduction of **gate arrays** and **standard cells** brought comparable benefits to IC designers. A gate array is an array of transistors and routing channels which can be configured into an IC through a metalization process during semiconductor fabrication. The metalization phases are used for cell definition, such as defining a NOR cell, and for interconnecting the cells. The electrical characteristics of cells after metalization have been carefully defined and are embodied in a databook. Standard cells are combinational and sequential logic gates whose electrical characteristics have been carefully defined and embodied in a library. Standard cells are similar to gate arrays in that they are precharacterized in a databook, but they offer additional degrees of freedom since they go through all the mask steps of semiconductor processing.

Logic gates of VLSI circuits, especially for ASICs, are usually restricted to be implemented by selections from a *technology library* of *gates*. A **gate** is a primitive element available in a particular implementation technology; a **technology library** is a collection of these gates. A technology library is assumed to consist of a finite collection of gates. For example, the gates in a static CMOS gate-array (or standard-cell) design typically include inverters, NAND gates, NOR gates, and a variety of complex gates, whereas the gates in an ***emitter-coupled logic*** (ECL) gate-array are typically NOR gates and XOR gates.

These libraries are typically composed of a few hundred gates and sequential elements like latches and flip-flops for which highly optimized layouts have been manually designed for a particular technology. Each gate is assigned a number of values associated with the different cost functions under which it will be optimized. For example, each gate is assigned a value called the *area* of the gate representing the physical area occupied by the gate. The logic designers are then restricted to using these gates in their logic circuits.

Example 6.67 The combinational subset of a very simple library is shown in Figure 6.26. The library cell names, associated area costs, their functions, and their representations in terms of two-input NAND (NAND2) gates and inverters (INV's) are shown.

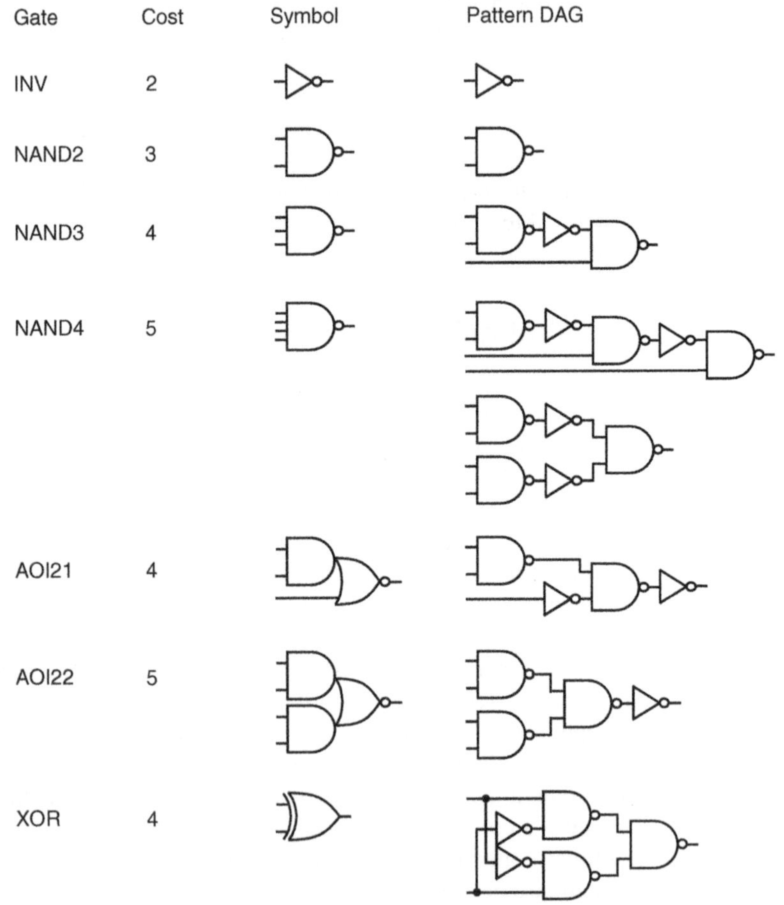

Gate	Cost	Symbol	Pattern DAG
INV	2		
NAND2	3		
NAND3	4		
NAND4	5		
AOI21	4		
AOI22	5		
XOR	4		

FIGURE 6.26

Gate library.

Given a technology library, the problem of technology mapping is finding a multilevel circuit equivalent to the given Boolean network such that it is comprised of gates in the library and has minimum cost, which could be the area, delay, testability, or power consumption of the resulting circuit.

6.4.2 **Graph covering**

A systematic approach to technology mapping is based on the notion of graph covering. With this formulation, the technology mapping problem can be viewed as the optimization problem of finding a minimum cost covering of the **subject graph** by choosing from the collection of **pattern graphs** for all gates in the library. A **cover** is a collection of pattern graphs such that every node of the subject graph is contained in one or more of the pattern graphs. Moreover, one restriction of any cover is that the inputs of one pattern in the covering must be the outputs of some other pattern in the covering. Otherwise it would imply that the inputs of one pattern come from internal nodes in another pattern. As these internal signal values are not visible outside the pattern, any covering without such a restriction would not be meaningful.

Example 6.68 The cover shown in Figure 6.27a is legitimate while that in Figure 6.27b is not.

In graph covering, the Boolean network to be covered is often represented in a special form, where each gate is either of a NAND2 or an INV. It is termed the **subject graph**, or **subject DAG**. In addition to the Boolean network to be covered, each library gate is also represented in this special form. Each realization is termed a **pattern graph**, or **pattern DAG**. Note that a gate may have more than one associated pattern DAG.

Example 6.69 In Figure 6.26, the pattern DAGs of the library cells are shown. The NAND4 gate has more than one pattern DAGs.

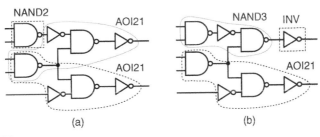

(a)   (b)

FIGURE 6.27

Graph coverings: (a) legal and (b) illegal.

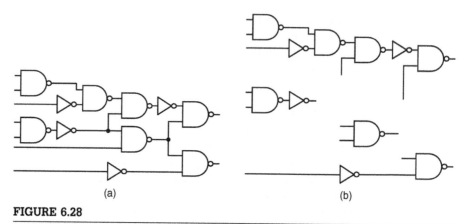

FIGURE 6.28

(a) Subject DAG example. (b) Subject DAG decomposed into a forest of trees.

Example 6.70 Figure 6.28a shows a subject DAG example.

The optimization problem of technology mapping can now be stated as: Find a minimum cost covering of the subject DAG by the pattern DAGs.

6.4.3 Choice of atomic pattern set

The choice of which atomic patterns to use for the subject and pattern graphs is an important consideration for graph covering algorithms. This decision influences the range of solutions for the covering problem and the number of patterns needed.

Why subject and pattern graphs are in terms of NAND2 and INV is motivated by the following observation. Adding additional functions such as a NOR2 gate, an AND2 gate, or an OR2 gate cannot provide higher-quality solutions; likewise, adding NAND, NOR, AND, or OR gates with more than two inputs cannot provide higher-quality solutions. This observation is based on the fact that given a cover for a subject graph using a larger set of functions, it is possible to show an equivalent cover where each function is replaced by an equivalent set of NAND2 gates and inverters.

Restricting ourselves to only a NAND2 gate and inverter does come at the price of increasing the number of patterns needed to represent some logic functions, as can be seen from the following example. Experience has shown that the increase in the number of patterns (and hence the increase in the memory and time required for technology mapping) is not significant.

Example 6.71 The logic function

$$f = \overline{a \cdot b \cdot c \cdot d + e \cdot f \cdot g \cdot h + i \cdot j \cdot k \cdot l + m \cdot n \cdot o \cdot p}$$

requires only one pattern corresponding to a tree of five NAND4 gates. However, representing all patterns for this same function using NAND2 gates and INV's requires 18 patterns.

6.4.4 **Tree covering approximation**

One technique (following the paradigm established in the domain of code generation [Aho 1976]) for solving the graph covering problem is to partition the subject graph into a forest of trees and solve the covering problem on each of the trees. A tree is a DAG where every node (including primary inputs) has a single fanout. The tree necessarily has a single sink (primary output) called the *root* and the sources (primary inputs) of the tree are called the *leaves* of the tree.

Example 6.72 The subject DAG of Figure 6.28a can be partitioned into a forest of trees as shown in Figure 6.28b.

The motivation for looking at the problem of tree covering is the existence of an efficient algorithm for the optimal tree covering problem [Keutzer 1987].

The application of the tree covering to technology mapping proceeds as follows. The first step is to convert the Boolean network into the NAND2-INV form, that is, every logic gate after the conversion is of type either NAND2 or INV. This subject DAG is then partitioned into a forest of trees by cutting the graph at each multiple-fanout stem. The resulting trees are optimally covered one tree at a time. Finding the optimum covering of a tree is done by generating the complete set of matches for each node in the tree (that is, the set of tree patterns which are candidates for covering a particular node) and then selecting the optimum match from among the candidates using a dynamic programming algorithm.

Example 6.73 Consider a Boolean network given by

$$
\begin{aligned}
Z &= X + \bar{Y} + h \\
Y &= W \cdot \bar{d} \\
X &= e \cdot f \cdot g, \text{ and} \\
W &= a \cdot b + c
\end{aligned}
$$

A NAND2-INV representation of the Boolean network is given in Figure 6.29a. The trivial covering of the subject DAG by pattern DAGs from the library of Figure 6.26 is also illustrated in Figure 6.29a. The cost of this trivial covering corresponds to the cost for seven NAND2 gates and five INV's, giving a cost of 31. A substantially better covering that exploits the larger gates in the library is shown in Figure 6.29b. The cost of this covering is the cost of two INV's, two NAND2's, one NAND3, and one NAND4 for a total cost of 19. A covering which utilizes an AOI gate with a lower cost of 17 is shown in Figure 6.29c.

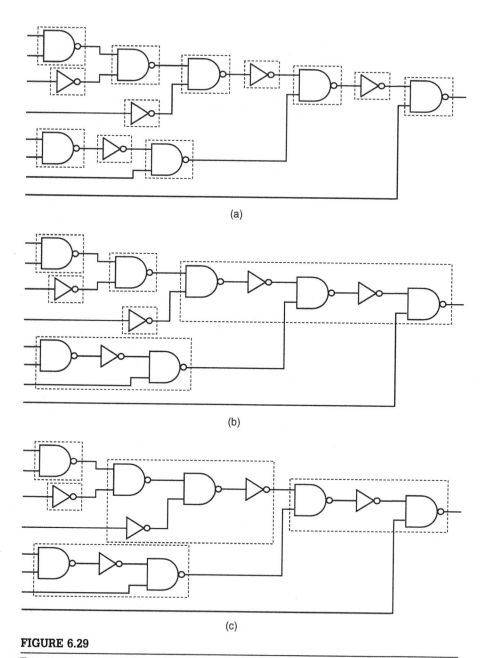

(a)

(b)

(c)

FIGURE 6.29

Tree coverings: (a) Trivial covering. (b) Better covering. (c) Optimum covering.

6.4.5 **Optimal tree covering**

A solution to establishing the initial set of candidate matches for a tree is to attempt to match each pattern at each node in the tree. If there are p patterns in the pattern set and n nodes in the subject graph, then this approach has complexity $O(n \cdot p)$.

Having generated a set of candidate matches for each node in the subject graph, an optimal tree cover must then be selected from among the candidates. Dynamic programming can be used for this purpose. Dynamic programming is a general technique for algorithm design which can be applied when the solution to a problem can be built from the solutions of a number of sub-problems.

Consider the problem of finding a minimum area cover for a subject tree T. A scalar cost is assigned to each tree pattern, and the cost for a cover is the sum of the costs for each pattern in the cover. The key observation is that the minimum-area cover for a tree T can be derived from the minimum-area covers for every node below the root of T. This is the *principle of optimality* for tree covering and is used as follows to find an optimal cover for T. For every match at the root of the tree the cost of an optimal cover containing that match equals the sum of the cost of the corresponding gate and the sum of the costs of the optimal covers for the nodes which are inputs to the match.[2] Note that the optimal covers for each input to the match at the root can be computed once and stored; it is not necessary to recompute the optimal cover for each input of each match.

Because each node in the tree is visited only once, the complexity of this algorithm is proportional to the number of nodes in the subject tree times the maximum number of matches at any node in the subject tree. The maximum number of matches is a function of the library size and is therefore a constant independent of the subject tree size. As a result the covering algorithm has linear complexity in the size of the subject tree, and the memory requirements are also linear in the size of the subject tree.

Example 6.74 We illustrate the optimum covering algorithm on the tree of Figure 6.30. We walk from the primary inputs to the primary output of the tree and determine the best match at each gate output. At each gate output, the match selected for the sub-tree whose root is the gate output has been shown along with the total cost of the optimal cover for this sub-tree. For the first-level gates, only NAND2 and INV matches are possible. At the output of gate 2 the only match is with a NAND2, and therefore the total cost is 8. At the output of gate 12 two matches are possible, with a NAND2 or with a NAND3. The former will result in a cost of 8, so we pick the latter which has a cost of 4. At the output of gate 4 the best match corresponds to an AOI gate with a cost of 9. The final cost at the primary output is 17. The optimum covering corresponds to that of Figure 6.29c.

---

[2] Recall the rules for legal coverings stated in Section 6.4.2.

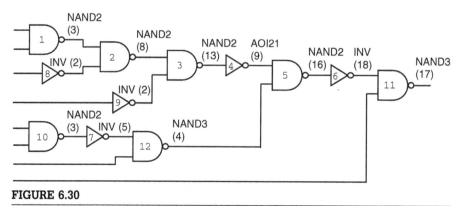

FIGURE 6.30

Dynamic programming for optimum tree covering.

6.4.6 **Improvement by inverter-pair insertion**

A simple way to improve the quality of circuits produced by the tree covering algorithm is by inserting inverter pairs. Redundant inverters are added to each tree to improve the number of patterns which can match at each node. This leads to an examination of more possible covers for each tree, leading directly to an improvement in the optimization quality.

The technique works as follows. Each edge in the subject tree and each edge in a pattern which connects two NAND gates is replaced with a pair of inverters. An extra pattern consisting of a pair of inverters is added to the matching patterns. This extra pattern is given zero area cost and zero delay cost. The tree covering algorithm is then applied unmodified.

Because of the optimality of the tree covering algorithm adding these extra inverters cannot lead to a cover with a greater cost. Each pair of inverters can be covered by the inverter-pair pattern, which leads to the solution which existed before the inverters were added. However, the advantage is that the tree covering algorithm is able to make the optimal choice between covering the extra inverters with the inverter-pair pattern at no cost or splitting the inverters between two patterns if this leads to a cover with less cost. The only disadvantage is that the number of nodes in the subject tree and the pattern trees has increased. The increase in the number of nodes is bounded by a factor of three (two extra inverter nodes for each node in the subject tree); however, the actual increase is typically less because redundant inverters are added only at the output of a NAND gate and not at the output of each inverter in the subject tree.

6.4.7 **Extension to non-tree patterns**

Some gates in a technology library cannot be represented in a tree form. Common examples are the XOR gate shown at the bottom of Figure 6.26, a two-to-one multiplexor, and a three-input majority gate (logic function $f = a \cdot b + a \cdot c + b \cdot c$). However, a simple extension allows these patterns to be included.

A leaf-DAG is a DAG where the only nodes with fanout greater than one are the primary inputs. Patterns which are trees, and patterns which are leaf-DAGs can be used directly by the tree covering algorithm. Hence the leaf-DAG patterns may include the XOR pattern shown in Figure 6.26. Note, however, that because of the multiple-fanout of one of these matches, the XOR gate must match at the leaves of the tree.

6.4.8 Advanced subjects

The success of the graph covering formulation has helped formulate the logic synthesis and optimization problem as an integration of technology-independent and technology-dependent portions. Graph covering based technology mapping is able to address a morass of technology specific issues, such as technology libraries and their area and timing characterization, which would significantly complicate higher level optimizations. The major limitation of graph covering, however, is its dependence on the structure of the given subject graph. This limitation was overcome in [Lehman 1997], where logic decomposition during technology mapping is proposed as a way of bridging the gap between technology-independent optimization and technology mapping. The approach was further developed in [Chatterjee 2006].

In our discussion, we focused on standard cell technology mapping. As the mapping algorithms heavily depend on the target implementation technology, different design styles may need different technology mapping methods. For instance, technology mapping for FPGAs [Scholl 2001], and even for standard cells [Kravets 2001], can be formulated very differently.

6.5 TIMING ANALYSIS

After correct logical functioning, the speed of an integrated circuit is one of the most important design characteristics. Timing optimization is thus an important aspect of logic synthesis. Any optimization system is only as good as the models that guide it, and as a result good timing optimization is entirely dependent on accurate timing analysis. For these reasons we spend a good deal of attention on techniques for accurate timing estimation of synchronous sequential circuits.

Accurate timing estimation relies on **component delay calculation** and **circuit delay calculation**. Component delay calculation is the method used for actually calculating the delay of individual components, such as gates and wires, within a circuit. In calculating gate delays, timing data such as the **inertial** and **propagation delays** of gates are typically gathered from extensive transistor-level and/or device-level simulation of the circuit components. In calculating wire delays, timing data arising from the parasitic capacitances and resistances of wires can be estimated through simulation or can be back-annotated from the final circuit layout. In our discussion we are mainly concerned about gate delays as wire delays can be embedded into the gate delays by the delay model to be introduced.

If we view a circuit as a graph, then the method used for delay calculation at the vertices of the graph is gate delay calculation while circuit delay calculation is the model used for calculating delay for the entire graph.

Below we present a simple gate delay model and then focus on the topic of circuit delay calculation, which is the most challenging and relevant problem in timing estimation for the developer of a logic optimization system.

Gate Delay Model. A popular (CMOS) gate delay model is a simple linear model [Sutherland 1999]: The delay T_d of a gate g is given by the equation

$$T_d = T_p + T_e \times \frac{C_{out}}{C_{in}} \qquad (6.19)$$

where T_p is the **parasitic delay** of the gate, T_e is the **logical effort**, C_{in} is the input capacitance, and C_{out} is the capacitive load at the gate output. It does not consider more refined details such as the effect of slow rising or falling transitions on the transistors associated with this gate. In this model, parameters T_p, T_e, and C_{in} are fixed constants for a standard cell whereas C_{out} varies depending on the fanout load of a gate (which may include wiring capacitances).

Gate delay calculations are performed extensively in timing analysis and logic optimization, and as a result tradeoffs have evolved between the accuracy of a model and the runtime of calculation. Although Equation (6.19) is a simple approximation, it is good enough for logic optimization purposes. More accurate nonlinear models are possible and often stored as look-up tables. Delay calculation often depends on the circuit implementation method.

Circuit Delay Calculation. We explain how to use gate delay calculation to compute the delay of an entire synchronous circuit. A simple implementation model of a clocked, or synchronous, sequential circuit is shown in Figure 6.31, where a clocked memory element (register), *e.g.*, an edge-triggered flip-flop, is used. At each active clock edge the next state is loaded into the flip-flops and becomes the current state.

Registers have a **propagation delay** associated with the interval between a clock edge and valid outputs. In order to guarantee that an input is not sampled when invalid, a period of validity extending slightly before and after the active edge is specified. Specification of a **setup time** t_s and **hold time** t_h dictates that the register inputs must be valid and stable during a period that begins t_s before the active clock edge and ends t_h after the edge.

Given a sufficiently long clock period and appropriate constraints on the timing of transitions on the inputs, the inputs to the flip-flops can be guaranteed to be stable at each active clock edge, ensuring correct operation. Correct operation depends on the assumptions that:

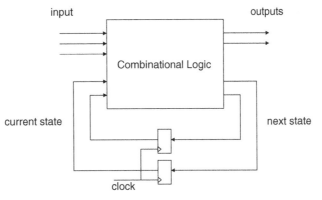

FIGURE 6.31

Clocked model for a sequential circuit.

1. The clock period is longer than the sum of the maximum propagation delay through the combinational logic, the setup time of the registers, and the maximum propagation delay through the registers.
2. The circuit's input signals are stable and valid for a sufficient period surrounding each active clock edge to accommodate both the maximum propagation delay through the combinational logic and the setup time of the registers.
3. The minimum propagation delay through the combinational logic exceeds the hold time requirement of the registers.

The most important constraint above is the first one. The length of the clock period of a sequential circuit is directly related to the maximum propagation delay through the combinational logic of the circuit.

Given that the delay calculation of the sequential circuit primarily depends on the delay of the combinational logic, we will focus on the problem of correctly computing the maximum propagation delay of a multilevel combinational circuit. We will show in the next section how to optimize a circuit so as to minimize the delay through the circuit.

For some time the most common approach to estimating and validating the delay of a synchronous circuit was **timing simulation**. The approach is diminishing in utility because of the incompleteness and excessiveness of input stimuli required to accurately determine circuit performance. Instead, **timing verification** is being used for validating the timing of circuits, and we will focus exclusively on using timing verification for estimating and validating the timing of a synchronous circuit.

Terminology. Before delving into timing analysis, we introduce terminology that will allow us to discuss timing issues. A combinational circuit can be viewed as a DAG $G = (V, E)$ where vertices or nodes V in the graph correspond to gates in the circuit and edges E correspond to connections in the circuit. Primary inputs are **sources** $\subseteq V$ while primary outputs are **sinks** $\subseteq V$. A **path** in a combinational circuit is an alternating sequence of vertices and edges, $\{v_0, e_0, \ldots, v_n, e_n, v_{n+1}\}$, where edge $e_i = (v_i, v_{i+1})$, $1 \leq i \leq n$, connects the output of vertex v_i to an input of vertex v_{i+1}. For $1 \leq i \leq n$, v_i is a gate g_i, v_0 is a primary input, and v_{n+1} is a primary output. Each e_i is a wire (or a two-terminal net) in the actual circuit.

Let $p = \{v_0, e_0, \ldots, v_n, e_n, v_{n+1}\}$ be a path. The inputs of v_i other than e_{i-1} are referred to as the **side-inputs** to p, that is, the set of signals not on p but feeding to the gates on p.

Each gate g_i (or wire e_i) is assumed to have a delay which can be a fixed quantity under the **fixed delay model** or can vary in a given range under the **monotone speedup delay model**.

A **controlling value** at a gate input is the value that determines the value at the output of the gate independent of the other inputs. For example, 0 is a controlling value for an AND gate. A **non-controlling value** at a gate input is the value which is not a controlling value for the gate. For example, 1 is a non-controlling value for an AND gate. We say that a gate g has the **controlled value** if one of its inputs has a controlling value; otherwise, we say that g has the **non-controlled value.**

Path sensitization studies the conditions under which signals can propagate from the primary inputs to the primary outputs of a combinational circuit. The conditions depend on the delay models and modes of operation assumed for the circuit.

We will precisely characterize the delay of a multilevel logic circuit, and see that the delay of a multilevel circuit depends on various assumptions relating to the mode of operation of the circuit and the delay model chosen. We begin with the simplest **topological timing analysis**, which is conservative but sound. The complexity of the analysis is linear in the circuit size. We will then introduce **functional timing analysis**, which is accurate at the cost of computation overhead.

6.5.1 Topological timing analysis

Most timing analyzers fall into *the topological timing analysis* category, where the topologically longest path in the circuit is assumed to dictate the critical delay of the circuit. We describe a topological timing analyzer that determines the longest path in the circuit without regard to the Boolean functionality of the circuit.

Circuit speed is measured by most optimization systems using a fixed delay model, where each gate and wire in the network has a given and fixed delay. Typically, a worst-case design methodology is followed, where the given delay for the gate is an upper bound on the actual delay of the fabricated gate.

The **arrival time** of a signal s, denoted A_s, is the time at which the signal settles to its steady state value. For a given circuit, using the arrival times of the primary inputs we can compute the arrival time of every signal in the circuit. For a gate in the circuit, the arrival time of the gate output equals the maximum among the arrival times of the gate inputs plus the gate delay. That is, the arrival time of the output signal o of a gate g with gate delay d can be computed by

$$A_o = \max_{i \in FI(g)} \{A_i\} + d$$

where $FI(g)$ denotes the set of fanin signals of g.

The **required time** of a signal s, denoted R_s, is the time at which the signal is required to be stable. For a given circuit, using the required times of the primary outputs we can compute the required time of every signal in the circuit. For a gate in the circuit, the required time of any input of the gate equals the minimum among the required times of the gate outputs minus the gate delay. That is, the required time of any input signal i of a gate g with gate delay d can be computed by

$$R_i = \min_{o \in FO(g)} \{R_o\} - d$$

where $FO(g)$ denotes the set of fanout signals of g.

The **slack time** of a signal s, denoted S_s, is the difference between its required time and arrival time, *i.e.*,

$$S_s = R_s - A_s$$

The slack value of a signal measures its looseness in terms of timing criticality. Negative slack values indicate timing violation.

Starting with the primary input arrival times, we can compute the arrival time for every signal in a **topological order** from primary inputs to primary outputs. Similarly, using the primary output required times, we can compute the required times for every signal in a reverse topological order from primary outputs to primary inputs. Thus the slack at each node can be obtained as well.

Example 6.75 The arrival time, required time, and slack of each signal in Figure 6.32 are shown as a 3-tuple. We are given the arrival times for the four primary inputs and the required time for the output. The delay of each node is indicated within the node. The arrival time of signal e is the maximum of the arrival times of primary inputs a and b (= 1) plus the delay of the node (= 1), equaling 2. Similarly the arrival times of the other signals can be calculated. On the other hand, given a required time of 8 at output h, the required times for signals f and g can be computed as 8 minus the delay of the output node (= 2), equaling 6. However, given the required time of 6 at f, the required times at signals e and g are calculated to be 4. The required time for signal g is the minimum of the computed required times, namely 4. This is intuitive because, if g does not stabilize by time 4, f will not stabilize by time 6 and the output h will not stabilize by time 8. Similarly, the required times at the other signals can be calculated.

The **topologically longest path** of a circuit is a path where each signal has the minimum slack. Static timing analyzers assume that the **critical delay** of the circuit is the delay of the topologically longest path. Under this (pessimistic) assumption the longest path is also called the **critical path**.

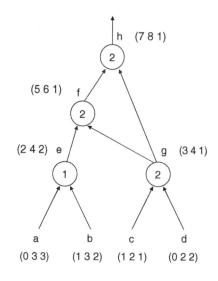

FIGURE 6.32

Topological timing analysis.

6.5.2 **Functional timing analysis**

The problem with topological analysis of a circuit is that not all critical paths in a circuit need be responsible for the circuit delay. Critical paths in a circuit can be **false**, *i.e.*, not responsible for the delay of a circuit. The **critical delay** of a circuit is defined as the delay of the longest **true** path in the circuit. Thus, if the topologically longest path in a circuit is false, then the critical delay of the circuit will be less than the delay of the longest path. The critical delay of a combinational logic circuit is dependent on not only the topological interconnection of gates and wires, but also the Boolean functionality of each node in the circuit. Topological analysis only gives a conservative upper bound on the circuit delay.

Example 6.76 Assume the fixed delay model, and consider the carry bypass circuit of Figure 6.33. The circuit uses a conventional ripple-carry adder (the output of gate 11 is the ripple-carry output) with an extra AND gate (gate 10) and an additional multiplexor. If the propagate signals p0 and p1 (the outputs of gates 1 and 3, respectively) are high, then the carry-out of the block c2 is equal to the carry-in of the block c0. Otherwise it is equal to the output of the ripple-carry adder. The multiplexor thus allows the carry to skip the ripple-carry chain when all the propagate bits are high. A carry-bypass adder of arbitrary size can be constructed by cascading a set of individual carry-bypass adder blocks, such as those of Figure 6.33.

Assume the primary input c0 arrives at time $t = 5$ and all the other primary inputs arrive at time $t = 0$. Let us assign a gate delay of 1 for AND and OR gates and gate delays of 2 for the XOR gates and the multiplexor. The longest path including the late

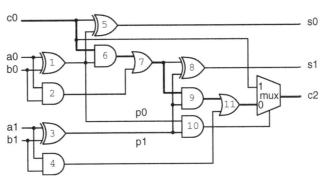

FIGURE 6.33

2-bit carry-bypass adder.

arriving input in the circuit is the path shown in bold, call it P, from c0 to c2 through gates 6, 7, 9, 11, and the multiplexor (the delay of this path is 11). A transition can never propagate down this path to the output because in order for that to happen the propagate signals have to be high, in which case the transition propagates along the bypass path from c0 through the multiplexor to the output. This path is false since it cannot be responsible for the delay of the circuit.

For this circuit, the path that determines the worst-case delay of c2 is the path from a0 to c2 through gates 1, 6, 7, 9, 11, and the multiplexor. The output of this critical path is available after 8 gate delays. The critical delay of the circuit is 8 and is less than the longest path delay of 11.

6.5.2.1 *Delay models and modes of operation*

Whether a path is a true or false delay path closely depends on the **delay model** and the **mode of operation** of a circuit.

In the commonly used **fixed delay model**, the delay of a gate is assumed to be a fixed number d, which is typically an upper bound on the delay of the component in the fabricated circuit. In contrast, the **monotone speedup delay model** takes into account the fact that the delay of each gate can vary. It specifies the delays as an interval $[0, d]$, with the lower bound 0 and upper bound d on the actual delay.

Consider the operation of a circuit over the period of application of two consecutive input vectors υ_1 and υ_2. In the **transition mode** of operation, the circuit nodes are assumed to be ideal capacitors and retain their values set by υ_1 until υ_2 forces the voltage to change. Thus, the timing response for υ_2 is also a function of υ_1 (and possibly other previously applied vectors). In contrast, in the **floating mode** of operation the nodes are not assumed to be ideal capacitors, and hence their state is unknown until it is set by υ_2. Thus, the timing behavior for υ_2 is independent of υ_1.

Transition Mode and Monotone Speedup. In our analysis of the carry-bypass adder we assumed fixed delays for the different gates in the circuit and applied a vector pair to the primary inputs. It was clear that an event (a signal transition, either $0 \rightarrow 1$ or $1 \rightarrow 0$) could not propagate down the longest path in the circuit. A precise characterization is that the path cannot be **sensitized,** and thus false, under the transition mode of operation and under (the given) fixed gate delays. Varying the gate delays in Figure 6.33 does not change the sensitizability of the path shown in bold.

False path analysis under the fixed delay model and the transition mode of operation, however, may be problematic as seen from the following example.

Example 6.77 Consider the circuit of Figure 6.34a, taken from [McGeer 1989]. The delays of each of the gates are given inside the gates. In order to determine the critical delay of the circuit we will have to simulate the two vector pairs corresponding to a, making a $0 \rightarrow 1$ transition and a $1 \rightarrow 0$ transition. Applying $0 \rightarrow 1$ and $1 \rightarrow 0$ transitions on a does not change the output f from 0. Thus, one can conclude that the circuit has critical delay 0 under the transition mode of operation for the given fixed gate delays.

Now consider the circuit of Figure 6.34b which is identical to the circuit of Figure 6.34a except that the buffer at the input to the NOR gate has been sped up from 2 to 0. We might expect that speeding up a gate in a circuit would not increase the critical delay of a circuit. However, for the $0 \rightarrow 1$ transition on a, the output f switches both at time 5 and time 6, and the critical delay of the circuit is 6.

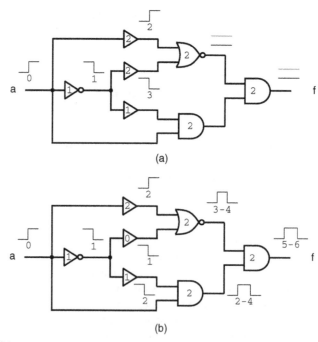

(a)

(b)

FIGURE 6.34

Transition mode with fixed delays.

This example shows that a sensitization condition based on transition mode and fixed gate delays is unacceptable in the **worst-case design methodology**, where we are given the upper bounds on the gate delays and are required to report the (worst-case) critical path in the circuit. Unfortunately, if we use only the upper bounds of gate delays under the transition mode of operation, an erroneous critical delay may be computed.

To obtain a useful sensitization condition, one strategy is to use the transition mode of operation and monotone speedup as the following example illustrates.

Example 6.78 Consider the circuit of Figure 6.35, which is identical to the circuit of Figure 6.34a, except that each gate delay can vary from 0 to its given upper bound. As before, in order to determine the critical delay of the circuit, we will have to simulate the two vector pairs corresponding to a making a $0 \rightarrow 1$ transition and a $1 \rightarrow 0$ transition. However, the process of simulating the circuit is much more complicated since the transitions at the internal gates may occur at varying times. In the figure, the possible combinations of waveforms that appear at the outputs of each gate are given for the $0 \rightarrow 1$ transition on a. For instance, the NOR gate can either stay at 0 or make a $0 \rightarrow 1 \rightarrow 0$ transition, where the transitions can occur between [0, 3] and [0, 4], respectively. In order to determine the critical delay of the circuit, we scan all the possible waveforms at output f and find the time at which the last transition occurs over all the waveforms. This analysis provides us with a critical delay of 6.

Timing analysis for a worst-case design methodology can use the above strategy of monotone speedup delay simulation under the transition mode of operation. The strategy however has several disadvantages. Firstly, the search space is 2^{2n} where n is the number of primary inputs to the circuit, since we may have to simulate each possible vector pair. Secondly, monotone speedup delay simulation is significantly more complicated than fixed delay simulation. These difficulties have motivated delay computation under the floating mode of operation.

Floating Mode and Monotone Speedup. Under floating mode, the delay is determined by a single vector. As compared to transition mode, critical delay under floating mode is significantly easier to compute for the fixed or monotone speedup delay model because large sets of possible waveforms do not need

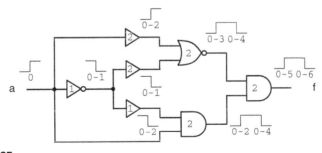

FIGURE 6.35

Transition mode with monotone speedup.

to be stored at each gate. Single-vector analysis and floating mode operation, by definition, make pessimistic assumptions regarding the previous state of nodes in the circuit. The assumptions made in floating mode operation make the fixed delay model and the monotone speedup delay model equivalent.[3]

6.5.2.2 *True floating mode delay*

The necessary and sufficient condition for a path to be responsible for circuit delay under the floating mode of operation is a delay-dependent condition.

The fundamental assumptions made in single-vector delay-dependent analysis are illustrated in Figure 6.36. Consider the AND gate of Figure 6.36a. Assume that the AND gate has delay d and is embedded in a larger circuit, and a vector pair $\langle \upsilon_1, \upsilon_2 \rangle$ is applied to the circuit inputs, resulting in a rising transition occurring at time t_1 on the first input to the AND gate and a rising transition at time t_2 on the second input. The output of the gate rises at a time given by $\max\{t_1, t_2\} + d$. The abstraction under floating mode of operation only shows the value of υ_2. In this case a 1 arrives at the first and second inputs to the AND gate at times t_1 and t_2, respectively, and a 1 appears at the output at time $\max\{t_1, t_2\} + d$. Similarly, in Figure 6.36b two falling transitions at the AND gate inputs result in a falling transition at the output at a time that is the minimum of the input arrival times plus the delay of the gate.

Now consider Figure 6.36c, where a rising transition occurs at time t_1 on the first input to the AND gate and a falling transition occurs at time t_2 on the second input. Depending on the relationship between t_1 and t_2 the output will either stay at 0 (for $t_1 \geq t_2$) or glitch to a 1 (for $t_1 < t_2$). It is possible to accurately determine whether the AND gate output is going to glitch or not if a simulation is carried out to determine the range of values that t_1 and t_2 can have on $\langle \upsilon_1, \upsilon_2 \rangle$. (This was illustrated in Figure 6.35.) However, under the floating mode of operation we only have the vector υ_2. The 1 at the first input to the AND gate arrives at time t_1, and the 0 at the second input arrives at time t_2. The output of the AND on υ_2 obviously settles to 0 on υ_2, but at what time does it settle? If $t_1 \geq t_2$, then the output of the gate is always 0, and the 0 effectively arrives at time

---

[3]To understand this effect, consider a circuit C with fixed values on its gate delays. Let p be a path through C and υ be a vector applied to C. In order to determine if p is responsible for the delay of C on υ, we inspect the side-inputs of p. At any gate g on p, the side-inputs have to be at non-controlling values when the controlling or non-controlling value propagates along p through g. If the value at a side-input i to g is non-controlling on υ, monotone speedup (under the transition or floating mode) allows us to disregard the time that the non-controlling value arrives, since we can always assume that it arrives before the value along p. Let the delay of all paths from the primary inputs to i be greater than the delay of the sub-path corresponding to p ending at g. Under monotone speedup, we can speed up all the paths to i, ensuring that the non-controlling value arrives in time. Under floating mode with fixed delays we cannot change the delays of the paths to i, but we can assume that υ_1, the vector applied before υ, was providing a non-controlling value! We do not have to wait for υ to provide the non-controlling value. In either case, the arrival time of non-controlling values on side-inputs does not matter.

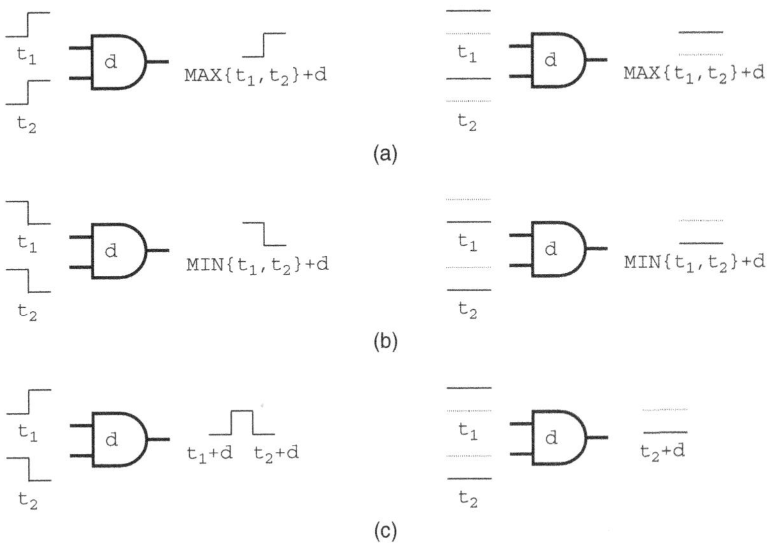

FIGURE 6.36

Fundamental assumptions made in floating mode operation.

0. If $t_1 < t_2$, then the gate output becomes 0 at $t_2 + d$. In order not to underestimate the critical delay of a circuit all single-vector sensitization conditions *have* to assume that the 1 (the non-controlling value for the AND gate) arrives before the 0 (the controlling value for the AND gate), *i.e.*, that $t_1 < t_2$. Under the floating mode of operation this corresponds to assuming that the values on the previous vector υ_1 were non-controlling. (The above assumption also captures the essence of transition mode delay under the monotone speedup delay model. Given that the AND gate is embedded in a circuit, under the monotone speedup model the sub-circuit that is driving the first input can be sped up to cause the rising transition to arrive before the falling transition.)

The rules in Figure 6.36 represent a timed calculus for single-vector simulation with delay values that can be used to determine the correct floating mode delay of a circuit under an applied vector υ_2 (assuming pessimistic unknown values for υ_1) and the paths that are responsible for the delay under υ_2. The rules can be generalized as follows:

1. If the gate output is at a controlling value, pick the minimum among the delays of the controlling values at the gate inputs. (There has to be at least one input with a controlling value. The non-controlling values are ignored.) Add the gate delay to the chosen value to obtain the delay at the gate output.

2. If the gate output is at a non-controlling value, pick the maximum of all the delays at the gate inputs. (All the gate inputs have to be at non-controlling values.) Add the gate delay to the chosen value to obtain the delay at the gate output.

To determine whether a path is responsible for floating mode delay under a vector υ_2, we simulate υ_2 on the circuit using the timed calculus. As shown in [Chen 1991], a path is responsible for the floating mode delay of a circuit on υ_2 if and only if for each gate along the path:

1. If the gate output is at a controlling value, then the input to the gate corresponding to the path has to be at a controlling value and furthermore has to have a delay no greater than the delays of the other inputs with controlling values.
2. If the gate output is at a non-controlling value, then the input to the gate corresponding to the path has to have a delay no smaller than the delays at the other inputs.

Let us apply the above conditions to determine the delay of the following circuits.

Example 6.79 Consider the circuit of Figure 6.34a reproduced in Figure 6.37. Applying the vector $a = 1$ sensitizes the path of length 6 shown in bold, illustrating that the sensitization condition takes into account monotone speedup (unlike transition mode fixed delay simulation). Each wire has both a logical value and a delay value (in parentheses) under the applied vector.

Example 6.80 Consider the circuit of Figure 6.38. Applying the vector $(a, b, c) = (0, 0, 0)$ gives a floating mode delay of 3. The paths {a, d, f, g} and {b, d, f, g} can be seen to be responsible for the delay of the circuit.

Example 6.81 Consider the circuit of Figure 6.39. Applying $a = 0$ and $a = 1$ results in a floating mode delay of 5.

We presented informal arguments justifying the single-vector abstractions of Figure 6.36 to show that the derived sensitization condition is necessary and sufficient for a path to be responsible for the delay of the circuit under the floating mode of operation. For a topologically oriented formal proof of the necessity and sufficiency of the derived condition, see [Chen 1991].

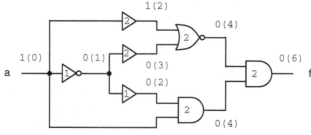

FIGURE 6.37

First example of floating mode delay computation on a circuit.

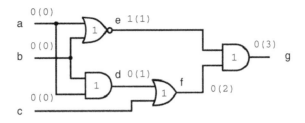

FIGURE 6.38

Second example of floating mode delay computation on a circuit.

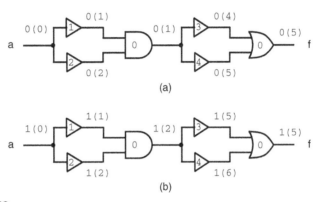

FIGURE 6.39

Third example of floating mode delay computation on a circuit.

6.5.3 **Advanced subjects**

There has been significant research done in an effort to arrive at the correct sensitization criterion in the late 1980s and early 1990s. A detailed history may be found in [McGeer 1991]. The computation of true critical delay of a circuit can be formulated with satisfiability solving [McGeer 1991; Guerra E Silva 2002] or **timed automatic test pattern generation** [Devadas 1992].

As for sequential circuit timing analysis, depending on the register types (*e.g.*, edge-triggered flip-flops and level-sensitive latches) and the number of clock phases used, their timing correctness requires careful analysis and verification. On the other hand, for IC manufacturing in the nanometer regime, process variations may cause substantial variations in circuit performance. This fabrication imperfection has motivated the development of **statistical static timing analysis** in replacement of the traditional (worst-case) static timing analysis (*i.e.*, the presented topological timing analysis). A good introduction to sequential circuit timing analysis and statistical static timing analysis can be found in [Sapatnekar 2004].

6.6 **TIMING OPTIMIZATION**

Being able to meet timing requirements is absolutely essential in synthesizing logic circuits. Timing optimization of combinational circuits can be performed both at the technology-independent level and during technology mapping. We consider the restructuring operations used in logic synthesis systems to improve circuit speed. We give an overview of basic restructuring methods that take into account timing constraints specified as input-arrival times of the primary inputs and output-required times of the primary outputs. The goal is to meet the timing constraints while keeping the area increase to a minimum. The methods use topological timing analysis, described in Section 6.5.1, to compute arrival times, required times, and slack times. Topological timing analysis is typically deployed in timing optimization tools due to its simple and fast calculation; functional timing analysis, in contrast, is mostly used for timing verification purposes instead due to its expensive computation cost.

6.6.1 **Technology-independent timing optimization**

For a given circuit to be delay minimized, the timing constraints are specified as the arrival times at the primary inputs and required times at the primary outputs. The optimization algorithm manipulates the network topology to achieve improved speed until the timing constraints are satisfied or no further decrease in the delay can be achieved.

The **critical section** of a Boolean network is composed of all the critical paths from primary inputs to primary outputs. Given a critical path, the total delay on the path can be reduced if any section of the path is sped up. **Collapsing** and **redecomposition** are the basic steps taken in restructuring. The nodes along the critical paths chosen to be collapsed and redecomposed form the **redecomposition region**.

Example 6.82 In Figure 6.40a we have a critical path $\{a, x, y\}$. The critical path can be reduced by first collapsing x and y and then redecomposing y in a different way to minimize the critical path as shown in Figure 6.40b.

Since a critical section usually consists of several overlapping critical paths, we select a minimum set of subsections, called **redecomposition points**, which when sped up will reduce the delays on all of the critical paths. (Note that it is not always possible to do so.) A weight is assigned to each candidate redecomposition point to account for possible area increase and for the total number of redecomposition points required. The goal is to select a set of points which cut all the critical paths and have the minimum total weight.

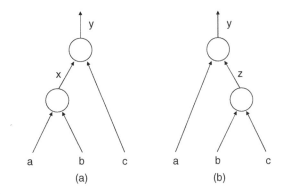

FIGURE 6.40

Collapsing and redecomposition.

Once the redecomposition points are chosen, they are sped up by the collapsing-decomposing procedure as described in Section 6.3.3. Since in a multi-level network we can reduce the area by sharing common functions, we first attempt to extract area saving divisors that do not contain critical signals. After all such divisors have been extracted, we decompose the node into a tree and place late arriving signals closer to the outputs, thus making them pass through a smaller number of gates.

Example 6.83 In Figure 6.41, the critical paths in the original network are shown in bold and begin from signals c and d. Node f is collapsed, and a divisor k is selected which has the desired property that substituting k into f, places the critical signals c and d closer to the output.

Note that the critical paths in the decomposed network may have changed. The collapsing-decomposing procedure can be iterated by identifying a new

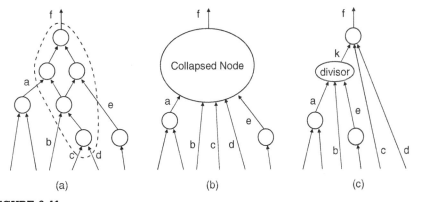

FIGURE 6.41

Basic idea of timing decomposition.

critical section. The algorithm proceeds until the requirement is satisfied or no improvement in delay can be made. A detailed exposition of speed optimization algorithms can be found in, *e.g.*, [Singh 1992; Devadas 1994].

6.6.2 **Timing-driven technology mapping**

Technology-independent delay optimization algorithms cannot estimate the delay of a circuit accurately, largely due to the lack of accurate technology-independent delay models. Therefore, such algorithms are not guaranteed to produce faster circuits, when circuit speed is measured after technology mapping and physical design. We will present a more accurate approach to delay optimization during technology mapping. The tree covering algorithm presented in Section 6.4.5, in the context of technology mapping for minimum area, will be modified to target circuit speed.

The most accurate estimation of the delay of a gate in a circuit can only be obtained after the entire circuit has been placed and routed. Since technology mapping has to be performed before placement and routing, an approximate delay model with reasonable accuracy has to be used. We adopt the linear delay model of Equation (6.19) of Section 6.5 in the following discussion.

6.6.2.1 *Delay optimization using tree covering*

The tree covering algorithm of Section 6.4.5 can only be used if the cost of a match at a gate can be determined by examining the cost of the match and the cost of the inputs to the match (for which the cost has already been determined). For area optimization the cost of a gate depends on the area cost of the match and the area cost of the inputs of the match. For delay optimization, the cost is signal arrival time at the output of the match. Therefore, the cost of a match for delay optimization depends not only on the structure of the tree beneath the gate, but also on the capacitive load seen by the match. This load cannot be determined at the time of the selection of the match as it depends on the unmapped portion of the tree. Several attempts have been made to generalize tree covering to produce minimum delay implementations [Rudell 1989; Touati 1990; Chaudhary 1992].

Load-Independent Tree Covering. The tree covering algorithm of Section 6.4.5 can be used to produce a minimum delay implementation of a circuit provided the loads of all the gates in the circuit are the same. Under the assumption that the delay of a gate is independent of the fanout of the gate, the tree covering algorithm provides the minimum arrival time cover, if we compute and store the arrival time at each node and choose the minimum arrival time match at each node.

Example 6.84 Consider the technology library shown in Figure 6.42 and the circuit shown in Figure 6.43a. For each gate in the library, its name, area, symbol, and pattern DAG are presented. In addition, the delay parameters for our delay model are shown. By Equation (6.19), the

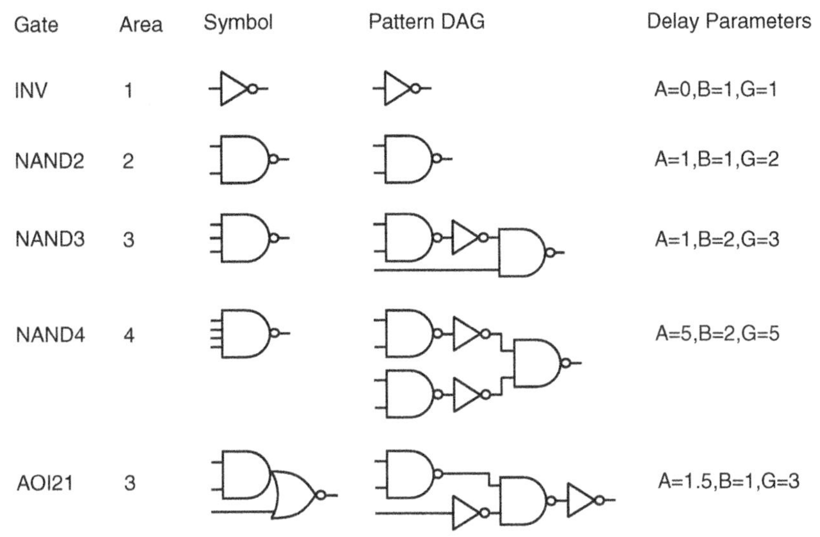

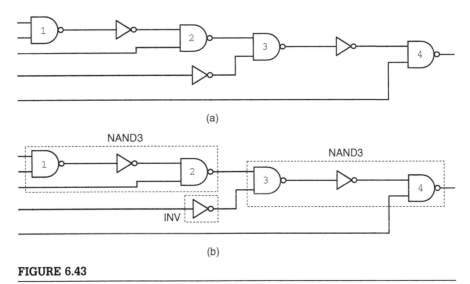

FIGURE 6.42

Gate library.

FIGURE 6.43

Circuit and its mapped implementation.

intrinsic delay, T_p, is denoted by A, the load dependent coefficient T_e/C_{in} is denoted by B, and the load C_{in} presented by the gate to any input gate is denoted by G. Note that in order to calculate the delay of a gate using Equation (6.19), we will use A and B for the gate and sum up the G values for all its fanout gates.

If the load of each gate in the circuit is considered to be 1, then the perfect match at each gate can be determined in one bottom-up pass, as in Section 6.4.5. For gate 1, this corresponds to a 2-input NAND gate with a delay of 2. The best match at gate 2 is a 3-input NAND gate with a delay of 3. The best covering for this circuit under the fixed load assumption is shown in Figure 6.43b.

Load-Dependent Tree Covering. The above load-independent tree covering does not necessarily produce the optimal solution because the load of all gates is not the same. As can be seen from the library in Figure 6.42, different gates provide different load values to their inputs.

An algorithm, originally presented in [Rudell 1989], can be used to take into account the effect of different loads. The first step of the algorithm is a pre-processing step over the technology library in order to create n load bins and quantize the load values for all the pins in the library. For each load bin, a representative load value is selected, and the remaining load values are mapped to their closest value in the chosen set. The value of n determines the accuracy and the run time of the algorithm. If n is equal to the number of distinct loads in the library, then the algorithm is most accurate. However, the larger the value of n, the more computation will be required. Instead of quantizing load values *a priori* based on the library information, a better way is to adapt the quantization intervals to each gate. In one pre-computation phase, we can determine all possible load values at a gate by examining all the possible matches at the gate. These load values can then be used to determine the values of the quantization intervals.

For a match at a gate, an array of costs (one for each load value) is calculated. The cost is the arrival time of the signal at the output of the gate. For each bin or load value, the match that gives the minimum arrival time is stored. For each input i of the match, the optimum match for driving the pin load of pin i of the match is assumed, and the arrival time for that match is used. This calculation can be done by traversing the tree once forward from the leaves of the tree to its root. The tree is then traversed backward from the root to the leaves, whereby the load values are propagated down and, for each gate, the best match at the gate is selected depending on the value of the load seen at the gate.

Example 6.85 We illustrate the algorithm using the circuit of Figure 6.43a and the library of Figure 6.42. Consider the best matches shown in Figure 6.44. Since the number of distinct load values in our example is only four, four bins are considered. For gate 1 the only match is a NAND2 gate. For each load value, the delay of this gate then gives the arrival time at the output of the match (assuming zero arrival time at the inputs). For the inverter at the output of this NAND gate, the only match is that of an inverter. Since the inverter presents a load of 1 to the NAND gate, the arrival time at the input of the inverter is the arrival time corresponding to the first bin of the NAND gate. Using this arrival time, the arrival times at the output of the inverter for all possible load values are computed and are shown in the figure.

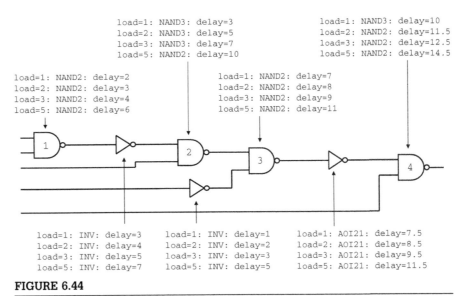

FIGURE 6.44

Technology mapping considering load values.

At gate 2, there are two possible matches corresponding to 2-input and 3-input NAND gates. If we consider the NAND2 gate, the two arrival times at the inputs of the match are 0 (corresponding to the primary input connection to gate 2) and 4 (corresponding to the inverter connection to gate 2 seeing a load of 2). The maximum arrival time at the inputs is 4. The arrival times at the output of the gate for the four different load values are 6, 7, 8, and 10. *E.g.*, for a load value of 5, a NAND2 gate has a delay $1 + 1 \times 5 = 6$. This delay added to the arrival time of 4 at the input of the NAND gate produces an arrival time of 10 at the output. For the NAND3 gate, the arrival times of all inputs are 0, and therefore the arrival times at the output are 3, 5, 7, and 11. Therefore, for the first three load values, the NAND3 is a better choice, while for the last load value the NAND2 is a better choice.

The final mapping is determined during backward traversal and depends on the load seen by gate 4. Assuming a load of 1, the best match at gate 4 is a NAND3 gate. This gate presents a load of 3 to its inputs, implying that the best match for a load value of 3 at gate 2 has to be chosen. This match is another NAND3 gate. The resulting mapping is shown in Figure 6.45a, which is coincidentally the same mapping obtained assuming constant load (Figure 6.43b). However, if the load is greater than 1, then the mapping of Figure 6.45b is better.

To improve the computation, we may apply adaptive quantization of load values. For instance, for gate 1 in the circuit of Figure 6.44, only a load value of 1 has to be considered because all possible matches at the inverter consist of only an inverter; for gate 2, load values of 2 and 3 have to be considered. This type of adaptive quantization produces results close to the optimum within reasonable amounts of computation time.

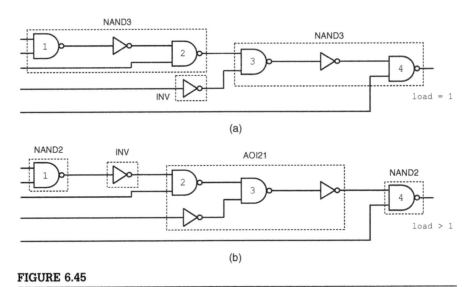

(a)

(b)

FIGURE 6.45

Two different implementations of the circuit depending on load value.

Note that, under the more general linear delay model, the principle of optimality of tree covering does not apply.

6.6.2.2 *Area minimization under delay constraints*

The tree covering algorithm used above can be generalized to minimize the area under a delay constraint. It may not be necessary to obtain the fastest circuit, but instead we may want to obtain a circuit that meets certain timing constraints and has the minimum possible area. This timing constraint is expressed as a required time at the root of the tree and can be propagated down the tree together with load values during backward traversal. In this case the cost of a match at a gate includes not only the arrival time but also the area of a match. During backward traversal the minimum area solution that meets the required timing constraint is chosen. If no such solution is available, then the minimum delay solution is chosen. Since not all of the sub-trees need to be maximally fast, the area of the circuit can be minimized.

Example 6.86 Consider the mapping shown in Figure 6.46a. The circuit has been mapped for minimum delay, and the arrival time at the output of gate 7 is 7. However, the required time at the output of this gate is 9, and the other match at gate 7 has an arrival time of 9 but a smaller area. Selecting this match gives us a circuit with the same delay but a smaller area, as shown in Figure 6.46b.

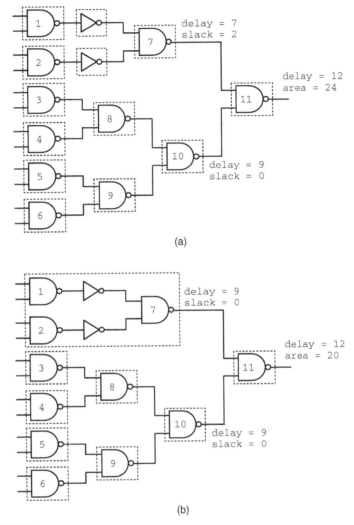

FIGURE 6.46

Example illustrating area recovery.

6.6.3 **Advanced subjects**

Fanout Optimization. Tree covering alone does not generate good quality solutions because most circuits are not trees but DAGs. In such circuits, a signal may feed two or more destinations. Due to the large amount of capacitance that has to be driven, the delay through the gate that drives this signal could be large. The optimization of this delay is called fanout optimization. Buffer insertion and gate sizing, among other techniques, are important approaches to fanout optimization. A survey on fanout optimization can be found in [Hassoun 2002].

Sequential Circuit Timing Optimization. In addition to logic restructuring, we may exploit optimization techniques special for sequential circuits. Promising sequential timing optimization methods include, for instance, retiming [Leiserson 1983, 1991] and clock skew scheduling. See, *e.g.*, [Sapatnekar 2004] for introduction.

6.7 CONCLUDING REMARKS

This chapter presents some important classic problems in combinational logic synthesis and basic techniques to solve them. Since logic synthesis has become very broad and continues to evolve, many important developments cannot be covered and only a few of them are mentioned here.

To invite and motivate future investigations, we list some logic synthesis trends:

Scalable Logic Synthesis. The capacity of logic synthesis tools is constantly being challenged by the ever-increasing complexity of modern industrial designs commonly consisting of millions of gates. The data structures and algorithms of logic synthesis tools must be effective and robust enough in order to handle large problem instances. It is interesting to note that every capacity leap in the history of logic synthesis can be attributed to some data structure revolution, *e.g.*, from truth tables to covers, from covers to BDDs, and from BDDs to AIGs and SAT. As SAT solvers have become much faster in recent years, a paradigm shift is taking place in logic synthesis. More and more SAT-based algorithms emerge in replacement of BDD-based ones. Searching for new effective data structures may transform logic synthesis tools.

Verifiable Logic Synthesis. As noted earlier, due to the hardness of verification, industrial synthesis methodologies are often conservative and mostly conduct only combinational optimization, despite the existence of practical sequential synthesis techniques.[4] This phenomenon is changing because progressive optimization methods are necessary to meet more stringent timing constraints, and also verification techniques are made more effective, especially for circuits optimized in particular ways [Jiang 2007]. To completely overcome the verification barrier, a general consensus is that essential synthesis information should be revealed to verifiers. Verifiable logic synthesis sets forth the criterion that whatever can be synthesized can be verified effectively [Brayton 2007].

Parallelizable Logic Synthesis. One way to speed up logic synthesis algorithms is to take advantage of hardware and software technologies. As multi-core computers support more and more parallelism, EDA tools can benefit from this technology advancement. How to utilize parallelism in logic synthesis algorithms is a challenge for EDA companies.

---

[4]One exception is FPGA synthesis, where sequential optimization methods find wide applications. The reconfigurability of FPGAs makes verification not as critical as general ASIC designs because incorrect logic transformations can be rectified later through reconfiguration.

Statistical Logic Synthesis. The continuous miniaturization of semiconductor devices imposes serious threats to circuit design robust against process variations and environmental fluctuations. Various uncertainties appear in both pre- and post-design phases. How to synthesize a robust circuit optimal in a statistical sense with respect to design constraints is an important challenge that needs to be addressed.

Physically Aware Logic Synthesis. Logic synthesis and physical design are traditionally separated to enable a divide-and-conquer approach to VLSI design automation. This separation becomes problematic when interconnect becomes the dominating factor of circuit delays. Lacking wiring information, logic synthesis cannot produce accurate timing estimation and precise timing optimization; lacking logic information, physical design cannot exploit logic flexibility and has limited optimization power. Therefore, before timing constraints are met, often several iterations of logic synthesis and physical design are performed in order to reach timing closure. Unfortunately there is no guarantee that the process will converge. This phenomenon leads to a serious design closure problem, which slows down design cycles and therefore time to market. Even though there are approaches to timing closure, such as gain-based synthesis, incremental placement and resynthesis, etc., there is still plenty of room for improvement.

Logic Synthesis for Emerging Technologies. As the miniaturization of electronic devices approaches physical limits, Moore's Law is expected to be broken sooner or later. Alternatives to silicon-based computation devices are actively being researched. For the next computation model, we might need very different logic synthesis tools, perhaps even beyond propositional logic and Boolean algebra.

6.8 EXERCISES

6.1. (Commutativity between Cofactor and Boolean Operations) Given two Boolean functions f and g and a Boolean variable v, prove or disprove the following equalities:

(a) $(\neg f)_v = \neg(f_v)$

(b) $(f \langle op \rangle g)_v = (f_v) \langle op \rangle (g_v)$ for $\langle op \rangle = \{\wedge, \oplus\}$

6.2. (Boolean Difference) Let $f(x, y, z) = h(g(x, y, z), y, z)$. Prove or disprove the following equalities:

(a) $\dfrac{\partial^2 f(x,y,z)}{\partial x \partial y} = \dfrac{\partial^2 f(x,y,z)}{\partial y \partial x}$

(b) $\dfrac{\partial f(x,y,z)}{\partial x} = \dfrac{\partial h(u,y,z)}{\partial u} \dfrac{\partial g(x,y,z)}{\partial x}$

(c) $\dfrac{\partial f(x,y,z)}{\partial y} = \dfrac{\partial b(u,y,z)}{\partial u}\dfrac{\partial g(x,y,z)}{\partial y} \oplus \dfrac{\partial^2 b(u,y,z)}{\partial u \partial y}\dfrac{\partial g(x,y,z)}{\partial y}\dfrac{\partial y}{\partial y}$

6.3. (Quantified Boolean Formula) For Boolean functions f and g, show that

$$\textbf{(a)} \quad \neg(\exists x.f(x,y)) = \forall x.\neg f(x,y)$$

$$\textbf{(b)} \quad \neg(\forall x.f(x,y)) = \exists x.\neg f(x,y)$$

$$\textbf{(c)} \quad \exists x.(f(x,y) \wedge g(x,y)) \neq (\exists x.f(x,y)) \wedge (\exists x.g(x,y))$$

$$\textbf{(d)} \quad \neg\forall x, \exists z.(f(x,y) \wedge g(x,z)) = \exists x.(\neg f(x,y) \vee \forall z.\neg g(x,z))$$

6.4. (Boolean Function Bi-decomposition) For a given Boolean function $f(X_A, X_B)$ with non-empty variable sets X_A and X_B, with $X_A \cap X_B = \varnothing$, what is the condition on $f(X_A, X_B)$ such that the rewriting $f(X_A, X_B) = f_A(X_A) \wedge f_B(X_B)$ is possible for some $f_A(X_A)$ and $f_B(X_B)$ to exist? (Express the condition with a quantified Boolean formula.)

6.5. (Characteristic Functions) Let $f : \mathbb{B}^3 \to \mathbb{B}^2$ be the vector (f_1, f_2) of Boolean functions with $f_1 = x_1 \vee \neg x_1 x_2$ and $f_2 = x_3 \wedge (x_1 \vee \neg x_1 x_2)$; let $\chi_S = x_1 \vee x_2$ be a characteristic function representing a set $S \subseteq \mathbb{B}^3$.

(a) Write down the characteristic function $Img_f(S)$ (in terms of a quantified Boolean formula) of the *image* of S under the mapping of f, that is, the set $\{q \in \mathbb{B}^2 \mid q = f(p), p \in S\}$.

(b) Perform quantifier elimination to obtain a quantifier-free formula equivalent to $Img_f(S)$ in (a).

(c) Justify that the formula in (b) indeed represents the image of S under f by enumerating all the truth assignments of (x_1, x_2, x_3) and the corresponding valuations of χ_S and f.

6.6. (BDD APPLY) Let F and G be the ROBDDs of Boolean functions $f = abc$ and $g = bd + b'd$, respectively, under the variable ordering *index* $(a) < index(b) < index(c) < index(d)$.

(a) Draw F and G.
(b) Derive the ROBDD of $F \cdot G$ using the BDDAPPLY procedure.
(c) Derive the ROBDD of $F + G$ using the BDDAPPLY procedure.
(d) Derive the ROBDD of $F \oplus G$ using the BDDAPPLY procedure.

6.7. (ROBDD Variable Ordering) Let F be the ROBDD of an arbitrary Boolean function $f(a, b, c, d, e)$ under variable ordering $index(a) < index(b) < index(c) < index(d) < index(e)$.

Show that the new ROBDD $F^\dagger$ under variable ordering

$index(a) < index(b) < index(d) < index(c) < index(e),$

must have the same BDD structure as F except for the nodes controlled by variables c and d.

6.8. (ROBDD Variable Ordering) Consider the Boolean function

$f = a_1 b_1 + a_2 b_2 + \cdots + a_n b_n.$

(a) Show that the ROBDD under variable ordering

$index(a_1) < index(b_1) < \cdots < index(a_n) < index(b_n)$

has $2n + 2$ nodes.

(b) Show that the ROBDD under variable ordering

$index(a_1) < \cdots < index(a_n) < index(b_1) < \cdots < index(b_n)$

has 2^{n+1} nodes.

6.9. (ROBDDs of Symmetric Functions) Totally symmetric functions are characterized by the fact that the value of each such function is determined by the number of variables which are 1 under a truth assignment; it does not matter which particular variables are. For example, functions $f_1 = x_1 \wedge \cdots \wedge x_n$, $f_2 = x_1 \vee \cdots \vee x_n$, and $f_3 = x_1 \oplus \cdots \oplus x_n$ are totally symmetric. A totally symmetric function on n variables can be described by a set $S \subseteq \{0, 1, \ldots, n\}$ such that for a minterm $a \in \mathbb{B}^n$, $f(a) = 1$ iff the number of 1's in a is a member of S. Prove that the ROBDD of any n-ary totally symmetric function has at most $O(n^2)$ nodes under any variable ordering.

6.10. (Circuit-to-CNF Conversion) Convert each of the following circuits to a CNF formula representing the consistency condition. In each case, list the truth assignments to the input/output variables that make the CNF true.

(a) An inverter with input a and output b.

(b) An OR2 gate with inputs a, b and output c.

(c) An XOR gate with inputs a, b and output c.

6.11. (Global Function Derivation) Consider the AIG of Figure 6.16. Derive the global function of x_7 (in terms of primary inputs x_1, x_2, x_3) using the following two methods.

(a) Existentially quantify out the intermediate variables x_4, x_5, x_6 from its corresponding consistency CNF formula and then perform a positive cofactor with respect to the variable x_7.

(b) Derive the global function of x_7 by recursively substituting intermediate variables with their local functions.

Verify that the above two methods yield the same result. Explain why these two approaches are equivalent.

6.12. (SOP and Tautology) Show that the tautology checking of any SOP formula with at most 2 literals in each product term can be done with time complexity polynomial in the formula size.

(Remark: The dual problem is the 2SAT problem in computer science, which is checkable in polynomial time.)

6.13. (Prime and Irredundant Cubes) Let

$$C = \{a'c'd', abd', a'b'd', a'bc', ab'c', a'b'c, abc, a'bd\}$$

be a cover of a completely specified function f.

(a) For each cube in C, determine whether it is prime and/or irredundant.

(b) Can we delete all the redundant cubes at once without affecting the function of f? Which redundant cubes can we delete from C if we successively delete removable cubes from left to right? How about from right to left? (Assume the cubes listed in C is ordered.)

6.14. (Quine-McCluskey Two-Level Logic Minimization) Given function

$$f = w'x'y'z' + wx'z' + wxz + w'x'z$$

with don't care set

$$d = w'xyz' + wx'yz + w'xyz$$

minimize f using the Quine-McCluskey procedure.

6.15. (Column Covering) Column covering is an essential computation step in Quine-McCluskey procedure. It can be solved in different ways.

(a) Show that the column covering problem can be formulated as a CNF satisfiability problem. Give an algorithm that performs such conversion. (The so-derived covering need not be minimum.)

(b) Show that the *minimum* column covering problem can be formulated in term of ROBDD. Give a polynomial-time algorithm solving the problem.

6.16. (Number of Prime Implicants) Show that

$$\frac{C^n_{\lceil \frac{n-2}{3} \rceil} 2^{n - \lceil \frac{n-2}{3} \rceil}}{n - \lceil \frac{n-2}{3} \rceil}$$

is a lower bound on the number of prime implicants for any n-ary Boolean function.

6.17. (Node Value and Elimination) Recall that the value of a node represents the saved literal count due to the existence of the node rather than collapsing it into its fanouts. Given the Boolean network of Figure 6.47, what are the values of nodes 5 and 6? What is the new value of node 6 after collapsing node 5 into its fanouts? (Here we treat Boolean formulas as polynomials in an algebraic sense, and assume that Boolean simplifications, such as $x \wedge \neg x = 0$, $x \vee \neg x = 1$, $x \wedge x = x$, and $x \vee x = x$, are not involved.)

6.18. (Algebraic Division) Prove that algebraic division produces a unique quotient and remainder. (Note that by definition the remainder is made as few cubes as possible.)

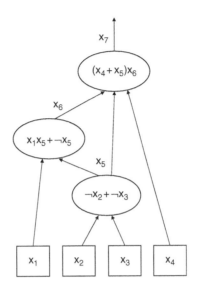

FIGURE 6.47

Boolean network.

6.19. (Kernels and Cokernels) Let expression

$F = aefh + aegh + aei + befh + begh + bei + cdefh + cdegh + cdei.$
Apply KERNEL1(0, F) to compute the kernels and corresponding cokernels of F. Identify which kernels are of level 0.

6.20. (Factoring) Continuing Exercise 6.19, apply GFACTOR to factor the function F. Use different level-0 kernels as divisors. What is the best factoring for F? For an arbitrary expression, can GFACTOR always produce a minimum-literal factoring with some proper level-0 kernels as divisors?

6.21. (Common Divisor Extraction) Let expressions

$F = ac + ad + bc + bd + adf + aef + ag + bcdf + bcef + bcg$, and
$G = ag + bcg + bcf + bcg + bdf + bdg + bef + beg$

(a) Iteratively reexpress F and G in terms of a common expression that yields the most reduction in literal count until no more common expressions exist. (A common expression can be a cube-free expression or a cube.)

(b) Extract an optimal common divisor of F and G by finding rectangles in the cokernel-cube matrix.

6.22. (Kernel Intersection) For two expressions F and G, suppose any kernel k_f of F and any k_g of G have at most one term in common. Show that F and G have no common algebraic divisor with more than one term.

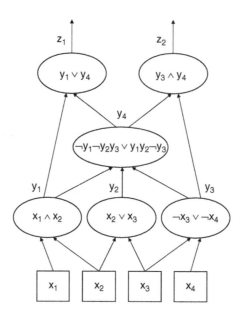

FIGURE 6.48

Boolean network.

6.23. (SDC and ODC) Consider the Boolean network of Figure 6.48.

   (a) Write down a Boolean formula representing the SDC of the entire circuit. That is, it represents the inconsistency condition of the circuit.

   (b) Write down a Boolean formula for the satisfiability don't cares SDC_4 of node 4 (with output y_4). Since SDC_4 is induced by the transitive fanins of node 4, the formula should depend on variables $x_1, \ldots, x_4, y_1, \ldots, y_3$. How can you make SDC_4 refer only to y_1, y_2, y_3 such that we can minimize node 4 directly?

   (c) Compute the observability don't cares ODC_4 of node 4.

6.24. (Don't Cares in Local Variables) Continuing Exercise 6.23, suppose the XDC for z_1 is $\neg x_1 \neg x_2 \neg x_3 \neg x_4$ and that for z_2 is $x_1 x_2 x_3 x_4$.

   (a) Compute the don't cares D_4 of node 4 in terms of its local input variables y_1, y_2, and y_3. (Note that in general the computation of ODC may be affected by XDC especially when there exist different XDCs for different primary outputs.)

   (b) Based on the computed don't cares, what is the best implementable function for node 4 (in terms of the literal count and cube count)?

6.25. (Complete Flexibility) Continuing Exercise 6.24, let $Y = \{y_1, y_2, y_3\}$ and $Z = \{z_1, z_2\}$.

    (a) Suppose the XDC for z_1 is $\neg x_1 \neg x_2 \neg x_3 \neg x_4$ and that for z_2 is $x_1 x_2 x_3 x_4$. Write down the specification relation $S(X,Z)$.

    (b) What is the influence relation $I_4(X, y_4, Z)$ of node 4?

    (c) What is the environment relation $E_4(X, Y)$ of node 4?

    (d) What is the complete flexibility $CF_4(Y, y_4)$ of node 4?

    (e) Is the previously computed don't care set D_4 of node 4 subsumed by CF_4?

6.26. (Static Timing Analysis) Given the circuit of Figure 6.49 with gate delays shown, assume the arrival times for the primary inputs are 0 except for input *b* with arrival time 1ns, and the required times for the primary output are 8ns. Compute the *arrival time, required time*, and *slack* of every net. Identify the critical path(s).

6.27. (Time Slack and Critical Path) Prove or disprove the following statement: The most critical path (with the smallest slack) must be a thorough path all the way from some primary input to some primary output.

6.28. (Arrival/Required Time Computation) Given a black box that computes arrival times for a Boolean network with specified gate delays and input arrival times, devise a way of reusing this black box to compute required times for a Boolean network.

6.29. (Tree Mapping) Decompose the subject DAG of Figure 6.50 into trees and perform dynamic programming to find optimum tree

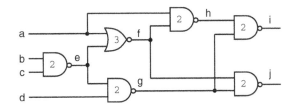

FIGURE 6.49

Circuit for timing analysis.

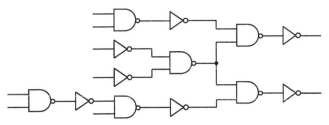

FIGURE 6.50

Subject graph.

mappings with respect to the pattern graphs of Figure 6.26. What is the optimum solution that you can get among different decomposition approaches?

6.30. **(DAG Mapping as SAT Solving)** Formulate the DAG mapping *feasibility* problem as a satisfiability problem. For the subject graph of Figure 6.50 and the pattern graphs of Figure 6.26, what is the CNF formula representing feasible DAG mappings?

ACKNOWLEDGMENTS

The authors are grateful to Professor Robert Brayton and Dr. Alan Mishchenko of the University of California at Berkeley, and Professor Jianwen Zhu of the University of Toronto for valuable feedback on the manuscript.

REFERENCES

R6.0 Books

[Brayton 1984] R. K. Brayton, G. Hachtel, C. McMullen, and A. Sangiovanni-Vincentelli, *Logic Minimization Algorithms for VLSI Synthesis,* Kluwer, 1984.

[Brown 2003] F. M. Brown, *Boolean Reasoning: The Logic of Boolean Equations,* Dover, 2003.

[Devadas 1994] S. Devadas, A. Ghosh, and K. Keutzer, *Logic Synthesis,* McGraw-Hill, 1994.

[Garey 1979] M. R. Garey and D. S. Johnson, *Computers and Intractability: A Guide to the Theory of NP-Completeness,* W. H. Freeman, 1979.

[Hassoun 2002] S. Hassoun and T. Sasao, *Logic Synthesis and Verification,* Kluwer, 2002.

[Kohavi 1978] Z. Kohavi, *Switching and Finite Automata Theory,* McGraw-Hill, 1978.

[McGeer 1991] P. McGeer and R. K. Brayton, *Integrating Functional and Temporal Domains in Logic Design,* Kluwer, 1991.

[Mo 2004] F. Mo and R. K. Brayton, *Regular Fabrics in Deep Sub-Micron Integrated-Circuit Design,* Kluwer, 2004.

[Minato 1996] S. Minato, *Binary Decision Diagrams and Applications to VLSI CAD,* Kluwer, 1996.

[Papadimitriou 1993] C. Papadimitriou, *Computational Complexity,* Addison Wesley, 1993.

[Sapatnekar 2004] S. Sapatnekar, *Timing,* Springer, 2004.

[Scholl 2001] C. Scholl, *Functional Decomposition with Applications to FPGA Synthesis,* Kluwer, 2001.

[Sutherland 1999] I. Sutherland, R. Sproull, and D. Harris, *Logical Effort: Designing Fast CMOS Circuits,* Margan Kaufmann, 1999.

[Villa 1997] T. Villa, T. Kam, R. K. Brayton, and A. Sangiovanni-Vincentelli, *Synthesis of Finite State Machines: Logic Optimization,* Kluwer, 1997.

R6.1 Introduction

[ABC 2005] Berkeley Logic Synthesis and Verification Group, ABC: A system for sequential synthesis and verification, http://www.eecs.berkeley.edu/~alanmi/abc/, 2005.

[Brayton 1987] R. K. Brayton, R. Rudell, A. Sangiovanni-Vincentelli, and A. Wang, MIS: Multiple-level interactive logic optimization system, *IEEE Trans. on Computer-Aided Design,* 6(6), pp. 1062–1081, November 1987.

[Gao 2002] M. Gao, J.-H. R. Jiang, Y. Jiang, Y. Li, A. Mishchenko, S. Sinha, T. Villa, and R. K. Brayton, Optimization of multi-valued multilevel networks, in *Proc. IEEE Int. Symp. on Multiple-Valued Logic*, pp. 168–177, May 2002.

[Rudell 1987] R. Rudell and A. Sangiovanni-Vincentelli, Multiple-valued minimization for PLA optimization, *IEEE Trans. on Computer-Aided Design*, 6(5), pp. 727–751, September 1987.

[Sentovich 1992] E. Sentovich, K. Singh, L. Lavagno, C. Moon, R. Murgai, A. Saldanha, H. Savoj, P. Stephan, R. K. Brayton, and A. Sangiovanni-Vincentelli, SIS: A system for sequential circuit synthesis, *Memo. UCB/ERL M92/41*, 1992.

R6.2 Data Structures for Boolean Representation and Reasoning

[ABC 2005] Berkeley Logic Synthesis and Verification Group, ABC: A system for sequential synthesis and verification, http://www.eecs.berkeley.edu/~alanmi/abc/, 2005.

[Akers 1978] S. B. Akers, Binary decision diagrams, *IEEE Trans. on Computers*, C-27(6), pp. 509–516, June 1978.

[Biere 2007] A. Biere, The AIGER and-inverter graph (AIG) format, http://fmv.jku.at/aiger/, 2007.

[Bryant 1986] R. E. Bryant, Graph-based algorithms for Boolean function manipulation, *IEEE Trans. on Computers*, C-35(8), pp. 677–691, August 1986.

[Bryant 1991] R. E. Bryant, On the complexity of VLSI implementations and graph representations of Boolean functions with application to integer multiplication, *IEEE Trans. on Computers*, C-40(2), pp. 205–213, February 1991.

[Bryant 1992] R. E. Bryant, Symbolic Boolean manipulation with ordered binary decision diagrams, *ACM Computg Surveys*, 24(3), pp. 293–318, September 1992.

[Kautz 1970] W. Kautz, The necessity of closed circuit loops in minimal combinational circuits, *IEEE Trans. on Computers*, C-19(2), pp. 162–164, February 1970.

[Kuehlmann 1997] A. Kuehlmann and F. Krohm, Equivalence checking using cuts and heaps, in *Proc. ACM/IEEE Design Automation Conf.*, pp. 263–268, June 1997.

[Lee 1959] C. Y. Lee, Representation of switching circuits by binary-decision programs, *Bell Systems Techncal J.*, 38(4), pp. 985–999, July 1959.

[Madre 1988] J.-C. Madre and J.-P. Billon, Proving circuit correctness using formal comparison between expected and extracted behaviour, in *Proc. ACM/IEEE Design Automation Conf.*, pp. 205–210, June 1988.

[Rudell 1990] R. L. Rudell, K. S. Brace, and R. E. Bryant, Efficient implementation of a BDD package, in *Proc. ACM/IEEE Design Automation Conf.*, pp. 40–45, June 1990.

[Rudell 1993] R. Rudell, Dynamic variable ordering for binary decision diagrams, in *Proc. IEEE/ACM Int. Conf. on Computer-Aided Design*, pp. 42–47, November 1993.

[Rudell 1987] R. Rudell and A. Sangiovanni-Vincentelli, Multiple-valued minimization for PLA optimization, *IEEE Trans. on Computer-Aided Design*, 6(5), pp. 727–751, September 1987.

[Tseitin 1970] G. S. Tseitin, On the complexity of derivation in propositional calculus, *Studies in Constructive Mathematics and Mathematical Logic*, Part II, (A. O. Slisenko, editors), pp. 115–125. Consultants Bureau, New York, 1970.

R6.3 Combinational Logic Minimization

[ABC 2005] Berkeley Logic Synthesis and Verification Group, ABC: A system for sequential synthesis and verification, http://www.eecs.berkeley.edu/~alanmi/abc/, 2005.

[Bartlett 1986] K. Bartlett, W. Cohen, A. J. De Geus, and G. D. Hachtel, Synthesis of multilevel logic under timing constraints, *IEEE Trans. on Computer-Aided Design*, CAD-5(4), pp. 582–595, October 1986.

[Bjesse 2004] P. Bjesse and A. Boralv, DAG-aware circuit compression for formal verification, in *Proc. IEEE/ACM Int. Conf. on Computer-Aided Design*, pp. 42–49, November 2004.

[Bostick 1987] D. Bostick, G. D. Hachtel, R. Jacoby, M. R. Lightner, P. Moceyunas, C. R. Morrison, and D. Ravenscroft, The Boulder optimal logic design system, in *Proc. IEEE/ACM Int. Conf. on Computer-Aided Design*, pp. 62–65, November 1987.

[Brayton 1982] R. K. Brayton and C. McMullen, The decomposition and factorization of Boolean expressions, in *Proc. IEEE Int. Symp. on Circuits and Systems*, pp. 49–54, May 1982.

[Brayton 1984] R. K. Brayton and C. McMullen, Synthesis and optimization of multistage logic, in *Proc. IEEE Int. Conf. on Computer Design*, pp. 23–28, October 1984.

[Brayton 1987] R. K. Brayton, R. Rudell, A. Sangiovanni-Vincentelli, and A. Wang, MIS: Multiple-level interactive logic optimization system, *IEEE Trans. on Computer-Aided Design*, 6(6), pp. 1062–1081, November 1987.

[Brayton 1990] R. K. Brayton, G. D. Hachtel, and A. Sangiovanni-Vincentelli, Multilevel logic synthesis, *Proceedgs of the IEEE*, 78(2), pp. 264–300, February 1990.

[Coudert 1994] O. Coudert, Two-level logic minimization: An overview, *Integrato*, 17(2), pp. 97–140, October 1994.

[Coudert 1995] O. Coudert, Doing two-level logic minimization 100 times faster, in *Proc. ACM/SIAM Symp. on Discrete Algorithms*, pp. 112–121, January 1995.

[Dagenais 1986] M. Dagenais, V. K. Agarwal, and N. Rumin, McBOOLE: A procedure for exact Boolean minimization, *IEEE Trans. on Computer-Aided Design*, CAD-5(1), pp. 229–237, January 1986.

[Darringer 1981] J. Darringer, W. Joyner, L. Berman, and L. Trevillyan, Logic synthesis through local transformations, *IBM J. of Research and Development*, 25(4), pp. 272–280, July 1981.

[Darringer 1984] J. Darringer, D. Brand, J. Gerbi, W. Joyner, and L. Trevillyan, LSS: A system for production logic synthesis, *IBM J. of Research and Development*, 28(5), pp. 537–545, September 1984.

[Devadas 1989] S. Devadas, A. R. Wang, A. R. Newton, and A. Sangiovanni-Vincentelli, Boolean decomposition in multilevel logic optimization, *IEEE J. of Solid State Circuits*, 24(2), pp. 399–408, April 1989.

[Fleisher 1975] H. Fleisher and L. I. Maissel, An introduction to array logic, *IBM J. of Research and Development*, 19(3), pp. 98–109, March 1975.

[Hong 1974] S. J. Hong, R. G. Cain, and D. L. Ostapko, MINI: A heuristic approach for logic minimization, *IBM J. of Research and Development*, 18(4), pp. 443–458, September 1974.

[Jiang 2004] J.-H. R. Jiang and R. K. Brayton, Functional dependency for verification reduction, in *Proc. Int. Conf. on Computer Aided Verification*, pp. 268–280, July 2004.

[Jiang 2006] J.-H. R. Jiang and R. K. Brayton, Retiming and resynthesis: A complexity perspective, *IEEE Trans. on Computer-Aided Design*, 25(12), pp. 2674–2686, December 2006.

[Keutzer 1987] K. Keutzer, DAGON: Technology mapping and local optimization, in *Proc. ACM/IEEE Design Automation Conf.*, pp. 341–347, June 1987.

[Kuehlmann 1997] A. Kuehlmann and F. Krohm, Equivalence checking using cuts and heaps, in *Proc. ACM/IEEE Design Automation Conf.*, pp. 263–268, June 1997.

[Lee 2007] C.-C. Lee, J.-H. R. Jiang, C.-Y. Huang, and A. Mishchenko, Scalable exploration of functional dependency by interpolation and incremental SAT solving, in *Proc. IEEE/ACM Int. Conf. on Computer-Aided Design*, pp. 227–233, November 2007.

[Leiserson 1983] C. Leiserson and J. Saxe, Optimizing synchronous systems, *J. of VLSI and Computer Systems*, 1(1), pp. 41–67, Spring 1983.

[Leiserson 1991] C. Leiserson and J. Saxe, Retiming synchronous circuitry, *Algorithmica*, 6(1), pp. 5–35, December 1991.

[Ling 2007] A. Ling, J. Zhu, and S. Brown, BddCut: Towards scalable symbolic cut enumeration, in *Proc. Asia and South Pacific Design Automation Conf.*, pp. 408–413, January 2007.

[Malik 1991] S. Malik, E. Sentovich, R. K. Brayton, and A. Sangiovanni-Vincentelli, Retiming and resynthesis: Optimizing sequential networks with combinational techniques, *IEEE Transactions on Computer-Aided Design*, 10(1), pp. 74–84, 1991.

[McCluskey 1956] E. J. McCluskey, Minimization of Boolean functions, *Bell Systems Technical J.*, 35(6), pp. 1417-1444, November 1956.

[McGeer 1987] P. C. McGeer and R. K. Brayton, Efficient, stable algebraic operations on logic expressions, in *Proc. IFIP Int. Conf. on Very Large Scale Integration*, August 1987.

[Mishchenko 2002] A. Mishchenko and R. K. Brayton, Simplification of non-deterministic multi-valued networks, in *Proc. IEEE/ACM Int. Conf. on Computer-Aided Design*, pp. 557-562, November 2002.

[Mishchenko 2005] A. Mishchenko, R. K. Brayton, J.-H. R. Jiang, T. Villa, and N. Yevtushenko, Efficient solution of language equations using partitioned representations, in *Proc. Design Automation and Test in Europe*, pp. 418-423, March 2005.

[Mishchenko 2006a] A. Mishchenko and R. K. Brayton, A theory of non-deterministic networks, *IEEE Trans. on Computer-Aided Design*, 25(6), pp. 977-999, June 2006.

[Mishchenko 2006b] A. Mishchenko, S. Chatterjee, and R. K. Brayton, DAG-aware AIG rewriting: A fresh look at combinational logic synthesis, in *Proc. ACM/IEEE Design Automation Conf.*, pp. 532-536, June 2006.

[Mishchenko 2007a] A. Mishchenko, R. K. Brayton, J.-H. R. Jiang, and S. Jang, SAT-based logic optimization and resynthesis, in *Proc. Int. Workshop on Logic Synthesis*, pp. 358-364, May 2007.

[Mishchenko 2007b] A. Mishchenko, S. Cho, S. Chatterjee, and R. K. Brayton, Combinational and sequential mapping with priority cuts, in *Proc. IEEE/ACM Int. Conf. on Computer-Aided Design*, pp. 354-361, November 2007.

[Rudell 1987] R. Rudell and A. Sangiovanni-Vincentelli, Multiple-valued minimization for PLA optimization, *IEEE Trans. on Computer-Aided Design*, 6(5), pp. 727-751, September 1987.

[Rudell 1989] R. Rudell, *Logic Synthesis for VLSI Design*, Ph.D. dissertation, University of California, Berkeley, 1989.

[Yevtushenko 2001] N. Yevtushenko, T. Villa, R. K. Brayton, A. Petrenko, and A. Sangiovanni-Vincentelli, Solution of parallel language equations for logic synthesis, in *Proc. IEEE/ACM Int. Conf. on Computer-Aided Design*, pp. 103-110, November 2001.

R6.4 Technology Mapping

[Aho 1976] A. Aho and S. Johnson, Optimal code generation for expression trees, *J. of the ACM*, 23(2), pp. 488-501, July 1976.

[Chatterjee 2006] S. Chatterjee, A. Mishchenko, R. K. Brayton, X. Wang, and T. Kam, Reducing structural bias in technology mapping, *IEEE Trans. on Computer-Aided Design*, 25(12), pp. 2894-2903, December 2006.

[Keutzer 1987] K. Keutzer, DAGON: Technology mapping and local optimization, in *Proc. ACM/IEEE Design Automation Conf.*, pp. 341-347, June 1987.

[Kravets 2001] V. Kravets, *Constructive Multilevel Synthesis by Way of Functional Properties*, Ph.D. dissertation, University of Michigan, Ann Arbor, 2001.

[Lehman 1997] E. Lehman, Y. Watanabe, J. Grodstein, and H. Harkness, Logic decomposition during technology mapping, *IEEE Trans. on Computer-Aided Design*, 16(8), pp. 813-834, August 1997.

R6.5 Timing Analysis

[Chen 1991] H.-C. Chen and D. H. Du, Path sensitization in critical path problem, in *Proc. IEEE/ACM Int. Conf. on Computer-Aided Design*, pp. 208-211, November 1991.

[Devadas 1992] S. Devadas, K. Keutzer, S. Malik, and A. Wang, Computation of floating mode delay in combinational logic circuits: Practice and implementation, in *Proc. Int. Symp. on Logic Synthesis and Microprocessor Architecture*, pp. 68-75, July 1992.

[Guerra E Silva 2002] L. Guerra E Silva, J. Marques-Silva, L. Silveira, and K. Sakallah, Satisfiability models and algorithms for circuit delay computation, *ACM Trans. on Design Automation of Electronic Systems*, 7(1), pp. 137–158, January 2002.

[McGeer 1989] P. McGeer and R. K. Brayton, Efficient algorithms for computing the longest viable path in a combinational network, in *Proc. ACM/IEEE Design Automation Conf.*, pp. 561–567, June 1989.

[McGeer 1991] P. McGeer, A. Saldanha, P. Stephan, R. K. Brayton, and A. Sangiovanni-Vincentelli, Timing analysis and delay-fault test generation using path-recursive functions, in *Proc. IEEE/ACM Int. Conf. on Computer Aided-Design*, pp. 180–183, November 1991.

R6.6 Timing Optimization

[Chaudhary 1992] K. Chaudhary and M. Pedram, A near optimal algorithm for technology mapping minimizing area under delay constraints, in *Proc. ACM/IEEE Design Automation Conf.*, pp. 492–498, June 1992.

[Leiserson 1983] C. Leiserson and J. Saxe, Optimizing synchronous systems, *J. of VLSI and Computer Systems*, 1(1), pp. 41–67, Spring 1983.

[Leiserson 1991] C. Leiserson and J. Saxe, Retiming synchronous circuitry, *Algorithmica*, 6(1), pp. 5–35, December 1991.

[Rudell 1989] R. Rudell, *Logic Synthesis for VLSI Design*, Ph.D. dissertation, University of California, Berkeley, 1989.

[Singh 1992] K. Singh, *Performance Optimization of Digital Circuits*, Ph.D. dissertation, University of California, Berkeley, 1992.

[Touati 1990] H. Touati, *Performance-Oriented Technology Mapping*, Ph.D. dissertation, University of California, Berkeley, 1990.

R6.7 Trends in Logic Synthesis

[Brayton 2007] R. K. Brayton, The synergy between logic synthesis and equivalence checking, in *Proc. Formal Methods in Computer Aided Design*, (Tutorial), November 2007.

[Jiang 2007] J.-H. R. Jiang and W.-L. Hung, Inductive equivalence checking under retiming and resynthesis, in *Proc. IEEE/ACM Int. Conf. on Computer-Aided Design*, pp. 326–333, November 2007.

Test synthesis

7

Laung-Terng (L.-T.) Wang
SynTest Technologies, Inc., Sunnyvale, California

Xiaoqing Wen
Kyushu Institute of Technology, Fukuoka, Japan

Shianling Wu
*SynTest Technologies, Inc., Princeton Junction,
New Jersey*

ABOUT THIS CHAPTER

Test synthesis is an important step in VLSI testing for automating the process of producing testable VLSI designs. The test synthesis flow typically includes testability rule checking and repair in the beginning to guarantee that the design has complied with all given testability rules. Once all rules are met, test synthesis is then performed to automatically insert test logic into the design. The test logic can include ***design-for-testability*** (DFT) circuitry used for scan design, logic ***built-in self-test*** (BIST), and/or test compression. Depending on the test requirements, additional circuitries for ***design-for-debug-and-diagnosis*** (DFD) and ***design-for-reliability*** (DFR) could also be inserted. These circuitries altogether are intended for improving the quality and reducing the test cost of the manufactured devices as well as simplifying the test, debug, and diagnosis tasks.

The focus of this chapter is on widely used scan synthesis flow and an emerging BIST synthesis flow. A set of scan design rules used in the scan synthesis flow, with which a design must comply, is described first. This is followed by the gate-level scan synthesis flow. Discussion then moves to new design rules required for logic BIST in the emerging BIST synthesis flow. This is followed by a BIST design example with all necessary steps involved in designing the logic BIST system, verifying its correctness, and further improving its fault coverage. The chapter concludes with a discussion of the scan design flow at the ***register-transfer level*** (RTL) which helps further reduce DFT design iterations and test development time. The RTL scan design flow can be readily extended for logic synthesis and integrated with other advanced DFT, DFD, and DFR features implemented at the RTL.

405

7.1 **INTRODUCTION**

Test synthesis is a **test automation** process to audit whether a digital design complies with all **testability rules** and then insert the required test logic into the digital design. In modern test synthesis tools, the automatic repair function, which automatically modifies the digital design to make it comply with the **testability rules**, is often provided. The digital design can be specified at the gate level or at the *register-transfer level* (RTL). After all testability rule violations are identified and repaired either manually or automatically, the resulting digital design is often referred to as a **testable design** or **test-ready core**.

By using the scan method, the test logic to be incorporated into the testable design will require conversion of internal storage elements, such as D flip-flops or D latches, into **scan cells**, such as *muxed-D scan cells* or *LSSD scan cells*. The resulting testable design is often called a **scan design**. In addition to being the dominant DFT architecture used for detecting manufacturing defects, scan design has become the backbone for more advanced DFT techniques, such as logic BIST and test compression [Wang 2006a]. Furthermore, as designs continue to move toward the nanometer scale, scan design is being used as a design feature to facilitate silicon debug, fault diagnosis, failure analysis, and reliability enhancement against **soft errors** [SIA 2005, 2006; Gizopoulos 2006; Wang 2007].

Recently, *design for testability* (DFT) has started to migrate from the gate level to the RTL. The motivation for this migration is to allow integration of additional DFT features, such as logic BIST and test compression, at the RTL to reduce test development time and to create reusable and testable RTL cores. This further allows the integrated RTL DFT design to be processed as part of synthesis-based optimization as to reduce test power, performance degradation, and area overhead.

Figure 7.1 shows a typical logic BIST system that has been synthesized in a scan design with embedded logic BIST and test compression circuitry. The logic BIST circuitry is composed of a logic BIST controller, a *test pattern generator* (TPG), and an *output response analyzer* (ORA). The test compression circuitry includes a **decompressor** and a **compactor**. Depending on the nature of the *circuit under test* (CUT) or scan design, the decompressor and the compactor can be a part of the TPG and the ORA, respectively. However, for the purpose of illustration, we separate the logic BIST circuitry from the test compression circuitry. Generally speaking, the logic BIST system will, at the minimum configuration, operate in five modes: normal mode, logic BIST mode, ATPG compression mode, ATPG mode, and serial debug and diagnosis mode.

In normal mode, all test logic will be disabled, and the CUT will function as intended. When the logic BIST system operates in BIST mode, the TPG will bypass the decompressor and automatically generate test patterns for application to the CUT (refer to Figure 3.13 in Chapter 3). The compactor will be bypassed, and the ORA will automatically compact the output responses of the CUT into a *signature*. Specific BIST timing control signals, including scan enable signals and clocks, are generated by the logic BIST controller for

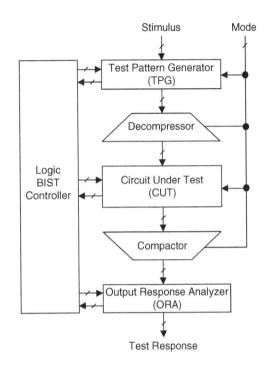

FIGURE 7.1

A typical logic BIST system with test compression circuitry.

coordinating the BIST operation among the TPG, CUT, and ORA. The logic BIST controller provides a pass/fail indication once the BIST operation is complete. It includes comparison logic to compare the *final signature* with an embedded expected or *golden signature*, and often comprises diagnostic logic for fault diagnosis. Because the ORA usually cannot tolerate unknown (X) values during output response analysis, it is required that all storage elements in the TPG, CUT, and ORA be initialized to known states before self-test, and no unknown (X) values be allowed to propagate from the CUT to the ORA. In other words, the CUT must comply with additional **BIST-specific design rules**.

When the system operates in ATPG compression mode, the TPG will be bypassed, and the decompressor will take over and decompress external compressed scan patterns (compressed stimuli) from *automatic test equipment* (ATE) for application to the scan data inputs of the CUT. At the same time, the compactor will compact the scan data outputs of the CUT, bypass the ORA, and shift the compressed test responses out to the ATE for comparison.

When the system operates in ATPG mode, both logic BIST and test compression circuitry will be bypassed. Scan patterns (stimuli) are directly shifted in from the ATE to the CUT, and the test responses are immediately shifted out to the ATE for comparison. In contrast to the ATPG mode, the serial debug and diagnosis mode requires that all scan chains be concatenated into one serial scan chain so test responses from the CUT can be shifted out bit-by-bit to a target computer or the ATE for analysis. This mode of operation is especially important for on-board or in-system diagnosis when the chip operates in the field.

7.2 SCAN DESIGN

Scan design is currently the most widely used structured DFT approach. As discussed in Chapter 3, it is implemented by connecting selected storage elements of a design into one or more shift registers, called **scan chains**, to provide external access to the storage elements. Scan design is obtained by replacing all selected storage elements with scan cells, each scan cell having one additional **scan input** (SI) port and one shared or additional **scan output** (SO) port. By connecting the SO port of one scan cell to the SI port of the next scan cell, one or more scan chains are created.

For a scan design to achieve the desired **defective parts per million** (PPM) goal, specific circuit structures and design practices that may affect fault coverage must be identified and repaired. This requires compiling a set of **scan design rules** that the design must adhere to. Hence, it is required to identify and fix scan design rule violations in the design before inserting or synthesizing scan chains into the design and generating test patterns for the scan design.

7.2.1 Scan design rules

To implement scan into a design, the design must comply with a set of scan design rules [Cheung 1997]. In addition, certain design styles must be avoided, because they may limit the fault coverage that can be achieved. A number of scan design rules that are required to successfully use scan and achieve the target fault coverage goal are listed in Table 7.1. In this table, a possible solution is recommended for each scan design rule violation. Scan design rules that are labeled "avoid" must be repaired throughout the shift and capture operations. Scan design rules that are labeled "avoid during the shift" must be fixed only during the shift operation. Detailed descriptions are provided for some critical scan design rules.

7.2.1.1 *Tristate buses*

Bus contention occurs when two bus drivers force opposite logic values onto a tristate bus, which can damage the chip. Bus contention is designed not to happen during the normal operation and is typically avoided during the capture operation, because advanced ATPG programs can generate test patterns that guarantee only one bus driver controls a bus. However, during the shift operation, no such guarantees can be made; therefore, certain modifications must be made to each tristate bus to ensure that only one driver controls the bus. For example, for the tristate bus shown in Figure 7.2a, which has three bus drivers (D_1, D_2, and D_3), circuit modification can be made as shown in Figure 7.2b,

Table 7.1 Typical Scan Design Rules

Design Style	Scan Design Rule	Recommended Solution
Tristate buses	Avoid during shift	Fix bus contention during shift
Bidirectional I/O ports	Avoid during shift	Force to input or output mode during shift
Gated clocks (muxed-D full-scan)	Avoid during shift	Enable clocks during shift
Derived clocks (muxed-D full-scan)	Avoid	Bypass clocks
Combinational feedback loops	Avoid	Break the loops
Asynchronous set/ reset signals	Avoid	Use external pin(s)
Clocks driving data	Avoid	Block clocks to the data portion
Floating buses	Avoid	Add bus keepers
Floating inputs	Not recommended	Tie to V_{DD} or V_{SS}
Cross-coupled NAND/NOR gates	Not recommended	Use standard cells
Non-scan storage elements	Not recommended for full-scan design	Initialize to known states, bypass, or make transparent

where EN_1 is forced to 1 to enable the D_1 bus driver, whereas EN_2 and EN_3 are set to 0 to disable both D_2 and D_3 bus drivers, when $SE = 1$.

In addition to bus contention, a bus without a pull-up, pull-down, or bus keeper may result in fault coverage loss. The reason is that the value of a floating bus is unpredictable, which makes it difficult to test for a stuck-at-1 fault at the enable signal of a bus driver. To solve this problem, a pull-up, pull-down, or bus keeper can be added. The bus keeper added in Figure 7.2b is an example of fixing this problem by forcing the bus to preserve the logic value driven onto it prior to when the bus becomes floating.

7.2.1.2 *Bidirectional I/O ports*

Bidirectional I/O ports are used in many designs to increase the data transfer bandwidth. During the capture operation, a bidirectional I/O port is usually specified as being either input or output; however, conflicts may occur at a bidirectional I/O port during the shift operation. An example is shown in Figure 7.3a, in which a bidirectional I/O port is used as an input, and the direction control is provided by the scan cell. Because the output value of the scan cell can vary

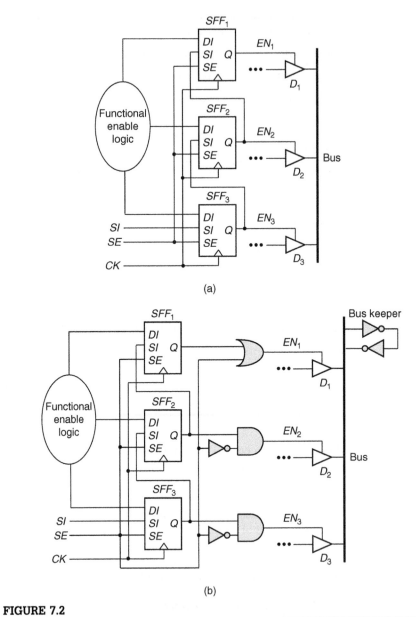

FIGURE 7.2

Fixing bus contention: (a) Original circuit. (b) Modified circuit.

during the shift operation, the output tristate buffer may become active, resulting in a conflict if BO and the I/O port driven by the tester have opposite logic values. Figure 7.3b shows an example of how to fix this problem by forcing the tristate buffer to be inactive when $SE = 1$, and the tester is used to drive

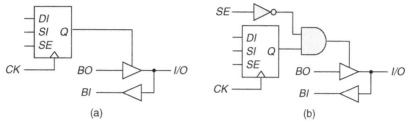

FIGURE 7.3

Fixing bidirectional I/O ports: (a) Original circuit. (b) Modified circuit.

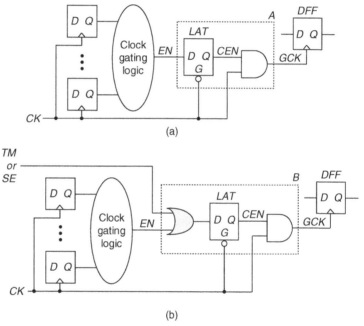

FIGURE 7.4

Fixing gated clocks: (a) Original circuit. (b) Modified circuit.

the I/O port during the shift operation. During the capture operation, the applied test vector determines whether a bidirectional I/O port is used as input or output and controls the tester appropriately.

7.2.1.3 *Gated clocks*

Clock gating is a widely used design technique for reducing power by eliminating unnecessary switching activity at storage elements. An example is shown in Figure 7.4a. The clock enable signal (*EN*) is generated at the rising edge of *CK* and is loaded into the latch *LAT* at the failing edge of *CK* to become *CEN*. *CEN* is then used to enable or disable clocking for the flip-flop *DFF*. Although clock gating is a good approach for reducing power consumption, it prevents the clock ports of some flip-flops from being directly controlled by primary inputs. As a result, modifications are necessary to allow the scan shift operation to be conducted on these storage elements.

The clock gating function should be disabled, at least during the shift operation. Figure 7.4b shows how the clock gating can be disabled. In this example, an OR gate is used to force *CEN* to 1 with either the test mode signal *TM* or the scan enable signal *SE*. If *TM* is used, *CEN* will be held at 1 during the entire scan test operation (including the capture operation). This will make it impossible to detect faults in the clock gating logic, causing fault coverage loss. If *SE* is used, *CEN* will be held at 1 only during the shift operation but will be released during the capture operation; hence, higher fault coverage can be achieved but at the expense of increased test generation complexity.

7.2.1.4 *Derived clocks*

A derived clock is a clock signal generated internally from a storage element or a clock generator, such as *phase-locked loop* (PLL), frequency divider, or pulse generator. Because derived clocks are not directly controllable from primary inputs, to test the logic driven by these derived clocks, these clock signals must be bypassed during the entire test operation. An example is illustrated in Figure 7.5a, in which the derived clock *ICK* drives the flip-flops DFF_1 and DFF_2. In Figure 7.5b, a multiplexer selects *CK*, which is a clock directly controllable from a primary input, to drive DFF_1 and DFF_2 during the entire test operation when $TM = 1$.

7.2.1.5 *Combinational feedback loops*

Depending on whether the number of inversions on a combinational feedback loop is even or odd, it can introduce either sequential behavior or oscillation into a design. Because the value stored in the loop cannot be controlled or determined

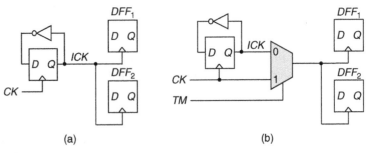

(a)   (b)

FIGURE 7.5

Fixing derived clocks: (a) Original circuit. (b) Modified circuit.

during test, this can lead to an increase in test generation complexity or fault coverage loss. Because combinational feedback loops are not a recommended design practice, the best way to fix this problem is to rewrite the RTL code generating the loop. In cases where this is not possible, a combinational feedback loop, as shown in Figure 7.6a, can be fixed by using a test mode signal *TM*. This signal permanently disables the loop throughout the entire shift and capture operations by inserting a scan point (*i.e.*, a combination of control and observation points) to break the loop, as shown in Figure 7.6b.

7.2.1.6 *Asynchronous set/reset signals*

Asynchronous set/reset signals of scan cells that are not directly controlled from primary inputs can prevent scan chains from shifting data properly. To avoid this problem, it is required that these asynchronous set/reset signals be forced to an inactive state during the shift operation. These asynchronous set/reset signals are typically referred to as being sequentially controlled. An example of a sequentially controlled reset signal *RL* is shown in Figure 7.7a. A method for

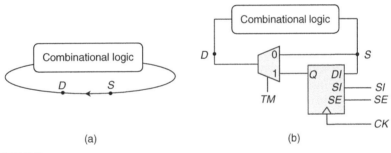

(a)   (b)

FIGURE 7.6

Fixing combinational feedback loops: (a) Original circuit. (b) Modified circuit.

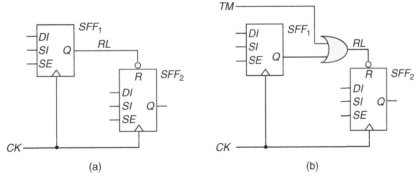

(a)   (b)

FIGURE 7.7

Fixing combinational feedback loops: (a) Original circuit. (b) Modified circuit.

fixing this asynchronous reset problem with an OR gate with an input tied to the test mode signal *TM* is shown in Figure 7.7b. When *TM* = 1, the asynchronous reset signal *RL* of scan cell SFF_2 is permanently disabled during the entire test operation.

The disadvantage of using the test mode signal *TM* to disable asynchronous set/reset signals is that faults within the asynchronous set/reset logic cannot be tested. The use of the scan enable signal *SE* instead of *TM* makes it possible to detect faults within the asynchronous set/reset logic, because during the capture operation (*SE* = 0), these asynchronous set/reset signals are not forced to the inactive state. However, this might result in mismatches because of race conditions between the clock and asynchronous set/reset ports of the scan cells. A better solution is to use an independent reset enable signal *RE* to replace *TM* and to conduct test generation in two phases. In the first phase, *RE* is set to 1 during both shift and capture operations to test data faults through the *DI* port of the scan cells while all asynchronous set/reset signals are held inactive. In the second phase, *RE* is set to 1 during the shift operation and 0 during the capture operation without applying any clocks to test faults within the asynchronous set/reset logic.

7.2.2 Scan design flow

Although conceptually scan design is not difficult to understand, the practice of inserting scan into a design to turn it into a scan design requires careful planning. This often requires many circuit modifications for which care must be taken not to disrupt the normal functionality of the circuit. In addition, many physical implementation details must be taken into consideration to guarantee that scan testing can be performed successfully. Finally, a good understanding of scan design, with respect to which scan cell design and scan architecture to use, is required to better plan in advance which scan design rules must be complied with and which debug and diagnose features must be included to facilitate simulation, debug, and fault diagnosis [Gizopoulos 2006; Wang 2007].

The shift operation and the capture operation are the two key scan operations in which care needs to be taken to guarantee that the scan design can operate properly. The shift operation, which is common to all scan designs, must be designed to perform successfully, regardless of the clock skew that exists within the same clock domain and between different clock domains. The capture operation is also common to all scan designs, albeit with more stringent scan design rules in some scan designs compared with others. It must be designed such that the ATPG tool is able to correctly and deterministically predict the expected responses of the generated test patterns. This requires a basic understanding of the logic simulation and fault models used for ATPG, as well as the clocking scheme used during the capture operation.

A typical design flow for implementing scan in a sequential circuit is shown in Figure 7.8. In this figure, scan design rule checking and repair are first

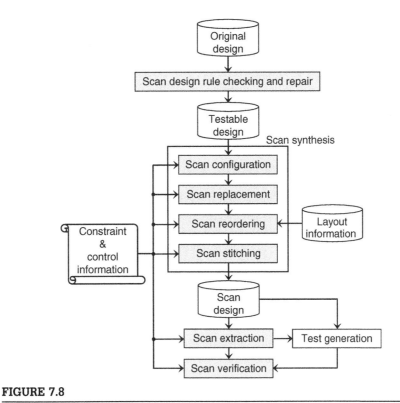

FIGURE 7.8

Typical scan design flow.

performed on a pre-synthesis RTL design or on a post-synthesis gate-level design, typically referred to as a **netlist**. The resulting design after scan repair is referred to as a **testable design**. Once all scan design rule violations are identified and repaired, scan synthesis is performed to convert the testable design into a scan design. The scan design now includes one or more scan chains for scan testing. A scan extraction step is used to further verify the integrity of the scan chains and to extract the final scan architecture of the scan chains for ATPG. Finally, scan verification is performed on both shift and capture operations to verify that the expected responses predicted by the zero-delay simulator used in test generation or fault simulation match with the full-timing behavior of the circuit under test. The steps shown in the scan design flow are described in the following subsections in more detail.

7.2.2.1 *Scan design rule checking and repair*

The first step in implementing a scan design is to identify and repair all scan design rule violations to convert the original design into a testable design.

Repairing these violations allows the testable design to meet target fault coverage requirements and guarantees that the scan design will operate correctly. These scan design rules were described in the previous section. In addition to these scan design rules, certain clock control structures may have to be added for at-speed delay testing. Typically, scan design rule checking is also performed on the scan design after scan synthesis to confirm that no new violations occur.

On successful completion of this step, the testable design must guarantee the correct shift and capture operations. During the shift operation, all clocks controlling scan cells of the design are directly controllable from external pins. The clock skew between adjacent scan cells must be properly managed so as to avoid any shift failure. During the capture operation, fixing all scan design rule violations should guarantee correctness for data paths that originate and terminate within the same clock domain. For data paths that originate and terminate in different clock domains, additional care must be taken in terms of the way the clocks are applied to guarantee the success of the capture operation. This is mainly because the clock skew between different clock domains is typically large. A data path originating in one clock domain and terminating in another might result in a mismatch when both clocks are applied simultaneously, and the clock skew between the two clocks is larger than the data path delay from the originating clock domain to the terminating clock domain. To avoid the mismatch, the timing governing the relationship of such a data path shown in the following equation must be observed:

$$\textit{Clock skew} < \textit{Data path delay} + \textit{Clock-to-Q delay (originating clock)}$$

If this is not the case, a mismatch may occur during the capture operation. To prevent this from happening, clocks belonging to different clock domains can be applied sequentially (with the **staggered clocking** scheme), as opposed to simultaneously, such that any clock skew that exists between the clock domains can be tolerated during the test generation process. It is also possible to apply only one clock during each capture operation with the **one-hot clocking** scheme. On the other hand, a design typically contains a number of noninteracting clock domains. In this case, these clocks can be applied simultaneously, which can reduce the complexity and the final pattern count of the pattern generation and fault simulation process. **Clock grouping** is a process used to identify all independent or noninteracting clocks that can be grouped and applied simultaneously.

An example of the clock grouping process is shown in Figure 7.9. This example shows the results of performing a circuit analysis operation on a testable design to identify all clock interactions, marked with an arrow, where a data transfer from one clock domain to a different clock domain occurs. As seen in Figure 7.9, the circuit in this example has seven clock domains ($CD_1 \sim CD_7$) and five cross-clock-domain data paths ($CCD_1 \sim CCD_5$). From this example, it can be seen that CD_2 and CD_3 are independent from each other; hence, their related clocks can be applied simultaneously during test as CK_2. Similarly, clock

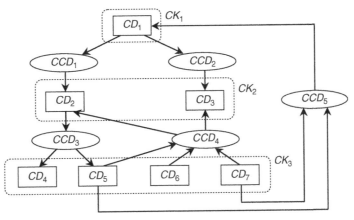

FIGURE 7.9

Clock grouping example.

domains CD_4 through CD_7 can also be applied simultaneously during test as CK_3. Therefore, in this example, three grouped clocks instead of seven individual clocks can be used to test the circuit during the capture operation.

7.2.2.2 *Scan synthesis*

When all the repairs of scan design rule violations have been made to the circuit, the scan synthesis flow is commenced. The scan synthesis flow converts a testable design into a scan design without affecting the functionality of the original design. Static analysis tools and equivalence checkers, which compare the logic circuitry of two circuits under given constraints, are typically used to verify that this is, indeed, the case. Depending on the types of scan cells used and the types of scan architecture implemented, minor modifications to the scan synthesis flow shown in Figure 7.8 may be necessary.

During the 1990s, this scan synthesis operation was typically performed with a separate set of scan synthesis tools, which were applied after the logic synthesis tool had synthesized a gate-level netlist out of an RTL description of the design. Recently, these scan synthesis features are being integrated into the logic synthesis tools, and scan designs are synthesized automatically from the RTL. The process of performing scan synthesis during logic synthesis is often referred to as **one-pass synthesis** or **single-pass synthesis**.

The scan synthesis flow shown in Figure 7.8 includes four separate steps: (1) scan configuration, (2) scan replacement, (3) scan reordering, and (4) scan stitching. Each of these steps is described in the following in more detail.

7.2.2.2.1 Scan configuration

Scan configuration describes the initial step in scan chain planning, in which the general structure of the scan design is determined. The main decisions that are made at this stage include: (1) the number of scan chains used; (2) the types of scan cells used to implement these scan chains; (3) storage elements to be excluded from the scan synthesis process; and (4) the way the scan cells are arranged within the scan chains.

The number of scan chains used is typically determined by analyzing the input and output pins of the circuit to determine how many pins can be allocated for the scan use. So as not to increase the number of pins of the circuit, which is typically limited by the size of the die, scan inputs and outputs are shared with existing pins during scan testing. In general, the larger the number of scan chains used, the shorter the time to perform a test on the circuit. This is because the maximum length of the scan chains dictates the overall test application time required to run each test pattern. One limitation that can preclude many scan chains from being used is the presence of high-speed I/O pads. The addition of any wire load to the high-speed I/O pads may adversely affect the timing of the design. An additional limitation is the number of tester channels available for scan testing.

The second issue regarding the types of scan cells to use typically depends on the process library. In general, for each type of storage element used, most process libraries have a corresponding scan cell type that closely resembles the functionality and timing of the storage element during the normal operation.

The third issue relates to which storage elements to exclude from scan synthesis. This is typically determined by investigating parts of the design where replacing storage elements with functionally equivalent scan cells may adversely affect timing. Therefore, storage elements lying on the critical paths of a design where the timing margin is very tight are often excluded from the scan replacement step to guarantee that the manufactured device will meet the restricted timing. In addition, certain parts of a design may be excluded from the scan for many different reasons, including security reasons (*e.g.*, parts of a circuit that deal with encryption). In these cases, individual storage element types, individual storage element instances, or a complete section of the design can be specified as "don't scan."

The remaining issue is to determine how the storage elements are arranged within the scan chains. This typically depends on how the number of clock domains relates to the number of scan chains in the design. In general, a scan chain is formed out of scan cells belonging to a single clock domain. For clock domains that contain a large number of scan cells, several scan chains are constructed, and a scan-chain balancing operation is performed on the clock domain to reduce the maximum scan-chain length. Oftentimes, a clock domain will include both negative-edge and positive-edge scan cells. If the number of negative-edge scan cells in a clock domain is large enough to construct a separate scan chain, these scan cells can be allocated as such. In cases where a scan chain has to include both negative-edge and positive-edge scan cells, all negative-edge scan cells are arranged in the scan chains such that they precede all

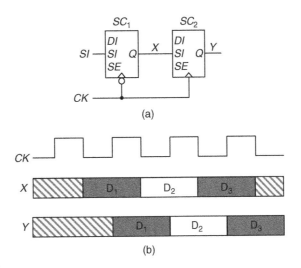

FIGURE 7.10

Mixing negative-edge and positive-edge scan cells in a scan chain: (a) Circuit structure. (b) Timing diagram.

positive-edge scan cells to guarantee that the shift operation can be performed correctly.

Figure 7.10a shows an example of a circuit structure made up of a negative-edge scan cell followed by a positive-edge scan cell. The associated timing diagram, shown in Figure 7.10b, illustrates the correct shift timing of the circuit structure. During each shift clock cycle, Y will first take on the state X at the rising CK edge before X is loaded with the SI value at the falling CK edge. If we accidentally place the positive-edge scan cell before the negative-edge scan cell, both scan cells will always incorrectly contain the same value at the end of each shift clock cycle.

In cases where scan chains must include scan cells from several different clock domains, a lock-up latch is inserted between adjacent cross-clock-domain scan cells to guarantee that any clock skew between the clocks can be tolerated. Clock skew between different clock domains is expected, because clock skew is controlled within a clock domain to remain below a certain threshold but not controlled across different clock domains. As a result, a race caused by a hold time violation could occur between these two scan cells if a lock-up latch is not inserted.

Figure 7.11a shows an example of a circuit structure having a scan cell SC_p belonging to clock domain CK_1 driving a scan cell SC_q belonging to clock domain CK_2 through a lock-up latch. The associated timing diagram is shown in Figure 7.11b, where CK_2 arrives after CK_1, to demonstrate the effect of clock skew on cross-clock-domain scan cells. During each shift clock cycle, X will first take on the SI value at the rising CK_1 edge, then Z will take on the Y value at

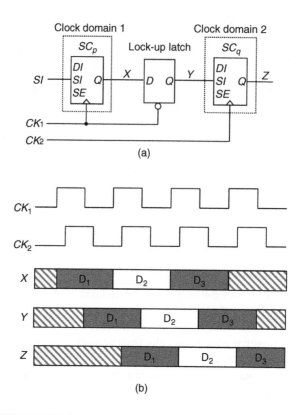

FIGURE 7.11

Adding a lock-up latch between cross-clock-domain scan cells: (a) Circuit structure. (b) Timing diagram.

the rising CK_2 edge. Finally, the new X value is transferred to Y at the falling CK_1 edge to store the SC_p contents. If CK_2 arrives earlier than CK_1, Z will first take on the Y value at the rising CK_2 edge. Then, X will take on the SI value at the rising CK_1 edge. Finally, the new X value is transferred to Y at the falling CK_1 edge to store the SC_p contents. In both cases, the lock-up latch design in Figure 7.11a allows correct shift operation regardless of whether CK_2 arrives earlier or later than CK_1. It is important to note that this scheme works only when the clock skew between CK_1 and CK_2 is less than the width (duty cycle) of the clock pulse. If this is not the case, then slowing down the shift clock frequency or enlarging the duty cycle of the shift clock can guarantee that this approach will work for any amount of clock skew. Other lock-up latch and lock-up flip-flop designs can also be used.

Once the clock structure of the scan chains is determined, it is still necessary to determine which scan cells should be stitched together into one scan chain and the order in which these scan cells should be placed. In some scan synthesis flows, a preliminary layout placement is used to allocate scan cells to different scan chains belonging to the same clock domain. Then, the best order in which to stitch these scan cells within the scan chains is determined to minimize the scan routing required to connect the output of each scan cell to the scan input of the next scan cell. In cases where a preliminary placement is not available, scan cells can be assigned to different scan chains on the basis of an initial floorplan of the testable design by grouping scan cells in proximate regions of the design together. Once the final placement is determined, the scan chains can then be reordered and stitched, and the scan design is modified on the basis of the new scan chain order.

7.2.2.2.2 Scan replacement

After scan configuration is complete, **scan replacement** replaces all original storage elements in the testable design with their functionally equivalent scan cells. The testable design after scan replacement is often referred to as a **scan-ready design**. Functionally equivalent scan cells are the scan cells that most closely match power, speed, and area characteristics of the original storage elements. The scan inputs of these scan cells are often tied to the scan outputs of the same scan cell to prevent floating inputs from being present in the circuit. These connections are later removed during the scan-stitching step. In cases where one-pass or single-pass synthesis is used, scan replacement is transparent to tool users. Recently, some RTL scan-synthesis tools have implemented scan replacement at the RTL, even before going to the logic/scan synthesis tool, to reflect the scan design changes in the original RTL design.

7.2.2.2.3 Scan reordering

Scan reordering refers to the process of reordering scan cells in scan chains on the basis of the physical scan cell locations to minimize the number of interconnect wires used to implement the scan chains. During design implementation, if the physical location of each scan cell instance is not available, a "random" scan order based purely on the module-level and bus-level connectivity of the testable design can be used. However, if a preliminary placement is available, scan cells can be assigned to different scan chains on the basis of the initial floorplan of the design. Only after the final placement process of the physical implementation is performed on this testable design is the physical location of each scan cell instance taken into consideration. During the routing process of the physical implementation, scan reordering can be performed with *intrascan-chain reordering*, *interscan-chain reordering*, or a combination of both. **Intrascan-chain reordering**, in which scan cells are reordered only within their respective scan chains, does not reorder any scan cells across clock or clock-polarity boundaries. **Interscan-chain reordering**, in which scan cells are reordered among different scan chains, must make sure that the clock structure of the scan chains is preserved. In both intrascan-chain reordering and interscan-chain reordering, care must also be taken to limit the minimum distance between scan cells to avoid timing violations that can destroy the integrity of the shift operation.

Advanced techniques have also been proposed to further reduce routing congestion while avoiding timing violations during the shift operation [Duggirala 2002, 2004]. For deep submicron circuits, the capacitance of the scan chain interconnect must also be taken into account to guarantee correct shift operation [Barbagallo 1996].

7.2.2.2.4 Scan stitching

Finally, the **scan-stitching** step is performed to stitch all scan cells together to form scan chains. Scan stitching refers to the process of connecting the output of each scan cell to the scan input of the next scan cell on the basis of the scan order specified previously. An additional step is also performed by connecting the scan input of the first scan cell of each scan chain to the appropriate scan chain input port and the scan output of the last scan cell of each scan chain to the appropriate scan chain output port to make the scan chains externally accessible. In cases where a shared I/O port is used to connect to the scan chain input or the scan chain output, additional signals must be connected to the shared I/O port to guarantee that it always behaves as either input or output, respectively, throughout the shift operation. As mentioned earlier, it is important to avoid the use of high-speed I/O ports as scan chain inputs or outputs, because the additional loading could result in a degradation of the maximum speed at which the device can be operated. In addition to stitching the existing scan cells, lock-up latches or lock-up flip-flops are often inserted during the scan-stitching step for adjacent scan cells where clock skew may occur. These lock-up latches or lock-up flip-flops are then stitched between adjacent scan cells.

7.2.2.3 *Scan extraction*

When the scan stitching step is complete, the scan synthesis process is complete. The original design has now been converted into a scan design; however, an additional step is often performed to verify the integrity of the scan chains, especially if any design changes are made to the scan design. **Scan extraction** is the process used for extracting all scan cell instances from all scan chains specified in the scan design. This procedure is performed by tracing the design for each scan chain to verify that all the connections are intact when the design is placed in shift mode. Scan extraction can also be used to prepare for the test generation process to identify the scan architecture of the design in cases where this information is not otherwise available.

7.2.2.4 *Scan verification*

When the physical implementation of the scan design is completed, including placement and routing of all the cells of the design, a timing file in *standard delay format* (SDF) is generated. This timing file resembles the timing behavior of the manufactured device. This is then used to verify that scan testing can be successfully performed on the manufactured scan design.

Other than the trivial problems of scan chains being incorrectly stitched, verification errors during the shift operation are typical because of hold time violations between adjacent scan cells, where the data path delay from the output of a driving scan cell to the scan input of the following scan cell is smaller than the clock skew that exists between the clocks driving the two scan cells. In cases where the two scan cells are driven by the same clock, this may indicate a failure of the *clock tree synthesis* (CTS) process in guaranteeing that the clock skew between scan cells belonging to the same clock domain be kept at a minimum. In cases where the two scan cells are driven by different clocks, this may indicate a failure of inserting a required lock-up latch between the scan cells of the two different clock domains.

Apart from clock skew problems, other scan shift problems may occur. Often, they stem from (1) an incorrect scan initialization sequence that fails to put the design into test mode; (2) incomplete scan design rule checking and repair in which the asynchronous set/reset signals of some scan cells are not disabled during the shift operation or the gated/generated clocks for some scan cells are not properly enabled or disabled; or (3) incorrect scan synthesis in which positive-edge scan cells are placed before negative-edge scan cells.

Scan capture problems typically occur because of mismatches between the zero-delay model used in test generation and fault simulation tools and the full-timing behavior of the real device. In these cases, care must be taken during the scan design and test application process to (1) provide enough clock delay between the supplied clocks such that the clock capture order becomes deterministic, and (2) prevent simultaneous clock and data switching events from occurring. Failure to take clock events into proper consideration can easily result in a breakdown of the zero-delay (cycle-based) simulator used in the test generation and fault simulation process. More detailed information regarding scan verification of the shift and capture operations is described in the following.

7.2.2.4.1 Verifying the scan shift operation

Verifying the scan shift operation involves performing **flush tests** with a full-timing logic simulator during the shift operation. A flush test is a shift test in which a selected flush pattern is shifted all the way through the scan chains to verify that the same flush pattern arrives at the end of the scan chains at the correct clock cycle. For example, a scan chain containing 1000 scan cells requires 1000 shift cycles to be applied to the scan chain for the selected flush pattern to begin arriving at the scan output. If the data arrive early by a number of shift cycles, this may indicate that a similar number of hold time problems exist in the circuit.

To detect clock skew problems between adjacent scan cells, the selected flush pattern is typically a pattern that is capable of providing both 0-to-1 and

1-to-0 transitions to each scan cell. To ensure that a 0-to-0 or 1-to-1 transition of a scan cell does not corrupt the data, the selected flush pattern is further extended to provide these transitions. A typical flush pattern used for testing the shift operation is "01100," which includes all four possible transitions. Different flush patterns can also be used for debugging different problems, such as the all-zero and all-one flush patterns used for debugging stuck-at faults in the scan chain.

Because observing the arrival of the data on the scan chain output cannot pinpoint the exact location of any shift error in a faulty scan chain, flush testbenches are typically created to observe the values at all internal scan cells to identify the locations at which the shift errors exist. By use of this technique, the faulty scan chain can be easily and quickly diagnosed and fixed during the scan shift verification process; for example:

- Scan hold time problems that exist between scan cells belonging to different clock domains indicate that a lock-up latch may be missing. Lock-up latches should be inserted between these adjacent scan cells.
- Scan hold time and setup time problems that exist between scan cells belonging to the same clock domain indicate that the CTS process was not performed correctly. In this case, either the CTS has to be redone or additional buffers need to be inserted between the failing scan cells to slow down the path.
- Scan hold time problems caused by positive-edge scan cells followed by negative-edge scan cells indicate that the scan chain order was not performed correctly. Lock-up flip-flops rather than lock-up latches can be inserted between these adjacent scan cells or the scan chains may have to be reordered by placing all negative-edge scan cells before all positive-edge scan cells.

An additional approach to scan shift verification that has become more popular in recent years involves performing *static timing analysis* (STA) on the shift path in shift mode. In this case, the STA tool can immediately identify the locations of all adjacent scan cells that fail to meet timing. The same solutions mentioned earlier are then used to fix problems identified by the STA tool.

7.2.2.4.2 Verifying the scan capture operation

Verifying the scan capture operation involves simulating the scan design with a full-timing logic simulator during the capture operation. This is used to identify the location of any failing scan cells in which the captured response does not match the expected response predicted by the zero-delay logic simulator used in test generation or fault simulation. To reduce simulation time, a **broadside-load testbench** is often used, in which a test pattern is loaded directly into all scan cells in the scan chains and only the capture cycle is simulated. Because the **broadside-load test** does not involve any shift cycle in the test pattern, broadside-load testbenches often include at least one shift cycle in the capture verification testbench to ensure that each test pattern can at least shift once.

This requires loading the test pattern into the outputs of the previous scan cells rather than directly into the outputs of the current scan cells. In addition, verifying the scan capture operation often includes a *serial simulation*, in which a limited number of test patterns, typically three to five or as many as can be simulated within a reasonable time, are simulated. In this serial simulation, a test pattern is simulated exactly the same as how it would be applied on the tester by shifting in each pattern serially through the scan chain inputs. Next, a capture cycle is applied. The captured response is then shifted out serially to verify that the complete scan chain operation can be performed successfully.

As mentioned before, mismatches in the capture cycle indicate that the zero-delay simulation model used by the test generator and the fault simulator failed to capture all the details of the actual timing occurring in the device. Debugging these types of failures is tedious and may involve observing all signals of the mismatching scan cells and signal lines (also called nets) driving these scan cells. One brute-force method commonly used by designers for removing these mismatches is to mask off the locations by changing the expected response of the mismatching location into an unknown (*X*) value. A new approach that has become more popular is to use the static timing analysis tool for both scan shift and scan capture verification.

7.3 LOGIC BUILT-IN SELF-TEST (BIST) DESIGN

Compared with scan synthesis, considerable efforts are required to implement or synthesize a BIST design. Because logic BIST is mostly scan-based, the BIST design must first comply with all scan design rules. In addition, because BIST designs usually cannot tolerate unknown (*X*) values propagated to the output response analyzers, **BIST-specific design rules** are required to deal with unknown sources originating from analog blocks, memories, non-scan storage elements, asynchronous set/reset signals, tristate buses, false paths, and multiple-cycle paths, to name a few. The need to implement a logic BIST controller that automatically coordinates BIST pattern generation and response analysis for at-speed testing of delay faults further complicates the process. Last, because pseudo-random patterns (as opposed to deterministic patterns) are commonly used for BIST pattern generation, additional test points (including control points and observation points) may have to be added to improve the circuit's fault coverage.

7.3.1 BIST design rules

Because logic BIST requires many more stringent design restrictions than conventional scan, many *scan design rules* discussed in Section 7.2 that are optional for scan designs become mandatory for BIST designs. The major logic BIST design restriction relates to the propagation of unknown (*X*) values. Because any unknown (*X*) value that propagates directly or indirectly to the *output response analyzer* (ORA) will corrupt the *signature* and cause the BIST signature to become useless, no unknown (*X*) values can be tolerated. This is different from scan designs in which unknown (*X*) values present in a scan design only result in fault coverage degradation. Therefore, when designing a logic BIST system, it is essential that the *circuit under test* (CUT) meet all scan design rules and BIST-specific design rules, called **BIST design rules**. The process of taking a scan-based design and making it meet all additional BIST-specific design rules turns the design into a **BIST-ready core**.

7.3.1.1 *Unknown source blocking*

There are many unknown (*X*) sources in a CUT or BIST-ready core. Any unknown (*X*) source in the BIST-ready core, which is capable of propagating its unknown (*X*) value to the ORA directly or indirectly, must be blocked and fixed with a DFT repair approach often called **X-bounding** or **X-blocking**. Figure 7.12 shows a few of the more typically used X-bounding methods for blocking an unknown (*X*) source: The **0-control point** forces an *X* source to 0; the **1-control point** controls the *X* source to 1; the **bypass logic** allows the output of the *X* source to receive both 0 and 1 from a *primary input* (PI) or an internal node; the **control-only scan point** drives both 0 and 1 through a storage element, such as D flip-flop; and finally, the **scan point** can capture the *X*-source value and drive both 0 and 1 through a scan cell, such as scan D flip-flop or *level-sensitive scan design* (LSSD) *shift register latch* (SRL) [Eichelberger 1977].

Depending on the nature of each unknown (*X*) source, several *X*-bounding methods can be appropriate for use. The most common problems inherent in these approaches include: (1) that they might increase the area of the design, and (2) that they might impact timing.

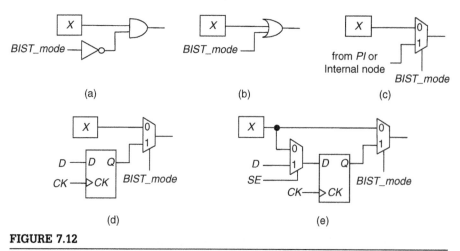

(a)   (b)   (c)

(d)   (e)

FIGURE 7.12

Typical X-bounding methods for blocking an unknown (*X*) source: (a) 0-control point. (b) 1-control point. (c) Bypass logic. (d) Control-only scan point. (e) Scan point.

7.3.1.1.1 Analog blocks

Examples of analog blocks are *analog-to-digital* **converters** (ADCs). Any analog block output that can exhibit unknown (X) behavior during a test has to be forced to a known value. This can be accomplished by adding a 0-control point, a 1-control point, bypass logic, or a control-only scan point. We recommend the latter two approaches, because they yield higher fault coverage than the former two approaches.

7.3.1.1.2 Memories and non-scan storage elements

Examples of memories are *dynamic random-access memories* (DRAMs), *static random-access memories* (SRAMs), or flash memories. Examples of non-scan storage elements are D flip-flops or D latches. Bypass logic is typically used to block each unknown (X) value originating from a memory or non-scan storage element. Another approach is to use an initialization sequence to set a memory or non-scan storage element to a known state. This is typically done to avoid adding delay to critical (functional) paths. Care must be taken to ensure that the stored state is not corrupted throughout the BIST operation.

7.3.1.1.3 Combinational feedback loops

All combinational feedback loops must be avoided. If they are unavoidable, then each loop must be broken with a 0-control point, a 1-control point, or a scan point. We recommend adding scan points because they yield higher fault coverage than the other approaches.

7.3.1.1.4 Asynchronous set/reset signals

As indicated in the preceding section, asynchronous set or reset can destroy the data during the shift operation if a pattern causes the set/reset signal to become active. The asynchronous set or reset can be disabled with an external set/reset disable (RE) pin given in Figure 7.7. This set/reset disable pin must be set to 1 during the shift operation. This may become cumbersome for BIST applications in which there is a need to use the pin for other purposes. Thus, we recommend using the existing scan enable (SE) signal to protect each shift operation and adding a set/reset clock point ($SRCK$) on each set/reset signal to test the set/reset circuitry as illustrated in Figure 7.13.

In addition, we recommend testing all data and set/reset faults with two separate BIST sessions as shown in Figure 7.14. The timing diagram in this figure is used for testing a circuit having one system clock (CK) and one added set/reset clock. To test data faults in the functional logic, a clock pulse C_1 is triggered from CK while $SRCK$ is held inactive in one capture window. Similarly, to test set/reset faults in the set/reset circuitry, C_2 is enabled while CK is held inactive in another capture window. By use of this approach, we can avoid races and hazards and prevent data in scan cells from being destroyed by the set/reset signals.

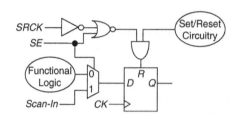

FIGURE 7.13

Set/reset clock point for testing a set/reset-type scan cell.

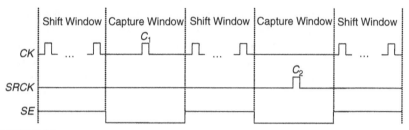

FIGURE 7.14

Example timing control diagram for testing data and set/reset faults.

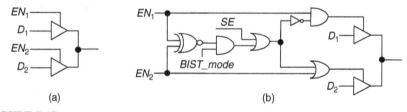

FIGURE 7.15

A one-hot decoder for testing a tristate bus with two drivers: (a) A tristate bus. (b) A one-hot decoder.

7.3.1.1.5 Tristate buses

Bus contention occurs when two drivers force different values on the same bus that can damage the chip; hence, it is important to prevent bus conflicts during the normal operation and the shift operation [Cheung 1997]. For BIST applications, because pseudo-random patterns are commonly used, it is also crucial to prevent bus contention from happening during the capture operation [Al-Yamani 2002]. To avoid potential bus contention, it is best to resynthesize each bus with multiplexers. If this is impractical, make sure only one tristate driver is enabled at any given time. The **one-hot decoder** shown in Figure 7.15 is an example of a circuit that can ensure only one driver is selected during each shift or capture operation.

7.3.1.1.6 False paths

False paths are not normal functional paths. They do no make any harm to the chip during the normal operation; however, for delay fault testing, a pseudorandom pattern might adversely attempt to test a selected false path. Because false paths are not exercised during the normal circuit operation, they typically do not meet timing specifications, which can result in a mismatch during logic BIST delay fault testing. To avoid this potential problem, we recommend adding a 0-control point or 1-control point to each false path.

7.3.1.1.7 Critical paths

Critical paths are timing-sensitive functional paths. Because the timing of such a path is critical, no extra gates are allowed to be added to the path to prevent increasing the delay of the critical path. To remove an unknown (X) value from a critical path, we recommend adding an extra input pin to a selected combinational gate, such as an inverter, NAND gate, or NOR gate, on the critical path to minimize the added delay. The combinational gate is then converted to an embedded 0-control point or embedded 1-control point as shown in Figure 7.16, where an inverter is selected for adding the extra input.

7.3.1.1.8 Multiple-cycle paths

Multiple-cycle paths are normal functional paths, but data are expected to arrive after two or more cycles. Similar to false paths, they can cause mismatches if exercised during delay fault testing, because they are intended to be tested in one cycle. To avoid this potential problem, we recommend adding a 0-control point or 1-control point to each multiple-cycle path or holding certain scan cell output states to avoid those multiple-cycle paths.

7.3.1.1.9 Floating ports

Neither *primary inputs* (PIs) nor *primary outputs* (POs) can be floating. These ports must have a proper connection to power (V_{DD}) or ground (V_{SS}). Also, floating inputs to any internal modules must be avoided. This has a potential chance to propagate unknown (X) values to the ORA.

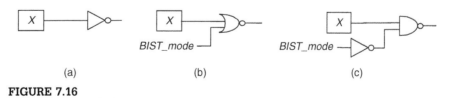

(a)   (b)   (c)

FIGURE 7.16

Embedded control points for testing a critical path having an inverter: (a) An inverter. (b) Embedded 0-control point. (c) Embedded 1-control point.

7.3.1.1.10 Bidirectional I/O ports

Bidirectional I/O ports are commonly used in a design. For BIST operations, the direction of each bidirectional I/O port should be forced to either input or output mode. Figure 7.17 shows an example of forcing a bidirectional I/O port to output mode.

7.3.1.2 *Re-timing*

Because the TPG and the ORA are typically placed far from the CUT, races and hazards caused by clock skews may occur between the TPG and the (scan chain) inputs of the CUT and between the (scan chain) outputs of the CUT and the ORA. To avoid these potential problems and ease physical implementation, we recommend adding **re-timing logic** between the TPG and the CUT and between the CUT and the ORA. The re-timing logic should consist of at least one negative-edge **pipelining register** (D flip-flop) and one positive-edge pipelining register (D flip-flop). Figure 7.18 shows an example re-timing logic among the TPG, CUT, and ORA that uses two pipelining registers on each end. Note that the three clocks (CK_1, CK_2, and CK_3) could belong to one clock tree.

7.3.2 **BIST design example**

In this subsection, we show an example of designing a logic BIST system for testing a scan-based design (core) composed of two clock domains with s38417 and s38584. The two clock domains are taken from the ISCAS-1989 benchmark circuits [Brglez 1989], and their statistics are shown in Table 7.2. The design we consider is described at the *register-transfer level* (RTL). We

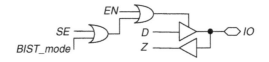

FIGURE 7.17

Forcing a bidirectional port to output mode.

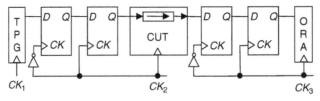

FIGURE 7.18

Re-timing logic among the TPG, CUT, and ORA.

Table 7.2 Design Statistics

Clock Domain	No. of PIs	No. of POs	No. of Flip-Flops	No. of Gates
CD_1 (s38417)	28	106	1,636	22,179
CD_2 (s38584)	12	278	1,452	19,253

show all the necessary steps to arrive at the logic BIST system design, verify its correctness, and improve its fault coverage.

7.3.2.1 *BIST rule checking and violation repair*

The first step is to perform logic BIST design rule checking on the RTL design. All DFT rule violations of the *scan design rules* and *BIST-specific design rules* provided in preceding sections must be repaired. Once all DFT rule violations are repaired, the design should meet all scan and logic BIST design rules. In addition, we should be aware of the following design parameters:

- The number of test clocks present in the design, each used for controlling one clock domain.
- The number of set/reset clocks present in the design to be used for breaking all asynchronous set/reset loops.

In the preceding example, the design contains two test clocks and does not require any additional set/reset clock. The new RTL design (core) after BIST rule repair is performed is referred to as an *RTL BIST-ready core*.

7.3.2.2 *Logic BIST system design*

The second step is to design the logic BIST system at the RTL. The decisions that need to be made at this stage include:

- The type of logic BIST architecture to adopt.
- The number of PRPG-MISR (or PEPG-MISR) pairs to use
- The length of each PRPG-MISR (or PEPG-MISR) pair.
- The faults to be tested and BIST timing control diagrams to be used for testing these faults.
- The types of optional logic to be added to ease physical implementation and facilitate debug and diagnosis, as well as improve the circuit's fault coverage.

7.3.2.2.1 Logic BIST architecture

We choose to implement **STUMPS**-based logic BIST architecture, because it is easy to integrate with scan/ATPG and is the architecture widely used in the industry. We recommend the use of one PRPG-MISR pair for each clock domain, whenever possible, because the resulting BIST architecture is easier to debug.

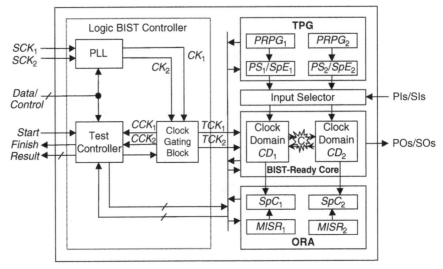

FIGURE 7.19

A logic BIST system for testing a design with two cores.

In addition, the use of one PRPG-MISR pair for each clock domain can eliminate the need for additional design efforts for managing clock skews between interacting clock domains, even when they operate at the same frequency. If it is required to use a single PRPG-MISR pair to test multiple clock domains, these clock domains should be placed within physical proximity to simplify physical implementation. An example of logic BIST system based on the STUMPS architecture for testing the design given in Table 7.2 is shown in Figure 7.19.

The BIST architecture used for testing the BIST-ready core consists of a TPG for generating test stimuli, an input selector for providing pseudo-random or ATPG patterns to the core-under-test, an ORA for compacting the test responses, and a logic BIST controller for coordinating the overall BIST operation. The logic BIST controller consists of a test controller and a clock-gating block. The test controller initiates the BIST operation on receiving a *Start* signal, issues a *Finish* signal once the BIST operation is complete, and reports the pass/fail status of the test through the *Result* bus. The clock-gating block accepts internal PLL clocks (CK_1 and CK_2) derived from external functional clocks (SCK_1 and SCK_2), and generates the required test clocks (TCK_1 and TCK_2) and controller clocks (CCK_1 and CCK_2) for controlling the BIST-ready core and test controller, respectively. During normal functional operation, both CK_1 and CK_2 can run faster or slower than SCK_1 and SCK_2, respectively.

7.3.2.2.2 TPG and ORA

Next, we need to determine the length of each PRPG-MISR pair. The use of a separate PRPG-MISR pair for each clock domain allows us to reduce the length of each PRPG and MISR. In the example shown in Figure 7.19, the linear phase shifters, PS_1 and PS_2, and space expanders, SpE_1 and SpE_2, can be used to further reduce the length of the PRPGs, whereas the space compactors, SpC_1 and SpC_2, can be used to further reduce the length of the MISRs. Each space expander or space compactor typically consists of an XOR-gate tree.

Now, suppose we decide to (1) synthesize the two clock domains, CD_1 and CD_2, each with 20 balanced scan chains; (2) run 100,000 pseudo-random patterns to obtain very high BIST fault coverage by adding additional test points; and (3) perform top-up ATPG after BIST to further increase the circuit's fault coverage. Because CD_1 has 28 PIs, a logical conclusion would be to expect the length of the $PRPG_1$ to be 48 for the use of a 48-stage PRPG to drive 28 PIs and 20 scan chains. Because we plan to perform top-up ATPG, which requires sharing 20 out of the 28 PIs with **scan inputs** (SIs), and another 20 POs with **scan outputs** (SOs), another possible length for the $PRPG_1$ would be 28. What we need to determine is whether a 28-stage PRPG, constructed from a maximum-length LFSR or **cellular automata** (CA), is adequate for generating the required 100,000 pseudo-random patterns.

For a CD_1 with 20 balanced scan chains, 82 shift clock pulses are required (1636 flip-flops/20 scan chains) to scan in a single pseudo-random pattern. This means that a total of 8.2 million shift clock pulses are required to scan in all 100,000 patterns. This number is much smaller than the 256 million ($2^{28}-1$) patterns generated with a 28-stage maximum-length LFSR or CA for the $PRPG_1$. From Table 3.5 given in Chapter 3, we chose a 28-stage maximum-length LFSR with characteristic polynomial, $f(x) = 1 + x^3 + x^{28}$.

A similar analysis applies for CD_2. The main difference is that CD_2 has 12 PIs. Suppose we pick 10 out of the 12 PIs to share with 10 SIs for top-up ATPG. We will need to use a 10-to-20 space expander (SpE_2) for driving the 20 scan chains and a 20-to-10 space compactor (SpC_2) for driving the 10 SOs. Because testing this clock domain requires a total of 7.3 million (1452/20 × 00,000) shift clock pulses, we need to use at least a 23-stage maximum-length LFSR or CA as $PRPG_2$ to drive the 12 PIs. From Table 3.5 given in Chapter 3, we chose a 25-stage maximum-length LFSR with characteristic polynomial, $f(x) = 1 + x^3 + x^{25}$.

As indicated in Section 3.4.2.3, each MISR can cause an *aliasing* problem, but the problem is of less concern when the MISR length is greater than 20. Because CD_1 and CD_2 both have 106 and 278 POs, we choose a 106-to-27 space compactor (SpC_1) and a 278-to-35 space compactor (SpC_2), respectively. Thus, we will use a 47-stage MISR and a 45-stage MISR to compact the test responses from both CD_1 and CD_2, respectively, where $47 = 27$ (shared POs) + 20 (SOs) and $45 = 35$ (shared POs) + 10 (SOs). From Table 3.5 given in Chapter 3, we choose to implement the 47-stage MISR with $f(x) = 1 + x^5 + x^{47}$, and the 45-stage MISR with $f(x) = 1 + x + x^3 + x^4 + x^{45}$. Table 7.3 shows the decisions made for each PRPG-MISR pair so far.

Table 7.3 PRPG-MISR Choices

Clock Domain	No. of Scan Chains	No. of Shared SIs or SOs	Max. Scan Chain Length	PRPG Length	MISR Length
CD_1 (s38417)	20	20	82	28	47
CD_2 (s38584)	20	10	73	25	45

7.3.2.2.3 Test controller

The test controller plays a central role in coordinating the overall BIST operation. In general, a *finite-state machine* written at the RTL is used to implement the test controller for interfacing with all external signals, such as *Start*, *Finish*, and *Result*, and generating the required timing control signals for controlling each PRPG-MISR pair and the BIST-ready core. Comparison logic is included in the test controller to compare the *final signature* with an embedded *golden signature*.

Often, these interface signals are controlled through an IEEE 1149.1 boundary-scan-standard-based **test access port (TAP) controller** [IEEE 1149.1-2001]. In this case, all signals can be assessed through the TAP: TDI (*Test Data In*), TDO (*Test Data Out*), TCK (*Test Clock*), and TMS (*Test Mode Select*). Optionally, an IEEE 1500 standard-based **wrapper** may be also used to isolate each selected clock domain [IEEE 1500-2005].

To test structural faults in the BIST-ready core, we chose the *staggered single-capture* approach rather than the *one-hot single-capture* approach. The slow-speed timing control diagram is shown in Figure 7.20, where test clocks TCK_1 and TCK_2 are staggered and generated by the clock-gating block shown in Figure 7.19.

To test delay faults in the BIST-ready core, we chose the *staggered double-capture* approach if CD_1 and CD_2 are asynchronous or the *aligned double-capture* approach if they are synchronous. This is because either approach

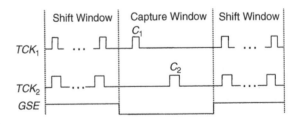

FIGURE 7.20

Slow-speed timing control using staggered single-capture.

allows us to operate a *global scan enable (GSE)* signal at slowspeed for driving all clock domains simultaneously in both BIST and scan ATPG modes. The at-speed timing control diagrams with the *staggered double-capture* and *launch aligned double-capture* schemes are shown in Figures 7.21 and 7.22, respectively.

7.3.2.2.4 Clock gating block

To generate an ordered sequence of *single-capture* or *double-capture* clocks, *clock suppression* [Rajski 2003], *daisy-chain clock-triggering*, or *token-ring clock-enabling* [Wang 2005a] can be used. The clock suppression scheme typically requires the use of a reference clock operating at the highest frequency. Daisy-chain clock-triggering means that a completion of one event automatically triggers the next event as the arrows shown in Figure 7.23. The only difference

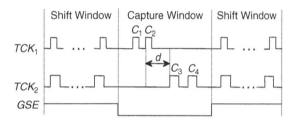

FIGURE 7.21

At-speed timing control using staggered double-capture.

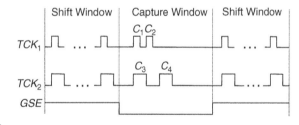

FIGURE 7.22

At-speed timing control using launch aligned double-capture.

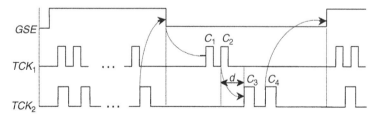

FIGURE 7.23

Daisy-chain clock-triggering.

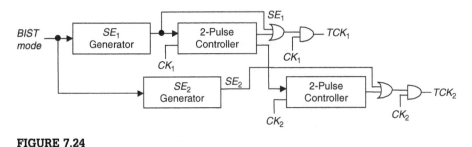

FIGURE 7.24

A daisy-chain clock-triggering circuit for generating the waveform given in Figure 7.23.

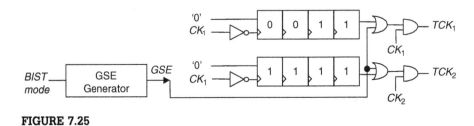

FIGURE 7.25

A clock suppression circuit for generating the waveform given in Figure 7.22.

between daisy-chain clock-triggering and token-ring clock-enabling is that the former uses a clock edge to trigger the next event, whereas the latter uses a signal level to enable the next event.

Figure 7.24 shows a daisy-chain clock-triggering circuit for generating the *staggered double-capture* waveform given in Figure 7.23. When the BIST mode is activated, the SE_1/SE_2 generators and 2-pulse controllers will generate the required scan enable and double-capture clock pulses, per the arrows shown in Figure 7.23. Each SE_1/SE_2 can be treated as a *GSE* signal for CD_1/CD_2.

Figure 7.25 shows a clock suppression circuit for generating the *launch aligned double-capture* waveform given in Figure 7.22. This circuit uses a reference clock (CK_1) to program the capture window. The contents of the 8-bit shift register are preset to {0011,1111} during each shift window. Because of its programmability, the approach can also be used to generate timing waveforms for testing asynchronous designs. One major requirement is that we guarantee that the delay measured by the number of reference clock pulses be longer than delay d between C_2 and C_3, as shown in Figure 7.21.

7.3.2.2.5 Re-timing logic

The main difference between ATE-based scan testing and logic BIST is that the latter requires that more complex BIST circuitry be implemented on the functional circuitry. Successfully completing the physical implementation of the functional circuitry of a high-speed and high-performance design is a challenge in itself. If the BIST circuitry adds a large number of timing critical signals and requires strict clock-skew management, the physical implementation of logic BIST can become extremely difficult. Therefore, we recommend adding two pipelining registers (see Figure 7.18) between each PRPG and the BIST-ready core and two additional pipelining registers between the BIST-ready core and each MISR. In this case, the maximum scan chain length for each clock domain, CD_1 or CD_2, is effectively increased by 2, not 4.

7.3.2.2.6 Fault coverage enhancing logic and diagnostic logic

The drawback of using pseudo-random patterns is that the circuit may not meet the target fault coverage goal. To improve the circuit's fault coverage, we recommend adding extra test points and additional logic for top-up ATPG support at the RTL. A general rule of thumb is to add one extra test point for every 1000 gates. For top-up ATPG support, the inserted logic includes an input selector for selecting test patterns either from the PRPGs or PIs/SIs, as shown in Figure 7.19, as well as circuitry for reconfiguring the scan chains to perform top-up ATPG in (1) ATPG mode or (2) ATPG compression mode, which is discussed in more detail in Chapter 3.

We also recommend including **diagnostic logic** in the RTL BIST code to facilitate debug and diagnosis [Wang 2006b]. One simple approach is to connect all PRPG-MISR pairs (and all scan chains) as a serial scan chain and make them externally accessible. (Refer to Chapter 7 [Wang 2006a] for more advanced BIST diagnosis techniques.) Table 7.4 summarizes all possible test modes of the logic BIST system along with the effective scan chain counts for each test mode.

7.3.2.3 *RTL BIST synthesis*

Once all decisions regarding the logic BIST architecture are made, it is time to create the RTL logic BIST code. At this stage, it is possible to either design the

Table 7.4 Example Test Modes Supported by the Logic BIST System

Test Mode	CD1 Effective Chain Count	CD2 Effective Chain Count
Normal	0	0
BIST	20	20
ATPG	20	10
ATPG compression	20	20
Serial debug and diagnosis	1	1

438 CHAPTER 7 Test synthesis

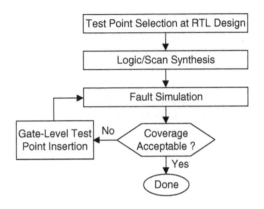

FIGURE 7.26

Fault simulation and test point insertion flow.

logic BIST system by hand or generate the RTL code automatically with an RTL logic BIST tool (commercially available). In either case, the number of scan chains for each clock domain should be specified along with the names of their associated scan inputs and scan outputs without inserting the actual scan chains into the circuit. The scan synthesis task can be handled as part of the general synthesis task, implemented with any commercially available synthesis tool for converting the RTL BIST-ready core and the logic BIST system into a gate-level netlist.

7.3.2.4 *Design verification and fault coverage enhancement*

Finally, the synthesized netlist needs to be verified with functional or timing verification to ensure that the logic BIST system functions as intended. If any pattern mismatch occurs, the problem must be identified and resolved. Next, fault simulation must be performed on the pseudo-random patterns generated by the TPG to determine the circuit's fault coverage. If the circuit does not reach the target fault coverage goal, additional test points should be inserted or top-up ATPG should be used. The extra test points that were added in advance at the RTL design should allow you to achieve the target fault coverage goal; otherwise, the test point insertion and fault simulation process may have to be repeated until the final fault coverage goal is reached. Once this process is complete, the *golden signature* can be either recorded to be compared externally or hard-coded into the comparison logic. The fault simulation and test point insertion flow is illustrated in Figure 7.26.

7.4 RTL DESIGN FOR TESTABILITY

During the 1990s, the testability of a circuit was primarily assessed and improved at the gate level. The reason was because the circuits were not too large that the logic/scan synthesis process took an unreasonable amount of time. As device size grows toward tens to hundreds of millions of transistors, tight timing, potential yield loss, and low power issues begin to pose serious challenges. When combined with increased core reusability and time-to-market pressure, it is becoming imperative that most, if not all, testability issues be addressed at the RTL. This allows the logic/scan synthesis tool and the physical synthesis tool, which takes physical layout information into consideration, to optimize area, power, and timing after DFT repairs are made. Fixing DFT problems at the RTL also allows designers to create testable RTL cores that can be reused without having to repeat the DFT checking and repair process for a number of times.

Figure 7.27 shows a design flow for performing testability repair at the gate level. It is clear that performing testability repair at the gate level introduces a loop in the design flow that requires repeating the time-consuming logic synthesis process every time testability repair is made. This makes it logical to attempt to perform testability checking and repair at the RTL instead, so testability violations can be detected and fixed at the RTL, as shown in Figure 7.28, without having to repeat the logic synthesis process.

An additional benefit of performing testability repair at the RTL is that it allows scan to be more easily integrated with other advanced DFT features implemented at the RTL, such as memory BIST, logic BIST, test compression, boundary scan, and *analog and mixed-signal* (AMS) BIST. This makes it possible to perform all testability integration at the RTL as opposed to the current practices of integrating the advanced DFT features at the RTL and later integrating them with scan at the gate level. In the following, we describe the RTL DFT problems by focusing mainly on scan design.

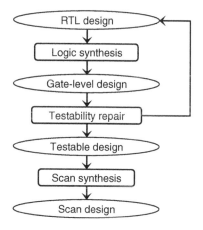

FIGURE 7.27

Gate-level testability repair design flow.

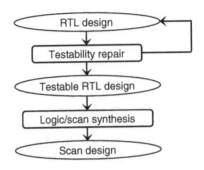

FIGURE 7.28

RTL testability repair design flow.

Some modern synthesis tools now incorporate testability repair and scan synthesis as part of the logic synthesis process, such that a testable design free of scan rule violations is generated automatically. In this case, if the DFT fixes made are acceptable and do not have to be incorporated into the RTL, the flow can proceed directly to test generation and scan verification.

7.4.1 RTL scan design rule checking and repair

To perform scan design rule checking and repair at the RTL, a **fast synthesis** step of the RTL is usually performed first. In fast synthesis, combinational RTL code is mapped onto combinational primitives and high-level models, such as adders and multipliers. This allows us to identify all possible scan design rule violations and infer all storage elements in the RTL design.

Static solutions for identifying testability problems at the RTL without having to perform any test vector simulation or dynamic solutions that simulate the structure of the design through the RTL have been developed. These solutions make it possible to identify almost all testability problems at the RTL. Although a few testability problems remain that can be identified only at the gate level, this approach does reduce the number of iterations involving logic synthesis, as shown in Figure 7.28. In addition, it has become common to add scan design rules as part of RTL "*lint*" tools that check for good coding and reusability styles, as well as user-defined coding style rules [Keating 1999]. To further optimize testability results, *clock grouping* can also be performed at the RTL as part of scan design rule checking [Wang 2005b].

Automatic methods for repairing RTL testability problems have also been developed [Wang 2005b]. An example is shown in Figure 7.29. The RTL code shown in Figure 7.29a, which is written in the Verilog *Hardware Description Language* (HDL) [IEEE 1463-2001], represents a generated clock. In this example, a flip-flop *clk_15* can be inferred, whose value is driven to 1 when a counter value q is equal to "1111." The output of this flip-flop is then used to trigger the second "always" statement, where an additional flip-flop can be

```
always @(posedge clk)
  if (q == 4'b1111)
    clk_15 <= 1;
  else
    begin
      clk_15 <= 0;
      q <= q + 1;
    end
always @(posedge clk_15)
  d <= start;
```

(a)

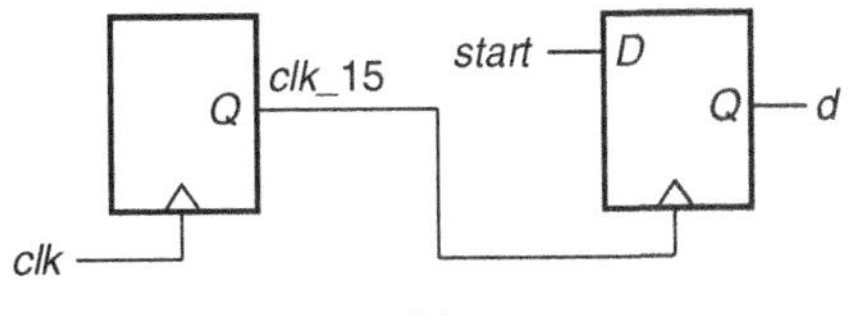

(b)

```
always @(posedge clk)
  if (q == 4'b1111)
    clk_15 <= 1;
  else
    begin
      clk_15 <= 0;
      q <= q + 1;
    end
assign clk_test = (TM) ? clk : clk_15;
always @(posedge clk_test)
  d <= start;
```

(c)

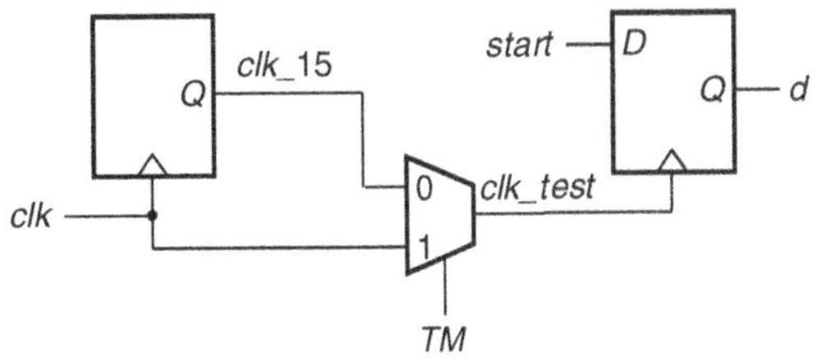

(d)

FIGURE 7.29

Automatic repair of a generated clock violation at the RTL: (a) Generated clock (RTL code). (b) Generated clock (schematic). (c) Generated clock repair (RTL code). (d) Generated clock repair (schematic).

inferred. Figure 7.29b shows a schematic of the flip-flop generating the *clk*_15 signal, as well as the flip-flop driven by the generated clock, which is likely to be the structure synthesized out of the RTL with a logic synthesis tool. This scan design rule violation can be fixed with the test mode signal *TM* by modifying the RTL code as shown in Figure 7.29c. The schematic for the modified RTL code is shown in Figure 7.29d.

7.4.2 **RTL scan synthesis**

When storage elements have been identified during RTL scan design rule checking, either **RTL scan synthesis** or **pseudo RTL scan synthesis** can be performed. In RTL scan synthesis, the scan synthesis step as described in Section 7.2.2.2 is performed. The only difference is that the scan equivalent of each storage element does not refer to a library cell but to an RTL structure that is equivalent to the original storage element in normal mode. In this case, the scan chains are inserted into the RTL design. In pseudo RTL scan synthesis, the scan synthesis step is not performed; only pseudo primary inputs and pseudo primary outputs are specified and stitched to primary inputs and primary outputs, respectively. This approach is becoming more appealing to designers now, because it can cope with many advanced DFT structures, such as logic BIST and test compression, where scan chains are driven internally by additional test structures synthesized at the RTL. Once all advanced DFT structures are inserted at the RTL, a one-pass or single-pass synthesis step is performed with the RTL design flow as shown in Figure 7.28.

Several additional steps are actually performed to identify the storage elements in the RTL design. First, all clocks are identified, either explicitly by tracing from specified clock signal names or implicitly by analyzing the sensitivity list of all "always" blocks. When the clocks have been identified, all registers, each consisting of one or more storage elements in the RTL design, are inferred by analyzing all "assign" statements to determine which assignments can be mapped onto a register while keeping track of the clock domain to which each register belongs. In addition, the clock polarity of each register is determined.

After all registers have been identified and each converted into its scan equivalent at the RTL, the next step is to stitch these individual scan cells into one or more scan chains. One approach is to allocate scan cells to different scan chains on the basis of the driving clocks and to stitch all scan cells within a scan chain in a random fashion [Aktouf 2000]. Although this approach is simple and straightforward, it can introduce wiring congestion as well as high interconnect area overhead. To solve these issues, it is better to take full advantage of the rich functional information available at the RTL [Roy 2000; Huang 2001]. Because storage elements are identified as registers as opposed to a large number of unrelated individual storage elements, it is beneficial to connect the scan cells (which are scan equivalent of these storage elements) belonging to the same register sequentially in a scan chain. This has been found to be able to dramatically reduce wiring congestion and interconnect area overhead.

7.4.3 **RTL scan extraction and scan verification**

To verify the scan-inserted RTL design (also called *RTL scan design*), both scan extraction and scan verification must be performed. Scan extraction relies on performing fast synthesis on the RTL scan design. This generates a software model where scan extraction can be performed by tracing the scan connections of each scan chain in a similar manner as scan extraction from a *gate-level scan design*. Scan verification relies on a flush testbench that is used to simulate flush tests on the RTL scan design. Because the inputs and outputs of the RTL scan design should match the inputs and outputs of its gate-level scan design, the same flush testbench can be used to verify the scan operation for both RTL and gate-level designs. It is also possible to apply broadside-load tests for verifying the scan capture operation at the RTL. In this case, either random test patterns or deterministic test patterns generated at the RTL can be used [Ghosh 2001; Ravi 2001; Zhang 2003].

7.5 CONCLUDING REMARKS

This chapter has discussed the design rules and test synthesis steps required to implement the basic *design-for-testability* (DFT) techniques presented in Chapter 3 into modern digital circuits. By modeling the circuit at the gate level or the *register-transfer level* (RTL), modern test synthesis programs can perform design rule checking and repair before scan synthesis (including synthesis of test compression logic) or logic *built-in self-test* (BIST) synthesis. Modern RTL (logic) synthesis programs can further incorporate test synthesis into the logic synthesis flow.

In this chapter, we have presented a comprehensive discussion of scan synthesis. This includes scan design rules and a typical scan design flow. We have also provided a comprehensive description of scan-based logic BIST synthesis. This includes BIST-specific design rules and a BIST design example. These BIST-specific design rules are mandatory for logic BIST in addition to following all scan design rules. The RTL DFT techniques that include RTL scan design rule checking and RTL scan synthesis were briefly touched on at the end of the chapter; these techniques were used to enable DFT integration at the RTL.

Implementing testability logic in a design could require many dedicated test pins. Modern test synthesis programs have incorporated support of a few IEEE-endorsed standards into the test synthesis flow to reduce additional pin count or to facilitate test, debug, and diagnosis. The most popular standard supported is the IEEE 1149.1 boundary-scan standard [IEEE 1149.1-2001], because it requires only four or five dedicated pins regardless of what testability logic is to be implemented. A few others that start to gain popularity include the IEEE 1149.6 boundary-scan standard for advanced digital networks [IEEE 1149.6-2003] and the IEEE 1500 embedded core-test standard [IEEE 1500-2005]. For more information on these emerging standards, refer to [Wang 2006a, 2007].

7.6 EXERCISES

7.1. (Lock-Up Latch) Suppose that a scan chain is configured as $SI \rightarrow SFF_1 \rightarrow SFF_2 \rightarrow SFF_3 \rightarrow SFF_4 \rightarrow SFF_5 \rightarrow SO$, where SFF_1 through SFF_5 are muxed-D scan cells, and SI and SO are the scan input pin and scan output pin, respectively. Suppose that this scan chain fails scan shift verification, in which the flush test sequence $<t_1\ t_2\ t_3\ t_4\ t_5> = <01010>$ is applied but the response sequence is $<r_1\ r_2\ r_3\ r_4\ r_5> = <01100>$. Identify the scan flip-flops that may have caused this failure, and show how to fix this problem by use of a lock-up latch.

7.2. (Lock-Up Latch) A scan chain may contain both positive-edge-triggered and negative-edge-triggered muxed-D scan cells. If, by accident, all positive-edge-triggered scan cells are placed before all negative-edge-triggered muxed-D scan cells, show how to stitch them into one single scan chain. (*Hint*: Positive-edge–triggered muxed-D scan cells and negative-edge–triggered muxed-D scan cells should be placed into two separate sections.)

7.3. **(Lock-Up Latch)** Refer to Figure 7.11. The scheme works only when the clock skew between CK_1 and CK_2 is less than the width (duty cycle) of the clock pulse. If CK_2 is delayed more than the duty cycle of CK_1 (*i.e.*, CK_1 and CK_2 become nonoverlapping), show whether or not it is possible to stitch the two cross-clock-domain scan cells into one single scan chain with a lock-up latch. If not, can it be done with a lock-up flip-flop instead?

7.4. **(Scan Stitching)** Use examples to show why a scan chain may not be able to perform the shift operation properly if two neighboring scan cells in the scan chain are too close to or too far from each other. Also describe how to solve these problems.

7.5. **(Test Signal)** Describe the difference between the test mode signal *TM* and the scan enable signal *SE* used in scan testing.

7.6. **(Clock Grouping)** Show an algorithm to find the smallest number of clock groups in clocking grouping.

7.7. **(BIST Design Rules)** A scan design can contain many asynchronous set/reset signals that may require adding two or more set/reset clock points to break all ripple set/reset loops. A ripple set/reset loop is a combinational feedback loop. Assume that the design now contains two system clocks (CK_1 and CK_2) and two set/reset clocks ($SRCK_1$ and $SRCK_2$). Derive two BIST timing control diagrams, including a scan enable (*SE*) signal, to test all data faults and set/reset faults controlled by these four clocks. Explain which timing control diagram can detect more faults.

7.8. **(BIST Design Rules)** Design a one-hot decoder for testing a tristate bus with four independent tristate drivers in BIST mode.

7.9. **(BIST Design Rules)** Design an X-bounding circuit for improving the fault coverage of a bidirectional I/O port by forcing it to input mode during BIST operation.

7.10. **(Aligned Skewed-Load *versus* Aligned Double-Capture)** Assume there are four synchronous clock domains each controlled by a capture clock CK_1, CK_2, CK_3, or CK_4, and each operated at a frequency $F_1 = 2 \times F_2 = 4 \times F_3 = 8 \times F_4$. Derive BIST timing control diagrams with aligned skewed-load and aligned double-capture to test all intra-clock-domain and inter-clock-domain delay faults. Specify by arrows the delay faults that can be detected in the diagram.

7.11. **(Staggered Skewed-Load *versus* Staggered Double-Capture)** Assume there are four asynchronous clock domains each controlled by a capture clock CK_1, CK_2, CK_3, or CK_4, and each operated at a frequency $F_1 > F_2 > F_3 > F_4$. Derive BIST timing control diagrams with staggered skewed-load and staggered double-capture to test all intra-clock-domain and inter-clock-domain delay faults. Specify by arrows the delay faults that can be detected in the diagram.

7.12. **(Hybrid Double-Capture)** Assume there are four mixed synchronous and asynchronous clock domains controlled by a capture clock, CK_1, CK_2, CK_3, and CK_4, operating at $F_1 = 100$ MHz, $F_2 = 50$ MHz, $F_3 = 60$ MHz, and $F_4 = 30$ MHz, respectively. Derive a BIST timing control diagram with a hybrid double-capture scheme composed of staggered double-capture and aligned double-capture to test all intra-clock-domain and inter-clock-domain delay faults. Specify by arrows the delay faults that can be detected in the diagram.

7.13. **(RTL Testability Enhancement)** Read the following Verilog HDL code and draw its schematic. Then determine whether there is any scan design rule violation. If there is any violation, then modify the RTL code to fix the problem, and draw the schematic of the modified RTL code.

```
reg [3:0] tri_en;
always @(posedge clk)
begin
  case (bus_sel)
  0: tri_en[0] = 1'b1;
  1: tri_en[1] = 1'b1;
  2: tri_en[2] = 1'b1;
  3: tri_en[3] = 1'b1;
  endcase
end
assign dbus = (tri_en[0])? d1 : 8'bz;
assign dbus = (tri_en[1])? d2 : 8'bz;
assign dbus = (tri_en[2])? d3 : 8'bz;
assign dbus = (tri_en[3])? d4 : 8'bz;
```

7.14. **(A Design Practice)** Use the scan design rule checking programs and user's manuals provided on the companion Web site to show whether you can detect any asynchronous set/reset signal violation and bus contention. Try to redesign a Verilog circuit to include such violations. Then, fix the violations by hand, and see whether the problems have disappeared.

7.15. **(A Design Practice)** Use the scan synthesis programs and user's manuals provided on the companion Web site to convert the two ISCAS-1989 benchmark circuits s27 and s38417 [Brglez 1989] into scan designs. Perform scan extraction and then run Verilog flush tests and broadside-load tests on the scan designs to verify whether the generated testbenches pass Verilog simulation.

7.16. (A Design Practice) Use the logic BIST programs and user's manuals provided on the companion Web site to design the logic BIST system with staggered double-capture for the circuit given in Section 7.3.2. Report the circuit's BIST fault coverage at every 10,000 increments up to 100,000 pseudo-random patterns.

7.17. (A Design Practice) Repeat Exercise 7.16, but instead implement the two pseudo-random pattern generators, PRPG1 and PRPG2, with a 28-stage CA and a 25-stage CA, respectively, with the construction rules given in Table 3.6. Explain why the CA-based logic BIST system can or cannot reach higher BIST fault coverage than the LFSR-based logic BIST system given in Exercise 7.16.

7.18. (A Design Practice) Use the ATPG programs and user's manuals provided on the companion Web site to report the circuit's ATPG fault coverage when the logic BIST system is reconfigured in ATPG mode. If the BIST fault coverage in Exercise 7.16 is lower than the ATPG fault coverage, insert as many test points as needed in the logic BIST system to reach the ATPG fault coverage; alternately, run top-up ATPG in both ATPG compression and ATPG modes and report the circuit's final fault coverage.

ACKNOWLEDGMENTS

We thank Khader S. Abdel-Hafez of Synopsys and formerly of SynTest Technologies for providing a portion of the materials in the Scan Design Flow section, Professor Wen-Ben Jone of the University of Cincinnati and Dr. Ravi Apte of SynTest Technologies for reviewing the chapter, and Teresa Chang of SynTest Technologies for drawing most of the figures.

REFERENCES

R7.0 Books

[Gizopoulos 2006] D. Gizopoulos, editor, *Advances in Electronic Testing: Challenges and Methodologies,* Morgan Kaufmann, San Francisco, 2006.

[Keating 1999] M. Keating and P. Bricaud, *Reuse Methodology Manual for System-on-a-Chip Designs,* Springer, Boston, 1999.

[Wang 2006a] L.-T. Wang, C.-W. Wu, and X. Wen, editors, *VLSI Test Principles and Architectures: Design for Testability,* Morgan Kaufmann, San Francisco, 2006.

[Wang 2007] L.-T. Wang, C. E. Stroud, and N. A. Touba, editors, *System-on-Chip Test Architectures: Nanometer Design for Testability,* Morgan Kaufmann, San Francisco, 2007.

R7.1 Introduction

[SIA 2005] SIA, *The International Technology Roadmap for Semiconductors: 2005 Edition,* Semiconductor Industry Association, San Jose, CA, http://public.itrs.net, 2005.

[SIA 2006] SIA, *The International Technology Roadmap for Semiconductors: 2006 Update*, Semiconductor Industry Association, San Jose, CA, http://public.itrs.net, 2006.

R7.2 Scan Design

[Barbagallo 1996] S. Barbagallo, M. Bodoni, D. Medina, F. Corno, P. Prinetto, and M. Sonza Reorda, Scan insertion criteria for low design impact, in *Proc. IEEE VLSI Test Symp.*, pp. 26–31, April 1996.

[Cheung 1997] B. Cheung and L.-T. Wang, The seven deadly sins of scan-based designs, *Integrated System Design*, www.eetimes.com/editorial/1997/test9708.html, August 1997.

[Duggirala 2002] S. Duggirala, R. Kapur, and T. W. Williams, System and Method for High-Level Test Planning for Layout, U.S. Patent No. 6,434,733, August 13, 2002.

[Duggirala 2004] S. Duggirala, R. Kapur, and T. W. Williams, System and Method for High-Level Test Planning for Layout, U.S. Patent No. 6,766,501, July 20, 2004.

R7.3 Logic Built-In Self-Test (BIST) Design

[Al-Yamani 2002] A. A. Al-Yamani, S. Mitra, and E. J. McCluskey, *Avoiding Illegal States in Pseudorandom Testing of Digital Circuits*, Center for Reliable Computing, Technical Report (CRC TR) No. 02-2, Stanford University, December 2002.

[Brglez 1989] F. Brglez, D. Bryan, and K. Kozminski, Combinational profiles of sequential benchmark circuits, in *Proc. IEEE Int. Symp. on Circuits and Systems*, pp. 1929–1934, August 1989.

[Cheung 1997] B. Cheung and L.-T. Wang The seven deadly sins of scan-based designs, *Integrated System Design*, www.eetimes.com/editorial/1997/test9708.html, August 1997.

[Eichelberger 1977] E. B. Eichelberger and T. W. Williams, A logic design structure for LSI testability, in *Proc. ACM/IEEE Design Automation Conf.*, pp. 462–468, June 1977.

[IEEE 1149.1-2001] IEEE Std. 1149.1-2001, *IEEE Standard Test Access Port and Boundary Scan Architecture*, IEEE Press, New York, 2001.

[IEEE 1500-2005] IEEE Std. 1500-2005, *IEEE Standard for Embedded Core Test*, IEEE Press, New York, 2005.

[Rajski 2003] J. Rajski, A. Hassan, R. Thompson, and N. Tamarapalli, Method and Apparatus for At-Speed Testing of Digital Circuits, U.S. Patent Application No. 20030097614, May 22, 2003.

[Wang 2005a] L.-T. Wang, X. Wen, P.-C. Hsu, S. Wu, and J. Guo, At-speed logic BIST architecture for multi-clock designs, in *Proc. IEEE Int. Conf. on Computer Design*, pp. 475–478, October 2005.

[Wang 2006b] L.-T. Wang, X. Wen, K. S. Abdel-Hafez, S.-H. Lin, H.-P. Wang, M.-T. Chang, P.-C. Hsu, S.-C. Kao, M.-C. Lin, and C.-C. Hsu, Method and Apparatus for Unifying Self-Test with Scan-Test during Prototype Debug and Production Test, European Patent No. 1,364,436, May 24, 2006.

R7.4 RTL Design for Testability

[Aktouf 2000] C. Aktouf, H. Fleury, and C. Robach, Inserting scan at the behavioral level, *IEEE Design & Test of Computers*, 17(3), pp. 34–42, July 2000.

[Ghosh 2001] I. Ghosh and M. Fujita, Automatic test pattern generation for functional register-transfer level circuits using assignment decision diagrams, *IEEE Trans. on Computer-Aided Design*, 20(3), pp. 402–415, March 2001.

[Huang 2001] Y. Huang, C. C. Tsai, N. Mukherjee, O. Samoan, W.-T. Cheng, and S. M. Reddy, On RTL Scan Design, in *Proc. IEEE Int. Test Conf.*, pp. 728–737, November 2001.

[IEEE 1463-2001] IEEE Std. 1463-2001, IEEE Standard Description Language Based on the Verilog Hardware Description Language, IEEE Press, New York, 2001.

[Ravi 2001] S. Ravi and N. Jha, Fast test generation for circuits with RTL and gate-level views, in *Proc. IEEE Int. Test Conf.*, pp. 1068–1077, November 2001.

[Roy 2000] S. Roy, G. Guner, and K.-T. Cheng, Efficient test mode selection and insertion for RTL-BIST, in *Proc. IEEE Int. Test Conf.*, pp. 263–272, October 2000.

[Wang 2005b] L.-T. Wang, A. Kifli, F.-S. Hsu, S.-C. Kao, X. Wen, S.-H. Lin, and H.-P. Wang, Computer-Aided Design System to Automate Scan Synthesis at Register-Transfer Level Test, U.S. Patent No. 6,957,403, October 18, 2005.

[Zhang 2003] L. Zhang, I. Ghosh, and M. S. Hsiao, Efficient sequential ATPG for functional RTL circuits, in *Proc. IEEE Int. Test Conf.*, pp. 290–298, October 2003.

R7.5 Concluding Remarks

[IEEE 1149.1-2001] IEEE Std. 1149.1-2001, *IEEE Standard Test Access Port and Boundary Scan Architecture*, IEEE Press, New York, 2001.

[IEEE 1149.6-2003] IEEE Std. 1149.6-2003, *IEEE Standard for Boundary Scan Testing of Advanced Digital Networks*, IEEE Press, New York, 2003.

[IEEE 1500-2005] IEEE Std. 1500-2005, *IEEE Standard for Embedded Core-Test*, IEEE Press, New York, 2005.

Logic and circuit simulation

8

Jiun-Lang Huang
National Taiwan University, Taipei, Taiwan

Cheng-Kok Koh
Purdue University, West Lafayette, Indiana

Stephen F. Cauley
Purdue University, West Lafayette, Indiana

ABOUT THIS CHAPTER

Logic simulation and circuit simulation are typically used in conjunction with functional verification to verify the correctness of an integrated circuit. During the logic design stage, designers rely on logic simulation to verify whether the design meets its specifications and contains any design errors. During the circuit design stage, designers use circuit simulation to test and characterize digital cell libraries, memory models, and ***analog and mixed-signal*** (AMS) circuits that require detailed timing analysis to ensure the correct operation of these circuits.

This chapter begins with a discussion of logic simulation. After an introduction to the logic circuit models, the popular compiled-code and event-driven logic simulation techniques are described. This is followed by hardware-accelerated logic simulation that is commonly referred to as hardware emulation and is intended to bridge the growing gap between circuit complexity and software simulator efficiency. Commonly used hardware emulation techniques are introduced first, followed by a description of the two crucial ingredients of emulators: reconfigurable computing units and interconnection architectures. The second half of the chapter is devoted to circuit-level simulation. After describing the circuit simulation models and essential numerical methods, the chapter explains the procedures required to simulate ***very large-scale integration*** (VLSI) circuits with interconnects and nonlinear devices. By working through this chapter, the reader will learn about the major logic simulation, hardware emulation, and circuit simulation techniques. This background will be valuable in selecting the simulation method that best meets the design needs.

8.1 **INTRODUCTION**

Simulation empowers a designer to predict a design's behavior without physically implementing it. In the design phase, the main purpose of simulation is **design verification**. Figure 8.1 depicts the flow of using simulation for design verification. During each design stage, the functional specification documents the required functionality and performance for the design and a corresponding circuit description is generated in conformance with the given specification. To ensure conformance, verification testbenches consisting of a set of input stimuli and expected output responses are created. The simulator then takes the circuit description and the input stimuli as inputs and produces the simulated responses. Any discrepancy between the simulated and expected responses (detected by the response analysis process) indicates that redesign or modification is necessary. Once the circuit has been verified to an acceptable confidence level, the design process advances to the next design stage.

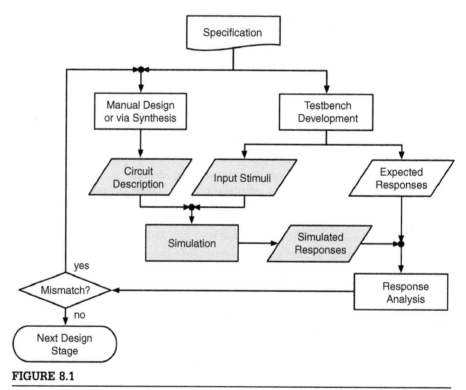

FIGURE 8.1

Simulation for design verification.

Table 8.1 Design Descriptions at Different Circuit Abstraction Levels

Design Stage	Circuit Description
Behavioral Level or Electronic System Level (ESL)	C/C++, SystemC [SystemC 2008] SystemVerilog [SystemVerilog 2008]
Register-Transfer Level (RTL)	Verilog [IEEE 1463–2001] [Thomas 2002] VHDL [IEEE 1076–2002]
Gate Level	Gate-Level Netlists
Circuit Level	Transistor-Level Schematics

8.1.1 Logic simulation

Digital circuit simulation can be performed at different abstraction levels—from the highest behavioral level to the lowest device level. At each level, a suitable description language that captures the required functional specification is used to describe the design. Table 8.1 lists the commonly used abstraction levels and the corresponding circuit descriptions.

In general, design verification begins at the behavioral level or *electronic-system level* (ESL) where the algorithm correctness and system throughput are the major concerns. Then, at the *register-transfer level* (RTL), the design is described in terms of blocks such as registers, counters, data processing units, and controllers, as well as the data/control flow between these blocks. Because ESL/RTL verification usually does not involve detailed timing analysis, ESL or RTL design verification is also referred to as **functional verification** [Wile 2005].

Logic/scan synthesis comes into play after the RTL design stage. The gate-level netlist corresponding to the RTL design, which includes scan cells, is synthesized with logic elements provided in a cell library. Finally, the transistor-level description provides the most accurate model of the design. However, because it is much slower than gate-level simulation, transistor-level simulation is usually used only for characterizing timing critical paths and library cells.

This chapter first discusses gate-level logic simulation. Although digital circuits use two-valued Boolean algebra as the underlying mathematics, most logic simulators include two more values, unknown (u) and high-impedance (Z), to handle the inevitable uncertainties in practical circuits. Understanding the capabilities and limitations of the 4-valued logic system prevents the users from incorrectly interpreting the simulation results. Furthermore, to deal with timing, delay information must be incorporated into the logic element descriptions. Thus, both gate and wire delays must be taken into account for modern designs. The two major logic simulation techniques are **compiled-code simulation** and **event-driven simulation**. Grounded in distinct principles, each of them has its own advantages and drawbacks and finds applications in different areas of the design process.

8.1.2 **Hardware-accelerated logic simulation**

As the circuit complexity continues growing, logic simulation becomes the bottle-neck of design verification—available logic simulators are too slow for practical *system-on-chip* (SOC) designs or *hardware/software* (HW/SW) **co-simulation** applications. Several types of **hardware-accelerated logic simulation** techniques have been proposed, including **simulation acceleration, (in-circuit) emulation**, and **hardware prototyping**, each of which has its advantages and shortcomings. A modern emulator may be a hybrid of the preceding types or be able to execute several types to meet the requirements of different design stages.

Most emulator systems consist of arrays of reconfigurable logic computing units that are directly or indirectly interconnected. Although *field program-mable gate array* (FPGA) is a natural choice for the computing unit, the emulation system performance is severely limited by the available *input/output* (I/O) pins. Indirect interconnect architectures such as full and partial crossbars and time-multiplexed I/O are possible solutions to improve the inter-chip data bandwidth. Other approaches include exploring different use models of FPGA and the use of programmable processors as the reconfigurable computing units.

8.1.3 **Circuit simulation**

Circuit simulation is an increasingly indispensable tool for the design of *integrated circuits* (ICs). The turnaround time and cost of fabrication, along with the sheer number of design parameters under consideration, prevent circuit designers from relying on intuition and extensive experimentation to meet their design specifications. Instead, designers can use an understanding of the dynamic behavior for their circuits, learned through circuit simulation, to save both time and resources for fabrication.

The simulation of ICs is a very structured area that is grounded in the first principles of **current** and **voltage** relationships. The process of simulating a circuit begins with the "modeling" of each element from the circuit in terms of basic building blocks such as current and voltage sources, resistors, capacitors, and inductors. The parameters for each element in the model may be time-varying or time-invariant. The goal of these models is to accurately mimic the dynamic behavior of the elements while providing the simplest possible representation. Specifically, with these models, the designer can easily construct a set of current and voltage relationships that describe the behavior of the entire circuit.

There are several different representations of circuit equations that primarily rely on the *Kirchhoff's voltage law* (KVL) and *Kirchhoff's current law* (KCL) in the formulations. Although the behavior of some very simple circuits can be described analytically, we must investigate general numerical techniques for even modest sized ICs. The scalability of these techniques is crucial when considering the growth of modern designs. Thus, in practice, the modeling of circuit elements and the subsequent formulation of circuit equations should be performed while keeping in mind the computation load required for the resulting numerical techniques.

8.2 **LOGIC SIMULATION MODELS**

In this section, we discuss the gate-level simulation models for combinational and sequential networks, which have widespread acceptance in the integrated circuit community.

8.2.1 **Logic symbols and operations**

In addition to 1 and 0, logic simulators often include two more symbols: u (unknown) and Z (high-impedance); the former represents the uncertain circuit behavior, and the latter helps resolve the behavior of tristate logic. For cases in which 0, 1, u, and Z are insufficient to meet the required simulation accuracy, intermediate logic states that incorporate both value and strength may be used.

8.2.1.1 *"1" and "0"*

The basic mathematics for most digital systems is the two-valued Boolean algebra. In two-valued Boolean algebra, a variable can assume only one of the two values, *true* or *false*, which are represented by the two symbols 1 and 0, respectively. Note that 1 and 0 here do not represent numerical quantities. Physical representations of the two symbols depend on the logic family of choice. Consider the most popular CMOS logic as an example; the two symbols 1 and 0 represent two distinct voltage levels, *power* (V_{DD}) and *ground* (V_{SS}), respectively. Here we assume positive logic is used. Whether a signal's value is 1 or 0 depends on which voltage source it is connected to.

8.2.1.2 *The unknown value* u

Almost all practical digital circuits contain memory elements (*e.g.*, flip-flops and memories) to store the circuit state; however, when these circuits are powered up, the initial states of their memory elements are usually unknown. To handle such situations, the logic symbol u is introduced to indicate an *unknown* logic value. By associating u with a signal, we mean that the signal is 1 or 0, but we are not sure which one is the actual value.

8.2.1.3 *The high-impedance state* Z

Until now, the logic signal states that we have discussed are 1 and 0, indicating that the signal is connected to either V_{DD} or V_{SS}. (The unknown symbol indicates uncertainty; however, the signal of interest is still 1 or 0.) In addition to 1 or 0, tristate gates have a third, high-impedance state, denoted by logic symbol Z. Tristate gates permit several gates to time-share a common wire, called a *bus*. A signal is in the Z state if it is connected to neither V_{DD} nor V_{SS}.

Figure 8.2 depicts a typical bus application. In this example, three bus drivers (G_1, G_2, and G_3) drive the bus wire y. Each driver G_i is controlled by an *enable* signal e_i, and its output o_i is determined as follows:

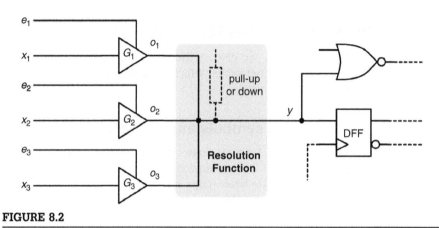

FIGURE 8.2

A tristate circuit example.

$$o_i = \begin{cases} x_i & \text{if } e_i = 1 \\ Z & \text{if } e_i = 0 \end{cases}$$

When $o_i = Z$, G_i has no effect on the bus wire y, leaving the control to other drivers.

A **bus conflict** occurs if at least two drivers drive the bus wire to opposite binary values. Such situations may cause the circuit to be permanently damaged. In addition to design errors, abnormal bus states could occur during testing when the circuit is not in its normal operating environment and may receive illegal input sequences. On the other hand, if no driver is activated, the bus is in a **floating state,** because it is not connected to V_{DD} or V_{SS}. A pull-up or pull-down network that connects the bus to V_{DD} or V_{SS} by means of a resistor may be added to provide a default 1 or 0 logic value (Figure 8.2); otherwise, the bus wire will retain its previous value as a result of trapped charge in the parasitic wire capacitance. Because the charge could decay in time, a **bus keeper** consisting of two weak inverters is usually added to the bus wire to keep its previous value as long as the bus is in the floating state.

8.2.1.4 *Basic logic operations*

The input/output relationships of the three basic logic operations (AND, OR, and NOT) that use the four logic symbols (0, 1, u, and Z) are summarized in Table 8.2. It is worth noting that:

1. Simulation results based on these truth tables are pessimistic (*i.e.*, a signal may be reported as unknown even though its value can be uniquely determined as 0 or 1 [Breuer 1972]).
2. The four symbols are insufficient to handle intermediate signal values (*e.g.*, values in the middle of V_{DD} and V_{SS}, which may occur in tristate buses, switch-level networks, or defective circuits). To enhance the

Table 8.2 Truth Tables of AND, OR, and NOT

AND	0	1	*u*	*z*	OR	0	1	*u*	*z*	NOT	0	1	*u*	*z*
0	0	0	*u*	0	**0**	0	1	*u*	*u*		1	0	*u*	*u*
1	0	1	*u*	*u*	**1**	1	1	1	1					
u	0	*u*	*u*	*u*	***u***	*u*	1	*u*	*u*					
z	0	*u*	*u*	*u*	***z***	*u*	1	*u*	*u*					

```
table
    //  select      a       b       :       out
    //  0           0       ?       :       0
    //  0           1       ?       :       1
    //  1           ?       0       :       0
    //  1           ?       1       :       1
    //  ?           0       0       :       0
    //  ?           1       1       :       1
endtable
```

FIGURE 8.3

The truth table of a two-input multiplexer.

simulation resolution, intermediate logic states that incorporate both signal value and strength may be used [Miczo 2003].

To support customized logic elements with complex sequential or combinational behavior, modern simulation tools support ***user-defined primitives*** (UDPs). Figure 8.3 shows how Verilog models a two-input multiplexer with a truth table. Here, *a* and *b* are the data inputs, *select* is the control input, and *out* is the multiplexer output. The symbol "?" is a shorthand notation for 0, 1, or *u*.

8.2.2 Timing models

Delay is a fact of life for all electrical components, including logic gates and interconnection wires. In this section, we discuss the commonly used gate and wire delay models.

8.2.2.1 *Transport delay*

The **transport delay** refers to the time duration it takes for the effect of gate input changes to appear at gate outputs. Several transport delay models characterize this phenomenon from different aspects.

The **nominal delay** model specifies the same delay value for the output rising and falling transitions. Consider the AND gate G in Figure 8.4 as an example. Here B is fixed at 1; thus, the output of G is only affected by A. Assuming that G has a nominal delay of $d_N = 2$ ns and A is pulsed to 1 for 1 ns, the

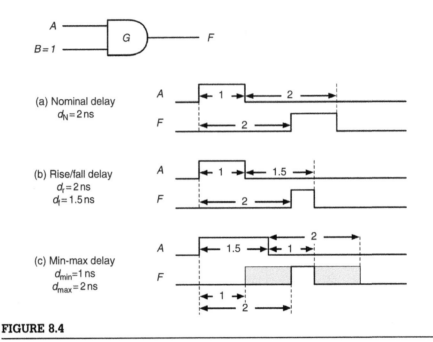

FIGURE 8.4

Transport delay models.

corresponding simulation result is shown in Figure 8.4a. Under the nominal delay model, the output waveform at F is simply a version of A delayed by 2 ns.

For cases in which the rising and falling times are different (*e.g.*, the pull-up and pull-down transistors of the gate have different driving strengths), one may opt for the **rise/fall delay** model. In Figure 8.4b, the setup is the same as that in Figure 8.4a except that the rise/fall delay model is used instead; the rise and fall delays are $d_r = 2$ ns and $d_f = 1.5$ ns, respectively. Because of the difference between the two delays, the duration of the output pulse shrinks from 1 to 0.5 ns.

If the gate transport delay cannot be uniquely determined (*e.g.*, because of process variations), one may use the **min–max delay** model. In the min-max delay model, the minimum and maximum gate delays (d_{min} and d_{max}) are specified to represent the ambiguous time interval in which the output change may occur. In Figure 8.4c, the minimum and maximum delays are 1 and 2 ns, respectively, and a 1.5-ns pulse is applied at A. In response to the delay uncertainty, two ambiguous intervals (the shaded regions), corresponding to the rising and falling transitions, are observed at output F. Within the two ambiguous intervals, the exact output value is unknown.

8.2.2.2 *Inertial delay*

The **inertial delay** is defined as the minimum input pulse duration necessary for the output to switch states. Pulses shorter than the inertial delay cannot pass

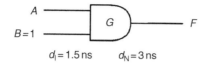

$d_I = 1.5\,\text{ns}$ $d_N = 3\,\text{ns}$

(a) Pulse duration less than d_I

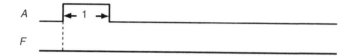

(a) Pulse duration longer than d_I

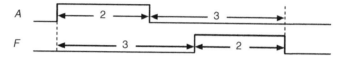

FIGURE 8.5

Inertial delay.

through the circuit element. The inertial delay models the limited bandwidth of logic gates. Figure 8.5 illustrates this filtering effect. Assume that the AND gate has an inertial delay (d_I) of 1.5 ns and a nominal delay of 3 ns. Let us fix B at 1 and apply a pulse on A. In Figure 8.5a, the 1-ns pulse is filtered and the output remains at a constant 0. In Figure 8.5b, the pulse is long enough (2 ns) and an output pulse is observed 3 ns later.

8.2.2.3 *Functional element delay model*

Functional elements, such as flip-flops, have more complicated behaviors than simple logic gates and require more sophisticated timing models. In Table 8.3, the I/O delay model of the positive-edge-triggered D flip-flop is depicted. Take the asynchronous preset operation (second row) as an example. Regardless of the *Clock* and *D* values, if the current flip-flop state (q) is 0 and *ClearB* remains 1, changing *PresetB* from 1 to 0 (denoted by the down arrow) will cause output transitions at Q and QB after 1.6 and 1.8 ns, respectively. Besides the input-to-output transport delay, the flip-flop timing model usually contains timing constraints, such as setup/hold times and inertial delays for each input.

8.2.2.4 *Wire delay*

Figure 8.6a illustrates the distributed RLC model of a metal wire. In the presence of the passive components, it takes finite time, called the **propagation delay**, for a signal to travel from point p to point q.

In general, wire delays are specified for each connected gate output and gate input pair, because the physical distances and thus the propagation delays between the driver and receiver gates vary. In Figure 8.6b, the inverter output

Table 8.3 The D Flip-Flop I/O Delay Model

Input Condition				Present State	Outputs		Delay (ns)		
D	Clock	PresetB	ClearB	q	Q	QB	to Q	to QB	Comments
X	X	↓	1	0	↑	↓	1.6	1.8	Asynchronous preset
X	X	1	↓	1	↓	↑	1.8	1.6	Asynchronous clear
1	↑	1	1	0	↑	↓	2	3	Q: 0→1
0	↑	1	1	1	↓	↑	3	2	Q: 1→0

Note: X indicates "don't care."

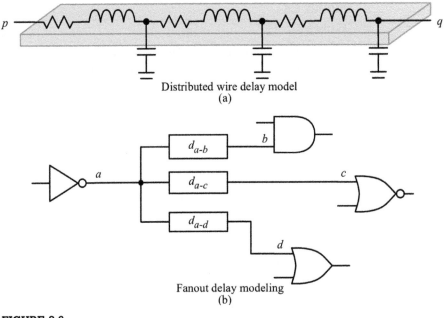

Distributed wire delay model
(a)

Fanout delay modeling
(b)

FIGURE 8.6

Wire delay model.

a branches out to drive three gates. To model the wire delays associated with the three signal paths, one may insert delay elements d_{a-b}, d_{a-c}, and d_{a-d} into the fanout branches. For convenience, wire delays may also be viewed as the receiver gate input delays and become part of the receiver gate delay model.

8.3 LOGIC SIMULATION TECHNIQUES

The general model of a gate-level or RTL network consists of the combinational network and memory elements as depicted in Figure 8.7. In this figure, X and Z denote the *primary inputs* (PI) and *primary outputs* (PO), and Q and Q^+ denote the present and next states of the flip-flops. Q and Q^+ are also called the *pseudo primary inputs* (PPI) and *pseudo primary outputs* (PPO), respectively.

For ease of illustration, we assume that a single clock and D-type flip-flops are used. The simplified synchronous sequential circuit simulation flow is depicted in Figure 8.8. At the beginning of each clock cycle, the simulator evaluates the flip-flops and then makes the necessary updates (*i.e.*, $Q \leftarrow Q^+$). Then, the input vector is read in and the combinational part is evaluated. The simulation continues until the specified simulation time has been reached or there is no more input vector left.

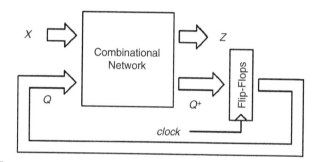

FIGURE 8.7

A general sequential circuit model.

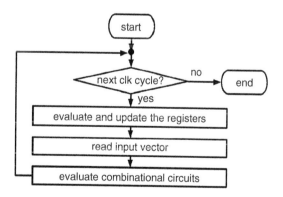

FIGURE 8.8

The simplified synchronous circuit simulation flow.

In this section, we will discuss two commonly used logic simulation methods: *compiled-code simulation* and *event-driven simulation*. Although the circuits are combinational in the discussions, these techniques can be easily extended to deal with sequential circuits.

8.3.1 Compiled-code simulation

The idea of *compiled-code simulation* is to translate the digital circuit into a series of machine instructions that model the functions of individual gates and the interconnects between them.

8.3.1.1 *Preprocessing*

In practice, logic optimization and levelization are performed before the actual code generation process. The purpose of **logic optimization** is to enhance the simulation efficiency. A typical optimization process consists of the transformations illustrated in Figure 8.9 [Wang 1987]. Because each gate corresponds to one or more statements in the compiled code, logic optimization reduces the program size and execution time.

To avoid unnecessary computations, logic gates must be evaluated in an order such that a gate will not be evaluated until all its driving gates have been evaluated. The logic levelization algorithm starts by assigning all the PI's and

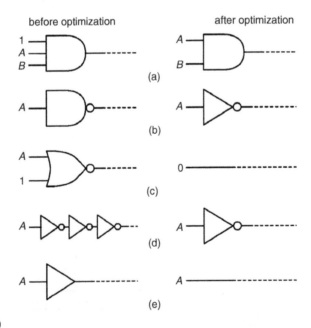

FIGURE 8.9

Logic optimization for compiled-code simulation.

PPI's level 0. For each logic element, its level is equal to the maximum of its driving elements' levels plus 1.

8.3.1.2 *Code generation*

Depending on performance, portability, and maintainability needs, different code generation techniques may be used [Wang 1987]. The three approaches are interpreted code, high-level programming language source code, and native machine code.

In the interpreted code approach, the target machine is a software emulator. During simulation, the instructions are interpreted and executed one at a time. This approach offers the best portability and maintainability at the cost of reduced performance. Mapping the simulated digital network to a high-level programming language such as C enhances performance and is portable to any target machine that has a C compiler. The compilation time could be a severe limitation for fault simulators that require recompilation for each faulty circuit. The native machine code approach generates the native machine code directly without the need of compilation; this makes it a more viable solution to fault simulation. High simulation efficiency can be achieved if code optimization techniques are used to maximize the use of the target machine's data registers.

Take the network in Figure 8.10 as an example. The generated pseudocode is shown in the following. In the actual implementation, each statement is replaced with the corresponding language constructs or machine instructions.

```
while true
  read(A, B, C);
  E = OR(B, C);
  H = AND(A, E);
  J = NOT(E);
  K = NOR(H, J);
  output(K);
end
```

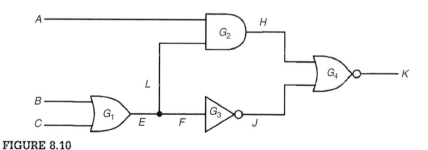

FIGURE 8.10

An example network for native machine code generation.

8.3.1.3 *Applications*

The main limitation of compiled-code simulation is that it is incapable of timing modeling. As a result, it fails to detect timing problems such as glitches and race conditions. Despite the limitations, compiled-code simulation finds its application in **cycle-based simulation,** where only the logic function correctness is of interest and zero-delay model is used. Compiled-code simulation is most effective when binary logic simulation suffices. In such cases, machine instructions are readily available for Boolean operations (*e.g.*, AND, OR, and NOT). Further speedup is possible with the bit-wise logic operation instructions that allow concurrent simulation of independent vectors or vector sequences. With 3 or more logic symbols, which is usually the case, the logic evaluation processes are more complicated but still manageable.

8.3.2 **Event-driven simulation**

In contrast to compiled-code simulation, *event-driven simulation* exhibits high simulation efficiency by performing gate evaluations only when necessary. We will use Figure 8.11 to illustrate the event-driven simulation concept. In this example, two consecutive input patterns $ABC = 001$ and 111 are applied to the circuit, and the corresponding circuit values are shown. Note that the application of the second vector does not change the input of G_3, so G_3 is not evaluated for the second vector. In event-driven simulation, the switching of a signal's value is called an **event**, and an event-driven simulator monitors the occurrences of events to determine which logic elements to evaluate.

8.3.2.1 *Zero-delay event-driven simulation*

Figure 8.12 depicts the zero-delay event-driven simulation flow. (A zero-delay simulation is one in which gates and interconnections are assumed to have zero delay.) At the beginning of the simulation flow, the initial signal values, which may be given or simply unknown, are read in and assigned. Then, a new input vector is loaded and the primary inputs at which events occur (called active PIs) are identified. To propagate the events toward primary outputs, gates driven by

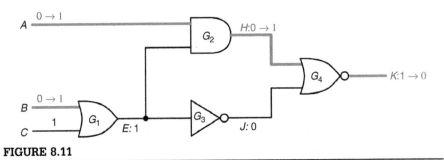

FIGURE 8.11

Signal transitions between consecutive input vectors.

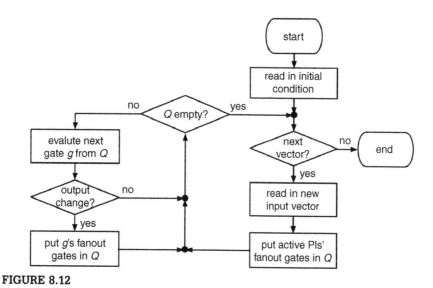

FIGURE 8.12

The zero-delay event-driven simulation flow.

active primary inputs are put in the event queue Q, which stores the gates to be evaluated. As long as Q is not empty, a gate g is dequeued from Q and evaluated. If the output of g changes (*i.e.*, a new event occurs), the fanout gates of g are placed in Q. When Q becomes empty, the simulation for the current input vector is finished, and the simulator proceeds to process the next input vector.

8.3.2.2 *Nominal-delay event-driven simulation*

The greatest advantage of event-driven simulation over compiled-code simulation is that it can handle any delay model. A sophisticated **event scheduler** keeps track of event occurrences and schedules the necessary gate evaluations at the proper time points. Because events must be evaluated in chronological order, the scheduler is implemented as a priority queue.

Figure 8.13 depicts one possible priority queue implementation for a nominal delay event-driven simulator. In the priority queue, the vertical list is an ordered list that stores the time stamps when events occur. Attached to each time stamp t_i is a horizontal list of events that occur at time t_i. During simulation, a new event that will occur at time t_i is appended to the event list of time stamp t_i. For example, in Figure 8.13, the value of signal w will switch to v_w^+ at t_i. If t_i is not in the time stamp list yet, the scheduler will first place it in the list according to the chronological order.

For the priority queue scheduler in Figure 8.13, the time needed to locate a time stamp to insert an event grows with the circuit size. To improve the event scheduler efficiency, one may use, instead of a linked list, an array of evenly spaced time stamps. Although some entries in the array may have empty event

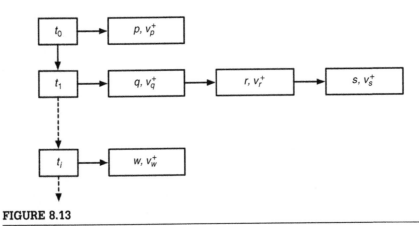

FIGURE 8.13

Priority queue event scheduler.

lists, the overall search time is reduced, because the target time stamp can be indexed by its value. Further enhancement is possible with the concept of **timing wheel** [Ulrich 1969]. Let the time resolution be one time unit and the array size M. A time stamp that is d time units ahead of current simulation time (with array index i) is stored in the array and indexed by $(i + d)$ modulo M if d is less than M; otherwise, it is stored in an overflow remote event list similar to that shown in Figure 8.13. The array is referred to as the timing wheel because of the modulo-M-induced circular structure. Remote event lists are brought into the timing wheel once their time stamps are within M-1 time units from current simulation time.

A two-pass strategy for nominal delay event-driven simulation is depicted in Figure 8.14. When there are still events with future time stamps to process, the event list L_E of next time stamp t is retrieved. L_E is processed in a two-pass manner. In pass one (the left shaded box), the simulator determines the set of gates to be evaluated. The notation (g, v_g^+) indicates that the output of gate g is to become v_g^+. For each event (g, v_g^+), if v_g^+ is the same as g's current value v_g, this event is false and is discarded. On the other hand, if $v_g^+ \neq v_g$, i.e., (g, v_g^+) is a valid event, then v_g is updated to v_g^+, and the fanout gates of g are appended to the activity list L_A. In the second pass (the right shaded box), gates are evaluated and new events are scheduled. As long as the activity list L_A is non-empty, a gate g is retrieved and evaluated. Let the evaluation result be v_g^+. The scheduler will schedule the new event (g, v_g^+) at time stamp $t + \text{delay}(g)$, where $\text{delay}(g)$ denotes the nominal delay of gate g. The two-pass strategy avoids repeated evaluation of gates with events on multiple inputs.

In the following, we will use the circuit in Figure 8.10 to demonstrate the two-pass event-driven strategy. In this example, the nominal delays for G_1, G_2, G_3, and G_4 are 8, 8, 4, and 6 ns, respectively, and there are four input events: $(A, 1, 0)$, $(C, 0, 2)$, $(B, 0, 4)$, and $(A, 0, 8)$, where the notation (w, v_w^+, t)

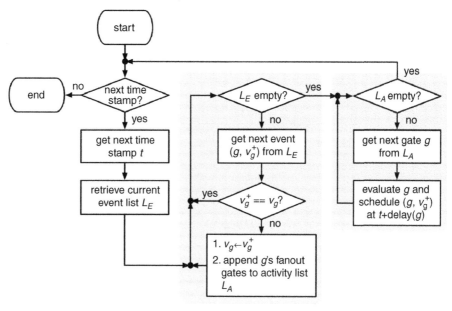

FIGURE 8.14

The two-pass nominal-delay event-driven simulation flow.

represents the event that signal w switches to v_w^+ at time t. The simulation progress is shown in Table 8.4. At time 0, there is only one primary input event $(A, 1)$. Because A drives G_2, G_2 is added to activity list L_A. Evaluation of G_2 returns $H = 1$; therefore, the event $(H, 1)$ is scheduled at time 8 (*i.e.*, 8 ns, the delay of G_2 after the current time). At time stamps 2 and 4, the two input events at C and B are processed in the same way. There are two events at time 8: the input event $(A, 0)$ and the scheduled event $(H, 1)$ from time stamp 0. Because both events are valid, the two affected gates, G_2 and G_4, are put in L_A for evaluation. The corresponding events $(H, 0)$ and $(K, 0)$ are scheduled at times 16 and 14, respectively. Note that the event $(E, 1)$ at time 10 is false, because it does not cause a signal transition; therefore, no gate evaluation is performed.

8.4 HARDWARE-ACCELERATED LOGIC SIMULATION

As the IC density and complexity continue growing, verifying the correctness of a new design before its first silicon has become the key to success. Although versatile and accurate, logic simulation is too slow for large designs, not to mention SOC designs that necessitate *hardware/software* (HW/SW) co-simulation.

Various hardware-acceleration techniques have been developed to bridge the gap between the IC complexity and logic simulation efficiency. A simplified block diagram of an FPGA-based hardware emulator is illustrated in Figure 8.15.

Table 8.4 A Two-Pass Event-Driven Simulation Example

Time	L_E	L_A	Scheduled Events
0	{(A,1)}	{G_2}	{(H,1,8)}
2	{(C,0)}	{G_1}	{(E,1,10)}
4	{(B,0)}	{G_1}	{(E,0,12)}
8	{(A,0),(H,1)}	{G_2,G_4}	{(H,0,16),(K,0,14)}
10	{(E,1)}		
12	{(E,0)}	{G_2,G_3}	{(H,0,20),(J,1,16)}
14	{(K,0)}		
16	{(H,0),(J,1)}	{G_4}	{(K,0,22)}
20	{(H,0)}		
22	{(K,0)}		

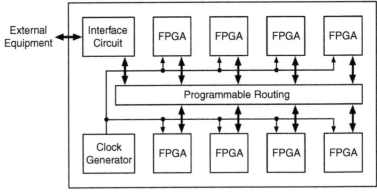

FIGURE 8.15

A simplified hardware emulator block diagram.

(In some hardware emulators, programmable ***application-specific integrated circuits*** [ASICs] are used instead of FPGAs.) The hardware emulator (called emulator hereafter) includes a circuit board holding a set of FPGAs, a routing system, and an interface circuit. For the emulator to emulate an IC, an external host computer programs each FPGA to emulate a portion of the IC and programs the routing system to route inter-FPGA signals. External test equipment or the target system board can then verify the emulated IC by supplying test signals to the FPGAs and monitoring the output responses from the FPGAs by means of the interface circuit.

The fundamental differences between software simulators and hardware emulators are as follows.

1. A logic simulator executes the RTL code or evaluates the logic elements serially. An emulator, on the other hand, executes the whole design concurrently or in massive parallelism.
2. Logic simulators are more flexible in terms of supported logic symbols and timing models. Emulators are basically 2-state machines and are more suitable for verifying logic correctness (*e.g.*, cycle-based simulation).
3. Logic simulators support a rich set of debugging capabilities. For example, one can stop the design at the middle of a cycle or even return to a previous state if stored. Emulators in general provide limited signal observability, although 100% visibility is possible at the cost of execution speed and hardware resources.

This section will introduce the commonly used hardware acceleration methods and the supporting technologies.

8.4.1 **Types of hardware acceleration**

Among the different methods to imitate a logic design, logic simulation and silicon implementation represent the two extremes in terms of resemblance to the final silicon. In between, various hardware acceleration techniques have been proposed to provide different balances among simulation speed, debugging capability, compilation time, and cost.

Figure 8.16a depicts a typical verification setup in which the "verification environment" provides the input vectors to the "design under verification" and analyzes the design's output responses. As illustrated in Figure 8.16b, in "simulation acceleration," the synthesizable part of the design under verification is mapped into hardware and executed in the emulator, whereas the remaining portions, in general the verification environment and the behavior code of the design, are executed in the workstation. High-speed channels exist between the workstation and the emulator to transport the simulation vectors and the responses. Because the simulator on the workstation is slower than the emulator, it becomes the bottleneck. However, with optimized testbench and simulation techniques, the high-speed channels could become the bottleneck.

In emulation (Figure 8.16c), the verification environment is also executed on the emulator to further increase the simulation speed. The communication between the workstation and the emulator is on demand, for example, to display the execution process. To do so, both the verification environment and the design itself must be synthesizable, implying a restricted coding style to the synthesizable subset.

In-circuit emulation (ICE) shown in Figure 8.16d relies on external hardware, which is usually the target system board, to provide "live-stimuli" and thus provides a more realistic verification environment. The target system may be

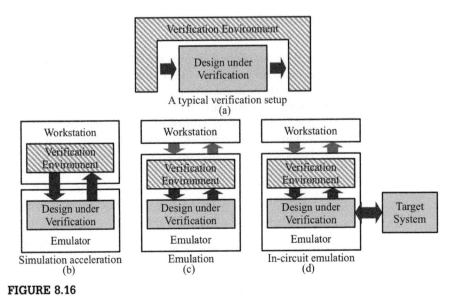

FIGURE 8.16

Simulation acceleration, emulation, and in-circuit emulation.

"static" or "dynamic." In the former case, the emulator supplies clock to the target system. Thus, one may stop and resume or slow down the simulation without causing problems. In the latter case, care must be taken to make sure that the whole system functions correctly (*e.g.*, to handle the differences in data and clock rate). As a result, the execution cannot be arbitrarily stopped, which limits debugging capability.

Hardware prototyping with FPGA is a well-known and widely adopted approach to verify the design correctness. The designers use the tool set provided by the FPGA vendors, including synthesis, placement, and routing, to map the design onto their FPGA technology. For small and medium-size ASIC's that can fit into a single FPGA, this approach is inexpensive and offers good simulation speed. However, as the number of FPGAs needed to realize the design increases, the partitioning and mapping process becomes cumbersome and error-prone.

The main ingredients of an emulator are the reconfigurable computing units programmed to perform the assigned tasks after design partitioning and the interconnection network that joins these computing units. They are discussed in the following sections.

8.4.2 Reconfigurable computing units

Today, the reconfigurable computing units for emulators are generally in the form of arrays of FPGAs or customized ASICs.

For an FPGA-based emulator, there are two FPGA use models. In one use model, the RTL design is partitioned into sub-circuits that can be fit into an FPGA. Then, each sub-circuit is synthesized at the gate level and mapped onto the target FPGA technology (Figure 8.17a). Because the gate count to I/O pin count ratios of the sub-circuits are very often greater than those of available commercial FPGAs, the FPGA logic is usually underused. As a result, the I/O pin resource becomes a severe limitation on this use model. Time-multiplexed interconnection schemes (Section 8.4.3.3) help relieve this I/O pin resource limitation.

In the second FPGA use model, the design is compiled into RCC (ReConfigurable Computing) elements [Lin 2002] (Figure 8.17b). Each RCC element is a small compact processor dedicated to perform one function (*e.g.*, Boolean expression, addition, multiplication, and case statement) at the RTL or gate level. This way, the design can be verified at different levels of abstraction, and the user does not have to debug at the gate level.

In [Beausoleil 1996], customized ASICs consisting of bit-wise Boolean processors are used as the computing units. The Boolean processors are designed to evaluate any n-input Boolean operations, and their inputs are selected with multiplexers. In this method, a circuit simulation cycle is divided into several emulator cycles, and the processors can perform different tasks and receive inputs from different sources in each emulator cycle according to the stored instructions. In Figure 8.18, the combinational network is simulated with three processors in 7 emulator cycles. The scheduling table shows one of the possible schedules. Processor 1, for example, samples A, B, and C in the first three emulator cycles and performs NOR(A, B, C) in the fourth emulator cycle.

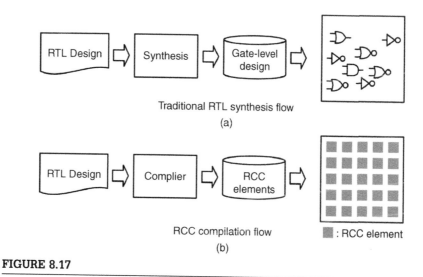

Traditional RTL synthesis flow
(a)

RCC compilation flow
(b)

FIGURE 8.17

Traditional and RCC synthesis/compilation flows.

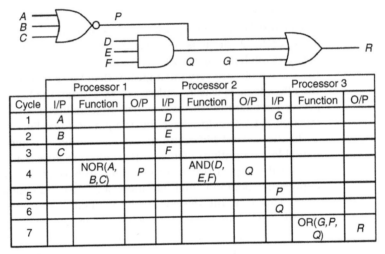

	Processor 1			Processor 2			Processor 3		
Cycle	I/P	Function	O/P	I/P	Function	O/P	I/P	Function	O/P
1	A			D			G		
2	B			E					
3	C			F					
4		NOR(A, B,C)	P		AND(D, E,F)	Q			
5							P		
6							Q		
7								OR(G,P, Q)	R

FIGURE 8.18

A processor scheduling example.

8.4.3 Interconnection architectures

The interconnection architectures can be divided into two categories: direct and indirect. In the former architecture, FPGA or ASIC chips are connected to each other directly through a fixed set of physical wires. In the latter architecture, dedicated routing chips are used to connect the chips, which relieves the computing units of inter-chip routing.

8.4.3.1 *Direct interconnection*

Because the computing units are usually arranged in a 2D array structure, the 2D-mesh-type direct interconnection architecture is an apparent choice. However, for the simple 2D-mesh interconnection scheme in Figure 8.19a, the valuable I/O pins may be used up by global interconnects routed through several computing units. In [Lin 2002], another mesh-type direct connection architecture (Figure 8.19b) is proposed. In this scheme, two "hops" and "jumps" are sufficient for any type of net; however, because each chip is used for routing as well as computing, the I/O pin resource limitation still exists.

In the direct interconnection architecture, the available I/O pin resource severely limits the logic use and the system routability. Because the chip I/O pin count grows at a slower rate than the gate count, either some logic resources must be used to route signals or some logic resources are wasted. Both indirect/time-multiplexed interconnection schemes and dynamically reconfigurable or programmable computing units help relieve this problem.

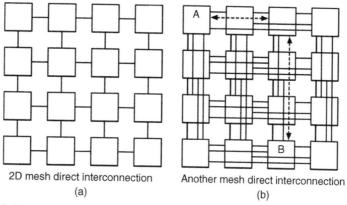

FIGURE 8.19

Direct interconnection architectures.

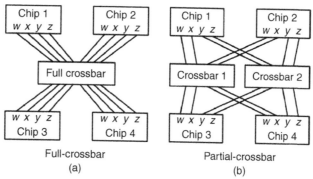

FIGURE 8.20

The full and partial-crossbar schemes.

8.4.3.2 *Indirect interconnect*

The crossbar-based indirect interconnection scheme can be used to relieve the computing units of inter-chip routing. In the full-crossbar configuration (Figure 8.20a), one full-crossbar chip, after programming, is able to make connections between any two pins; however, this scheme soon runs out of steam, because the crossbar chip size increases quadratically with the total pin count. In the partial-crossbar configuration [Varghese 1993], the I/O pins are divided into groups, and one crossbar is assigned to connect pins of the same group. In Figure 8.20b, pins are divided into two groups (w and x in one group; y and z in the other), and two crossbars are used to realize intragroup connections. To further reduce the crossbar size, one may use the multilevel partial crossbar configuration.

8.4.3.3 *Time-multiplexed interconnect*

Alternatively, time-multiplexed interconnection schemes are used to overcome the pin resource limitation by dividing the pin bandwidth among several interchip logical signals.

The virtual wire technique in [Babb 1997] exploits the fact that logic values between circuit partitions only need to be transmitted once and that circuit communication patterns repeat in a predictable fashion.[1] As shown in Figure 8.21a, a physical wire is multiplexed among pipelined shift registers. Each register in the pipeline carries a single bit of information from one logical output to the corresponding logical input in the neighboring FPGA. Figure 8.21b illustrates the phase-based virtual wire operating principle. The simulation clock is divided into micro cycles; micro cycles are grouped into sequential phases to support combinational paths that extend across multiple chips. Each phase consists of the evaluation and communication time spans. In the former, logic outputs are evaluated according to the inputs; in the latter, the evaluation results are transferred to the other combinational logic partition.

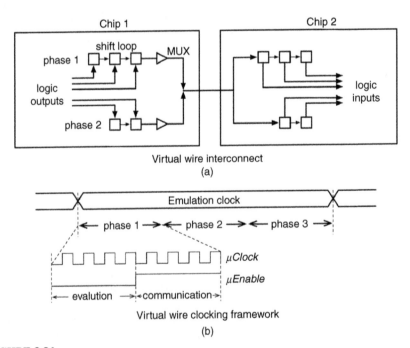

FIGURE 8.21

The virtual wire interconnection scheme.

---

[1]Assume that the circuit is synchronous and has no combinational loops.

In [Li 1998], a dynamic field programmable interconnect device (FPID), which consists of several layers of reprogrammable switching networks, is proposed. In Figure 8.22, there are n such reprogrammable switching networks, each of which can be a full or a partial crossbar. Here, n is the ratio of the speed of the physical wires and the switching network to that of the computing units. The n select lines are used to activate only one switching network at one time; thus, the I/O pin connections can be dynamically configured as expected.

The time-multiplexed interconnect scheme may be combined with the partial crossbar architecture as in Figure 8.23 [Sample 1999]. If n-to-1 multiplexers and de-multiplexers are used, the number of required pins can be reduced by a factor of n. In Figure 8.24, two signals share one I/O pin (*i.e.*, $n = 2$). Figure 8.24 depicts an example timing diagram for the wire *pq*. The *mux_clock* is divided by 2 to produce the divided clock *div_clock*, and the clock divider is reset by the *sync* signal to ensure that all clock dividers in the system are synchronized. *div_clock* is used to sample internal signal p at the rising edge and q at the falling

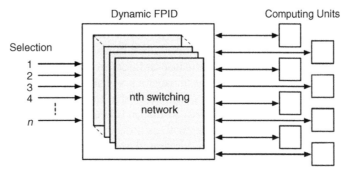

FIGURE 8.22

The dynamic FPID architecture.

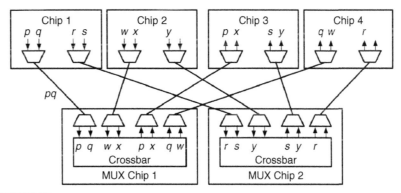

FIGURE 8.23

An interconnect scheme that combines TDI and partial crossbar.

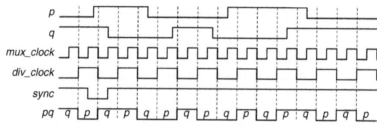

FIGURE 8.24

An example timing diagram of *pq* in Figure 8.23.

edge. Both *p* and *q* are stored in a flip-flop or a latch. When *div_clock* is high, *p* is transferred to the output; when *div_clock* is low, *q* is transferred to the output.

8.4.4 Timing issues

Achieving timing closure is a challenging task for emulators. Although many timing issues can be resolved by lowering the emulation frequency, this practice not only sacrifices the emulation throughput but also is insufficient for SOC designs that use multiple clock domains and different clock phases.

Consider the emulator architecture in Figure 8.15. The on-board clock generator supplies one or more clock signals to the FPGAs through PCB traces. These traces are carefully designed so that the clock edges arrive at the FPGAs concurrently; the clock trees inside each FPGA then forward these clock signals to the flip-flops and latches with as little skew as possible. However, things get complicated for internally generated secondary clocks. For example, the secondary clock signal *clk_2* in Figure 8.25a is derived from the primary clock *clk_1*; it is generated in FPGA 1 but is also shared by FPGA 2. Because of the large inter-FPGA signal path delay, *clk_2* signal can exhibit excessive skew. One solution to this problem is to duplicate the clock logic for each FPGA that requires *clk_2* as shown in Figure 8.25b—each FPGA has its own copy of *clk_2* (*i.e.*, *clk_2'* and *clk_2''*, respectively).

Another timing issue is related to the hold-time error. In Figure 8.26, d_1 denotes the delay from *clk_1* to D1's clock input, d_2 denotes the total delay from *clk_1* to D2's clock input, and d_3 denotes the delay of the logic block. For flip-flop D2 to capture the correct logic block output, d_2 must be no less than the sum of d_1 and d_3. However, if d_2 gets too long, the next *clk_1* edge may have arrived and cause the logic block output to change value before D2 can capture it; this leads to hold-time error. In an emulator, d_1, d_2, and d_3 depends on the emulator architecture; it is very difficult to adjust their values to ensure that hold-time errors do not occur. In [Wang 2006], the circuit to be emulated is modified so that the emulator can adequately control the clock edge timing relationship; correct operation of the modified circuit can be ensured by adjusting the clock frequency. The readers may refer to

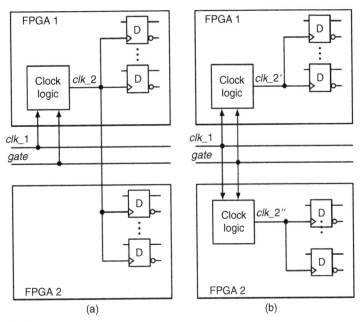

FIGURE 8.25

Duplicate clock logic to reduce clock skew.

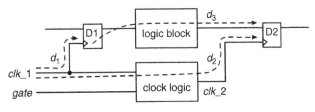

FIGURE 8.26

Hold-time error.

[Dai 1995], [Selvidge 1997], and [Tseng 2001] for other emulation techniques to eliminate hold-time errors.

8.5 CIRCUIT SIMULATION MODELS

In order to understand the modeling and simulation of ICs, we begin by visiting several basic circuit elements, namely, voltage and current sources, resistors, capacitors, and inductors. We also highlight several fundamental concepts from linear circuit analysis, specifically, the governing relationships for current and voltage throughout a circuit consisting of these basic elements. Next, we discuss the formulation of structured representations for circuit equations. In general, using analytical expressions for the solutions of these circuit equations would be computationally impractical for large-scale analysis. Hence, we will explore alternate numerical techniques for their solution.

8.5.1 Ideal voltage and current sources

An ideal voltage source is a circuit element where the voltage across the device is independent of the current flowing through it. There can be two kinds of ideal voltage sources.

- Independent voltage source: The voltage across the terminals is independent of any other variables in the circuit.
- Dependent voltage source: The voltage across the terminals is a function of some other variables in the circuit.

In a similar fashion, we can describe an ideal current source. An ideal current source is a circuit element where the current flowing through the device is independent of the voltage across it. Again, there are two kinds of ideal current sources.

- Independent current source: The current flowing through the device is independent of any other variables in the circuit.
- Dependent current source: The current flowing through the device is a function of some other variables in the circuit.

8.5.2 Resistors, capacitors, and inductors

Resistors, capacitors, and inductors are passive circuit elements. The amount of current flowing through a resistor is proportional to the voltage supplied across it. Ohm's law gives us the relationship between branch voltage (V), branch current (I), and resistance (R), which is measured in ohms (Ω):

$$V = IR$$

The charge Q stored on a capacitor is proportional to the voltage applied across it, and is given by

$$Q = CV$$

where C denotes the capacitance value measured in farads (F). Differentiating the preceding equation with respect to time t gives the following circuit equation for a capacitor:

$$I = C\frac{dV}{dt}$$

It is a well understood physical phenomenon that an electric current flowing through an inductor of inductance value L, which is measured in henrys (H), produces a magnetic flux Φ as follows:

$$\Phi = LI$$

Differentiating the preceding equation with respect to time gives the following circuit equation for an inductor:

$$V = L\frac{dI}{dt}$$

8.5.3 Kirchhoff's voltage and current laws

Kirchhoff's current law (KCL) states that at each node in a circuit, the sum of currents entering it is zero. This law is derived from the principle of conservation of charge. Consider the example circuit as shown in Figure 8.27. Applying KCL at node A gives $I_s - I_{l1} = 0$. Applying KCL at node B gives $I_{l1} - I_{l2} - I_{c1} = 0$. Applying KCL at node C gives $I_{l2} - I_{c2} = 0$.

Kirchhoff's voltage law (KVL) states that the voltage drop across every loop in a circuit is zero. This law is derived from the principle of conservation of energy. For the circuit shown in Figure 8.27, applying KVL on loop 1 gives $V_A - I_{l1}R_1 - L_1 \, (dI_{l1}/dt) - V_B = 0$. Applying KVL across loop 2 gives $V_B - I_{l2}R_2 - L_2 \, (dI_{l2}/dt) - V_C = 0$.

8.5.4 Modified nodal analysis

Through the combination of the branch equations and Kirchhoff's laws, governing equations can be systematically constructed to describe the dynamic behavior of the circuit. These formulations are often designed to exploit specific numerical methods for their solution. In addition, different formulations may be used in order to calculate quantities of interest. Specifically, the tableau formulation preserves all of the branch currents, branch voltages, and nodal voltages for the circuit. Alternatively, the nodal analysis (NA) formulation is used to solve only the nodal equations. Here, the nodal voltages are available during

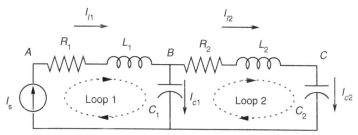

FIGURE 8.27

Example circuit.

simulation, which is the information most relevant to digital integrated circuit designers. However, the NA formulation cannot be used directly when the branch equations are dependent on branch currents (*e.g.*, when inductors and voltage sources are included). The ***modified nodal analysis*** (MNA) formulation also includes branch currents as unknown variables for these types of circuit elements [Ho 1975]. Due to the generality and computational efficiency of the MNA formulation, it is of broad interest to the circuit community, and we will focus on its development.

Given a linear circuit, we first define the adjacency matrix, A_J, for the system as follows:

$$A_J(i, j) = \begin{cases} +1 & \text{if node } j \text{ is the source of branch } i, \\ -1 & \text{if node } j \text{ is the sink of branch } i, \\ 0 & \text{otherwise.} \end{cases}$$

The adjacency matrix can be decomposed into sub-matrices according to the types of the branches:

$$A_J = \begin{pmatrix} A_i \\ A_g \\ A_c \\ A_l \end{pmatrix}$$

Here, A_i, A_g, A_c, and A_l, respectively, correspond to the adjacency matrices for current sources, resistances, capacitances, and inductances. We also define the corresponding branch voltages and branch currents,

$$v_b = \begin{pmatrix} v_i \\ v_g \\ v_c \\ v_l \end{pmatrix}, \quad i_b = \begin{pmatrix} i_i \\ i_g \\ i_c \\ i_l \end{pmatrix}$$

which are related as follows:

$$i_i = I_s, \quad i_g = R^{-1}v_g, \quad i_c = C\dot{v}_c, \quad v_l = L\dot{i}_l$$

R is the resistance matrix and C is the capacitance matrix, both of which are diagonal. L is the inductance matrix. The corresponding adjacency matrices for R, L, and C are A_g, A_l, and A_c, respectively. I_s is the current source vector with adjacency matrix A_i. Here, we consider only current sources in the formulation due to the ability to convert voltage sources into their Norton equivalent circuits.

Now, we introduce v_n, the node voltages, which are related to v_g, v_c, and v_l through KVL as follows:

$$v_g = A_g v_n, \quad v_c = A_c v_n, \quad v_l = A_l v_n$$

Applying KCL, $A_J^T i_b = 0$ and eliminating most branch currents except for those flowing through inductors, we obtain the MNA formulation, which allows for a structured accumulation of current–voltage relationships in the form of matrix equations as follows:

$$\tilde{G}x + \tilde{C}\dot{x} = b \tag{8.1}$$

where

$$\tilde{G} = \begin{bmatrix} \mathcal{G} & A_l^T \\ -A_l & 0 \end{bmatrix}, \ \tilde{C} = \begin{bmatrix} \hat{C} & 0 \\ 0 & L \end{bmatrix}, \ x = \begin{bmatrix} v_n \\ i_l \end{bmatrix}$$

$$b = \begin{bmatrix} -A_I^T I_s \\ 0 \end{bmatrix}, \ \mathcal{G} = A_g^T R^{-1} A_g, \ \text{and} \ \hat{C} = A_c^T C A_c$$

Example 8.1 For the circuit shown in Figure 8.27, the adjacency matrix is illustrated below:

$$A_J = \left(\frac{\frac{A_I}{A_g}}{\frac{A_c}{A_l}} \right) = \begin{pmatrix} 1 & -1 & 0 & 0 & 0 & 0 \\ 0 & 1 & -1 & 0 & 0 & 0 \\ 0 & 0 & 0 & 1 & -1 & 0 \\ -1 & 0 & 0 & 1 & 0 & 0 \\ -1 & 0 & 0 & 0 & 0 & 1 \\ 0 & 0 & 1 & -1 & 0 & 0 \\ 0 & 0 & 0 & 0 & 1 & -1 \end{pmatrix}$$

The nodes, which correspond to the columns in A_J, are *GND*, *A*, *AB*, *B*, *BC*, and *C*, where nodes *AB* and *BC* are the nodes to the right of R_1 and R_2, respectively. As *GND* is the reference node in this example, we eliminate that column from A_J before constructing the matrices $\hat{G}$ and $\hat{C}$ as follows:

$$\tilde{G} = \begin{bmatrix} 1/R_1 & -1/R_1 & 0 & 0 & 0 & 0 & 0 \\ -1/R_1 & 1/R_1 & 0 & 0 & 0 & 1 & 0 \\ 0 & 0 & 1/R_2 & -1/R_2 & 0 & -1 & 0 \\ 0 & 0 & -1/R_2 & 1/R_2 & 0 & 0 & 1 \\ 0 & 0 & 0 & 0 & 0 & 0 & -1 \\ 0 & -1 & 1 & 0 & 0 & 0 & 0 \\ 0 & 0 & 0 & -1 & 1 & 0 & 0 \end{bmatrix}$$

$$\tilde{C} = \begin{bmatrix} 0 & 0 & 0 & 0 & 0 & 0 & 0 \\ 0 & 0 & 0 & 0 & 0 & 0 & 0 \\ 0 & 0 & C_1 & 0 & 0 & 0 & 0 \\ 0 & 0 & 0 & 0 & 0 & 0 & 0 \\ 0 & 0 & 0 & 0 & C_2 & 0 & 0 \\ 0 & 0 & 0 & 0 & 0 & L_1 & 0 \\ 0 & 0 & 0 & 0 & 0 & 0 & L_2 \end{bmatrix}$$

where

$$R = \begin{bmatrix} R_1 & 0 \\ 0 & R_2 \end{bmatrix}, \ L = \begin{bmatrix} L_1 & 0 \\ 0 & L_2 \end{bmatrix}, \ C = \begin{bmatrix} C_1 & 0 \\ 0 & C_2 \end{bmatrix}$$

The corresponding input vector b and the vector of voltage and current variables x are as follows:

$$b = [I_s \ 0 \ 0 \ 0 \ 0 \ 0 \ 0]^T$$
$$x = [V_A \ V_{AB} \ V_B \ V_{BC} \ V_C \ I_{l1} \ I_{l2}]^T$$

Equation (8.1) can be separated into two smaller matrix differential equations as follows [Chen 2003]:

$$\begin{aligned} \mathcal{G}\,v_n + A_l^T i_l + \hat{C}\dot{v}_n &= -A_i^T I_s \\ -A_l v_n + L\dot{i}_l &= 0 \end{aligned} \tag{8.2}$$

Although an analytic solution for Equation (8.1) or (8.2) exists, the evaluation of the solution is computationally expensive. Therefore, we shall explore numerical methods for solving ordinary differential equations of this type.

8.6 NUMERICAL METHODS FOR TRANSIENT ANALYSIS

Ordinary differential equations, or ODEs, are encountered when modeling the behavior of numerous physical systems. In order to construct a procedure for solving ODEs, we first have to understand general approximation methods for functions. This will serve as a foundation for the development of numerical techniques for integration. Finally, the application of these techniques to the solution of ODEs similar to the MNA Equations (8.1) and (8.2) will be addressed.

8.6.1 Approximation methods and numerical integration

First, we examine a general integration problem

$$I(f) = \int_a^b f(x)\,dx$$

where $[a, b]$ is a closed and bounded interval. If no explicit form of the antiderivative exists or the evaluation time is prohibitively slow, the use of approximation methods becomes necessary. Let $\tilde{f}(x)$ be an approximation of $f(x)$ such that the approximation of our integral $I(\tilde{f}) = \int_a^b \tilde{f}(x)\,dx$ can be solved more readily or more efficiently.

In order to analyze how well $I(\tilde{f})$ can be used to approximate the original integral $I(f)$, we define the error for the estimate as follows:

$$\begin{aligned} |E(f)| &= \left| I(f) - I(\tilde{f}) \right| \\ &= \left| \int_a^b [f(x) - \tilde{f}(x)]\,dx \right| \\ &\leq \int_a^b |f(x) - \tilde{f}(x)|\,dx \\ &\leq (b - a)\sup_{a \leq x \leq b} |f(x) - \tilde{f}(x)| \end{aligned}$$

In general, we would like the error to be as small as possible. In the following, we define a set of functions: $\{\tilde{f}_n | n \geq 1\}$ for the increasingly more accurate approximation of our integral, where

$$I_n(f) = \int_a^b \tilde{f}_n(x)dx = I(\tilde{f}_n)$$

We would like the error $|E_n(f)| = |I(f) - I_n(\tilde{f})| = \left| \int_a^b [f(x) - \tilde{f}_n(x)]dx \right|$ to diminish as n increases, that is, $\sup_{a \leq x \leq b} |f(x) - \tilde{f}_n(x)| \to 0$ as $n \to \infty$.

We choose $\{\tilde{f}_n | n \geq 1\}$ such that

$$I_n(f) = \sum_{j=0}^{n} w_{j,n} f(x_{j,n}) \tag{8.3}$$

where $w_{j,n}$ are called the weights (or quadrature weights) and $x_{j,n}$ are the integration nodes. For convenience of representation, we will leave out the n dependency.

If we choose to approximate f on $[a, b]$ by a straight line (see Figure 8.28) joining the points $(a, f(a))$ and $(b, f(b))$, then the single approximating function would be

$$\tilde{f}_1(x) = f(a) + \frac{(x-a)}{(b-a)}[f(b) - f(a)]$$

with the integral being of the form:

$$I_1(x) = \int_a^b \tilde{f}_1(x)dx = \left(\frac{b-a}{2}\right)[f(a) + f(b)]$$

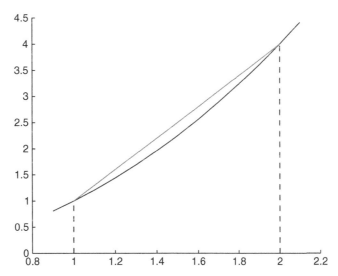

FIGURE 8.28

Approximation of the integral of $y = x^2$, $1 \leq x \leq 2$.

If we assume that f is twice differentiable, the error for our estimate is

$$E_1(f) = \frac{(b-a)^3}{12} f''(\eta), \quad \eta \in [a,b]$$

Example 8.2 $y = x^2$, $1 \leq x \leq 2$.

$$I(f) = \int_1^2 x^2 dx = \frac{x^3}{3} \Big|_1^2$$

$$= \frac{8}{3} - \frac{1}{3}$$

$$= \frac{7}{3}$$

$$E_1(f) = -\frac{(2-1)^3}{12} f''(\eta)$$

$$= -\frac{1}{12}(2)$$

$$= -\frac{1}{6}$$

$$I_1(f) = \left(\frac{2-1}{2}\right)[f(1) + f(2)]$$

$$= \left(\frac{1}{2}\right)[1 + 4]$$

$$= \frac{5}{2}$$

$$I(f) - I_1(f) = \frac{7}{3} - \frac{5}{2}$$

$$= \frac{14 - 15}{6}$$

$$= -\frac{1}{6} = E_1(f)$$

As $E_1(f) \propto (b-a)^3$, this estimate is inaccurate for large intervals. The general framework in Equation (8.3) allows us to break up the integral into smaller sub-regions as shown in Figure 8.29.

Specifically, we can form the *composite rule* as follows:

$$n \geq 1, \quad h = (b-a)/n, \quad x_j = a + jh, \quad j = 0, 1, \ldots, n$$

Assigning a linear function to approximate the function between the boundary points of each sub-region, we arrive at what is classically referred to as the *Trapezoidal Rule*:

$$I(f) = \int_a^b f(x)dx = \sum_{j=1}^n \int_{x_{j-1}}^{x_j} f(x)dx$$

$$= \sum_{j=1}^n \left\{ \left(\frac{h}{2}\right)[f(x_{j-1}) + f(x_j)] - \frac{h^3}{12}f''(\eta_j) \right\}, \quad x_{j-1} \leq \eta_j \leq x_j \tag{8.4}$$

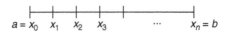

$$a = x_0 \quad x_1 \quad x_2 \quad x_3 \quad \cdots \quad x_n = b$$

FIGURE 8.29

Defining subregions for the composite rule.

Therefore, we can write the approximating integral as:

$$I_n(f) = b\left[\frac{1}{2}f(x_0) + f(x_1) + \ldots + f(x_{n-1}) + \frac{1}{2}f(x_n)\right]$$

with the error now quadratically dependent on the width of the sub-regions:

$$\begin{aligned}
E_n(f) &= I(f) - I_n(f) \\
&= \sum_{j=1}^{n} -\frac{b^3}{12}f''(\eta_j) \\
&= -\frac{(b-a)b^2}{12}f''(\eta), a \leq \eta \leq b
\end{aligned} \tag{8.5}$$

In order to further reduce the error, quadratic polynomials can be used to approximate the function, a method known as *Simpson's Rule*. Furthermore, the use of higher order approximations has been developed, and these approximations are called the *Newton-Cotes formulas* [Atkinson 1989].

8.6.2 Initial value problems

In general, we are interested in studying the initial value problem:

$$y' = f(x, y), \quad y(x_0) = Y_0 \tag{8.6}$$

where the function $f(x, y)$ is continuous for all (x, y) in some domain D, and Y_0 describes the initial condition for the differential equation. One of the simplest and most popular approaches to solving problems of this type is the *finite difference method*. Here we approximate the solution of Equation (8.6) at a discrete set of equally spaced grid points:

$$x_0 < x_1 < x_2 < \ldots < x_n < \ldots$$
$$x_j = x_0 + jb, \quad j = 0, 1, \ldots$$

where the true solution $Y(x)$ satisfies Equation (8.6), and we will denote our approximation at each grid point $y_j = y(x_j), j \geq 0$. *Euler's method* gives us a first order method for relating future values of our approximate solution to past values, given the underlying differential equation and initial condition:

$$y_{n+1} = y_n + bf(x_n, y_n), \quad n = 0, 1, \ldots$$
$$y_0 = Y_0$$

Example 8.3 $y' = 3x^2 + x + 1, y_0 = 0.$
The solution is of the form:

$$Y(x) = x^3 + \frac{1}{2}x^2 + x + c$$

where c is a constant. Applying initial condition gives $Y(0) = 0^3 + \frac{1}{2}0^2 + 0 + c = c$. As $y_0 = 0$, we have $c = 0$. Therefore,

$$Y(x) = x^3 + \frac{1}{2}x^2 + x$$

On the other hand, Euler's method gives the following approximation:

$$y_{n+1} = y_n + h(3x_n^2 + x_n + 1)$$

The following table gives the errors $Y(x_j) - y_j$ across varying step sizes:

	h = 0.01	**h = 0.1**	**h = 0.25**	**h = 0.5**
x = 0.0	0	0	0	0
x = 0.5	−0.006275	−0.065	−0.171875	−0.375
x = 1.0	−0.02005	−0.205	−0.53125	−1.125
x = 1.5	−0.041325	−0.42	−1.078125	−2.25
x = 2.0	−0.0701	−0.71	−1.8125	−3.75

Finally, we conclude by examining the relationship between numerical integration and the solution of ODEs. We first integrate Equation (8.6) across a region $[x_k, x_{k+1}]$ to obtain

$$Y(x_{k+1}) = Y(x_k) + \int_{x_k}^{x_{k+1}} f(x, Y(x))dx$$

Then, we apply Equation (8.4), the trapezoidal rule for integration, to reduce the preceding equation to

$$Y(x_{k+1}) = Y(x_k) + \frac{h}{2}[f(x_k, Y(x_k)) + f(x_{k+1}, Y(x_{k+1}))]$$

$$- \frac{h^3}{12}Y^{(3)}(\eta_j), \qquad x_k \leq \eta_j \leq x_{k+1}$$

which gives us the standard recursion for the trapezoidal method:

$$y_{k+1} = y_k + \frac{h}{2}[f(x_k, y_k) + f(x_{k+1}, y_{k+1})], \quad k \geq 0 \tag{8.7}$$

Consequently, the trapezoidal method has $O(h^2)$ convergence rate, which it inherits from the $O(h^2)$ error bound of the trapezoidal rule for integration in

Equation (8.5). It is also important to note that Equation (8.7) contains the unknown quantity y_{k+1} on both sides of the equality.

We can directly construct a numerical solution for the MNA circuit equations in Equation (8.1) by reformulating the problem in the form of Equation (8.6):

$$y' = \tilde{C}\dot{x}$$
$$f(t, x(t, y)) = -\tilde{G}x + b$$
$$x(t_0, y) = x_0$$

From here we can use the trapezoidal method in Equation (8.7), assuming a uniform discretization of the time axis with resolution h:

$$y_{k+1} = \tilde{C}x_{k+1}$$

$$= y_k + \frac{h}{2}[f(t_k, x(t_k, y)) + f(t_{k+1}, x(t_{k+1}, y))]$$

$$= \tilde{C}x_k + \frac{h}{2}[(-\tilde{G}x_k + b_k) + (-\tilde{G}x_{k+1} + b_{k+1})]$$

Thus, we solve the following recursion for each time step of our simulation:

$$\left(\frac{\tilde{G}}{2} + \frac{\tilde{C}}{h}\right)x_{k+1} = -\left(\frac{\tilde{G}}{2} - \frac{\tilde{C}}{h}\right)x_k + \frac{b_{k+1} + b_k}{2} \qquad (8.8)$$

8.7 SIMULATION OF VLSI INTERCONNECTS

On-chip interconnects, such as that shown in Figure 8.30, introduce capacitive, resistive, and inductive effects that can have a dominant impact on the circuit

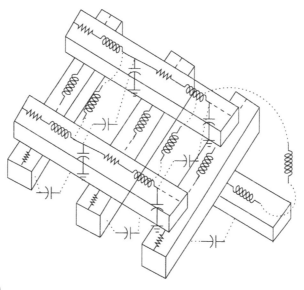

FIGURE 8.30

Distributed model of a typical three-dimensional VLSI interconnect structure.

operation and performance. In modern designs, the delay of global interconnects, possibly larger than the clock period, often dominates the gate delay. Moreover, coupling among interconnects can exacerbate the problem of signal integrity throughout the circuit. In this section, we will focus on the modeling and simulation of interconnects. The inclusion of nonlinear devices in a transient simulation procedure will be covered in the next section.

The capacitance of a wire is a typical component in nearly all interconnect models. The inclusion of resistive and inductive parameters in the modeling of interconnects is a recent trend. The scaling of devices reduces the resistance of a transistor. However, wire resistance increases significantly due to the scaling of the cross-sectional dimensions of interconnects and the increase in global interconnect lengths. As the relative significance of wire resistance increases, accurate models for the wire resistances must be employed. At higher frequencies, the impedance of an interconnect is mainly due to the inductance, which should be accounted for in a full-fledged interconnect model. In this section, we will examine a few commonly used models [Rabaey 1996, 2003; Cheng 1999], beginning with an analysis of the three physical quantities that are the building blocks for the models.

8.7.1 Wire resistance

The resistance of a rectangular wire with uniform cross-section, as shown in Figure 8.31, is given by:

$$R = \frac{\rho l}{wh} \tag{8.9}$$

where ρ is the resistivity of the material, and h, w, and l are, respectively, the height, width, and length of the wire. It is important to note that at lower frequencies, current flowing through a conductor can be assumed to be distributed uniformly throughout the cross-section of the conductor. Therefore, the resistance is accurately given by Equation (8.9). For higher frequencies, however, it is important to consider an electromagnetic induction phenomenon known as the *skin effect*. At high frequencies, current tends to crowd near the surface of the conductor, resulting in non-uniform distribution of current in a conductor, as shown in Figure 8.32, where a darker region indicates a higher

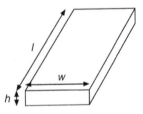

FIGURE 8.31

Rectangular wire.

FIGURE 8.32

Non-uniform current distribution in a conductor caused by the skin effect.

current density. As most of the current is carried by a small portion of the cross-section of the conductor, the effective resistance increases.

To formalize this concept, we introduce the current density of a conductor as a function of depth d from the surface:

$$J = J_s e^{-d/\delta}$$

Skin depth δ is defined as the depth at which current density is attenuated to e^{-1} of its value at the surface:

$$\delta = \sqrt{\frac{2\rho}{\omega\mu}}$$

where $\omega = 2\pi f$ is the angular frequency and μ is the permeability of the conductor. It is important to note that for higher frequencies and higher conductivity values, there is less penetration of current into the conductor.

8.7.2 Wire capacitance

Traditionally, the capacitance of interconnects is the most influential parasitic parameter that the designer of a CMOS circuit has to consider, especially before the emergence of sub-micron technologies. Moreover, this methodology remains adequate when considering short interconnects that are not part of a critical path for the circuit.

In order to accurately model the behavior of interconnects we have to consider several capacitive components. Specifically, we examine *area*, *fringing*, and *coupling* capacitance components for the model. Both the area capacitance and the fringing capacitance are considered to be the *grounded* capacitance between the interconnect and the substrate. The coupling capacitance is the parasitic effect due to interactions between two neighboring interconnects. When compared to the grounded capacitance, we consider the coupling capacitance to be *floating*.

A simple capacitive model for a typical wire is given in Figure 8.33. The parallel plate capacitance is a result of the electrical field, shown normal to the surface of the conductor terminating at the ground plate, and is described by the following equation:

$$C = \frac{\varepsilon w l}{t_{\text{ox}}}$$

where w and l are, respectively, the width and length of the wire, ε is the permitivity of the insulating material between the plates, and t_{ox} is the thickness of the insulator.

As the ratio w/b for the conductors decreases, which is typical when we scale the dimensions of interconnects, the parallel plate model becomes increasingly inaccurate. In this case, the capacitance between the side walls of the wires and the substrate (described by the fringing capacitance component) becomes a dominant contributor to the overall grounded capacitance. When approximating the total grounded capacitance, a typical approach treats the wires as rectangular sections with two hemi-spherical end caps, as described in [Yuan 1982] and illustrated in Figure 8.34. For this model, the total grounded capacitance is the sum of two components: a parallel-plate capacitance between a wire of width $w - b/2$ and the ground plane and a fringing capacitance modeled by a cylindrical wire with a radius of $b/2$. The interconnect grounded capacitance C_{grounded} can be calculated as follows:

$$C_{\text{grounded}} = \varepsilon \cdot l \cdot \left[\frac{w - \frac{b}{2}}{t_{\text{ox}}} + \frac{2\pi}{\ln \left\{ 1 + \frac{2t_{\text{ox}}}{b} + \sqrt{\frac{2t_{\text{ox}}}{b} \cdot \left(\frac{2t_{\text{ox}}}{b} + 2\right)} \right\}} \right]$$

The preceding discussion concerns the capacitance of a single interconnect. It is typical that every wire on a chip is surrounded by a number of neighboring

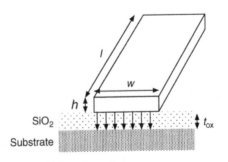

FIGURE 8.33

Parallel plate capacitance model of a wire.

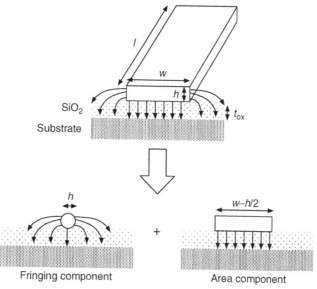

FIGURE 8.34

Fringing and area capacitance of a wire.

wires. These neighboring wires may completely or partially shield the wire of interest from the ground plane. As a result, the area capacitance and the fringing capacitance depend on the neighborhood configuration or the spacings from the neighboring wires as well.

While neighboring wires may reduce the grounded capacitance of a wire, they contribute to the coupling capacitances, which is due to electrical fields between adjacent wires that reside not only in the same layer, but also in different layers. Typically, three components of the coupling capacitance are considered. First, the area component of coupling capacitance is due to the parallel plates formed by the overlapping surfaces of wires in different routing layers. Second, the fringing component of the coupling capacitance is formed between the side-wall of one wire and the surface of a second wire above or below. Third, the lateral component of the coupling capacitance is due to the parallel plates formed by the side-walls of neighboring wires on the same layer. While there exist approximate models, fields-based solvers such as FastCap [Nabors 1991] are usually used to extract capacitive parameters of 3-D interconnect structures.

8.7.3 **Wire inductance**

With both the increase in clock frequencies and the decrease in signal transition times, on-chip inductance of interconnect wires has become a concern for circuit designers. The modeling of inductance effects is necessary when analyzing signal overshoot/undershoot and crosstalk noise. The self-inductance of a rectangular wire with uniform cross-section, as shown in Figure 8.31, can be approximated by the following equation [Keiser 1979]:

$$L = \frac{\mu_0}{2\pi} \left[l \ln\left(\frac{2l}{w+b}\right) + \frac{l}{2} + 0.2235(w+b) \right]$$

where μ_0 is the permeability of free space.

Exact formulas for computing the self-inductance and mutual inductance of rectangular wires are available in [Hoer 1965; Wu 1992; Zhong 2003]. As each of these formulas easily takes up more than half a page, we omit them in this textbook. These closed-form formulas are valid only at low frequencies, when the current distribution varies very little in the cross-sections and can be assumed to be uniform throughout the conductors. Consequently, the self and mutual inductances can be respectively computed one-by-one and pairwise even for a multi-conductor system.

At high frequencies, the current in a conductor is not uniformly distributed due to the skin effect. Moreover, the presence of neighboring wires also causes uneven distribution of current within a conductor, as shown in Figure 8.35. In this example, the currents in the two conductors flow in opposite directions. As current tends to flow in the path with the least loop impedance, the currents in the two conductors tend to crowd near the two closest surfaces of the conductors. This is known as the *proximity effect*. As the current distributions in the conductors affect each other, the inductive parameters of the whole system must be extracted at the same time, as is the extraction of capacitive parameters

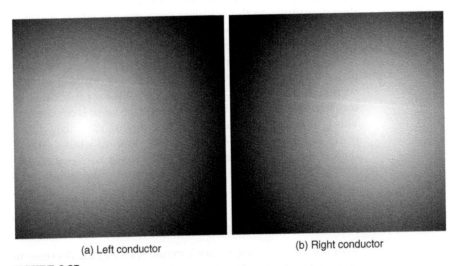

(a) Left conductor   (b) Right conductor

FIGURE 8.35

Non-uniform current distribution in two parallel conductors caused by the proximity effect.

of a multi-conductor system. The representative work is FastHenry [Kamon 1994], a parallel of FastCap [Nabors 1991].

8.7.4 Lumped and distributed models

The simplest model for an interconnect wire is a lumped capacitor model. As the resistive component of the wire becomes more significant, a ***resistive-capacitive*** (RC) model has to be adopted. The lumped RC model for a wire is shown in Figure 8.36. The use of a simple lumped RC model offers the potential to greatly simplify the analysis and optimization of interconnects, albeit with less accuracy.

The resistive and capacitive parasitics of a wire are in reality distributed along its length. In order to address the inaccuracy associated with a lumped model, especially when considering long interconnect wires, a distributed model can be formed by dividing a long wire into several segments, each of length ΔL. Let r and c be the unit-length wire resistance and capacitance, respectively. Each of the segments can be viewed as a lumped RC element with resistance $r\Delta L$ and capacitance $c\Delta L$. The distributed RC model for a wire is shown in Figure 8.37a.

In order to further improve upon the model, inductance can incorporated into the distributed framework; the distributed RLC line model is shown in Figure 8.37b, where l denotes the unit-length wire inductance. It is important to recall that in practice, there often exists significant coupling between parallel groups of interconnect wires. Therefore, in general we will consider both inductive and capacitive coupling between neighboring wire segments when we examine the dynamic behavior of the circuit as a whole. Consequently, the capacitance and inductance matrices are usually of large sizes. While capacitance matrices are in general sparse, the inductance matrices are dense.

8.7.5 Simulation procedure for interconnects

If we consider the structure of the MNA equations from Equation (8.2) in more detail, we arrive at the ***Nodal Analysis*** (NA) equations, used in [Chen 2003]. Specifically, using the trapezoidal method shown in Equation (8.7), we can formulate the following recursions:

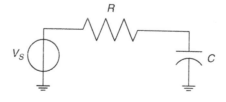

FIGURE 8.36

Lumped RC model of a wire.

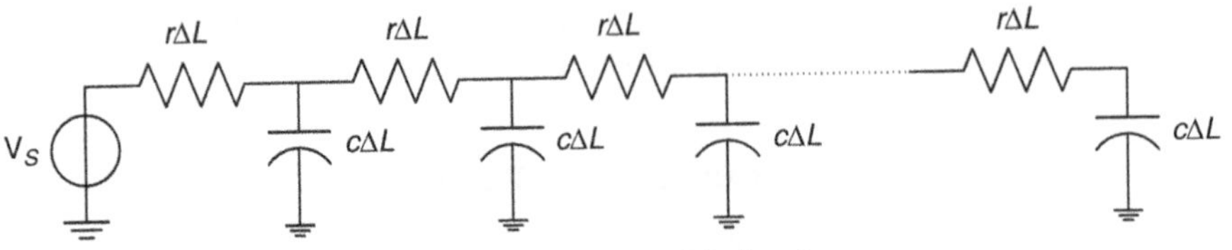

(a) Distributed RC model of a wire.

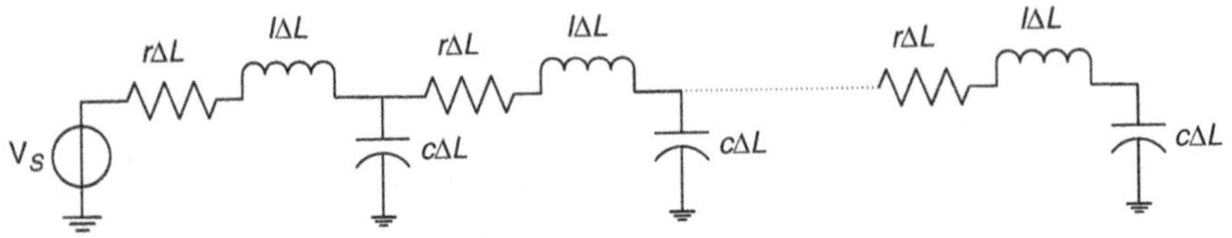

(b) Distributed RLC model of a wire.

FIGURE 8.37

Distributed wire models.

$$\underbrace{\left(\mathcal{G}+\frac{2}{b}\hat{C}+\frac{b}{2}S\right)}_{U} v_n^{k+1} = \underbrace{\left(-\mathcal{G}+\frac{2}{b}\hat{C}-\frac{b}{2}S\right)}_{V} v_n^k - 2A_l^T i_l^k - A_l^T\left(I_s^{k+1}+I_s^k\right) \qquad (8.10)$$

and

$$2A_l^T i_l^{k+1} = 2A_l^T i_l^k + bS\left(v_n^{k+1}+v_n^k\right) \qquad (8.11)$$

where $S = A_l^T L^{-1} A_l$. In order to avoid overloaded subscripts for both node voltages and branch currents, the time steps are embedded in the superscripts of these variables in Equations (8.10) and (8.11). In contrast, we use subscripts to denote the time steps in Equation (8.8).

Equation (8.10) allows for the determination of the voltages v_n^{k+1} at each time step, which can then be used for the calculation of the currents i_l^{k+1} in Equation (8.11). The voltage and current variables can then be carried back to be used for solving the voltage equation in the next time step. The first advantage of the NA equations over the MNA representation is the reduced problem size. Separating the equations as shown in Equation (8.2) allows for the solution of a smaller pair of equations, which yields computational savings. In addition, the NA representation allows us to observe specific structures or sparsity in the matrices, thereby improving efficiency of simulation.

When large numbers of interconnects are modeled using a distributed framework, the inductance matrix L will be extremely large and dense (all entries are non-zero). The computation load involved in the simulation of even modestly sized problems using direct numerical techniques will be prohibitive. However, if we examine the inverse of the inductance matrix, we see specific structures that can potentially be exploited.

Consider, for example, a group of 16 parallel wires. Each row of the inductance matrix describes the mutual coupling between all of the conductors and a specific conductor (corresponding to the row being examined). In particular, each entry relates how the rate of change of current in an conductor, say j, contributes to the voltage of the conductor being examined, say k:

$$v_k = \sum_{j=1}^{16} L_{kj} \frac{di_j}{dt}$$

Now, let us take a look at the inverse relationship due to L^{-1}:

$$\frac{di_k}{dt} = \sum_{j=1}^{16} L_{kj}^{-1} v_j$$

Each entry in the inverse of the inductance matrix relates how the voltage change in a conductor affects the current in the conductor being examined. This is, in fact, what we are more concerned with in a digital circuit—how does the voltage switching in a conductor affect the signal delay or signal integrity of other conductors. From Figure 8.38 we can clearly see a substantial decrease in the magnitude of the entries for the first row of the inverse of the inductance matrix. This is due to the fact that the conductor number here corresponds directly to their spatial location, that is, the conductors are placed in rows beginning with the number with 1 and finishing with 16. Therefore, the farther away the conductors are from conductor 1, the less coupled they are. This is

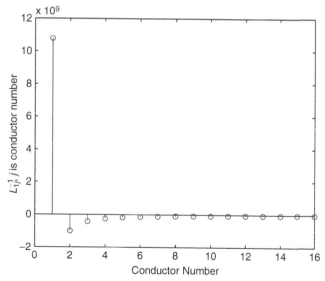

FIGURE 8.38

First row of the inverse of the inductance matrix for 16 parallel wires.

known as the locality effect or shielding effect. In other words, only conductors in close proximity of the conductor being examined are strongly coupled to the conductor of interest. These tightly coupled conductors shield faraway conductors from the conductor of interest.

This serves as the foundation for several approximation methods that reduce the amount of computation for simulation, while aiming to preserve accuracy. Several of these techniques rely on the fact that through the use of a threshold most of the entries in L^{-1} can be truncated (assumed to be identically zero), and all mathematical operations for the simulation can be performed with the resulting "sparse" L^{-1} matrices [Chen 2003]. Matrices U and V in Equation (8.10) are made up of sparse symmetric matrices G, $\hat{C}$, and S. Cholesky factorization of sparse U can, therefore, be computed efficiently. The stored Cholesky factor of U can then be used for the repeated solves of v_n^{k+1}.

An alternate formulation that also allows us to exploit the inherent structure in matrices in the MNA equations can be constructed by rewriting the separated MNA Equation (8.2) as follows [Jain 2004]:

$$A_l i_l + \hat{C}_t v_n = -A_i^T I_s,$$
$$-A_{gl} v_n + R i_l + L \dot{i}_l = 0 \tag{8.12}$$

where A_{gl} is the adjacency matrix formed by combining A_g and A_l and removing any zero columns created (from non-zero columns) as a result of inductor to resistor connections. Similarly, $\hat{C}_t$ is a truncated version of $\hat{C}$ obtained by removing the zero rows and columns in $\hat{C}$, which correspond to inductor-to-resistor connections. Typical VLSI interconnect has both resistance and inductance; this formulation relies on, as well as takes advantage of this property. For the example seen in Figure 8.27 we would have:

$$A_g + A_l = \begin{pmatrix} 1 & 0 & -1 & 0 & 0 \\ 0 & 0 & 1 & 0 & -1 \end{pmatrix}$$

As the third and fifth columns from $A_g + A_l$ are created from non-zero columns, we remove them to form A_{gl}:

$$A_{gl} = \begin{pmatrix} 1 & -1 & 0 \\ 0 & 1 & -1 \end{pmatrix}$$

In this example, $\hat{C}_t = \hat{C}$.

Using the trapezoidal method shown in Equation (8.7) we can formulate the following recursions

$$\underbrace{\left(\frac{L}{b} + \frac{R}{2} + \frac{b}{4} A_{gl} P A_{gl}^T\right)}_{X} i_l^{k+1} = \underbrace{\left(\frac{L}{b} + \frac{R}{2} + \frac{b}{4} A_{gl} P A_{gl}^T\right)}_{Y} i_l^k + A_{gl} v_n^k + \frac{b}{4} A_{gl} P \left(I_s^{k+1} + I_s^k\right),$$

$$\tag{8.13}$$

and

$$v_n^{k+1} = v_n^k - \frac{b}{2} P A_{gl}^T \left(i_l^{k+1} + i_l^k \right) + \frac{b}{2} P \left(I_s^{k+1} + I_s^k \right) \qquad (8.14)$$

where P is the inverse of capacitance matrix, that is, $P = \hat{C}_t^{-1}$. The first equation allows for the determination of the currents i_l^{k+1} at each time step, which can then be used to solve for the voltages v_n^{k+1}. The voltage can then be carried back to be used for the next time step of the current equation.

The matrix X is dense as it is made up of dense matrices $P = \hat{C}_t^{-1}$ and L. As both $P^{-1} = \hat{C}_t$ and L^{-1} exhibit sparsity structure due to locality and shielding effects, X^{-1} can be expected to be sparse as well. Indeed, for a three-dimensional interconnect topology, we might find the significant entries for X^{-1} to have a pattern similar to that shown in Figure 8.39. Therefore, a truncated X^{-1} helps to reduce the computational complexity required of the repeated solves of i_l^{k+1}.

8.8 SIMULATION OF NONLINEAR DEVICES

Metal-oxide-semiconductor field-effect transistors (MOSFETs) and parasitic diodes constitute the basic circuit elements of modern digital circuits. In order to describe the dynamic behavior of a digital circuit, it is essential to construct

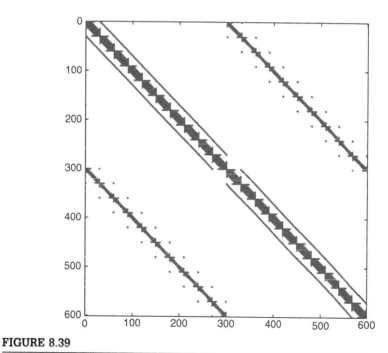

FIGURE 8.39

Significant entries for X^{-1} of three-dimensional interconnect structure.

models for these circuit building blocks. In general, the choice for a model is governed by both the accuracy in terms of describing the behavior of an actual circuit element in practice and the complexity (analysis time) associated with that model. Different models for devices exist based on the type of simulation performed by the designer. In this section, we limit our focus to device models used for transient simulation.

8.8.1 The diode

The diode, the schematic symbol of which is shown in Figure 8.40, is an important modern circuit element that is found in every MOSFET, the workhorse of modern digital circuits. Each source or drain diffusion region of a MOSFET naturally forms a pn-junction diode with the well in which the MOSFET resides.

The current-voltage characteristics or I–V characteristics for a typical diode can be divided into two regions: *forward-biased* and *reverse-biased*. When the voltage difference across the diode is smaller than a certain threshold voltage, it offers a very high resistance to the active current. Specifically, in the situation where no current is allowed to flow across the device we say that the diode is in the reverse-bias mode. As the potential drop is increased, the current can flow across the diode and the diode is said to be in the forward-bias mode. The I–V characteristics of a typical diode are shown in Figure 8.41. We can formulate the I–V characteristics of an ideal diode through the following equation:

$$I_D = I_S\left(e^{\frac{qV_D}{nkT}} - 1\right) \tag{8.15}$$

Here I_S is the reverse saturation current, q is the charge carried by an electron, k is Boltzmann's constant, T is the temperature in Kelvins, and n is the emission coefficient. The equivalent circuit model that provides us with Equation (8.15) is shown alongside the schematic symbol in Figure 8.40. The resistor R_s captures the series resistance due to the neutral regions on both sides of the junction. The nonlinear capacitance C_D, which is voltage-dependent, has two components, namely the *junction capacitance* and *diffusion capacitance*.

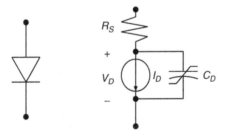

FIGURE 8.40

Circuit symbol for a diode and its equivalent circuit for transient analysis.

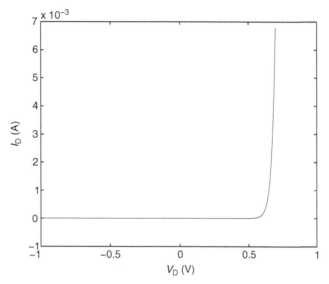

FIGURE 8.41

I–V characteristics of a diode.

The charge variation in the depletion region of a *pn*-junction diode due to variation in the potential difference across the junction is modeled as a nonlinear junction capacitance. The junction capacitance can be seen as the parallel-plate capacitance between the *n*-regions and *p*-regions of a *pn*-junction diode, and is given by the following expression [Rabaey 1996]:

$$C_j = \frac{dQ_j}{dV_D} = A_D \sqrt{\frac{q\varepsilon_{si}}{2} \frac{N_A N_D}{N_A + N_D} (\phi_0 - V_D)^{-1}}$$

where A_D is the junction area, ε_{si} is the permittivity of silicon, ϕ_0 is the zero bias potential across the junction, and N_A and N_D are the acceptor and donor concentrations, respectively. The preceding equation is valid only for an abrupt junction that has an instantaneous transition from *p*-material to *n*-material. For a linearly graded junction, a variation of the preceding equation can be used to model the junction capacitance [Rabaey 1996].

Under forward bias, excess carriers are stored at the boundaries of the depletion region. This effect is modeled by the diffusion capacitance, which is approximated by the following expression [Rabaey 1996]:

$$C_d = \frac{dQ_d}{dV_D} = \frac{q\tau_T I_S}{kT} e^{qV_D/nkT}$$

where τ_T is the mean transit time for the charge to flow across the diode.

8.8.2 **The field-effect transistor**

The metal-oxide-semiconductor field-effect-transistor (MOSFET) is the key component in present-day VLSI circuits. There are several existing models with varying degrees of sophistication that have been presented in literature [Tsividis 1987]. In this section, we will concentrate on an NMOS transistor, whose schematic symbol is shown in Figure 8.42. The behavior of a MOS transistor can be separated into three modes of operation depending on the voltages applied across its terminals: gate (G), source (S), drain (D), as well as body (B) or bulk. For simplicity, we assume that the body is tied to ground. The *I–V* characteristics of a *long-channel* NMOS transistor for each of the three modes can be described by the following equations:

- Cutoff Region ($V_{GS} \leq V_T$), where V_T is the threshold voltage of the transistor:

$$I_{DS} = 0$$

- Linear Region ($V_{DS} < V_{GS} - V_T$):

$$I_{DS} = \mu_n C_{ox} \frac{W}{L} \left((V_{GS} - V_T)V_{DS} - \frac{V_{DS}^2}{2} \right)$$

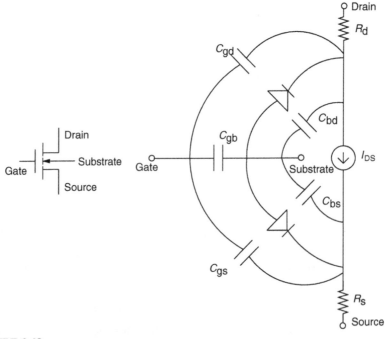

FIGURE 8.42

Circuit symbol of an NMOS transistor and its equivalent circuit for transient analysis.

Here, μ_n is the mobility of the transistor, C_{ox} is the per unit area capacitance of the oxide, and W and L are, respectively, the width and length of the transistor.

- Saturation Region ($V_{DS} \geq V_{GS} - V_T$)

$$I_{DS} = \mu_n C_{ox} \frac{W}{L} \frac{(V_{GS} - V_T)^2}{2}$$

In the preceding equation, we assume that in saturation mode the transistor will act like a perfect current source. However, the applied voltage at the drain would shorten the channel length. To account for that, we must include some dependence of the actual effective channel length on V_{DS}. This is accomplished through the inclusion of a channel-length modulation factor λ:

$$I_{DS} = \mu_n C_{ox} \frac{W}{L} \frac{(V_{GS} - V_T)^2}{2} (1 + \lambda V_{DS})$$

The *I-V* characteristics for a typical long-channel NMOS transistor are shown in Figure 8.43a.

Short-channel devices exhibit current-voltage characteristics that are considerably different from the long-channel devices. In particular, we have to incorporate the velocity-saturation effect of carriers in the equations for I_D. In a long-channel device, we assume that the velocity of carriers is proportional to the electrical field. For a short-channel device, the velocity remains constant or saturates at v_{sat} when the electrical field reaches a critical value, ξ_c. An in-depth analysis of this effect is outside the scope of this book. Interested readers are encouraged to explore [Rabaey 2003]. For short-channel devices, the equation for I_D in the linear region is

$$I_{DS} = \mu_n C_{ox} \frac{W}{L} \left((V_{GS} - V_T) V_{DS} - \frac{V_{DS}^2}{2} \right) \kappa(V_{DS})$$

where

$$\kappa(V) = \frac{1}{1 + V/(\xi_c L)}$$

When the velocity of carriers saturates, the transistor operates in the saturation region, and the equation for I_D becomes

$$I_{DS} = v_{sat} C_{ox} W (V_{GS} - V_T - V_{DSAT})$$

where $V_{DSAT} = (V_{GS} - V_T)\kappa(V_{GS} - V_T)$ is the drain-source voltage at which velocity saturation occurs. Again, the accuracy of the model can be further improved by including the channel length modulation factor. The *I-V* characteristics for a short-channel NMOS transistor is shown in Figure 8.43b.

While it is fine to assume $I_{DS} = 0$ when $V_{GS} \leq V_T$ for long-channel devices, sub-threshold leakage is no longer negligible for short-channel devices. The current in the sub-threshold region can be approximated as follows [Rabaey 2003]:

$$I_{DS} = I_S e^{\frac{qV_{GS}}{nkT}} \left(1 - e^{-\frac{qV_{DS}}{kT}} \right) (1 + \lambda V_{DS})$$

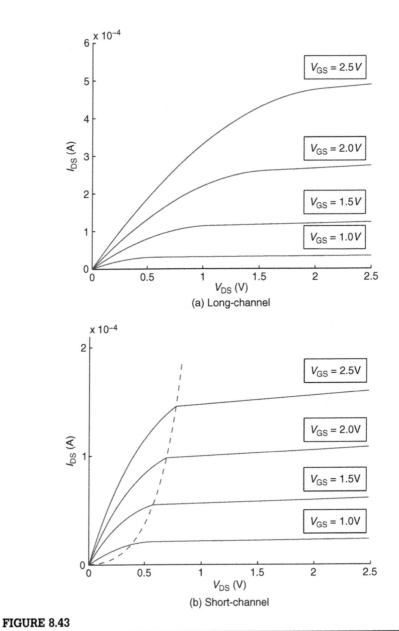

FIGURE 8.43

I–V characteristics of an NMOS transistor.

where I_s and n are empirical parameters, with $n \approx 1.5$.

We now describe the different capacitances associated with an MOS transistor. First, due to lateral diffusion under the poly gate during the ion implanation of source and drain, the gate overlaps with source and drain by x_d. Consequently, a transistor of gate length L has an effective length of $L_{eff} = L - 2x_d$. The overlap capacitance at the source or drain is $C_{ox}Wx_d$, where W denotes the channel width of the transistor. The gate-to-channel capacitance of a MOSFET can be divided into three capacitances, with C_{gs}, C_{gb}, C_{gb} being the capacitances between the gate and source, drain, and body respectively. These capacitances vary as the operating condition varies. An approximation of these capacitance values are given in Table 8.5.

Alternatively, a more accurate piecewise-linear approximation is typically used to make these capacitances continuous, resulting in a more stable simulation. For example, we can represent the gate-to-drain capacitance as:

$$C_{gd} = -\frac{dQ_g}{dV_D} \tag{8.16}$$

where the rate of change for both the gate charge Q_g and drain voltage V_D are obtained from the estimated increases over a fixed period of time. While it should be clear that the drain voltage can be determined directly from KVL-type equations, there are several different models available to approximate the charges seen at different areas of the MOSFET. If we consider the BSIM4 model [Dunga 2007], which is often provided as part of commercially available simulation packages, the amounts of charge observed at different terminals (*i.e.*, gate, body, drain and source) of a MOSFET in the saturation region will be as follows:

$$Q_g = C_{oxc}W_{active}L_{active}\left[V_{gs} - V_{FB} - \Phi_s - \frac{V_{DSAT}}{3}\right]$$

$$Q_b = -C_{oxe}W_{active}L_{active}\left[V_{FB} + \Phi_s - V_T + \frac{(1 - A'_{bulk})V_{DSAT}}{3}\right] \tag{8.17}$$

$$Q_d = -\frac{4}{15}C_{oxe}W_{active}L_{active}(V_{GS} - V_T)$$

$$Q_s = -(Q_g + Q_d + Q_b)$$

Table 8.5 Approximation of gate capacitances

Mode	C_{gs}	C_{gd}	C_{gb}
Cutoff	$C_{ox}WL_{eff}$	0	0
Linear	0	$\frac{C_{ox}WL_{eff}}{2}$	$\frac{C_{ox}WL_{eff}}{2}$
Saturation	0	$\frac{2C_{ox}WL_{eff}}{3}$	0

The model shown above is dependent on several parameters such as the effective gate oxide capacitance, C_{oxe}, the effective length and width of transistor, L_{active} and W_{active}, the surface potential Φ_s, the flat-band voltage V_{FB}, as well as A'_{bulk}, a parameter related to the bulk charge effect. Default values for these and other parameters can be found in resources similar to [Dunga 2007], along with a complete description of the charge relationships in all operating regions for the device.

The other two capacitances C_{bs} and C_{bd} shown in Figure 8.42 are the diffusion capacitances, contributed by the reverse biased source-bulk and drain-bulk pn junctions, respectively. They can be modeled by the nonlinear capacitance C_D associated with a pn-junction diode, as described in Section 8.8.1.

8.8.3 Simulation procedure for nonlinear devices

It is often the case that simulation problems involve nonlinear circuit device components. In this context, the nonlinearity refers to the nonlinear current-voltage relationship over time. Many nonlinear devices can be described using the simple building blocks of voltage/current sources, resistors, capacitors, and inductors. It is important to note that both the values of the voltage/current sources and the parasitic values for the model may change with time. This is due to the fact that the dynamic behavior for nonlinear devices is governed by the voltages seen at the terminals of the device. For example, if we consider the model of the MOSFET it should be clear that the drain current I_{DS} is dependent on the voltages seen at terminals, for example, V_{GS} and V_{DS}. In addition, from Figure 8.42 we can see that the capacitance values between the terminals, for example, C_{gd} and C_{gs}, would again be dependent on the potential difference seen across the terminals. Therefore, we should consider a simple "black box" approach to analyzing these nonlinear devices. Here, we will assume for our simulation a device that is dependent on both the current and voltage values seen at the terminals over time.

For example, the current seen at the output terminal can be described by some nonlinear function g of the voltage seen at each of the terminals across time:

$$i_{out} = g(v_{in}, v_{out}, t)$$

If we place this black box inside a simple circuit consisting of a voltage source and a load capacitance, shown in Figure 8.44, we can write down the voltage current relationships. For simplicity of illustration we will consider the nonlinear device to be the diode described in Section 8.8.1. Moreover, we ignore the nonlinear capacitance C_D associated with the diode. Here the current across the diode is given as:

$$i_D = I_S \left[e^{\frac{q}{nkT}(v_s - v_1)} - 1 \right]$$

and the current across the capacitor is:

$$i_C = C\frac{dv_1}{dt}$$

Combining the equations we get the ODE:

$$v_1' = \frac{I_s}{C}\left[e^{\frac{q}{nkT}(v_s-v_1)} - 1\right] = g(t, v_1)$$

Using the integration technique described in Section 8.6.2, we are left with a nonlinear expression

$$v_1^{k+1} = v_1^k + \frac{b}{2}\left[g(t_k, v_1^k) + g(t_{k+1}, v_1^{k+1})\right]$$

or

$$v_1^{k+1} - v_1^k - \frac{b}{2}\left[g(t_k, v_1^k) + g(t_{k+1}, v_1^{k+1})\right] = 0 \qquad (8.18)$$

In this formulation, we know the value of v_1^k and we are interested in solving for the value of v_1^{k+1}, that is, the voltage at the next instant in time. There are several classical methods that can be used to find the "zero" of this nonlinear equation. We will focus on Newton's method, which offers a computationally efficient and numerically stable approach to solve for a zero. If we denote the unknown variable to be $x = v_1^{k+1}$, and the left-hand side of Equation (8.18) to be $z(x)$, then we are attempting to determine a solution x^* such that our nonlinear equation $z(x^*) = 0$. This is accomplished by forming a sequence of iterates that will converge to this point x^*, starting with an initial guess x_0:

$$x_{n+1} = x_n - \frac{z(x_n)}{z'(x_n)}, \quad n = 0, 1, \ldots \qquad (8.19)$$

where $x_n \to x^*$ as $n \to \infty$. Note here that the subscripts in the preceding equation refer to the iteration numbers of the iterative procedure. In general, this procedure can be followed exactly when dealing with the inclusion of nonlinear devices into the MNA framework. Specifically, the effect of the nonlinear devices can be explicitly seen in the variable b below:

$$\tilde{G}x + \tilde{C}\dot{x} = b \qquad (8.20)$$

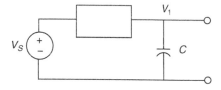

FIGURE 8.44

Circuit example including nonlinear device model.

where

$$\tilde{G} = \begin{bmatrix} \mathcal{G} & A_l^T \\ -A_l & 0 \end{bmatrix}, \quad \tilde{C} = \begin{bmatrix} \hat{C} & 0 \\ 0 & L \end{bmatrix}, \quad x = \begin{bmatrix} v_n \\ i_l \end{bmatrix}$$

$$b = \begin{bmatrix} -A_l^T I_s + I_{nl} \\ 0 \end{bmatrix}, \quad \mathcal{G} = A_g^T R^{-1} A_g, \quad \text{and } \hat{C} = A_c^T C A_c$$

The vector I_{nl} captures the effect of nonlinear loads and depends on the node voltages as $I_{nl} = f(v_n)$. Here, f is a function that varies depending on the load characteristics and in general, can be a nonlinear function. In addition, it is important to note that the values of the capacitance matrix C will change as the simulation progresses. This can clearly be seen by revisiting the piece-wise-linear approximations which were used for the junction capacitances in Equation (8.16). As the node voltages v_n change with time, we will see a corresponding change in the capacitance values reflected by changes to the matrices seen in Equation (8.20).

The result is a voltage-dependent system of linear equations that must be solved at each time step of the simulation process. At first glance, this would numerically lend itself to the use of an iterative solver. We can, for example, apply Equation (8.19) iteratively to solve for each x^k at the kth time step. This is due to the fact any preprocessing or initial factorization would not directly be of benefit from one time step to the next as the matrices change over time. However, the matrix equations observed during the trapezoidal integration and Newton's method steps described above have a very special structure when combining nonlinear devices with various interconnect topologies. Specifically, the matrices can be grouped into those that remain constant with respect to time, and those components that are time-dependent or voltage-dependent.

Based upon a nodal analysis scheme, it should be clear that these two sets are not independent, that is, there exist branches or coupling between the nodes, and they must be eventually solved together. However, the special nature of the underlying matrix elements lends itself to the use of different numerical techniques for efficient solution. Specifically, the linear portion is time-independent and a direct factorization (e.g., LU or Cholesky) can be reused for each time-step. The nonlinear, time-dependent portion will not benefit from this preprocessing; however, a preconditioned iterative method can be employed (see [Jain 2006] and [Zhu 2007], for details relating to this procedure).

8.9 CONCLUDING REMARKS

We have presented two classes of fundamental techniques, logic simulation and circuit simulation, which are useful for predicting the behavior of a design before its physical implementation is built. During this design stage, designers heavily rely on simulation to verify and debug their designs. Oftentimes, simulation is combined with functional verification, which is a major topic in Chapter 9, to further ensure the correctness of their designs. Commercial simulation tools have been available for modern *system-on-chip* (SOC) designs modeled at the behavioral level down to the lowest transistor level. This chapter covered the fundamental logic simulation, hardware-accelerated logic simulation, and circuit simulation techniques at the gate level and the transistor level.

For logic simulation, event-driven simulation that can take timing (delay) models and sequential circuit behavior into consideration is the technique most widely used in commercially available logic simulators. Examples of logic simulators include Verilog-XL, NC-Verilog (both from Cadence [Cadence 2008]), ModelSim (from Mentor Graphics [Mentor 2008]), and VCS (from Synopsys [Synopsys 2008]). These logic simulators can accept gate-level models as well as RTL and behavioral descriptions of the circuits written in hardware description languages, such as Verilog and VHDL, both of which are IEEE standards. HDLs are beyond the scope of this book but are important topics for digital designers to learn. More detailed descriptions of both languages can be found in books or Web sites, such as [Palnitkar 1996], http://www.verilog.com, and http://www.verilog.net.

Although flexible and low-cost, software simulators are becoming too slow for modern SOC designs and hardware/software co-simulation applications. Hardware-accelerated logic simulation techniques have been developed to bridge the growing gap. Emulators are differentiated by their interconnection architectures and the types and use models of the reconfigurable computing units. Each combination has its advantages and disadvantages and finds its applications in different verification environments. A few popular commercially available emulators include Incisive Acceleration and Emulation (from Cadence [Cadence 2008]), ZeBu (from EVE [EVE 2008]), and Veloce (from Mentor Graphics [Mentor 2008]).

Circuit simulation (commonly referred to as SPICE simulation [Nagel 1975]) at the transistor level, although too slow for practical designs, is important when the circuit's dynamic behavior or accurate timing information is desired. In general, circuit simulation is used to characterize the cell library, memory models, and the timing critical portion of the circuit. A few popular circuit simulators include Hspice (from Synopsys [Synopsys 2008]) and Spectre (from Cadence [Cadence 2008]).

As the design complexity continues growing and has reached two billion transistors [Stackhouse 2008], verifying the correctness of these designs has become a much more challenging task than ever. As a result, it is becoming imperative that advanced techniques for both logic simulation and circuit simulation, either hardware-accelerated or pure software-based, be developed to address the high-performance and high-capacity issues.

8.10 **EXERCISES**

8.1. (3-Valued Logic Simulation) By use of 3-valued logic simulation, what is the output value of circuit M in Figure 8.45 if the input pattern is $ABCD = 1u0u$? Show that there is information loss in this example.

8.2. (Timing Models) For circuit M shown in Figure 8.45, complete the timing diagram in Figure 8.46 with respect to each timing model given below:

a. *Nominal delay*—Two-input gate, 10 ns; three-input gate, 12 ns; inverter, 8 ns.
Inertial delay—All gates, 4 ns.

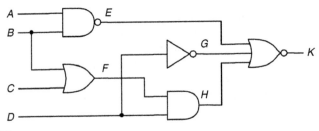

FIGURE 8.45

The example circuit M.

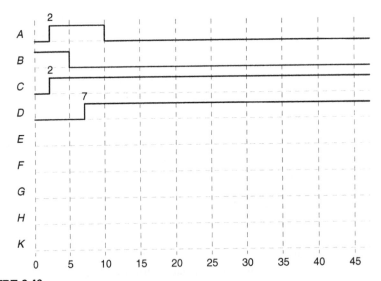

FIGURE 8.46

The timing diagram.

    b. *Rise delay*—Two-input gate, 8 ns; three-input gate, 10 ns; inverter, 6 ns.

    Fall delay—Two-input gate, 6 ns; three-input gate, 8 ns; inverter, 4 ns.

    c. *Minimum delay*—Two-input gate, 8 ns; three-input gate, 10 ns; inverter, 6 ns.

    Maximum delay—Two-input gate, 10 ns; three-input gate, 12 ns; inverter, 8 ns.

8.3. **(Compiled-Code Simulation)** One approach to speed up compiled-code simulation is to use the bit-wise logic operations. If two-valued logic is used and the host computer's data word width is 32-bit, one can store in a single word 32 copies of a signal (with respect to different input vectors) and process them at the same time. In this problem, we consider a logic simulator with four logic symbols (0, 1, u, and Z) that are encoded as follows: $v_0 = (00)$, $v_1 = (11)$, $v_u = (01)$, and $v_Z = (10)$. To simulate w input vectors in parallel, two words (X_1 and X_2) are allocated for each signal X to store the first and second bits of the logic symbol codes, respectively.

    a. Derive the gate evaluation procedures for AND, OR, and NOT operations.

    b. Derive the evaluation procedures for 2-to-1 multiplexer, XOR, and tristate buffer.

8.4. **(Event-Driven Simulation)** Redo Problem 8.2a, using the nominal-delay event-driven simulation technique. Show the event and activity lists of each time stamp.

8.5. **(Interconnection Architectures)** It is known that the full-crossbar chip complexity (Figure 8.20a) grows quadratically with the total pin count. How about the partial-crossbar solution (Figure 8.20b)?

8.6. **(Numerical Integration)** Show that the degree of precision for Simpon's rule is 3. Begin first by considering the integration formula:

$$I_3(f) = \sum_{j=0}^{2} \alpha_j f(x_j)$$

Assuming a uniform time step h, show that the formula is exact for the functions $f(x) = 1$, x, x^2 across the points $x_0 = -h$, $x_1 = 0$, and $x_2 = h$ (this should involve solving a system of three equations for the coefficients α_j). Finally, try these coefficients for higher order polynomials, *i.e.*, $f(x) = x^3$, x^4 to determine the error scaling as a function of h.

8.7. **(Numerical Integration)** With the formula for I_3 derived above, write a simple script to evaluate

$$I_3\,(f) = \int\limits_{-5}^{5} 5x^4 + 3x^2 - 1$$

Use three different values of the time step h and compare with the expected error scaling determined in Exercise 8.6.

8.8. (Modified Nodal Analysis) Formulate the MNA equation representation for the circuit shown in Figure 8.47. The equations should be in terms of the parameters $(v_s, R_1, R_2, C_1, C_2)$ and the unknown voltages and currents $(v_1, v_2, i_1, i_2, i_3, i_4)$.

8.9. (Modified Nodal Analysis) Given resistance values: $R_1 = R_2 = 10\ \Omega$ and capacitance values: $C_1 = C_2 = 10^{-10}$ F, use the MNA equations constructed in Exercise 8.8 to plot $v_2(t)$ for a step input from the source v_s.

8.10. (Wire Capacitance) Assuming the following dimensions for the two parallel wires shown in Figure 8.48: $h = 1\ \mu m$, $w/h = 2$, $l = 100\ \mu m$, $d = 2\ \mu m$, and $t = 0.75\ \mu m$, compute both the total and coupling capacitances of the wires. Assume that the field terminates as shown

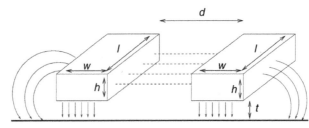

FIGURE 8.47

RLC circuit example.

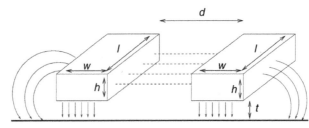

FIGURE 8.48

Parallel wires.

in Figure 8.48. Now, assume that the ratio w/h is doubled. How does this affect the capacitances? Finally, assume instead that the distance between the wires is doubled. How does this affect the capacitances?

8.11. (Newton's Method) Write a simple program that implements the Newton's method seen in Equation (8.19). Follow the pseudocode shown in Algorithm 8.1.

Algorithm 8.1 Newton's Method

Newton (f, x_0, ϵ)
1. $n = 0$;
2. $x_{n+1} = x_n - \frac{f(x_n)}{f'(x_n)}$;
3. **if** $f(x_{n+1}) \leq \epsilon$ **then**
4.   **return** $x^* = x_{n+1}$;
5. **else** $n = n + 1$; **goto** step 2;

8.12. (Newton's Method) Use the routine from Exercise 8.11 to solve the nonlinear equation (8.18) for the three consecutive time points assuming:

$$b = 10^{-11}\,\text{s},$$
$$C = 1\,\text{F},$$
$$R = 1\,\Omega,$$
$$I_S = 1\,\text{A},$$
$$n = 1,$$
$$V_S = 2\,\text{V},$$
$$V_1 = 1\,\text{V},$$
$$T = 300\,\text{K}.$$

Provide the number of iterations and tolerance assumed.

ACKNOWLEDGMENTS

We thank Dr. Jitesh Jain of Purdue University for his invaluable contribution to the development of Sections 8.5 to 8.8. We also thank Dr. Tsung-Hao Chen of Mentor Graphics and Professor Yu-Min Lee of National Chiao Tung University, Jensen Tsai of SpringSoft, Professor Ren-Song Tsay of National Tsing Hua University, Dr. Ming-Yang Wang of Fortelink, and Professor Duncan M. (Hank) Walker of Texas A&M University for reviewing this chapter.

REFERENCES

R8.0 Books

[Atkinson 1989] K. Atkinson, *An Introduction to Numerical Analysis*, John Wiley & Sons, New York, 1989.

[Cheng 1999] C.-K. Cheng, J. Lillis, S. Lin, and N. Chang, *Interconnect Analysis and Synthesis*, John Wiley & Sons, New York, 1999.

[Dunga 2007] M. V. Dunga, W. M. Yang, X. J. Xi, J. He, W. Liu, M. Cao, X. Jin, J. J. Ou, M. Chan, A. M. Niknejad, and C. Hu, *BSIM4.6.1 MOSFET Model—User's Manual*, University of California, Berkeley, CA, 2007.

[IEEE 1076-2002] *IEEE Standard, VHDL Language Reference Manual (IEEE Std. 1076-2002)*, IEEE Press, New York, 2002.

[IEEE 1463-2001] *IEEE Standard Description Language Based on the Verilog Hardware Description Language (IEEE Std. 1463-2001)*, IEEE Press, New York, 2001.

[Keiser 1979] B. E. Keiser, *Principles of Electromagnetic Compatibility*, Artech House, Dedham, MA, 1979.

[Miczo 2003] A. Miczo, *Digital Logic Testing and Simulation*, 2nd ed., John Wiley & Sons, Hoboken, New Jersey, 2003.

[Palnitkar 1996] S. Palnitkar, *Verilog HDL: A Guide to Digital Design and Synthesis*, Sunsoft, Mountain View, CA, 1996.

[Rabaey 1996] J. M. Rabaey, *Digital Integrated Circuits: A Design Perspective*, Prentice-Hall, Upper Saddle River, NJ, 1996.

[Rabaey 2003] J. M. Rabaey, A. Chandrakasan, and B. Nikoli'c, *Digital Integrated Circuits: A Design Perspective*, 2nd ed., Prentice-Hall, Upper Saddle River, NJ, 2003.

[Thomas 2002] D. E. Thomas and P. R. Moorby, *The Verilog Hardware Description Language*, Springer Science, New York, 2002.

[Tsividis 1987] Y. Tsividis, *Operation and Modeling of the MOS Transistor*, McGraw-Hill, New York, 1987.

[Wile 2005] B. Wile, J. C. Goss, and W. Roesner, *Comprehensive Functional Verification*, Morgan Kaufmann, San Francisco, 2005.

R8.1 Introduction

[SystemC 2008] SystemC, http://www.systemc.org, 2008.

[SystemVerilog 2008] SystemVerilog, http://systemverilog.org, 2008.

R8.2 Logic Simulation Models

[Breuer 1972] M. A. Breuer, A note on three valued logic simulation, *IEEE Trans. on Computers*, C-21(4), pp. 399–402, April 1972.

R8.3 Logic Simulation Techniques

[Ulrich 1969] E. G. Ulrich, Exclusive simulation of activity in digital networks, *Communications of the ACM*, 12(2), pp. 102–110, February 1969.

[Wang 1987] L.-T. Wang, N. E. Hoover, E. H. Porter, and J. J. Zasio, SSIM: A software levelized compiled-code simulator, in *Proc. ACM/IEEE Design Automatic Conf.*, pp. 2–8, June 1987.

R8.4 Hardware-Accelerated Logic Simulation

[Babb 1997] J. Babb, R. Tessier, M. Dahl, S. Z. Hanono, D. M. Hoki, and A. Agarwal, Logic emulation with virtual wires, *IEEE Trans. on Computer-Aided Design*, 16(6), pp. 609–626, June 1997.

[Beausoleil 1996] W. F. Beausoleil, T.-K. Ng, and H. R. Palmer, Multiprocessor for Hardware Emulation, U.S. Patent No. 5,551,013. August 27, 1996.

[Dai 1995] W.-J. Dai, L. Galbiati, III, J. Varghese, and D. V. BuiS. P. Sample, Method of Removing Gated Clocks from the Clock Nets of a Netlist for Timing Sensitive Implementation of the Netlist in a Hardware Emulation System, U.S. Patent No. 5,452,239, September 19, 1995.

[Li 1998] J. Li and C.-K. Cheng, Routability improvement using dynamic interconnect architecture, *IEEE Trans. on Very Large Scale Integration Systems*, 6(3), pp. 498–501, September 1998.

[Lin 2002] S. S.-P. Lin, and P.-S. Tseng, Converification System and Method, U.S. Patent No. 6,389,379, May 14, 2002.

[Sample 1999] S. P. Sample, M. Bershteyn, M. R. Butts, and J. R. Bauer, Emulation System with Time-Multiplexed Interconnect, U.S. Patent No. 5,960,191, September 28, 1999.

[Selvidge 1997] C. W. Selvidge and M. L. Dahl, Transition Analysis and Circuit Resynthesis Method and Device for Digital Circuit Modeling, U.S. Patent No. 5,649,176, July 15, 1997.

[Tseng 2001] P.-S. Tseng, S. S.-P. Lin, and Q. K.-H. Shen, Timing-Insensitive Glitch-Free Logic System and Method, U.S. Patent No. 6,321,366, November 20, 2001.

[Varghese 1993] J. Varghese, M. Butts, and J. Batcheller, An efficient logic emulation system, *IEEE Trans. on Very Large Scale Integration Systems*, 1(2), pp. 171–174, June 1993.

[Wang 2006] M. Y. Wang, S. Shei, and V. Chiu, Clock Distribution in a Circuit Emulator, U.S. Patent No. 7,117,143, October 3, 2006.

R8.5 Circuit Simulation Models

[Chen 2003] T.-H. Chen, C. Luk, and C. C.-P. Chen, INDUCTWISE: Inductance-wise interconnect simulator and extractor, *IEEE Trans. on Computer-Aided Design*, 22(7), pp. 884–894, July 2003.

[Ho 1975] C. W. Ho, A. E. Ruehli, and P. A. Brennan, The modified nodal approach to network analysis, *IEEE Trans. on Circuits and Systems*, 22(6), pp. 504–509, June 1975.

R8.7 Simulation of VLSI Interconnects

[Chen 2003] T.-H. Chen, C. Luk, and C. C.-P. Chen, INDUCTWISE: Inductance-wise interconnect simulator and extractor, *IEEE Trans. on Computer-Aided Design*, 22(7), pp. 884–894, July 2003.

[Hoer 1965] C. Hoer and C. Love, Exact inductance equations for rectangular conductors with applications to more complicated geometries, *J. of Research of the National Bureau of Standards*, 69C, pp. 127–137, April-June 1965.

[Jain 2004] J. Jain, C.-K. Koh, and V. Balakrishnan, Fast simulation of VLSI interconnects, in *Proc. IEEE/ACM Int. Conf. on Computer-Aided Design*, pp. 93–98, November 2004.

[Kamon 1994] M. Kamon, M. J. Tsuk, and J. K. White, FASTHENRY: A multipole-accelerated 3-D inductance extraction program, *IEEE Trans. on Microwave Theory and Techniques*, 42, pp. 1750–1758, September 1994.

[Nabors 1991] K. Nabors and J. White, Fastcap: A multipole accelerated 3-D capacitance extraction program, *IEEE Trans. on Computer-Aided Design*, 10(11), pp. 1447–1459, November 1991.

[Wu 1992] R.-B. Wu, C.-N. Kuo, and K. K. Chang, Inductance and resistance computations for three-dimensional multiconductor interconnect structures, *IEEE Trans. on Microwave Theory and Techniques*, 40, pp. 263–270, February 1992.

[Yuan 1982] C. P. Yuan and T. N. Trick, A simple formula for the estimation of the capacitance of two-dimensional interconnects in VLSI circuits, *IEEE Electronic Device Letters*, EDL-3(12), pp. 391–393, December 1982.

[Zhong 2003] G. Zhong and C.-K. Koh, Exact closed form formula for partial mutual inductances of on-chip interconnects, *IEEE Trans. on Circuits and Systems I: Fundamental Theory and Applications*, 50(10), pp. 1349–1353, October 2003.

R8.8 Simulation of Nonlinear Devices

[Jain 2006] J. Jain, S. Cauley, C.-K. Koh, and V. Balakrishnan, SASIMI: Sparsity-aware simulation of interconnect-dominated circuits with nonlinear devices, in *Proc. IEEE/ACM Asia and South Pacific Design Automation Conf.*, pp. 422–427, January 2006.

[Zhu 2007] Z. Zhu, H. Peng, K. Rouz, M. Borah, C. K. Cheng, and E. S. Kuh, Two-stage Newton-Raphson method for transistor level simulation, *IEEE Trans. on Computer-Aided Design*, 26(5), pp. 881–895, May 2007.

R8.9 Concluding Remarks

[Cadence 2008] Cadence Design Systems, http://www.cadence.com, 2008.

[EVE 2008] EVE, http://eve-team.com, 2008.

[Mentor 2008] Mentor Graphics, http://www.mentor.com, 2008.

[Nagel 1975] L. W. Nagel, *SPICE2: A computer program to simulate semiconductor circuits*, Technical Report ERL-M520, University of California, Berkeley, May 1975.

[Stackhouse 2008] B. Stackhouse, A 65nm 2-billion-transistor quad-core itanium processor, in *Proc. IEEE International Solid-State Circuits Conference*, pp. 592–598, February 2008.

[Synopsys 2008] Synopsys, http://www.synopsys.com, 2008.

Functional verification

Hung-Pin (Charles) Wen
National Chiao-Tung University, Taiwan

Li-C. Wang
University of California, Santa Barbara, California

Kwang-Ting (Tim) Cheng
University of California, Santa Barbara, California

ABOUT THIS CHAPTER

In a typical **integrated circuit** (IC) design flow, functional verification ensures that the implementation conforms to the specification. Because of the rapid growth of both design size and complexity, functional verification has become one of the key bottlenecks in the design process. For example, it has been reported in [Bailey 2002] that the functional verification process consumes more than 70% of the design effort, and this number might continue to increase. Functional verification is critical, because an undetected bug in a design may result in significant financial loss for a company. The Pentium recall for the famous FDIV bug, for example, cost Intel more than $450 million in 1995. Therefore, effective verification strategies and techniques have become indispensable to the design flow to ensure high verification quality.

This chapter starts with an overview of the basic concepts of functional verification and its general flow. Current challenges are explained to help readers to understand the complexity of functional verification. Meanwhile, modern designs usually follow the principle of hierarchism by decomposing a complex system into multiple components. Each decomposition boundary is referred to as a **level**. A brief discussion of verification at each of these levels is introduced.

To assess the verification quality, coverage metrics are developed for measuring the extent of an intended verification task. Coverage metrics can be divided into two categories: structural and functional. Structural coverage metrics calculate a coverage number on the basis of specific structural representations, such as lines and branches, in the hardware description model and are the most popular measures. Functional metrics, on the other hand, focus on the semantics or the design intent of the hardware description model. In this chapter, various structural coverage metrics will be reviewed in detail.

513

Simulation-based verification is the most widely used approach in functional verification. Simulation is based on testbenches. In a typical verification task, testbenches accompanied with a design description model are developed and include input stimuli and expected output responses by the design. The efficiency of the simulation determines the efficiency of the verification, and, hence, having compact and high-quality stimuli is critical to this approach. An alternative to simulation-based verification is **formal verification**. Formal verification relies on mathematical reasoning techniques to verify a design. There can be two types of formal verification methods, one to prove specific properties of a design and the other to prove that two models of a design are equivalent. The former is called property checking, and the later is often referred to as equivalence checking. At the end of this chapter, some of these formal verification techniques will be introduced as supplemental materials.

9.1 INTRODUCTION

Verification processes happen everywhere in our daily life. One general definition of verification given in [ANSI/ASQC 1978] is "the act of reviewing, inspecting, testing, checking, auditing, or otherwise establishing and documenting whether or not items, processes, services or documents conform to specified requirements." Within the context of design automation of IC design, shown in Figure 9.1, functional verification is the step to ensure that the specifications

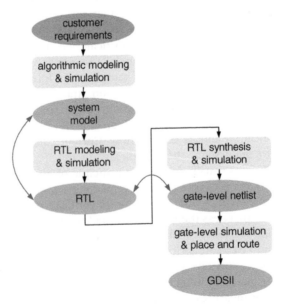

FIGURE 9.1

Typical design flow overview.

and/or the implementations of the design at various abstraction levels are in accord with the design intent.

In a typical design flow, representations for a design at different abstraction levels often contain thousands of lines or more of **_Hardware Description Language_** (HDL) code. These representations are error-prone because of the high complexity of the design. Verification plays an important role in identifying various kinds of problems that may have occurred at different design stages. For many medium-scale to large-scale processors, **_application-specific integrated circuits_** (ASICs), or **_system-on-chips_** (SOCs), functional verification can consume more than 70% of the total labor effort in the design process [Piziali 2006]. The difficulty inherent in functional verification is a result of the following three issues:

1. **Ambiguous specifications:** Customer requirements are often written colloquially into the specification. It may be difficult to precisely specify the requirements with a natural language such as English. Moreover, a specification is often described at the system level. When verifying a unit or block inside a system, a clear specification for the unit or the block usually is not available.

2. **Complexity explosion:** In general, the complexity of a Boolean circuit can grow exponentially in terms of both the number of inputs and the number of internal states. Exhaustive simulation (of all input value combinations and/or state combinations) is simply infeasible for any nontrivial design.

3. **Quality concerns**: Ensuring highest-quality verification with limited engineering resources and within limited time is the challenge to every verification task. To effectively use resources and time, one needs coverage metrics to guide the spending of verification effort. Although various coverage metrics exist to measure verification coverage, none of these metrics have been shown to be the golden metric that can reliably and accurately reflect the verification quality. As a result, signing off a design with respect to functional verification can become a managerial decision that heavily depends on one's experience and is often influenced by time-to-market pressure as well.

9.2 VERIFICATION HIERARCHY

Modern IC designs typically follow a **top-down** implementation flow in which a system is hierarchically partitioned into components. Each partitioning boundary defines the level of the design components. Within the hierarchy, verification tasks need to be performed before individual components are assembled. The V diagram in Figure 9.2 illustrates the design, verification, and integration

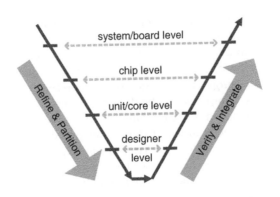

FIGURE 9.2

V diagram of design, verification, and integration.

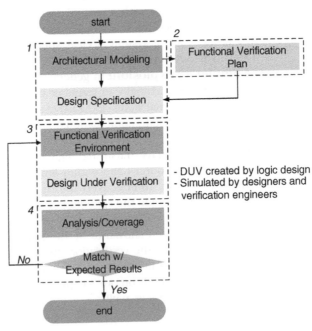

FIGURE 9.3

Generic design verification flow.

flow starting from the system/board level, through the chip and core/unit levels, to the designer level.

A generic verification flow [Palnitkar 2003a] for each level consists of several steps, as shown in Figure 9.3. In Step 1, architects need to prepare a design specification for the best architecture on the basis of analysis of simulation result. In Step 2, a functional verification plan is created to define the basic parameters that are used later in the functional verification environment. Test vectors and testbenches are either generated manually or automatically by tools during Step 3. A software simulator applies these test vectors and testbenches to the ***design under verification*** (DUV) and collects the related information after simulation. In Step 4, the output data are analyzed and checked against the expected results to calculate verification coverage. If the desired coverage goal is not achieved, Step 3 is repeated to generate more test vectors to improve the coverage. After the coverage goal is met, optional steps of hardware-accelerated simulation, emulation, and assertion-based verification could be applied to further improve verification quality and to reduce the risk of needing a future re-spin.

9.2.1 Designer-level verification

In the top-down implementation flow shown in Figure 9.2, the designer level is the lowest level that defines the smallest of the RTL modules such as an arbiter or a **first-in first-out** (FIFO) that one designer can be in charge of in a project. Designer-level blocks are usually verified individually to ensure that the basic functionalities of the blocks understood by the designer from the system specification are correctly implemented. As the tasks involved in verifying a designer-level block do not require interaction with other blocks, the designer is given full control of the block, and thus a high standard of verification is expected at this level.

During the early phase of a design project, the functionality of a block would not be completely fixed and likely will be modified frequently. For example, part of a block's functionality may need to move across the interface to other blocks for better unit/core/chip optimization. It is, therefore, not uncommon to repeat the designer-level verification process multiple times.

A variety of verification techniques are available at this level. Testbench development is relatively easy because the block inputs and outputs are treated as primary inputs and outputs at this stage. The designers often explore most of or even the entire input space of the target block by simulation. Formal methods such as property checking can also be applied relatively easily at this level because of the small design size. It is important to note that, for designer-level verification, the main challenge is not in verifying the block itself as an independent design, but in verifying the block in the context of the environment in which it will be placed. For example, a property may not be verified as always true if the block operates independently. However, under specific constraints imposed by the environment surrounding the block, the property could become always true. Establishing proper environmental constraints for designer-level verification is, therefore, an important (and usually not trivial) task.

9.2.2 **Unit-level verification**

A complex design is usually divided into several logical components that are referred to as **units**. The units intercommunicate through buses following pre-specified protocols. Figure 9.4 shows an AMBA bus-based SOC design. Memory, UART, Bridge, and Arbiter are among the units created from many different designer-level components. In this example, the communications between units go through two PCI buses. [Scafidi 2004] reported that even when the full-chip model of Intel's Itanium-2 processor was close to the tape-out quality, unit-level verification still uncovered additional bugs.

The functionality at the unit level is specified more clearly, and usually the specification is more stable than that at the designer level. Each unit usually has a precise specification where its physical and timing characteristics will abide by the requirements of the bus protocol. Each unit implements a set of specified operations. Therefore, the goal of unit-level verification is to guarantee that each operation performed by the unit conforms to the desired functionality and satisfies the bus interface's communication constraints.

Because of the high accessibility of units through buses, high-quality verification that guarantees each unit correctly meets its formal specification is usually achievable. In an ideal situation, once the unit-level verification is completed, bugs residing within these units can be excluded from the list of candidates. When performing verification at the next level, only those bugs originated from the communication and physical interfaces need to be considered.

9.2.3 **Core-level verification**

In the example of Figure 9.4, units such as the ARM processor core, the DMA core, and the third-party IPs are initially designed for general purpose use and are equipped with more generalized functionalities. They are incorporated into

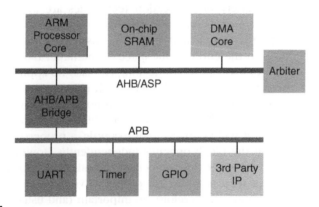

FIGURE 9.4

AMBA bus-based SOC.

an SOC design to avoid the need for developing dedicated logic, which often requires only a subset of the original functionalities. Such reusable components are referred to as **cores** and can be either acquired from other companies or developed internally in a company. In modern SOC designs, a core is often used multiple times within a system or across different systems. For core providers, it is necessary to thoroughly verify the functionality of the core before it is delivered to the core integrators.

Cores are often designed as a stand-alone component in the first place. In addition to core-specific functionalities, standardized bus protocols and/or physical interface standards are then incorporated to offer core reusability. The corresponding verification components used to stimulate and monitor these standard buses or interfaces can, therefore, be reused and shared among cores by use of the same bus protocols or physical interfaces.

Even if a core has its own stand-alone specification, this specification can change because of bug fixing or functionality enhancement, either of which may alter the original functionality. Therefore, it is necessary to re-ensure that operations defined in a previous version of the core will still work correctly in a subsequent version. This requirement is called **backward compatibility**. To meet this requirement, a **regression test suite** is commonly used. Such a test suite is developed by collecting interesting and useful tests from verification conducted on previous versions of the design. A new version must pass these tests to ensure backward compatibility. Note that if a bug exists in old versions of the design, we should not expect regression tests to capture this bug in the new version even if a fix to the bug has been inserted. For that purpose, new tests are required to verify the correctness of the inserted fix.

9.2.4 **Chip-level verification**

A chip-level design consists of multiple units/cores that have complete RTL and bus functional models with well-defined I/O boundaries. At this level, the specification usually does not change significantly from its initial architecture. Hence, the verification requirement is usually well defined.

The aim of chip-level verification is to ensure that the components are properly connected through the interfaces and the entire design abides by the specification. For a regular interface structure such as a bus protocol, only a restricted set of sequences of control and data signals, typically called **transactions**, are permitted. On the basis of the specified interactions between the units, transaction-based tests can be developed to verify the interfaces.

A transaction-based test usually consists of one top-level RTL file that includes all units and bus interfaces and one testbench file that produces transactions to propagate events from one unit to another through the bus interface. Responses at the primary I/Os and/or memory contents are monitored to check the overall behavior of the system.

9.2.5 **System-/board-level verification**

System-level integration is a complex task that requires many tools for design creation, simulation, and analysis. In [Bailey 2007], system-level verification is defined as "the utilization of appropriate abstractions to increase comprehension about a system, and to enhance the probability of a successful implementation of functionality in a cost-effective manner."

Verification at this level involves checking the integration through the interconnections between different chips on the board. The functionality at the lower levels is assumed to have been fully verified. Often, the application software is applied at this stage to verify the entire system.

Verification engineers frequently use programmable logic devices, such as *field programmable gate arrays* (FPGAs), to emulate the design. With the design implemented in programmable devices, the testbenches can be executed directly on such emulated implementations, which is significantly faster than executing the testbenches with a software simulator.

9.3 **MEASURING VERIFICATION QUALITY**

"When can one claim that the verification is complete?" This is a perpetual and still unanswerable question. Even if a verification team performs all the scheduled tasks, and even if no more new bugs can be discovered over an extended verification period, say a few weeks, there is no guarantee that additional simulation would not discover a new bug. The total space to be verified is well beyond what can be exhaustively simulated. Considering a logic block with 64-bit inputs, the combinatorial possibilities for its input space reach 16×10^{18} billion. If simulating one instance takes one nanosecond, then simulating all of them will take 5.07 centuries. Obviously, some modeling, analysis, and optimization techniques need to be used to avoid simulating all tests exhaustively. Various measures are developed to guide the selection of tests for simulation. These measures are typically referred to as **coverage metrics**. Rather than simulating all tests, the idea is to simulate just enough tests to reach a desired coverage goal on the basis of the given metric. The assumption is that achieving the coverage goal implies that a sufficient verification quality has been accomplished.

In this section, we will first introduce the concept of random testing followed by the coverage-driven verification paradigm to outline the concept of coverage in verification. We will also introduce a classification of verification metrics and common coverage metrics within each category.

9.3.1 **Random testing**

Random testing is the most intuitive verification approach. A test generation program is used to generate random tests according to a set of test templates along with a seed. Multiple random instances of each test template are generated and applied to exercise a variety of scenarios for exploring various design corners. A refinement of this approach, called **constrained random verification**, relies on a collection of additional constraints to guide the generation of tests. Figure 9.5 illustrates the concept of the random testing approach.

Random test generation requires two types of inputs to **constrain** the test generation process: (1) a template that serves as the skeleton of the test case, which contains a set of unknown input fields, and (2) a set of arguments for which the values can be set during the generation process. Instead of hand-crafting tests directly, users specify these arguments for input fields within their legal ranges. Multiple instances of physical test cases are then automatically generated from each template by specifying values in the input fields. Templates, along with the changeable arguments, provide an abstract mechanism for hiding the structural details from users while simultaneously satisfying all architectural constraints.

Take microprocessor verification as an example. Its test template is an assembly program with a set of predefined bias arguments. On the basis of these parameters, one can create arguments to:

1. Select an instruction,
2. Select the next instruction on the basis of the current one,
3. Select an operand,
4. Use branch and jump,
5. Cause an overflow or underflow,
6. Interrupt to cause an exception.

However, all the preceding arguments must conform to the architectural constraints, such as, for instance, 32 registers (20 general-purpose, 12 special-purpose), 24-bit addressing, and indirect addressing.

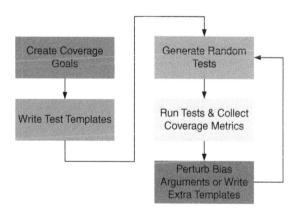

FIGURE 9.5

Flow of random testing.

One corresponding template may look like the following:

MUL < random R1-R4 >< random R4-R8 >< random R8-R20 >

or

< Pr(**ADD**) = 90% & Pr(**SUB**) = 10% > R3 R5 < random R4-R7 >

In the first template, the instruction is designated to be **MUL** (multiplication), and its three operands can be selected from different registers. In the second example, the actual instruction is decided with a probability, where 90% is to be an **ADD** (addition) and 10% is a **SUB** (subtraction) where the third operand is randomly selected from registers R4 through R7.

Random testing is usually applied at the beginning of the verification process for modern designs. Random tests are applied to randomly exercise the design space that often can cover some nontrivial cases and some corner cases. Advanced constrained random test generation uses architecture knowledge of the design and past experience to better guide the test generation process. Both templates and bias arguments help hide the detailed information from users while still being able to generate legal tests that conform to the architectural constraints of the design.

9.3.2 **Coverage-driven verification**

Storing information during simulation is necessary to identify those scenarios that have been previously verified. Such a task is called **functional coverage analysis**. The stored information facilitates the generation of new test cases. *Coverage-driven verification* (CDV) represents such a method. It measures the current verification progress [James 2003] and then guides the development of new strategies for uncovering any missing features or scenarios.

CDV uses a single test stimulus to explore multiple scenarios automatically. Inheriting the characteristics of random testing, CDV can also discover corner cases that might occur beyond a user's expectations. **Coverage points** such as **assertions** are often placed in the environment to collect data for analysis. After collecting and analyzing the data, the constraints for guiding test generation can be modified, either automatically or manually, to target the missing features or scenarios before the next round of test generation is called. This iterative test generation process is known as *coverage-directed generation* (CDG). Figure 9.6 illustrates a typical coverage-driven verification design flow.

CDV [Benjamin 1999; Bergeron 2000; Verisity 2001; Gluska 2003; Palnitkar 2003b] is more effective than constrained random verification and thus achieves verification closure faster. Figure 9.7 illustrates the effectiveness comparison of these two approaches.

Coverage is created to identify the error-prone areas in which bugs may reside. It originates from software testing, which provides a means of assessing the thoroughness of software development. A general definition of coverage is a

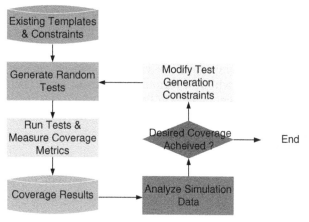

FIGURE 9.6

Coverage-driven verification design flow.

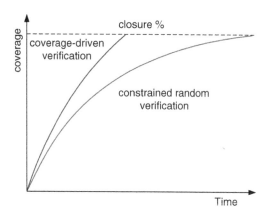

FIGURE 9.7

Effectiveness comparison between coverage-driven verification and constrained random verification approaches.

measure of the extent to which the features and scenarios of the design under verification are covered.

Coverage metrics can be classified into two categories—**functional coverage** and **structural coverage**—according to the verification intent. Functional coverage checks the concordance of the **semantic** design intent with the designer's implementation, and it is measured by the number of features and scenarios defined in the design specification that are exercised by the test set. **Structural coverage** aims at measuring the degree of confidence for **syntactic** correctness of the physical implementation that the test set achieves.

9.3.3 **Structural coverage metrics**

Structural coverage measure is also referred to as **code coverage metric,** because the objective is to evaluate whether various kinds of elements in the HDL implementation are exercised by a given test set. Because code coverage metric ties with test vectors and physical representation in the hardware description language, simulation engines can be easily modified to provide the coverage information. Code coverage comes in many forms. The following describes a few among the commonly used metrics.

9.3.3.1 *Line coverage (a.k.a. statement coverage)*

This metric takes the syntactical HDL implementation and counts the number of lines exercised during the simulation run. The line coverage is defined as:

$$\text{Line Coverage} = \frac{\#\text{ of exercised lines in HDL}}{\text{Total}\#\text{ of lines in HDL}} \times 100\%$$

Consider the following Verilog HDL code in Box 9.1:

BOX 9.1
```
1. always @(in or reset) begin
2.    out = in;
3. if (reset)
4.    out = 0;
5. en = 1;
6. end
```

If the testbench exercises lines 1, 2, 3, 5, and 6, the line coverage would be 5/6 = 83.3%. The line coverage is easy to comprehend, and the missed line explicitly indicates the absence of signal activities. One obvious drawback of line coverage is its lack of a clear connection between the number of exercised lines and the correctness of design intent.

9.3.3.2 *Toggle coverage*

This metric checks whether signals in the design change their values during simulation. It helps verify the quality of the test set and locate the unexercised areas. Signals that fail to be initialized or to toggle by the test cases can be easily identified. Box 9.2 is a sample toggle coverage report.

BOX 9.2
```
1. //net toggle coverage
2. //name      Toggle    0→1    1→0
3. clk         Yes
```

---

4. reset	No	Yes	No
5. start	Yes		
6. state[6:0]	Yes		
7. state[9:7]	No	No	No
8. op[2:0]	Yes		
9. op[3]	No	No	Yes
10. op[4]	Yes		
11. op[5]	No	No	Yes
12. round[1:0]	Yes		
13. src1[63:0]	Yes		
14. src2[63:0]	Yes		

Although the toggle coverage is easy to compute, it has similar drawbacks to the line coverage in that it does not provide any insight about the design intent from the toggle events.

9.3.3.3 *Branch/path coverage*

This metric evaluates the control flow, such as *if* and *case*, in RTL statements. It counts the number of branches at decision points that are exercised during simulation. The branch coverage is defined as:

$$\text{Branch Coverage} = \frac{\#\,\text{of exercised branches}}{\text{Total}\,\#\,\text{of possible branches}} \times 100\%$$

The path coverage refines the branch coverage concept. It does not look at decision points independently. Instead, it considers the whole sequence of decision points, called a path, which could possibly be involved in one clock cycle. Note that when *if* or *case* statements are nested, the total number of possible paths may grow exponentially. Therefore, reaching a 100% path coverage may become difficult.

Consider the preceding exemplar Verilog HDL code in the discussion of line coverage. Assume the signal *reset* is always 1. Then, for the *if* statement, only the *reset* = 1 branch is exercised. Thus, the branch coverage is 1/2 = 50%. Now consider another example:

BOX 9.3
```
1. if (x != y)
2.    z = 0;
3. w = z;
```

In Figure 9.8, the RTL code is represented in two flowcharts — each of which is from the line and branch coverage viewpoints, respectively. Assume the values of signal *x* are never equal to those of *y* during simulation. Then line 2 will be exercised, resulting in a final line coverage of 100%. But the branch ($x == y$),

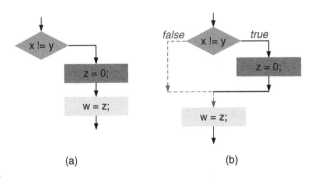

FIGURE 9.8

(a) Flowchart for line coverage. (b) Flowchart for branch coverage.

represented by the dotted line in Figure 9.8b, is never exercised, resulting in a branch coverage of only 50%.

Note that designers can implement the branch condition implicitly without the use of *if* or *case* statements. For example, an *if-else* condition can be implemented by a multiplexer that uses AND or AND-NOT operations. Hence, it may not be always apparent to know exactly where to collect the branch statistics to calculate a branch coverage. In many situations, a branch not explicitly implemented by use of *if* or *case* statements may not be accounted for in the coverage.

9.3.3.4 *Expression coverage*

The expression coverage enhances the line and branch coverages and provides more information about concurrent signal assignments. It focuses the analysis on the expression in the right-hand side of an assignment or the expression in a condition statement.

Typically, one expression can be recursively decomposed into multiple **sub-expressions**, which are either a single variable or two variables connected by a logical operator. These sub-expressions are monitored individually during simulation. An expression is fully covered if all of the sub-expressions are exercised. Otherwise, the expression coverage for a line is calculated by deriving the ratio of the total number of exercised cases to the total number of possible cases among all of its sub-expressions.

$$\text{Expression Coverage} = \frac{\sum_{i=1}^{k} \# \text{of exercised cases for sub-expressions } i}{\sum_{i=1}^{k} \# \text{of possible cases for sub-expressions } i} \times 100\%$$

The expression coverage can be further classified into three categories: **multiple sub-condition**, **basic sub-condition**, and **focused expression coverages** [Dempster 2002].

The *multiple sub-condition coverage* (MSC) is the most popular and straightforward one. It enumerates all possible combinations of the sub-expressions. That is, if there are N sub-expressions, then 2^N cases need to be covered to achieve a 100% multiple sub-condition coverage. Consider the following expression in Box 9.4:

BOX 9.4

1. **if** ((A == 0) || ((B == 1) && (C == 0)))

The participating sub-expressions are (A == 0), (B == 1), and (C == 0). Thus, the test vectors have to cover all $2^3 = 8$ possible cases to achieve a 100% multiple sub-condition coverage.

The **basic sub-condition coverage** (BSC) checks both the true and false states of each sub-expression during simulation. For the preceding example, there are six possible cases: (A == 0) is true, (A == 0) is false, (B == 1) is true, (B == 1) is false, (C == 0) is true, and (C == 0) is false. A sample report, after the basic sub-condition coverage is derived, is listed in Box 9.5:

BOX 9.5

	Count	Sub-expression	Outcome
1.	Count	Sub-expression	Outcome
2.	4	A == 0	true
3.	6	A == 0	false
4.	8	B == 1	true
5.	2	B == 1	false
6.	0	C == 0	true
7.	10	C == 0	false

In this report, because the condition "(C == 0) is true" has never been exercised during simulation, the basic sub-condition coverage is (5/6) = 83.33%.

An expression is a function of the participating variables combined with Boolean operators. If one variable in focus can control the result of the expression, there should be a pair of variable assignments for which the values at all other variables, except the focused variable, are the same so that one assignment evaluates the expression to be true and the other assignment to be false. On the basis of this notion, the **focused expression coverage** (FEC) is developed, which helps identify the minimum set of tests required for verifying a complicated branching expression. To achieve a 100% FEC for an expression, for each participating variable in the expression, the test set must include a pair of vectors that assign identical values to all other variables except the target variable, and these two vectors evaluate the expression to different values.

To illustrate this notion, consider the expression in Box 9.6:

BOX 9.6

1. **if** (A && B)

The focused expression coverage criteria for variable A are [A, B] = [0, 1] and [A, B] = [1, 1]. Note that in both cases, B has to be 1 for the effect of changing

A to be observed. Similarly, the criteria for variable B are [A, B] = [1, 0] and [A, B] = [1, 1]. Because [A, B] = [1, 1] is a common assignment, it would require only three assignments to fully validate expression (A && B).

Now consider the following example in Box 9.7:

BOX 9.7

1. **if** (((X == 1) && (Y == 0)) || (Z == 0))

The three sub-expressions are expr_1 = (X == 1), expr_2 = (Y == 0), and expr_3 = (Z == 0). To achieve a 100% FEC, the test set must include the following tests:

- To target expr_1, [expr_1, expr_2, expr_3] = [0, 1, 0] and [expr_1, expr_2, expr_3] = [1, 1, 0] are required. Note that expr_2 has to be 1 because it is **AND**ed with expr_2. Similarly, expr_3 has to be 0 because it is **OR**ed with the rest of the expression. The result is that (X, Y, Z) = (0, 0, 1) and (1, 0, 1) must be covered.
- To target expr_2, [expr_1, expr_2, expr_3] = [1, 0, 0] and [expr_1, expr_2, expr_3] = [1, 1, 0] are required. Therefore, (X, Y, Z) = (1, 1, 1) and (1, 0, 1) must be covered.
- To target expr_3, there are three different ways to ensure expr_3 controlling the overall expression: [expr_1, expr_2] = [0, 0], [0, 1] and [1, 0] respectively. Therefore, one of following three pairs, (X, Y, Z) = {(0, 1, 1), (0, 1, 0)}, {(0, 0, 1), (0, 0, 0)}, and {(1, 1, 1), (1, 1, 0)} must be included in the test set.

Combining these three requirements, the minimum test set for a 100% FEC includes 4 tests which are either {(0, 0, 1), (1, 0, 1), (1, 1, 1), (0, 0, 0)} or {(0, 0, 1), (1, 0, 1), (1, 1, 1), (1, 1, 0)}.

Suppose a given test set contains only two tests, (X, Y, Z) = (1, 0, 1) and (X, Y, Z) = (1, 0, 0), which evaluate [expr_1, expr_2, expr_3] to [1, 1, 0] and [1, 1, 1], respectively. With respect to the focused expression notion, none of the three sub-expressions is satisfied by these two tests and, thus, its focused expression coverage is 0%.

9.3.3.5 *Trigger coverage (a.k.a. event coverage)*

This metric simply measures the number of exercised variables in the sensitivity list. Consider the example given in Box 9.8:

BOX 9.8

1. **always** @(a **or** b **or** c)
2. **begin**
3.   ...
4. **end**

Signals a, b, and c are monitored throughout the simulation. If only b and c change values during simulation, then the trigger coverage would be $2/3 = 66.67\%$.

9.3.3.6 *Finite state machine (FSM) coverage*

The FSM coverage plays an important role in verifying the control unit of a design. As its name implies, this metric is tied to the HDL structure of finite state machines in a design and can be divided into three sub-classes. The **state coverage** reports the states that are visited and their frequencies during simulation. The **arc coverage** records the state transitions that are traversed during simulation. Even if 100% *state* and *arc coverages* are achieved, there is no guarantee that the FSM is bug-free. Therefore, the third class of FSM coverage, called **sequential arc coverage** (*a.k.a.* **transition coverage**), was designed. The metric measures the coverage on the basis of an increased sequential depth of state visitation or arc traversal. It also identifies the fundamental cyclic sequences in various lengths. Figure 9.9 shows an FSM example and the arc sequences starting from s_1 for calculating the sequential arc coverage. For example, $\{s_1 \rightarrow s_2 \rightarrow s_2\}$ is a *2-arc* transition starting from $s1$ to be monitored for the sequential arc coverage.

In calculating the coverage, the conventional FSM coverage interprets the RTL code syntactically. That is, it treats each state as a unique state and its state transition to any other state as a unique arc. Although each state has a unique state code, it is common that a group of states have identical or very similar behavior. Therefore, interpreting the FSM syntactically may result in many unnecessary checks. Consider the following partial RTL code of a 4-bit binary counter with *reset* and *load* signals in Box 9.9:

```
BOX 9.9
1. always @(posedge clk) begin
2.    if (reset) count = 0;
3.    if (load) count = in;
4.    else if (count == 15)
5.       count = 0;
6.    else
7.       count = count + 1;
8. end
```

The implementation has 16 states. Because any state can go to any other state including itself (either through incrementing the count variable or through loading a new state value in), each state has 16 outgoing arcs, resulting in a total of 256 arcs. Figure 9.10a illustrates this conventional interpretation of the FSM. If the counter is 8-bit, the total number of states will increase to 256 states with 65,526 arcs.

To represent the design as an FSM, it is better to interpret it semantically, which defines the states on the basis of the unique actions taken during the

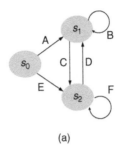

length of transition	1-arc	2-arc
	$s_1 \rightarrow s_1$	$s_1 \rightarrow s_1 \rightarrow s_1$
sequence of states from s_1	$s_1 \rightarrow s_2$	$s_1 \rightarrow s_1 \rightarrow s_2$
		$s_1 \rightarrow s_2 \rightarrow s_1$
		$s_1 \rightarrow s_2 \rightarrow s_2$

(a) (b)

FIGURE 9.9

(a) FSM example. (b) Transition sequences from s_1.

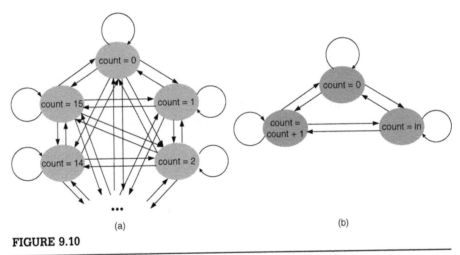

(a) (b)

FIGURE 9.10

Illustration of (a) the conventional FSM coverage. (b) the semantic FSM coverage.

operation. For the preceding example, there are only three different actions: count = 0, count = in, and count = count + 1. Figure 9.10b shows the FSM of this interpretation, which consists of only three states and nine arcs. The **semantic FSM coverage**, calculated on the basis of this representation, can greatly reduce the number of tests required for achieving a high coverage.

9.3.3.7 *More on structural coverage*

Different metrics for structural coverage can be associated with different HDL structures at different design stages. In general, during the behavioral-level design stage, only line, branch, condition, path, trigger, and FSM coverage can be measured. Toggle coverage is often applied to gate-level designs only. The RTL-level design stage has the broadest possible coverage spectrum, and all types of metrics can be applied.

Table 9.1 Typical Coverage Targets for Different Metrics

Metric	Coverage Goal (%)
Line	100
Branch	100
Condition	60~100
Path	>50
Trigger	100
Toggle	100
FSM (state and arc)	100

Because these metrics are simple and straightforward, it is often desirable to achieve a high structural coverage. The typical coverage goals for various metrics are listed in Table 9.1 [Dempster 2002].

Even if the desired coverage for these metrics is achieved, it does not guarantee a bug-free design. None of these metrics — or even were we to combine them all — can be guaranteed to cover all the possible erroneous scenarios.

The structural coverage attempts to explore the design space from the implementation perspective. Although the targets of the structural coverage do not necessarily have direct correlation to functional bugs, achieving a high structural coverage can likely increase the chance of bug discovery. A bug may be revealed by a new test that was designed to detect a not-yet-covered structural target.

9.3.4 Functional coverage metrics

Functional coverage metrics guide test generation and verification from a semantic perspective. They supplement the deficiencies in the code coverage and help improve verification quality. Some companies have stated that functional coverage would be an important component of their next-generation verification methods [Drucker 2002].

Functional coverage metrics usually involve the interpretation of functionality and the related measurements from the specification, and require domain knowledge and instrumentation from the designer and/or verification engineers. Therefore, an automated means of creating functional coverage models does not exist. Typically, verification engineers need to manually develop a list of target functionalities to be verified and to devise different strategies to exercise each case in the list. A functional bug is claimed to be found if the design does not behave as expected with respect to the functional specification after exercising the related verification scenarios.

The verification method based on the functional coverage includes four major tasks:

1. Determining the coverage events to be verified
2. Preparing stimuli to exercise the target events
3. Collecting data from the design under verification
4. Analyzing results to quantify the coverage and identify missing events

Basically, it is the designer's job to determine the functions to be covered. Verification engineers are required to create a verification plan on the basis of their understanding of the design's functional specification. In addition to enumerating the functions under verification, external resource expenditures, including verification time, manpower, and related software and tool costs, should also be carefully considered.

The verification plan forms the basis for developing the corresponding test programs. Random testing techniques are often used at the transaction level to facilitate test program development. For the AMBA APB part of the example in Figure 9.4, transactions considered for functional coverage could be based on either a simple operation, like a Read/Write to RAM, or a complicated operation, like a sequence of back-to-back Reads to the same address in RAM.

9.4 SIMULATION-BASED APPROACH

[Bergeron 2000] introduced a *re-convergence model* for the general design and verification process. Figure 9.11 illustrates the application of this model to functional verification. The designer's effort is dedicated to transforming the functional specification into an implementation in HDL, whereas the verification effort ensures that the transformation is as intended without misinterpreting any functionality.

The functional verification process is typically associated with the concept of *testbench,* which refers to the environment used to apply the predetermined sequence of input vectors to the ***design under verification*** (DUV) and to observe the responses. Figure 9.12 illustrates a DUV surrounded by a testbench. No external communication is required in this system. The testbench models certain aspects of the design intent and is responsible for delivering the input sequences to the DUV and for receiving the output responses for subsequent analysis.

FIGURE 9.11

Re-convergence model for the design and verification process.

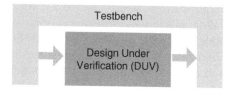

FIGURE 9.12

Generic structure of the design under verification and its testbench.

9.4.1 Testbench and simulation environment development

In general, the testbench is an HDL description used to create a closed system on top of the design under verification. A testbench consists of three fundamental components: a **stimuli driver**, a **monitor**, and a **checker**.

The **stimuli driver** is responsible for providing stimuli to the DUV. The stimuli can be either predetermined or generated during simulation. The purpose of the stimuli driver is not to mimic the behavior of the entire neighboring blocks but to maintain the interface coherence to the DUV.

The **monitor** is used to observe signal at the inputs, outputs, and any internal wires of interest on the DUV. The values at the input and output signals must be consistent with the interface protocol, and the monitor will issue an error if any exception occurs.

A **checker** can be viewed as a special type of monitor for checking the functionality of the design intent. Traditionally, designers create the functionality checkers manually and use them to compare the responses from the design with the specification. As designs become more complicated, the need to automate the development of such checkers increases.

On the basis of the coverage metrics, verification engineers try to prepare a set of test cases to cover the target functional events. In developing such test cases, experience plays a crucial role. Creating meaningful test cases for some specific events often rely heavily on a designer's knowledge and interpretation of the specifications.

Consider a 16-bit one-hot encoding bus protocol. To achieve an optimal coverage for all scenarios, the test cases would require each bit taking a turn to be 1 with others being 0. In deriving the test cases, it could be difficult to observe the regularity solely from the structure of a design implementation. However, having knowledge of the functionality of the protocol would help capture the regularity and similarity for each bit that make test generation easier and more efficient.

Enumerating deterministic test cases to cover all functions is tedious. An alternative is to convert a design specification into an HDL model to automate the checking. Such a testbench is called a *self-checking* testbench, because checking instrumentation is no longer needed. The *self-checking* testbench paradigms can be divided into three types: **checking with golden vectors**, **checking against a reference model**, and **transaction-based checking**.

Checking with golden vectors is the most widely used approach among the three. Given coverage metrics, the verification engineers search for test cases at inputs and derive the corresponding output responses manually or by use of an auxiliary program. Such combinations of input and output vectors are called the golden vectors. After the testbench applies the input vectors to the DUV, the actual responses are captured and compared with the golden vectors. A bug is found when a mismatch occurs between the golden and the actual responses. Figure 9.13 shows the components of this method.

The **checking-against-a-reference-model** paradigm uses a reference model that captures all functions in the specification. The reference model is typically implemented at a more abstract level with either a high-level programming or a verification language. All input vectors are applied to both the reference model and the DUV, and their responses are evaluated and compared. If the comparison takes place at the end of each cycle, the reference model must be cycle-accurate. The checker compares the responses from both the DUV and the reference model, as illustrated in Figure 9.14. If the specifications change, the reference model would need to be modified accordingly. This modification effort is usually much lower than the effort of reproducing all golden vectors required for the *checking-with-golden-vectors* paradigm.

Transaction-based checking is applicable to the DUV that can correspond to commands and data in a transaction. It uses a scoreboard to record the verified command and data. The checker is used to query the scoreboard. It issues an error if the identifier cannot match any transaction in the scoreboard or if the

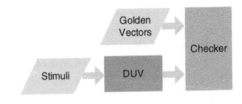

FIGURE 9.13

Self-checking testbench with golden vectors.

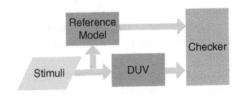

FIGURE 9.14

Self-checking testbench with a reference model.

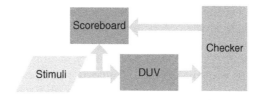

FIGURE 9.15

Transaction-based self-checking testbench.

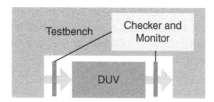

FIGURE 9.16

Black-box verification.

command and data are not the expected values given by the scoreboard. This concept is illustrated in Figure 9.15.

9.4.2 **Methods of observation points**

As we can see in the preceding, the monitor and checker in one testbench are tightly tied to the concept of observation of signal changes in the DUV. Such observation approaches will also determine the strategy used for generating stimuli. The three common verification paradigms regarding the observation points are the **black-box**, the **white-box**, and the **grey-box** methods.

The **black-box** method assumes the internal signals of the DUV are not accessible during verification. Only the external input/output interfaces are directly controllable and observable. The verification plan, including the test-bench development, is developed based only on input/output functionality. Figure 9.16 illustrates this method.

The major advantages of black-box verification are its simplicity and independence from specific implementation information. Of all the verification methods, it requires the least amount of knowledge about the DUV. Even if the design's HDL code is not ready, the verification process can be started, and stimuli can be developed as long as a reliable specification for the DUV becomes available. Whether the DUV is realized as an ASIC, an FPGA, a circuit board, or a software program is irrelevant. The black-box method only aims at verifying the functionality defined with respect to the design boundaries.

On the other hand, without any structural information, black-box verification lacks the observability and controllability internal to the DUV, which sometimes might be required to determine whether the DUV passes or fails a specific test. It is challenging to precisely identify what and where a problem is in the DUV with this method. It may not be feasible for the black-box verification to check for DUV's low-level features and structural changes. Black-box verification may not be suitable for designer-level blocks, because many interesting corner cases may be observed only when implementation details are provided.

In short, the black-box method requires no implementation knowledge and demands only design specification to complete the testbench development. Being independent of the implementation makes the generated stimuli more reusable for different realizations, but it also makes the stimuli generation process more difficult because of the lack of observability and controllability internal to the DUV.

The **white-box** method, which is illustrated in Figure 9.17, represents another extreme scenario. Here, the full observability and controllability internal to the DUV is assumed to be available. For controllability, verification engineers can easily derive stimuli for the desired events by setting up the required internal states and justifying these states backward toward the inputs. Likewise, regarding observability, any changes in internal signals can be directly observed. Therefore, the white-box method can pinpoint the problematic area in the DUV once a mismatch from the expected value is observed.

Low-level features and implementation changes can be incorporated in the white-box approach, because such verification is tied to a specific implementation. Therefore, the generated test cases may only be valid for the specific implementation. Modification to the generated test cases would be necessary if the implementation changes. Therefore, the maintenance efforts required for the white-box method would be much greater than those for the black-box method.

White-box verification can ensure that implementation-dependent features are verified. For example, it becomes feasible to generate test cases to exercise a timing-critical path when the full observability and controllability to the internal structure of the DUV is available.

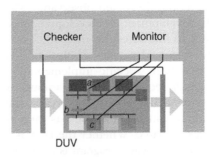

FIGURE 9.17

White-box verification.

The **grey-box** approach is a compromise between the black-box and the white-box approaches, which inherits the advantages from both methods. This approach intends to exercise only those significant features associated with the implementation.

The general architecture of the DUV is assumed to be known by the verification engineers, and only a limited number of internal points are accessible. These observation points are often located in the inter-block interface and adhere to specific communication protocols. In other words, the grey-box verification method observes only a select set of important internal signals, which are typically located at the boundary of a building block. Therefore, for the illustration in Figure 9.17, a grey-box method would preclude observation of the monitor *c* but would include the other two observation points.

Similar to the white-box approach, the grey-box approach could exercise a desired event by applying a test case directly at inter-block interfaces. Even if the implementation of the components changes, as long as the interfaces between the components within the DUV remain unchanged, the generated test cases can be reused.

9.4.3 **Assertion-based verification**

Assertion-based verification is becoming popular in the industry and has drawn much attention in the recent literature [Foster 2004]. This method embeds a set of assertions in various parts of the implementation for monitoring design properties. Assertion-based verification can be viewed as a variant of the white-box method.

The concept of assertions is originated from software testing. An assertion is a line in the program that checks the validity of an expression. A correct program must guarantee that such expressions are always true; otherwise, a warning or exit signal should be issued. Software engineers frequently write assertions to check the possible existence of unexpected scenarios. Many high-level programming languages such as C/C++, Java, and Eiffel support assertions by the use of a system library or by the use of the language definition itself. Actually, the first standardization of VHDL defined its language constructs to support simple assertions, as shown in Figure 9.18.

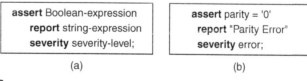

assert Boolean-expression	**assert** parity = '0'
report string-expression	**report** "Parity Error"
severity severity-level;	**severity** error;

(a)   (b)

FIGURE 9.18

(a) Syntax. (b) example of an even-parity assertion in VHDL.

Similar to software testing, assertions in hardware design are also expressed as part of the design description in the HDL code. Many contemporary *hardware verification languages* (HVLs), such as **SystemVerilog** [Accellera 2002a] and **OpenVera** [Synopsys 2001], were developed to facilitate the writing of assertions in conjunction with the design itself. Another flavor of practical solutions is to use an auxiliary specification language. Several different proprietary formats of specification languages exist, such as **PSL/Sugar** [Accellera 2002b]. Assertions can be written in the specification language with a proper interface to the design.

The use of assertions in verification has various advantages. In black-box verification, for example, assertions can be used to replace the original monitors for the purpose of collecting coverage data. In white-box verification, the origin of an assertion failure could be confined to a limited area to facilitate the debugging process. It is also a good practice to use assertions as formal comments in place of comments in natural language. Meanwhile, assertions can be reused as part of the verification IP associated with the IP core delivered to the customers. Moreover, because assertions are placed in the HDL code, they can be directly used as properties to be checked for the use of formal methods.

9.4.3.1 *Assertion coverage and classification*

The term *assertion coverage* has a variety of definitions. It could be used to indicate the ratio of the number of assertions to the number of HDL code lines. However, **assertion density**, suggested in [Piziali 2004], is considered a better term for this definition. The better definition for assertion coverage should be similar to that of functional coverage, which is defined on the basis of the number of exercised scenarios over the total number of scenarios to be covered. *Assertion coverage* counts the number of exercised assertions to the total number of assertions extracted from the design implementation.

Assertions can be classified into two types: *static* and *temporal*.

- **Static assertions** dictate those legal scenarios that are not related to time, and, as such, they are required to be held for all time. These scenarios can be described by the first-order logic. The one-hot encoding bus is an example. Only one bit in such a bus can be one, and the rest should be zero. A static assertion monitors the bus during the course of simulation and sends an error message whenever this rule is violated.
- **Temporal assertions** extend the capability of static assertions to temporal logic. The consequent statement needs to be evaluated during the specified period of time after which the antecedent condition is triggered. Consider the following SystemVerilog example in Box 9.10:

BOX 9.10
```
1. @(posedge clk)
2.    init_event |=> abort_event;
```

where |=> denotes the non-overlapping implication operator. This example states that once an antecedent condition, **init_event**, successfully completes, a consequent statement, **abort_event**, will occur in the next clock cycle.

The behavior of temporal assertions can be illustrated by a finite state machine, as shown in Figure 9.19. In the *idle* state, the assertion moves to the *evaluate* state when its antecedent condition is triggered. The *evaluate* state repeatedly checks the consequent statement before a Pass/Fail result is issued. Once there is a result, either an error signal is generated or the system moves back to the *idle* state.

To illustrate a *SystemVerilog Assertion* (SVA) example, assume that the intended property in a design is the following: "after the request signal is asserted, the acknowledge signal must be generated from 1 to 3 cycles later." Figure 9.20 shows its timing diagram and the corresponding code in SVA.

9.4.3.2 *Use of assertions*

For different types of properties, assertions can be divided into two categories: **coverage assertions** and **checker assertions**. Coverage assertions primarily record the occurrence frequency of a specified event. Such assertions usually monitor events defined in the functional coverage metrics. For the example of a 16-bit one-hot coded bus, the assertion defines all possible combinations of 16 one-hot cases and records the case(s) exercised during simulation.

Checker assertions function as sentinels. They watch the violation of static or temporal properties. At the module level, in white-box verification, assertions

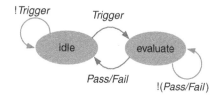

FIGURE 9.19

Finite state machine for generic assertions.

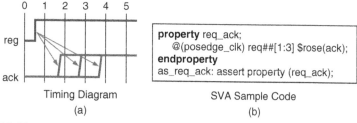

Timing Diagram
(a)

SVA Sample Code
(b)

FIGURE 9.20

Example of a temporal assertion in SVA.

can check implementation details, whereas in black-box verification, assertions check against the specification through both module inputs and outputs. For higher-level verification, checker assertions are used to monitor the interfaces across components. Because the interfaces must abide by their corresponding protocols, checker assertions signal errors once unexpected scenarios occur. A two-hot message in a one-hot coded bus is such an example.

9.4.3.3 *Writing assertions*

One of the most frequently asked questions in assertion-based verification is "Who should write the assertions?" In practice, this job is shared by the entire design and verification team. At different levels of the design abstraction, different properties are converted into assertions. It may be difficult to ask a designer responsible for designing a small block and lacking a system-level view to write high-level assertions.

At the architectural level, a design is described by use of the input/output functions of each component and the interface protocols that connect them, without implementation detail. Assertions at this level model high-level relationships and ensure that system-level behavior is consistent with the system-level specification. Also at this level, observation points are located at inputs and outputs of the components and at bus interfaces only.

Assertions try to capture one's understanding of the design intent. Once a design component is created, the designer can write assertions for it on the basis of the functionality from the specification and the implementation he or she chooses. At this level, assertions are frequently used for debugging and for measuring coverage.

If applicable, verification engineers may use formal methods to prove assertions to complement the deficiencies of simulation-based methods. Also, assertions accompanied with IP cores from IP providers would need to be integrated into the verification plan.

9.5 FORMAL APPROACHES

Advances in modern simulators allow full-chip simulation to be efficiently conducted. Nevertheless, the success of simulation-based verification remains dependent largely on the quality of the stimuli. The stimuli exercise a *design under verification* (DUV) and traverse its state space. Verification can be considered as a process of exploring reachable state space of the design. Modern designs rapidly increase in size and complexity, and, consequently, their reachable state space can grow exponentially. As a result, it becomes difficult to exhaust all reachable states for complete verification by use of only simulation.

Formal approaches aim to make complete verification possible, where completeness is in the sense that all reachable states are explored. The underlying idea is to infer the design properties by reasoning without explicitly simulating stimuli. A property models certain aspects of design behavior associated with all or a subset of reachable states. Proving design properties with formal approaches requires the use of efficient search or reasoning engines, many of which have been developed over the years. Significant advances have been achieved in recent years.

The remainder of this chapter provides an overview of modern formal verification approaches. Three major types of formal approaches are introduced: *model checking*, *equivalence checking*, and *theorem proving*. For each approach, we explain the underlying theory, illustrate its use, give examples, and discuss the advantages and disadvantages. Finally, we include a brief review of advanced research topics in the area.

9.5.1 Equivalence checking

Modern VLSI design flow is partitioned into a number of synthesis steps that take the idea from system specification into GDSII. This results in descriptions at different abstraction levels, which include behavioral, RTL, gate, and switch levels. Ensuring equivalence between two alternative descriptions of the same design is a commonly encountered problem in a design process. This task is referred to as *equivalence checking*. Although such a general concept can be applied to detect any mismatch from two descriptions given at any level, commercially available equivalence checking tools typically address the equivalence between the design's RTL code and its various gate-level netlists, as shown in Figure 9.21. That is the focus of this section.

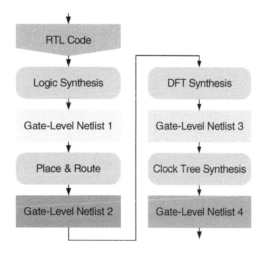

FIGURE 9.21

RTL to gate-level design flow.

Boolean circuits, in general, can be viewed as *finite state machines* (FSMs), and, therefore, *Boolean equivalence checking* (BEC) over two circuits, FSM_1 *and* FSM_2, can be formulated as the problem of checking for the output of the **miter circuit**, as shown in Figure 9.22, being constant 0 or not. FSM_1 consists of combinational logic C_1 and a state-holding element set, S_1, whereas FSM_2 consists of combinational logic C_2 and a state-holding element set, S_2. Both primary inputs are m bits and primary outputs are n bits. PPO_1 (PPO_2) denotes the pseudo-primary outputs from to C_1 (C_2) to S_1 (S_2). Note that the number of state-holding elements can be different in the two FSMs. Each pair of corresponding primary output bits — one from C_1 and the other from C_2 — connects to an XOR gate. If any XOR output becomes 1 with respect to any input vector or sequence, these two FSMs are not equivalent.

A simplified version of the BEC problem is *combinational equivalence checking* (CEC). This problem assumes that FSM_1 and FSM_2 have a complete, one-to-one mapping between the state-holding elements and that they start with the same initial state. The assumption is also made that PPO_1 always has the same value as PPO_2. Hence, the original miter circuit can be recast as that shown in Figure 9.23; here, we only focus on the comparison between combinational logic C_1 and C_2 without any sequential elements. The combinational equivalence checking problem is thus formulated as the following: Given two combinational Boolean netlists C_1 and C_2, check whether the corresponding outputs of C_1 and C_2 are equal for all input combinations. There are two types of approaches for solving the CEC problem: *functional equivalence* and *structural equivalence*.

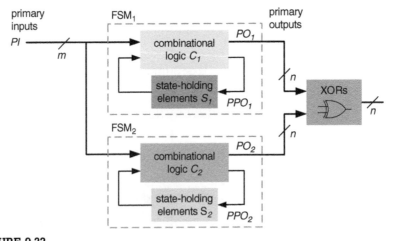

FIGURE 9.22

Miter circuit for checking equivalence of two FSMs.

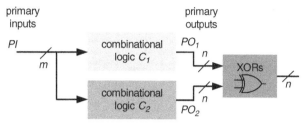

FIGURE 9.23

Combinational equivalence checking.

9.5.1.1 *Checking based on functional equivalence*

The first step of functional CEC is to translate the combinational circuits into a *canonical representation*. A representation of a Boolean function is *canonical* if the representation for each function is unique and independent of the implementation of the function. A *truth table* is one example of a canonical representation for Boolean functions. Equivalence can be determined by directly comparing the two canonical representations. Among all canonical representations, the reduced **ordered binary decision diagram** (OBDD), introduced in Chapter 4, is the most prevalent, because OBDD yields a more compact representation than other representations. The CEC problem can be resolved by building the OBDDs for the outputs of the circuits on the basis of their primary inputs. Two circuits are equivalent if the OBDDs from each pair of corresponding outputs are graphically isomorphic.

9.5.1.2 *Checking based on structural search*

A structural search approach checks to see whether any vector exists at primary inputs that would cause a mismatch between the two circuits at their primary outputs. If no such input vector can be found, the two circuits are proven equivalent. The **satisfiability** (SAT) solvers, introduced in Chapter 4, can be used as the structural search engine for checking equivalence. A SAT solver can be used to check if an assignment at PIs exists to satisfy a 1 at the miter's output. An UNSAT answer from the solver proves the equivalence of the two circuits. An ATPG tool developed for generating manufacturing tests for stuck-at faults can also be used for checking structural equivalence. As illustrated in Figure 9.24, if the stuck-at-0 fault at the XOR output is proven a redundant fault by an ATPG tool, the two circuits are equivalent. A thorough treatment of ATPG techniques will be provided in Chapter 14.

For complex circuits, directly applying SAT solving at the miter's output signal may result in an exponential number of backtracks, which makes the approach inefficient. Structural similarity between the two circuits under checking can be explored to improve its efficiency, which attempts to solve the structural equivalence problem by incrementally solving a sequence of easier sub-problems

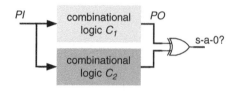

FIGURE 9.24

Checking structural inequivalence by generating a test for XOR output stuck-at-0 fault.

[Brand 1993; Kunz 1993; Goldberg 2000; Huang 2000]. On the basis of a divide-and-conquer strategy, various heuristics have been developed to identify internal equivalent points from the two circuits under checking. For example, when two signals are proved to be equivalent, the equivalence of the two signals can be encoded as a SAT clause and added back to the SAT formulation of the problem. Such equivalence clauses can then help to speed up the SAT search, as shown in [Lu 2003].

For the **sequential equivalence checking** (SEC) problem, shown in Figure 9.22, *state traversal* techniques are often used. The most common state traversal technique is *reachability analysis*. Note that two FSMs, M_1 and M_2, are equivalent if, and only if, the output of the *miter circuit* $M_{1\otimes2}$ is constant 0 under all combinations of input assignments for all *reachable states* of $M_{1\otimes2}$. Therefore, checking sequential equivalence would require the ability of deriving the set of states reachable from a given initial state set I for a given FSM M. An intuitive approach that explicitly enumerates state transitions over the state graph of the FSM is not scalable to large design and, thus, is often impractical. Practical solutions usually adopt a *symbolic* technique implemented by OBDD that implicitly derives the reachable state set by use of *transition functions*.

Symbolic reachable analysis consists of two steps: (1) encoding the FSM symbolically and (2) performing reachability analysis iteratively. Given FSM $M_1 = (Q_1, I_1, \sum_1, \Omega_1, \delta_1, \lambda_1)$ and FSM $M_2 = (Q_2, I_2, \sum_2, \Omega_2, \delta_2, \lambda_2)$, where Q_i's, I_i's, $\sum_i$'s, Ω_i's, δ_i's, λ_i's denote the state spaces, the initial state sets, the input and output alphabets, transition functions, and output functions, respectively, the FSM $M_{1\otimes2} = (Q_m, I_m, \sum_m, \Omega_m, \delta_m, \lambda_m)$ for the miter circuit can be constructed as follows:

- The state space $Q_m = Q_1 \times Q_2$
- The initial state set $I_m = I_1 \times I_2$
- $\sum_m$ and Ω_m are the same input and output alphabet sets as in M_1 and M_2 (that is, $\sum_m = \sum_1 = \sum_2$ and $\Omega_m = \Omega_1 = \Omega_2$)
- The transition function $\delta_m(s, a)$: $\sum_m \times Q_m \rightarrow Q_m$, where s and a represent for one state in Q_m and one input vector in Σ, respectively
- The output function $\lambda_m(s, x)$: $\sum_m \times Q_m \rightarrow \Omega_m$

We define a new function, called *transition relation*, which is denoted as $R(x, s, s')$: $(\sum_m \times Q_m) \times Q_m \rightarrow \{0, 1\}$. $R(a, p, q) = 1$ if there exists a transition from the state p to the state q under an input vector a for $M_{1\otimes2}$; otherwise, $R(a, p, q) = 0$. Assume given an input vector set $x = (x_1, x_2, \ldots, x_k)$ with the corresponding sequence of state transitions $\delta_x = (\delta_1, \delta_2, \ldots, \delta_k)$, the transition relation from the state s to the state s' can be formulated as:

$$R(x, s, s') = (s_1' \equiv \delta_1(s,\ x)) \wedge (s_2' \equiv \delta_2(s,x)) \wedge \ldots \wedge (s_k' \equiv \delta_k(s,x)) = \Pi_i(s_i' \equiv \delta_i(s,x))$$

Therefore, if the input vector set x can bring the finite state machine from the state s to the state s', then $R(x, s, s') = 1$; otherwise, $R(x, s, s') = 0$.

We then annotate the existential quantification operator $\exists$ to the transition relation R. A pair of states $(p, q) \in R_\exists$ if, and only if, there exists an input vector x such that the machine transitions from state p to state q after applying x. Applying the *existential quantification* notation $\exists$ to the preceding *transition relation* results in $R_\exists(s, s')$. Such a notation is called **quantified transition relation** and represented as:

$$R_\exists(s, s') = \exists x.(s_1' \equiv \delta_1(s,x)) \wedge (s_2' \equiv \delta_2(s,x)) \wedge \ldots \wedge (s_k' \equiv \delta_k(s,x)))$$
$$= \exists x.\ \Pi_i(s_i' \equiv \delta_i(s,x))$$

Given $M_{1\otimes2} = (Q_m, I_m, \sum_m, \Omega_m, \delta_m, \lambda_m)$ and its *quantified transition relation*, we can apply $R_\exists$ to derive all reachable states. Such a process is called reachability analysis and can be done by the **image computation** denoted as $Img(S, R_\exists)$, where S is a set of given states and $R_\exists$ is the *quantified transition relation* defined by $M_{1\otimes2}$. The output of $Img(S, R_\exists)$ is the set of states reachable from S in one clock cycle. One approach to reachability analysis is to iteratively perform image computation starting from the initial state set I_m. Such an approach is called *forward* reachability analysis, and the generic pseudocode is outlined as follows:

Algorithm 9.1 Forward_Reachability

---

1. $i := 0$	// counter for looping
2. $Q^i := I$	// i-th set of reachable states
3. **do** {	
4. $Q_{new} := Img(Q^i, R_\exists)$;	// compute image from current states
5. $Q^{i+1} := Q^i \vee Q_{new}$;	// update the state set for next iteration
6. $i := i + 1$;	// counter increments
7. } **until** $(Q^{i+1} \equiv Q^i)$	// stop when state set is stable
8.	
9. **return** Q^{i+1}	

Consider the 7-state FSM shown in Figure 9.25 for which state 0 is the only initial state. The forward reachability algorithm derives all reachable states from state 0 as follows in Table 9.2:

The iterative process stops at iteration 4 for which the current set of reachable states is equivalent to the next set of reachable states. Therefore, the set of reachable states from state 0 is {0, 1, 2, 3}. From this analysis, we find that states

Table 9.2 Reachable States by Forward Reachability Algorithm

Iteration	1	2	3	4
Q^j	{0}	{0, 1, 2}	{0, 1, 2, 3}	{0, 1, 2, 3}
Q_{new}	{0}	{1, 2}	{1, 3}	{0, 1, 3}
Q^{j+1}	{1, 2}	{1, 3}	{0, 1, 3}	{0, 1, 2, 3}

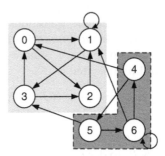

FIGURE 9.25

Example of forward reachability analysis.

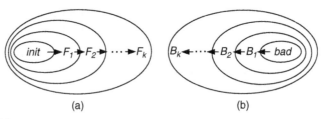

FIGURE 9.26

Intuitions behind forward and backward reachability analysis.

4, 5, and 6, which are surrounded by the dotted line in Figure 9.25, can never be reached from the initial state 0. These states form the set of unreachable states for state 0.

Reachability analysis can be also conducted through a *background* traversal of the state space [Abdulla 2000]. For a target final state (which could be a state that causes non-equivalence of the two FSMs), the search attempts to compute the set of previous states that can transition into this target state. If the backward reachability analysis can eventually reach an initial state, the search stops, and the two FSMs are proven not equivalent. Intuitions behind the forward and backward reachability analysis are illustrated in Figure 9.26a and Figure 9.26b, respectively.

Image computation may suffer from too many iterations and/or memory explosion. Several techniques that attempt to avoid memory explosion, such as the use of SAT solving instead of BDD-based techniques [Abdulla 2000], have been proposed.

Boolean equivalence checking has been widely accepted and incorporated into industrial design flows. Most leading EDA vendors offer BEC tools that include Encounter Conformal from Cadence and Formality from Synopsys. Combinational equivalence checkers have enjoyed tremendous success, partially thanks to the recent advances in SAT solving, which help to improve both performance and scalability of CECs. Sequential equivalence checking has also made significant progress in recent years. SEC tools such as SLEC from Calypto [Calypto 2008] are also commercially available.

9.5.2 Model checking (property checking)

Given a property and a design, a model checking tool allows a user to check whether the property holds true on the design. To develop such a tool, one needs to ask two basic questions: how to specify or describe a property and how to efficiently prove that a property holds true or is violated. The first question concerns the language used to express properties. Such a language determines what properties can be described and what properties cannot be described and, hence, limits the applicability of a model checking tool. The second question concerns the computation engine used to prove properties. Like equivalence checking described previously, OBDD and SAT are two prevalent methods that are used to implement the core computation engine of a model-checking tool. In this section, we begin by introducing the (formal) languages used to describe properties, followed by a brief review of how OBDD and SAT can be used to implement a model checking tool.

Temporal logic, introduced by Arthur Prior in 1960s [Prior 1957] and initially known as **Tense Logic**, provides a formal system for qualitatively describing and inferring how the values of statements for properties vary over *time* in a system. In temporal logic, a statement's truth value can change over time. In contrast, in traditional predicate logic, a statement's truth value is either true or false, which does not change over time. Application of temporal logic in verification started to receive attention in 1980s.

Temporal logic consists of two types of formulas: (1) *state formulas*, a form of **atomic propositions** (AP) that indicate the validity of specific states; and (2) *path formulas*, in which the property of a path holds constant. Note that a path here refers to a sequence of states. According to the views taken with respect to the underlying nature of time, temporal logic can be classified into (1) *linear temporal logic* (LTL), where the future value can only be derived along its linear computation path; and (2) *branching time temporal logic* (BTTL), which is a tree-like structure that allows quantifications over many different futures at each moment. Whether LTL or BTTL is more suitable for model checking depends on the property and the design being checking [Emerson 1990].

LTL allows applications to reason about the nondeterministic behavior. It models time as a sequence of discrete states starting from an initial moment with no predecessors and extending infinitely into the future. Such a sequence of states is known as either a computation path or an execution path. LTL derives the change over time with a linear time model $M = (S, \rightarrow, L)$, which is also known as a **Kripke** structure [Kripke 1963]. Here,

> S: a set of state formulas $\{s_0, s_1, \ldots\}$
> $\rightarrow$: the transition relation where $\forall s \in S$, $\exists s' \in S$, s.t. $s \rightarrow s'$
> L: a labeling function $L:S \rightarrow P(AP)$ in which each state is labeled with a set of atomic propositions from AP.

Figure 9.27 shows a simple example of a linear time model, M_1, where

> $S = \{s_0, s_1, s_2, s_3\}$
> $\rightarrow = \{(s_0, s_1), (s_0, s_2), (s_1, s_0), (s_1, s_3), (s_2, s_3), (s_3, s_0), (s_3, s_3)\}$
> $L = \{(s_0, \{p,q\}), (s_1, \{r,t\}), (s_2, \{q,t\}), (s_3, \{r\})\}$

A path π in $M = (S, \rightarrow, L)$ is an infinite sequence of ordered states $\{s_i \in S\}$ such that for each $i \geq 1$, $s_i \rightarrow s_{i+1}$. Therefore, path π can be expressed as $\pi = \{s_1 \rightarrow s_2 \rightarrow \ldots \rightarrow s_i \rightarrow \ldots\}$. Particularly, π^k denotes the suffix of a path starting from the kth state. For example, $\pi^3 = \{s_3 \rightarrow s_4 \rightarrow \ldots\}$. The notations $\models$ and $\not\models$ denote the *satisfaction* relation and the *unsatisfaction* relation, respectively. Given a *Kripke* structure $M = (S, \rightarrow, L)$, $\pi \models \phi$ denotes that the formula ϕ holds true (*i.e.*, is satisfied by the system) at the starting point of the path π in M. Let $I(s_1)$ be the set of formulae that hold true at the starting point of path π. Then, "$\pi \models \phi$" means "$\phi \in I(s_1)$."

LTL is built up from a set of propositional variables $p_1, p_2, \ldots, \top$ (true) and $\bot$ (false), the usual logic connectives $\neg$(negation), $\vee$(disjunction), $\wedge$(conjunction), $\rightarrow$(imply), and the following temporal modal operators: **X**(*Next*), **G**(*Always*), **F**(*Finally*), **U**(*Until*), and **R**(*Release*):

- *Next* (**X**) operator is unary and specifies that a formula holds at the *second* state on the path π:

  $\pi \models \mathbf{X}\phi$ *iff* $\pi^2 \models \phi$

- *Always* (**G**) operator is unary and specifies that a formula holds along *every* state on the path π:

  $\pi \models \mathbf{G}\phi$ *iff* $\forall i \geq 1, \pi^i \models \phi$

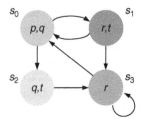

FIGURE 9.27

Example of an LTL model.

- *Finally* (**F**) operator is unary and specifies that a formula holds at *some* future state on the path π:

$$\pi \models \mathbf{F}\phi \ \textit{iff} \ \exists i \geq 1, \pi^i \models \phi$$

- *Until* (**U**) operator is binary and specifies that for some $i \geq 1$, π^0 to π^{k-1} satisfies the first formula ϕ and π^k satisfies the second formula ψ:

$$\pi \models \phi\mathbf{U}\psi \ \textit{iff} \ \exists i \geq 1, \text{s.t.} \ \pi^i \models \psi \text{ and } \forall j < i, \pi^j \models \phi$$

- *Release* (**R**) operator is binary and specifies that for some $i \geq 1$, we have either there exists $j < k$ such that π^j satisfies the first formula ϕ or π^k satisfies the second formula ψ:

$$\pi \models \phi\mathbf{R}\psi \ \textit{iff} \ \text{either } \exists i > 1, \text{s.t.} \ \pi^i \models \phi \text{ and } \forall j \leq i, \pi^j \models \psi$$

$$\text{or}$$

$$\forall k \geq 1, \pi^k \models \psi$$

Figure 9.28 illustrates examples for the semantics of various LTL operators assuming that all examples show on a path π in $\boldsymbol{M} = (\boldsymbol{S}, \rightarrow, \boldsymbol{L})$. We can apply LTL to the *Kripke* structure $\boldsymbol{M}_1$ in Figure 9.27 and derive the following formulas:

1. $s_0 \models \mathbf{X}t$ for all path π, and $s_0 \not\models \mathbf{X}(q \wedge r)$ because the next state of s_0 can not satisfy both q and r.

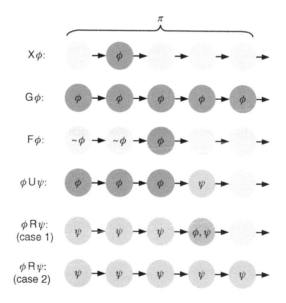

FIGURE 9.28

Examples for semantics of LTL operations.

2. $s_0 \models \mathbf{G}\neg(p \wedge t)$ and $s_3 \models \mathbf{G}r$ because $\mathbf{M}_1$ can loop at s_3 forever.
3. $s_0 \not\models \mathbf{GF}p$ denotes that not every path starting from s_0 can finally hold the formula p. $\pi = \{s_0 \rightarrow s_1 \rightarrow s_3 \rightarrow s_3\ldots\}$ is such one example.
4. $s_0 \models \mathbf{GF}p \rightarrow \mathbf{GF}r$ denotes that every path starting from s_0 which satisfies the formula p will always satisfy the formula r, but not for the case $s_0 \not\models \mathbf{GF}r \rightarrow \mathbf{GF}p$.
5. $\forall s \in S$ in $\mathbf{M}_1$, $s \models \mathbf{X}(q \vee r) \rightarrow \mathbf{F}r$ denotes that the next state of one path starting from every state in $\mathbf{M}_1$ can be q or r, and then the formula r will also hold on the path finally.

The expressive power of LTL is limited and implicitly quantifies *universally* over paths. An LTL formula can be satisfied if, and only if, all paths starting from the given state satisfy such a formula. A LTL system cannot decide whether one specific formula can be satisfied along some paths in $\mathbf{M}$. Therefore, ***computation tree logic*** (CTL), one type of BTTL, is evaluated over a branching-time structure and it quantifies the paths explicitly by introducing both the *existential* operator (**E**) and the *universal* operator (**A**) over paths.

The *Existential* (**E**) operator is defined as follows:

- $\mathbf{EX}\phi$ specifies that there is a path such that ϕ holds at the next state:

$$s \models \mathbf{EX}\phi \text{ } \textit{iff} \text{ } \exists \pi = \{v_1 \rightarrow v_2 \rightarrow \ldots v_i \rightarrow \ldots|_{v_1=s}\} \text{ s.t. } v_2 \models \phi$$

- $\mathbf{EG}\phi$ specifies that there is a path along which ϕ holds at every state:

$$s \models \mathbf{EG}\phi \text{ } \textit{iff} \text{ } \exists \pi = \{v_1 \rightarrow v_2 \rightarrow \ldots v_i \rightarrow \ldots|_{v_1=s}\} \text{ s.t. } \forall v_i, v_i \models \phi$$

- $\mathbf{EF}\phi$ specifies that there is a path along which ϕ holds finally:

$$s \models \mathbf{EF}\phi \text{ } \textit{iff} \text{ } \exists \pi = \{v_1 \rightarrow v_2 \rightarrow \ldots v_i \rightarrow \ldots|_{v_1=s}\} \text{ s.t. } \exists v_i, v_i \models \phi$$

- $\mathbf{E}[\phi\mathbf{U}\psi]$ specifies that there is a path along which ϕ holds until ψ holds:

$$s \models \mathbf{E}[\phi\mathbf{U}\psi] \text{ } \textit{iff} \text{ } \exists \pi = \{v_1 \rightarrow v_2 \rightarrow \ldots v_i \rightarrow \ldots|_{v_1=s}\} \text{ s.t. } \pi \models \phi\mathbf{U}\psi$$

The Universal (**A**) operator is defined as follows:

- $\mathbf{AX}\phi$ specifies that for all paths, ϕ holds at the next state:

$$s \models \mathbf{AX}\phi \text{ } \textit{iff} \text{ } \forall \pi = \{v_1 \rightarrow v_2 \rightarrow \ldots v_i \rightarrow \ldots|_{v_1=s}\} \text{ s.t. } v_2 \models \phi$$

- $\mathbf{AG}\phi$ specifies that for all paths, ϕ holds at every state of the path:

$$s \models \mathbf{AG}\phi \text{ } \textit{iff} \text{ } \forall \pi = \{v_1 \rightarrow v_2 \rightarrow \ldots v_i \rightarrow \ldots|_{v_1=s}\} \text{ s.t. } \forall v_i, v_i \models \phi$$

- $\mathbf{AF}\phi$ specifies that for all paths, ϕ holds finally:

$$s \models \mathbf{AF}\phi \text{ } \textit{iff} \text{ } \forall \pi = \{v_1 \rightarrow v_2 \rightarrow \ldots v_i \rightarrow \ldots|_{v_1=s}\} \text{ s.t. } \exists v_i, v_i \models \phi$$

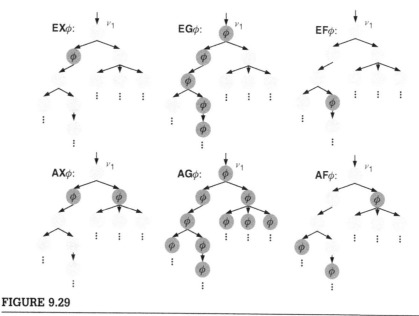

FIGURE 9.29

Illustrations for CTL *Existential* and *Universal* operations.

- $\mathbf{A}[\phi\mathbf{U}\psi]$ specifies that for all paths, ϕ holds until ψ holds:

$$s \models \mathbf{A}[\phi\mathbf{U}\psi]\textit{iff } \forall\pi = \left\{v_1 \rightarrow v_2 \rightarrow \ldots v_i \rightarrow \ldots|_{v_1=s}\right\} \text{ s.t. } \pi \models \phi\mathbf{U}\psi$$

Figure 9.29 illustrates partial examples for the *Existential* and *Universal* operations according to the preceding definitions.

CTL is capable of specifying branching behaviors such as $\mathbf{AG(EF}f)$, which is also known as **resetability**—meaning there is always a path back to f. This property cannot be modeled by LTL because of the lack of the path quantifier **E**. Likewise, there exists some LTL formulas that cannot be expressed in CTL. For example, $\mathbf{FG}\phi$ in LTL means that the formula ϕ will finally hold along every path from the given point. Its semantic should be expressible as $\mathbf{A(FG}\phi)$. However, in CTL, every temporal operator (**F** and **G**) must be preceded by a path quantifier (**E** or **A**). Hence, CTL cannot express $\mathbf{A(FG}\phi)$. CTL* extends the expressiveness from both LTL and CTL and primarily allows a path quantifier to be used followed by an arbitrary LTL formula. The relationships between the expressiveness of LTL, CTL, and CTL* can be viewed as LTL $\cup$ CTL $\subset$ CTL*, which are illustrated in Figure 9.30. Particularly, there is a set $\{\phi_4\}$ of CTL* formulas that can be expressed neither in CTL nor in LTL. $\mathbf{E(GF}\phi)$ is such an example, saying that there is a path where from one certain state, ϕ's holds through arbitrarily many states to the end [Huth 2004].

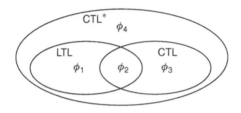

FIGURE 9.30

Relationships between the expressiveness of LTL, CTL, and CTL*.

The properties of design systems can be divided into two types [Owicki 1982]:

1. **Safety properties** that indicate that some bad event will never happen. For a sequential program, safety guarantees that no incorrect outcome will be produced by the program. For a finite state machine, safety checking denotes those properties whose violation can always find a finite trace. Another typical example of safety is a mutual exclusive property that states that having more than one process in the critical section will never occur.
2. **Liveness properties** that indicate that some good event will eventually happen. For a sequential program, the program will terminate as it produces a legal outcome. For a finite state machine, those properties that may be violated will never have a finite witness. CTL can model the simple liveness for the phrase "The light will turn green" as $light \models \mathbf{AF}(green)$. "Any request will eventually be satisfied" is another example semantic phrase that can be expressed and the corresponding CTL expression is $\mathbf{AG}(Req) \Rightarrow \mathbf{AF}(Sat)$. Liveness focuses on a slice in the tree structure and may incur the witness as a computation path of infinite steps.

To illustrate the safety and liveness properties, consider a two-input Muller C-element used for asynchronous circuit connections. Figure 9.31a shows its gate-level netlist with two Boolean inputs (x, y) and one output (z). The corresponding dynamic behavior is represented by the state transition graph in Figure 9.31b.

A safety property of the C-element is that if all inputs and outputs are equal, then the output z will not change its value until all inputs flip their values. There are two situations: all values are 0 and all values are 1.

- $\mathbf{AG}((x = 0 \wedge y = 0 \wedge z = 0) \Rightarrow \mathbf{A}(z = 0 \; \mathbf{U} \; (x = 1 \wedge y = 1)))$
- $\mathbf{AG}((x = 1 \wedge y = 1 \wedge z = 1) \Rightarrow \mathbf{A}(z = 1 \; \mathbf{U} \; (x = 0 \wedge y = 0)))$

A liveness property of the C-element is that if both inputs become equal, then the output z will eventually change to the corresponding value. There are two situations: both input values are 0 and both input values are 1.

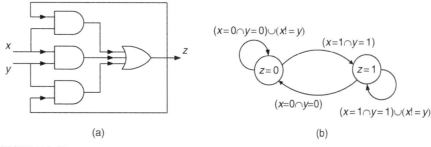

FIGURE 9.31

(a) Gate-level netlist. (b) state transition graph of a C-element.

- **AG(A**($x = 0 \wedge y = 0$) **U** ($z = 0 \vee x = 1 \vee y = 1$))
- **AG(A**($x = 1 \wedge y = 1$) **U** ($z = 1 \vee x = 0 \vee y = 0$))

9.5.2.1 *Model checking with temporal logic*

Let a *Kripke* structure $M = (S, \rightarrow, L)$ represent a finite state concurrent system. The model-checking problem can be formulated as: given a model M, a property p specified as a temporal formula, and a state s, does $s \models p$ hold in M? The corresponding result is either (1) yes, $s \models p$ in M, or (2) no, $s \not\models p$ in M. Especially for the latter case, such a result is derived from finding a counterexample that invalidates p in M. Therefore, the modeling checking problem can be addressed by computing the state set S_p that satisfies p in M.

The **labeling algorithm,** proposed by E. Clarke, E. Emerson, and A. Sistla [Clarke 1986], is a basic algorithm for the model checking problem. Given a CTL formula, the labeling algorithm labels the set of states in which the target formula p holds, which is denoted as $[[p]] \triangleq \{\forall s \in S \text{ in } M, s \models p\}$, and called the **denotation** of p. Deriving $[[p]]$ starts by decomposing p into a set of sub-formulas in a *bottom up* manner. Because $\{\bot, \neg, \wedge\}$ and $\{\textbf{AF}, \textbf{EX}, \textbf{EU}\}$ can form an adequate set of connectives for CTL [Martin 2004], and all other propositional and temporal connectives can be written in terms of this set, a preprocessing step to convert the target formula p into an equivalent form in terms of this adequate set is first invoked and then followed by labeling states in M for $[[p]]$. Later, the denotation $[[p]]$ is compared with the set S_{init} of all initial states to check whether $S_{init} \subseteq [[p]]$.

The labeling algorithm explicitly enumerates the states in the model whose size often grows exponentially in terms of the numbers of variables in the system. This problem is typically referred to as the *state explosion problem*. To overcome this issue, a more efficient technique called **fix-point computation** is proposed, which incorporates OBDD for symbolic computation and implicit representation of states. Model checking with OBDDs is often referred to as **symbolic model checking** [Burch 1990], and **SMV,** developed at Carnegie Mellon University, is one such verifier [McMillan 1992].

Fix-point computation finds the set of states that satisfies the specific *global* CTL formula. A function $x_{i+1} = f(x_i)$ is called a **fix-point** if $\exists x_k$, where $k \geq 0$, s.t. $x_{k+1} = f(x_k) = x_k$. Given a starting value x_0, a **fix-point** can be found by iteratively mapping f to x_i until $f(x_k) = x_k$. To help calculate the fix-points on a *Kripke* structure $M = (S, \rightarrow, L)$, we define a function τ called a *predicate transformer*, which takes a subset of S and outputs another subset. In other words, the function τ is defined on the basis of the power set $P(S)$, which is the set of all subsets of S. $\tau^i(S')$ denotes i applications of τ to the given subset $S' \subseteq S$. That is,

$$\tau^i(S') = \underbrace{\tau(\tau(\ldots(\tau(S'))\cdots)}_{i \text{ times}}$$

τ is *monotonic*, provided that for any two subsets of S, P, and Q, if $P \subseteq Q \subseteq P(S)$, then $\tau(P) \subseteq \tau(Q)$. Note that because τ is monotonic, by starting from a subset of S and continuously applying τ, a fixed point can always be reached.

Let τ be monotonic, $\emptyset$ be the empty set, and U be a finite set $\{s_0, s_1, \ldots, s_n\} \subseteq P(S)$ of n elements in M, then $\exists l$, s.t. $\tau^l(\emptyset) = \tau^{l+1}(\emptyset)$ and $\exists u$, s.t. $\tau^u(U) = \tau^{u+1}(U)$. $\tau^l(\emptyset)$ and $\tau^u(U)$ are called the *least* and *greatest* fix-points of τ, which are denoted by fp_{min} and fp_{max}, respectively. Each basic CTL* operator can be further represented by either fp_{min} or fp_{max} over an appropriate predicate transformer. For a complete treatment of the underlying theory and proof, please refer to [Granas 2003].

Suppose that we would like to apply the fix-point computation to check $\mathbf{AG}(\phi \Rightarrow \mathbf{AF}\psi)$, then sub-annotations will be computed in a bottom-up manner. That is then $[[\psi]]$, $[[\mathbf{AF}\psi]]$, $[[\phi]]$, $[[\phi \Rightarrow \mathbf{AF}\psi]]$, and $[[\mathbf{AG}(\phi \Rightarrow \mathbf{AF}\psi)]]$ in this example. Assuming $\psi = p$ and $\phi = q$, let's check the process of calculating the formula on the basis of the example given in Figure 9.27.

- $[[\psi]] = [[r]] = \{s_3\}$
- $[[\mathbf{AF}\psi]] = [[\mathbf{AF}r]] = \{s_0, s_1, s_2, s_3\}$ can be computed as the union of
  - $[[\psi]] = [[r]] = \{s_3\}$
  - $[[r \vee \mathbf{AX}r]] = \{s_3\}\mathrm{U}\{s_1, s_2\} = \{s_1, s_2, s_3\}$
  - $[[r \vee \mathbf{AX}(r \vee \mathbf{AX}r)]] = \{s_3\}\mathrm{U}\{s_0, s_1, s_2, s_3\} = \{s_0, s_1, s_2, s_3\}$
  - no need to repeat since $\{s_0, s_1, s_2, s_3\}$ converges
- $[[\phi]] = [[p]] = \{s_0\}$
- $[[\phi \Rightarrow \mathbf{AF}\psi]] = [[\neg\phi \vee (\mathbf{AF}\psi)]] = [[\neg p \vee (\mathbf{AF}r)]]$
  - $[[\neg p \vee (\mathbf{AF}r)]] = \{s_1, s_2, s_3\}\mathrm{U}\{s_0, s_1, s_2, s_3\} = \{s_0, s_1, s_2, s_3\}$
- $[[\mathbf{AG}\mu]]$ can be computed as the intersection of $[[\mu]]$, $[[\mu \wedge \mathbf{AX}\mu]]$, $[[\mu \wedge \mathbf{AX}(\mu \wedge \mathbf{AX}\mu)]]$, and etc. Therefore, $[[\mathbf{AG}(\phi \Rightarrow \mathbf{AF}\psi)]]$ can be obtained from the following and result in $\{s_0, s_1, s_2, s_3\}$:
  - $[[\mu]] = [[\phi \Rightarrow \mathbf{AF}\psi]] = \{s_0, s_1, s_2, s_3\}$
  - $[[\mu \wedge \mathbf{AX}\mu]] = \{s_0, s_1, s_2, s_3\} \cap \{s_0, s_1, s_2, s_3\} = \{s_0, s_1, s_2, s_3\}$
  - $[[\mu \wedge \mathbf{AX}(\mu \wedge \mathbf{AX}\mu)]] = \{s_0, s_1, s_2, s_3\}$
  - ... all remaining computations converge to $\{s_0, s_1, s_2, s_3\}$

Because every state belongs to $[[\mathbf{AG}(\phi \Rightarrow \mathbf{AF}\psi)]] = \{s_0, s_1, s_2, s_3\}$, the *Kripke* structure $M = (S, \rightarrow, L)$ satisfies this property. As we can see, computing the state set for propositional connectives is straightforward. The computation for temporal connectives such as $\mathbf{EX}\phi$ is relatively sophisticated and requires applying the temporal operations over the current state set repeatedly until there is no change.

Symbolic model checking is often limited by the sizes of corresponding OBDDs used in the computation. Typically, a good variable ordering is crucial for minimizing OBDD size. However, finding optimal ordering is a proven NP-complete problem. In some cases, even with the best ordering, the OBDD size is still larger than the available computation resource. To address this problem, an alternative method, called *bounded model checking* (BMC), was proposed, which only tries to find counterexamples for properties within a bounded number of clock cycles (state transitions). Most of the bounded model checkers use a propositional decision (SAT) procedure [Biere 1999]. Several efficient satisfiability solvers have been developed in recent years that are capable of solving problems with more than thousands of variables. Bounded model checking can find minimal length counterexamples as the propositional decision procedure traverses the state-transition graph step by step. This feature can also make users easily understand counterexamples and consequently facilitate the debugging process.

Given the *Kripke* structure $M = (S, \rightarrow, L)$ and a safety property ϕ, by use of BMC we can determine whether a length-k execution path of M that satisfies ϕ exists. That is, $M \models_k \mathbf{E}\phi$. Let a propositional formula $T(s,s')$ define the relationship of the state transition in M and let $I(s)$, a predicate over the state variables, define the initial states. The BMC problem is equivalent to the satisfiability problem of a Boolean formula $[[M, \phi]]_k = [[M]]_k \wedge [[\phi]]_k$ where $[[M]]_k$ and $[[\phi]]_k$, respectively, encode the set of length-k execution paths of M and the set of length-k paths that satisfy ϕ in M.

For a valid length-k path $\pi = \{s_0 \rightarrow s_1 \rightarrow s_2 \rightarrow \ldots \rightarrow s_k\}$, $[[M]]_k$ can be defined as

$$[[M]]_k = I(s_0) \wedge T(s_0, s_1) \wedge T(s_1, s_2) \wedge \ldots \wedge T(s_{k-1}, s_k) = I(s_0) \wedge \prod_{i=0}^{k-1} T(s_i, s_{i+1})$$

The core of encoding for a formula ϕ with k steps depends on whether M contains any loop that starts at s_l and ends at s_k. Therefore, $[[\phi]]_k$ can be computed as the disjunction of two cases:

1. **Without loopback in M:** $[[\phi]]_k \triangleq \left(\neg \left(\prod_{l=0}^{k} T(s_l, s_k) \wedge [[\phi]]_k^0 \right) \right)$, where for every $[[.]]_k^i$, k is the length of the prefix of the path and i is the current position in this prefix.

2. **With a loopback in M:** $[[\phi]]_k \triangleq \prod_{l=0}^{k} \left(T(s_l, s_k) \wedge_l [[\phi]]_k^0 \right)$, where for every $_l[[.]]_k^i$, i is the current position in the path π, k is the length of the prefix of this path, and l is the position where the loop starts.

For example, given a formula $\phi = \mathbf{F}p$, $M \models_k \phi$ is used to check whether any reachable state in which a property p holds in M within k steps exists. Bounded model checking will first derive $[[M, \phi]]_k = I(s_0) \wedge \prod_{i=0}^{k-1} T(s_i, s_{i+1}) \wedge \prod_{j=0}^{k} p(s_j)$, where $p(s_j) = 1$ if the property p holds on s_j, otherwise $p(s_j) = 0$. This satisfiability problem can be solved with an SAT solver. It will return 1 if such a path is found. To check whether any reachable state that satisfies p, provided that q holds infinitely (*i.e.*, $\phi = \mathbf{GF}q \wedge \mathbf{F}p$) exists, modeling the loopback behavior in M is required. That is,

$$[[M, \phi]]_k = I(s_0) \wedge \prod_{i=0}^{k-1} T(s_i, s_{i+1}) \wedge \prod_{j=0}^{k} p(s_j) \wedge \prod_{l=0}^{k} \left(T(s_l, s_k) \wedge_l [[q]]_k^0 \right)$$

Although bounded model checking with the propositional decision (**SAT**) procedure can handle larger circuits, it is an incomplete technique. If the checking formulas are unsatisfiable (*i.e.*, the property holds true over a bounded length k of checking, there is no guarantee that the property will hold or not over a length greater than k.

9.5.3 **Theorem proving**

We have introduced how propositional and temporal logic can be automated to compare two representations in equivalence checking and to validate properties from the specifications against a given model in model checking. The effectiveness of both equivalence checking and model-checking techniques is often limited by the capacity and performance of the underlying engines used such as OBDD and SAT. Sometimes, the complexity of a verification task for an arithmetic circuit, such as a data path or a signal processing unit, can be reduced if a more general mathematical formulation of the circuit, with a better abstraction of the word-level information, is provided. Theorem proving techniques are applied for such purposes.

Theorem proving is the process for determining whether a given implementation satisfies the target specification by means of mathematical reasoning, as shown in Figure 9.32. Both the implementation and specification need to be transformed into formulas in a formal logic system. The relationships between implementation and specification are regarded as theorems in logic. The conformance is then established by proving the theorems either from implementation

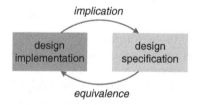

FIGURE 9.32

Verification by theorem proving.

to specification, denoted by the *implication* arrow in the figure, or from specification to implementation, denoted by the *equivalence* arrow.

A *proof system* (or *calculus*) **S** consists of:

1. *Expressions* of **S**: a finite sequence of symbols
2. *Well-formed formulas* of **S**: a subset of the expressions of **S**
3. *Axioms* of **S**: a finite set of the well-formed formulas of **S**
4. *Inference rules* of **S**: a finite set of derivation rules from a given finite set of well-formed formulas to a new well-formed formula

The general form of an inference rule is $\frac{\alpha_1, \alpha_2, \dots, \alpha_k}{\beta}$, where the well-formed formulas $\alpha_1, \alpha_2, \dots, \alpha_k$ are called the *premises* of the rule, whereas the well-formed formula β is called the *conclusion*.

In such a proof system **S**, a *proof* is a finite sequence of formulas, $\phi_1, \phi_2, \dots, \phi_n$ in which ϕ_i can be either an *axiom* or else *derived* from applying an inference rule of **S** over $\{\phi_1, \phi_2, \dots, \phi_{i-1}\}$, which is denoted as $\{\phi_1, \phi_2, \dots, \phi_{i-1}\} \vdash \phi_i$. The last formula ϕ_n is the goal of the proof, which is known as a *theorem* of **S**. Sometimes, proofs may require supplementary *assumptions*, such as $\Gamma = \{\psi_1, \psi_2, \dots, \psi_{i-1}\}$ from the domain specific axioms. The term $\Gamma \vdash \phi$ asserts that the formula ϕ is valid if all assumptions in Γ are true. If Γ is empty, we write this as $\vdash \phi$.

Many modern theorem proving systems are publicly available. These include Coq [Coq 2003], *Z/Eves* [Saaltink 1999], **High-Order Logic** (HOL) [Nipkow 2002], *PVS* [Owre 1992], and *ACL2* [Kaufmann 2002]. To illustrate the deduction process involved in theorem proving, we use HOL, developed at the University of Cambridge [Gordon 1993], for the remainder of the discussion. HOL supports the use of standard predicate operators, five axioms, and eight primitive inference rules, which are listed in Table 9.3, for expressing most ordinary mathematical theories.

The first step of the proof method in HOL is to formalize both the specification and the implementation into the formal logic used in the proof system. Then, the formulation of a proof goal can be achieved by either proof of implication (forward) or proof of equivalence (backward) with the inference rules. In the forward manner, a theorem prover starts with simple lemmas that can be proven directly to develop new rules. Rules are successively combined into more difficult lemmas until the target theorem is proven. Figure 9.33 shows an example for such an HOL theorem proving. The functional specification of the underlying black-box, shown in Figure 9.33a, is an NOR function denoted by $f = \bar{x} \times \bar{y}$. Its formal specification can be expressed as $SPEC(x, y, z) \triangleq z = (\neg x \land \neg y)$. And the implementation, which is shown in Figure 9.33b, may use only primitive AND, OR, and NOT gates. The corresponding descriptions of these gates in formal logic are:

- AND$(i_1, i_2, out) \triangleq out = (i_1 \land i_2)$, where i_1 and i_2 are input ports and *out* is an output port.

Table 9.3 Base Rules of Higher Order Logics Used in HOL

Name	Explanation	Rule	Remark
ASSUME	Assumption introduction	$\dfrac{\quad}{t \vdash t}$	
REFL	Reflexivity	$\dfrac{\quad}{\vdash t = t}$	
ABS	Abstraction	$\dfrac{\Gamma \vdash t_1 = t_2}{\Gamma \vdash (\lambda x.t_1) = (\lambda x.t_2)}$	If x is not free in Γ, where $(\lambda x.t_i)$ denotes the function defined by $f(x) = t_i$
BETA_CONV	Beta-conversion	$\dfrac{\quad}{\vdash (\lambda x.t_1)t_2 = t_1[t_2/x]}$	$t_1[t_2/x]$ substitutes t_2 for x in t_1 with the restriction that no free variables in t_2 become bound after substitution into t_1
SUBST	Substitution	$\dfrac{\Gamma_1 \vdash t_1 = t_2 \mid \Gamma_2 \vdash t[t_1]}{\Gamma_1 \cup \Gamma_2 \vdash t[t_2]}$	$t[t_i]$ denotes a term t containing a subterm t_i
INST_TYPE	Type instantiation	$\dfrac{\Gamma \vdash t}{\Gamma \vdash t[\sigma_1, \ldots, \sigma_n / v_1, \ldots, v_n]}$	$t[\sigma_1, \ldots, \sigma_n / v_1, \ldots, v_n]$ substitutes in parallel the types $\sigma_1, \ldots, \sigma_n$ for the variables $v_1, \ldots, v_n$ in t
DISCH	Assumption discharging	$\dfrac{\Gamma \vdash t_2}{\Gamma - \{t_1\} \vdash t_1 \Rightarrow t_2}$	$\Gamma - \{t_1\}$ denotes the set subtracting $\{t_1\}$ from Γ
MP	Modus ponens	$\dfrac{\Gamma_1 \vdash t_1 \Rightarrow t_2 \mid \Gamma_2 \vdash t_1}{\Gamma_1 \cup \Gamma_2 \vdash t_2}$	

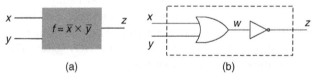

(a) (b)

FIGURE 9.33

Example of theorem proving by HOL.

- $OR(i_1, i_2, out) \triangleq out = (i_1 \vee i_2)$, where i_1 and i_2 are input ports and *out* is an output port.
- $NOT(i, out) \triangleq out = (\neg i)$, where i is an input port and *out* is an output port.

Therefore, the formal definition for the implementation in Figure 9.33b is $IMPL(x, y, z) \triangleq \exists w.OR(x, y, w) \wedge NOT(z, w)$. The goal of this proof is to derive *SPEC(x, y, z)* from *IMPL(x, y, z)* by applying the inference rules specified in Table 9.3. The proof — given step-by-step — is as follows in Table 9.4. Please note that the actual process executed with HOL software may not look exactly the same though. However, it should be similar to what it is shown below.

Table 9.4 Step-by-step Proof for an NOR Function

Proof

$IMPL(x, y, z)$	{ from the circuit diagram }
$\vdash \exists w.OR(x, y, w) \wedge NOT(z, w)$	{ by definition of the implementation }
$\vdash OR(x, y, w) \wedge NOT(z, w)$	{ strip off $\exists w$ }
$\vdash (w = x \vee y) \wedge NOT(z, w)$	{ by formal definition of OR gate }
$\vdash (w = x \vee y) \wedge (z = \neg w)$	{ by formal definition of NOT gate }
$\vdash (z = \neg(x \vee y))$	{ substitute w with $x \vee y$ }
$\vdash (z = \neg x \wedge \neg y)$	{ distribute $\neg$ over $x \vee y$ }
$\vdash SPEC(x, y, z)$	{ by definition of the specification }
$\vdash IMPL(x, y, z) \Rightarrow SPEC(x, y, z)$	

Q.E.D.

Theorem proving can be applied to verify implementations described at different levels of abstraction. The formal specification of the behavior of a transistor-level CMOS inverter, for example, can be expressed by $SPEC(x, y) \triangleq y = (\neg x)$ [Gordon 1992]. Consider the network structure shown in Figure 9.34. The implementation is built on basic modules and includes a power cell, a ground cell, a P-type transistor, and an N-type transistor which are denoted as *VDD(p)*, *GND(q)*, *PTran(x, p, y)*, and *NTran(x, y, q)*, respectively. The behaviors of these basic modules can be formally defined as:

- $VDD(p) \triangleq (p = \top(\text{true}))$
- $GND(q) \triangleq (q = \bot(\text{flase}))$
- $PTran(x, p, y) \triangleq (\neg x \Rightarrow (p = y))$
- $NTran(x, y, q) \triangleq (x \Rightarrow (y = q))$

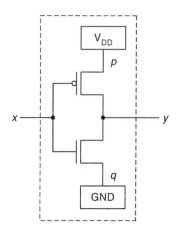

FIGURE 9.34

CMOS inverter.

Then, the entire network structure can be formulated as:

$$IMPL(x,y) \triangleq \exists p, q. V_{DD}(p) \wedge PTran(x,p,y) \wedge Ntran(x,y,q) \wedge GND(q)$$

Again, the proof goal is to derive $SPEC(x, y)$ from $IMPL(x, y)$ by applying inference rules. The step-by-step proof process is as follows.

Proof

$IMPL(x,y)$ {from the network structure   }

$\vdash \exists p, q. VDD(p) \wedge PTran(x,p,y) \wedge NTran(x,y,q) \wedge GND(q)$
                {by definition of the implementation   }

$\vdash VDD(p) \wedge PTran(x,p,y) \wedge NTran(x,y,q) \wedge GND(q)$
                {strip off $\exists p, q$ }

$\vdash (p = \top) \wedge PTran(x,p,y) \wedge NTran(x,y,q) \wedge (q = \bot)$
                {by definition of VDD and GND cells   }

$\vdash (p = \top) \wedge PTran(x,\top,y) \wedge NTran(x,y,\bot) \wedge (q = \bot)$
                {substitute p in $PTran$, q in $NTran$ }

$\vdash (\exists p.p = \top) \wedge PTran(x,\top,y) \wedge NTran(x,y,\bot) \wedge (\exists q.q = \bot)$
                {use $\exists a.t_1 \wedge t_2 = (\exists a.t_1) \wedge t_2$ if a is free in t_2 }

$\vdash (\top) \wedge PTran(x,\top,y) \wedge NTran(x,y,\bot) \wedge (\top)$
                {use$(\exists a.a = \top) = \top$ and $(\exists a.a = \bot) = \top$ }

$\vdash PTran(x,\top,y) \wedge NTran(x,y,\bot)$
                {use $(x \wedge \top) = x$ }

$\vdash (\neg x \Rightarrow (\top = y)) \wedge (x \Rightarrow (y = \bot))$
                {by definition of $PTran$ and $NTran$ cells   }

$\vdash (x \vee (\top = y)) \wedge (\neg x \vee (y = \bot))$
                {by$(a \Rightarrow b) = (\neg a \vee b)$ }

$$\vdash (x \land \neg x) \lor (x \land (y = \bot)) \lor ((\top = y) \land \neg x) \lor ((\top = y) \land (y = \bot))$$

$$\vdash (\bot) \lor (x \land (y = \bot)) \lor ((\top = y) \land \neg x)(\bot)$$

$$\vdash (x \land (y = \bot)) \lor ((\top = y) \land \neg x)$$

$$\{\text{apply Boolean simplification} \quad \}$$

$$\vdash y = (\neg x)$$

$$\{\text{if } x = \top \Rightarrow (y = \bot) \text{ and if } x = \bot \Rightarrow (y = \top) \quad \}$$

$$\vdash IMPL(x,y) \Rightarrow SPEC(x,y)$$

Q.E.D.

Theorem proving has been successfully applied to the verification of hardware designs, such as the TAMARACK microprocessor [Joyce 1986] and the Viper microprocessor [Cohn 1988]. Its strength is its ability to support the expressiveness of higher order logics, to relate circuit behaviors at different levels of abstraction [Melham 1988], and to provide many effective reasoning utilities. Moreover, the design hierarchy and regularity can be exploited by theorem provers, which enable users to be in full control of the verification process. Higher order logics can specify and verify generic and parameterized hardware designs. One such example would be a channel encoder with words in n-bit width. Also, tactics of inference rules can continuously evolve during the deduction process. Particularly frequent and useful theories/theorems can be customized and retained for future proofs.

Verification by theorem proving requires users to familiarize themselves with the proof system and to spend a considerable amount of effort toward developing the formal models for both the specification and the implementation. This is one of the major disadvantages of the approach. Moreover, because of the lack of sound proof systems for higher order logic, the derivation of inference rules may require a great deal of human intervention, especially for complex and large theorems. For these reasons, the application of theorem proving has been limited and not widely used for industrial design projects.

9.6 ADVANCED RESEARCH

Simulation remains the mainstream verification approach in the industry. Its scalability, along with its easy applicability to designs at almost any abstraction level, makes it attractive for complex verification tasks. When used as a stand-alone technique, simulation can detect simple and easy-to-find bugs. Its effectiveness in finding corner-case, hard-to-detect bugs can be limited because of the availability of high-quality stimuli that can cover a wide range of the corner cases and can activate and reveal the subtle bugs. Although traditional formal techniques—broadly speaking, model checking and theorem proving—can, in principle, analyze and find subtle bugs, their applicability can be limited by their runtime inefficiency and/or difficulty in use.

For simulation-based approaches, measuring the coverage and preparing the test vectors are the two most important things in the verification plan. The coverage-driven verification (CDV) flow, shown in Figure 9.6, links these two together and can be automated if the test generation constraints can be modified automatically [Bai 2003; Chen 2003; Wen 2006, 2007]. Such improvements can substantially save the amount of manual efforts needed for coverage analysis and test preparation. The improvements in coverage-driven verification can be divided into two categories: feedback-based coverage-driven verification and coverage-driven verification by construction.

Feedback-based coverage-driven verification modifies the biases and seeds to direct the automatic test generation. A generic algorithm [Bose 2001] can be applied to resynthesize test cases for optimizing the coverage. The authors in [Tasiran 2001] represent the DUV as a Markov chain model and analyze the feedback data to modify the model's parameters. The authors in [Fine 2003] cast the coverage-driven test generation in a statistical inference framework by modeling the relationship between coverage information and the directives to the test generation as Bayesian networks. A machine-learning-based technique in [Fine 2006] was later proposed to provide enhanced coverage through automatically learning the relationship between the initial state and vector generation success.

Coverage-driven verification by construction derives an abstract model that can capture the logical constraints in the DUV and assemble the new directives to correctly hit the uncovered events. [Ur 1999] abstract the processor control as a set of FSMs and use them to automate the verification tasks. A physical test case is derived from a sequential trace of the state traversal in the FSM. The works in [Chen 2003] and [Bai 2003] generate tests to target stuck-at and crosstalk faults in processors and use a *virtual constraint circuit* (VCC) for assisting the module-level test generation process. The application is for *software-based self-test* (SBST) [Lai 2000]. A data-mining approach based on simulation data was proposed in [Wen 2006, 2007] to approximate the functionality of the DUV as BDDs that can then be used to better guide the test generation process.

Although the capacity and performance of formal methods has improved significantly over the past decade, such improvements barely kept pace with the growth in design complexity. The search for new solutions resulted in some powerful hybrid techniques that combined formal and informal approaches. These hybrid techniques attempt to address verification bottlenecks by enhancing coverage of the state space traversed.

Researchers who investigated formal methods have widely recognized the importance of providing a way to combine disparate tools. Joyce and Seger experimented with combining *trajectory evaluation* with theorem proving. They used trajectory evaluation as a decision procedure for the *higher-order logic* (HOL) system [Joyce 1993]. A proposal called interface logics [Guttman 1991] discusses the idea of combining different *theorem provers* by defining a single logic such that the logic of each individual tool can be viewed as its sublogics. [Jang 1997] used CTL model checking to verify a set of properties of embedded microcontrollers, and the proof of the top-level specification was achieved through a compositional argument by use of the properties instead of through a theorem prover. A hybrid of two model-checking techniques, called MIST [Hazelhurst 2002], enables a handshake between *symbolic trajectory evaluation* and *symbolic model checking*.

Generally speaking, **hybrid methods** combining formal and informal techniques aim to increase the design space coverage and, thus, the probability of finding design errors. These types of methods include *control space exploration, directed functional test generation, combining ATPG with formal techniques*, and *heuristic-based traversal*. Control space exploration addresses the problem of finding bugs and increases space coverage by exploring control logic [Iwashita 1994; Ho 1995; Geist 1996; Moondanos 1998]. Directed functional test generation leverages the strengths of both formal verification and simulation techniques to generate functional tests [Sumners 2000; Ganai 2001; Mishra 2005]. Because ATPG can avoid state space explosion by use of dual justification and propagation techniques to localize the search, adding formal techniques can compensate for the inherent incompleteness of ATPG, making the combination a more complete and effective verification approach [Boppana 1999; Huang 2001; Vedula 2004]. Heuristic-based traversal tackles the need to efficiently traverse state space by an extensive use of heuristics [Yang 1998; Wagner 2005; Shyam 2006]. Note that because of the inherent incompleteness of informal techniques, any method that combines an informal technique with another is also an incomplete verification method.

9.7 CONCLUDING REMARKS

This chapter reviews the basic concepts of functional verification and the challenges associated with it. Different levels of the verification hierarchy, including the designer level, unit level, core level, chip level, and system/board level, are explained. Various coverage metrics used for measuring the explored extent of verification are provided. The simulation-based approach is currently the most pervasive form of verification. Key components such as testbench and simulation environment development are reviewed. The emerging assertion-based verification method is explained in detail. To compensate for the incompleteness of simulation-based verification, formal methods built on mathematical theories were developed. Basic concepts in equivalence checking, model checking, and theorem proving are reviewed. Current research efforts toward advancing functional verification are summarized to conclude this chapter.

9.8 EXERCISES

9.1. (Line Coverage) Suppose that the module in Box 9.11 was specified in your Verilog HDL design:

BOX 9.11
```
 1. module test;
 2. reg X, Y, Z;
 3. initial
 4. begin
 5.    X = 1'b0;
 6.    Y = 1'b1;
 7. if (X)
 8.    Z = Y;
 9. else
10.    Z = ~Y;
11. end
12. endmodule
```

Calculate the line coverage after simulation and identify the line or lines that has/have not been covered.

9.2. (Toggle Coverage) Suppose that the following module in Box 9.12 was specified in your Verilog HDL design:

BOX 9.12
```
 1. module test;
 2. reg [2:0] X;
 3. initial
 4. begin
 5.    X = 3'b000;
 6.    #100;
 7.    X = 3'b110;
 8.    #100;
 9.    X = 3'b010;
10.    #100;
11. end
12. endmodule
```

After simulation, the register would have achieved a total toggle percentage of 50%. Please identify which toggles are missing.

9.3. (**Expression Coverage**) Suppose that the following module was specified in your Verilog HDL design:

BOX 9.13
1. **module** test;
2. **reg** X, Y;
3. **wire** Z;
4. **assign** Z = X|Y;
5. **initial**
6. **begin**
7.   X = 1'b0;
8.   Y = 1'b0;
9.   #50;
10.   X = 1'b1;
11.   #50;
12.   Y = 1'b1;
13.   #50;
14. **end**
15. **endmodule**

This module consists of only one expression: X|Y. Calculate the expression coverage after simulation and identify those cases that are not covered.

9.4. (**FSM Coverage**) Suppose that the module in Box 9.14 was specified in your Verilog HDL design:

BOX 9.14
1. **module** test;
2. **reg** [1:0] D;
3. **wire** W, X, Y, Z;
4.
5. **assign** Y = D[1] ∧ D[0];
6. **assign** Z = X ∧ Y;
7. **assign** W = ~Z;
8. **always** @(**posedge** clk) **begin**
9.   D[1] = W;
10.   D[0] = Z;
11. **end**
12.
13. **always** #50 clk = ~clk;
14. **initial**
15. **begin**

```
16.    clk = 0;
17.    D = 2'b00;
18.    # 100 X = 1'b1;
19.    # 100 X = 1'b0;
20.    # 100 X = 1'b1;
21.    # 100 X = 1'b0;
22. end
23. endmodule
```

Please first draw the corresponding finite state machine and then calculate both the state and the arc coverage from the simulation.

9.5. (Equivalence Checking) Determine whether the following two combinational circuits are functionally equivalent. If not, produce a counterexample.

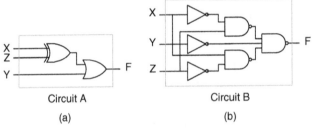

Circuit A   Circuit B

(a)   (b)

FIGURE 9.35

Gate-level schematics for the two circuits in Exercise 9.5.

9.6. (Equivalence Checking) Determine whether the following two sequential circuits are functionally equivalent. If not, produce a counterexample. Note that the initial states of all flip-flops are zero.

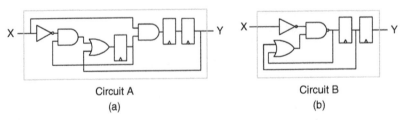

Circuit A   Circuit B

(a)   (b)

FIGURE 9.36

Gate-level schematics for the two circuits in Exercise 9.6.

9.7. (**Kripke Structure**) Derive the Kripke structure for the following circuit.

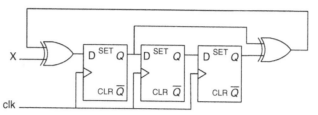

FIGURE 9.37

Gate-level schematic used for Exercise 9.7.

9.8. (**Kripke Structure**) Derive the Kripke structure for the following circuit.

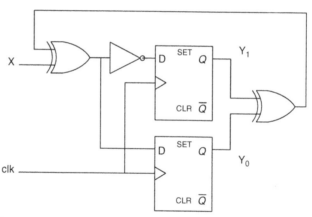

FIGURE 9.38

Gate-level schematic used for Exercise 9.8.

9.9. (**Model Checking**) Assume that ϕ, ψ, and γ are atomic propositions. Please use LTL to describe the following design properties:

  (a) If ψ occurs, γ never occurs in the future.

  (b) Always if ϕ occurs, then eventually ψ occurs immediately followed by γ.

  (c) Any occurrence of ϕ is followed eventually by an occurrence of ψ. Furthermore, γ never occurs between ϕ and ψ.

9.10. (**Model Checking**) Prove or disprove the following equivalences of all LTL formulas:

  (a) $\phi \mathbf{W} \psi \equiv \phi \mathbf{U} \psi \vee \mathbf{G} \phi$

(b) $\phi \mathbf{R} \psi \equiv \phi \mathbf{W}(\phi \vee \psi)$
(c) $\phi \mathbf{U} \psi \equiv \psi \mathbf{R}(\phi \vee \psi)$

9.11. (Model Checking) Prove the following equivalences of all CTL formulas:

(a) $\mathbf{AG}\phi \equiv \phi \wedge \mathbf{AXAG}\phi$
(b) $\mathbf{EF}\psi \equiv \psi \vee \mathbf{EXEF}\psi$
(c) $\mathbf{E}[\phi \mathbf{U} \psi] \equiv \psi \vee (\phi \wedge \mathbf{EXE}[\phi \mathbf{U} \psi])$

9.12. (Model Checking) Consider the model M in Figure 9.39. Please check whether $s_0 \models \phi$ and $s_3 \models \phi$ hold the following CTL formulas ϕ's in M:

(a) $\mathbf{AG(AF}a)$
(b) $\mathbf{EX(EX}c)$
(c) $\mathbf{AG(EF}(c \vee d))$

9.13. (Model Checking) Assume that ϕ is an atomic proposition. Please prove or disprove that the formula $\mathbf{EGF}\phi$ in CTL* is equivalent to the formula $\mathbf{EGEF}\phi$ in CTL.

9.14. (Theorem Proving) The exclusive-or function XOR can be defined as $f = x \otimes y = \bar{x}y + x\bar{y}$ in Figure 9.40a, and its implementation is shown in Figure 9.40b. Please derive $SPEC(x, y, z)$ from $IMPL(x, y, z)$ by applying the inference rules specified in Table 9.3.

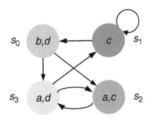

FIGURE 9.39

Finite state machine for the model M used for Exercise 9.12.

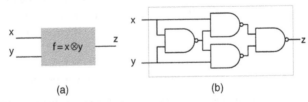

FIGURE 9.40

Specification and implementation views in Exercise 9.14. (a) $SPEC(x,y,z)$. (b) $IMPL(x,y,z)$.

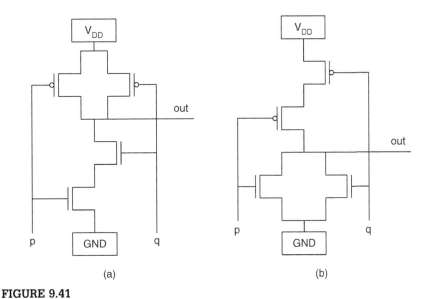

FIGURE 9.41

Transistor schematics for NAND and NOR gates in Exercise 9.16. (a) a NAND gate. (b) a NOR gate.

9.15. (**Theorem Proving**) Given i_1, i_2 as input ports and *out* as an output port, the formal specifications for NAND and XOR gates can be represented as:

- NAND(i_1,i_2,*out*) ≜ $out = \neg(i_1 \wedge i_2)$
- XOR(i_1,i_2,*out*) ≜ $out = (i_1 \wedge \neg i_2) \vee (\neg i_1 \wedge i_2)$

(a) Derive the formal descriptions for the two circuits in Exercise 9.5.

(b) Prove that the two circuits are equivalent by applying inference rules specified in Table 9.3.

9.16. (**Theorem Proving**) Given i_1 and i_2 as input ports and *out* as an output port, the formal specifications for NAND and NOR gates are:

- NAND(i_1, i_2, *out*) ≜ $out = \neg(i_1 \wedge i_2)$
- NOR(i_1, i_2, *out*) ≜ $out = \neg(i_1 \vee i_2)$

(a) Derive the formal specifications for a NAND gate from the CMOS implementation in Figure 9.41a.

(b) Derive the formal specifications for a NOR gate from the CMOS implementation in Figure 9.41b.

ACKNOWLEDGMENTS

We thank Professor Michael S. Hsiao of Virginia Tech, Professor Jing-Yang Jou of National Chiao Tung University, and Professor Jie-Hong (Roland) Jiang of National Taiwan University for reviewing the text and providing helpful comments.

REFERENCES
R9.0 Books

[Bailey 2007] G. Bailey, G. Martin, and A. Piziali, *ESL Design and Verification: A Prescription for Electronic System Level Methodology,* Morgan Kaufmann, San Francisco, February 2007.

[Bergeron 2000] J. Bergeron, *Writing Testbenches, Function Verification of HDL Models,* Second edition, Kluwer Academic Publishers, New York, February 2003.

[Dempster 2002] D. Dempster and M. Stuart, *Verification Methodology Manual: Techniques for Verifying HDL Designs,* Third Edition, Teamwork International, Hampshire, UK, June 2002.

[Foster 2004] H. Foster, A. Krolnik, and D. Lacey, *Assertion-Based Design,* Second Edition, Kluwer Academic Publishers, New York, May 2004.

[Gorden 1993] M. J. C. Gorden and T. F. Melham, *Introduction to HOL: A Theorem Proving Environment for Higher Order Logic,* Cambridge University Press, London, June 1993.

[Granas 2003] A. Granas and J. Dugundji, *Fixed Point Theory,* Springer, New York, June 2003.

[Huth 2004] M. Huth and M. Ryan, *Logic in Computer Science: Modelling and Reasoning about Systems,* Second Edition, Cambridge University Press, New York, June 2004.

[James 2003] P. James, *Verification Plans: The Five-Day Verification Strategy for Modern Hardware Verification Languages,* Kluwer Academic Publishers, New York, October 2003.

[Nipkow 2002] T. Nipkow, L. C. Paulson, and M. Wenzel, *Isabelle/HOL: A Proof Assistant for Higher-Order Logic,* Springer-Verlag, Berlin Heidelberg, May 2002.

[Palnitkar 2003a] S. Palnitkar, *Verilog® HDL: A Guide to Digital Design and Synthesis,* Second Edition, Prentice Hall PTR, New Jersey, March 2003.

[Palnitkar 2003b] S. Palnitkar, *Design Verification with e,* Prentice Hall PTR, New Jersey, October 2003.

[Piziali 2004] A. Piziali, *Functional Verification Coverage Measurement and Analysis,* Springer, New York, October 2004.

[Prior 1957] A. N. Prior, *Time and Modality,* Clarendon Press, Oxford, 1957.

R9.1 Introduction

[ANSI/ASQC 1978] ANSI/ASQC A3, Quality systems terminology. *American Society for Quality Control,* Milwaukee, WI, 1978.

[Bailey 2002] B. Bailey, The wake of the sleeping giant-verification, *Scalable Verification Technical Publications,* http://www.mentor.com, April 2002.

[Piziali 2006] A. Piziali, Verification planning to functional closure of processor-based SoCs, in *Proc. DesignCon,* 3-TP2, February 2006.

R9.2 Verification Hierarchy

[Scafidi 2004] C. Scafidi, J. D. Gibson, and R. Bhatia, Validating the Itanium 2 exception control unit: A unit-level approach, *IEEE Design & Test of Computers,* 21(2), pp. 94–101, March 2004.

R9.3 Measuring Verification Quality

[Benjamin 1999] M. Benjamin, D. Geist, A. Hartman, Y. Wolfsthal, G. Mas, and R. Smeets, A study in coverage-driven test generation, in *Proc. ACM/IEEE Design Automation Conf.*, pp. 970-975, June 1999.

[Drucker 2002] L. Drucker, Functional coverage metrics—the next frontier, *EETimes*, http://www.eetimes.com, August 2002.

[Gluska 2003] A. Gluska, Coverage-oriented verification of Banias, in *Proc. ACM/IEEE Design Automation Conf.*, pp. 280-284, June 2003.

[Verisity 2001] Verisity Design Inc., *Coverage-Driven Functional Verification*, White Paper, http://www.verisity.com, 2001.

R9.4 Simulation-Based Approach

[Accellera 2002a] Accellera, http://www.systemverilog.org, 2002

[Accellera 2002b] Accellera, http://www.accellera.org, 2002

[Synopsys 2001] Synopsys, http://www.open-vera.com, 2001

R9.5 Formal Approaches

[Abdulla 2000] P. A. Abdulla, P. Bjesse, and N. Eén, Symbolic reachability analysis based on SAT solvers, in *Proc. 6th Int. Conf. on Tools and Algorithms for the Construction and Analysis of Systems*, pp. 411-425, March 2000.

[Biere 1999] A. Biere, A. Cimatti, E. Clarke, and Y. Zhu, Symbolic model checking without BDDs, in *Proc. Workshop on Tools and Algorithms for the Construction and Analysis of Systems*, pp. 193-207, March 1999.

[Brand 1993] D. Brand, Verification of large synthesized designs, in *Proc. IEEE/ACM Int. Conf. on Computer-Aided Designs*, pp. 534-537, November 1993.

[Burch 1990] J. R. Burch, E. M. Clarke, K. L. McMillan, D. L. Dill, and L.-J. Hwang, Symbolic model checking: 10^{20} states and beyond, in *Proc. IEEE Symp. on Logic in Computer Science*, pp. 1-33, June 1990.

[Calypto 2008] Calypto Design Systems, *SLEC System,* http://www.calypto.com, 2008.

[Clarke 1986] E. M. Clarke, E. A. Emersion, and A. P. Sistla, Automatic verification of finite state concurrent system using temporal logic specifications, *ACM Trans. on Programming Languages and System*, 8(2), pp. 144-163, April 1986.

[Cohn 1988] A. Cohn, Correctness properties of the VIPER block model: The second level, *Technical Report No. 134*, University of Cambridge, Computer Laboratory, May 1988.

[Coq 2003] The Coq Development Team, *The Coq Proof Assistant Reference Manual*, version 7.4, INRIA, http://coq.inria.fr/doc/main.html, February 2003 .

[Emerson 1990] E. A. Emerson, Temporal and modal logic, in *Handbook of Theoretical Computer Science*, Vol. B, Elsevier, pp. 996-1072, 1990.

[Goldberg 2000] E. Goldberg, M. Prasad, and R. Brayton, Using SAT for combinational equivalence checking, in *Proc. Int. Workshop on Logic Synthesis*, pp. 185-191, May 2000.

[Huang 2000] S.-Y. Huang, K.-T. Cheng, K.-C. Chen, C.-Y. Huang, and F. Brewer, AQUILA: An Equivalence Checking System for Large Sequential Designs, *IEEE Trans. on Computers*, 49(5), pp. 443-464, May 2000.

[Joyce 1986] J. J. Joyce, G. Birtwistle, and M. Gordon, Proving a computer correct in higher order logic, *Technical Report No. 134*, University of Cambridge, Computer Laboratory, 1986.

[Kaufmann 2002] M. Kaufmann and J. Moore, A computational logic for applicative common lisp, in *A Companion to Philosophical Logic*, pp. 724-741, Blackwell Publishers, 2002.

[Kripke 1963] S. A. Kripke, Semantic consideration on modal logic, in *Proc. A Colloquium: Model and Many Valued Logic, Acta Philosophica Fennica*, 16, pp. 83–94, August 1963.

[Kunz 1993] W. Kunz, HANNIBAL: An efficient tool for logic verification based on recursive learning, in *Proc. IEEE/ACM Int. Conf. on Computer-Aided Design*, pp. 538–543, November 1993.

[Lu 2003] F. Lu, L.-C. Wang, K.-T. Cheng, and R. C.-Y. Huang. A circuit SAT solver with signal correlation guided learning, in *Proc. IEEE/ACM Design, Automation and Test in Europe Conf.*, pp. 892–897, March 2003.

[Martin 2004] A. Martin, Adequate sets of temporal connectives in CTL, *Elsevier Electronic Notes in Theoretical Computer Science*, 52(1), pp. 1–11, January 2004.

[McMillan 1992] K. L. McMillan, *Symbolic Model Checking—An Approach to the State Explosion Problem*, PhD thesis, SCS, Carnegie Mellon University, 1992.

[Melham 1988] T. F. Melham, Abstraction mechanisms for hardware verification, in *VLSI Specification, Verification, and Synthesis*, pp. 129–157, Kluwer Academic Publishers, Boston, 1988.

[Owicki 1982] S. Owicki and L. Lamport, Proving liveness properties of concurrent programs, *ACM Trans. on Programming Languages and Systems*, 4(3), pp. 455–495, July 1982.

[Owre 1992] S. Owre, J. M. Rushby, and N. Shankar, PVS: A prototype verification system, in *Proc. 11th Int. Conf. on Automated Deduction (CADE)*, pp. 748–752, June 1992.

[Saaltink 1999] M. Saaltink, The Z/EVES Users Guide, *Technical Report TR-97-5493-06*, ORA, Canada, 1999.

R9.6 Advanced Research

[Bai 2003] X. Bai, L. Chen, and S. Dey, Software-based self-test for crosstalk in processors, in *Proc. Int. Workshop on High Level Design Validation and Test*, pp. 11–16, November 2003.

[Bose 2001] M. Bose, J. Shin, E. M. Rudnick, T. Dukes, and M. Abadir, A genetic approach to automatic bias generation for biased random instruction generation, in *Proc. 2001 Congress on Evolutionary Computation*, pp. 442–448, May 2001.

[Chen 2003] L. Chen, S. Ravi, A. Raghunathan, and S. Dey, A scalable software-based self-test methodology for programmable processors, in *Proc. ACM/IEEE Design Automation Conf.*, pp. 548–553, June 2003.

[Fine 2003] S. Fine and A. Ziv, Coverage directed test generation for functional verification using Bayesian networks, in *Proc. ACM/IEEE Design Automation Conf.*, pp. 286–291, June 2003.

[Fine 2006] S. Fine, A. Freund, I. Jaeger, Y. Mansour, Y. Naveh, and A. Ziv, Harnessing machine learning to improve the success rate of stimuli generation, *IEEE Trans. on Computers*, 55(11), pp. 1344–1355, November 2006.

[Ganai 2001] M. Ganai, P. Yalagandula, A. Aziz, A. Kuehlmann, and V. Singhal, SIVA: A system for coverage-directed state space search, *J. of Electronic Testing: Theory and Applications*, 17(1), pp. 11–27, February 2001.

[Geist 1996] D. Geist, M. Farkas, A. Landver, Y. Lichtenstein, S. Ur, and Y. Wolfsthal, Coverage-directed test generation using symbolic techniques, in *Proc. Int. Conf. on Formal Methods in Computer-Aided Design*, pp. 143–158, November 1996.

[Guttman 1991] J. D. Guttman, A proposed interface logic for verification environments, *Technical Report M91-19*, the MITRE Corporation, March 1991.

[Hazelhurst 2002] S. Hazelhurst, G. Kamhi, O. Weissberg, and L. Fix, A hybrid verification approach: Getting deep into the design, in *Proc. ACM/IEEE Design Automation Conf.*, pp. 111–116, June 2002.

[Ho 1995] R. C. Ho, C. H. Yang, M. A. Horowitz, and D. L. Dill, Architecture validation for processors, in *Proc. Int. Symp. on Computer Architecture*, pp. 404–413, May 1995.

[Huang 2001] C.-Y. Huang and K.-T. Cheng, Using word-level ATPG and modular arithmetic constraint-solving techniques, *IEEE Trans. on Computer-Aided Design*, 20(3), pp. 381–391, March 2001.

[Iwashita 1994] H. Iwashita, S. Kowatari, T. Nakata, and F. Hirose, Automatic test program generation for pipelined processors, in *Proc. IEEE/ACM Int. Conf. on Computer-Aided Design*, pp. 580-583, November 1994.

[Jang 1997] J.-Y. Jang, S. Qadeer, M. Kaufmann, and C. Pixley, Formal verification of FIRE: A case study, in *Proc. ACM/IEEE Design Automation Conf.*, pp. 173-177, June 1997.

[Joyce 1993] J. J. Joyce and C. H. Seger, Linking BDD-based symbolic evaluation to interactive theorem-proving, in *Proc. ACM/IEEE Design Automation Conf.*, pp. 469-474, June 1993.

[Lai 2000] W.-C. Lai, A. Krstic, and K.-T. Cheng, Functionally testable path delay faults on a microprocessor, *IEEE Design & Test of Computers*, 17(4), pp. 6-14, October 2000.

[Mishra 2005] P. Mishra and N. Dutt, Functional coverage driven test generation for validation of pipelined processors, in *Proc. IEEE/ACM Design, Automation and Test in Europe Conf.*, pp. 678-683, March 2005.

[Moondanos 1998] D. Moondanos, J. A. Abraham, and Y. V. Hoskote, Abstraction techniques for validation coverage analysis and test generation, *IEEE Trans. on Computers*, 47(1), pp. 2-14, January 1998.

[Shyam 2006] S. Shyam and V. Bertacco, Distance-guided hybrid verification with GUIDO, in *Proc. IEEE/ACM Design, Automation and Test in Europe Conf.*, pp. 1211-1216, March 2006.

[Tasiran 2001] S. Tasiran, F. Fallah, D. G. Chinnery, S. J. Weber, and K. Keutzer, A functional validation technique: Biased-random simulation guided by observability-based coverage, in *Proc. IEEE/ACM Int. Conf. on Computer-Aided Design*, pp. 82-88, September 2001.

[Ur 1999] S. Ur and Y. Yadin, Micro architecture coverage directed generation of test programs, in *Proc. ACM/IEEE Design Automation Conf.*, pp. 175-180, June 1999.

[Vedula 2004] V. M. Vedula, W. J. Townhead, and J. A. Abraham, Program slicing for ATPG-based property checking, in *Proc. Int. Conf. on VLSI Design*, pp. 591-596, January 2004.

[Wagner 2005] I. Wagner, V. Bertacco, and T. Austin, StressTest: An automatic approach to test generation via activity monitors, in *Proc. ACM/IEEE Design Automation Conf.*, pp. 783-788, June 2005.

[Wen 2006] H.-P. Wen, L.-C. Wang, and K.-T. Cheng, Simulation-based functional test generation for embedded processors, *IEEE Trans. on Computers*, 55(11), pp. 1-9, November 2006.

[Wen 2007] H.-P. Wen, L.-C. Wang, and J. Bhadra, An incremental learning framework for estimating signal controllability in unit-level verification, in *Proc. IEEE/ACM Int. Conf. on Computer-Aided Design*, pp. 250-257, November 2007.

[Yang 1998] C. H. Yang and D. Dill, Validation with guided search of the state space, in *Proc. ACM/IEEE Design Automation Conf.*, pp. 599-604, June 1998.

Floorplanning

Tung-Chieh Chen
National Taiwan University, Taipei, Taiwan

Yao-Wen Chang
National Taiwan University, Taipei, Taiwan

ABOUT THIS CHAPTER

Floorplanning is an essential design step for hierarchical, building-module design methodology. Floorplanning provides early feedback that evaluates architectural decisions, estimates chip areas, and estimates delay and congestion caused by wiring. As technology advances, design complexity is increasing and the circuit size is getting larger. To cope with the increasing design complexity, hierarchical design and *intellectual property* (IP) modules are widely used. This trend makes floorplanning much more critical to the quality of a *very large-scale integration* (VLSI) design than ever.

This chapter starts with the formulation of the floorplanning problem. After the problem formulation, the two most popular approaches to floorplanning, simulated annealing and analytical formulations, are discussed. On the basis of simulated annealing, three popular floorplan representations, normalized Polished expression, B*-tree, and sequence pair, are further covered and compared. Some modern floorplanning issues such as soft modules, fixed-outline constraints, and large-scale designs are also addressed.

10.1 INTRODUCTION

In Chapter 1, we introduced the electronic design automation flow. Floorplanning is the first major step in physical design; it is particularly important because the resulting floorplan affects all the subsequent steps in physical design, such as placement and routing that are discussed in Chapters 11 and 12, respectively.

10.1.1 Floorplanning basics

Two popular approaches to floorplanning, **simulated annealing** and **analytical formulation**, are typically used to solve the floorplanning problem [Sait 1999; **575**

Sherwani 1999]. Basically, simulated annealing-based floorplanning relies on the representation of the geometric relationship among modules, whereas an analytical approach usually captures the absolute relationship directly. The topological representation profoundly affects the operations of modules and the complexity of a simulated annealing-based floorplan design process. In this chapter, three popular floorplan representations, **normalized Polish expression** [Wong 1986], **B*-tree** [Chang 2000], and **Sequence Pair** [Murata 1995], are introduced. In general, these representations are efficient, flexible, and effective in modeling **geometric relationships** (*e.g.*, left, right, above, and below relationships) among modules for floorplan designs. The simulated annealing-based floorplanning is concluded with the comparisons of popular floorplan representations in the recent literature.

The analytical approach applies **mathematical programming** that is composed of an objective function and a set of constraints. The objective function models the cost metric (*e.g.*, area and wirelength) for floorplan optimization, whereas the constraints capture the geometric and dimensional restrictions among modules (*e.g.*, the nonoverlapping and aspect ratio constraints). Specifically, this chapter introduces **mixed integer linear programming** (ILP) for the floorplanning problem [Sutanthavibul 1990]. For the mixed ILP formulation, an approximated area is modeled by a linear cost function, whereas the nonoverlapping and aspect ratio constraints are modeled by a set of linear equations. To handle the expensive time complexity of mixed ILPs, a **successive augmentation** method that solves a partial problem at each step to reduce the floorplanning complexity is also introduced.

In addition to chip area minimization, modern VLSI floorplanning also needs to handle some important issues such as **soft modules** and **fixed-outline constraints**. Unlike a **hard module** that has a fixed dimension (width and height), the shape of a soft module is to be decided during floorplanning, although its area is fixed. Therefore, a floorplanner needs to find a desired aspect ratio for each soft module to optimize the floorplan cost. As pointed out by [Kahng 2000], modern VLSI design is based on a fixed-die (fixed-outline) floorplan, rather than a variable-die one. An area-optimized floorplan without considering the fixed-outline constraint may be useless, because it might not fit into the given outline. Therefore, modern floorplanning should address the fixed-outline consideration.

As the transistor feature size scales down, design complexity is increasing drastically. To cope with the increasing design complexity, *intellectual property* (IP) modules are widely reused for large-scale designs. Consequently, a modern VLSI design often consists of large-scale functional modules, and designs with billions of transistors are already in production. Therefore, efficient and effective design methods and tools capable of placing and optimizing large-scale modules are essential for modern chip designs. In addition to the enhancement in floorplanning tools, the floorplanning frameworks are evolving to tackle the challenges in design complexity. This chapter also addresses the **multilevel frameworks** for large-scale building module designs.

10.1.2 **Problem statement**

The floorplanning problem can be stated as follows: Let $B = \{b_1, b_2, \ldots, b_m\}$ be a set of m rectangular modules whose respective width, height, and area are denoted by w_i, h_i, and a_i, $1 \le i \le m$. Each module is free to rotate. Let (x_i, y_i) denote the coordinate of the bottom-left corner of module b_i, $1 \le i \le m$, on a chip. A floorplan F is an assignment of (x_i, y_i) for each b_i, $1 \le i \le m$, such that no two modules overlap with each other. The goal of floorplanning is to optimize a predefined cost metric such as a combination of the area (*i.e.*, the minimum bounding rectangle of F) and wirelength (*i.e.*, the sum of all interconnection lengths) induced by a floorplan. For modern floorplan designs, other costs such as routability, power, and thermal might also need to be considered.

10.1.3 **Floorplanning model**

We can classify floorplans into two categories for discussions: (1) **slicing floorplans** and (2) **non-slicing floorplans**. A slicing floorplan can be obtained by repetitively cutting the floorplan horizontally or vertically, whereas a non-slicing floorplan cannot. The given dimension of each hard module must be kept. All modules are free of rotation; if a module is rotated, its width and height are exchanged.

Example 10.1 Two example floorplans are shown in Figure 10.1. Consider the non-slicing floorplan with five modules shown in Figure 10.1a first. The five modules can be rearranged to form a slicing floorplan given in Figure 10.1b, in which each module can be extracted by repetitively cutting the floorplan horizontally or vertically.

10.1.3.1 *Slicing floorplans*

On the basis of the slicing property of a slicing floorplan, we can use a binary tree to represent a slicing floorplan [Otten 1982]. A **slicing tree** is a binary tree with modules at the leaves and cut types at the internal nodes. There are two cut types, H and V. The H cut divides the floorplan horizontally, and the left (right) child represents the bottom (top) sub-floorplan. Similarly, the V cut divides the floorplan vertically, and the left (right) child represents the left (right) sub-floorplan.

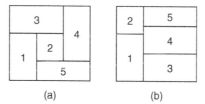

(a)   (b)

FIGURE 10.1

Examples of (a) a non-slicing floorplan. (b) a slicing floorplan.

Note that a slicing floorplan may correspond to more than one slicing tree, because the order of the cut-line selections may be different. This representation duplication might incur a larger solution space and complicate the optimization process. Therefore, it is desirable to prune such redundancies to facilitate floorplan design. As such, we refer to a slicing tree as a **skewed slicing tree** if it does not contain a node of the same cut type as its right child.

Example 10.2 Figure 10.2 shows the slicing floorplan from Figure 10.1b and its corresponding slicing tree. The tree root, V, represents the vertical cut-line that divides the floorplan into the left sub-floorplan (modules 1 and 2) and the right sub-floorplan (modules 3, 4, and 5). The left child of the root is a node H, which horizontally divides the left sub-floorplan into the bottom sub-floorplan (module 1) and the top sub-floorplan (module 2). Similarly, the right child of the root horizontally cuts the sub-floorplan into the bottom (module 3) and the top (modules 4 and 5) sub-floorplans, and the top sub-floorplan is further divided into the bottom (module 4) and the top (module 5) sub-floorplans.

There are two slicing trees corresponding to the floorplan in Figure 10.2a. Figure 10.3a and Figure 10.3b show the two slicing trees. The slicing tree in Figure 10.3a is a non-skewed slicing tree, because there is a node H as the right child of anther node H (see the dashed box). The slicing tree in Figure 10.3b gives a skewed one.

10.1.3.2 *Non-slicing floorplans*

The non-slicing floorplan is more general than the slicing floorplan. However, because of its non-slicing structure, we cannot use a slicing tree to model it. Instead, we can use a *horizontal constraint graph* (HCG) and a *vertical constraint graph* (VCG) to model a non-slicing floorplan. The horizontal constraint graph defines the horizontal relations of modules, and the vertical constraint graph defines the vertical ones. In a constraint graph, a node represents a module. If there is an edge from node A to node B in the HCG (VCG), then module A is at the left (bottom) of module B.

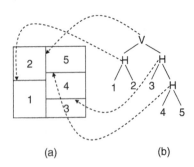

(a)   (b)

FIGURE 10.2

An example of (a) a slicing floorplan. (b) a slicing tree modeling.

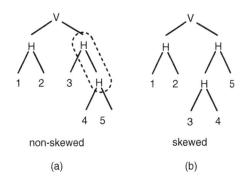

FIGURE 10.3

An example of (a) a non-skewed slicing tree. (b) a skewed slicing tree. Both slicing trees represent the same floorplan shown in Figure 10.2a.

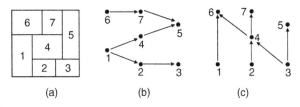

FIGURE 10.4

An example of (a) a non-slicing floorplan. (b) The horizontal constraint graph. (c) the vertical constraint graph.

Example 10.3 Consider the example non-slicing floorplan given in Figure 10.4a. Figure 10.4b and Figure 10.4c show the corresponding horizontal and vertical constraint graphs of the floorplan, respectively. Because module 1 is at the left of module 4, we add a directed edge from node 1 to node 4 in the horizontal constraint graph. Similarly, module 2 is below module 4, and thus we add a directed edge from node 2 to node 4 in the vertical constraint graph. On the basis of the left/right and above/below relationships, we can construct a horizontal and a vertical constraint graph corresponding to a floorplan.

10.1.4 **Floorplanning cost**

The goal of floorplanning is to optimize a predefined cost function, such as the area of a resulting floorplan given by the minimum bounding rectangle of the floorplan region. The floorplan area directly correlates to the chip silicon cost. The larger the area, the higher the silicon cost. The space in the floorplan bounding rectangle uncovered by any module is called **white space** or **dead space**.

Other floorplanning cost, such as wirelength, will also be considered. Shorter wirelength not only can reduce signal delay but also can facilitate wire interconnection at the routing stage. The floorplanning objective can also be a combined cost, such as area plus wirelength.

Example 10.4 Consider the two different floorplans for the same seven modules given in Figure 10.5. The left figure (see Figure 10.5a) is an optimal floorplan in terms of area, because there is no wasted area among modules. The right figure (see Figure 10.5b) illustrates a non-optimal floorplan. It is clear that the area of the right floorplan is larger than that of the left one, because there are white spaces in the floorplan shown in Figure 10.5b.

There are two popular approaches to find a desired floorplan: simulated annealing and the analytical approach. The two approaches are discussed in the following two sections.

10.2 SIMULATED ANNEALING APPROACH

Simulated annealing (SA) is probably the most popular method for floorplan optimization [Kirpatrick 1983]. It has the significant advantage of easily incorporating an optimizing goal into the objective function. To apply simulated annealing for floorplan design, it needs to first encode a floorplan as a solution, called a **floorplan representation**, which models the geometric relation of modules in a **floorplan**. A floorplan representation not only induces a solution space that contains all feasible solutions defined by the representation but also induces a unique solution structure that guides the search of simulated annealing to find a desired floorplan in the solution space. In this section, we detail three popular floorplan representations, **Normalized Polish Expression** [Wong 1986], **B*-tree** [Chang 2000], and **Sequence Pair** [Murata 1995], and summarize the properties of some popular recent representations in the literature.

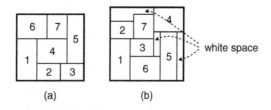

(a)   (b)

FIGURE 10.5

An example of (a) an optimal floorplan in terms of area. (b) a non-optimal floorplan.

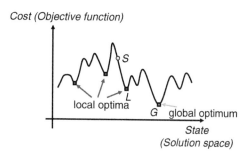

FIGURE 10.6

Illustration of simulated annealing.

10.2.1 **Simulated annealing basics**

Figure 10.6 illustrates the simulated annealing process. In Figure 10.6, the **cost** of a floorplan (represented by the vertical axis) is plotted as a function of the **state** of a floorplan configuration (represented by the horizontal axis). Given an initial solution S, it tries to search for a globally optimal floorplan solution with the lowest cost. For a greedy approach, it might iteratively search for a neighboring solution with a lower cost than that of the current solution until such a neighboring solution cannot be found. For this greedy mechanism, it is very likely that we get stuck in a locally optimal solution, such as L in Figure 10.6. Once it gets stuck at L, it is impossible to escape from this locally optimal solution, because all neighboring solutions have higher costs than L. Unlike the greedy approach, simulated annealing adopts a **hill-climbing** technique to escape from the locally optimal solution L. Given a solution, simulated annealing provides a non-zero probability to move from the current solution to a neighboring one even with a higher cost. With this **uphill move** capability, it is possible to reach the globally optimal solution G no matter where the initial solution starts. The probability of accepting a neighboring solution depends on two factors: (1) the magnitude of the uphill move and (2) the search time. To implement this idea, the probability of accepting a new solution S' is defined by:

$$\text{Prob}(S \rightarrow S') = \begin{cases} 1 & \text{if } \Delta C \leq 0 \quad \text{(down-hill move)} \\ e^{\Delta C/T} & \text{if } \Delta C > 0 \quad \text{(up-hill move)} \end{cases}$$

where $\Delta C = \text{cost}(S') - \text{cost}(S)$, and T is the current temperature. Every down-hill move is accepted, and the probability of accepting an uphill move depends on the magnitude of the move (cost difference) and the search time (annealing temperature). Initially, we are assigned a high temperature. As the annealing process goes by, the temperature is typically decreased by a fixed ratio, say 0.9. For example, a simple annealing schedule is given by $T = T_0, T_1, T_2, \ldots$, and $T_i = r_i T_{i-1}$, $r < 1$. At each temperature, we perturb the current solutions to search for a number of neighboring solutions for k times, where k is a user-defined value, and keep the best solution found so far. This process continues until the temperature is reduced to a "frozen" state or a predefined termination condition is reached. Then, the best-found solution is reported. It is clear that the probability of accepting an "inferior" solution is higher if the cost increase is smaller and/or the current temperature T is higher. In practice, simulated annealing is often quite effective in searching for a desired solution. The whole simulated annealing process is an analogy of annealing an iron to produce a craft work, and it is where the name simulated annealing comes from. At a high temperature, the atoms of iron get more energy and thus have more freedom to move around. As the temperature reduces gradually, the atoms reach an equilibrium state step by step. As a result, the iron forms a desired shape. Algorithm 10.1 summarizes the generic algorithm of simulated annealing.

In addition to simulated annealing, we could also apply the **iterative method** (or **greedy search**) for floorplan designs. For this method, all uphill moves are rejected. The implementation is easier, and it typically can converge to a solution faster. However, it is very likely that this method gets stuck at some local optimum (see Figure 10.6 for an illustration). Consequently, its solution quality highly depends on the selected initial solution, and thus this approach is not as popular as simulated annealing for floorplan designs.

Algorithm 10.1 Simulated Annealing Algorithm for Floorplanning

---

1. Get an initial floorplan S; $S_{Best} = S$;
2. Get an initial temperature $T > 0$;
3. **while** not "frozen" **do**
4.   **for** $i = 1$ to k **do**
5.   Perturb the floorplan to get a neighboring S' from S;
6. $\Delta C = \text{cost}(S') - \text{cost}(S)$;
7.   **if** $\Delta C \leq 0$ **then**   // down-hill move
8             $S = S'$;
9.   **else**   // uphill move
10. $S = S'$ with the probability $e^{-\Delta C/T}$;
11.   **end if**
12.   **if** $\text{cost}(S_{Best}) > \text{cost}(S)$ **then**
13. $(S_{Best}) = S$;
14.   **end if**
15.   **end for**
16. $T = rT$;   // reduce temperature
17. **end while**
18. **return** S_{Best};

There are four basic ingredients for simulated annealing: (1) **solution space**, (2) **neighborhood structure**, (3) **cost function (objective function)**, and (4) **annealing schedule**. A floorplan representation defines the solution space and the neighborhood structure, the cost function is defined by the optimization goal, and the annealing schedule captures the temperature change during the annealing process.

10.2.2 Normalized Polish expression for slicing floorplans

In Section 10.1.3, a binary tree is used to model a slicing floorplan. We can record the binary tree by use of a **Polish expression** $E = e_1 e_2 \ldots e_{2n-1}$ where $e_i \in \{1, 2, \ldots, n, H, V\}$. Here, each number denotes a module and H (V) represents a horizontal (vertical) cut in the slicing floorplan. The Polish expression is the **postfix ordering** of a binary tree, which can be obtained from the post-order traversal on a binary tree given in Algorithm 10.2. The length of E is $2n-1$, where n is the number of modules.

Because the Polish expression is the postfix order of a slicing tree, it has the **balloting property:** for every sub-expression $E_i = e_1 \ldots e_i$, $1 \leq i \leq 2n-1$, the number of operands is always larger than the number of operators.

As shown in Section 10.1.3, a slicing floorplan might induce multiple slicing trees, resulting in significant redundancies. Such redundancies will enlarge the solution space and complicate the search for a desired solution. Therefore, it is desired to prune such redundant solutions. As such, the **normalized Polish expression** E corresponding to the skewed slicing tree T is defined [Wong 1986]; a skewed slicing tree does not contain any node of the same cut type as its right child, and so does a normalized Polished expression, which does not contain consecutive operators of the same type, which is "HH" or "VV," in E. This results in a 1-1 correspondence between a set of skewed slicing trees with n modules and the corresponding set of normalized Polish expressions of length $2n-1$.

Algorithm 10.2 PostOrderTraversal(T)

---

1. /* Post-order traversal of a binary tree */
2. **if** root(T) ! $=$ NULL **then**
3.   PostOrderTraversal(LeftSubtree(T));
4.   PostOrderTraversal(RightSubtree(T));
5.   Visit(root(T));
6. **end if**
7. **return**;

To transform a normalized Polish expression E to its corresponding floorplan F, we can use a bottom-up method to recursively combine the slicing sub-floorplans on the basis of E. There are two binary operators, H and V. If a and b are two modules or sub-floorplans, the expression abH implies to place a below b, and abV implies to place a to the left of b, as illustrated in Figure 10.7. The packing is performed by a post-order procedure. Each time, we combine two slicing sub-floorplans according to the operator type. For example, $E = 12$H implies that module 1 is placed

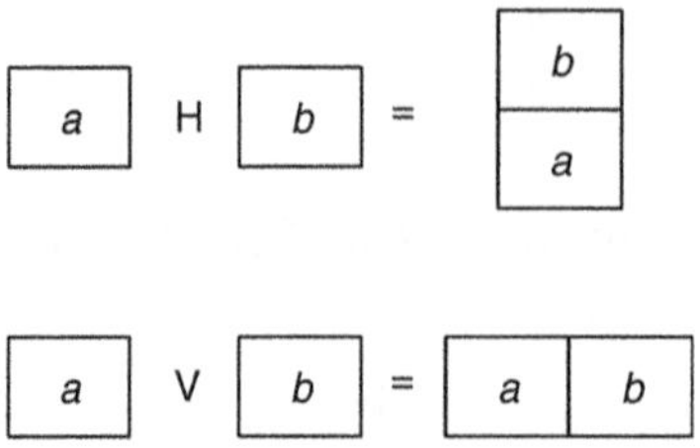

FIGURE 10.7

Binary operators, H and V, for slicing floorplans.

below module 2, and $E = 34V$ implies that module 3 is to the left of module 4. For $E = 123H\ldots$, because E is in the postfix expression, the operator H takes the operands 2 and 3, and module 2 is to the left of module 3.

Because E is in the postfix form, we can use a stack to facilitate the packing procedure (see Algorithm 10.3). Each time, we check an operand or an operator from E. If it is an operand, we push it into the stack; if it is an operator, we pop two operands from the stack and derive the new sub-floorplan based on the two operands and the operator. Then, the resulting sub-floorplan is treated as an operand and is pushed into the stack. This procedure continues until all operands/operators in E are processed, and the final floorplan is popped out from the stack.

Algorithm 10.3 Polish Expression Evaluation (E)

1. stack s;
2. **for** $i = 1$ to $2n$-1 **do**
3.   **if** e_i is an operand **then** s.push(e_i);
4.   **if** e_i is an operator **then**
5. $a = s$.pop( ); $b = s$.pop( ); $c = a\ e_i\ b$;
6. s.push(c);
7.   **end if**
8. **end for**
9. **return** s.pop( );

Example 10.5 Given a binary tree shown in Figure 10.8, we can construct a Polish expression $E = $ 12H34H5HV based on the post-order traversal.

The balloting property is verified in the following table. For each column, the number of operands is always larger than that of operators. Further, $E = $ 12H34H5HV is a normalized Polish expression because there are no consecutive operators of the same type.

	1	2	H	3	4	H	5	H	V
No. of operands	1	2	2	3	4	4	5	5	5
No. of operators	0	0	1	1	1	2	2	3	4

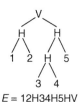

$E = 12H34H5HV$

FIGURE 10.8

A binary tree and its Polish expression.

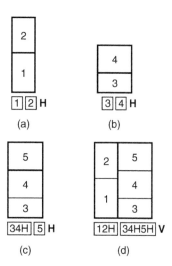

(a)   (b)

(c)   (d)

FIGURE 10.9

The packing process of the normalized Polish expression $E =$ 12H34H5HV.

After obtaining the normalized Polish expression, $E =$ 12H34H5HV from the given slicing tree, Figure 10.9 gives steps to construct the corresponding floorplan from E. In step (a), we place module 1 below module 2 to obtain a slicing floorplan 12H. In step (b), we place module 3 below module 4 to obtain a slicing floorplan 34H. In step (c), we place the slicing floorplan 34H below module 5 to obtain the slicing floorplan 34H5H. In the final step, we place the sub-floorplan 12H to the left of the sub-floorplan 34H5H to obtain the final floorplan 12H34H5HV in Figure 10.9d.

10.2.2.1 *Solution space*

The set of all normalized Polish expressions forms the solution space. Given a normalized Polish expression with n operands (modules) and n-1 operators, the total number of combinations can be computed by the number of unlabeled binary trees with $2n$-1 nodes and the permutation of n labels. The permutation of n labels is n! From [Hilton 1991], the counting of an unlabeled p-ary tree with n node is given by

$$\frac{1}{(p-1)n+1}\binom{pn}{n}$$

Applying **Stirling's approximation**

$$n! = \Theta\left(\sqrt{2\pi n}\left(\frac{n}{e}\right)^n\right)$$

and setting p to 2, we have the following asymptotic form

$$O\left(\frac{1}{(2-1)n+1}\binom{2n}{n}\right)$$

$$= O\left(\frac{(2n)!}{(n+1)n!(2n-n)!}\right)$$

$$= O\left(\frac{\sqrt{4\pi n}(2n/e)^{2n}}{(n+1)2\pi n(n/e)^{2n}}\right)$$

$$= O\left(\frac{2^{2n}}{n^{1.5}}\right)$$

With the H/V label on internal nodes (2^{n-1}) and the permutation on external nodes (n!), the total number of possible skewed slicing floorplans (normalized Polish expressions) with n modules is

$$O\left(n!2^{n-1}\frac{2^{2n}}{n^{1.5}}\right)$$

$$= O\left(n!\frac{2^{3n}}{n^{1.5}}\right)$$

Note that the upper bound is not tight. The reader can refer to [Shen 2003] for the derivation of the tighter bound of $\Theta(n!\ 2^{2.543n}/n^{1.5})$ for the total number of skewed slicing floorplans.

10.2.2.2 *Neighborhood structure*

Given a solution, we can perturb it to obtain a "neighboring" solution. The perturbation plays an important role in the search for a desired solution. For a normalized Polish expression, two operands are said to be **adjacent** if there is no operand between them. Two operators are said to be adjacent if there is no operand or operator between them. An operand and an operator are said to be adjacent if they are next to each other in E. A **chain** is a sequence of adjacent operators. For a normalized Polish expression, no consecutive operators of the same type are allowed, and thus there are only two types of chains, HVHVH... or VHVHV.... In other words, no chain can be HH... or VV....

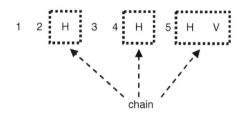

FIGURE 10.10

An example of the chains in a normalized Polish expression.

Example 10.6 In Figure 10.10, the operands 1 and 2 are adjacent, and so are the operands 3 and 4; the operand 3 and the operator H are also adjacent. There are three chains H, H, and HV in Figure 10.10.

We define three types of operations to perturb one normalized Polish expression to another.

Op1: Swap two adjacent operands.
Op2: Invert a chain by changing V to H and H to V.
Op3: Swap two adjacent operands and operators.

Performing Op1 and Op2 on a normalized Polish expression always produces a legal normalized Polish expression. However, Op3 could make the number of operands not greater than that of operators, which violates the balloting property, or it could generate two identical consecutive operators, which violates the property of a normalized Polish expression. As a result, we will only accept those Op3 operations that result in legal normalized Polish expressions. It turns out not to be difficult to check the legality of Op3. Assume that Op3 swaps the operand e_i and the operator e_{i+1}, $1 \leq i \leq k-1$. Then, the swap will not violate the balloting property if and only if $2N_{i+1} < i$, where N_k is the number of operators in the expression $E = e_1\, e_2 \dots\, e_k$, $1 \leq k \leq 2n-1$.

Two normalized Polish expressions are said to be **neighboring** if one can be perturbed to another by use of one of the three operations. Furthermore, these three operations are sufficient to generate any normalized Polish expression from a given normalized Polish expression by a sequence of the preceding operations.

Example 10.7 Figure 10.11 gives an example of applying the three types of operations on the normalized Polish expression E to obtain its corresponding slicing floorplans. Given $E = 12H4H35VH$ and the modules' dimensions, the initial floorplan is shown in Figure 10.11a. Applying Op1 to swap the two adjacent operands 4 and 3, we obtain $E' = 12H3H45VH$ and the resulting floorplan as shown in Figure 10.11b. We then apply Op2 to invert the last chain VH to obtain $E'' = 12H3H45HV$ and its resulting floorplan

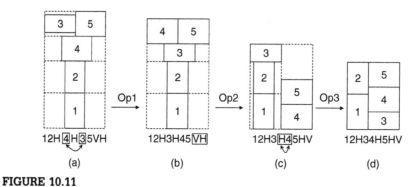

FIGURE 10.11

Illustrations of the perturbations in Example 10.7.

as shown in Figure 10.11c. We further apply Op3 to swap the adjacent operand and operator H4 to obtain $E''' = $ 12H34H5VH. Consequently, we obtain the final floorplan as shown in Figure 10.11d, which has zero dead space.

10.2.2.3 *Cost function*

Area and wirelength are perhaps the two most commonly used costs for floorplan design. We can adopt the following cost function for simulated annealing:

$$Cost = \alpha \frac{A}{A_{norm}} + (1 - \alpha) \frac{W}{W_{norm}}$$

where A is the floorplan area, A_{norm} is the average area, W is the total wirelength, W_{norm} is the average wirelength, and α controls the weight between area and wirelength. To compute A_{norm} and W_{norm}, we can perturb the initial normalized Polish expression by use of the three operations for m times to obtain m floorplans and compute the average area A_{norm} and the average wirelength W_{norm} of these floorplans. The value m is proportional to the problem size.

To illustrate the area evaluation for a normalized Polish expression, we first construct the corresponding skewed slicing tree. All feasible floorplan implementations with the minimum areas are recorded in the corresponding nodes. Because module rotation is allowed, we might have two possible floorplan implementations, (w, h) and (h, w), for a module of the dimension (w, h). Of course, we have only one possible implementation for a square module. The floorplan size can be obtained in a bottom-up manner. Consider a non-leaf node with its left child of the dimension (w_1, h_1) and its right child of the dimension (w_2, h_2). If the cut type is H, the resulting implementation is $(\max(w_1, w_2), h_1 + h_2)$; if the cut type is V, the resulting implementation is $(w_1 + w_2, \max(h_1, h_2))$. For a node with two possible implementations in its left child and its right child each, it may generate four resulting implementations. For any two implementations $m_i = (w_i, h_i)$ and $m_j = (w_j, h_j)$, m_i **dominates** m_j if $w_i \leq w_j$ and $h_i \leq h_j$. In other words, the implementation m_j is **redundant,** because the implementation m_i gives a

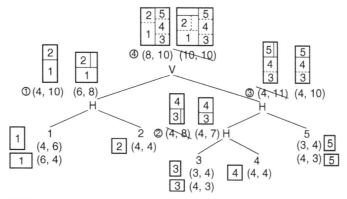

FIGURE 10.12

Dimension computation based on the slicing tree.

Table 10.1 The Dimensions of Modules in Example 10.8.

Module No.	Width	Height
1	4	6
2	4	4
3	3	4
4	4	4
5	3	4

floorplan solution of smaller area. By doing so, we can prune redundant implementations. Stockmeyer shows that the number of resulting irredundant slicing floorplan implementations after combining two nodes grows only linearly [Stockmeyer 1983]. (See Exercise 10.5.) Consequently, the area cost for a normalized Polish expression can be computed efficiently in polynomial time.

To compute the wirelength, we can only resort to approximation, because actual wiring is not performed yet at the floorplan stage. A popular wirelength approximation for a net is to measure its ***half-perimeter wirelength*** (HPWL). The HPWL is the half-perimeter length of the smallest bounding box that encloses all pins. If pin positions are not given, we can compute HPWL by use of the centers of modules.

Example 10.8 Find a floorplan implementation with the minimum area for the normalized Polish expression E = 12H34H5HV. The module dimensions are listed in Table 10.1.

First, we construct the corresponding skewed slicing tree and record the dimension candidates for each module with its corresponding leaf node. Because modules 2 and 4 are square, they have only one dimension candidate. The dimension candidates

Table 10.2 Area Evaluation for Example 10.8.

Step	Operator	Left child	Right child	Results
1	H	(4, 6) (6, 4)	(4, 4)	(4, 10) (6, 8)
2	H	(3, 4) (4, 3)	(4, 4)	~~(4, 8)~~ (4, 7)
3	H	(4, 7)	(3, 4) (4, 3)	~~(4, 11)~~ (4, 10)
4	V	(4, 10) (6, 8)	(4, 10)	(8, 10) ~~(10, 10)~~

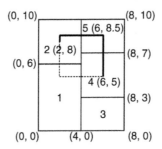

FIGURE 10.13

An example of the HPWL computation.

(irredundant implementations) for each internal node are updated in a post-order traversal order. See the steps in Table 10.2 and Figure 10.12. The resulting implementations that are dominated by others are crossed out.

Finally, the only irredundant implementation for the root is (8, 10), implying that the resulting floorplan has width = 8, height = 10, and area = 8 * 10 = 80. This floorplan has the minimum area for the normalized Polish expression E = 12H34H5HV with the given module dimensions. In case that we have more than one irredundant implementation for the root, we would pick the one with the minimum area.

Example 10.9 Given the floorplan in Figure 10.13, compute the HPWL of a net connecting modules 2, 4, and 5. The center coordinates of modules 2, 4, and 5 are (2, 8), (6, 5), and (6, 8.5), respectively. The height of the minimum bounding box is 3.5 and the width is 4. So the HPWL of this net is 3.5 + 4 = 7.5.

10.2.2.4 *Annealing schedule*

We can apply the **classical simulated annealing** algorithm for floorplanning described in Algorithm 10.4. The annealing schedule is $T = T_0, T_1, T_2, \ldots$, and $T_i = r_i T_{i-1}$, $r < 1$. The initial temperature is set to a high value so that the probability of accepting all perturbation is close to 1.0. We can compute the initial temperature as follows. Before the simulated annealing process starts, we perturb the initial normalized Polish expression for a certain time to compute the average of all positive (uphill) cost change Δ_{avg}. Then, T_0 is initialized as $T_0 = -\Delta_{avg}/\ln P$, where P is the initial probability of accepting an uphill solution. We can set P very close to 1.0 (but certainly not 1.0). The temperature is reduced by a fixed ratio r at each iteration of annealing. The value 0.85 is recommended by most previous works. The larger the r, the longer the annealing time; however, a larger r often results in a better floorplan solution.

Algorithm 10.4 Wong-Liu Floorplanning (P, ε, r, k)

---

1. $E = 12V3V4V \ldots nV$;   // initial solution
2. $E_{Best} = E$; $T = -\Delta_{avg} / \ln P$;
3. **do**
4. $reject = 0$;
5.   **for** $ite = 0$ to k **do**
6.   SelectOperation(Op);
7.   **Case** Op
8. Op_1: Select two adjacent operands e_i and e_j; $E' = \text{Swap}(E, e_i, e_j)$;
9. Op_2: Select a nonzero length chain C; $E' = \text{Complement}(E, C)$;
10. Op_3: done = FALSE
11.   **while** not (done) **do**
12.   Choice 1: Select two adjacent operand e_i and operator e_{i+1};
13.   **if** $(e_{i-1} \neq e_{i+1})$ and $(2N_{i+1} < i)$ **then** done = TRUE;
14.   Choice 2: Select two adjacent operator e_i and operand e_{i+1};
15.   **if** $(e_i \neq e_{i+2})$ **then** done = TRUE;
16.   **end while**
17. $E' = \text{Swap}(E, e_i, e_{i+1})$;
18.   **end case**
19. $\Delta\text{Cost} = \text{cost}(E') - \text{cost}(E)$;
20.   **if** $(\Delta\text{Cost} \leq 0)$ **or** $(Random < e^{-\Delta Cost/T})$ **then**
21. $E = E'$;
22.   **if** $\text{cost}(E) < \text{cost}(E_{Best})$ **then** $E_{Best} = E$;
23.   **else**
24. $reject = reject + 1$;
25.   **end if**
26. **end for**
27. $T = rT$;   // reduce temperature
28. **until** $(reject/k > 0.95)$ **or** $(T < \varepsilon)$ **or** (OutOfTime)

The excessive running time, however, is a significant drawback of the classical SA algorithm. To improve the efficiency of SA for searching for desired solutions, several annealing schemes of controlling the temperature changes during the annealing process have been proposed in the literature. The annealing schedule used in TimberWolf [Sechen 1986, 1988] is one of the most popular schemes. It increases r gradually from its lowest value (0.8) to its highest value (approximately 0.95) and then gradually decreases r back to its lowest value.

Recently, a *fast-simulated annealing* (Fast-SA) scheme was proposed in [Chen 2005b]. The motivation is to reduce the number of uphill moves in the beginning, because most of the uphill moves at this stage lead to inferior solutions. Because it is not efficient and effective to accept too many uphill moves in the beginning, a greedy search can be applied to find a local optimum faster.

Starting with the local optimum, we then switch to normal SA. By doing so, it can save time for searching for desired solutions. To implement the preceding scheme, Fast-SA consists of three stages: (1) the high-temperature random search stage, (2) the pseudo-greedy local search stage, and (3) the hill-climbing search stage. At the first stage, the temperature is set to a very large value, and thus the probability of accepting an inferior solution approaches 1. This can avoid getting trapped in a local optimum in the very beginning. At the second stage, the temperature is set approaching zero to accept only a small number of inferior solutions. At the third stage, the temperature is raised to facilitate hill climbing to search for better solutions. The temperature reduces gradually, and very likely it finally converges to a desired solution. See Figure 10.14a and Figure 10.14b for the respective temperature changes over the search time with classical SA and Fast-SA.

Because the Fast-SA scheme saves significant iterations to explore the solution space, it could devote more time to finding better solutions in the hill-climbing stage. This makes the annealing much more efficient and effective. To implement this annealing scheme, we derive the temperature updating function T of Fast-SA by the following equations:

$$
T_r = \begin{cases} \dfrac{-\Delta_{avg}}{\ln P} & r = 1 \\[2ex] \dfrac{T_1 \langle \Delta_{cost} \rangle}{rc} & 2 \le r \le k \\[2ex] \dfrac{T_1 \langle \Delta_{cost} \rangle}{r} & r > k \end{cases}
$$

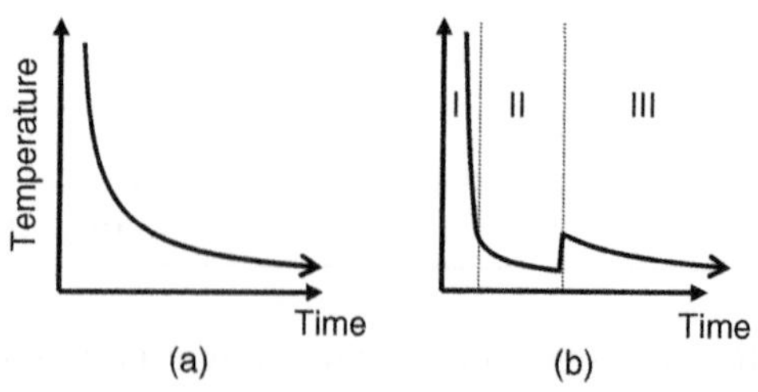

FIGURE 10.14

Temperature versus search time for (a) classical SA. (b) Fast-SA.

Here, r is the number of iterations, Δ_{avg} is the average uphill cost, P is the initial probability for accepting uphill solutions, $<\Delta_{cost}>$ is the average cost change (new cost − old cost) for the current temperature, and c and k are user-specified parameters. At the first iteration, the temperature is set according to the given initial probability P and the average uphill cost Δ_{avg}. Because P is usually set close to 1, it performs a random search to find a good solution. Then, it enters the pseudo-greedy local search stage until the kth iteration. Here, c is a user-defined parameter to control how low the temperature is in the second stage. We usually choose a large c to make $T \rightarrow 0$ such that it only accepts good solutions to perform pseudo-greedy searches. After k iterations, the temperature jumps up to further improve the solution quality. The value of $<\Delta_{cost}>$ affects the reduction rate of the temperature. If the cost of a neighboring solution changes significantly, $<\Delta_{cost}>$ is larger and thus the temperature reduces slower. In contrast, if $<\Delta_{cost}>$ is smaller, it implies that the cost of the neighboring solution only changes a little; in this case, we reduce the temperature more to reduce the number of iterations. Because the cost function is normalized to 1, this implies that $<\Delta_{cost}>$ is less than 1, and it ensures that the temperature is decreased. The number of iterations in the second stage can be determined by the problem size. The smaller the problem size, the smaller the k value. We can set $c = 100$ and $k = 7$ for typical floorplanning problems. Note that the initial temperature for Fast-SA is the same as that for the classical SA (*i.e.*, $T_1 = -\Delta_{avg}/\ln P$). The initial temperature T_1 needs to be kept high to avoid getting trapped into a local optimum in the very beginning.

10.2.3 **B*-tree for compacted floorplans**

B*-trees are based on ordered binary trees to model **compacted floorplans**. In a compacted floorplan, no modules can be moved toward left or bottom in the floorplan [Chang 2000]. Consequently, an area-optimal floorplan always corresponds to some B*-tree. Inheriting from the nice properties of ordered binary trees, B*-trees are very easy for implementation and can perform the three primitive tree operations **search**, **insertion**, and **deletion** in constant, constant, and linear time, respectively.

Unlike the slicing floorplan and its corresponding slicing tree(s), there exists a unique correspondence between a compacted floorplan and its induced B*-tree. In other words, given a compacted floorplan F, we can construct a unique B*-tree corresponding to F, and the packing corresponding to the B*-tree is the same as F. The nice property of the unique correspondence between a compacted floorplan and its induced B*-tree prevents the search space from being enlarged with redundant solutions and guarantees that an area-optimal placement can be found by searching on B*-trees. In the following, we describe the procedures for the transformation between a floorplan and a B*-tree.

10.2.3.1 *From a floorplan to its B*-tree*

First, we compact all modules to the left and bottom to obtain a compacted floorplan, because the B*-tree can only model compacted floorplans. A B*-tree is an ordered binary tree whose root corresponds to the module on the bottom-left corner. Similar to the ***depth-first-search*** (DFS) procedure, we construct a B*-tree T for a compacted floorplan in a recursive fashion: starting from the root, we first recursively construct the left sub-tree and then the right sub-tree. Let R_i be the set of modules located on the right-hand side and adjacent to module i. The left child of the node n_i corresponds to the lowest, unvisited module in R_i. The right child of n_i represents the lowest module located above and with the same x-coordinate as that of module i.

Example 10.10 Figure 10.15 shows an example of constructing a B*-tree from the floorplan given in Figure 10.15a. First, we compact all modules to the left and bottom to obtain a compacted floorplan as shown in Figure 10.15b. Module 1 is the root of the B*-tree, because it locates at the bottom-left corner. Module 2 is the lowest unvisited adjacent module on the right of module 1, so we make node n_2 the left child of node n_1. Module 3 has the same x-coordinate as that of module 1, and it is the lowest unvisited module above module 1, so we make node n_3 the right child of node n_1. Similarly, node n_4 is the left child of node n_3, node n_6 is the right child of node n_3, and node n_5 is the left child of node n_4.

10.2.3.2 *From a B*-tree to its floorplan*

Given a B*-tree T, its root represents the module on the bottom-left corner, and thus the coordinate of the module is $(x_{root}, y_{root}) = (0, 0)$. If node n_j is the left child of node n_i, module j is placed on the right-hand side and adjacent to module i, *i.e.*, $x_j = x_i + w_i$. Otherwise, if node n_j is the right child of n_i, module j is placed above module i, with the same x-coordinate as that of module j, *i.e.*, $x_j = x_i$. Therefore, given a B*-tree, the x-coordinates of all modules can be determined by traversing the tree once in linear time.

To efficiently compute the y-coordinates from a B*-tree, a **contour data structure** [Guo 2001] is used to facilitate the operations on modules. The

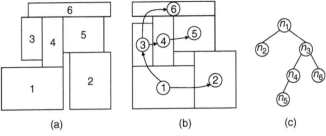

(a)   (b)   (c)

FIGURE 10.15

An example of (a) a given floorplan. (b) A compacted floorplan. (c) the corresponding B*-tree.

contour structure is a double-linked list storing the coordinates of the contour curve in the current compaction direction. A horizontal contour can reduce the running time for computing the y-coordinate of a newly inserted module. Without the contour, the running time for determining the y-coordinate of a newly inserted module would be linear to the number of modules. By maintaining the contour structure, however, the y-coordinate of a module can be computed in amortized $O(1)$ time.

Example 10.11   Find the floorplan corresponding to the B*-tree given in Figure 10.15c. The module dimensions are given in Table 10.3.

At first, there is no module. Therefore, we initialize the contour data structure $C = <(0,0)\ (\infty,0)>$. On the basis of the **depth-first order**, we pack modules one by one in six steps. The detailed processing is explained below, summarized in Table 10.4, and illustrated in Figure 10.16.

Step (a): Because n_1 is the **root**, the coordinate of module 1 is (0, 0). Inserting module 1 introduces three more contour points $C_{new} = <(0, 6), (9, 6), (9, 0)>$. To generate the new contour list, we need to find two sub-contour lists that are before and after the x-span [0, 9] of module 1: $C_{before} = <(0, 0)>$ and $C_{after} = <(\infty, 0)>$. The resulting contour $C = <C_{before}, C_{new}, C_{after}> = <(0, 0), (0, 6), (9, 6), (9, 0), (\infty, 0)>$.

Step (b): n_2 is the **left** child of n_1. Therefore, the x-coordinate of module 2 is $x_2 = x_1 + w_1 = 9$. To determine the y-coordinate of module 2, we search the contour to find the maximum y-coordinate between the x-span $[x_2, x_2 + w_2] = [9, 15]$. The maximum y-coordinate is 0, so we have $y_2 = 0$. Inserting module 2 introduces three more contour points $C_{new} = <(9, 8), (15, 8), (15, 0)>$. Again, we need to find two sub-contour lists that are before and after the x-span of module 1, [9, 15], to generate the new contour list: $C_{before} = <(0, 0), (0, 6), (9, 6)>$ and $C_{after} = <(\infty, 0)>$. The resulting contour $C = <C_{before}, C_{new}, C_{after}> = <(0, 0), (0, 6), (9, 6), (9, 8), (15, 8), (15, 0), (\infty, 0)>$.

Step (c): n_3 is the **right** child of n_1. Therefore, module 3 has the same x-coordinate as module 1. To determine the y-coordinate of module 3, we search the contour to

---

Table 10.3   The Dimensions of Modules in Example 10.11

Module No.	Width	Height
1	9	6
2	6	8
3	3	6
4	3	7
5	6	5
6	12	2

find the maximum y-coordinate in the x-span $[x_3, x_3 + w_3] = [0, 3]$. Because the maximum y-coordinate is 6, we have $y_3 = 6$. Inserting module 3 introduces three more contour points $C_{new} = <(0, 12), (3, 12), (3, 6)>$. We have the two sub-contour lists that are before and after the x-span of module 3, $[0, 3]$: $C_{before} = <(0, 0)>$ and $C_{after} = <(9, 6), (9, 8), (15, 8), (15, 0), (\infty, 0)>$. So the resulting contour $C = <C_{before}, C_{new}, C_{after}> = <(0, 0), (0, 12), (3, 12), (3, 6), (9, 6), (9, 8), (15, 8), (15, 0), (\infty, 0)>$.

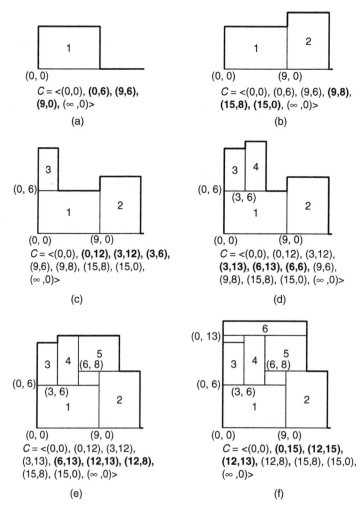

FIGURE 10.16

The B*-tree packing process. The double linked list C of the contour is shown below each figure. The horizontal contour lines are in bold.

Table 10.4 B*-Tree Packing for Example 10.11.

Step	Mod.	x-Coordinate	y-Coordinate	Contour C
Contour initialization				$<(0, 0), (\infty, 0)>$
(a)	1	$x1 = 0$	$y1 = 0$	$<(0, 0), (0, 6), (9, 6),(9, 0), (\infty, 0)>$
(b)	2	$x2 = x1 + w1 = 9$	$y2 = 0$	$<(0, 0), (0, 6), (9, 6), (9, 8), (15, 8), (15, 0), (\infty, 0)>$
(c)	3	$x3 = x1 = 0$	$y3 = \max(y1 + h1) = 6$	$<(0, 0), (0, 12), (3, 12), (3, 6), (9, 6), (9, 8), (15, 8), (15, 0), (\infty, 0)>$
(d)	4	$x4 = x3 + w3 = 3$	$y4 = \max(y1 + h1) = 6$	$<(0, 0), (0, 12), (3, 12), (3, 13), (6, 13), (6, 6), (9, 6), (9, 8), (15, 8), (15, 0), (\infty, 0)>$
(e)	5	$x5 = x4 + w4 = 6$	$y5 = \max(y1 + h1, y2 + h2) = 8$	$<(0, 0), (0, 12), (3, 12), (3, 13), (6, 13), (12, 13), (12, 8), (15, 8), (15, 0), (\infty, 0)>$
(f)	6	$x6 = x3 = 0$	$y6 = \max(y3 + h3, y4 + h4, y5 + h5) = 13$	$<(0, 0), (0, 15), (12, 15), (12, 13), (12, 8), (15, 8), (15, 0), (\infty, 0)>$

Step (d): n_4 is the **left** child of n_3. Therefore, $x_4 = x_3 + w_3 = 3$. To determine the y-coordinate of module 4, we search the contour to find the maximum y-coordinate between the x-span $[x_4, x_4 + w_4] = [3, 6]$. Because the maximum y-coordinate is 6, we have $y_4 = 6$. Inserting module 4 introduces three more contour points $C_{new} = <(3, 13), (6, 13), (6, 6)>$ and two sub-contour lists that are before and after the x-span of module 4, $[3, 6]$: $C_{before} = <(0, 0), (0, 12), (3, 12)>$ and $C_{after} = <(9, 6), (9, 8), (15, 8), (15, 0), (\infty, 0)>$. Consequently, the resulting contour $C = <C_{before}, C_{new}, C_{after}> = <(0, 0), (0, 12), (3, 12), (3, 13), (6, 13), (6, 6), (9, 6), (9, 8), (15, 8), (15, 0), (\infty, 0)>$.

Step (e): n_5 is the **left** child of n_4. Therefore, $x_5 = x_4 + w_4 = 6$. The maximum y-coordinate in the x-span $[x_5, x_5 + w_5] = [6, 12]$ is 8, and so is $y_5 = 8$. Inserting module 5 introduces the three contour points $C_{new} = <(6, 13), (12, 13), (12, 8)>$ and the two sub-contour lists before and after the x-span of module 5, $[6, 12]$: $C_{before} = <(0, 0), (0, 12), (3, 12), (3, 13)>$ and $C_{after} = <(15, 8), (15, 0), (\infty, 0)>$. The resulting contour $C = <C_{before}, C_{new}, C_{after}> = <(0, 0), (0, 12), (3, 12), (3, 13), (6, 13), (12, 13), (12, 8), (15, 8), (15, 0), (\infty, 0)>$.

Step (f): n_6 is the **right** child of n_3. Therefore, $x_6 = x_3 = 0$. The maximum y-coordinate in the x-span $[x_6, x_6 + w_6] = [0, 12]$ is 13, and so is $y_6 = 13$. Inserting module 6 introduces the three contour points $C_{new} = <(0, 15), (12, 15), (12, 13)>$ and the two sub-contour lists before and after the x-span of module 6, $[0, 12]$: $C_{before} = <(0, 0)>$ and $C_{after} = <(12, 8), (15, 8), (15, 0), (\infty, 0)>$. The resulting contour $C = <C_{before}, C_{new}, C_{after}> = <(0, 0), (0, 15), (12, 15), (12, 13), (12, 8), (15, 8), (15, 0), (\infty, 0)>$.

10.2.3.3 *Solution space*

The total number of B*-trees can be computed by the number of unlabeled binary trees and the permutation of n labels. The permutation of n labels is $n!$. From [Hilton 1991], the counting of an unlabeled p-ary tree with n node is

$$\frac{1}{(p-1)n+1}\binom{pn}{n}$$

Applying **Stirling's approximation**, we have

$$n! = \Theta\left(\sqrt{2\pi n}\left(\frac{n}{e}\right)^n\right)$$

Setting p to 2, we have the following asymptotic form:

$$O\left(\frac{2^{2n}}{n^{1.5}}\right)$$

Thus, the total number of possible floorplans for a B*-tree with n nodes is

$$O\left(n!\frac{2^{2n}}{n^{1.5}}\right)$$

10.2.3.4 *Neighborhood structure*

Each B*-tree corresponds to a floorplan. Therefore, the solution space consists of all B*-trees with the given nodes (modules). To find a neighboring solution, we can perturb a B*-tree to get another B*-tree by the following operations:

Op1: Rotate a module.
Op2: Move a node to another place.
Op3: Swap two nodes.

For Op1, we rotate a module for a B*-tree node, which does not affect the B*-tree structure. For Op2, we move a node to another place in the B*-tree. Op2 consists of two steps, **deletion** and **insertion**, which will be explained later. For Op3, we swap two nodes in the B*-tree. After packing for a B*-tree, we obtain a resulting floorplan.

There are three cases for the deletion operation. Note that in Cases 2 and 3, the relative positions of the modules might be changed after the operation, and thus we might need to reconstruct a corresponding floorplan for further processing.

- Case 1: A leaf node. We can simply delete the target leaf node.
- Case 2: A node with one child. We remove the target node and then place its only child at the position of the removed node. The tree update can be performed in $O(1)$ time.
- Case 3: A node with two children. We replace the target node n_t by either its right child or its left child n_c. Then we move a child of n_c, if any, to the original position of n_c. This process proceeds until the corresponding leaf node is handled. It is obvious that such a deletion operation requires $O(h)$ time, where h is the height of the B*-tree.

We can insert a new node into either an internal or an external position as follows.

- **Internal position:** A position between two nodes in a B*-tree.
- **External position:** A position pointed by a NULL pointer. Each node has two pointers, the left child and the right child. When the node has no left or right child, it points NULL.

Example 10.12 Figure 10.17 gives an example of the three types of operations on the B*-tree and its corresponding floorplans. (a) Rotate module 3 (Op1). It does not affect the B*-tree structure. (b) Move n_4 to the left child of n_6 (Op2). First, we delete n_4. Because n_4 has the only left child n_5, we attach n_5 to the left child of n_3. Then, we insert n_4 to the left child of n_6. (c) Swap n_1 and n_2 (Op3). Finally, we obtain the B*-tree and the corresponding floorplan in (d).

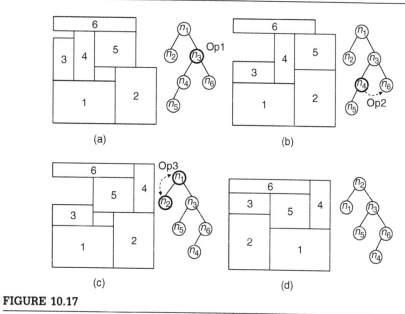

(a)   (b)

(c)   (d)

FIGURE 10.17

Illustration of the B*-tree perturbations.

10.2.3.5 *Cost function*

Similar to the normalized Polish expression, we can use the floorplan area and the total wirelength as the cost function of the simulated annealing:

$$Cost = \alpha \frac{A}{A_{norm}} + (1 - \alpha) \frac{W}{W_{norm}}$$

where A is the floorplan area, A_{norm} is the average area, W is the total wire-length, W_{norm} is the average wirelength, and α controls the weight between area and wirelength. (See Section 10.2.2 for the computation of A_{norm} and W_{norm}.) The area for a B*-tree can be computed by its width/height of the resulting floorplan, and HPWL can be used to evaluate the wirelength.

10.2.3.6 *Annealing schedule*

We can apply either classical SA or Fast-SA to B*-trees to find a desired floorplan. See Section 10.2.2 for more information on classical SA and Fast-SA.

10.2.4 Sequence pair for general floorplans

Sequence pair (SP) is a flexible representation to model a general floorplan [Murata 1995]. A sequence pair consists of an ordered pair of module name sequences. For example, (124536, 326145) can represent a floorplan of the six modules 1, 2,..., 6. In the following, we describe the procedures for the transformation between a floorplan and a sequence pair.

10.2.4.1 *From a floorplan to its sequence pair*

Given six modules shown in Figure 10.18a, we first stretch modules one by one to obtain **rooms**, each room containing only one module. Figure 10.18b shows the floorplan F with rooms derived from Figure 10.18a.

The following procedure encodes F by a pair of the module name sequences. For each module i, we draw two rectilinear curves, **right-up locus** and **left-down locus**. The right-up locus of module i is initially located at the center of module i and starts to move rightward. It turns its direction up and right alternately when it hits (1) the sides of rooms, (2) previously drawn lines, or (3) the boundary of the chip. The locus goes until it reaches the upper-right corner. The union of the two loci of module i forms the **positive locus** of module i. Figure 10.19a shows an example of positive loci. With the construction of positive loci, no two positive loci cross each other. Therefore, these positive loci can be linearly ordered, as well as the corresponding modules. Here we order the positive loci from left to right. Let Γ_+ be the module name sequence in this order. In Figure 10.19a, the first sequence $\Gamma_+ = 124536$ is obtained.

Negative loci can be obtained similar to the positive loci. The difference is that a negative locus is the union of the **up-left locus and down-right locus.** Let Γ_- be the module name sequence in the order of the negative loci from left

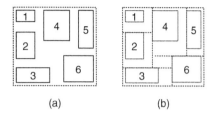

FIGURE 10.18

(a) Given modules. (b) A floorplan of the "rooms."

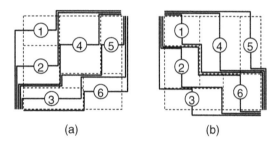

FIGURE 10.19

(a) Positive loci. (b) Negative loci.

to right. An example of negative loci is shown in Figure 10.19b. As a result, we have the negative loci from left to right and obtain $\Gamma_- = 326145$. Finally, the sequence pair $(\Gamma_+, \Gamma_-) = (124536, 326145)$ is obtained.

10.2.4.2 *From a sequence pair to its floorplan*

Given a sequence pair (Γ_+, Γ_-), the geometric relation of modules can be derived from the sequence pair as follows:

Rule 1 (horizontal constraint): module i is left to module j if i appears before j in both Γ_+ and Γ_- $(\ldots i \ldots j \ldots, \ldots i \ldots j \ldots)$.

Rule 2 (vertical constraint): module i is below module j if i appears after j in Γ_+ and i appears before j in Γ_- $(\ldots i \ldots j \ldots, \ldots i \ldots j \ldots)$.

The following steps describe a procedure to transform a sequence pair to its floorplan. Consider an $n \times n$ grid, where n is the number of modules. Label the horizontal grid lines and the vertical grid lines with module names along Γ_+ and Γ_- from top and from left, respectively. A cross point of the horizontal grid line of label i and the vertical grid line of label j is referred to by (i, j). Then, rotate the resulting grid counterclockwise by 45 degrees to get an oblique grid. (See Figure 10.20.) Place each module i with its center being on (i, i). See Figure 10.20 for an example.

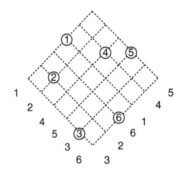

FIGURE 10.20

An oblique grid for $(\Gamma+, \Gamma-) = (124536, 326145)$.

On the basis of the preceding constraints, we can create a **horizontal-constraint graph** with a source and a sink, and a node-weighted directed acyclic graph $G_H (V, E)$, where V is the set of nodes, and E is the set of edges as follows:

- V: source s, sink t, and n nodes labeled with module names.
- E: (s, t) and (i, t) for each module i, and (i, j) if and only if module i is on the left of module j (horizontal constraint).
- Nodes weight: zero for s and t, width of module i for node i.

Similarly, a **vertical-constraint graph** $G_V (V, E)$ can be constructed on the basis of the vertical constraints and the height of each module. Note that both the horizontal and the vertical constraint graphs are acyclic. If two modules i and j are in horizontal relation, then there is an edge between nodes i and j in G_H, and thus they do not overlap horizontally in the resulting floorplan. Similarly, if modules i and j are in vertical relation, they do not overlap vertically. Because any pair of modules is either in horizontal or vertical relation, no modules overlap with each other in the resulting floorplan.

The module locations can be obtained from the constraint graphs. The x-coordinate of module i is given by the longest path length from the source s to node i in G_H. Similarly, the y-coordinate of module i can be computed on G_V. Consequently, the width and the height of the resulting floorplan can be computed by the longest path length between the source and the sink in G_H and G_V, respectively. The longest path length computation on each node-weighted directed acyclic graph, G_H or G_V, can be performed in $O(n^2)$ time by applying the well-known **longest path algorithm** [Lawler 1976], where n is the number of modules. In other words, given a sequence pair (Γ_+, Γ_-), the area-optimal packing can be obtained in quadratic time.

For the example sequence pair $(\Gamma_+, \Gamma_-) = (124536, 326145)$, we can construct the corresponding G_H and G_V shown in Figure 10.21. Table 10.5 lists the module dimensions. The weight and the width (height) of each module are indicated in each node of G_H (G_V). As mentioned earlier, the x-coordinates

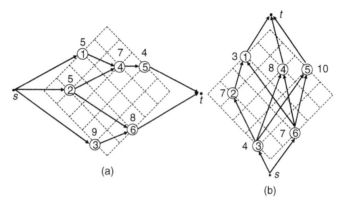

FIGURE 10.21

The constraint graphs with the source s and sink t induced from the sequence pair (124536, 326145): (a) The horizontal constraint graph G_H. (b) The vertical constraint graph G_V. (Note that the existence of the edges (a, b) and (b, c) implies that the transitive edge (a, c) is also in the constraint graph. Transitive edges are not shown in both graphs for simplicity.)

Table 10.5   The Dimensions of the Modules in Figure 10.21

Module No.	Width	Height
1	5	3
2	5	7
3	9	4
4	7	8
5	4	10
6	8	7

can be computed by the longest path length from the source in G_H, and thus we have

Module 1: $x_1 = 0$
Module 2: $x_2 = 0$
Module 3: $x_3 = 0$
Module 4: $x_4 = \max(x_1 + w_1, x_2 + w_2) = \max(0 + 5, 0 + 5) = 5$
Module 5: $x_5 = x_4 + w_4 = 5 + 7 = 12$
Module 6: $x_6 = \max(x_2 + w_2, x_3 + w_3) = \max(0 + 5, 0 + 9) = 9$
Floorplan width $= \max(x_5 + w_5, x_6 + x_6) = \max(12 + 4, 9 + 8) = 17$.

The y-coordinates can be computed from G_V similarly as follows:

Module 1: $y_1 = \max(y_2 + h_2, y_6 + h_6) = \max(4 + 7, 0 + 7) = 11$

Module 2: $y_2 = y_3 + b_3 = 4$
Module 3: $y_3 = 0$
Module 4: $y_4 = \max(y_3 + b_3, y_6 + b_6) = \max(0 + 4, 0 + 7) = 7$
Module 5: $y_5 = \max(y_3 + b_3, y_6 + b_6) = \max(0 + 4, 0 + 7) = 7$
Module 6: $y_6 = 0$
Floorplan height $= \max(y_1 + b_1, y_4 + b_4, y_5 + b_5) = \max(11 + 3, 7 + 8,$
$\qquad 7 + 10) = 17.$

The resulting floorplan is shown in Figure 10.22, and the coordinate of each module and the resulting floorplan dimension are as follows:

Module 1 = (0, 11)
Module 2 = (0, 4)
Module 3 = (0, 0)
Module 4 = (5, 7)
Module 5 = (12, 7)
Module 6 = (9, 17)
Floorplan dimension = (17, 17)

10.2.4.3 *Solution space*

Each permutation of Γ_+ and Γ_- gives a floorplan solution. For n modules, the lengths of Γ_+ and Γ_- are both n, and thus each of Γ_+ and Γ_- have $n!$ permutations. Consequently, there are $(n!)^2$ total permutations for a sequence pair with n modules.

10.2.4.4 *Neighborhood structure*

To search for a desired floorplan solution, we can use the following three types of operations to perturb a sequence pair to another:

Op1: Rotate a module.
Op2: Swap two module names in only one sequence.
Op3: Swap two module names in both sequences.

Each of the three operations results in a legal sequence pair and floorplan solution. Furthermore, these three operations are sufficient to generate any sequence pair from a given sequence pair by a sequence of operations.

FIGURE 10.22

The minimal area packing result of $(\Gamma+, \Gamma-) = (124536, 326145)$.

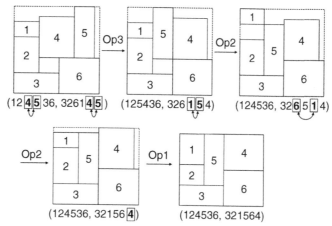

FIGURE 10.23

Effects of the perturbations in sequence pairs.

Figure 10.23 gives an example of these types of operations on the sequence pair and its resulting floorplans.

10.2.4.5 *Cost function*

Again, we can use the floorplan area and the total wirelength as the cost function of the simulated annealing:

$$Cost = \alpha \frac{A}{A_{norm}} + (1 - \alpha) \frac{W}{W_{norm}}$$

where A is the floorplan area, A_{norm} is the average area, W is the total wirelength, W_{norm} is the average wirelength, and α controls the weight between area and wirelength. (See Section 10.2.2 for the computation of A_{norm} and W_{norm}.) The area for a sequence pair can be computed by its width/height (longest path length in G_H/G_V) of the corresponding floorplan, and HPWL can be used to evaluate the wirelength.

10.2.4.6 *Annealing schedule*

Again, we can apply either classical SA or Fast-SA based on sequence pair to find a desired floorplan solution. See Section 10.2.2 for classical SA or Fast-SA.

10.2.5 **Floorplan representation comparison**

In addition to normalized Polish expression, B*-tree, and sequence pair, there are quite a few popular floorplan representations, such as **BSG** [Nakata 1996], **O-tree** [Guo 1999], **Corner Block List** (CBL) [Hong 2000], **Transitive Closure Graph** (TCG) [Lin 2001], **TCG-S** [Lin 2002], **Corner Sequence** (CS) [Lin 2003], **Twin Binary Sequences** (TBS) [Young 2003], **Adjacent Constraint Graph** (ACG) [Zhou 2004], etc. Some representations are closely related. For example, B*-tree is equivalent to O-tree, yet with faster operations,

Table 10.6 Comparison among Floorplan Representations

Representation	Solution Space	Packing Time	Flexibility
Normalized Polish Expression	$O(n!2^{3n}/n^{1.5})$	$O(n)$	Slicing
Corner Block List	$O(n!2^{3n})$	$O(n)$	Mosaic
Twin Binary Sequence	$O(n!2^{3n}/n^{1.5})$	$O(n)$	Mosaic
O-tree	$O(n!2^{2n}/n^{1.5})$	$O(n)$	Compacted
B*-tree	$O(n!2^{2n}/n^{1.5})$	$O(n)$	Compacted
Corner Sequence	$\leq (n!)^2$	$O(n)$	Compacted
Sequence Pair	$(n!)^2$	$O(n^2)$	General
BSG	$O(n!C(n^2, n))$	$O(n^2)$	General
Transitive Closure Graph	$(n!)^2$	$O(n^2)$	General
TCG-S	$(n!)^2$	$O(n \lg n)$	General
Adjacent Constraint Graph	$O((n!)^2)$	$O(n^2)$	General

simpler data structures, and higher flexibility in handling various placement constraints. TCG and sequence pair are also equivalent but their induced operations are significantly different.

Table 10.6 summarizes the sizes of the solution spaces, packing times, and flexibility of the popular floorplan representations. Among the representations, sequence pair, TCG, TCG-S, and ACG can represent general floorplans; O-tree, B*-tree, and corner sequence can model only compacted floorplans; CBL and TBS model the floorplan with each room containing exactly one module, called the mosaic floorplan; and normalized Polish is restricted to slicing floorplans. The general floorplan has the highest flexibility, followed by the compacted floorplan, then followed by the mosaic floorplan, and the slicing floorplan has the least flexibility. (For tighter bounds of slicing, mosaic, and general floorplans, please see [Shen 2003].)

For the packing time, sequence pair, TCG, and ACG require $O(n^2)$ time to generate a floorplan, where n is the number of modules. Note that sequence pair can reduce its packing time to $O(n \lg \lg n)$ time based on the longest common subsequence technique [Tang 2001]. TCG-S needs $O(n \lg n)$ time for packing. For O-tree, B*-tree, corner sequence, and the normalized Polish expression, the packing time is only linear time mainly because they keep relatively simpler information in their data structures.

As a remark for floorplan representations, the evaluation of a floorplan representation should be made based on at least the following three criteria: (1) the definition/properties of the representation, (2) its induced solution structure (not merely the size of its solution space), and (3) its induced operations. We shall avoid the pitfall that judges a floorplan representation by only one of the aforementioned three criteria alone; for example, claiming a floorplan representation A is superior to another floorplan representation B simply because A has a smaller solution space and a faster packing time. Here is an analogy: the representation itself is like the body of an automobile, the induced operations are like the wheels of the automobile, and the solution structure is like the highway network. An automobile with its body alone can go nowhere. For a comprehensive study of floorplan representations, similarly, we shall evaluate them from at least all the aforementioned three criteria.

10.3 ANALYTICAL APPROACH

In addition to simulated annealing, we can resort to the **analytical approach** to floorplan designs [Sutanthavibul 1991]. The analytical approach is a mathematical programming formulation that includes an objective function and a set of constraints. For the floorplanning problem, we need to consider two sets of basic constraints: (1) the **module nonoverlapping constraint** and (2) the **dimension constraint**.

Two modules, i and j are said to be **nonoverlapping**, if at least one of the following cases (linear constraints) is satisfied:

		p_{ij}	q_{ij}
Case 1: i to the left of j	$x_i + w_i \leq x_j$	0	0
Case 2: i below j	$y_i + h_i \leq y_j$	0	1
Case 3: i to the right of j	$x_i - w_j \geq x_j$	1	0
Case 4: i above j	$y_i - h_j \geq y_j$	1	1

where two binary variables, p_{ij} and q_{ij}, are introduced to denote that one of the above inequalities is enforced. For example, when $p_{ij} = 0$ and $q_{ij} = 1$, the inequality equation $y_i + h_i \leq y_j$ is enforced.

Let W and H be upper bounds of the width and height of the floorplan, respectively. We have the following linear constraints for module nonoverlap:

$$x_i + w_i \leq x_j + W\left(p_{ij} + q_{ij}\right)$$

$$y_i + h_i \leq y_j + H\left(1 + p_{ij} - q_{ij}\right)$$

$$x_i - w_j \geq x_j - W\left(1 - p_{ij} + q_{ij}\right)$$

$$y_i - h_j \geq y_j - H\left(2 - p_{ij} - q_{ij}\right)$$

where

$$1 \leq i \leq j \leq n$$

For the dimension constraint, each module must be enclosed within a rectangle of the width W and the height H of the floorplan. Specifically, we have

$$x_i + w_i \leq W$$

$$y_i + h_i \leq H$$

where

$$1 \leq i \leq j \leq n$$

Our objective is to minimize the floorplan area, xy, where x and y are the width and height of the resulting floorplan, respectively. Notice that the area xy is non-linear, and it is much harder to solve a non-linear system than a linear one. To transform the original non-linear objective into a linear one, we can approximate the problem by fixing the floorplan width W and minimizing the height y. As a result, we need to modify the dimension constraints to $x_i + w_i \leq W$ and $y_i + h_i \leq y$, where $1 \leq i \leq n$ and y is the height of the current floorplan. In summary, we have the following four types of constraints:

1. There is no overlap between any two modules $(\forall i, j : 1 \leq i < j \leq n)$.
2. Each module is enclosed within a rectangle of width W and height H $(x_i + w_i \leq W, y_i + h_i \leq y, 1 \leq i \leq n)$. Here, w_i and h_i are known.
3. $x_i \geq 0, y_i \geq 0, 1 \leq i \leq n$
4. $p_{ij}, q_{ij} \in \{0, 1\}$

On the basis of the above discussions, we can formulate the floorplan designs as the following **mixed integer linear program** (**MILP**). Note that both the objective function and all constraints are linear. Our floorplan design problem is to minimize the height y for a given bound of the floorplan width W, subject to the following system of inequality constraints:

Minimize	y	
subject to	$x_i + w_i \leq W$	$1 \leq i \leq n$
	$y_i + h_i \leq y$	$1 \leq i \leq n$
	$x_i + w_i \leq x_j + W(p_{ij} + q_{ij})$	$1 \leq i < j \leq n$
	$y_i + h_i \leq y_j + H(1 + p_{ij} - q_{ij})$	$1 \leq i < j \leq n$
	$x_i - w_j \geq x_j - W(1 - p_{ij} + q_{ij})$	$1 \leq i < j \leq n$
	$y_i - h_j \geq y_j - H(2 - p_{ij} - q_{ij})$	$1 \leq i < j \leq n$
	$x_i, y_i \geq 0$	$1 \leq i \leq n$
	$p_{ij}, q_{ij} \in \{0, 1\}$	$1 \leq i < j \leq n$

For the size of the mixed ILP for n modules, the number of continuous variables is $O(n)$, the number of integer variables is $O(n^2)$, and the number of linear constraints is $O(n^2)$. There are a few popular mixed ILP solvers, such as **GLPK**

[GLPK 2008], ***CPLEX*** [ILOG 2008], ***LINDO*** [LINDO 2008], ***lp_solve*** [lp_solve 2008], etc. ILP has the exponential time complexity in the worst case, and thus it is time-consuming for problems of large sizes. To cope with problems of large sizes, methods such as the divide-and-conquer and the progressive approaches are often used. We will elaborate on this issue later.

The preceding formulation does not consider the rotation of modules. We can extend the aforementioned MILP formulation by introducing a new binary variable r_i to consider the rotation of the module i. When $r_i = 0$, module i is not rotated (*i.e.*, rotated by 0 degree); when $r_i = 1$, module i is rotated by 90 degrees. The system of inequality constraints now becomes

$x_i + r_i h_i + (1 - r_i)w_i \leq W$	$1 \leq i \leq n$
$y_i + r_i w_i + (1 - r_i)h_i \leq y$	$1 \leq i \leq n$
$x_i + r_i h_i + (1 - r_i)w_i \leq x_j + M(p_{ij} + q_{ij})$	$1 \leq i < j \leq n$
$y_i + r_i w_i + (1 - r_i)h_i \leq y_j + M(1 + p_{ij} - q_{ij})$	$1 \leq i < j \leq n$
$x_i - r_j h_j - (1 - r_j)w_j \geq x_j - M(1 - p_{ij} + q_{ij})$	$1 \leq i < j \leq n$
$y_i - r_j w_j - (1 - r_j)h_j \geq y_j - M(2 - p_{ij} - q_{ij})$	$1 \leq i < j \leq n$
$x_i, y_i \geq 0$	$1 \leq i \leq n$
$r_i, p_{ij}, q_{ij} \in \{0, 1\}$	$1 \leq i < j \leq n$

where $M = \max\{W, H\}$. The following gives an example of the MILP formulation for floorplan design. (The preceding formulation considers only the area optimization. If wirelength also needs to be considered, we need to modify the objective function to minimize the total wirelength.)

Example 10.13 Given the dimensions of modules listed in Table 10.7. The total module area is 8 * 6 + 8 * 5 + 11 * 2 = 110. Because a square floorplan is often desired, the square root of 110 is approximately 10. Therefore, we set $W = 10$ and $M = \max\{8, 6\} + \max\{8, 5\} + \max\{11, 2\} = 27$ to find a floorplan with width less than 10. We can use the publicly available lp_solve program to solve this problem. The Figure 10.24 shows the input file in the lp-format. The objective is to minimize the floorplan height (y). The constraints c_1 to c_6 define the bounding box of the floorplan, and the constraints c_7 to c_{10}, c_{11} to c_{14},

Table 10.7 The Dimensions of Modules in Example 10.13

Module No.	Width (w_i)	Height (h_i)
1	8	6
2	8	5
3	11	2

```
min: y;

c1:  x1 +  6 r1 +  8 -  8 r1 <= 10;
c2:  x2 +  5 r2 +  8 -  8 r2 <= 10;
c3:  x3 +  2 r3 + 11 - 11 r3 <= 10;
c4:  y1 +  8 r1 +  6 -  6 r1 <= y;
c5:  y2 +  8 r2 +  5 -  5 r2 <= y;
c6:  y3 + 11 r3 +  2 -  2 r3 <= y;

c7:   x1 + 6 r1 + 8 - 8 r1 <= x2 + 27 p12 + 27 q12;
c8:   y1 + 8 r1 + 6 - 6 r1 <= y2 + 27 + 27 p12 - 27 q12;
c9:   x1 - 5 r2 - 8 + 8 r2 >= x2 - 27 + 27 p12 + 27 q12;
c10:  y1 - 8 r2 - 5 + 5 r2 >= y2 - 54 + 27 p12 + 27 q12;

c11:  x1 + 6 r1 +  8 -  8 r1 <= x3 + 27 p13 + 27 q13;
c12:  y1 + 8 r1 +  6 -  6 r1 <= y3 + 27 + 27 p13 - 27 q13;
c13:  x1 - 2 r3 - 11 + 11 r3 >= x3 - 27 + 27 p13 + 27 q13;
c14:  y1 - 11 r3 - 2 + 2 r3 >= y3 - 54 + 27 p13 + 27 q13;

c15:  x2 + 5 r2 +  8 -  8 r2 <= x3 + 27 p23 + 27 q23;
c16:  y2 + 8 r2 +  5 -  5 r2 <= y3 + 27 + p23 - 27 q23;
c17:  x2 - 2 r3 - 11 + 11 r3 >= x3 - 27 + 27 p23 + 27 q23;
c18:  y2 - 11 r3 - 2 + 2 r3 >= y3 - 54 + 27 p23 + 27 q23;

r1 <= 1;
r2 <= 1;
r3 <= 1;
p12 <= 1;
q12 <= 1;
p13 <= 1;
q13 <= 1;
p23 <= 1;
q23 <= 1;

int r1, r2, r3, p12, q12, p13, q13, p23, q23;
```

FIGURE 10.24

The input file for lp_solve to minimize y in Example 10.13.

and c_{15} to c_{18} define the nonoverlapping relationship between modules 1 and 2, modules 1 and 3, and modules 2 and 3, respectively. The remaining constraints define the 0-1 integer variables.

Applying lp_solve to solve the preceding MILP program, we can obtain the outputs shown in Figure 10.25, which gives the co-ordinates of modules: Module 1 $(x_1, y_1) = (0, 5)$; module 2 $(x_2, y_2) = (0, 0)$; module 3 $(x_3, y_3) = (8, 0)$. Only module 3 is rotated $(r_3 = 1)$. The resulting floorplan height is 11 $(y = 11)$, and the final floorplan in shown in Figure 10.26.

As mentioned earlier, the time complexity of MILP is exponential, and thus it is prohibitively expensive for problems of large sizes. To cope with problems of large sizes, methods such as the progressive and the divide-and-conquer approaches are often used to reduce the problem sizes. We will examine a **progressive augmentation method** that solves a partial problem at each step to reduce the floorplanning complexity. Each time, we select a set of modules and place them into the current partial floorplan, as illustrated in Figure 10.27.

```
Value of objective function: 11

Actual values of the variables:
y                          11
x1                          0
r1                          0
x2                          0
r2                          0
x3                          8
r3                          1
y1                          5
y2                          0
y3                          0
p12                         1
q12                         1
p13                         0
q13                         0
p23                         0
q23                         0
```

FIGURE 10.25

The outputs from lp_solve in Example 10.13.

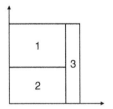

FIGURE 10.26

The resulting floorplan in Example 10.13.

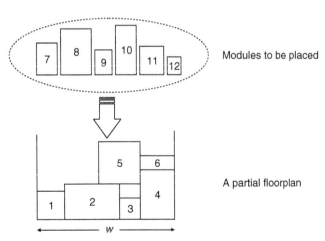

Modules to be placed

A partial floorplan

FIGURE 10.27

Floorplanning with an existing partial floorplan.

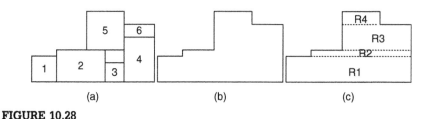

FIGURE 10.28

Reducing the problem size by a set of covering rectangles: (a) The original partial floorplan. (b) The outline of the partial floorplan. (c) A set of rectangles covering the partial floorplan.

To reduce the problem size, we limit the number of the modules to be placed at each step and also minimize the problem size of the current partial floorplan. We can replace the already placed modules by a set of covering rectangles. Figure 10.28 illustrates the procedure for obtaining these rectangles. First, we find the outline of the six placed modules, as shown in Figure 10.28b. The dead spaces among the placed modules are also enclosed in the outline, because it is impossible for the newly added modules to use them. Then, we horizontally dissect the outline into rectangles, R1, R2, R3, and R4. By doing so, the number of rectangles is usually much smaller than that of placed modules, and so are the number of variables and constraint in the MILP formulation.

Besides the preceding MILP-based floorplanning, a sophisticated analytical floorplanning method was proposed [Zhan 2006]. It first roughly determines the module positions by uniformly distributing modules. Then, the overlaps are gradually removed in the second stage to obtain the final floorplan. This approach has much better scalability to handle large-scale designs. However, this approach cannot guarantee a nonoverlap floorplan solution.

10.4 MODERN FLOORPLANNING CONSIDERATIONS

Increasing design complexity and new circuit properties and requirements have reshaped the modern floorplanning problem. The new considerations and challenges make the problem much more difficult. In this section, we will discuss such crucial considerations. Specifically, we will focus on (1) soft modules, (2) fixed-outline constraints, and (3) large-scale floorplanning, and then highlight other important issues for modern floorplanning.

10.4.1 Soft modules

Unlike hard modules with fixed heights and widths, **soft modules** can change their heights and widths while keeping the same module area. The aspect ratio bounds are given as inputs for each module. There are many techniques for the adjustment of soft-module dimensions. In the following, we introduce an effective and efficient heuristic that adjusts soft-module dimensions to optimize the chip area. The underlying concept of this sizing method is to align the module width/height to its adjacent module to reduce the dead space [Chang 2000; Chi 2003; Chen 2005a, 2006].

Given a set B of modules, we assume that module i's bottom-left coordinate is (x_i, y_i) and its top-right coordinate is $(x_i + w_i, y_i + h_i)$. Each soft module has four candidates for the dimensions (*i.e.*, shapes). The candidates are defined as follows:

- $R_i = x_a + w_a - x_i$, where $x_a + w_a = \min \{x_k + w_k \mid x_k + w_k > x_i + w_i, k \in B\}$
- $L_i = x_b + w_b - x_i$, where $x_b + w_b = \max \{x_k + w_k \mid x_k + w_k < x_i + w_i, k \in B\}$
- $T_i = x_c + h_c - y_i$, where $x_c + h_c = \min \{x_k + h_k \mid x_k + h_k > x_i + h_i, k \in B\}$
- $B_i = x_d + h_d - y_i$, where $x_d + h_d = \max \{x_k + h_k \mid x_k + h_k < x_i + h_i, k \in B\}$

Define the **aspect ratio** of a module as the ratio of the height over width of the module. After determining the candidates of the module shapes, we may change the shape of a soft module i by choosing one of the following five choices during simulated annealing:

1. Change the width of module i to R_i.
2. Change the width of module i to L_i.
3. Change the height of module i to T_i.
4. Change the height of module i to B_i.
5. Change the aspect ratio of module i to a random value in the range of the given soft aspect ratio constraint.

We can add the module resizing as one floorplan perturbation operation during simulated annealing so that the module shapes could be changed to obtain a more desired floorplan.

Example 10.14 Consider the example of soft-module resizing given in Figure 10.29. Module 4 has four shape candidates, R_4, L_4, T_4, and B_4, with four candidate lines being shown in Figure 10.29a. If we stretch the right boundary of module 4 to R_4 (the height is also changed correspondingly to maintain a fixed area), it can generate a more compacted floorplan as shown in Figure 10.29b.

The preceding soft-module sizing technique can be applied to any floorplan representations based on simulated annealing or iterative improvement. For the normalized Polish expression (slicing tree), we can use a more sophisticated method, **shape curve**, to handle soft modules. Because the area of a soft module is fixed, the shape function of a module is a hyperbola: $wh = A$, or $h = A/w$, where w is the width, h is the height, and A is the area of the

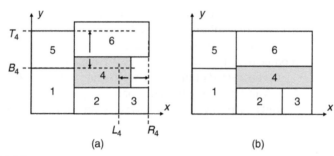

FIGURE 10.29

A soft-module resizing example: (a) the original floorplan with four shape candidates for resizing module 4. (b) a compacted floorplan by stretching the right boundary of module 4 to R_4.

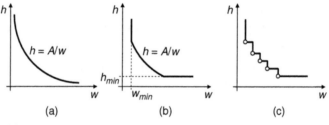

FIGURE 10.30

The shape curve of a module: (a) The shape curve in a hyperbola function. (b) The shape curve with the minimum width/height constraint. (c) The piecewise linear shape curve.

module. See Figure 10.30a for an example of the shape curve. Because module width and height are usually constrained to avoid very thin modules, $h \geq h_{\min}$ and $w \geq w_{\min}$; see Figure 10.30b for the resulting shape curve. In practice, we can use piecewise linear functions to record the shape curve for easier implementation. We only need to record the **corner points** of the shape curve, as shown in Figure 10.30c.

The shape curves can record not only the shapes of a basic soft module, but also that of a composite module formed by a set of basic modules (*i.e.,* a sub-floorplan). In a slicing tree, we first generate a shape curve for each module and record this shape curve with the corresponding leaf node, as shown in Figure 10.31. Then, the shape curve of a composite module can be derived from its children nodes and recorded in the corresponding internal node. By use of the bottom-up procedure, we can find the shape curve of the root node, which gives all possible shapes of the resulting floorplans.

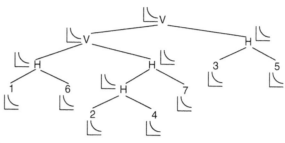

FIGURE 10.31

Shape curves in a slicing tree.

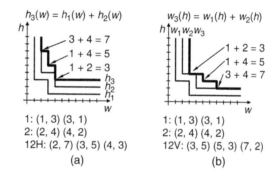

FIGURE 10.32

Examples of updating the shape curve: (a) The H operator. (b) The V operator.

Example 10.15 Given two piecewise linear shape curves for modules 1 and 2, derive the shape curves of the composite modules 12H and 12V. Figure 10.32a illustrates the derivation of the shape curve for the composite module 12H. For the H operator, two modules are merged vertically; we have $h_3(w) = h_1(w) + h_2(w)$, and the minimum width of the resulting floorplan cannot be smaller than $\max(\min w_1, \min w_2)$. See the bold lines in Figure 10.32a for the shape curve of 12H. Similarly, for the V operator, we have $w_3(h) = w_1(h) + w_2(h)$ and the height of the resulting floorplan cannot be smaller than $\max(\min h_1, \min h_2)$. The shape curve of 12V is represented by the bold lines in Figure 10.32b.

10.4.2 **Fixed-outline constraint**

Modern VLSI design is typically based on a **fixed-die (fixed-outline) floorplan** [Kahng 2000], rather than a **variable-die floorplan**. A floorplan with pure area minimization without any fixed-outline constraints may be useless, because it cannot fit into the given outline. Unlike classical floorplanning that usually handles only module packing to minimize silicon area, modern floorplanning should be formulated as **fixed-outline floorplanning**.

The fixed-outline constraint is given as follows. We first construct a fixed outline with the aspect ratio R^* (*i.e.*, height/width). For a collection of modules with the given total area A and the **maximum percentage of dead space** Γ, we have the chip area $= H^*W^* = (1 + \Gamma)A$ and the chip aspect ratio $= H^*/W^* = R^*$. Therefore, the new height H^* and width W^* of the outline are defined by the following equations [Adya 2001]:

$$H^* = \sqrt{(1 + \Gamma)AR^*}$$

$$W^* = \sqrt{(1 + \Gamma)A/R^*}$$

Example 10.16 Figure 10.33 gives three floorplan examples with different R^*'s and 's. The three floorplans contain the same modules. Figure 10.33a and Figure 10.33b have the same maximum percentage of dead spaces, 15%, yet with different outline ratios, 2.0 and 1.0, respectively. Figure 10.33b and Figure 10.33c have the same outline ratio, 1.0, yet with different maximum percentages of dead spaces, 15% and 50%, respectively.

To handle the fixed-outline constraint, we will modify the cost function for simulated annealing. In addition to the wirelength/area objective, we may add an aspect ratio penalty to the cost function [Chen 2005a, 2006]. The rationale is that if the aspect ratio of the floorplan is similar to that of the outline, and the dead space of the floorplan is smaller than the maximum percentage of the dead space Γ, then the floorplan can fit into the outline. Suppose that the current aspect ratio of the floorplan is R. We define the cost function Φ for a floorplan solution F by the following equation:

$$\Phi(F) = \alpha A + \beta W + (1 - \alpha - \beta)(R^* - R)^2$$

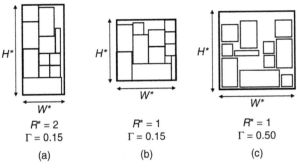

$R^* = 2$	$R^* = 1$	$R^* = 1$
$\Gamma = 0.15$	$\Gamma = 0.15$	$\Gamma = 0.50$
(a)	(b)	(c)

FIGURE 10.33

Three floorplans with different outline ratios (R^*) and the maximum percentages of dead spaces (Γ) based on the same modules.

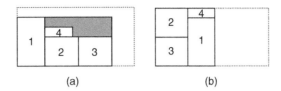

FIGURE 10.34

Examples of: (a) a floorplan with the aspect ratio the same as the one of the outline. (b) the optimal floorplan with a different aspect ratio from the aspect ratio of the outline.

where A is the floorplan area, W is the wirelength, R is the floorplan aspect ratio, R^* is the desired floorplan aspect ratio, and α and β are user-defined parameters.

The best aspect ratio of the floorplan in the fixed outline may not be the same as that of the outline, as shown in Figure 10.34. In this case, we will decrease the weight of the aspect ratio penalty to concentrate more on the wirelength/area optimization. We can use an adaptive method to control the weights in the cost function according to the most recent floorplans found in [Chen 2005a]. If there are more feasible floorplans in most recent floorplans found during simulated annealing, it implies that this instance is easier to be fit into the floorplan outline, and thus we will reduce the weight of the aspect ratio penalty to focus more on the wirelength/area optimization. Figure 10.35 shows the resulting floorplans for the MCNC circuit ami49 with various aspect ratios. There are 49 modules in this circuit [Chen 2005a].

In addition to the objective function adjustment, new perturbations can also be applied to better guide a local search for fixed-outline floorplanning on the basis of sequence pair or normalized Polish expression [Adya 2003; Lin 2004a]. However, unlike the objective function adjustment that can be applied to all floorplan representations, the new perturbations are specific to the target floorplan representation. On the basis of the generalized slicing tree, DeFer is proposed to handle fixed-outline floorplanning efficiently and effectively [Yan 2008]. DeFer generates a collection of possible floorplan solutions and chooses the best one that can fit into the fixed outline with the smallest wirelength at the last stage.

10.4.3 Floorplanning for large-scale circuits

As technology advances, the number of modules in a chip becomes larger. Simulated annealing alone cannot handle large-scale floorplanning instances effectively and efficiently. To cope with the **scalability** problem, **hierarchical** floorplanning is proposed. The hierarchical approach recursively divides a floorplan region into a set of sub-regions and solves those sub-problems independently. Patoma is a fast hierarchical floorplanner based on recursive bipartitioning [Cong 2005]. It partitions a floorplan and uses ***row-oriented block*** (ROB) packing and

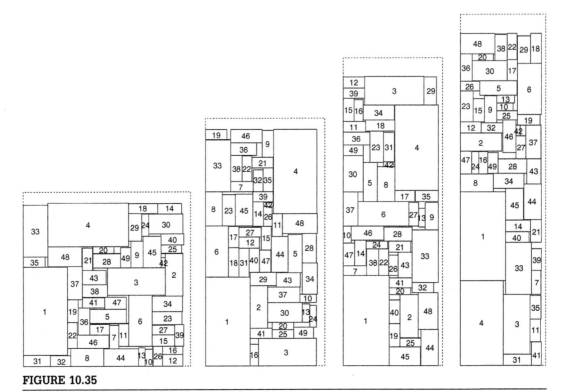

FIGURE 10.35

The floorplans of the MCNC circuit ami49 with fixed-outline ratios 1, 2, 3, and 4.

zero-dead space (ZDS) floorplanning to find legal sub-floorplans. The top-down, hierarchical technique is efficient in handling large-scale problems. Nevertheless, a significant drawback of the hierarchical approach is that it might lack the global information for the floorplanning interactions among different sub-regions, because each sub-region is processed independently. As a result, the hierarchical approach might not find desired solutions.

To remedy the deficiency, **multilevel floorplanning** is proposed to find a better trade-off between the scalability and solution quality. The multilevel framework applies a two-stage technique, **bottom-up coarsening** and **top-down uncoarsening**. We take the MB*-tree [Lee 2003] as an example to explain the concept of multilevel floorplanning. Figure 10.36 shows the MB*-tree multilevel framework based on a two-stage technique of bottom-up **clustering** (coarsening) followed by top-down **declustering** (uncoarsening). It should be noted that although we use the MB*-tree as an example to explain the multilevel floorplanning framework, this framework itself is general to all floorplan representations.

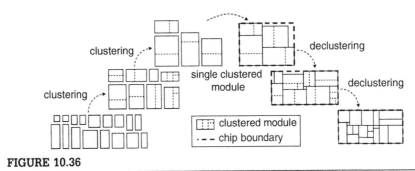

FIGURE 10.36

Multilevel floorplanning that uses recursive clustering and declustering.

The **clustering** stage iteratively groups a set of (primitive or composite) modules (say, two modules) on the basis of a cost metric defined by area utilization, wirelength, and connectivity among modules, and at the same time establishes the geometric relations among the newly clustered modules by constructing a corresponding **B*-subtree**. The clustering procedure repeats until a single cluster containing all modules is formed (or the number of modules is smaller than a predefined threshold that can be handled by a classical floorplanner), denoted by a one-node B*-tree that records the entire clustering scheme. During clustering, we will record how two modules i and j are clustered into a new composite module k. The relation for each pair of modules in a cluster is established and recorded in the corresponding B*-subtree during clustering. It will be used for determining how to expand a node into a corresponding B*-subtree during declustering.

The **declustering** stage iteratively ungroups a set of previously clustered modules (*i.e.*, expanding a node into a subtree according to the B*-tree topology constructed at the clustering stage) and then refines the floorplan solution on the basis of a simulated annealing scheme. The refinement should lead to a "better" B*-tree structure that guides the declustering at the next level. It is important to note that we always keep only one B*-tree for processing at each iteration, and the multilevel B*-tree–based floorplanner preserves the geometric relations among modules during declustering (*i.e.*, the tree expansion), which makes the B*-tree an ideal data structure for the multilevel floorplanning framework.

The MB*-tree algorithm is summarized in Algorithm 10.5. We first perform clustering to reduce the problem size level by level and then enter the declustering stage. In the declustering stage, we perform floorplanning for the modules at each level with simulated annealing.

Algorithm 10.5 MB*-tree Floorplanning

Input: A set of modules and a set of nets.
Output: A final area-optimized floorplan.
Stage I: Clustering

1. **while** the number of modules/clusters is still large
2.   Cluster modules according to their dimensions and connectivity;
3. **end while**;

Stage II: Declustering
4. **while** still having clusters
5.   Decluster a set of clusters;
6.   Perform simulated annealing to refine the floorplan;
7. **end while**;
8. **return** the final floorplan;

Figure 10.37 illustrates the MB*-tree algorithm. For easier explanation, we cluster three modules each time in Figure 10.37. Figure 10.37a lists seven modules to be packed, i's, $1 \leq i \leq 7$. Figure 10.37b to Figure 10.37d illustrate the clustering process. Figure 10.37b shows the resulting configuration after clustering modules 5, 6, and 7 into a new cluster module 8 (*i.e.*, the clustering scheme of 8 is {{5, 6}, 7}); note that the B*-tree for the packing of modules 5, 6, and 7 is recorded with module 8. Similarly, we cluster modules 1, 2, and 4 into module 9 with the clustering scheme {{2, 4}, 1} and record the B*-tree with module 9 for packing modules 1, 2, and 4. Finally, we cluster modules 3, 8, and 9 into module 10 by use of the clustering scheme {{3, 8}, 9} and record a one node B*-tree for module 10. The clustering stage is thus done, and the declustering stage begins, in which simulated annealing is applied to the floorplanning. In Figure 10.37e, we first decluster module 10 into modules 3, 8, and 9 (*i.e.*, expand the node n_{10} into the B*-subtree illustrated in Figure 10.37e). We then refine the solution by moving module 8 to the top of module 9 (perform Op2 on n_8) during simulated annealing (see Figure 10.37f). As shown in Figure 10.37g, we further decluster module 9 into modules 1, 2, and 4, and then rotate module 2 and move module 3 on top of module 2 (perform Op1 on n_2 and Op2 on n_3), resulting in the configuration shown in Figure 10.37h. Finally, we decluster module 8 shown in Figure 10.37i to modules 5, 6, and 7, and move module 4 to the right of module 3 (perform Op2 on n_4), which results in the area optimal floorplan shown in Figure 10.37j.

Figure 10.38 shows the layout for the circuit ami49_200 with 9800 modules and 81,600 nets (not shown in the layout) [Lee 2003] obtained by MB*-tree. It has a dead space of only 3.44%. Without the use of the multilevel approach,

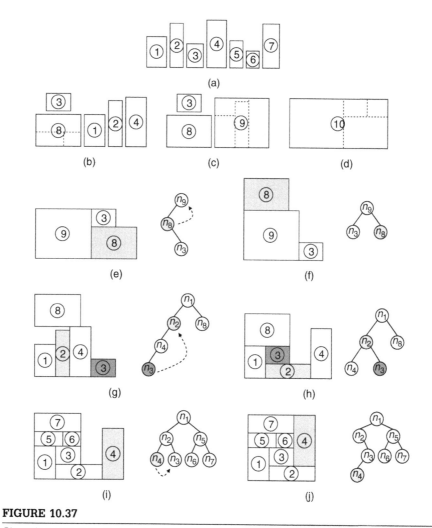

FIGURE 10.37

Clustering and declustering: (a) given seven *i*s, $1 \leq i \leq 7$. (b) Cluster modules 5, 6, and 7 into 8. (c) Cluster modules 1, 2, and 4 into 9. (d) Cluster modules 3, 8, and 9 into 10. (e) Decluster module 10 into modules 3, 8, and 9. (f) Perform Op2 on module 8. (g) Decluster module 9 into modules 1, 2, and 4. (h) Perform Op1 and Op2 on modules 2 and 3, respectively. (i) Decluster module 8 into module 5, 6, and 7. (j) Perform Op2 on module 4 to obtain the final floorplan.

the flat floorplanning method could not handle large circuits of this magnitude effectively.

The MB*-tree approach is referred to the ∧-**shaped multilevel framework**, because it starts with bottom-up coarsening (clustering) followed by top-down

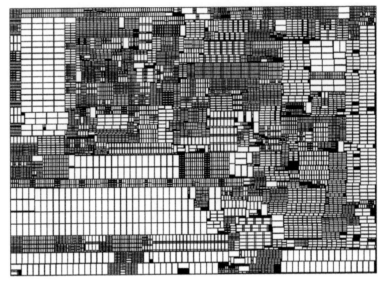

FIGURE 10.38

The resulting floorplan for the circuit ami49_200 with 9800 modules and 81,600 nets; the resulting dead space (the dark regions) is only 3.44%.

uncoarsening (declustering). In contrast, the **V-shaped multilevel framework** works from top-down uncoarsening (partitioning) followed by bottom-up coarsening (merging) [Chen 2005b]. The V-shaped multilevel framework often outperforms the ∧-shaped one in the optimization of global circuit effects, such as interconnection optimization, because the V-shaped framework considers the global configuration first and then processes down to local ones level by level and thus the global effects can be handled at earlier stages.

10.4.4 **Other considerations and topics**

In addition to the aforementioned modern floorplanning considerations, there are many other issues that might need to be considered. In the following, we briefly describe these issues.

Modern circuit designs often need to integrate analog and digital circuits on a single chip and thus may suffer from **substrate noise coupling**. A pioneering work along this direction was proposed in [Cho 2006]. With the continued increase in system frequency and design complexity, existing techniques for reducing substrate noise may need to be enhanced substantially. Considering substrate noise in early floorplanning is now desirable.

For nanometer VLSI designs, **interconnect** dominates overall circuit performance. However, the conventional design flow often deals with interconnect optimization at the routing or post-routing stages. When the interconnect complexity grows drastically, it is often too late to perform aggressive interconnect optimization during or after routing, because most silicon and routing resources are occupied. Therefore, it is desirable to optimize interconnect as early as possible. Many techniques have been proposed for interconnect optimization. Some examples are wiring topology construction, buffer/repeater insertion and sizing, and wire sizing and spacing [Cong 1997]. Among these interconnect optimization techniques, **buffer insertion** is generally considered the most effective and popular technique to interconnect delay reduction, especially for global signals [Alpert 1997]. With so many buffers being added, the buffer positions should be planned as early as possible to ensure timing closure and design convergence; in particular, current VLSI designs often do not allow buffers to be inserted inside a circuit module, because they consume silicon resources and require connections to the power/ground network. Consequently, buffers are placed in channels and dead spaces of the current floorplan and are often clustered to form buffer blocks between existing circuit modules of the floorplan, which inevitably increases the chip area [Cong 1999]. It is thus desirable to carefully plan for the buffers during/after floorplanning to minimize the area overhead and facilitate routing, which is referred to as the **buffer block planning**. Furthermore, long interconnects affect microarchitecture designs very much, because multiple cycles are necessary to communicate global signals across the chip. As a result, it is desirable to handle **microarchitecture aware floorplanning** considering interconnect pipelining to improve the performance of microarchitecture designs [Jagannathan 2005; Ma 2007].

Because interconnection on the chip becomes more congested as technology advances, bus routing becomes a challenging task. Because buses have different widths and go through multiple modules, the positions of the modules greatly affect the bus routing. To make the bus routing easier, bus planning should be considered at the floorplanning stage, which is called **bus-driven floorplanning**. The feasibility conditions of bus-driven floorplanning for sequence pair [Xiang 2003] and B*-tree [Chen 2005a] are studied to reduce their solution spaces and find the desired floorplans efficiently. When the number of modules through which a bus goes is large, multi-bend bus structure can be used to find better solutions [Law 2005].

As technology advances, the metal width decreases, whereas the global wirelength increases. This trend makes the resistance of the power wire increase substantially. Therefore, floorplanning considering **voltage (IR) drop** in the ***power/ground* (P/G) network** becomes important. Because of IR-drop, supply voltage in logic may not be an ideal reference. An important problem of P/G network synthesis is to use the minimum amount of wiring area for a P/G network under the power integrity constraints such as IR drop and electromigration. As the design complexity increases dramatically, it is necessary to handle the IR-drop problem earlier in the design cycle for better design convergence. Most existing commercial tools deal with the IR-drop problem at the post-layout stage, when entire chip design is completed and detailed layout and current information are known. It is, however, often very difficult and computationally expensive to fix the P/G network synthesis at the post-layout stage. Therefore, researchers started to consider the P/G network analysis at an earlier design stage [Yim 1999; Wu 2004; Lin 2007].

Recently, **3-D floorplanning** was developed to handle dynamically reconfigurable *field programmable gate arrays* (FPGAs) to improve logic capacity by time-sharing. We may use the 3-D space (x, y, t) to model a dynamically reconfigurable system. The x and y coordinates represent the 2-D plane of FPGA resources (spatial dimension), whereas the t coordinate represents the time axis (temporal dimension). Each "task" [*Reconfigurable Functional Unit Operation* (RFUOP), the execution unit in a reconfigurable FPGA] is modeled by a rectangular box (module). We may denote each module as a 3-D box with the spatial dimensions x and y and the temporal dimension t. Figure 10.39a shows a program with four parts of codes to be mapped into RFUOPs. Because of the capacity constraint, we may not load all modules into the device at the same time. Therefore, it is desirable to consider the 3-D floorplanning problem of placing these modules into the *Reconfigurable Functional Unit* (RFU) (see Figure 10.39b). The objective is to allocate modules to optimize the area and execution time and to satisfy specified constraints. To deal with the 3-D floorplanning problem, a few 3-D floorplan representations extending the 2-D floorplan ones are proposed. For example, **Sequence Triple** [Yamazaki 2003] and **Sequence Quintuple** [Yamazaki 2003] are extensions of sequence pair for 2-D packing. **K-tree** [Kawai 2005], **T-tree** [Yuh 2004a], and **3D-subTCG**

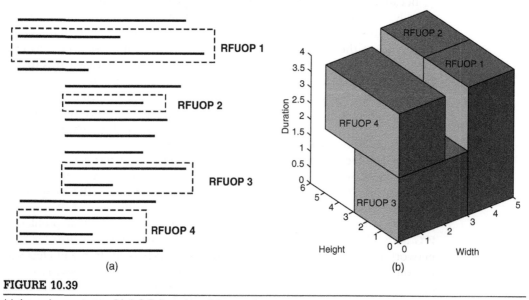

(a)   (b)

FIGURE 10.39

(a) A running program. (b) A 3-D floorplan of the running program.

[Yuh 2004b] are extensions of O-tree [Guo 1999], B*-tree [Chang 2000], and **Transitive Closure Graph** (TCG) [Lin 2001] for 2-D packing, respectively. Furthermore, heat dissipation is the most critical challenge of system-in-package design, sometimes called 2.5-D IC's (discrete layers are added into the traditional x and y spatial dimensions). Layer partitioning followed by 2-D floorplanning is often adopted to handle the thermal constraints for the 2.5-D IC designs [Cong 2004].

In addition to the floorplanning for VLSI modules, the floorplanning techniques can also be applied to other problems, such as system-on-chip test scheduling [Wu 2005] and digital microfluidic biochip placement [Yuh 2006].

10.5 CONCLUDING REMARKS

Floorplanning is an essential design step for hierarchical, building-module design methodology. It provides valuable insights into the hardware decisions and estimation of various costs. The most popular floorplanning method resorts to the modeling of the floorplan structure and then optimizes the floorplan solutions with simulated annealing. There exist many floorplan representations in the literature. Yet, normalized Polish expression, B*-tree, and sequence pair have been recognized as the most valuable representations because of their superior simplicity, effectiveness, efficiency, and flexibility.

In additional to simulated annealing, analytical floorplanning approaches have shown their advantage in the effective wirelength optimization [Zhan 2006]; however, it is harder to handle the module overlaps and the fixed-outline constraint for such an approach. Floorplanning considering both hard and soft modules is also more challenging for the analytical approach.

After floorplanning, all hard modules are fixed. For each soft module, we might need to further place standard cells inside the module. The placement problem will be introduced in Chapter 11. Once the positions of all hard modules and standard cells are decided, we need to route all signal and power/ground nets, which will be introduced in Chapters 12 and 13, respectively.

10.6 EXERCISES

10.1. (Polish Expression and B*-tree) Given the following Polish expression, $E = 12V34HVH5$,

   (a) Does the above expression have the balloting property? Justify your answer.

   (b) Is E a normalized Polish expression? If not, exchange an operator and an operand to transform E into a normalized Polish expression E'.

(c) Give the slicing tree that corresponds to the Polish expression E. Also, give the slicing tree corresponding to the "resulting" normalized Polish expression E', if E is not a normalized Polish expression.

(d) Assume that modules 1, 2, 3, 4 and 5 have the sizes and shapes listed in Table 10.8. If all modules are rigid (hard) and rotation is allowed, what will be the size of the smallest bounding rectangle corresponding to the "resulting" normalized Polish expression E'? Show all steps that lead to your answer.

(e) Give a B*-tree for the floorplan derived in (d).

(f) Show all steps for computing the coordinates of the modules from the resulting B*-tree of (e).

10.2. (Polish Expression and B*-tree) Given the following Polish expression, $E = 12V3H4V$,

(a) Give a slicing tree corresponding to the expression E.

(b) Assume modules 1, 2, 3, and 4 have the sizes and shapes indicated in Table 10.9. If all modules are rigid and rotation is allowed, what will be the size of the smallest bounding rectangle corresponding to the Polish expression E? Show all steps that lead to your answer.

(c) Give a B*-tree for the floorplan derived in (b).

(d) Show all steps for computing the coordinates of the modules from the resulting B*-tree of (c).

Table 10.8. The Dimensions of Modules in Exercise 10.1

Module No.	Width	Height
1	2	3
2	2	2
3	5	3
4	3	3
5	1	3

Table 10.9. The Dimensions of Modules in Exercise 10.2

Module No.	Width	Height
1	2	2
2	3	2
3	2	4
4	5	3

10.3. **(Sequence Pair and B*-tree)** Consider the floorplan of five modules 1, 2, 3, 4, and 5 and their dimensions shown in Table 10.10 and Figure 10.40.

   (a) Derive the sequence pair $S = (\Gamma_+, \Gamma_-)$ for the floorplan. Show your procedure.

   (b) Show all steps on the sequence pair to evaluate the area cost for the $S = (\Gamma_+, \Gamma_-)$-packing. What is the area cost?

   (c) Derive the B*-tree for the floorplan shown in the figure.

   (d) Show all steps on the B*-tree for evaluating the cost efficiently. What is the area cost?

10.4. **(Rectilinear Modules)** In this chapter, we assume all modules are rectangular. However, in real-world application, some modules may be of a rectilinear shape. Show how to extend B*-tree and sequence pair to handle rectilinear modules. (*Hint: Dissect a rectilinear module into rectangular submodules.*)

10.5. **(Shape Curve Candidates)** Consider two lists, $A = \{(p_1, q_1), \ldots, (p_m, q_m)\}$ and $B = \{(x_1, y_1), \ldots, (x_n, y_n)\}$, with $p_i \leq p_{i+1}$, $q_i \geq q_{i+1}$, $x_i \leq x_{i+1}$, and $y_i \geq y_{i+1}$. Combine A and B by considering each element (p_i, q_i) of A and each element (x_j, y_j) of B to produce an element of a list C: $(p_i + x_j, \max\{q_i, y_j\})$. Thus C has $m \times n$ elements. If there are two elements (c_i, d_i) and (c_j, d_j) in C with $c_i \leq c_j$ and $d_i \leq d_j$, then delete

Table 10.10 The Dimensions of Modules in Exercise 10.3

Module No.	Width	Height
1	2	2
2	2	3
3	3	3
4	3	3
5	4	2

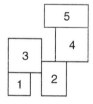

FIGURE 10.40

The floorplan for Exercise 10.3.

(c_j, d_j) from C. Prove that the resulting list C has at most a linear function of m and n elements. Find the linear function.

10.6. **(Multilevel Framework)** Give the strengths and weaknesses of the $\wedge$- and V-shaped multilevel frameworks. Here, the $\wedge$-shaped multilevel framework consists of two stages of bottom-up processing followed by top-down processing, whereas the V-shaped one uses top-down processing followed by bottom-up processing.

10.7. **(Boundary Constraints on B*-tree)** It is often useful to identify the modules being placed along a chip boundary because those modules are closet to the I/O pads in a traditional chip package with peripheral I/Os. Given a B*-tree, you are asked to derive the feasibility of the B*-tree for the boundary conditions.

    (a) For the nodes corresponding to the modules along the bottom boundary of a floorplan, what are their positions in the B*-tree?

    (b) For the nodes corresponding to the modules along the left boundary of a floorplan, what are their positions in the B*-tree?

    (c) For the nodes corresponding to the modules along the right boundary of a floorplan, what are their positions in the B*-tree?

    (d) For the nodes corresponding to the modules along the top boundary of a floorplan, what are their positions in the B*-tree?

10.8. **(Programming)** This programming assignment asks you to write a chip floorplanner that can handle hard macros and provide *graphic user interface* (GUI) to show the floorplanning result with interconnections (center-to-center connection for each net). The evaluation is based on the resulting floorplan area, wirelength, and running time.

(1) Input/Output specification

Input format

Each test case has two input files, *problem_no.mac* and *problem_no. net*. The first file defines chip and macro information includes chip width and chip height. The later includes name, area, and aspect ratio constraints of a macro and lists all nets. For example, there are two input files, problem1.mac and problem1.net. The first file format is as follows:

```
.chip_bbox (width, height)
//the lower-left corner of this bounding box is (0, 0)
.module name width height
.module name width height
... More modules
```

The format of the second file (netlist) is:

```
.net net_name module_name1 module_name2 ...
.net net_name module_name1 module_name2 ...
... More nets
// one line defines a net
// for example, if net N1 connects macro A, B, and C, the definition is
// .net N1 A B C
```

Output format

The output file consists of three parts: (1) bounding box for each macro (specified by the coordinates of the lower-left and upper-right corners), (2) total wirelength estimated by the *half-perimeter wirelength* (HPWL) of all nets, and (3) area (it may be smaller than the chip bounding box). The area can be obtained by $X * Y$, where X (Y) is the difference between rightmost (topmost) edge and leftmost (bottommost) edge among all modules. The report file format is as follows:

```
.module module_name (x1, y1) (x2, y2)
.module module_name (x1, y1) (x2, y2)
// (x1, y1): lower-left corner, (x2, y2): upper-right corner
... More modules
.wire total_wire_length
.area chip_area
// area = (max_x2 – min_x1) * (max_y2 – min_y1)
```

(2) Problem statement

Given (1) a set of rectangular modules and (2) a set of nets interconnecting these modules, the floorplanner places all modules within a specified fixed-outline (*i.e.*, a rectangular bounding box). We assume that the lower-left corner of this bounding box is the origin (0, 0) and no space (channel) is needed between two modules. The main objective is to minimize the total wirelength. The net terminals are assumed to be at the center of their corresponding module. The second objective is to minimize the chip area. Figure 10.41 illustrates an example of all input/output files.

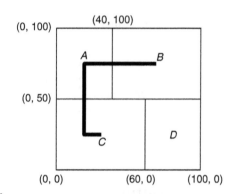

FIGURE 10.41

A floorplan problem and its solution, the bold line representing net N1.

Input files

[PROBLEM1.MAC]
.chip_bbox (100, 100)
.module A 50 40
.module B 60 50
.module C 60 50
.module D 50 40

[PROBLEM1.NET]
.net N1 A B C

Output file

[PROBLEM1.RPT]
.module A (0, 50) (40, 100)
.module B (40, 50) (100, 100)
.module C (0, 0) (60, 50)
.module D (60, 0) (100, 50)
.wire 100
.area 10000

ACKNOWLEDGMENTS

We thank Dr. Laung-Terng Wang of SynTest Technologies Inc., Professor Chris Chu of Iowa State University, Professor Cheng-Kok Koh of Purdue University, Professor Evangeline F.-Y. Young of the Chinese University of Hong Kong, Professor Martin D. F. Wong of the University of Illinois at Urbana-Champaign, and the National Taiwan University students in the physical design class for their very careful review of this chapter. We also thank SpringSoft Inc. for providing the programming assignment.

REFERENCES

R10.0 Books

[Lawler 1976] E. L. Lawler, *Combinatorial Optimization: Networks and Matroids,* Holt, Rinehart, and Winston, New York, 1976.

[Sait 1999] S. M Sait and H. Youssef, *VLSI Physical Design Automation: Theory and Practice,* World Scientific, Singapore, 1999.

[Sherwani 1999] N. Sherwani, *Algorithms for VLSI Physical Design Automation,* Third Edition, Kluwer Academic, Boston, 1999.

[Wong 1988] D. F. Wong, H. W. Leong, and C. L. Liu, *Simulated Annealing for VLSI Design,* Kluwer Academic, Boston, 1988.

R10.1 Introduction

[Kahng 2000] A. B. Kahng, Classical floorplanning harmful?, *in Proc. ACM Int. Symp. on Physical Design,* pp. 207–213, April 2000.

[Otten 1982] R. H. J. M. Otten, Automatic floorplan design, *in Proc. ACM/IEEE Design Automation Conf.,* pp. 261–267, June 1982.

R10.2 Simulated Annealing Approach

[Chang 2000] Y.-C. Chang, Y.-W. Chang, G.-M. Wu, and S.-W. Wu, B*-trees: a new representation for non-slicing floorplans, *in Proc. ACM/IEEE Design Automation Conf.,* pp. 458–463, June 2000.

[Guo 1999] P.-N. Guo, C.-K. Cheng, and T. Yoshimura, An O-tree representation of non-slicing floorplan and its applications, *in Proc. ACM/IEEE Design Automation Conf.,* pp. 268–273, June 1999.

[Hilton 1991] P. Hilton and J. Pederson, Catalan numbers, their generalization, and their uses, *Math. Intelligencer,* 13(2), pp. 64–75, February 1991.

[Hong 2000] X. Hong, G. Huang, Y. Cai, J. Gu, S. Dong, C.-K. Cheng, and J. Gu, Corner block list: An effective and efficient topological representation of non-slicing floorplan, *in Proc. IEEE/ACM Int. Conf. on Computer-Aided Design,* pp. 8–13, November 2000.

[Kirpatrick 1983] S. Kirpatrick, C. D. Gelatt, and M. P. Vecchi, Optimization by simulated annealing, *Science,* 220(4598), pp. 671–680, May 13, 1983.

[Lin 2001] J.-M. Lin and Y.-W. Chang, TCG: A transitive closure graph based representation for general floorplans, *in Proc. ACM/IEEE Design Automation Conf.,* pp. 764–769, June 2001.

[Lin 2002] J.-M. Lin and Y.-W. Chang, TCG-S: Orthogonal coupling of P*-admissible representations for general floorplans, *in Proc. ACM/IEEE Design Automation Conf.,* pp. 842–847, June 2002.

[Lin 2003] J.-M. Lin, Y.-W. Chang, and S.-P. Lin, Corner sequence: A P-admissible floorplan representation with a worst case linear-time packing scheme, *IEEE Trans. on Very Large Scale Integration Systems,* 11(4), pp. 679–686, August 2003.

[Murata 1995] H. Murata, K. Fujiyoshi, S. Nakatake, and Y. Kajatani, Rectangle packing based module placement, *in Proc. IEEE/ACM Int. Conf. on Computer-Aided Design*, pp. 472–479, November 1995.

[Otten 1982] R. H. J. M. Otten, Automatic floorplan design, *in Proc. ACM/IEEE Design Automation Conf.*, pp. 261–267, June 1982.

[Otten 1983] R. H. J. M. Otten, Efficient floorplan optimization, *in Proc. IEEE Int. Conf. on Computer Design*, pp. 499–502, November 1983.

[Sechen 1986] C. Sechen and A. Sangiovanni-Vincentelli, TimberWolf3.2: A new standard cell placement and global routing package, *in Proc. IEEE/ACM Design Automation Conf.*, pp. 432–439, June 1986.

[Sechen 1988] C. Sechen, Chip-planning, placement, and global routing of macro/custom cell integrated circuits using simulated annealing, *in Proc. IEEE/ACM Design Automation Conf.*, pp. 73–80, June 1988.

[Shen 2003] C. Shen and C. Chu, Bounds on the number of slicing, mosaic and general floorplans, *IEEE Trans. on Computer-Aided Design*, 22(10), pp. 1354–1361, October 2003.

[Stockmeyer 1983] L. Stockmeyer, Optimal orientations of cells in slicing floorplan designs, *Information and Control*, 57(2-3), pp. 91–101, May/June 1983.

[Tang 2001] X. Tang and D. F. Wong, FAST-SP: A fast algorithm for block placement based on sequence pair, *in Proc. IEEE/ACM Asia South Pacific Design Automation Conf.*, pp. 521–526, January 2001.

[Wong 1986] D. F. Wong and C. L. Liu, A new algorithm for floorplan design, *in Proc. ACM/IEEE Design Automation Conf.*, pp. 101–107, June 1986.

[Young 2003] E. F.-Y. Young, C. C.-N. Chu, and Z. C. Shen, Twin binary sequences: A non-redundant representation for general non-slicing floorplan, *IEEE Trans. on Computer-Aided Design*, 22(4), pp. 457–469, April 2003.

[Zhou 2004] H. Zhou and J. Wang, ACG-adjacent constraint graph for general floorplans *in Proc. IEEE Int. Conf. on Computer Design*, pp. 572–575, October 2004.

R10.3 Analytical Approach

[GLPK 2008] *GLPK* (GNU Linear Programming Kit), http://www.gnu.org/software/glpk/, 2008.

[ILOG 2008] ILOG CPLEX, http://www.ilog.com/products/cplex/, 2008.

[LINDO 2008] LINDO System Inc., http://www.lindo.com/, 2008.

[lp_solve] lp_solve, http://tech.groups.yahoo.com/group/lp_solve/, 2008.

[Sutanthavibul 1991] S. Sutanthavibul, E. Shragowitz, and J. B. Rosen, An analytical approach to floorplan design and optimization, *IEEE Trans. on Computer-Aided Design of Integrated Circuits an Systems*, 10(6), pp. 761–769, June 1991.

[Zhan 2006] Y. Zhan, Y. Feng, and S. S. Sapatnekar, A fixed-die floorplanning algorithm using an analytical approach, *in Proc. IEEE/ACM Asia South Pacific Design Automation Conf.*, pp. 771–776, January 2006.

R10.4 Modern Floorplanning Considerations

[Adya 2003] S. N. Adya and I. Markov, Fixed-outline floorplanning: enabling hierarchical design, *IEEE Trans. on Very Large Scale Integration Systems*, 11(6), pp. 1120–1135, December 2003.

[Alpert 1997] C. J. Alpert and A. Devgan, Wire segmenting for improved buffer insertion, *in Proc. ACM/IEEE Design Automation Conf.*, pp. 588–593, June 1997.

[Chang 2000] Y.-C. Chang, Y.-W. Chang, G.-M. Wu, and S.-W. Wu, B*-trees: A new representation for non-slicing floorplans, *in Proc. ACM/IEEE Design Automation Conf.*, pp. 458–463, June 2000.

[Chen 2005a] T.-C. Chen and Y.-W. Chang, Modern floorplanning based on fast simulated annealing, *in Proc. ACM Int. Symp. on Physical Design*, pp. 104–112, April 2005.

[Chen 2005b] T.-C. Chen, Y.-W. Chang, and S.-C. Lin, IMF: interconnect-driven multilevel floorplanning for large-scale building-module designs, *in Proc. IEEE/ACM Int. Conf. on Computer-Aided Design*, pp. 159–164, November 2005.

[Chen 2006] T.-C. Chen and Y.-W. Chang, Modern floorplanning based on B*-tree and fast simulated annealing, *IEEE Trans. on Computer-Aided Design*, 25(4), pp. 637–650, April 2006.

[Chi 2003] J.-C. Chi and M. C. Chi, A block placement algorithm for VLSI circuits, *Chung Yuan Journal*, 31(1), pp. 69–75, March 2003.

[Cho 2006] M. Cho, H. Shin, and D. Z. Pan, Fast substrate noise-aware floorplanning with preference directed graph for mixed-signal SOCs, *in Proc. IEEE/ACM Asia South Pacific Design Automation Conf.*, pp. 765–770, January 2006.

[Cong 1997] J. Cong, L. He, K.-Y. Khoo, C.-K. Koh, and Z. Pan, Interconnect design for deep submicron ICs, *in Proc. IEEE/ACM Int. Conf. on Computer-Aided Design*, pp. 478–485, November 1997.

[Cong 1999] J. Cong, T. Kong, and D. Z. Pan, Buffer block planning for interconnect-driven floorplanning, *in Proc. IEEE/ACM Int. Conf. on Computer-Aided Design*, pp. 358–363, November 1999.

[Cong 2004] J. Cong, J. Wei, and Y. Zhang, A thermal-driven floorplanning algorithm for 3D ICs, *in Proc. Int. Conf. on Computer-Aided Design*, pp. 306–313, November 2004.

[Cong 2005] J. Cong, M. Romesis, and J. R. Shinnerl, Fast floorplanning by look-ahead enabled recursive bipartitioning, *in Proc. IEEE/ACM Asia South Pacific Design Automation Conf.*, pp. 1119–1122, January 2005.

[Jagannathan 2005] A. Jagannathan, H. H. Yang, K. Konigsfeld, D. Milliron, M. Mohan, Mi. Romesis, G. Reinman, and J. Cong, Microarchitecture evaluation with floorplanning and interconnect pipelining, *in Proc. ACM/IEEE Asia South Pacific Design Automation Conf*, pp. 8–15, January 2005.

[Kawai 2005] H. Kawai and K. Fujiyoshi, 3D-block packing using a tree representation, *in Proc. Workshop on Circuits and Systems in Karuizawa*, pp. 199–204, April 2005.

[Law 2005] J. H. Y. Law and E. F. Y. Young, Multi-bend bus driven floorplanning, *in Proc. ACM Int. Symp. Physical Design*, pp. 113–120, April 2005.

[Lee 2003] H.-C. Lee, Y.-W. Chang, J.-M. Hsu, and H. H. Yang, Multilevel floorplanning/placement for large-scale modules using B*-trees, *in Proc. ACM/IEEE Design Automation Conf.*, pp. 812–817, June 2003.

[Lin 2001] S. Lin and N. Chang, Challenges in power-ground integrity, *in Proc. IEEE/ACM Int. Conf. on Computer-Aided Design*, pp. 651–654, November 2001.

[Lin 2004] C.-T. Lin, D.-S. Chen, and Y.-W. Wang, Robust fixed-outline floorplanning through evolutionary search, *in Proc. IEEE/ACM Asia and South Pacific Design Automation Conf.*, pp. 42–44, January 2004.

[Liu 2007] C.-W. Liu and Y.-W. Chang, Power/ground network and floorplan co-synthesis for fast design convergence, *IEEE Trans. on Computer-Aided Design*, 26(4), pp. 693–704, April 2007.

[Ma 2007] Y. Ma, Z. Li, J. Cong, X. Hong, G. Reinman, S. Dong, and Q. Zhou, Micro-architecture pipelining optimization with throughput-aware floorplanning, *in Proc. ACM/IEEE Asia South Pacific Design Automation Conf.*, pp. 920–925, January 2007.

[Nakatake 1996] S. Nakatake, K. Fujiyoshi, H. Murata, and Y. Kajatani, Module placement on BSG-structure and IC layout applications, *in Proc. IEEE/ACM Int. Conf. on Computer-Aided Design*, pp. 261–267, November 1996.

[Wu 2004] S.-W. Wu and Y.-W. Chang, Efficient power/ground network analysis for power integrity-driven design methodology, *in Proc. ACM/IEEE Design Automation Conf.*, pp. 177–180, June 2004.

[Wu 2005] J.-Y. Wu, T.-C. Chen, and Y.-W. Chang, SoC test scheduling using the B*-tree based floorplanning technique, *in Proc. ACM/IEEE Asia South Pacific Design Automation Conf.*, pp. 1188–1191, January 2005.

[Xiang 2003] H. Xiang, X. Tang, and M. D. F. Wong, Bus-driven floorplanning, *in Proc. IEEE/ACM Int. Conf. Computer-Aided Design*, pp. 66–73, November 2003.

[Yan 2008] J. Z. Yan and C. Chu, DeFer: Deferred decision making enabled fixed-outline floorplanner, *in Proc. IEEE/ACM Design Automation Conf.*, June 2008.

[Yamazaki 2003] H. Yamazaki, K. Sakanushi, S. Nakatake, and Y. Kajitani, The 3D-packing by meta data structure and packing heuristics, *IEICE Trans. on Fundamentals of Electronics, Communications and Computer*, E82-A(4), pp. 639–645, April 2003.

[Yuh 2004a] P.-H. Yuh, C.-L. Yang, and Y.-W. Chang, Temporal floorplanning using the T-tree representation, *in Proc. IEEE/ACM Int. Conf. on Computer-Aided Design*, pp. 300–305, November 2004.

[Yuh 2004b] P.-H. Yuh, C.-L. Yang, Y.-W. Chang, and H.-L. Chen, Temporal floorplanning using 3D-subTCG, *in Proc. IEEE Asia and South Pacific Conf. on Circuits and Systems*, pp. 725–730, January 2004.

[Yuh 2006] P.-H. Yuh, C.-L. Yang, and Y.-W. Chang, Placement of digital microfluidic biochips using the T-tree formulation, *in Proc of ACM/IEEE Design Automation Conf.* pp. 931–934, July 2006.

[Yim 1999] J.-S. Yim, S.-O. Bae, and C.-M. Kyung, A floorplan-based planning methodology for power and clock distribution in ASICs, *in Proc. ACM/IEEE Design Automation Conf.*, pp. 766–771, June 1999.

[Zhou 2004] H. Zhou and J. Wang, ACG–adjacent constraint graph for general floorplans, *in Proc. IEEE Int. Conf. on Computer Design*, pp. 572–575, October 2004.

R10.5 Concluding Remarks

[Zhan 2006] Y. Zhan, Feng, and S. S. Sapatnekar, A fixed-die floorplanning algorithm using an analytical approach, *in Proc. IEEE/ACM Asia South Pacific Design Automation Conf.*, pp. 771–776, January 2006.

Placement   **11**

Chris Chu
Iowa State University, Ames, Iowa

ABOUT THIS CHAPTER

Placement is the process of determining the locations of circuit devices on a die surface. It is an important stage in the VLSI design flow, because it affects routability, performance, heat distribution, and to a less extent, power consumption of a design. Traditionally, it is applied after the logic synthesis stage and before the routing stage. Since the advent of deep submicron process technology around the mid-1990s, interconnect delay, which is largely determined by placement, has become the dominating component of circuit delay. As a result, placement information is essential, even in early design stages, to achieve better circuit performance. In recent years, placement techniques have been integrated into the logic synthesis stage to perform physical synthesis and into the architecture design stage to perform physical-aware architecture design.

This chapter begins with an introduction to the placement stage. Next, various placement problem formulations are discussed. Then, partitioning-based approach, simulated annealing approach, and analytical approach for global placement are presented. After that, legalization and detail placement algorithms are described. The chapter concludes with a discussion of other placement approaches and useful resources to placement research.

11.1 INTRODUCTION

Traditionally, placement is the design stage after logic synthesis and before routing in the VLSI design flow. In logic synthesis, a netlist is generated. Then in placement, the locations of the circuit modules in the netlist are determined. After placement, routing is performed to lay out the nets in the netlist.

Placement is a critical step in the VLSI design flow mainly for the following four reasons. First, placement is a key factor in determining the performance of a circuit. Placement largely determines the length and, hence, the delay of interconnect wires. As feature size in advanced VLSI technology continues to reduce, interconnect delay has become the determining factor of circuit performance. **635**

Interconnect delay can consume as much as 75% of clock cycle in advanced design. Therefore, a good placement solution can substantially improve the performance of a circuit. Second, placement determines the routability of a design. A well-constructed placement solution will have less routing demand (*i.e.*, shorter total wirelength) and will distribute the routing demand more evenly to avoid routing hot spots. Third, placement decides the distribution of heat on a die surface. An uneven temperature profile can lead to reliability and timing problems. Fourth, power consumption is also affected by placement. A good placement solution can reduce the capacitive load because of the wires (by having shorter wires and larger separation between adjacent wires). Hence the switching power consumption can be reduced.

In recent years, it has become essential for the logic synthesis stage to incorporate placement techniques to perform physical design aware logic synthesis (*i.e.*, physical synthesis). The reason is that without some placement information, it is impossible to estimate the delay of interconnect wires. Hence, given the significance of interconnect delay, logic synthesis will not have any meaningful timing information to guide the synthesis process. As a result, the synthesized netlists will have poor performance after placement. For the same reason, consideration of placement information during architecture design is also increasingly common.

Placement is a computationally difficult problem. Even the simple case of placing a circuit with only unit-size modules and 2-pin nets along a straight line to minimize total wirelength is NP-complete [Garey 1974]. The VLSI placement problem is much more complicated. The circuit may contain modules of different sizes and may have multi-pin nets. The placement region is two-dimensional. Other cost functions may be used rather than total wirelength. There may also be different constraints for different design styles. (Details of problem formulations can be found in Section 11.2.) As designs with millions of modules are now common, it is a major challenge to design efficient placement algorithms to produce high-quality placement solutions.

One way to overcome the complexity issue is to perform placement in several manageable steps. One common flow is as follows.

1. **Global placement.** Global placement aims at generating a rough placement solution that may violate some placement constraints (*e.g.*, there may be overlaps among modules) while maintaining a global view of the whole netlist.

2. **Legalization.** Legalization makes the rough solution from global placement legal (*i.e.*, no placement constraint violation) by moving modules around locally.

3. **Detailed placement.** Detailed placement further improves the legalized placement solution in an iterative manner by rearranging a small group of modules in a local region while keeping all other modules fixed.

The global placement step is the most important one of the three. It has the most impact on placement solution quality and runtime, and has been the focus of most prior research works. After global placement, the placement solution is almost completely determined. In legalization and detailed placement, only local changes in module locations will be made. Therefore, the main emphasis of this chapter is the global placement step. The most commonly used global placement approaches are **partitioning-based approach, simulated annealing approach**, and **analytical approach**. The analytical approach will be presented with the most details, because it is currently the best approach in both quality and runtime.

11.2 PROBLEM FORMULATIONS

The input to the placement problem is a placement region, a set of modules, and a set of nets. The widths and heights of the placement region and all modules are given. The locations of I/O pins on the placement region and on all modules are fixed. Sometimes, some input modules (*e.g.*, buffer bays, I/O modules, IP blocks) are preplaced by designers, and, hence, their locations are also fixed before placement. Each net specifies a collection of pins in the placement region and/or in some modules that are connected. Basically, placement is to find a position for each module within the placement region so that there is no overlap among the modules and some objective is optimized. Many variations in the placement problem formulation exist, because different designs may require different objectives and different design styles may introduce different constraints. The placement problems for common design styles and objectives are presented in the following.

11.2.1 Placement for different design styles

11.2.1.1 *Standard-cell placement*

In a standard-cell design, all modules have the same height. The placement of standard cells has to be aligned with some prespecified standard-cell rows in the placement region. Because of the popularity of standard-cell design, most placement algorithms assume a standard-cell design style.

11.2.1.2 *Gate array/FPGA placement*

In gate array or FPGA design, the modules can only be placed at some predefined sites that are arranged in a regular array.

11.2.1.3 *Macro block placement*

In macro block placement, each module is a macro block of fixed shape and orientation. The macro blocks have to be placed within the placement region without overlap among them. The macro block placement problem is similar to the fixed-outline floorplanning problem. However, in floorplanning, the shape and the orientation of the macro blocks are usually assumed to be changeable. Macro block placement can be considered to be a special case of fixed-outline floorplanning. If the number of modules is small, it can be solved by floorplanning techniques (please refer to Chapter 10).

11.2.1.4 *Mixed-size placement*

Mixed-size placement places both macro blocks and standard cells in a circuit. Modern designs often contain a large number of macro blocks together with a huge number of standard cells. As a result, mixed-size placement is a common formulation in recent years. Because macro blocks are typically orders of magnitude larger than standard cells, the handling of the nonoverlapping constraints among the modules presents a unique challenge.

11.2.2 Placement objectives

11.2.2.1 *Total wirelength*

Total wirelength is the most commonly used objective in placement formulations. Minimization of total wirelength indirectly optimizes several other objectives. First, routability can be improved by less routing demand. Second, timing can be better because shorter wires have less delay. Third, power consumption can be reduced because shorter wires also introduce less capacitive load. Notice that total wirelength minimization is only a heuristic in optimizing these other objectives. Even if the *total* wirelength is reduced, nets in the most congested region, along the timing critical paths, or with the highest switching activities may not be shorter. To specify the importance of different nets in optimizing another objective, a weight can be assigned to each net. Then the total weighted wirelength will have a much better correlation to the other objective.

It is difficult to predict during placement the wirelength of a net after routing, because it is router-dependent. Several approaches estimate the routed wirelength for a given placement. The most widely used approach is *half-perimeter wirelength* (HPWL). The HPWL of a net is equal to half of the perimeter of the smallest bounding rectangle that encloses all the pins of the net. For example, for the 5-pin net in Figure 11.1a, the HPWL is $W + H$ as shown in Figure 11.1b. HPWL is popular because it can be computed in linear time and it can be written as a simple closed-form function of the coordinates of the pins (see Section 11.5.1). It also provides exact wirelength for optimally routed nets with two and three pins. However, HPWL can significantly underestimate the wirelength for nets with four or more pins.

Another approach of wirelength estimation is based on *rectilinear minimum spanning tree* (RMST). A RMST is a tree with minimum total wirelength in Manhattan distance to connect a given set of nodes. For example, the RMST of the net in Figure 11.1a is given in Figure 11.1c. The best time complexity for RMST construction is $O(n \log n)$ [Guibas 1983]. RMST wirelength is exact for nets with two pins, but it can overestimate the wirelength of optimally routed nets with three or more pins by up to 50% [Hwang 1976]. In practice, RMST

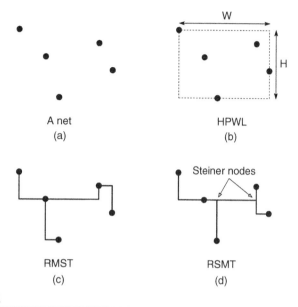

FIGURE 11.1

Wirelength estimation techniques.

can produce more accurate wirelength estimation than HPWL (especially for nets with high pin count) in a runtime several times more than that of HPWL.

A highly accurate approach is based on *rectilinear Steiner minimal tree* (RSMT). A RSMT is a tree with minimum total edge length in Manhattan distance to connect a given set of nodes possibly through some extra (*i.e.*, Steiner) nodes. For example, the RSMT of the net in Figure 11.1a is given in Figure 11.1d. If there is no routing congestion, RSMT is the preferred way to route a net, because it gives the minimum wirelength. Thus, RSMT wirelength has a very high correlation with routed wirelength unless the design is heavily congested and requires a lot of detour in routing. RSMT construction is NP-complete [Garey 1979]. In the past, all exact and heuristic RSMT algorithms were very time-consuming. Hence, traditionally, RSMT was rarely used for wirelength estimation during placement. Recently, a lookup–table based RSMT heuristic algorithm called FLUTE [Chu 2008] was introduced. It is more accurate than all previous RSMT heuristics, yet it is as fast as RMST algorithms. With FLUTE, it is feasible for placers to use the highly accurate RSMT wirelength in guiding the placement process.

11.2.2.2 *Routability*

Routability is the most basic requirement of a placement solution. Any placement solution is useless if the routing cannot be completed. However, the routability of a placement solution is very hard to evaluate. Routability is router-dependent. There is no objective measure of routability. Even for a specific router, the routability is still very hard to estimate because of the complicated behavior of a router. One way of routability estimation is to call the router to perform a rough routing (*e.g.*, global routing), but this way is computationally very expensive. A more popular way is to assume the routing of each net follows some probability distribution and then estimate the routing congestion of each edge in the routing grid by the expected number of crossing nets. However, this way is not accurate and is still quite expensive computationally.

Because of the high computational cost, routability estimation is rarely incorporated into the placement objective function in guiding the placement process in practice. Instead, routability is often indirectly optimized during placement by a **white space allocation approach**. Regions that are predicted to be congested are given more white space (*i.e.*, placed with a lower module density) to provide more routing tracks.

11.2.2.3 *Performance*

Placement has significant impact on the delay of interconnects, and hence the performance of circuits. Because interconnect delay becomes a more dominating component of circuit delay as feature size continues to decrease, performance-driven placement is increasingly important. However, the delay of a net heavily depends on other factors like routing, buffering, driver size, wire width, and wire spacing. It is computationally too expensive to perform those tasks during the placement process. In other words, it is basically impossible to obtain accurate timing information during placement. In practice, the delay of a net is heuristically controlled during placement by controlling the length of the net. The net delay can be either reduced by assigning a larger net weight or bounded by constraining the net length.

11.2.2.4 *Power*

For most circuits, the major component of power consumption is switching power, which is consumed whenever a gate switches (*i.e.*, the capacitive load driven by the gate is charged/discharged). The capacitance of a net is proportional to its wirelength. So the power minimization problem can be formulated as a wirelength minimization problem. As power consumption is also proportional to the amount of switching, the nets should be weighted by their switching activity factors. Note that the clock distribution network is a global net driving a huge load and switches every clock cycle. It consumes a significant portion of switching power. Therefore, special attention should be paid to the placement of clocked elements (including latches, flip-flops, memories, and dynamic gates) so that the capacitance of clock wires can be minimized without increasing clock skew or degrading timing.

11.2.2.5 *Heat distribution*

An uneven temperature profile on a chip may adversely affect the characteristics of temperature-sensitive circuits. It may also lead to reliability problems. Therefore, it is desirable to properly distribute the heat-generating elements of a circuit to achieve an even temperature profile.

Thermal-driven placement is difficult for several reasons. First, the heat generation of each element changes over time, because it depends on the operations performed by the circuit. Second, heat is generated by both transistors and wires, and the heat generation by wires is hard to predict during placement. Third, the temperature profile is determined by both generation and transfer of heat and is difficult to approximate without the use of time-consuming simulation.

A practical solution to the thermal-driven placement problem is to distribute the average heat generation of modules evenly in the placement region (by assuming the dynamic nature of heat generation, wire heat generation, and heat transfer are secondary effects). This module heat distribution problem is similar to the module area distribution problem in placement and can be solved by similar techniques.

11.2.3 A common placement formulation

Although there are many variations in placement problem formulations for different design styles and with different objectives, the underlying issues for the various formulations are the same. For placement of different design styles and for thermal-driven placement, the modules have to be properly distributed. For optimization of different objectives, the lengths of wires have to be reduced.

In this chapter, the focus is on the very popular problem of minimum-wirelength placement for standard-cell design. In particular, total HPWL is considered to be the objective function, because RSMT wirelength is traditionally expensive to compute and difficult to optimize. An algorithm for this formulation can usually be extended to handle other design styles and objectives. For other design styles (even with preplaced modules), as long as the module density can be properly controlled during global placement, minor placement constraint violations can be easily resolved during legalization. The thermal-driven and routability-driven placement problems can also be handled by controlling heat density and white space density, respectively, in addition to area density during global placement. The performance and power optimization problems can be formulated as a weighted wirelength minimization problem.

11.3 GLOBAL PLACEMENT: PARTITIONING-BASED APPROACH

Roughly speaking, the **partitioning problem** is to divide a circuit into several subcircuits of similar sizes such that the number of connections among subcircuits is minimized. A circuit placement can be generated by recursively applying a partitioning procedure. Such an approach is called **partitioning-based placement** or **min-cut placement**. In the following sections, the basics for partitioning will first be introduced. The application of partitioning to placement will then be presented.

11.3.1 **Basics for partitioning**

Most partitioning algorithms solve the **bipartitioning** (or **2-way partitioning**) problem, which is to divide a circuit into two subcircuits. The bipartitioning problem will be discussed in the following.

11.3.1.1 *Problem formulation*

The circuit bipartitioning problem can be formulated as a hypergraph bipartitioning problem by modeling a circuit as a hypergraph. A circuit module is represented by a vertex, and the area of the module is modeled as the vertex size. A net is represented by a hyperedge, and the criticality of the net is modeled as the hyperedge weight. For example, the circuit in Figure 11.2a can be modeled by the hypergraph in Figure 11.2b in which the set of vertices is {A, B, C, D, E} and the set of hyperedges is {{A, B}, {A, C}, {A, D}, {B, D}, {D, E}, {B, C, E}}.

Given a hypergraph $G(V, E)$, where each vertex $v \in V$ has a size $s(v)$ and each hyperedge $e \in E$ has a weight $w(e)$, the hypergraph bipartitioning problem is to divide the set V into two subsets V_1 and V_2 such that the total weight of hyperedges being cut (*i.e.*, spanning both subsets) is minimized and the total sizes of vertices in V_1 and in V_2 are close to some user-defined values. Formally, the cost function can be written as:

$$CutCost(V_1, V_2) = \sum_{e \in E \text{ s.t. } e \cap V_1 \neq \emptyset \wedge e \cap V_2 \neq \emptyset} \omega(e)$$

The size constraint on V_1 can be specified by use of a ratio parameter γ and a tolerance parameter ε as follows:

$$\gamma - \epsilon \leq \sum_{v \in V_1} s(v) / \sum_{v \in V} s(v) \leq \gamma + \epsilon$$

The size constraint on V_2 is indirectly specified as

$$\sum_{v \in V_2} s(v) = \sum_{v \in V} s(v) - \sum_{v \in V_1} s(v)$$

Hypergraph bipartitioning is NP-complete [Garey 1979].

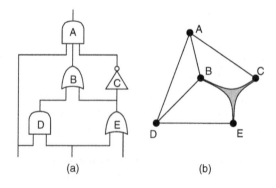

(a)   (b)

FIGURE 11.2

The hypergraph model of a circuit.

11.3.1.2 *The Fiduccia-Mattheyses algorithm*

A classical approach to solve the hypergraph bipartitioning problem is an iterative heuristic by Fiduccia and Mattheyses [Fiduccia 1982]. This heuristic is commonly called the FM algorithm. To explain the FM algorithm, the concept of the **gain** of a vertex has to be first introduced. Given a bipartitioning solution, the gain $g(v)$ of a vertex v is the reduction in the cut cost if vertex v is moved from its current partition to the other partition. For example, for the hypergraph in Figure 11.2b, assume all hyperedges have a weight of 1. In the initial bipartition as shown in Figure 11.3a, three hyperedges {A, C}, {B, C, E}, and {D, E} are cut. Hence the cut cost is 3. If vertex E is moved to the other partition as shown in Figure 11.3b, the cut cost becomes 2. Therefore, $g(E) = 3 - 2 = 1$. If vertex A is moved to the other partition as shown in Figure 11.3c, the cut cost becomes 4. Therefore, $g(A) = 3 - 4 = -1$.

The FM algorithm is described in Algorithm 11.1. The basic idea of the FM algorithm is to iteratively refine the current bipartition by greedily moving a vertex with maximum gain to the other side (step 6). Steps 3 to 12 are called a **pass**. In a pass, each vertex is tentatively moved to the other side one time (step 8). Once a vertex is moved, it will be locked (step 9) and will not be considered again until next pass. At the end of a pass, the sequence of first k moves that provides the best total gain G is made permanent (step 12). Note that g_j for some $j \in \{1, \ldots, k\}$ may be negative. That means moving v_j to the other side will make the cut worse. For a simple greedy algorithm that looks at the gain value of one vertex to decide a move, it will not proceed further. In other words, it gets trapped at a local minimum. For the FM algorithm, however, the move may still be taken, because moves subsequent to that may eventually improve the cut. Therefore, it can get out of the local minimum and generate bipartitions with better cut cost. The algorithm is repeated until there is no improvement in a pass (*i.e.*, $G = 0$).

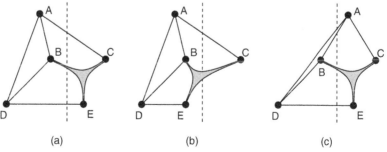

(a)	(b)	(c)

FIGURE 11.3

Illustration of the gain of a vertex.

Algorithm 11.1 The *Fiduccia and Mattheyses* (FM) Algorithm

---

1. Get an initial bipartition;
2. Repeat
3.   Unlock all vertices;
4.   For $i = 1$ to n where $n = |V|$
5.   Begin
6.   Select a vertex v_i with maximum gain among all unlocked vertices that will
          not violate the size constraints if moved to the other side;
7.   Set $g_i = g(v_i)$;
8.   Tentatively move vertex v_i to the other side;
9.   Lock vertex v_i;
10.   End
11.   Find k such that $G = \sum_{i=1}^{k} g_i$ is maximized;
12.   Make the moves for $v_1, \ldots, v_k$ permanent
        and discard the moves for $v_{k+1}, \ldots, v_n$;
13. Until $G = 0$;

Let n be the number of vertices of the hypergraph and p be the total number of pins in all hyperedges. For a given bipartition, the gain of all vertices can be computed in a straightforward manner described in the following. Let $g_e(v)$ be the contribution of hyperedge e to $g(v)$ for any $v \in V$ and $e \in E$. In other words, $g(v) = \sum_{e \in E} g_e(v)$. If $v \notin e$, then obviously $g_e(v) = 0$, because moving vertex v to the other partition does not affect whether or not hyperedge e is cut. Consider $v \in e$. Suppose the vertices in e are divided into two subsets e_1 and e_2 according to the bipartition. Assume without loss of generality that $v \in e_1$. If $e_1 = \{v\}$, then $g_e(v) = w(e)$, as e is cut in the given bipartition but will not be cut if v is moved to the other partition. If $e_2 = \emptyset$, then $g_e(v) = -w(e)$ as e is not cut in the given bipartition but will be cut if v is moved to the other partition. Otherwise, $g_e(v) = 0$ as e remains cut regardless of the partition v is in. For any $e \in E$, $g_e(v)$ for all $v \in e$ can be computed in $O(|e|)$time. Therefore, $g(v)$ for all $v \in V$ can be computed in $O(p)$ time. In other words, step 6 takes $O(p)$ time. Hence, each pass takes $O(np)$ time.

A technique based on incremental gain computation is presented in [Fiduccia 1982]. This technique reduces the runtime of each pass to $O(p)$ if all hyperedges have unit weight.[1] Instead of computing the gains from scratch in each pass, a **gain bucket** data structure is used to store the gain values. The gain bucket data structure is illustrated in Figure 11.4. It consists of a pointer array with index

---

[1]The technique can be easily extended for hyperedges with bounded integer weight.

ranging from $-p_{max}$ to p_{max}, where p_{max} is the maximum number of hyperedges connecting to a vertex. Indices of the array correspond to possible gain values. The entry with index g in the array points to a list of vertices with gain g. A MAX-GAIN index is maintained to keep track of the bucket for the vertices with maximum gain. Besides, a VERTEX array is kept to allow direct access to all vertices in the vertex lists.

In the FM algorithm, one gain bucket data structure is used to record the gains of the unlocked vertices for each partition. At the beginning of each pass, the gains for the current bipartition are computed, and the gain bucket data structures are constructed in $O(p)$ time. Then the vertices with the maximum gain can be obtained immediately. After a vertex is moved, the gains of the vertices connecting to it may be changed. With the help of the VERTEX array, the update of the gain bucket data structure for each vertex with changed gain takes only constant time. The number of vertices connecting to each vertex is equal to the number of pins in it. Therefore, the total update time for one pass is proportional to the total number of pins, *i.e.*, $O(p)$.

11.3.1.3 *A multilevel scheme*

The FM algorithm works well in practice for hypergraphs with up to hundreds of vertices. For larger hypergraphs, a bottom-up hierarchical scheme on the basis of recursive clustering, which is often called a **multilevel scheme**, can produce higher-quality bipartitions in a relatively small amount of runtime. The multilevel scheme consists of three phases, namely, **coarsening phase, initial partitioning phase**, and **uncoarsening and refinement phase**. The three phases are illustrated in Figure 11.5. During the coarsening phase, a sequence of successively smaller (coarser) hypergraphs is constructed by clustering heavily connected vertices together. During the initial partitioning phase, a bipartition of the coarsest hypergraph is computed by any bipartitioning algorithm. During the uncoarsening and refinement phase, the bipartition is

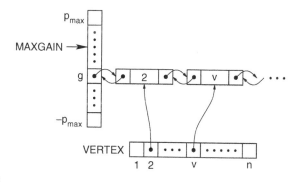

FIGURE 11.4

Gain bucket data structure of the FM algorithm.

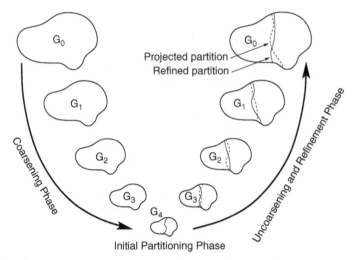

FIGURE 11.5

The three phases of multilevel bipartitioning [Karypis 1997].

successively projected to the next level finer hypergraph, and at each level an iterative refinement algorithm such as FM is used to further improve the bipartition.

hMetis [Karypis 1997] is one of the earliest and best multilevel hypergraph partitioning algorithms. In hMetis, three different coarsening techniques are applied. The initial bipartition is simply a random bipartition. The refinement of the uncoarsening hypergraphs is done by FM. Interested readers may refer to [Karypis 1997] for more details of hMetis and [Alpert 1997] for another example of high-quality multilevel partitioning algorithm. Note that the multilevel scheme can also be applied to placement as pointed out in Section 11.4.2 and discussed in Section 11.5.4.

11.3.2 **Placement by partitioning**

11.3.2.1 *The basic idea*

Partitioning algorithms can be used to perform placement. The partitioning-based placement approach is illustrated in Figure 11.6. Given a circuit and a placement region, the placement problem is to assign each circuit module to some specific location in the placement region. The approach starts by partitioning the circuit into two subcircuits A and B, and correspondingly, dividing the placement region by a **cutline** into two subregions A and B (see Figure 11.6b). The areas of subregions A and B should be bigger than the total module areas of subcircuits A and B, respectively. Then subcircuits A and B are assigned to subregions A and B, respectively. In other words, the placement of

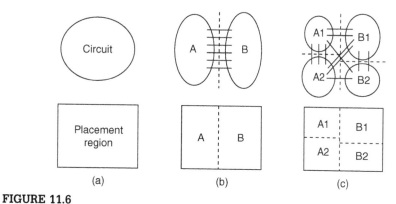

FIGURE 11.6

Partitioning-based placement approach.

each module is restricted to a smaller region. The locations of modules can be further restricted by recursively partitioning the subcircuits and dividing the subregions (as shown in Figure 11.6c). The process continues until each subcircuit contains a few modules that are assigned to a small subregion. After that, legalization is performed to pack all modules in each subcircuit into the corresponding subregion, and detailed placement is applied to further reduce the wirelength.

Note that cutlines of different directions (*i.e.*, horizontal or vertical) and locations may be used for each subregion. For example, in the division of subregion *A* in Figure 11.6c, the cutline could be vertical instead of horizontal. It could also be near the top instead of close to the middle. Different schemes of cutline selection are discussed in [Breuer 1977a, 1977b].

In partitioning-based placement, the minimization of the cut cost during partitioning can be considered to be an indirect way to minimize wirelength. The intuition is that to minimize wirelength, heavily connected modules should be placed close to one another. This can be achieved by minimizing the cut cost, because it will force heavily connected modules to be on the same side of a cut. Then, they will be placed in the same subregion. Alternately, cut cost minimization can be viewed as a way to minimize the routing congestion across the cutline.

11.3.2.2 *Terminal propagation technique*

One issue with recursive partitioning is that each subcircuit has not only nets internal to it but also nets connected to external modules. The effect of external nets should also be considered during the partitioning of a subcircuit. Dunlop and Kernighan [Dunlop 1985] developed a technique called **terminal propagation** to handle external nets. The technique is illustrated in Figure 11.7 and explained in the following.

Consider an example in Figure 11.7a in which subcircuit *A* has a net connecting module *p* to an external module *q*. To take into account the effect of

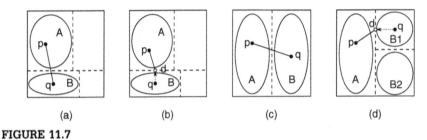

(a)   (b)   (c)   (d)

FIGURE 11.7

Terminal propagation.

this external net, assume module q is placed at the center of subregion B. The terminal in q is "propagated" to the closest point on the boundary of subregion A and is replaced by a **dummy terminal** d (as shown in Figure 11.7b). During the partitioning of A, the net between p and d is treated as an internal net and d is treated as a fixed terminal of subcircuit A. For this example, if subregion A is divided by a horizontal cutline, module p will be biased toward the lower partition because of the net between p and d. However, if subregion A is divided by a vertical cutline, d will be very near to the cutline. In this case, it is not clear whether the external net should bias p to the left or to the right partition. It is suggested in [Dunlop 1985] that dummy terminals that are within the middle third of the side should be ignored.

Another example is shown in Figure 11.7c. Assume both subregions A and B are divided by horizontal cutlines. Suppose subcircuit B is partitioned first. During the partitioning of B, the dummy terminal caused by p is too close to the cutline and hence is ignored. Suppose module q is assigned to the top subregion $B1$. Then during the partitioning of A, a dummy terminal d is introduced near to the top of subregion A (as shown in Figure 11.7d). Thus, it will bias p to the top.

11.3.3 Practical implementations

In the past, partitioning-based placement was generally perceived to be simple and efficient but not as good as simulated annealing or analytical approaches in terms of solution quality. In late-1990s, partitioning algorithms dramatically improved because of the multilevel scheme. Partitioning-based placement should also improve as a result. Capo [Caldwell 2000] and Fengshui [Yildiz 2001a, 2001b; Agnihotri 2003] are two placement algorithms that leverage this breakthrough in partitioning. They demonstrated that a careful implementation of the partitioning-based approach could produce very competitive wirelength with a relatively short runtime.

11.3.3.1 *The Capo algorithm*

In Capo, different bipartitioning techniques are applied to subcircuits of different sizes during recursive bipartitioning. For instances with more than 200 modules, a multilevel FM algorithm is used. For instances with 35 to 200 modules, the flat FM algorithm is used. Smaller instances are solved optimally with branch-and-bound.

In addition, much attention has been paid in Capo to the handling of partitioning tolerance. First, an "uncorking" technique is proposed to prevent large modules from being the first modules in a bucket of the FM algorithm. In an ordinary FM implementation, a large module at the head of a bucket may fail to move to the other partition without violating constraints on partition sizes. Smaller modules further in the same bucket will not be considered for moves, because the bucket is temporarily invalidated. This "corking" effect may degrade solution quality of FM. Second, repartitioning, which refers to a chained FM calls on the same partitioning instance, is presented. The first call is performed with a much larger tolerance than requested to ensure mobility of all modules. In subsequent calls, the tolerance gradually decreases to the original value. Third, a high tolerance of $\varepsilon = 20\%$ is used for vertical cutlines, because the cutline locations can be adjusted after partitioning according to sizes of partitions. However, this technique cannot be applied to horizontal cutlines, because their locations are more discrete (*i.e.*, aligned to standard-cell rows) and cannot be easily adjusted. Fourth, a formula is derived to determine the tolerance on the basis of the amount of white space in an instance. An instance with more white space will have a larger partitioning tolerance.

11.3.3.2 *The Fengshui algorithm*

The overall scheme of Fengshui is very similar to that of Capo. Several major differences are outlined in the following. In [Yildiz 2001a], instead of locally optimizing each subcircuit by bipartitioning, all subcircuits are partitioned simultaneously. The problem is formulated as multiway partitioning. Moreover, the partitioning cost function is HPWL rather than min-cut. In [Yildiz 2001b], a dynamic programming approach is presented to select a good sequence of cutlines. The sequence is optimal under certain assumptions. In [Agnihotri 2003], a fractional cut approach is introduced. In this approach, horizontal cutlines are not required to be aligned with standard-cell row boundaries. To handle the assignment of cells to rows that may be partially covered by a region, a dynamic programming–based legalization algorithm is developed. As fewer constraints are imposed on partitioning for horizontal cutlines, the wirelength can be reduced.

11.4 GLOBAL PLACEMENT: SIMULATED ANNEALING APPROACH

Simulated annealing is introduced in Section 4.4.4. It is an iterative heuristic for solving combinatorial optimization problems. The basic idea of simulated annealing is to search for a **configuration** with low cost by iteratively moving from the current configuration to a neighbor configuration. If the cost of the neighbor configuration is lower than that of the current configuration, the **move** will be taken. Otherwise (*i.e.*, the move causes an increase in the cost), the move may still be taken with a probability that is decreasing over time according to a **cooling schedule**. This probabilistic move helps the search procedure to get out of a local minimum.

Simulated annealing is very popular, because it is a very robust technique that can be easily applied to virtually any optimization problem. To design a simulated annealing based algorithm for a given problem, one simply needs to define the configuration space, several types of moves, the cooling schedule, and the cost function. However, simulated annealing based algorithms are usually comparatively slow, especially for large problem instances. The main challenge of designing a simulated annealing based algorithm is to make it efficient without compromising the solution quality. Two simulated annealing based placement algorithms will be described in the following. Readers may also find the discussions on simulated annealing based floorplanning algorithms in Chapter 10 useful.

11.4.1 The placement algorithm in TimberWolf

Simulated annealing based placement algorithm was popularized in the mid-1980s by the TimberWolf place-and-route package [Sechen 1986]. The Timber-Wolf standard-cell placement algorithm consists of two stages. Stage 1 allows overlaps among cells and movement of cells between rows. Stage 2 eliminates all overlaps and only performs interchange of adjacent cells. The details are described in the following.

11.4.1.1 *Stage 1*

In stage 1, a configuration is an arrangement of the cells into the standard-cell rows possibly with cell overlaps. Three moves are defined:

M1: Move a cell to a new location, which can be in a different row.
M2: Swap two cells, which can be in different rows.
M3: Mirror a cell's *x*-coordinates.

The three moves are selected randomly with unequal probability. In each step, a selection between M1 and M2 is first made, with M1 four times more likely than M2. If M1 is selected but the new configuration is rejected, then M3 will be attempted for the same cell with a probability of 1/10. The applicable range for M1 and M2 is specified by a rectangular window called range limiter. For M1, the window is centered at the center of the randomly selected cell. A random location within the window will be chosen as the destination of the cell. For M2, a swap will be attempted only if the window can be positioned such that it contains both centers of the two randomly selected cells. At the beginning of stage 1, the horizontal span and vertical span of the window are equal to twice the horizontal span and vertical span of the chip, respectively. (Therefore, if the center of the window is positioned at a corner of the chip, the window will still cover the entire chip.) During the annealing process, the horizontal span and vertical span of the window decrease slowly in proportion to the logarithm of the temperature.

The cost function has three components:

$$C = C_1 + C_2 + C_3$$

The first component, C_1, is an estimation of the total interconnect cost. For a net e, let w_e and h_e be the width and height of its bounding box, and β_e and γ_e be user-specified horizontal and vertical weights. Then,

$$C_1 = \sum_e (\beta_e \omega_e + \gamma_e h_e)$$

The second component, C_2, is an overlap penalty function. Let LinearOverlap(i, j) be the amount of overlap of cells i and j in the x-direction. Then,

$$C_2 = \sum_{i \neq j} \text{LinearOverlap}(i, j)^2$$

The third component, C_3, is a penalty function that serves to control the row lengths. For each row r, let $d(r)$ be the desired row length and $l(r)$ be the sum of the widths of the cells in row r. Then,

$$C_3 = \theta \times \sum_r |l(r) - d(r)|$$

where θ is a user-specified parameter.

11.4.1.2 *Stage 2*

When the vertical span of the range limiter window has been reduced to less than the center-to-center spacing between the rows, TimberWolf enters stage 2. At the beginning of stage 2, feed-through cells are inserted as required, and cell overlaps are eliminated by the following procedure. First, the cells in each row are sorted according to the x-coordinate of their centers. Then, they are re-placed side-by-side starting from the left edge of the row. After that, the simulated annealing continues.

In stage 2, the moves are more restrictive. M1 is not allowed. M2 considers swapping two adjacent cells only if they are in the same row. M3 is attempted only when M2 is attempted and rejected. In addition, the cost function is effectively just C_1. As there is no cell overlap, $C_2 = 0$. Because cells are not allowed to change rows, C_3 remains constant.

11.4.1.3 *Annealing schedule*

In the annealing schedule of TimberWolf, the initial temperature is 4,000,000. Then the temperature is decreased according to the following function:

$$T_{new} = \alpha(T_{old}) \times T_{old}$$

where α at different temperature is set according to a very specific table presented in [Sechen 1986]. Roughly, α starts at 0.8 when the temperature is high. Then it gradually increases as temperature decreases. It peaks at 0.94 when temperature is between 200 and 5000. After that, it steadily decreases to 0.7 as temperature drops. Finally, α is set to 0.1 when the temperature is below 1.5. The annealing process terminates when the temperature is less than 0.1. There are 117 temperature levels in this annealing schedule. At each temperature, a fixed number of moves are attempted. The number of moves per temperature is a function of the circuit size. For a 200-cell circuit, 100 moves per cell are recommended. Thus, 2.3×10^6 configurations have to be evaluated in total. For a 3000-cell circuit, 700 moves per cell are recommended. The number of configurations to be evaluated will increase dramatically to 245.7×10^6.

11.4.2 The Dragon placement algorithm

Simulated annealing based algorithms like TimberWolf can produce placement solutions of excellent quality for small circuits (with up to a few thousand cells). However, they tend to be increasingly inefficient for larger circuits. One way to improve their scalability is to perform placement in a hierarchical manner. A multilevel scheme (*i.e.*, bottom-up hierarchical scheme based on recursive clustering) is used in an improved version of TimberWolf [Sun 1995]. A top-down hierarchical scheme based on recursive partitioning is applied in the Dragon placement algorithm [Wang 2000]. The Dragon algorithm will be discussed in the following.

In fact, Dragon takes a hybrid approach that combines simulated annealing and partitioning. A hierarchy as shown in Figure 11.8 is formed by recursive quadrisectioning (*i.e.*, 4-way partitioning) by use of hMetis. At level b, the original circuit is partitioned into 4^b subcircuits. Correspondingly, the placement region is divided into a regular array of 4^b bins. Each subcircuit is assigned to one bin. Then low-temperature simulated annealing is applied to minimize the wirelength by swapping subcircuits among the bins. The recursive quadrisectioning terminates when a bin contains less than approximately 7 cells. Then low-temperature simulated annealing is again used to further reduce

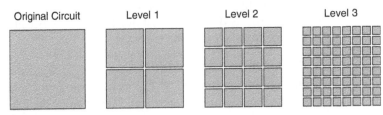

FIGURE 11.8

The top-down hierarchy used in Dragon.

wirelength by relocating a single cell to a different bin in each move. Finally, detailed placement is done by a greedy algorithm.

Compared with a flat annealing–based placement approach, annealing at high levels is swapping subcircuits among the bins. In other words, it is moving a large number of cells by a long distance in each move. It can be more efficient than swapping individual cells locally. Besides, quadrisectioning together with refinement at higher levels provides a good starting solution to simulated annealing and hence can shorten the annealing process significantly. Compared with a pure partitioning-based placement approach, the annealing-based swapping can correct wrong decisions made by quadrisectioning at higher levels. In addition, the swapping is based directly on wirelength rather than cut cost. Hence, it can generate higher-quality solutions than the partitioning-based approach alone.

11.5 GLOBAL PLACEMENT: ANALYTICAL APPROACH

The basic idea of the analytical approach is to express the cost function and the constraints as analytical functions of the coordinates of the modules. Then the placement problem is transformed into a mathematical program. To illustrate this approach, an exact, but impractical, formulation is first presented in Section 11.5.1. Practical analytical placement techniques can be classified as **quadratic** and **nonquadratic**, which are presented in Sections 11.5.2 and 11.5.3, respectively. Note that analytical placement techniques generally treat both standard cells and macros in the same manner. (Sometimes, special techniques are used to handle large macros to improve wirelength or to make legalization easier, but in theory, they are not essential.) Thus, the techniques presented in this section can be considered to be for mixed-size designs.

Some notations used in this section are introduced in the following. Consider a circuit with a set of modules V and a set of nets E is to be placed in a region of width W and height H. Assume the modules in V are indexed from 1 to n. For module $i \in V$, let w_i and h_i be its width and height, and let x_i and y_i be the x- and y-coordinates of its center. For each net $e \in E$, let c_e be its weight. Note that if a module i is a fixed module, then x_i and y_i are constants. Otherwise, they are variables. Assume for simplicity that all pins are located at the center of a module.

11.5.1 An exact formulation

It is instructive to first look at an exact formulation to the wirelength-minimized placement problem for mixed-size design. Then the rationales, pros, and cons of practical analytical placement techniques presented in Sections 11.5.2 and 11.5.3 can be better understood.

The HPWL of net $e \in E$ can be written as:

$$\text{HPWL}_e(x_1, \ldots, x_n, y_1, \ldots, y_n) = \left(\max_{i \in e}\{x_i\} - \min_{i \in e}\{x_i\}\right) + \left(\max_{i \in e}\{y_i\} - \min_{i \in e}\{y_i\}\right)$$

To express the nonoverlapping constraints among modules, it is necessary to introduce the following function:

$$\Theta([L1, R1], [L2, R2]) = [\min(R1, R2) - \max(L1, L2)]^+$$

where

$$[z]^+ = \begin{cases} z & \text{if } z > 0 \\ 0 & \text{if } z \leq 0 \end{cases}$$

$\Theta([L1, R1], [L2, R2])$ gives the length of the overlapping region of the intervals $[L1, R1]$ and $[L2, R2]$ as illustrated in Figure 11.9. Then the overlapping area between module i and module j is given by:

$$\text{Overlap}_{ij}(x_i, y_i, x_j, y_j) = \Theta\left(\left[x_i - \frac{\omega_i}{2}, x_i + \frac{\omega_i}{2}\right], \left[x_j - \frac{\omega_j}{2}, x_j + \frac{\omega_j}{2}\right]\right)$$

$$\Theta\left(\left[y_i - \frac{b_i}{2}, y_i + \frac{b_i}{2}\right], \left[y_j - \frac{b_j}{2}, y_j + \frac{b_j}{2}\right]\right)$$

The placement problem can be written as the following mathematical program:

Minimize $\quad \sum_{e \in E} c_e \times \text{HPWL}_e(x_1, \ldots, x_n, y_1, \ldots, y_n)$

Subject to $\quad \text{Overlap}_{ij}(x_i, y_i, x_j, y_j) = 0$ for all $i, j \in V$ s.t. $i \neq j$

$$0 \leq x_i - \frac{\omega_i}{2}, \; x_i + \frac{\omega_i}{2} \leq W \text{ for all } i \in V$$

$$0 \leq y_i - \frac{b_i}{2}, \; y_i + \frac{b_i}{2} \leq H \text{ for all } i \in V$$

This mathematical program is extremely difficult to handle. The functions $\text{Overlap}_{ij}(x_i, y_i, x_j, y_j)$ for $i, j \in V$ are highly nonconvex and not differentiable. The functions $\text{HPWL}_e(x_1, \ldots, x_n, y_1, \ldots, y_n)$ for $e \in E$, although convex, are

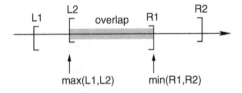

FIGURE 11.9

The overlapping region of the intervals $[L1, R1]$ and $[L2, R2]$.

not differentiable. Moreover, there are $O(n^2)$ constraints with the number of modules n being up to millions in modern designs. Therefore, this is not a practical formulation. In practice, the wirelength is approximated by some differentiable and convex functions. The nonoverlapping constraints are usually replaced by some simpler constraints to make the module distribution roughly even. Then legalization is performed to eliminate the module overlaps, and detailed placement is applied to refine the solution with a more accurate wirelength metric. Various techniques are presented in the remainder of this section.

11.5.2 Quadratic techniques

In quadratic placement techniques, the placement problem is transformed into a sequence of **convex quadratic programs**. Convex quadratic program is a mathematical program with a convex and quadratic objective function and linear constraints.

11.5.2.1 *Quadratic wirelength*

First, the way to express the placement cost function as a quadratic function is presented. Suppose for the time being that all nets in the circuit are 2-pin nets. (Multi-pin nets will be discussed later.) Consider a net $\{i, j\}$ (*i.e.*, connecting module i and module j). Its wirelength is given by the Manhattan distance between the modules:

$$L_{\{i, j\}} = |x_i - x_j| + |y_i - y_j|$$

This is usually referred to as linear wirelength. However, this function is not differentiable. So a common idea is to consider the squared Euclidean distance between the modules instead:

$$\tilde{L}_{\{i, j\}} = (x_i - x_j)^2 + (y_i - y_j)^2$$

This is usually called **quadratic wirelength**. To help visualize the functions, the x-components of $L_{\{i,j\}}$ and $\tilde{L}_{\{i,j\}}$ with a fixed value of $x_j = 2$ are plotted as functions of x_i in Figure 11.10.

In quadratic placement techniques, it is more convenient to set the cost function $\tilde{L}$ to be half[2] of the total weighted quadratic wirelength:

$$\tilde{L} = \frac{1}{2} \sum_{1 \leq i < j \leq n} c_{\{i, j\}} \times \left((x_i - x_j)^2 + (y_i - y_j)^2 \right)$$

---

[2]Half of the total weighted quadratic wirelength is used so that the derivatives will have simpler forms.

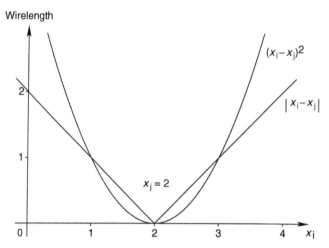

FIGURE 11.10

Comparison of quadratic wirelength and linear wirelength.

$\tilde{L}$ can be written succinctly in matrix form in terms of the coordinates of movable modules. In the following, assume modules 1 to r are movable and modules $r + 1$ to n are fixed. Therefore, $x_1, \ldots, x_r$ and $y_1, \ldots, y_r$ are variables, and $x_{r+1}, \ldots, x_n$ and $y_{r+1}, \ldots, y_n$ are constants. Let $x = (x_1 x_2 \cdots x_r)^T$ and $y = (y_1 y_2 \cdots y_r)^T$ be the vectors of x-coordinate and y-coordinate of movable modules, respectively. Let $D = (d_{ij})_{r \times r}$ be a diagonal matrix such that $d_{ii} = \sum_{j \in V} c_{\{i,j\}}$ for all $i \in \{1, \ldots, r\}$. Let $C = (c_{ij})_{r \times r}$ be the connectivity matrix among movable modules, *i.e.*, $c_{ij} = c_{ji} = c_{\{i,j\}}$ for all $i, j \in \{1, \ldots, r\}$. Let $Q = D - C$. Let $d_x = (d_{x_1} \cdots d_{x_r})^T$ such that $d_{x_i} = -\sum_j \in \{r+1, \ldots, n\} c_{ij} x_j$ and $d_y = (d_{y_1} \cdots d_{y_r})^T$ such that $d_{y_i} = -\sum_j \in \{r+1, \ldots, n\} c_{ij} y_j$. Then

$$\left[\tilde{L} = \frac{1}{2} x^T Q x + d_x^T x + \frac{1}{2} y^T Q y + d_y^T y + \text{constant terms} \right]$$

For example, for the circuit with 3 movable modules and 3 fixed modules as represented by the graph in Figure 11.11,

$$\tilde{L} = \frac{1}{2} (c_{12}((x_1 - x_2)^2 + (y_1 - y_2)^2)$$

$$+ (c_{13}((x_1 - x_3)^2 + (y_1 - y_3)^2)$$
$$+ (c_{14}((x_1 - x_4)^2 + (y_1 - y_4)^2)$$
$$+ (c_{24}((x_2 - x_4)^2 + (y_2 - y_4)^2)$$
$$+ (c_{25}((x_2 - x_5)^2 + (y_2 - y_5)^2)$$
$$+ (c_{36}((x_3 - x_6)^2 + (y_3 - y_6)^2)$$

$$= \frac{1}{2} x^T Q x + d_x^T x + \frac{1}{2} y^T Q y + d_y^T y$$

$$+ \frac{1}{2} ((c_{14} + c_{24}) x_4^2 + c_{25} x_5^2 + c_{36} x_6^2 + (c_{14} + c_{24}) y_4^2 + c_{25} y_5^2 + c_{36} y_6^2)$$

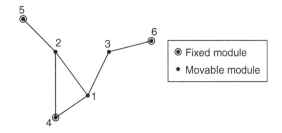

FIGURE 11.11

Connections of a circuit.

where

$$x = \begin{pmatrix} x_1 \\ x_2 \\ x_3 \end{pmatrix}, \ y = \begin{pmatrix} y_1 \\ y_2 \\ y_3 \end{pmatrix},$$

$$Q = \begin{pmatrix} c_{12} + c_{13} + c_{14} & 0 & 0 \\ 0 & c_{21} + c_{24} + c_{25} & 0 \\ 0 & 0 & c_{31} + c_{36} \end{pmatrix} - \begin{pmatrix} 0 & c_{12} & c_{13} \\ c_{21} & 0 & c_{23} \\ c_{31} & c_{32} & 0 \end{pmatrix}$$

$$d_x = \begin{pmatrix} -c_{14}x_4 \\ -c_{24}x_4 - c_{25}x_5 \\ -c_{36}x_6 \end{pmatrix} \text{ and } d_y = \begin{pmatrix} -c_{14}y_4 \\ -c_{24}y_4 - c_{25}y_5 \\ -c_{36}y_6 \end{pmatrix}$$

It is clear that $\tilde{L}_{\{i,j\}}$ for all $\{i, j\} \in E$ are convex and continuously differentiable functions. Hence, $\tilde{L}$, which is a weighted sum of $\tilde{L}_{\{i,j\}}$'s, is also convex and differentiable. So $\tilde{L}$ should be easy to minimize. In particular,

$$\frac{\partial \tilde{L}}{\partial x} = Qx + d_x \text{ and } \frac{\partial \tilde{L}}{\partial y} = Qy + d_y$$

Therefore, the placement with minimum wirelength is given by

$$Qx + d_x = 0 \text{ and } Qy + d_y = 0 \tag{11.1}$$

In other words, if nonoverlapping constraints are ignored, the quadratic placement problem is equivalent to solving a system of linear equations. If all movable modules are connected to fixed modules either directly or indirectly, Q is positive definite and thus invertible. This implies the existence of a unique global optimal solution. The simplicity of quadratic formulation is the main reason for its popularity. Note that x and y can be solved independently. For brevity's sake, sometimes only the x-component will be discussed from now on.

11.5.2.2 *Force interpretation of quadratic wirelength*

The problem of quadratic wirelength minimization can also be interpreted as a classical mechanics problem of finding the equilibrium configuration for a system of objects attached to zero-length springs. Consider each circuit module as an object and each 2-pin net $\{i, j\}$ as a stretched spring with spring constant $c_{\{i, j\}}$ connecting object i and object j. For the circuit represented by Figure 11.11, the corresponding spring system is shown in Figure 11.12.

The potential energy stored in spring $\{i, j\}$ is:

$$\varepsilon_{\{i, j\}} = \frac{1}{2} \times c_{\{i, j\}} \times (\text{Length of spring}\{i, \ j\})^2$$

$$= \frac{1}{2} \times c_{\{i, j\}} \times \left((x_i - x_j)^2 + (y_i - j_j)^2 \right)$$

Hence, the total potential energy of the spring system is equal to $\tilde{L}$. In other words, finding the minimum energy configuration for the spring system is equivalent to minimizing the quadratic wirelength in the quadratic placement formulation.

For a spring system, the minimum energy configuration is also the same as the force-equilibrium configuration. Note that the gradient of the total potential energy to the coordinates of an object gives the total force acting on the object. Therefore, the entries in the vectors $Qx + d_x$ and $Qy + d_y$ are the *x*-components and *y*-components of the total forces acting on the objects. In other words, another interpretation of the optimal placement conditions in Equation (11.1) is that all objects are in force equilibrium. For a nonequilibrium system (*i.e.*, a circuit placement with suboptimal quadratic wirelength), the total force on an object provides the best way of movement for the object to minimize the total energy (*i.e.*, the total weighted quadratic wirelength). Extra forces can also be added to influence the placement in a desirable manner (*e.g.*, to spread out the objects). Many quadratic placement algorithms use the guidance provided by springs/extra forces to optimize a placement solution (*e.g.*, [Quinn 1975]). Those algorithms are often called **force-directed placement algorithms**.

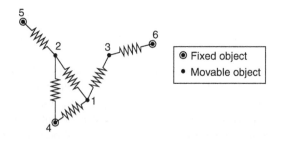

FIGURE 11.12

The spring system corresponding to the example in Figure 11.11.

The forces exerted by the springs are given by Hooke's law. The force exerted on object i by a spring connecting objects i and j (as illustrated in Figure 11.13) is:

$$\boldsymbol{F}_{ij} = c_{\{i,j\}} \times \text{Displacement from object } i \text{ to object } j$$

Its magnitude is:

$$\left| \boldsymbol{F}_{ij} \right| = c_{\{i,j\}} \times \sqrt{\left(x_i - x_j\right)^2 + \left(y_i - y_j\right)^2}$$

To find the total force exerted by several springs on an object, it is often more convenient to decompose the force by each spring into x- and y-components:

$$\left| \boldsymbol{F}_{ij}^x \right| = c_{\{i,j\}} \times \left| x_j - x_i \right| \text{ and}$$
$$\left| \boldsymbol{F}_{ij}^y \right| = c_{\{i,j\}} \times \left| y_j - y_i \right|$$

Then the x-component and y-component of the total force are the sum of the x-component and y-component of the forces by all springs, respectively.

11.5.2.3 *Net models for multi-pin nets*

Circuits typically contain both 2-pin nets and multi-pin nets. To place circuits with multi-pin nets by quadratic techniques, various models have been proposed to replace each net by a group of 2-pin nets.

The traditional model is to replace each net by a **clique** (*i.e.*, complete graph). For example, for the 5-pin net in Figure 11.14a, the clique model is shown in Figure 11.14b. The net weights of the 2-pin nets in the clique model should be set properly to balance the minimization of 2-pin nets and multi-pin nets. For a k-pin net with net weight c, the weight of each 2-pin net in the clique is usually set to either $c/(k-1)$ [Vygen 1997] or $2c/k$ [Kleinhans 1991; Eisenmann 1998].

Another model is the **star model** [Vygen 1997; Mo 2000] as illustrated in Figure 11.14c. In the star model, one extra dimensionless module called the star module is introduced for each net. The star module is placed together with other movable modules during placement. Therefore, two extra variables corresponding to the x- and y-coordinates of the star module are added to the placement problem.

It is proved in [Viswanathan 2004] that the clique model and the star model are equivalent in quadratic placement if the net weights are set properly.

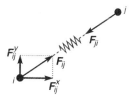

FIGURE 11.13

The forces by a stretched spring connecting objects i and j.

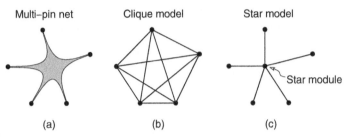

Multi–pin net Clique model Star model

Star module

(a) (b) (c)

FIGURE 11.14

Clique and star models for multi-pin nets.

Specifically, consider a k-pin net connecting modules $1, \ldots, k$. In the clique model, if the weight of each 2-pin net is set to c, then the total force on module i by the 2-pin nets in the clique is given by:

$$F_i^{clique} = c \times \sum_{j=1}^{k} (x_j - x_i)$$

In the star model, let x_s be the x-coordinate of the star module. If the weight of each 2-pin net is set to $k \times c$, then the total force on the star node is given by:

$$F_s = \sum_{j=1}^{k} k \times c \times (x_j - x_s)$$

By setting $F_s = 0$, the force-equilibrium position for the star node is:

$$x_s = \frac{1}{k} \sum_{j=1}^{k} x_j$$

The force on module i by the 2-pin net connecting to the star is given by:

$$
\begin{aligned}
F_i^{star} &= k \times c \times (x_s - x_i) \\
&= k \times c \times \left(\frac{1}{k} \sum_{j=1}^{k} x_j - x_i \right) \\
&= c \times \left(\sum_{j=1}^{k} x_j - k \times x_i \right) \\
&= c \times \sum_{j=1}^{k} (x_j - x_i) \\
&= F_i^{clique}
\end{aligned}
$$

Because the forces exerted are the same, the clique and the star models are equivalent, and they can be used interchangeably in quadratic placement.

On the basis of the equivalence of clique and star models, the **hybrid net model** [Viswanathan 2004] is a natural choice. In the hybrid net model, the clique model is used for nets with 2 to 3 pins, and the star model is used for nets with 4 or more pins. It has been shown empirically that the hybrid net model reduces the number of 2-pin nets by more than $10\times$ over the clique model for industrial circuits [Viswanathan 2007a]. It can also significantly reduce both the number of 2-pin nets and number of variables over the star model, because approximately 70% of nets in a typical circuit have 2 or 3 pins. Because the runtime to solve a quadratic placement problem is roughly proportional to the number of 2-pin nets and it increases slightly with the number of variables, the hybrid net model can speed up quadratic placement significantly.

11.5.2.4 *Linearization methods*

As shown in Figure 11.10, quadratic wirelength is a very rough approximation to linear wirelength. For small circuits, despite the inaccuracy of quadratic wirelength, quadratic placement techniques can still generate very competitive solutions. For larger circuits, however, this inaccuracy is a major bottleneck to the quality of quadratic placement solutions.

The authors in [Sigl 1991] presented a method to approximate the linear wirelength in a quadratic placement framework by iteratively adjusting the spring constant. Assume the star model is used to replace multi-pin nets. For a net e in the original circuit, let x_e be the coordinate of the associated star module. Then the total linear wirelength (for the circuit after applying the star model) can be written as:

$$L^{star} = \sum_{e \in E} \sum_{i \in e} |x_i - x_e|$$

Consider the function:

$$\tilde{L}^{star} = \sum_{e \in E} \sum_{i \in e} \frac{(x_i - x_e)^2}{g_{ie}}$$

If $g_{ie} = |x_i - x_e|$ then $\tilde{L}^{star} = L^{star}$. However, g_{ie}'s are set to constants so that $\tilde{L}^{star}$ would become a quadratic function. To approximate L^{star} with $\tilde{L}^{star}$, the function $\tilde{L}^{star}$ is optimized iteratively such that g_{ie} in current iteration is set according to the coordinates of previous iteration. Intuitively, $1/g_{ie}$ can be viewed as a variable spring constant that decreases with increasing spring length. The iterative process terminates when the g_{ie} factors no longer change significantly.

Notice that even L^{star} is just a rough approximation to the total HPWL objective function. Thus instead of setting g_{ie} to $|x_i - x_e|$, an experimentally-verified net specific factor is used:

$$g_{ie} = \sum_{i \in e} |x_i - x_e| \text{ for all } i \in e$$

662 | **CHAPTER 11** Placement

This choice has two advantages. First, the summation reduces the influence of nets with many connected modules and emphasizes most nets connecting only two or three modules. Second, the summation also prevents increasing the force on modules close to the star node by too much. According to HPWL, as long as a module is inside the net bounding box, it is not helpful to increase the force to pull it farther inside.

Nets becoming very short (*i.e.*, g_{ie} becoming very small) may cause numerical problems during the minimization of $\tilde{L}^{star}$. Therefore, g_{ie} is lower bounded (by the average module width for example) to ensure that g_{ie} will never be zero.

Spindler and Johannes [Spindler 2006] introduced a **BoundingBox net model**, which, when combined with the preceding wirelength linearization idea, can accurately model HPWL in a quadratic placement framework. In the BoundingBox net model, a multi-pin net is transformed into only a few characteristic 2-pin nets as illustrated in Figure 11.15a. It is different from the clique model in which all possible 2-pin nets are included as shown in Figure 11.15b. Consider a *k*-pin net. For a given placement, suppose the modules are indexed in ascending order of their *x*-coordinates. Therefore, module 1 is at the left boundary and module *k* is at the right boundary of the net's bounding box. All connections are joined to the two boundary modules. One 2-pin net is connecting modules 1 and *k*. Two 2-pin nets are connecting each of the remaining $k - 2$ inner modules to module 1 and module *k*, respectively. The total number of 2-pin nets is $1 + 2(k - 2)$. Let $N = \{\{1, k\}, \{1, 2\}, \{2, k\}, \{1, 3\}, \{3, k\}, \ldots, \{1, k - 1\}, \{k - 1, k\}\}$ be the set of 2-pin nets.

According to the BoundingBox net model, the wirelength $\tilde{L}^{BB}$ of the k-pin net is defined as:

$$\tilde{L}^{BB} = \frac{1}{2} \sum_{\{i, j\} \in N} \omega_{\{i, j\}} \times \left(x_i - x_j\right)^2$$

where

$$\omega_{\{i, j\}} = \frac{2}{k - 1} \times \frac{1}{l_{\{i, j\}}}$$

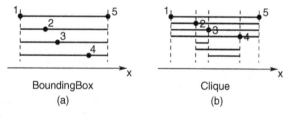

BoundingBox
(a)

Clique
(b)

FIGURE 11.15

The BoundingBox and the clique net models for a 5-pin net.

If $l_{\{i,\,j\}}$ is set to $|x_i - x_j|$ for all $\{i, j\} \in N$, then:

$$
\begin{aligned}
\tilde{L}^{BB} &= \frac{1}{2}\left(\sum_{\{i,\,j\}\in N}\frac{2}{k-1}\times\frac{1}{|x_i - x_j|}\times(x_i - x_j)^2\right)\\[4pt]
&= \frac{1}{k-1}\left(\sum_{\{i,\,j\}\in N}|x_i - x_j|\right)\\[4pt]
&= \frac{1}{k-1}\left(|x_1 - x_k| + \sum_{2\leq i\leq k-1}(|x_1 - x_i| + |x_i - x_k|)\right)\\[4pt]
&= \frac{1}{k-1}(|x_1 - x_k| + (k-2)\times|x_1 - x_k|)\\[4pt]
&= |x_1 - x_k|
\end{aligned}
$$

Thus, the BoundingBox net model gives the exact HPWL if the net weights are set appropriately.

Like the linearization method in [Sigl 1991], the correct net weights $w_{\{i,\,j\}}$ are searched by iteratively optimizing $\tilde{L}^{BB}$ such that $l_{\{i,j\}}$ in current iteration is set to $|x_i - x_j|$ of previous iteration. In each iteration, the $l_{\{i,\,j\}}$ factors are constants and hence $\tilde{L}^{BB}$ becomes a quadratic function of module coordinates.

For simplicity, only the x-component of the model is described previously. The y-component can be constructed similarly. However, because the boundary modules and module distances may be different in x- and y-directions, the set of 2-pin nets introduced and the $l_{\{i,\,j\}}$ factors are most likely different. Moreover, even for a given direction, the boundary modules may change from iteration to iteration. Hence, the set of 2-pin nets and the net weights have to be updated continually. The overhead associated with maintaining two copies of connectivity matrices and the need to search for the net weights by an iterative process are the main disadvantages of the BoundingBox model over the clique, star, and hybrid models.

The BoundingBox net model has several advantages. First, it allows quadratic techniques to perform placement with the HPWL metric. Second, unlike the clique or star models, no connection is introduced among the inner modules to pull them together.[3] The inner modules are able to move more freely as in the HPWL model. Third, this model introduces much fewer 2-pin nets than the clique model. It does introduce more 2-pin nets than the star/hybrid model for nets with 4 or more pins, but the difference is not very significant.

Another way to mitigate the inaccuracy of quadratic wirelength is to correct the mistakes by refining the placement solution with some linear metrics. Detailed placement can be viewed as one example of this approach. In detailed placement, as a simple problem of locally rearranging a few modules is considered, an accurate wirelength model (*e.g.*, HPWL or even RSMT wirelength) can usually be applied. However, because corrections are restricted by the local

---

[3]In the star model, the inner modules are pulled together indirectly through the star module.

nature and the legality requirement of module movements, the effectiveness of detailed placement in optimizing the linear cost function is limited.

A better technique called *iterative local refinement* (ILR) is proposed in FastPlace [Viswanathan 2004]. ILR can be applied to any global placement solution before legalization. It works in iterations. In each iteration, the placement region is divided into bins by a regular grid structure. The bin size is large at the beginning iterations and is gradually reduced to consider progressively finer module movements. After binning, the modules are examined one by one. For each module, it is tentatively moved from its original bin to its eight adjacent bins as shown in Figure 11.16. For each tentative move, one score is computed. The score is a weighted sum of a wirelength component and a density component. The wirelength component is the total change in HPWL of all nets connected to the module. The density component is a function of the module densities of the original bin and the target bin. It rewards movements from a dense bin to a sparse bin. If all eight scores are negative, the module will remain unmoved. Otherwise, the move with the highest score will be taken. This iterative process is repeated until there is no significant improvement in wirelength.

Because ILR is not constrained by the nonoverlapping requirement and can move modules by a relatively long distance, it is much more effective than detailed placement in correcting major problems in the placement solution. It also helps the spreading of modules. Besides, it is an extremely fast technique because of its simplicity.

11.5.2.5 *Handling nonoverlapping constraints*

In placement, the two primary goals are to minimize the wirelength and to avoid module overlaps. These two goals are in conflict with each other. Wirelength minimization brings modules together. Overlap avoidance requires modules to spread out. Note that if the nonoverlapping constraints are ignored, for a circuit

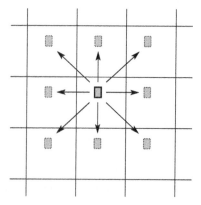

FIGURE 11.16

Eight tentative moves of iterative local refinement.

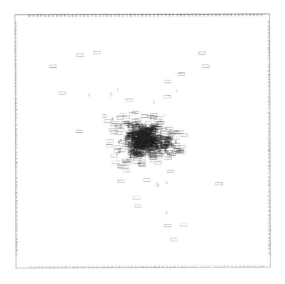

FIGURE 11.17

The placement solution for a circuit with fixed I/O pins at boundary when quadratic wirelength is minimized and nonoverlapping constraints are ignored.

without fixed modules, the optimal solution is to place all modules at the same location. This solution has zero wirelength but is meaningless because of the serious overlap issue. Even for a circuit with fixed modules (*e.g.*, I/O pins at boundary), which help pulling the movable modules away from each other, the wirelength-minimized placement without considering overlaps typically has a lot of overlaps at the center of the placement region as illustrated in Figure 11.17.

As pointed out in Section 11.5.1, instead of completely eliminating module overlaps in an analytical framework, an even distribution of modules is targeted in practice. In quadratic placement, there are basically two ways to make the module distribution more even. The first way is to add center-of-mass constraints to prevent modules from clustering together. The second way is to add forces to pull modules from dense regions to sparse regions. For both ways, the constraints/forces are added in an iterative manner to gradually spread out the modules. Note that even with the additional constraints/forces, the quadratic wirelength minimization problem can still be formulated as a convex quadratic program. Therefore, the circuit placement problem is transformed into a sequence of convex quadratic programs.

The technique of adding center-of-mass constraints is first introduced by GORDIAN [Kleinhans 1991]. Similar techniques are also used in BonnPlace [Brenner 2005] and hATP [Nam 2006]. The algorithm of GORDIAN is presented in the following. In GORDIAN, given an uneven quadratic placement solution, the module distribution can be improved by the following procedure. Assume the modules have to be spread horizontally. First, a vertical cutline is used to

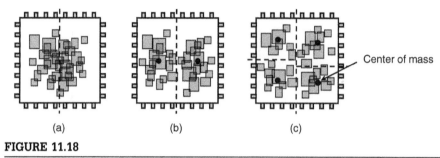

(a)   (b)   (c)

FIGURE 11.18

Module spreading by center-of-mass constraints in GORDIAN.

partition all modules into two subcircuits and the placement region into two subregions (see Figure 11.18a). Then, for each subcircuit, a constraint in the *x*-direction is added to force the center of mass of all its modules to be at the center of the corresponding region. Next, the placement problem with the two additional constraints is solved again. The center-of-mass constraints should pull the two subcircuits horizontally away from each other as shown in Figure 11.18b. This procedure is applied hierarchically to improve the distribution in each subregion (see Figure 11.18c) until each subcircuit contains less than a predefined number of modules. Note that at each hierarchical level, the placement of all subcircuits is considered together as a single global optimization problem.

The coordinates of the center of mass are the area weighted mean values (*i.e.*, linear functions) of the module coordinates. In other words, the center-of-mass constraints are linear equality constraints. Therefore, the global optimization problem at each hierarchical level is a convex quadratic program, which is equivalent to solving a system of linear equations.

Although the center-of-mass constraints help spreading, they hurt wirelength. For any two subcircuits belonging to the same parent in the hierarchy, the center-of-mass constraints draw them apart by a long distance. Hence the connections between them will become much longer. The wirelength impact can be minimized if the cut cost between the two subcircuits can be reduced. To avoid a large cut cost, both direction and position of the cutline are chosen carefully. To determine the cut cost of every possible partition by a vertical cutline, the cutline is scanning from left to right and the cut cost is updated whenever the cutline passes over a module. Only the partitions in which both subcircuits are at least 35% of the area of their parent are considered. The cut costs for horizontal cutline can be found similarly. The cutline with the smallest cut cost among all directions and positions is chosen. After that, the FM bipartitioning algorithm is optionally applied to refine the partition by moving modules that are close to the cutline. Moreover, after global optimization, if there are a lot of overlaps between two subcircuits, it indicates a bad bipartition, because many modules tend to migrate to the other region. In that case, they are repartitioned.

The technique of adding forces to spread modules in a quadratic placement framework was first introduced in Kraftwerk [Eisenmann 1998]. In Kraftwerk, density-based forces are derived to pull modules from high-density regions to low-density regions. However, constant forces are used, and the magnitude of the forces is set heuristically. As a result, the convergence is hard to control and the algorithm is not as fast as it should be. In the following, the improved technique in the new version of Kraftwerk [Spindler 2006] is presented. Note that the x-coordinates will be focused in the discussion following.

For a given placement, let x' be the vector of current module positions and x be the vector of variables representing module positions in a new placement to be determined. The additional force for each module is separated into two components: **hold force** and **move force**. The hold force vector F_x^{hold} is defined as:

$$F_x^{hold} = -(Qx' + d_x)$$

It is used to counterbalance the total forces by the nets of the circuit in the current placement x'. It makes sure that if the placement problem is solved again, all modules will be held in their current positions. Hold forces do not depend on x and hence are constant forces.

The move force is used to move a module toward less dense regions. Let $D(x, y)$ be the module density at location (x, y). It is defined as the number of modules that cover (x, y) minus the average density $(\sum_i w_i \times h_i)/(W \times H)$. The distribution $D(x, y)$ can be viewed as a charge distribution, which creates an electrostatic potential ϕ based on the Poisson equation:

$$\Delta \phi = -D(x, y)$$

The Poisson equation can be solved efficiently by geometric multigrid solvers (*e.g.*, [Kowarschik 2001]).

In the electrostatic formulation, the potential ϕ is high in regions where the distribution $D(x, y)$ is high, and *vice versa*. Hence the gradient of the potential $(\partial \phi/\partial x, \partial \phi/\partial y)^T$ can be used to move the modules away from high-density regions toward low-density regions and thereby reduce the overlaps among the modules. For each module i, its target position $\hat{x}_i$ is:

$$\hat{x}_i = x'_i - \left| \frac{\partial \Phi}{\partial x} \right|_{(x'_i, y'_i)}$$

Move forces are added to guide modules toward their target positions. They are generated by use of the **fixed-point** idea proposed in FAR [Hu 2002]. Each module i is connected to its target position (*i.e.*, a fixed point) by a spring with spring constant $\hat{c}_i$. So the move force vector F_x^{move} is given by:

$$F_x^{move} = \hat{Q}(x - \hat{x})$$

where $\hat{Q} = \text{diag}(\hat{c}_i)$. Note that spring forces rather than constant forces are used for move forces so that module movements are limited. Each module can be moved at most up to its target position. This helps the convergence of Kraft-werk significantly. The spring constants $\hat{c}_i$ control the tradeoff between rate of convergence and wirelength. If large $\hat{c}_i$ values are used, modules will be moved close to their target positions. Hence the placer will converge faster to an even density distribution. On the other hand, small $\hat{c}_i$ values allow module positions to be determined mostly by wirelength minimization.

In the new placement solution, the sum of net force, hold force, and move force on each module should be zero:

$$(\boldsymbol{Q}\boldsymbol{x} + \boldsymbol{d}_x) - (\boldsymbol{Q}\boldsymbol{x'} + \boldsymbol{d}_x) + \hat{\boldsymbol{Q}}(\boldsymbol{x} - \hat{\boldsymbol{x}}) = 0$$

Therefore, the new module positions $\boldsymbol{x}$ can be found by solving the following system of linear equations:

$$(\boldsymbol{Q} + \hat{\boldsymbol{Q}})(\boldsymbol{x} - \boldsymbol{x'}) = -\hat{\boldsymbol{Q}}(\boldsymbol{x'} - \hat{\boldsymbol{x}})$$

This spreading procedure is repeated until the placement density distribution is even enough. It can be proved that this procedure always converges to an overlap-free placement.

Note that besides this potential-based method, the target positions can also be computed by simpler heuristics like **cell shifting** [Viswanathan 2004] or **grid warping** [Xiu 2004]. All these methods try to equalize the placement density of nearby regions by locally moving modules. There is no proof of convergence for these methods, but they work well in practice.

To mitigate the negative effect of the additional forces on wirelength in force-directed spreading algorithms, a **force-vector modulation** technique is proposed in RQL [Viswanathan 2007b]. The technique is based on the observation that the spreading force is small for most modules but very huge for a few percent of all modules. A huge force implies that the corresponding module is pulled away from its natural position (*i.e.*, the force-equilibrium position if spreading force is removed) by a long distance. Thus, the nets connecting to the module become very long. The force-vector modulation technique nullifies the huge forces before the next quadratic optimization iteration. As a result, modules with nullified spreading forces can return to the minimum-wirelength positions. Hence, the total wirelength can be significantly improved. Because the spreading forces of only a few percent of modules are nullified, module spreading is not seriously affected.

11.5.3 **Nonquadratic techniques**

Another category of analytical approach is to formulate the placement problem as a single nonlinear program as in Section 11.5.1. However, instead of exact wirelength metric and exact nonoverlapping constraints, approximations are used. In particular, the placement region is divided into bins, and the nonoverlapping constraints are replaced by bin density constraints. Let $\boldsymbol{x}$ and $\boldsymbol{y}$ be the vectors of x- and y-coordinates of the modules, respectively. Then the problem can be formulated as follows:

$$\text{Minimize} \sum_{e \in E} c_e \times \text{WL}_e(\boldsymbol{x}, \boldsymbol{y})$$

$$\text{Subject to } D_b(\boldsymbol{x}, \boldsymbol{y}) = T_b \text{ for all bin } b$$

$\text{WL}_e()$ is a continuously differentiable function that may be more complicated than quadratic functions but may also be more accurate in approximating HPWL. T_b is the target density of bin b. $D_b()$ gives the density of bin b with respect to placement solution $\boldsymbol{x}$ and $\boldsymbol{y}$. The exact bin density function is a piece-wise linear function and hence is not differentiable. $D_b()$ is a smooth version of the exact one.

Examples of placers in this category are APlace [Kahng 2004], mPL [Chan 2005], and NTUPlace [Chen 2006]. To approximate the wirelength, APlace, mPL, and NTUPlace all use the **log-sum-exponential wirelength function** described in a patent by [Naylor 2001]. To smooth the density function, APlace and NTUPlace use a **bell-shaped function** proposed also by [Naylor 2001], and mPL uses **inverse Laplace transformation** [Evans 2002]. The wirelength approximation and density smoothing methods are described in the following.

11.5.3.1 *Log-sum-exponential wirelength function*

The log-sum-exponential function is defined as:

$$\text{LSE}_\alpha(z_1, \ldots, z_n) = \alpha \times \left(\log \left(\sum_{i=1}^n e^{z_i/\alpha} \right) \right)$$

It is an approximation of the maximum function:

$$\text{LSE}_\alpha(z_1, \ldots, z_n) \approx \max(z_1, \ldots, z_n)$$

α is a parameter controlling the accuracy of the approximation. As α converges to 0, the log-sum-exponential function converges to the maximum function. This is demonstrated in Figure 11.19, which shows for $\alpha = 0.1$ and for $\alpha = 2$, the log-sum-exponential function of two arguments.

The HPWL of a net e can be expressed in terms of the maximum function:

$$
\begin{aligned}
\text{HPWL}_e &(x_1, \ldots, x_n, \ y_1, \ldots, y_n) \\
&= \left(\max_{i \in e} \{x_i\} - \min_{i \in e} \{x_i\} \right) + \left(\max_{i \in e} \{y_i\} - \min_{i \in e} \{y_i\} \right) \\
&= \left(\max_{i \in e} \{x_i\} + \max_{i \in e} \{-x_i\} \right) + \left(\max_{i \in e} \{y_i\} + \max_{i \in e} \{-y_i\} \right)
\end{aligned}
$$

So HPWL can be approximated by the log-sum-exponential based function as follows:

$$\text{LSEWL}_{e,\alpha}(x_1, \ldots, x_n, \ y_1, \ldots, y_n)$$

$$= \alpha \times \left(\log \left(\sum_{i \in e} e^{x_i/\alpha} \right) + \log \left(\sum_{i \in e} e^{-x_i/\alpha} \right) + \log \left(\sum_{i \in e} e^{y_i/\alpha} \right) + \log \left(\sum_{i \in e} e^{-y_i/\alpha} \right) \right)$$

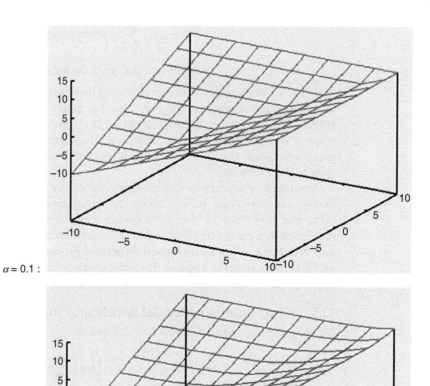

$\alpha = 0.1$:

$\alpha = 2$:

FIGURE 11.19

The log-sum-exponential function of two arguments for two different values of α.

$\text{LSEWL}_{e,\alpha}()$ is strictly convex, continuously differentiable, and converges to $\text{HPWL}_e()$ as α converges to 0 [Naylor 2001].

11.5.3.2 *Density constraint smoothing by bell-shaped function*

To illustrate the idea, assume for now that each module is much smaller than the bins and hence is considered to be a dot. Besides, assume each module has a unit area. For a bin b, let x_b be the x-coordinate of the center and w_b be the width of bin b. Then the overlap function $\Theta_x(b, i)$ in the x-direction

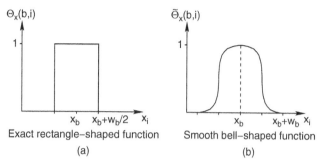

FIGURE 11.20

Overlap function between bin b and module i.

between bin b and module i is shown in Figure 11.20a. This function can be approximated by a smooth bell-shaped function $\tilde{\Theta}_x(b, i)$ as shown in Figure 11.20b. Let $d_x = |x_i - x_b|$. Then:

$$\tilde{\Theta}_x(b, i) = \begin{cases} 1 - 2 \times d_x^2/\omega_b^2 & \text{if } 0 \leq d_x \leq \omega_b/2 \\ 2 \times (d_x - \omega_b)^2/\omega_b^2 & \text{if } \omega_b/2 \leq d_x \leq \omega_b \\ 0 & \text{if } \omega_b \leq d_x \end{cases} \quad (11.2)$$

The overlap function $\tilde{\Theta}_y(b, i)$ in the y-direction is defined similarly.

The density function of bin b can be written as follows:

$$D_b(x, y) = \sum_{i \in V} C_i \times \tilde{\Theta}_x(b, i) \times \tilde{\Theta}_y(b, i)$$

where C_i is a normalization factor so that $\sum_b C_i \times \tilde{\Theta}_x(b, i) \times \tilde{\Theta}_y(b, i) = \omega_i \times h_i$ (*i.e.*, area of module i).

This idea can be extended to handle large modules [Kahng 2005]. For a module i with width w_i, the scope of the module in the x-direction is set to $w_b + w_i/2$ (*i.e.*, every bin within horizontal distance $w_b + w_i/2$ from the module's center is considered to be overlapping with the module). Then:

$$\tilde{\Theta}_x(b, i) = \begin{cases} 1 - a \times d_x^2 & \text{if } 0 \leq d_x \leq w_b/2 + w_i/2 \\ b \times (d_x - w_b - w_i/2)^2 & \text{if } w_b/2 + w_i/2 \leq d_x \leq w_b + w_i/2 \\ 0 & \text{if } w_b + w_i/2 \leq d_x \end{cases} \quad (11.3)$$

where

$$a = 4/((w_b + w_i)(2w_b + w_i))$$

$$b = 4/(w_b(2w_b + w_i))$$

so that the function is continuous when $d_x = w_b/2 + w_i/2$. Note that Equation (11.3) is the same as Equation (11.2) if $w_i = 0$.

11.5.3.3 *Density constraint smoothing by inverse Laplace transformation*

Inverse Laplace transformation is a commonly used method to smooth functions. For a given placement solution x and y and for any location (x, y) in bin b, let $d(x, y)$ be the density at (x, y), *i.e.*, $d(x, y) = D_b(x, y)$. The smoothing operator $\Delta_\epsilon^{-1} d(x, y)$ is defined by solving the Helmholtz equation:

$$\begin{cases} \Delta \Psi(x, y) - \epsilon \Psi(x, y) = d(x, y) & (x, y) \in R \\ \dfrac{\partial \Psi}{\partial v} = 0 & (x, y) \in \partial R \end{cases} \quad (11.4)$$

where $\psi(x, y)$ is a smoothed version of $d(x, y)$, $\epsilon > 0$ is a parameter controlling the smoothness, R is the placement region, ∂R is the boundary of R, v is the outer unit normal vector pointing outside the boundary, and $\Delta = \partial^2/\partial x^2 + \partial^2/\partial y^2$ is a differential operator.

The inverse operator $\Delta_\epsilon^{-1} d(x, y)$ is well defined as equation (11.4) has a unique solution for any $\epsilon > 0$. Because the solution of Equation (11.4) gains two more derivatives than $d(x, y)$, ψ is at least twice differentiable.

11.5.3.4 *Algorithms for nonlinear programs*

For nonquadratic techniques, as the objective function and all constraints are continuously differentiable, the resulting nonlinear program can be solved by any nonlinear programming algorithms. In APlace and NTUPlace, the nonlinear program is converted by the **quadratic penalty method** into a sequence of unconstrained minimization problems. Each unconstrained minimization problem has the following form:

$$\text{Minimize} \sum_{e \in E} c_e \times \text{WL}_e(x, y) + \beta \times \sum_b (D_b(x, y) - T_b)^2$$

The intuition of the quadratic penalty method is that any placement solution violating the density constraint for bin b will be charged a penalty of $(D_b(x, y) - T_b)^2$. β is a parameter to specify the importance of density constraints. Its value keeps increasing in the sequence of unconstrained problems to discourage uneven placement solutions. Each unconstrained problem is solved by the **conjugate gradient method** [Luenberger 1984].

In mPL, the nonlinear program is solved by the **Uzawa algorithm** [Arrow 1958] shown in Algorithm 11.2. In the Uzawa algorithm, x^k and y^k are the module locations at the k-th iteration, λ_b^k is the **Lagrange multiplier** at the k-th iteration, and α is a parameter to control the rate of convergence.

Algorithm 11.2 The Uzawa Algorithm

---

1. Initialize $\boldsymbol{x}^0$, $\boldsymbol{y}^0$, and λ_b^0 for all b
2. For $k = 0, 1, \ldots$
3.   Find $\boldsymbol{x}^{k+1}$ and $\boldsymbol{y}^{k+1}$ by solving the following equality:

$$\sum_{e \in E} c_e \times \nabla \mathrm{WL}_e\left(\boldsymbol{x}^{k+1}, \boldsymbol{y}^{k+1}\right) + \sum_b \lambda_b^k \times \nabla D_b\left(\boldsymbol{x}^k, \boldsymbol{y}^k\right) = 0$$

4.   Set $\lambda_b^{k+1} = \lambda_b^k + \alpha \times \left(D_b\left(\boldsymbol{x}^{k+1}, \boldsymbol{y}^{k+1}\right) - T_b\right)$ for all b

---

The nonquadratic techniques are elegant and comparable to the quadratic techniques in terms of wirelength. However, they are more complicated to implement and are usually more expensive computationally.

11.5.4 **Extension to multilevel**

To handle large-sized problems, a multilevel scheme is commonly used in analytical placement. The application of a multilevel scheme to placement is similar to its application to partitioning presented in Section 11.3.1. It consists of three phases. First, a hierarchy of coarser netlists is constructed by clustering heavily connected modules together. Second, an initial placement of the coarsest netlist is generated. Finally, the netlist is successively unclustered, and the placement at each level is refined. The multilevel scheme can improve both the runtime and the solution quality of analytical placement algorithms. Two popular clustering techniques for netlist coarsening are introduced in the following.

11.5.4.1 *First choice*

The **First Choice clustering technique** [Karypis 1997] first represents the netlist as a weighted graph by replacing the multi-pin nets with the clique model. The weight or **affinity** r_{ij} between any modules i and j in the graph is given by:

$$r_{ij} = \sum_{e \in E \wedge i, j \in e} \frac{c_e}{|e| - 1}$$

Then the modules are traversed in an arbitrary order. Each module i is clustered with an unclustered neighbor j with the largest r_{ij}. After all modules are traversed, the affinity graph is updated, and the clustering process is repeated until the number of modules has reached the target. The intuition behind First Choice is that modules with high affinity should stay close together in a good placement solution.

First Choice is originally proposed for multilevel partitioning. Several modifications of First Choice targeting placement are presented in [Chan 2005]. To reduce variation in cluster size, the affinity between modules i and j is redefined as:

$$r_{ij} = \sum_{e \in E \wedge i, j \in e} \frac{c_e}{(|e| - 1) \times \mathrm{area}(e)}$$

where area(e) is the total area of all modules in e. In addition, modules are visited in ascending order of module area (with preference to smaller module degree to break ties). This ordering is observed to balance the area of clusters better. If a good initial placement is provided, the proximity information between modules can be incorporated into the affinity as follows:

$$r_{ij} = \sum_{e \in E \land i, j \in e} \frac{c_e}{(|e| - 1) \times \text{area}(e) \times \text{dist}(i, j)}$$

where dist(i, j) is the Euclidean distance between i and j.

11.5.4.2 *Best choice*

In the **Best Choice clustering technique** [Alpert 2005], the affinity is defined as:

$$r_{ij} = \sum_{e \in E \land i, j \in e} \frac{c_e}{|e| \times (\text{area}(i) \times \text{area}(j))}$$

where area(i) and area(j) are the areas of modules i and j, respectively. In addition to the indirect control of the cluster size by the affinity, Best Choice imposes a hard upper limit for cluster size. Moreover, the pair of modules with the largest affinity among all pairs is clustered and, in principle, the netlist is immediately updated. In other words, Best Choice always selects the globally best pair for clustering. In practice, as the immediate update is time-consuming, a **lazy updating technique** is proposed to reduce the runtime. The idea is that instead of explicitly recomputing the affinities of module pairs affected by a given cluster, they are marked as invalid and are updated only after they have been selected for clustering. Because the pair selection is based on invalid affinities, lazy updating may incur some errors. However, the dramatic reduction in runtime outweighs the small errors.

As with the modified First Choice affinity, the proximity information of a good initial placement can be incorporated into the Best Choice affinity in the same way.

11.6 LEGALIZATION

Given an illegal placement, legalization is a process to eliminate all overlaps by perturbing the modules as little as possible. In the partitioning-based approach, the modules in each subcircuit at the lowest level have to be arranged in the corresponding subregion. In the simulated annealing approach, it is possible that overlaps are allowed (but penalized) in intermediate steps. In analytical placement, the nonoverlapping constraints are always replaced by density constraints. Therefore, legalization is required for all three approaches.

For standard-cell placement, the Tetris legalization algorithm [Hill 2002] is very commonly used. In this algorithm, modules are first sorted in ascending *x*-coordinate. Then the modules are packed to the left one at a time into the row that minimizes the total displacement for that module. This simple greedy algorithm is extremely fast. However, it sometimes may result in very uneven row lengths and hence may fail to pack all modules inside the placement region. This issue motivated a slight modification proposed in [Khatkhate 2004]. If the algorithm fails to pack all modules inside the placement region, the penalty for displacing a module in the vertical direction is gradually reduced. This modification encourages more even row lengths and so improves the chance of success. It is clear that the Tetris algorithm can also legalize mixed-size designs. It was suggested in [Khatkhate 2004] that the algorithm handles mixed-size designs well.

Because the Tetris algorithm is greedy in nature and attempts to pack modules to the left, it may perturb the original placement quite significantly. A more robust legalization method is proposed in [Ren 2005]. This method is based on a discrete approximation to a closed-form solution of the continuous diffusion equation. It generates a roughly legal placement. Then any legalizer can be applied to put modules onto rows without overlap. This diffusion-based method spreads the modules smoothly. Thus it helps preserve neighborhood characteristics of the original placement. As a result, the wirelength and timing of the resulting placement is better.

11.7 DETAILED PLACEMENT

Given a legalized placement solution, detailed placement further improves the wirelength (or other objectives) by locally rearranging the standard cells while maintaining legality. There may be significant room for wirelength improvement in a legalized global placement solution for several reasons. First, global placement typically uses inaccurate wirelength models (*e.g.*, cut cost, quadratic wirelength with clique net model, log-sum-exponential function). Second, global placement algorithms often place each cell into a subregion without paying much attention to the location of the cell within the subregion. Third, during legalization, the wirelength is likely to be worsened by the perturbations.

Simulated annealing can be easily adopted to perform detailed placement, but this technique is usually slow. Another technique is to iteratively consider different windows and use branch-and-bound to optimally rearrange the cells within each window [Caldwell 2000; Agnihotri 2003]. Because of the high computational complexity of branch-and-bound, a window can only contain up to 7 to 8 cells. Hence this technique is effective only in making very local modifications to the placement solution. Two detailed placement algorithms that work well in practice are presented in the following.

11.7.1 The Domino algorithm

A very high-quality yet efficient detailed placer is the Domino algorithm [Doll 1994]. Domino also uses a **sliding window approach** to iteratively refine a small region. For each region, the problem of assigning the cells to new locations is formulated as a **transportation problem**. To account for the different cell widths, each cell i with width w_i is divided into w_i unit-width subcells. Then the problem is to simultaneously transport the subcells to unit-width locations in an overlap-free manner that minimizes a cost function approximating wirelength. This problem can be transformed into a **minimum cost maximum flow problem** on a network as shown in Figure 11.21. This network consists of a source node S, a set of cell nodes i, a set of location nodes k, and a destination node D. The capacity of the arc between node S and cell node i is w_i. Because each location can hold at most one subcell, all capacities of arcs leading from location nodes to node D are set to one. The cost of assigning a subcell of cell i to location k is c_{ik}, which is determined by a net model described later.

By solving the network flow problem, the subcells are assigned to locations. For all subcells of each cell, as they are associated with the same transportation cost and they are pulled towards the cheapest location by the transportation algorithm, they tend to lie side by side. Each cell is placed in the row holding most of its subcells. The x-coordinate of the cell is determined by the center of gravity of the subcells. Finally, the cells in each row of the region are packed according to their x-coordinates to prevent overlap and unused space.

The cost c_{ik} of assigning a subcell of cell i to location k is the total HPWL of all nets connected to cell i. During the evaluation of the cost c_{ik}, cell i is assumed to be at location k. However, the locations of other cells in the region are still unknown. Hence the HPWL of nets are estimated according to the following net model. To estimate the HPWL of a net e connected to cell i, let e_I be the subset of cells connected to net e inside the region. Consider the following three cases:

- Case 1: $|e_I| = 1$.

  In this case, the only cell inside the region that net e connects to is cell i. Therefore, the HPWL of net e can be calculated exactly.

- Case 2: $1 < |e_I| < |e|$.

  In this case, net reconnects to some cell(s) other than cell i both inside and outside the region. The unknown locations of cells in $e_I - \{i\}$ are estimated by their coordinates in the current placement. Then the HPWL of net e can be calculated.

- Case 3: $|e_I| = |e|$.

  In this case, net e connects only to cells inside the region. Again, the locations of cells in $e_I - \{i\}$ are estimated by their coordinates in the

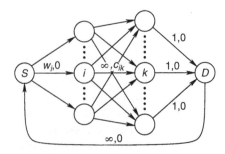

FIGURE 11.21

Transportation network in Domino. Arcs are labeled with "capacity, cost."

current placement. Besides, a virtual cell is introduced at the center of gravity of the cells in e_l with respect to the current placement. The HPWL of all cells in e together with the virtual cell is used as an estimate of the HPWL of net e.

An advantage of Domino over branch-and-bound based algorithms is that the network flow problem has a much lower computational complexity and hence much larger windows can be used. A larger window allows more cells to be placed simultaneously and potentially improves the wirelength. However, it also increases the runtime and results in a less accurate estimation of the HPWL in the cost function. In practice, a window size of approximately 20 to 30 cells per region yields a good tradeoff between wirelength and runtime.

The preceding description is a brief outline of the main ideas of the Domino algorithm. Another net model and theoretical analysis of the relations between the two net models and HPWL are presented in [Doll 1994]. Interested readers may refer to the original paper.

11.7.2 **The FastDP algorithm**

The FastDP algorithm [Pan 2005] is a greedy heuristic that can generate slightly better solutions than Domino and is an order of magnitude faster. The FastDP algorithm consists of four key techniques: **global swap, vertical swap, local reordering**, and **single-segment clustering**. The flow of FastDP is given in Algorithm 11.3.

Algorithm 11.3 The FastDP Detailed Placement Algorithm

---

1. Perform single-segment clustering
2. Repeat
3.   Perform global swap
4.   Perform vertical swap
5.   Perform local reordering
6. Until no significant improvement in wirelength
7. Repeat
8.   Perform single-segment clustering
9. Until no significant improvement in wirelength

Global swap is the technique that gives the most wirelength reduction. It examines all cells one by one. For each cell i, the goal is to move it to its **optimal region**. For a given placement, the optimal region of cell i is defined as the region such that if cell i is placed in it, the wirelength will be optimal. It can be determined on the basis of the median idea of [Goto 1981]. Let E_i be the set of nets connected to cell i. For each net $e \in E_i$, the bounding box excluding cell i

is computed. Let x_L^e and x_U^e be the x-coordinates of left and right boundaries, and y_L^e and y_U^e be the y-coordinates of lower and upper boundaries, respectively. Then the optimal x-coordinate for cell i is given by the median of the set of boundary coordinates $\{x_L^e : e \in E_i\} \cup \{x_U^e : e \in E_i\}$. In general, the optimal coordinate for cell i is a region rather than a single point as the number of elements in the set $\{x_L^e : e \in E_i\} \cup \{x_U^e : e \in E_i\}$ is even. Similarly, the optimal y-coordinate for cell i is given by the median of the set of boundary coordinates $\{y_L^e : e \in E_i\} \cup \{y_U^e : e \in E_i\}$. For example, the optimal region of a cell i connected to three nets $A = \{i, 1, 2, 3\}$, $B = \{i, 4, 5\}$, and $C = \{i, 6, 7\}$ is given in Figure 11.22.

Although it is desirable in terms of wirelength to move cell i to its optimal region, the optimal region may not have enough space to accommodate the cell. So in global swap, cell i attempts to swap with another cell or a space in the optimal region. A benefit function is computed for each cell and each space in the optimal region. If there exists a cell or a space with positive benefit, the one with the highest benefit will be swapped with cell i. The benefit function consists of two components. The first component is the improvement in total wirelength if the swap is performed. The second component is a penalty. Swapping cells of different sizes or swapping a big cell with a small space may create overlap. The overlap is resolved by shifting nearby cells away. The penalty is a function of the least amount of shifting required to resolve the overlap.

Vertical swap is similar to global swap. The difference is that a cell attempts to swap with a few nearby cells one row above/below its current position. Sometimes a cell fails to be moved by global swap, because there is no cell or space in the optimal region with a positive benefit. Vertical swap allows the cell to move toward its optimal region to reduce the vertical wirelength. In addition,

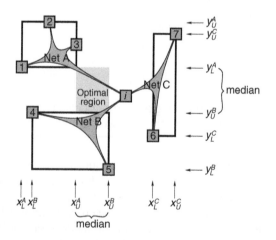

FIGURE 11.22

The optimal region of cell i.

vertical swap is much faster than global swap, because the number of candidates to be considered for swapping is much less.

Local reordering considers each possible group of n consecutive cells in a row. For each group, all possible left-right orderings of the cells are tried, and the one with the best wirelength is selected. Local reordering is a very inexpensive technique to locally minimize horizontal wirelength. In practice, n is set to 3. It is not necessary to use a larger n, because it is more efficient to fix nonlocal errors by global swap.

Single-segment clustering is a technique to minimize the horizontal wirelength by shifting the cells in a segment without changing the cell order. In FastDP, a segment is a maximal unbroken section of a standard cell row. Single-segment clustering examines each segment one by one. When a segment is considered, the locations of all cells in other segments are fixed. A very efficient algorithm based on a clustering idea is presented in [Pan 2005] to find the optimal nonoverlapping placement of the cells in a segment. Interested readers may refer to [Pan 2005] for the details.

11.8 CONCLUDING REMARKS

Placement is such a fundamental problem in VLSI design that there are so many different formulations and algorithms in the literature. This chapter is by no means a comprehensive survey on placement. In this chapter, total wirelength minimization for standard-cell design is focused. Algorithms for other design styles and other objectives are more or less extensions of the algorithms for this basic formulation.

Three popular approaches for global placement are presented. Other approaches that are not considered in this chapter are based on the eigenvalue method [Hall 1970], resistive network optimization [Cheng 1984], genetic algorithm [Cohoon 1987; Shahookar 1990], and artificial neural network [Yu 1989]. Those approaches are not successful in practice, but their insights may be intellectually interesting to readers.

A lot of useful and up-to-date information is collected in the book "Modern Circuit Placement" [Nam 2007]. That book contains descriptions of the underlying algorithms, implementation details, and latest experimental results for nine state-of-the-art academic placers. Another useful resource is a tutorial on large-scale circuit placement that summarizes results from recent optimality and scalability studies of placement tools and highlights recent techniques for optimization of wirelength, routability, and performance [Cong 2005]. A comprehensive survey of older placement techniques can be found in [Shahookar 1991].

Several placement benchmark suites are available in the public domain. The ISPD 2005 benchmarks [Nam 2005, 2007] and the ISPD 2006 benchmarks [Nam 2006, 2007] are derived from modern ASIC designs in IBM. They are mixed-size designs with both fixed and movable modules. The number of movable modules in the largest circuit is 2.2 million in the ISPD 2005 suite and 2.5 million in the ISPD

2006 suite. A major feature of the ISPD 2006 suite is that placement density targets are specified to address the routability concern; this feature is absent from the ISPD 2005 suite. The IBM-PLACE 2.0 [Yang 2002] and IBM-MSwPins [Adya 2004] benchmark suites are derived from smaller circuits released by IBM during ISPD 1998 [Alpert 1998]. The IBM-PLACE 2.0 benchmarks have standard cells only and have no connection to I/O pads, because all macros and the associated nets are removed from the netlists. They have exact pin locations. The IBM-MSwPins benchmarks contain large movable macros and many fixed pads distributed through the periphery. The Faraday Mixed-size benchmarks [Adya 2004] have sufficient routing information to run an industrial router on them after placement. Also, there are several variations of the PEKO benchmark suites [Chang 2003; Nam 2007], which are synthetic placement benchmarks with known optimal wirelength. They can be used to evaluate the optimality of placement algorithms.

These and many other benchmarks, as well as source codes/executables of many academic placers, can be downloaded from the "Wirelength-driven Standard-Cell Placement" slot of GSRC Bookshelf [Caldwell 2002].

11.9 EXERCISES

11.1. (**Introduction**) Consider a special placement problem that determines whether a standard-cell circuit without any net can be placed inside a given placement region. Prove that this problem is NP-complete.

11.2. (**Problem Formulations**) Prove that HPWL is a lower bound of RSMT wirelength. Prove that HPWL is the same as RSMT wirelength for 2- or 3-pin nets.

11.3. (**Problem Formulations**) Prove that RMST wirelength is an upper bound of RSMT wirelength. Give an example such that RMST wirelength overestimates RSMT wirelength by 50%.

11.4. (**Global Placement: Partitioning-Based Approach**) For the hypergraph in Figure 11.2b, assume all hyperedges have a weight of 1. What is the optimal bipartitioning if one partition should contain two vertices and the other should contain three?

11.5. (**Global Placement: Partitioning-Based Approach**) Draw the gain bucket data structures for the bipartition in Figure 11.3a.

11.6. (**Global Placement: Partitioning-Based Approach**) In the hMetis algorithm, a random bipartitioning is done in the initial partitioning phase. Would it be much better to replace the random bipartitioning by the FM algorithm?

11.7. (**Global Placement: Partitioning-Based Approach**) In Section 11.3, a placement approach based on bipartitioning is presented. In this question, a bipartitioning approach based on wirelength-minimized placement is considered. For a circuit with n modules, each module is first placed in one of the n integer coordinates $1, 2, \ldots, n$

on the *x*-axis. Then the modules in coordinates $1, \ldots, k$ form one subcircuit and those in coordinates $k + 1, \ldots, n$ form another subcircuit, where k is chosen according to the size constraints. Show by constructing an example that even if optimal placement can be found, the cut cost of the bipartitioning solution by this approach can be $O(n^2)$ times that of the optimal solution.

11.8. (Global Placement: Simulated Annealing Approach) Prove that the penalty function C_3 of TimberWolf in Section 11.4.1 is equivalent to the following function:

$$C_3' = 2\theta \times \sum_r [l(r) - d(r)]^+$$

$$\text{where } [z]^+ = \begin{cases} z & \text{if } z > 0 \\ 0 & \text{if } z \leq 0 \end{cases}$$

11.9. (Global Placement: Analytical Approach) Prove that the matrix Q defined in Section 11.5.2 is positive definite if all movable modules are connected to fixed modules either directly or indirectly.

11.10. (Global Placement: Analytical Approach) Consider the example in Figure 11.11. Assume all nets have a weight of 1. Assume module 4 is at (1, 0), module 5 is at (0, 3), and module 6 is at (4, 2).
  1. Determine the locations of the movable modules such that the total weighted quadratic wirelength is minimized.
  2. If module 1 is at (2, 2), module 2 is at (1, 3), and module 3 is at (3, 2), find the force vectors exerting on the movable modules.

11.11. (Global Placement: Analytical Approach) For each of the following net models, write the number of 2-pin nets and the number of extra variables introduced for a *k*-pin net as functions of *k*.

  1. Clique
  2. Star
  3. Hybrid
  4. BoundingBox

Plot the number of 2-pin nets for all 4 models in a graph for the range $2 \leq k \leq 15$.

11.12. (Global Placement: Analytical Approach) Prove that the function $\tilde{\Theta}_x(b, i)$ in Equation (11.3) is continuous when $d_x = w_b + w_i/2$. Is it differential when $d_x = w_b + w_i/2$?

11.13. (Global Placement: Analytical Approach) This question analyzes the error of the log-sum-exponential function defined in Section 11.5.3 as an approximation of the maximum function. The error function is defined as:

$$\text{err}_\alpha(z_1, \ldots, z_n) = \text{LSE}_\alpha(z_1, \ldots, z_n) - \max(z_1, \ldots, z_n)$$

Derive an upper bound and a lower bound of $\text{err}_\alpha(z_1, \ldots, z_n)$ over all possible values of $z_1, \ldots, z_n$ as functions of n and α.

11.14. (Detailed Placement) Consider the FastDP algorithm in Section 11.7. Prove that the optimal region is given by the medians of the sets of boundary coordinates.

ACKNOWLEDGMENTS

I thank Natarajan Viswanathan of Iowa State University, Professor Cheng-Kok Koh of Purdue University, Dr. Laung-Terng Wang of SynTest Technologies, Inc., and Professor Yao-Wen Chang of National Taiwan University for carefully reviewing the chapter.

REFERENCES

R11.0 Books

[Arrow 1958] K. Arrow, L. Huriwicz, and H. Uzawa, *Studies in Nonlinear Programming,* Stanford University Press, Stanford, CA, 1958.

[Evans 2002] L. C. Evans, *Partial Differential Equations,* American Mathematical Society, Providence, RI, 2002.

[Garey 1979] M. R. Garey and D. S. Johnson, *Computers and Intractability: A Guide to the Theory of NP-Completeness,* Freeman, New York, 1979.

[Luenberger 1984] D. G. Luenberger, *Linear and Nonlinear Programming,* second edition, Addison Wesley, Reading, MA, 1984.

[Nam 2007] G.-J. Nam and J. Cong, editors, *Modern Circuit Placement—Best Practices and Results,* Springer, Boston, 2007.

R11.1 Introduction

[Garey 1974] M. R. Garey, D. S. Johnson, and L. Stockmeyer, Some simplified NP-complete problems, in *Proc. ACM Symp. on Theory of Computing,* pp. 47-63, April-May 1974.

R11.2 Problem Formulations

[Chu 2008] C. Chu and Y.-C. Wong, FLUTE: Fast lookup table based rectilinear Steiner minimal tree algorithm for VLSI design, *IEEE Trans. on Computer-Aided Design,* 27(1), pp. 70-83, January 2008.

[Guibas 1983] L. J. Guibas and J. Stolfi, On computing all northeast nearest neighbors in the L1 metric, in *Information Processing Letters,* 17(4), pp. 219-223, April 1983.

[Hwang 1976] F. K. Hwang, On Steiner minimal trees with rectilinear distance, *SIAM J. of Applied Mathematics,* 30(1), pp. 104-114, January 1976.

R11.3 Global Placement: Partitioning-Based Approach

[Agnihotri 2003] A. Agnihotri, M. C. Yildiz, A. Khatkhate, A. Mathur, S. Ono, and P. H. Madden, Fractional cut: Improved recursive bisection placement, in *Proc. IEEE/ACM Int. Conf. on Computer-Aided Design*, pp. 307–310, November 2003.

[Alpert 1997] C. J. Alpert, J.-H. Huang, and A. B. Kahng, Multilevel circuit partitioning, in *Proc. ACM/IEEE Design Automation Conf.*, pp. 530–533, June 1997.

[Breuer 1977a] M. A. Breuer, A class of min-cut placement algorithms, in *Proc. ACM/IEEE Design Automation Conf.*, pp. 284–290, June 1977.

[Breuer 1977b] M. A. Breuer, Min-cut placement, *J. of Design Automation and Fault Tolerant Computing*, 1(4), pp. 343–382, October 1977.

[Caldwell 2000] A. E. Caldwell, A. B. Kahng, and I. L. Markov, Can recursive bisection produce routable placements, in *Proc. ACM/IEEE Design Automation Conf.*, pp. 477–482, June 2000.

[Dunlop 1985] A. E. Dunlop and B. W. Kernighan, A procedure for placement of standard-cell VLSI circuits, *IEEE Trans. on Computer-Aided Design*, 4(1), pp. 92–98, January 1985.

[Fiduccia 1982] C. M. Fiduccia and R. M. Mattheyses, A linear-time heuristic for improving network partitions, in *Proc. ACM/IEEE Design Automation Conf.*, pp. 175–181, June 1982.

[Karypis 1997] G. Karypis, R. Aggarwal, V. Kumar, and S. Shekhar, Multilevel hypergraph partitioning: Application in VLSI domain, in *Proc. ACM/IEEE Design Automation Conf.*, pp. 526–529, June 1997.

[Yildiz 2001a] M. C. Yildiz and P. H. Madden, Global objectives for standard cell placement, in *Proc. 11th ACM Great Lakes Symp. on VLSI*, pp. 68–72, March 2001.

[Yildiz 2001b] M. C. Yildiz and P. H. Madden, Improved cut sequences for partitioning based placement, in *Proc. ACM/IEEE Design Automation Conf.*, pp. 776–779, June 2001.

R11.4 Global Placement: Simulated Annealing Approach

[Sechen 1986] C. Sechen and A. L. Sangiovanni-Vincentelli, TimberWolf 3.2: A new standard cell placement and global routing package, in *Proc. ACM/IEEE Design Automation Conf.*, pp. 432–439, June 1986.

[Sun 1995] W.-J. Sun and C. Sechen, Efficient and effective placement for very large circuits, *IEEE Trans. on Computer-Aided Design*, 14(3), pp. 349–359, March 1995.

[Wang 2000] M. Wang, X. Yang, and M. Sarrafzadeh, Dragon2000: Standard-cell placement tool for large industry circuits, in *Proc. IEEE/ACM Int. Conf. on Computer-Aided Design*, pp. 260–263, November 2000.

R11.5 Global Placement: Analytical Approach

[Alpert 2005] C. Alpert, A. Kahng, G.-J. Nam, S. Reda, and P. Villarrubia, A semi-persistent clustering technique for VLSI circuit placement, in *Proc. ACM Int. Symp. on Physical Design*, pp. 200–207, April 2005.

[Brenner 2005] U. Brenner and M. Struzyna, Faster and better global placement by a new transportation algorithm, in *Proc. ACM/IEEE Design Automation Conf.*, pp. 591–596, June 2005.

[Chan 2005] T. Chan, J. Cong, and K. Sze, Multilevel generalized force-directed method for circuit placement, in *Proc. ACM Int. Symp. on Physical Design*, pp. 185–192, April 2005.

[Chen 2006] T.-C. Chen, Z.-W. Jiang, T.-C. Hsu, H.-C. Chen, and Y.-W. Chang, A high quality analytical placer considering preplaced blocks and density constraint, in *Proc. IEEE/ACM Int. Conf. on Computer-Aided Design*, pp. 187–192, November 2006.

[Eisenmann 1998] H. Eisenmann and F. Johannes, Generic global placement and floorplanning, in *Proc. ACM/IEEE Design Automation Conf.*, pp. 269–274, June 1998.

[Hu 2002] B. Hu and M. Marek-Sadowska, FAR: Fixed-points addition and relaxation based placement, in *Proc. ACM Int. Symp. on Physical Design*, pp. 161–166, April 2002.

[Kahng 2004] A. B. Kahng and Q. Wang, Implementation and extensibility of an analytical placer, in *Proc. ACM Int. Symp. on Physical Design*, pp. 18–25, April 2004.

[Kahng 2005] A. B. Kahng and Q. Wang, Implementation and extensibility of an analytic placer, *IEEE Trans. on Computer-Aided Design*, 24(5), pp. 734–747, May 2005.

[Karypis 1997] G. Karypis, R. Aggarwal, V. Kumar, and S. Shekhar, Multilevel hypergraph partitioning: Application in VLSI domain, in *Proc. ACM/IEEE Design Automation Conf.*, pp. 526–529, June 1997.

[Kleinhans 1991] J. Kleinhans, G. Sigl, F. Johannes, and K. Antreich, GORDIAN: VLSI placement by quadratic programming and slicing optimization, *IEEE Trans. on Computer-Aided Design*, 10(3), pp. 356–365, March 1991.

[Kowarschik 2001] M. Kowarschik and C. Weiß, DiMEPACK—a cache-optimized multigrid library, in *Proc. Int. Conf. on Parallel and Distributed Processing Techniques and Applications*, pp. 425–430, June 2001.

[Mo 2000] F. Mo, A. Tabbara, and R. Brayton, A force-directed macro-cell placer, in *Proc. IEEE/ACM Int. Conf. on Computer-Aided Design*, pp. 177–180, November 2000.

[Nam 2006] G.-J. Nam, S. Reda, C. J. Alpert, P. G. Villarrubia, and A. B. Kahng, A fast hierarchical quadratic placement algorithm, *IEEE Trans. on Computer-Aided Design*, 25(4), pp. 678–691, April 2006.

[Naylor 2001] W. Naylor, Non-linear Optimization System and Method for Wire Length and Delay Optimization for an Automatic Electric Circuit Placer, U.S. Patent No. 6,301,693. Oct. 9, 2001.

[Quinn 1975] N. R. Quinn, The placement problem as viewed from the physics of classical mechanics, in *Proc. ACM/IEEE Design Automation Conf.*, pp. 173–178, June 1975.

[Sigl 1991] G. Sigl, K. Doll, and F. M. Johannes, Analytical placement: A linear or a quadratic objective function, in *Proc. ACM/IEEE Design Automation Conf.*, pp. 427–431, June 1991.

[Spindler 2006] P. Spindler and F. Johannes, Fast and robust quadratic placement combined with an exact linear net model, in *Proc. IEEE/ACM Int. Conf. on Computer-Aided Design*, pp. 179–186, November 2006.

[Viswanathan 2004] N. Viswanathan and C. Chu, FastPlace: Efficient analytical placement by use of cell shifting, iterative local refinement and a hybrid net model, in *Proc. ACM Int. Symp. on Physical Design*, pp. 26–33, April 2004.

[Viswanathan 2007a] N. Viswanathan, M. Pan, and C. Chu, FastPlace: An efficient multilevel force-directed placement algorithm, in Gi-Joon Nam and Jason Cong, editors, *Modern Circuit Placement—Best Practices and Results*, pp. 193–228, Springer, Boston, 2007.

[Viswanathan 2007b] N. Viswanathan, G.-J. Nam, C. Alpert, P. Villarrubia, H. Ren, and C. Chu, RQL: Global placement via relaxed quadratic spreading and linearization, in *Proc. ACM/IEEE Design Automation Conf.*, pp. 453–458, June 2007.

[Vygen 1997] J. Vygen, Algorithms for large-scale flat placement, in *Proc. ACM/IEEE Design Automation Conf.*, pp. 746–751, June 1997.

[Xiu 2004] Z. Xiu, J. D. Ma, S. M. Fowler, and R. A. Rutenbar, Large-scale placement by grid-warping, in *Proc. ACM/IEEE Design Automation Conf.*, pp. 351–356, June 2004.

R11.6 Legalization

[Hill 2002] D. Hill, Method and System for High Speed Detailed Placement of Cells Within an Integrated Circuit Design, U.S. Patent No. 6,370,673. April 9, 2002.

[Khatkhate 2004] A. Khatkhate, C. Li, A. R. Agnihotri, M. C. Yildiz, S. Ono, C. K. Koh, and P. H. Madden, Recursive bisection based mixed block placement, in *Proc. ACM Int. Symp. on Physical Design*, pp. 84–89, April 2004.

[Ren 2005] H. Ren, D. Z. Pan, C. Alpert, and P. Villarrubia, Diffusion-based placement migration, in *Proc. ACM/IEEE Design Automation Conf.*, pp. 515–520, June 2005.

R11.7 Detailed Placement

[Agnihotri 2003] A. Agnihotri, M. C. Yildiz, A. Khatkhate, A. Mathur, S. Ono, and P. H. Madden, Fractional cut: Improved recursive bisection placement, in *Proc. IEEE/ACM Int. Conf. on Computer-Aided Design*, pp. 307-310, November 2003.

[Caldwell 2000] A. E. Caldwell, A. B. Kahng, and I. L. Markov, Optimal partitioners and end-case placers for standard-cell layout, *IEEE Trans. on Computer-Aided Design*, 19(11), pp. 1304-1314, November 2000.

[Doll 1994] K. Doll, F. M. Johannes, and K. J. Antreich, Iterative placement improvement by network flow methods, *IEEE Trans. on Computer-Aided Design*, 13(10), pp. 1189-1200, October 1994.

[Goto 1981] S. Goto, An efficient algorithm for the two-dimensional placement problem in electrical circuit layout, *IEEE Trans. on Circuits and Systems*, 28(1), pp. 12-18, January 1981.

[Pan 2005] M. Pan, N. Viswanathan, and C. Chu, An efficient and effective detailed placement algorithm, in *Proc. IEEE/ACM Int. Conf. on Computer-Aided Design*, pp. 48-55, November 2005.

R11.8 Concluding Remarks

[Adya 2004] S. N. Adya, S. Chaturvedi, J. A. Roy, D. A. Papa, and I. L. Markov, Unification of partitioning, floorplanning and placement, in *Proc. IEEE/ACM Int. Conf. on Computer-Aided Design*, pp. 550-557, November 2004.

[Alpert 1998] C. J. Alpert, The ISPD98 Circuit Benchmark Suite, in *Proc. ACM Int. Symp. on Physical Design*, pp. 80-85, April 1998. http://vlsicad.ucsd.edu/UCLAWeb/cheese/ispd98.html.

[Caldwell 2002] A. E. Caldwell, A. B. Kahng, and I. L. Markov, Toward CAD-IP reuse: The MARCO GSRC bookshelf of fundamental CAD algorithms, in *IEEE Design and Test*, pp. 72-81, 2002. http://www.gigascale.org/bookshelf/.

[Chang 2003] C.-C. Chang, J. Cong, and M. Xie, Optimality and scalability study of existing placement algorithms, in *Proc. IEEE/ACM Asia and South Pacific Design Automation Conf.*, pp. 621-627, January 2003.

[Cheng 1984] C. Cheng and E. Kuh, Module placement based on resistive network optimization, *IEEE Trans. on Computer-Aided Design*, 3(3), pp. 218-225, July 1984.

[Cohoon 1987] J. P. Cohoon and W. D. Paris, Genetic placement, *IEEE Trans. on Computer-Aided Design*, 6(6), pp. 956-964, November 1987.

[Cong 2005] J. Cong, J. R. Shinnerl, M. Xie, T. Kong, and X. Yuan, Large-scale circuit placement, in *ACM Trans. on Design Automation of Electronics Systems*, 10(2), pp. 389-430, April 2005.

[Hall 1970] K. M. Hall, An r-dimensional quadratic placement algorithm, *Management Science*, 17 (3), pp. 219-229, November 1970.

[Nam 2005] G.-J. Nam, C. J. Alpert, P. Villarubbia, B. Winter, and M. Yildiz, The ISPD2005 placement contest and benchmark suite, in *Proc. ACM Int. Symp. on Physical Design*, pp. 216-220, April 2005. http://www.sigda.org/ispd2005/contest.htm.

[Nam 2006] G.-J. Nam, The ISPD 2006 placement contest: Benchmark suite and results, in *Proc. ACM Int. Symp. on Physical Design*, p. 167, April 2006. http://www.sigda.org/ispd2006/contest.html.

[Shahookar 1990] K. Shahookar and P. Mazumder, A genetic approach to standard cell placement using metagenetic parameter optimization, *IEEE Trans. on Computer-Aided Design*, 9(5), pp. 500-511, May 1990.

[Shahookar 1991] K. Shahookar and P. Mazumder, VLSI cell placement techniques, *ACM Computing Surveys*, 23(2), pp. 143-220, June 1991.

[Yang 2002] X. Yang, B.-K. Choi, and M. Sarrafzadeh, Routability-driven white space allocation for fixed-die standard-cell placement, in *Proc. ACM Int. Symp. on Physical Design*, pp. 42-49, April 2002.

[Yu 1989] M. L. Yu, A study of the applicability of Hopfield decision neural nets to VLSI CAD, in *Proc. ACM/IEEE Design Automation Conf.*, pp. 412-417, June 1989.

Global and detailed routing

12

Huang-Yu Chen
National Taiwan University, Taipei, Taiwan

Yao-Wen Chang
National Taiwan University, Taipei, Taiwan

ABOUT THIS CHAPTER

After placement, the routing process determines the precise paths for nets on the chip layout to interconnect the pins on the circuit blocks or pads at the chip boundary. These precise paths of nets must satisfy the design rules provided by chip foundries to ensure that the designs can be correctly manufactured. The most important objective of routing is to complete all the required connections (*i.e.*, to achieve 100% routability); otherwise, the chip would not function well and may even fail. Other objectives, such as (1) reducing the routing wirelength and (2) ensuring each net to satisfy its required timing budget, have become essential for modern chip design. For modern large-scale circuit design, a chip may contain billions of transistors and millions of nets. To handle the high complexity, a routing algorithm often adopts the two-stage approach of global routing followed by detailed routing. Global routing first partitions the routing region into tiles and decides tile-to-tile paths for all nets, whereas detailed routing determines the exact tracks and vias for nets.

This chapter starts with a discussion of the routing problem. After introducing the problem definition, the techniques of general-purpose routing are described. This is followed by the introduction of popular global-routing algorithms that cover sequential and concurrent approaches. The second half of this chapter discusses detailed routing, for which channel and full-chip routing techniques are discussed, followed by modern routing techniques considering signal integrity and chip manufacture and yield. This chapter concludes with routing trends and future directions of routing. After reading through this chapter, the reader should have a clear picture about popular global and detailed routing algorithms. This background will be valuable in implementing/developing routing algorithms to meet the design needs.

12.1 INTRODUCTION

Routing is an important step in the design of ***integrated circuits*** (ICs). It generates wiring to interconnect pins of the same signal, while obeying the manufacturing **design rules**. As IC process advances to nanometer technology, foundries may fabricate billions of transistors in a single chip, and the number of transistors per die will still grow drastically in the near future. This increasing complexity imposes substantial challenges for physical design, especially for routing.

Research in VLSI routing has received much attention in the literature. Routing is typically a very complex combinatorial problem. To make it manageable, the routing problem is usually solved by use of a two-stage approach of **global routing** followed by **detailed routing**. Global routing first partitions the routing region into tiles and decides tile-to-tile paths for all nets while attempting to optimize some given objective function (*e.g.*, total wirelength and circuit timing). Then, guided by the paths obtained in global routing, detailed routing assigns actual tracks and vias for nets.

Figure 12.1 illustrates the process of global routing and detailed routing. After placement, we have a placed layout shown in Figure 12.1a, which contains the information about the exact locations of blocks, pins of blocks, and I/O pads at chip boundaries. We are also provided with a **netlist** that describes a list of connections by indicating which pins or pads should be electrically connected to form a set of nets. Figure 12.1b illustrates some global-routing paths. It first divides the routing region into **tiles** and then generates a "loose" route for each connection by finding the tile-to-tile paths to connect pins and/or pads. Figure 12.1c shows a result of detailed routing, which determines the exact route for each net by searching within the tile-to-tile path. Here, the exact route means a path specified by the actual geometric layout such as metal wires and vias.

In the following we formally give the problem definition of the routing problem and describe the routing model and constraints.

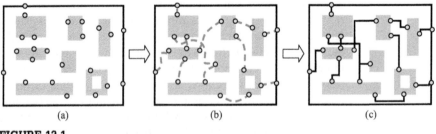

(a)	(b)	(c)

FIGURE 12.1

Routing problem: (a) A given placement result with fixed locations of blocks and pins. (b) Global routing. (c) Detailed routing.

12.2 PROBLEM DEFINITION

The problem definition for the general routing problem is as follows:
Inputs:

1. A placed layout with fixed locations of chip blocks, pins, and pads
2. A netlist
3. A timing budget for each critical net
4. A set of design rules for manufacturing process, such as resistance, capacitance, and the wire/via width and spacing of each layer

Output:

Wire connection for each net presented by actual geometric layout objects that meet the design rules and optimize the given objective, if specified.

12.2.1 Routing model

Routing in a modern chip is typically a very complex process, and it is thus usually hard to obtain solutions directly. Most routing algorithms are based on a graph-search technique guided by the congestion and timing information associated with routing regions and topologies [Saxena 2007]. A router assigns higher costs to route nets through congested areas to balance the net distribution among routing regions.

Applying the graph-search technique for routing requires modeling the routing resource as a graph where the graph topology can represent the chip structure. Figure 12.2 illustrates the graph modeling. For the modeling, a chip (routing region) is first partitioned into an array of rectangular tiles (or called **global-routing tiles**), each of which may accommodate tens of routing tracks in each dimension, as illustrated in Figure 12.2a. A node in the routing graph represents a tile in the chip, whereas an edge denotes the boundary between two adjacent tiles (see Figures 12.2b–c). Each edge is assigned a capacity according to the physical routing area or the number of tracks in a tile. This graph is called a **global-routing graph**.

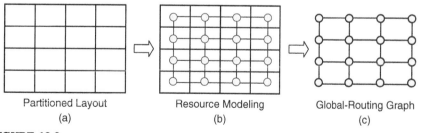

Partitioned Layout	Resource Modeling	Global-Routing Graph
(a)	(b)	(c)

FIGURE 12.2

The global-routing graph: (a) The chip (routing region) is partitioned into an array of rectangular tiles. (b) A node in the routing graph represents a tile in the chip, whereas an edge denotes the boundary between two adjacent tiles. (c) The final global-routing graph.

A global router finds tile-to-tile paths for all nets on the global-routing graph to guide the detailed router. The goal of global routing is to route as many nets as possible while meeting the capacity constraint of each edge and any other constraint, if specified. For example, for timing-driven routing, additional costs can be added to the routing topologies with longer critical path delays. For detailed routing, the router decides the actual physical interconnections of nets by allocating *wires* on each metal layer and *vias* for switching between metal layers.

Generally, there are two different layer models, the **reserved** and **unreserved layer models**. In the reserved layer model, each layer is allowed only one specific routing direction (*i.e.,* **preferred direction**). For example, the technology file may specify that the wires in the first metal layer are allowed to run only in the horizontal direction, the second metal layer contains only vertical wires, etc. A layer model is unreserved if it allows the placement of wires with any directions (*i.e.,* **non-preferred direction**). Most of the existing routers and design methodologies apply the reserved layer model, because it has lower complexity than the unreserved layer model and is much easier for implementation.

There are two kinds of detailed-routing models: the **grid-based** and **gridless** models. For grid-based routing, a routing **grid** is superimposed on the routing region, and then the detailed router finds routing paths in the grid, as shown in Figure 12.3a. The space between adjacent grid lines is called **wire pitch**, which is defined in the technology file and is larger than or equal to the sum of the minimum width and spacing of wires. Note that the router has to control the searching space such that the path in the horizontal/vertical layers can only

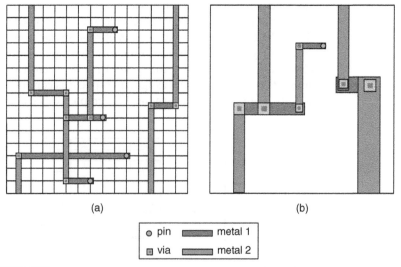

(a)   (b)

| ⊙ pin | ▬▬ metal 1 |
| ▣ via | ▬▬ metal 2 |

FIGURE 12.3

Two kinds of detailed-routing models: (a) Grid-based detailed routing. (b) Gridless detailed routing.

run horizontally/vertically for the reserved layer model, and switching from layer to layer is allowed only at the intersection of vertical and horizontal grid lines. In this way, the wires with the minimum width following the path in the grid would automatically satisfy the design rules. Therefore, grid-based detailed routing is much more efficient and easier for implementation.

The gridless detailed routing model (also called *shaped-based*) refers to any model that does not follow the grid-based model. A gridless detailed router does not follow the routing grid and thus can use different wire widths and spacing, as shown in the example in Figure 12.3b. Various gridless models have been proposed, such as the **connection graph** [Zheng 1996], the **implicit connection graph** [Cong 1999], the **implicit triple-line graph** [Chen 2007a], and **corner stitching** [Qusterhout 1984]. The main advantage of gridless routing lies in its greater flexibility; it can handle variable widths and spacing for wires and is, thus, more suitable for interconnect tuning optimization, such as wire sizing and perturbation. However, gridless detailed routing is generally much slower than the grid-based one because of its higher complexity.

Figure 12.4 illustrates an example of grid-based detailed routing for a two-pin net. After the global routing, we have a tile-to-tile global-routing path as shown in Figure 12.4a, and the detailed-routing graph is constructed only within the tiles of the global-routing path, as shown in Figure 12.4b. Then the final detailed-routing solution is found in the graph, as shown in Figure 12.4c. Constructing and searching the detailed-routing graph within the tiles of the global-routing path, the detailed router can substantially prune the searching space and thus reduce the routing time.

12.2.2 Routing constraints

The routing constraints can be classified into two major categories: (1) design-rule constraints and (2) performance constraints. The design-rule constraint is

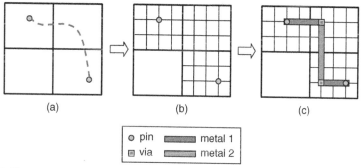

(a)   (b)   (c)

| ○ pin | ▬▬ metal 1 |
| ▣ via | ▬▬ metal 2 |

FIGURE 12.4

Detailed routing: (a) A tile-to-tile global-routing path connecting two pins on metal 1. (b) The detailed-routing graph is constructed within the tiles of the global-routing path. (c) A detailed-routing solution on the detailed-routing graph.

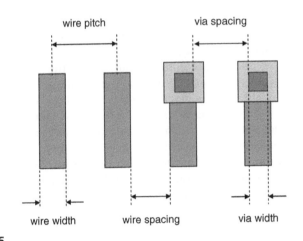

wire pitch   via spacing

wire width   wire spacing   via width

FIGURE 12.5

An example of design rules. Typical rules define wire width, wire spacing, wire pitch, via width, and via spacing on each layer.

often related with the manufacturing details during fabrication. To improve the manufacturing yield, connections of nets have to follow the rules provided by foundries. For example, in the 65-nm technology, the physical limitations of an optical lithography system would impose a constraint on a wire such that its width cannot be smaller than 65 nm.

Figure 12.5 illustrates a typical set of **design rules**. It defines the minimum widths of wires and vias, the minimum wire-to-wire spacing, and the minimum via-to-via spacing of a layer. The distance between two wires or routing tracks of the grid-based model is often called wire pitch. Other design rules of the manufacturing process, such as resistance and capacitance of each layer, are also included.

The objective of the performance constraint is to make the connections meet the performance specifications provided by chip designers. For example, the timing constraint is often the most important performance constraint for high-speed designs. The speed of a chip is limited by its **critical nets**, which have smaller timing budgets (or timing slacks) than others. To meet the performance constraint, it is desirable to carefully route these critical nets by proper routing topologies.

12.3 **GENERAL-PURPOSE ROUTING**

In Section 12.2.1, we modeled the routing resources by the global- and detailed-routing graphs. For global and detailed routing, we can perform a graph-search technique on these routing models. In the following, we introduce three popular graph-searching techniques, the **maze, line-search**, and **A*-search** routing algorithms. Note that these algorithms are **general-purpose routing algorithms**, because they can be applied to both global and detailed routing problems on the general routing structure.

12.3.1 **Maze routing**

Perhaps the most widely used algorithm for finding a path between two points is the **maze-routing algorithm** (also called **Lee's algorithm**) [Lee 1961], which is based on the *breadth-first-search* (BFS) technique.

Maze routing adopts a two-phase approach of **filling** followed by **retracing**. The filling phase works in the "wave propagation" manner. Starting from the source node S, the adjacent grid cells are progressively labeled one by one according to the distance of the "wavefront" from S until the target node T is reached. Figures 12.6a and b illustrates the "wave propagation" when the labels of "wavefronts" reach 2 and 3, respectively. Once the target node T is reached, a shortest path is then retraced from T to S with decreasing labels during the retracing phase. Note that any such a path with decreasing labels gives a shortest path. However, we often prefer the one with the least detours for other practical concerns such as the number of bends (vias). Figure 12.7 illustrates the two phases of Lee's algorithm.

A nice property of Lee's algorithm is that it *guarantees to* find a path between two points if such a path does exist, and the path is the *shortest* one, even with obstacles. In practice, however, Lee's algorithm is slow and memory consuming. It has the time and space complexity of $O(mn)$, where

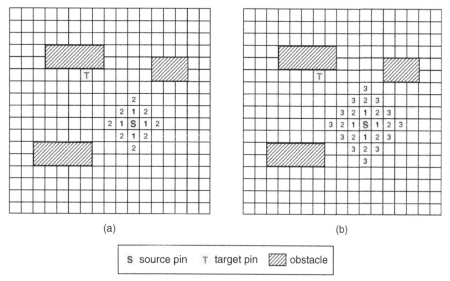

(a)   (b)

S source pin   T target pin   ▨ obstacle

FIGURE 12.6

An example of the filling process: (a) The filling (wave propagation) when labels of the "wavefront" reach 2. (b) The next filling step of (a) when labels of the "wavefront" reach 3.

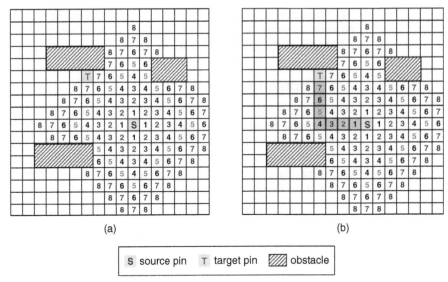

(a)   (b)

| S source pin | T target pin | ▨ obstacle |

FIGURE 12.7

Lee's maze-routing algorithm: (a) The wave propagation phase. (b) The retracing phase.

m and n are the respective numbers of horizontal and vertical grid cells. Consequently, it is difficult to apply for large-scale dense designs directly.

Because of the pervasive use of Lee's maze-routing algorithm and its high time and space complexity, many methods have been proposed to reduce its running time and memory requirements. These popular optimization methods can be classified into three major categories: (1) coding scheme, (2) search algorithm, and (3) search space.

12.3.1.1 *Coding scheme*

Akers observed that adjacent labels for k are either $k-1$ or $k+1$ [Akers 1967]. To retrace the path, it suffices to have a labeling scheme such that each label has its preceding label different from its succeeding label. With the observations, Akers developed a 2-bit **coding scheme** to reduce memory requirement. The coding scheme uses 1 bit for filling by labeling the grid cells with the sequence 0, 0, 1, 1, 0, 0, 1, 1, ... In this way, for each label, its preceding label is different from its succeeding one, and thus retracing can work correctly. This coding scheme requires another bit to indicate whether a node is blocked or not. Therefore, this coding scheme needs only two bits for each grid cell to perform the maze routing. Another economical coding scheme is from [Hadlock 1977]; it uses the *detour* numbers for the labeling to reduce the search space and runtime.

12.3.1.2 *Search algorithm*

Soukup combined BFS and the ***depth-first-search*** (DFS) approaches to propagate wavefronts [Soukup 1978]. Depth-first (**line**) search is first directed from the source S toward the target T until an obstacle or T is reached. BFS (as in Lee's algorithm) is then used to "bubble" around an obstacle if an obstacle is encountered. This algorithm has the same time and space complexity as that of Lee's algorithm, but is typically 10 to 50 times faster than Lee's algorithm. It can still find a path between S and T if such a path does exist, but it cannot guarantee a shortest path because of the DFS processing. Pure DFS (line-search) algorithms can further speed up the routing, at the cost of solution quality. See Section 12.3.2 for two line-search algorithms.

12.3.1.3 *Search space*

To reduce the running time for maze routing, techniques such as **starting point selection, double fanout**, and **framing** are in pervasive use [Sait 1999]. All the three techniques can substantially reduce the number of cells required to be labeled. The starting point selection is to choose the point closest to the chip boundary as the starting point for filling. In this way, we can discard more out-of-bound cells for labeling. Double fanout propagates waves from both the source and the target cells to reduce the area required for labeling. Framing searches only inside a rectangular region, say 10% larger than the bounding box formed by the source and the target. It needs to enlarge the rectangle and redo maze routing if the search fails. It is obvious that Lee's algorithm can no longer guarantee finding the shortest path with the framing heuristic.

12.3.2 **Line-search routing**

As mentioned earlier in Section 12.3.1, the major drawbacks of the maze-routing algorithm are the high memory use and long running time. The **line-search algorithm** alleviates these drawbacks by use of line segments to represent the routing space and paths at the cost of solution quality.

Mikami and Tabuchi proposed the first line-search algorithm (also called *line probe routing*) [Mikami 1968]. In contrast to the maze-routing algorithm, which mainly proceeds in a breadth-first manner, the line-search algorithm performs a depth-first search. The line-search algorithm initially sets the source S and the target T as **base points** and then generates four (two horizontal and two vertical) level-0 line segments passing through these base points. These line segments are extended until they hit the design boundary or obstacles. Then, each grid point of these line segments at level i are iteratively set as new base points, and a perpendicular line segment of level $i + 1$ is generated crossing each new base point. This process repeats until a segment generated from S intersects a segment generated from T, and a connection can then be found by tracing from this **intersection point** to both S and T. Figure 12.8 illustrates the Mikami-Tabuchi's line-search algorithm. The crossing points denote the base points, and the numbers denote the sequence of the search process. Like Lee's maze-routing algorithm, Mikami-Tabuchi's line-search algorithm also guarantees finding a path if one exists, but it may not always be the shortest. The line-search technique significantly reduces both memory requirements and execution times.

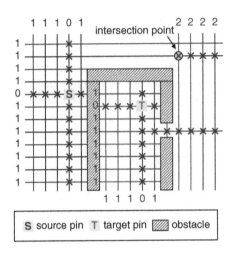

FIGURE 12.8

Mikami-Tabuchi's line-search algorithm.

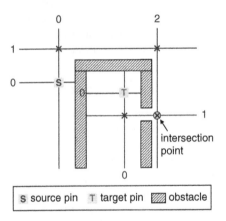

FIGURE 12.9

Hightower's line-search algorithm.

[Hightower 1969] proposed another line-search algorithm, which is similar to Mikami-Tabuchi's algorithm. The difference is that Hightower's algorithm only considers those line segments that are extendable beyond obstacles, and each line segment has at most two base points. Figure 12.9 illustrates Hightower's line-search algorithm. Because fewer line segments are considered, Hightower's algorithm has significantly more memory saving than Mikami-Tabuchi's algorithm. However, Hightower's algorithm might fail to find a path even if one exists. To remedy the deficiencies, it needs **backtracing** procedures to choose the right base points, and, therefore, the running time may not improve very much over Lee's maze-routing algorithm in practice.

12.3.3 **A*-search routing**

As discussed in Section 12.3.1, the maze routing that adopts the BFS searching is generally slow, although it guarantees finding a shortest path. In the searching field, the maze search is also called **blind search,** because it searches the routing region in a blind way without any prioritized choices. Intuitively, if a router does not need to consider points that are not likely to be on the routing path, the running time would be improved.

In [Hart 1968], a general graph search algorithm called **A*-search** was proposed, which uses the function $f(x) = g(x) + h(x)$ to evaluate the cost of a path x, where $g(x)$ is the cost from the source node to the current node of x, and $h(x)$ is the *estimated* (or predicted) cost from the current node of x to the target node. Every time the algorithm selects a node with the lowest path cost to propagate (*i.e.,* the lower $f(x)$), the higher the priority for propagation. As a result, the A*-search is also called the **best-first search**, because at each decision making it first searches the routes that are most likely to lead toward the target. Note that generally speaking, the BFS is a special case of A*-search algorithm, where $h(x) = 0$ for all x.

The A*-search has a good property that if $h(x)$ is *admissible*, meaning that it never overestimates the actual minimal cost from the current node to the target node, then A*-search is optimal. Therefore, for the Manhattan routing (*i.e.,* only allow horizontal and vertical connections), $h(x)$ might be set as the Manhattan distance from the current node to the target, because it is the smallest possible distance between any two points in the Manhattan space.

The A*-search algorithm has many applications, such as in the field of *artificial intelligence* (AI). The **A*-search routing** introduced by [Clow 1984] for VLSI routing and [McMurchie 1995] for FPGA routing are pervasive in modern routers [Chao 2007; Pan 2007; Roy 2007; Chang 2008; Hsu 2008].

12.4 **GLOBAL ROUTING**

Traditional routing algorithms adopt the flat framework that finds paths for nets in the whole routing region directly. These algorithms can be classified into sequential and concurrent approaches, which are based on the general-purpose routing for 2-pin nets mentioned in Section 12.3 or a Steiner-tree algorithm for the multi-pin nets to be introduced in Section 12.4.3.

12.4.1 **Sequential global routing**

Perhaps the most straightforward strategy for routing is to select a specific net order and then to route nets sequentially in that order. However, this sequential approach often leads to a poor routing result, because an earlier routed net might block the routing for its subsequent nets. Therefore, the quality of the routing solution greatly depends on the net ordering.

Figure 12.10a illustrates a simple one-layer routing instance with two two-pin nets *A* and *B*. If we arbitrarily choose the net ordering as routing *A* first followed by B, net B might be blocked by net *A* and thus requires more longer wirelength to complete the routing (see Figure 12.10b). In contrast, if we route *B* first and then *A*, we can get a better routing result with shorter total wirelength (see Figure 12.10c). Therefore, it is desired to find a good net-ordering scheme for general routing instances. Unfortunately, such a universally good scheme is hard to find. In an earlier study, Abel concluded that there is no single net-ordering scheme that performs better than any other ordering scheme in all routing problems [Abel 1972], and finding the optimal net ordering has proven to be NP-hard, meaning that most likely no polynomial-time algorithm exists to solve this problem.

To remedy the deficiencies, today's sequential routing often applies a heuristic net ordering and conducts a **rip-up and reroute** process to further refine the solution. Here we give some popular net-ordering schemes: (1) Order the nets in the ascending order of the number of pins within their bounding boxes. If there are more pins inside the bounding box of a net, this net would tend to block the nets inside this bounding box. (2) Order the nets in the ascending (descending) order of their lengths if routability (timing) is the most critical metric. Research shows that routing shorter nets first often leads to better routability, because they usually have less routing flexibility than the longer ones. In this way, the shorter and straight nets would be routed without excessive detours, and the routing resource would be used more efficiently. In contrast, longer nets should be routed earlier for high-performance designs because they typically determine the overall timing. (3) Order the nets on the basis of their timing criticality. In addition to the net-ordering schemes, we can first analyze the net distribution over the routing region, identify the congested regions, and then route nets in the most congested regions first.

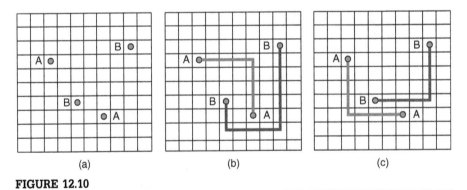

(a)   (b)   (c)

FIGURE 12.10

Routing based on different net orderings: (a) A one-layer routing instance with two two-pin nets *A* and *B*. (b) An inferior solution obtained by the net ordering of *A* followed by *B*. (c) A better solution resulted by the net ordering of *B* followed by *A*.

The rip-up and reroute process consists of two steps: (1) identify the bottle-neck regions and rip up some already routed nets and (2) route the blocked connections and reroute the ripped-up connections. The process often performs iteratively until all nets are routed or a time limit is exceeded. Generally, it can lead to more desirable routing solutions. As the example of Figure 12.10b, if the router has observed that net *B* is blocked or its length is substantially increased because of net *A*, it can rip up net A, and reroute *B* and then *A* to improve the solution. McMurchie and Ebeling developed a **negotiation-based rip-up and reroute** algorithm called *PathFinder* for *field-programmable gate array* (FPGA) [McMurchie 1995], which reveals its superiority in recent leading academic global routers, such as *BoxRouter* [Cho 2007], *FastRoute* [Pan 2007], *FGR* [Roy 2007], *NTHU-Route* [Chang 2008], and *NTUgr* [Hsu 2008].

Chen *et al.* and Kastner *et al.* developed **pattern-routing** schemes [Ho 1990; Chen 1999; Kastner 2002] that use patterns such as **L-shaped** (1-bend) or **Z-shaped** (2-bend) routes to make connections (see Figure 12.11). The pattern routing gives the shortest path length between two points and enjoys very high speed and less memory use, because the search space followed by patterns is much smaller than the maze-routing algorithm. As a result, pattern routing is pervasively used for global-routing applications.

12.4.2 Concurrent global routing

The major drawback of the sequential approach is that it suffers from the net-ordering problem. Under any net ordering, it is more difficult to route the nets that are considered later, because they are subject to more blockages. In addition, if the sequential routing fails to find a feasible solution, it is not clear whether this is because of no existing feasible solution or because of a poor selection of net order. Moreover, when the sequential routing does find a feasible solution, we do not know whether or not this solution is optimal or how far it is from the optimal solution. These questions may be answered if we solve the routing problem with the concurrent approach.

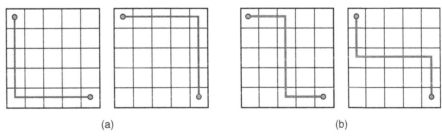

(a)   (b)

FIGURE 12.11

Pattern routing: (a) L-shaped (1-bend) routes. (b) Z-shaped (2-bend) routes.

One popular concurrent approach is to formulate global routing as a *0-1 integer linear programming* (0-1 ILP) problem. The layout is first modeled as a routing graph $G(V, E)$, where each node represents a tile and each edge denotes the boundary between two adjacent tiles. Each edge $e \in E$ is assigned a capacity, denoted by c_e, which represents the number of tracks crossing that boundary. Given a net, all of its possible routing patterns can be enumerated. Let the variable $x_{i,j} \in \{0,1\}$ indicate whether the routing pattern $r_{i,j}$ is selected from the set R_i of routing patterns of net n_i. Consequently, for a routing graph $G(V, E)$ with the netlist N, the congestion-driven global routing can be formulated as a 0-1 ILP problem as follows:

$$
\begin{aligned}
\text{Minimize} \quad & \lambda \\
\text{Subject to} \quad & \sum_{r_{i,j} \in R_i} x_{i,j} = 1, \quad && \forall n_i \in N \\
& x_{i,j} \in \{0,1\}, \quad && \forall n_i \in N, \forall r_{i,j} \in R_i \\
& \sum_{i,j:e \in r_{i,j}} x_{i,j} \le \lambda c_e, \quad && \forall e \in E
\end{aligned}
\tag{12.1}
$$

The first and the second constraints require that only one routing pattern can be chosen for each net, and the third constraint with the objective together ensure to minimize the maximum congestion. If a solution of $\lambda \le 1$ exists, a global-routing solution with the maximum congestion being minimized can be achieved.

Because the 0-1 ILP is NP-complete, the high time complexity greatly limits the feasible problem size. An alternate approach to this problem is to first solve the continuous *linear programming* (LP) relaxation, obtained by replacing the second constraint with the real variable $x_{i,j} \in [0,1]$, because LP problems can be solved in polynomial time. Then, the resulting fractional solution can be transformed to integer solutions through rounding such as randomized rounding [Raghavan 1987]. However, this approximation could inevitably lose the optimality.

In practice, the 0-1 ILP concurrent routing technique is often embedded into a larger overall global routing framework with a hierarchical, divide-and-conquer manner, such as solving a subproblem, in which the complexity of computing the optimal solution is manageable. Another approach to divide a routing region into subregions such that the routing problem can be handled subregion by subregion to reduce the problem size is BoxRouter [Cho 2006], which is based on box expansion to push the congestion outward progressively.

12.4.3 Steiner trees

The algorithms we have described so far are mainly for two-pin nets. If all nets are two-pin ones, we can apply a general-purpose routing algorithm to handle the problem, such as maze, line-search, and A*-search routing described in Section 12.3.

For three or more multi-pin nets, one naive approach is to *decompose* each net into a set of two-pin connections, and then route the connections one-by-one. One popular decomposition method is to find a *minimum spanning tree*

(MST) for pins of each net, which is a minimum-length tree of edges connecting all the pins. The MST can efficiently be computed in polynomial time by the **Kruskal** [Kruskal 1956] or **Prim-Dijkstra** [Prim 1957] algorithms. However, the routing result of this approach would depend on the decomposition and often leads to only suboptimal solutions. Figure 12.12 depicts an example 4-pin net decomposed by a rectilinear MST, where each segment runs horizontally or vertically.

A better and more natural method to route multi-pin nets is to adopt the **Steiner-tree**–based approach. Specifically, a ***minimum rectilinear Steiner tree*** (MRST) is used for routing a multi-pin net with the minimum wirelength. Given m points in the plane, an MRST connects all points by rectilinear lines, possibly via some extra points (called **Steiner points**), to achieve a minimum-wirelength tree of rectilinear edges. Let P and S denote the sets of original points and Steiner points, respectively. Then, we have the following relationship between MRST and MST.

$$\text{MRST}(P) = \text{MST}(P \cup S) \tag{12.2}$$

Figure 12.13b shows an example of the MRST with two Steiner points s_1 and s_2 for the four pins p_1, p_2, p_3, and p_4 in Figure 12.13a.

There could be an infinite number of Steiner points that need to be considered for the MRST construction. Fortunately, Hanan proved that for a set P of pins, there exists an MRST of P with *all* Steiner points chosen from the grid points of the **Hanan grid**, which is obtained by constructing vertical and horizontal lines through every pin in P. This is known as **Hanan's theorem** [Hanan 1966]. Figure 12.13c shows the Hanan grid for the four pins in Figure 12.13a. Both the Steiner points s_1 and s_2 of MRST in Figure 12.13b are on the grid points of the Hanan grid.

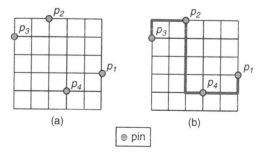

(a)   (b)

⬤ pin

FIGURE 12.12

A 4-pin net decomposed by a minimum rectilinear spanning tree: (a) A net consisting of four pins: p_1, p_2, p_3, and p_4. (b) An MST of (a), which decomposes the net into three two-pin connections.

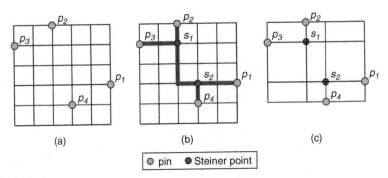

FIGURE 12.13

A minimum rectilinear Steiner tree (MRST) and its Hanan grid: (a) A net consisting of a set P of four pins: p_1, p_2, p_3, and p_4. (b) An MRST of (a) with the two Steiner points s_1 and s_2. (c) The Hanan grid of P. Note that all Steiner points s_1 and s_2 of MRST in (b) are chosen from the grid points on the Hanan grid.

The Hanan theorem greatly reduces the search space for the MRST construction from an infinite number of choices to only $m^2\text{-}m$ candidates for the Steiner points, where $m=|P|$. However, the MRST construction is still an NP-hard problem [Garey 1977]. Therefore, many heuristics have been developed.

The relationship between MST and MRST can be stated by **Hwang's theorem** [Hwang 1976] as follows:

$$\frac{Wirelength(MST(P))}{Wirelength(MRST(P))} \leq \frac{3}{2} \qquad (12.3)$$

Equation (12.3) gives a strong motivation for constructing an MRST by an MST-based approximation algorithm. Ho *et al.* constructed an MRST from an MST by maximizing monotonic (nondetour) edge (*e.g.*, L-shaped, Z-shaped) overlaps by dynamic programming [Ho 1990]. Kahng and Robins developed the **iterated 1-Steiner heuristic** [Kahng 1990] (see Algorithm 12.1). Starting with an MST, they iteratively select one Steiner point that can reduce the wirelength most and then add the Steiner point to the tree. The iterations continue until the wirelength cannot be further improved. Figures 12.14b–d illustrates the first, second, and third iterations after inserting Steiner points s_1, s_2, and s_3 into the initial MST in Figure 12.14a, respectively. Note that the iterated 1-Steiner heuristic may generate a "degenerate" Steiner point with the number of branches (degrees) $\leqq 2$, such as s_1 in Figure 12.14d. Therefore, we have to remove a degenerate Steiner point whenever it is created (see Figure 12.14e). Figure 12.14e shows the final MRST of Figure 12.14a.

On the basis of the **spanning graph** that contains an MRST in a sparse graph, [Zhou 2004] developed an efficient MRST algorithm with the worst-case time complexity of only $O(m \lg m)$ and solution quality close to that of the

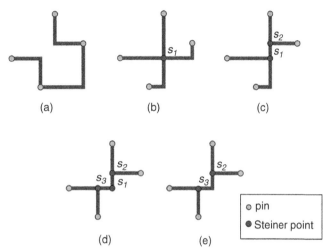

FIGURE 12.14

A step-by-step example of the iterated 1-Steiner heuristic for a 4-pin net: (a) The initial MST. (b) The MRST after the first iteration by inserting the Steiner point s_1. (c) The MRST after the second iteration by inserting the Steiner point s_2. (d) The MRST after the third iteration by inserting the Steiner point s_3. (e) The final MRST after removing the degenerate Steiner point s_1.

iterated 1-Steiner heuristic. [Chu 2004] developed the FLUTE package by use of precomputed lookup tables to efficiently and accurately estimate the wirelength for multi-pin nets. Lin *et al.* constructed a single-layer and a multi-layer obstacle-avoiding MRST to consider routing obstacles incurred from power networks, prerouted nets, IP blocks, and/or feature patterns for manufacturability/reliability improvements [Lin 2007, 2008]. Shi *et al.* constructed an obstacle-avoiding MRST based on a current-driven circuit model [Shi 2006].

Algorithm 12.1 Iterated 1-Steiner Algorithm

Input: P – a set of m pins.
Output: a Steiner tree on P.
1. $S \leftarrow \phi$;
   /*$H(P \cup S)$: set of Hanan points */
   /* $\Delta MST(A, B) = Wirelength(MST(A)) - Wirelength(MST(A \cup B))$ */
2. **while** $(Cand \leftarrow \{x \in H(P \cup S) \mid \Delta MST(P \cup S, \{x\}) > 0\} \neq \phi)$ **do**
3.   Find $x \in C$ *and* which maximizes $\Delta MST(P \cup S, \{x\})$;
4. $S \leftarrow S \cup \{x\}$;
5.   Remove points in S which have degree ≤ 2 in $MST(P \cup S)$;
6. **end while**
7. **Output** $MST(P \cup S)$;

12.5 **DETAILED ROUTING**

Given global-routing paths, detailed routing determines the exact tracks and vias for nets. Here, we discuss the two most popular types of detailed routing: **channel routing** and **full-chip routing**.

In earlier process technologies when the maximum number of available metal layers was only two or three, channel routing was pervasively used, because most wires were routed in the free space (*i.e.*, **routing channel**) between a pair of logic blocks (cell rows); see Figure 12.15. In modern technologies, a chip typically contains six to ten metal layers, and the number of available metal layers is expected to increase steadily in the near future. With more metal layers, routing over the logic block (cell rows) is common (*i.e.*, **over-the-cell routing**). As a result, routing regions become more like channel-less regions. This trend drives the need of a full-chip routing method.

12.5.1 **Channel routing**

Channel routing is a special case of the routing problem in which wires are connected within the routing channels. To apply channel routing, a routing region is usually decomposed into routing channels. Note that there are often various ways to decompose a routing region. For example, Figure 12.16 shows two ways of decomposition for the T-shaped routing region. The routing region shown in Figure 12.16a is decomposed into one horizontal channel (channel 1) and one vertical channel (channel 2), whereas that in Figure 12.16b is decomposed into two horizontal channels (channels 1 and 2) and one vertical channel (channel 3).

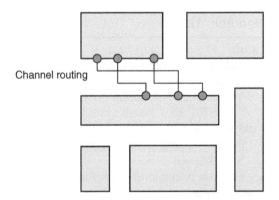

Channel routing

FIGURE 12.15

Channel routing between IC blocks.

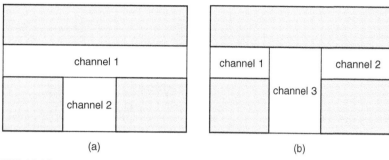

FIGURE 12.16

Two ways of routing region decomposition: (a) The routing region is decomposed into two channels. (b) The routing region is decomposed into three channels.

The order of routing regions significantly affects the channel-routing process. In Figure 12.16a, no conflicts occur in case of routing in the order of channel 2 and then channel 1. Instead, if channel 1 is routed first and all related wirings are fixed in the channel, channel 2 cannot be expanded if this channel cannot accommodate all the nets. In contrast, if channel 2 is routed first, we can still expand channel 1 for routing if needed. Note that it is not always possible to find a feasible channel ordering to avoid conflicts, for which we could resort to L-shaped channel routing to resolve the conflicts.

For modern chip routing, each routing layer typically has a preferred routing direction, either a horizontal or a vertical routing layer (*a.k.a.* **reserved routing model**). For example, the three-layer **HVH routing model** means that the preferred directions of the first, second, and third layers are horizontal, vertical, and horizontal, respectively. For the channel routing problem discussed in this section, we assume a two-layer HV routing model, unless stated otherwise.

We define some terminology of channel routing (see Figure 12.17 for an illustration). The inputs to a channel routing problem are two channel boundaries, the **upper boundary** and the **lower boundary**, with pin (terminal) numbers on columns of the channel boundaries. The pin number represents its unique net ID; pins of the same number belong to the same net and thus must be interconnected. The horizontal wire segments on the tracks are **trunks**, and the vertical wire segments connecting trunks to pins are **branches**. If the routing path of a net contains more than one trunk, this routing path is called a **dogleg**. The area of a routing channel is represented by the number of routing **tracks**, called **channel height**, inside the channel. Each column of a routing channel is associated with a **local density** to represent the total number of nets crossing the column. **Channel density**, the density of a routing channel, is then defined as the maximum local density inside the channel. It is obvious that channel density is a lower bound for the number of tracks required to complete the routing. The main objective of channel routing is to minimize the channel

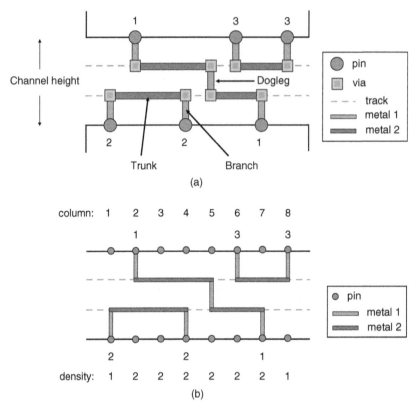

FIGURE 12.17

Channel routing illustration: (a) A channel routing configuration with two routing tracks. (b) A simplified illustration for (a).

height, which is directly related to the die size and thus the manufacturing cost. The general two-layer channel routing problem is NP-complete [Szymanski 1985], whereas some special cases of the problem can be solved optimally in polynomial time [Hashimoto 1971].

Figure 12.17a illustrates an example of two-layer channel routing that connects three nets with the pin numbers 1, 2, and 3, respectively. The channel height is two, and the connection of net 1 is a dogleg. For brevity, we would instead use the simplified illustration of Figure 12.17b throughout this chapter. As illustrated in Figure 12.17b, the routing channel contains eight columns with its local densities of 1, 2, 2, 2, 2, 2, 2, 1 for these columns (from left to right) and the channel density of 2.

To minimize the channel height, doglegs are commonly used to connect wire segments. For the same routing instance, the channel routing with doglegs shown in Figure 12.18b requires a channel height of only two tracks, whereas that without dogleg shown in Figure 12.18a needs four tracks to complete the routing.

In the following we introduce the **dogleg channel routing algorithm** [Deutsch 1976], which is an extension from the **constrained left-edge channel routing algorithm** [Hashimoto 1971]. The dogleg channel routing algorithm first decomposes multi-pin nets into two-pin connections and then assigns the trunk of each connection into a feasible track.

The dogleg channel routing algorithm contains three steps: (1) decompose each multi-pin net into 2-pin connections, (2) construct two constraint graphs to model the routing constraints, the *horizontal constraint graph* (*HCG*) and the *vertical constraint graph* (*VCG*), according to the locations of these connections, and (3) route each net without violating any constraints modeled in both HCG and VCG. As an example of the net decomposition, the 3-pin net 1 (represented by the interval [2, 7] because it spans from Column 2 to Column 7) is broken into two 2-pin connections, 1_a (interval [2, 5]) and 1_b (interval [5, 7]), as shown in Figure 12.19b.

The second step is to construct the *HCG* and *VCG* for the given routing instance. The *HCG* (*V, E*) is an *undirected* graph, where each node $v_i \in V$ represents a connection n_i, and an edge $(v_i, v_j) \in E$ exists if and only if a horizontal constraint exists between connections n_i and n_j (*i.e.*, the spans [intervals]) of n_i and n_j are overlapped) and thus n_i and n_j cannot share the same track or a circuit short would occur. In the example of Figure 12.19b, the spans of connections 2 and 4 ([1, 4] and [2, 4], respectively) are overlapped in the interval [2, 4], so there is a horizontal constraint in *HCG* between the nodes 2 and 4. Figure 12.19c depicts the *HCG* for the channel routing instance of Figure 12.19b. Note that there is no horizontal constraint between 1_a and 1_b, because they belong to the same net (net 1).

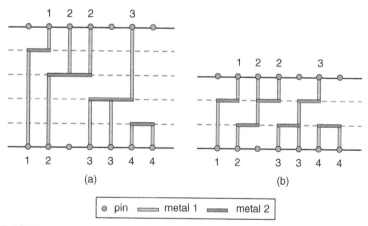

(a)   (b)

| ⊙ pin | ▭ metal 1 | ▭ metal 2 |

FIGURE 12.18

The effect of dogleg channel routing: (a) A channel routing solution without dogleg requires four tracks for routing completion. (b) A channel routing solution with dogleg only requires two tracks.

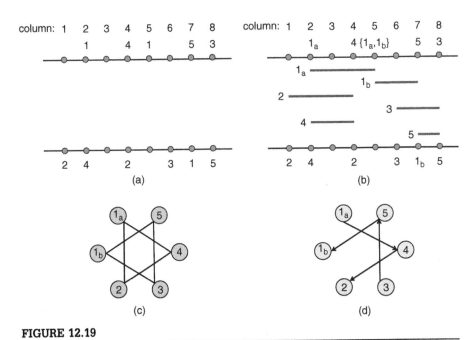

FIGURE 12.19

Constraint graph construction for dogleg channel routing: (a) A channel routing instance. (b) Multi-pin net decomposition. (c) The undirected horizontal constraint graph (HCG). (d) The directed vertical constraint graph (VCG).

The *VCG* (V, E) is a ***directed*** graph in which each node $v_i \in V$ represents a connection n_i, and a directed edge $(v_i, v_j) \in E$ exists if a vertical constraint exists between n_i and n_j (*i.e.*, the truck of n_i must be above that of n_j). The *VCG* can directly be constructed according to the pin locations in the upper and lower boundaries. For the example of Figure 12.19b, the pins in Column 4 of the upper and lower boundaries are 4 and 2, respectively; therefore, there is a directed edge (4, 2) in *VCG*. Figure 12.19d gives the *VCG* for the instance of Figure 12.19b.

The third step is to route each net under the constraints specified in both *HCG* and *VCG*. Suppose it routes nets to the routing tracks from top to bottom. In this step, the constrained left-edge algorithm [Hashimoto 1971] is applied. First, the algorithm treats each connection as an interval, and intervals are sorted according to their left-end *x*-coordinates. Then, the connections without any vertical constraint (*e.g.*, the nodes with zero in-degrees in the *VCG*) are routed one-by-one according to the order. For a connection, tracks in the channel are scanned from top to bottom, and the first track that can accommodate this connection is assigned to the connection. After all trunks (horizontal connections) are assigned to tracks, channel routing is completed by connecting the left ends and right ends of the trunks to the corresponding pins on the channel boundaries via branches. Note that the routing for a channel with no vertical

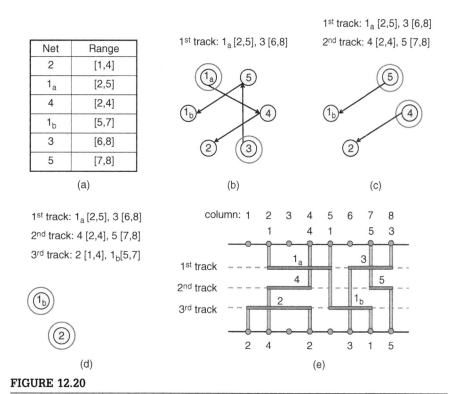

Net	Range
2	[1,4]
1_a	[2,5]
4	[2,4]
1_b	[5,7]
3	[6,8]
5	[7,8]

(a)

1st track: 1_a [2,5], 3 [6,8]

(b)

1st track: 1_a [2,5], 3 [6,8]
2nd track: 4 [2,4], 5 [7,8]

(c)

1st track: 1_a [2,5], 3 [6,8]
2nd track: 4 [2,4], 5 [7,8]
3rd track: 2 [1,4], 1_b[5,7]

(d)

(e)

FIGURE 12.20

Dogleg channel routing for the instance of Figure 12.19a (unconstrained connections in the VCG are circled): (a) Connections are sorted by the left-end coordinates. (b) Connections 1_a and 3 are assigned one-by-one to the first track. (c) Connections 4 and 5 are assigned one-by-one to the second track. (d) Connections 2 and 1_b are assigned one-by-one to the third track. (e) The final routing solution with three tracks.

constraints (see the instance shown in Figure 12.17 for an example) can be solved optimally in polynomial time by the left-edge algorithm [Hashimoto 1971].

Figure 12.20 illustrates dogleg channel routing for the instance of Figure 12.19a, which has the channel density of three. Connections are first sorted as $<2, 1_a, 4, 1_b, 3, 5>$ according to their left-end coordinates (see Figure 12.20a). As shown in Figure 12.20b, there are two unconstrained connections 1_a and 3 in the VCG, and according to the order, 1_a and 3 are routed one-by-one. Both 1_a and 3 are assigned to the first track. Then the VCG is updated by deleting nodes 1_a and 3 and related edges (see Figure 12.20c). The resulting unconstrained connections in the VCG are 4 and 5. Similarly, 4 and 5 are routed one-by-one, and both trunks of 4 and 5 are routed on the second track. The VCG is then updated by deleting the nodes 4 and 5 and related edges (see Figure 12.20d). The resulting unconstrained connections in the VCG are 1_b and 2. Finally, 2 and 1_b are routed one-by-one, and both trunks of 2 and 1_b are assigned to the third track. The final routing solution is then obtained (see Figure 12.20e) after connecting the left ends and right ends of each trunk to the pins on the corresponding channel boundaries via branches.

Note also that the dogleg channel routing algorithm introduced in [Deutsch 1976] applied two parameters to control the routing:

- Range: Determines the number of consecutive 2-pin connections of the same net that can be placed on the same track. This parameter would affect the number of doglegs and thus the number of vias.
- Routing sequence: Specifies the starting position and the direction of routing along the channel. The dogleg channel router assigns connections to the routing tracks from top to bottom, from bottom to top, or alternately with the two directions. Different routing sequences might result in different routing solutions. Note that the connections without any vertical constraint correspond to the nodes with zero out-degrees in the *VCG* if the routing sequence is from bottom to top.

12.5.2 **Full-chip routing**

Full-chip routing is typically a very complex combinatorial problem. To make it manageable, many routing algorithms adopt a two-stage technique of global routing followed by detailed routing. However, the continuously increasing design complexity imposes severe challenges for modern routers. The traditional **flat framework** does not scale well as the design size increases. A modern chip may contain billions of transistors and millions of nets. To cope with the scalability problem, routing frameworks are evolving, and the **hierarchical** and **multilevel frameworks** have become more and more popular for large-scale designs.

The hierarchical routing framework uses the divide-and-conquer approach by transforming a large and complicated routing problem into a series of smaller and simpler subproblems and then proceeds in a *top-down*, *bottom-up*, or *hybrid* manner, which can be applied to both global and detailed routing.

A top-down hierarchical global-routing framework has been proposed in [Burstein 1983]. The algorithm recursively divides the routing regions into successively smaller subregions, named **super cells**, and nets at each hierarchical level are routed sequentially or concurrently and are refined in the subsequent levels. Figure 12.21 illustrates an example of global routing for a 3-pin net by the top-down hierarchical approach, in which the routing region is recursively bisected into smaller super cells, and at each level, the net is routed in terms of these super cells at that level. This process is performed in a top-down manner until the sizes of super cells reduce to that of global-routing tiles.

A bottom-up hierarchical routing method is developed in [Marek-Sadowska 1984]. Initially, the routing region is partitioned into an array of super cells. At each hierarchical level, the routing is restrained within each super cell

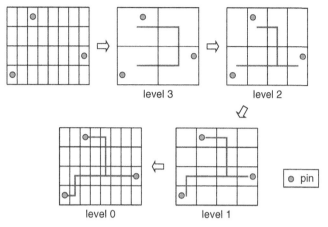

FIGURE 12.21

A level-by-level top-down hierarchical routing approach for a 3-pin net.

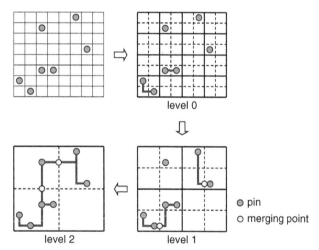

FIGURE 12.22

A level-by-level bottom-up hierarchical routing approach for a 7-pin net.

individually. When the routing at the current level is finished, every four super cells are merged to form a new larger super cell at the next higher level. This process continues until the top level containing the whole chip is reached. Figure 12.22 shows the process of bottom-up hierarchical routing for a 7-pin net, in which each solid rectangle represents a super cell, and the 2*2 dotted subregions of the previous level are merged together.

A major limitation in the top-down and the bottom-up hierarchical approaches is that the routing decision made at one hierarchical level may be suboptimal for subsequent levels. To alleviate this problem, Lin, Hsu, and Tsai proposed a hybrid hierarchical approach that combines the bounded maze-routing algorithm with both the top-down and bottom-up hierarchical methods into a unified routing framework [Lin 1990]. Their algorithm consists of three phases: (1) neighboring propagation, (2) preference partitioning, and (3) bounded routing.

Phase 1 performs bounded maze routing by propagating W circles of waves out of each pin, where W is a user-defined parameter. If the connection is not found, Phase 2 recursively maps the pins and blockages onto the adjacent upper level (see Figure 12.23a) and calls the bounded maze-routing algorithm until a path is found. Then, the connected path is mapped back to the lower level to *preferred regions* (see Figure 12.23b). Phase 3 finds a routing path in the preferred regions (see Figure 12.23c). Compared with pure top-down or bottom-up hierarchical routing, the hybrid hierarchical approach has more global information to generate better routing solutions.

Although the hierarchical routing approach can scale to larger designs, it has the significant drawbacks that the interactions among different routing subregions are lacking and the routing decision at a level is irreversible (*i.e.*, cannot be refined at later stages), thus limiting the solution quality. To remedy the deficiencies, researchers have proposed the **multilevel framework** to handle large-scale routing problems. The multilevel frameworks were first developed in [Cong 2001, 2002] for global routing and in [Lin 2002] and [Chang 2004] for both global and detailed routing. In the following, we introduce the routability-driven Λ-**shaped multilevel routing framework** [Chang 2004].

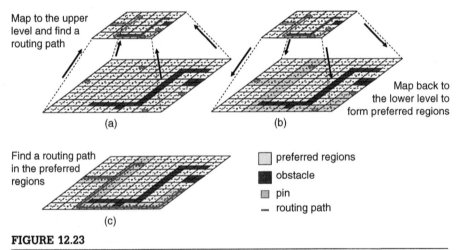

FIGURE 12.23

An example of global routing by use of the hybrid hierarchical approach: (a) Mapping pins and blockages up one level and then finding a routing path at the upper level. (b) Mapping the connection at the upper level to the lower level to form the preferred regions. (c) Finding a routing path in the preferred regions.

The multilevel routing framework models the routing resource as a **multilevel-routing graph**. At the beginning, the routing region is partitioned into an array of rectangular subregions, each of which may accommodate tens of routing tracks in each dimension (see Figure 12.24). These subregions are called *global cells* (*GCs*). A node in the routing graph represents a *GC* in the chip, whereas an edge denotes the boundary between two adjacent *GCs*. Each edge is assigned a capacity according to the physical area or the size of a *GC*. This routing graph is called the multilevel-routing graph of level 0, denoted by G_0, in which the subscript represents the level.

The Λ-shaped multilevel routing framework consists of bottom-up **coarsening** followed by top-down **uncoarsening**. The coarsening stage is a bottom-up approach that iteratively groups a set of *GCs* in the multilevel-routing graph. This process starts from the finest level (level 0) to the coarsest level; at each level k, four adjacent GC_k of G_k are merged into a larger GC_{k+1} of G_{k+1}, and at the same time it performs resource estimation for use at the $k+1$ level. Coarsening continues until the number of *GCs* at a level is below a threshold. In contrast, the uncoarsening stage iteratively ungroups a set of previously clustered *GCs* in a top-down manner. It proceeds from the coarsest level to the finest level; at each level k, a GC_k is decomposed into four smaller GC_{k+1}. Uncoarsening continues until the finest level is reached. Figure 12.25 illustrates the Λ-shaped multilevel framework.

Given a netlist, the multilevel routing first applies a minimum spanning tree (MST) algorithm to decompose each net into 2-pin connections. At each level k of the coarsening stage, global routing is first performed for the **local** 2-pin connections (those connections that entirely sit inside a GC_k), and then the detailed router is used to determine the exact wiring. Let the multilevel-routing graph of level 0 be $G_0 = (V_0, E_0)$, and the global-routing result for a local connection be $Re = \{e \in E_0 | e$ is the edge chosen for routing$\}$. For the congestion control, the cost function $\alpha : E_0 \to \mathfrak{R}$ is applied to guide the routing:

$$\alpha(R_e) = \sum_{e \in R_e} c_e \qquad (12.4)$$

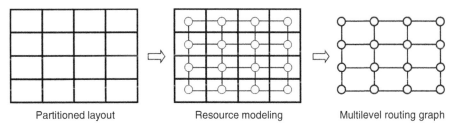

Partitioned layout · Resource modeling · Multilevel routing graph

FIGURE 12.24

The multilevel-routing graph.

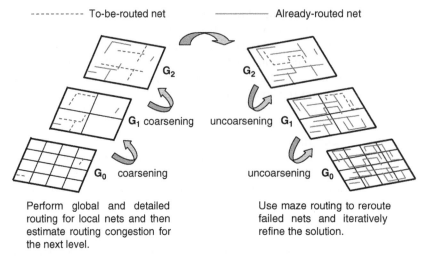

- - - - - - - - - - To-be-routed net   ——————— Already-routed net

G_2 G_2

G_1 coarsening   uncoarsening G_1

G_0 coarsening   uncoarsening G_0

Perform global and detailed routing for local nets and then estimate routing congestion for the next level.

Use maze routing to reroute failed nets and iteratively refine the solution.

FIGURE 12.25

The Λ-shaped multilevel routing framework.

where c_e is the congestion of edge e and is defined by

$$c_e = 1/2^{(p_e - d_e)} \qquad (12.5)$$

where p_e and d_e are the capacity and density associated with e, respectively. Note that we always search the shortest global-routing path between two pins in the coarsening stage therefore (*i.e.*, monotonic routes or no detours); therefore, the wirelength is the minimum, and thus the wirelength is not included in the cost function at the global routing stage. This cost function can guide the global router to select a path with smaller congestion.

After the global routing is completed, the detailed routing applies a ***simultaneous pathlength and via minimization*** (SPVM) algorithm to perform modified maze routing that simultaneously considers the pathlength and via minimization. For better circuit performance, it is desirable to minimize the number of vias used in a routing path, because vias typically have significantly larger RC delay than metal wires. The SPVM algorithm can find a shortest path with the minimum number of bends/vias, if such a path exists. It associates each *basic detailed routing region u* (could be a grid cell in grid-based routing or a basic routing region defined by the wire pitch in gridless routing) with two labels $d(u)$ and $b(u)$, where $d(u)$ is the distance of the shortest path from the source s to u, and $b(u)$ is the minimum number of bends/vias along the shortest path from s to u.

Initially, $d(s)$, $b(s) = 0$, and $d(u)$, $b(u) = \infty$, $\forall\ u \neq s$. In the filling phase of maze routing, the computation of label d is the same as the original maze-routing algorithm. Let u be a basic routing region on the wavefront of wave propagation and v a neighboring basic-routing region of u. The predecessor routing region of u is the region from which the wavefront was propagated for obtaining the minimum $b(u)$. The propagation direction of u is the direction from the predecessor routing region of u to u. The computation of $b(v)$ is shown in Algorithm 12.2.

Algorithm 12.2 Computation of $b(v)$ in the SPVM Algorithm

---

1. **if** $(d(v) \geq d(u) + 1)$ **do**
2.   **if** $((b(v) > b(u))$ **and**
      (v is along the propagation direction of u)) **do**
3. $b(v) \leftarrow b(u)$;
4.   Record u as the predecessor routing region of v;
5.   **end if**
6.   **if** $((b(v) > b(u) + 1)$ **and**
      (v is not along the propagation direction of u)) **do**
7. $b(v) \leftarrow b(u) + 1$;
8.   Record u as the predecessor routing region of v;
9.   **end if**
10. **end if**

The basic idea is to compare the distance label d first and then compare the bend/via number label b. The value $b(v)$ of a neighboring routing region v with $d(v) < d(u)$ remains unchanged, because the path from s through u to v is not the shortest path between s and v. The retracing phase is the same as that of the original maze-routing algorithm. Note that there may be several shortest paths with different numbers of bends/vias. The wave-propagation phase always keeps track of the shortest path with the minimum bend/via number to allow the retracing phase to find such a path.

When the global and detailed routing is performed at level k, four adjacent GC_k are merged into a larger GC_{k+1} and at the same time resource estimation is performed for use at the next level $k + 1$. Because the global routing, detailed routing, and resource estimation are integrated together at each level, the routing resource estimation is more accurate, thus facilitating the solution refinement (*e.g.*, the rip-up and reroute processes) at the uncoarsening stage. Algorithm 12.3 gives the algorithm of the Λ-shaped multilevel routing framework [Chang 2004].

12.6 MODERN ROUTING CONSIDERATIONS

As the process geometries scale down to the nanometer territory, the IC industry faces severe challenges in **signal integrity, manufacturability**, and **reliability**. In this section, we address the routing problems considering these issues. Specifically, we discuss **crosstalk** for signal integrity-aware routing,

---

Algorithm 12.3 Λ-Shaped Multilevel Routing Algorithm

Input: G – partitioned layout;
            N – netlist of multi-terminal nets.
Output: routing solutions for N on G
  1. partition the layout and build MST's for N;
     //coarsening stage
  2. **for** (each level at the coarsening stage) **do**
  3.   Choose a local net n;
  4.   **if** (n belongs to this level) **do**
  5.   Global_Pattern_Routing(n);
  6.   Detailed_Routing(n);
  7.   **end if**
  8. **end for**
     // uncoarsening stage
  9. **for** (each level at the uncoarsening stage) **do**
 10.   Choose a local net n;
 11.   Global_Maze_Routing(n);
 12.   Detailed_Routing(n);
 13. **end for**
 14. Output_Result();

---

optical proximity correction (OPC) and *chemical-mechanical polishing* (CMP) for manufacturability-aware routing, and **antenna effect** avoidance and **double-via insertion** for reliability-aware routing.

12.6.1 **Routing for signal integrity**

As the fabrication technology advances, on-chip minimum feature sizes continue to decrease, clock rates keep increasing, and devices and interconnection wires are placed in closer proximity to reduce interconnection delay and routing area. Consequently, increasing the aspect ratios of wires and decreasing interconnect spacing make the **coupling capacitance** larger than self-capacitance. In fact, the ratio of coupling capacitance is reported to be even as high as 70% to 80% of the total wiring capacitance, even in the 0.25-μm technology. As a result, **crosstalk** becomes a key issue for signal integrity.

12.6.1.1 *Crosstalk modeling*

Noise is an unwanted variation that makes the behavior of a manufactured circuit deviate from the expected response. The deleterious influences of noise can be classified into two categories. One is malfunctioning, which makes the logic values of gates differ from what we desire; the other is timing

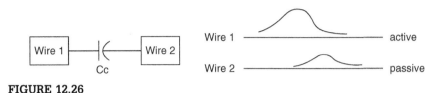

FIGURE 12.26

The crosstalk effect.

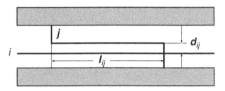

FIGURE 12.27

The capacitive crosstalk computation between two wires i and j.

change, which is caused by switching behavior. The main noise comes from the **crosstalk** effect, which is mostly caused by the coupling capacitance between interconnection wires. As an example shown in Figure 12.26, because of the coupling capacitance C_c between wires 1 and 2, wire 2 would induce an undesirable pulse when wire 1 is activated by a positive signal. If the unexpected pulse is larger than a threshold, the functionality of the circuit may fail. More precisely, the crosstalk between two wires switching in different directions would increase signal delays and decrease signal integrity; on the contrary, the crosstalk would decrease signal delays and increase signal integrity if the two wires switch in the same direction.

In general, the crosstalk between two wires is proportional to their coupling capacitance, which is determined by the relative positions of these wires. The coupling capacitance between orthogonal wires is negligible compared with that between adjacent parallel wires in current technology. Consequently, the crosstalk can be approximated by considering only adjacent parallel wires.

Figure 12.27 illustrates an instance with two wires i and j belonging to different nets. The coupling capacitance c_{ij} between i and j can be approximated as follows [Sakurai 1983]:

$$c_{ij} = a \frac{l_{ij}}{(d_{ij})^k} \quad (12.6)$$

where a is a technology-dependent constant, k is a constant between 1 and 2 (and close to 2), l_{ij} is the overlapping length of wires i and j, and d_{ij} is the distance between wires i and j. On the basis of Equation (12.6), we can see that the coupling capacitance between two parallel wires is proportional to their coupling length and is inversely proportional to the distance between them. More accurate crosstalk modeling can be found in [Vittal 1999; Jiang 2000; Cong 2001].

12.6.1.2 *Crosstalk-aware routing*

Routing with minimum crosstalk has been extensively studied in the literature [Gao 1996; Zhou 1998; Ho 2005, 2007]. Gao and Liu applied a mixed ILP (integer linear programming) formulation to permute the routing tracks in a given channel routing solution to minimize crosstalk [Gao 1996]. Zhou and Wong minimized crosstalk during global routing on the basis of a Steiner tree formulation and Lagrangian relaxation [Zhou 1998]. Chaudhary, Onozawa, and Kuh proposed a wire-spacing adjusting algorithm after detailed routing to reduce crosstalk [Chaudhary 1993]. However, it might not be easy to handle crosstalk during global routing or detailed routing. It might be too early to handle crosstalk during global routing, because the relative positions and ordering of nets are not determined at this stage; consequently, the best that one can possibly do is to use rough statistical estimators that discourage nets from entering unwanted proximity regions. Conversely, it might be too late for detailed routing to handle crosstalk, because detailed routers may encounter unsolvable rip-up/re-route problems when trying to embed a late-routing net into a dense region with conflicting aggressor or victim nets.

To address these problems, Ho *et al.* incorporated a **layer/track assignment** heuristic for crosstalk optimization in the intermediate stage of the Λ-shaped multilevel routing framework [Ho 2005], as shown in Figure 12.28.

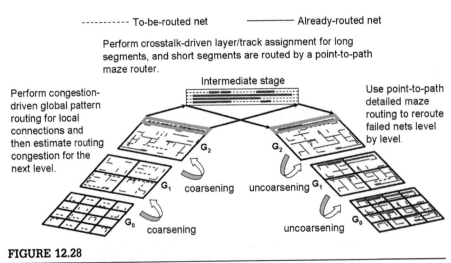

FIGURE 12.28

The Λ-shaped multilevel routing framework with an intermediate stage for crosstalk minimization.

The layer/track assigner works on a full row or column of the global cell array at a time, where a row (column) is called a ***panel***. In the layer/track assignment, the segments spanning more than one complete global cell in a row or a column are processed, and short segments are routed during detailed routing.

First, a horizontal constraint graph $HCG(V, E)$ is built for all segments in the panel. Each vertex $v \in V$ corresponds to a segment in the panel. Two vertices v_i and v_j are connected by an edge $e \in E$ if and only if these segments belong to two different nets and their spans overlap. The edge cost of $e = (v_i, v_j) \in E$ represents the coupling length if v_i and v_j are assigned to adjacent tracks. The crosstalk-driven layer assignment can be formulated as the max-cut, k-coloring (MC) problem. However, the general MC problem is NP-complete [Garey 1979]. Thus, a simple yet efficient heuristic is applied by constructing a maximum spanning tree of HCG followed by the k-coloring method to spread all segment into k layers. After k-coloring, the nodes are assigned to layers one-by-one in a decreasing order of their costs (coupling lengths).

After the crosstalk-driven layer assignment, the crosstalk-driven track assignment is applied. Let T be the set of tracks inside a panel. Each track $\tau \in T$ can be represented by the set of its constituent contiguous intervals. Denote these intervals by x_i. A segment $r \in S$ (set of segments) is said to be **assignable** to $\tau \in T$, $\tau \equiv \cup x_i$, if x_i is either a free interval or is an interval occupied by a segment of the same net.

After layer assignment, most of the edges with larger costs in an HCG are eliminated, and the HCG is decomposed into k subgraphs $subHCG_1$, $subHCG_2, \ldots, subHCG_k$ if there are k layers. Figure 12.29 shows an example of the track assignment problem for a $subHCG$, where $S = \{a, b, c, d, e, f\}$, $T = \{1, 2, 3, 4\}$, and obstacles on tracks are shaded in grey (*e.g.*, the two obstacles on tracks 3 and 4). A bipartite assignment graph is used to indicate the assignability of segments to tracks. For example, as shown in Figure 12.29b, edges between node a and nodes 1, 2, and 3 are introduced, because segment a can be assigned to track 1, 2, or 3, but not track 4. For easier implementation, the $subHCG$ and the bipartite assignment graph are merged into a combination graph, as shown in Figure 12.29c.

Because each vertex $v \in V$ corresponds to a segment and each edge $e \in E$ corresponds to the coupling cost in $HCG(V, E)$, the crosstalk-driven track assignment can be formulated as the Hamiltonian path problem which is NP-complete [Garey 1979]. Here is a heuristic for this problem. The heuristic starts by finding the maximal sets of conflicting segments. This is equivalent to finding the largest clique V_c in the subgraph $subHCG_i$. The algorithm first assigns one maximal subset of conflicting segments at a time by starting from the largest clique. Then the longest segment in the clique is chosen as the source s and assigned to the uppermost available track. Then, the minimum-cost edge (s, i) (and thus the minimal coupling) is chosen, and the segment associated with i is assigned to the first available track. If all tracks are occupied, the net associated with i is marked as a failed net that will be reconsidered at the uncoarsening stage.

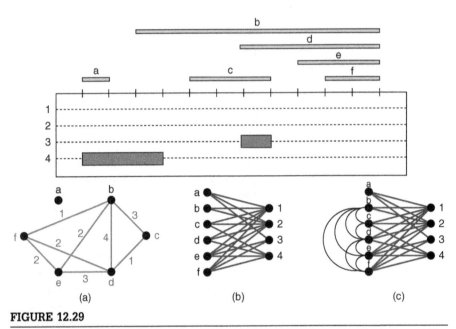

FIGURE 12.29

Constraint graph modeling for track assignment: (a) *SubHCG* for a given instance. (b) The corresponding bipartite assignment graph. (c) The combination graph.

The procedure is repeated by finding the minimum-cost edge (i, j) for further processing, where j is an unvisited node.

Figure 12.30 illustrates the track assignment process for the instance of Figure 12.29. The maximum clique in the *subHCG* is $\{b, d, e, f\}$, and the longest segment in the clique is b. Thus, the segment b is assigned to the uppermost available track, which is track 1. See Figure 12.30b for the updated combination graph after assigning b to track 1. Then, the heuristic makes b the source for constructing the Hamiltonian path for the clique. The minimum-cost edge $e = (b, f)$ incident on b is chosen, and f is assigned to the first available track. See Figure 12.30c for the updated combination graph after assigning f to track 2. The process is repeated until all nodes in the clique are visited. The final track assignment solution is shown in Figure 12.30a.

12.6.2 **Routing for manufacturability**

For manufacturability, OPC and CMP are two most important concerns for modern chip designs. The former adds or subtracts feature patterns to a **mask** to enhance the layout resolution and thus the **printability** of the mask patterns on the wafer, whereas the latter improves layout uniformity and chip planarization to achieve higher manufacturing yield.

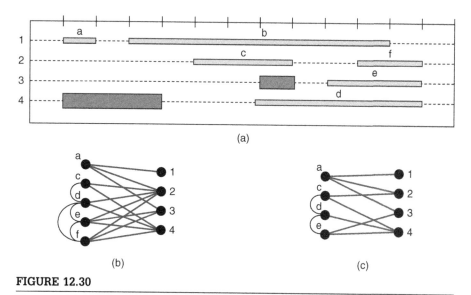

FIGURE 12.30

Process of track assignment: (a) Final track assignment for the instance of Figure 12.29. (b) The resulting combination graph after assigning *b* to track 1. (c) The resulting combination graph after assigning *f* to track 2.

12.6.2.1 *OPC-aware routing*

We will first introduce the manufacturing process. The process uses an **optical lithography system** and goes through many cycles of processing, each of which consists of two major steps: **exposure** followed by **etching**.

Figure 12.31 illustrates a basic optical lithography system. In the exposure step, it transfers the patterns on a **mask** to the light-sensitive **positive** or **negative photoresist** coated on the top of the wafer, which is performed by an intense ultraviolet light emitted from the light source through the apertures of the mask. Exposed by the light, the positive photoresist becomes soluble to the photoresist developer, whereas the negative photoresist becomes insoluble. This chemical change allows some of the photoresist to be removed by a special solution. In the etching step, a chemical agent removes the uppermost layer of the wafer in the areas that are not protected by photoresist to form the designed patterns on the wafer.

With the continuous shrinking of the minimum feature size, IC foundries have to use an optical lithography system with a larger wavelength of light to print a feature pattern with a much smaller size on a wafer, which is called the **sub-wavelength lithography gap** (see Figure 12.32). For the modern process technology, for example, we might need to print a 45-nm feature pattern by use of the light of 193-nm wavelength. The sub-wavelength lithography gap might lead to unwanted large shape *distortions* for the printed patterns on

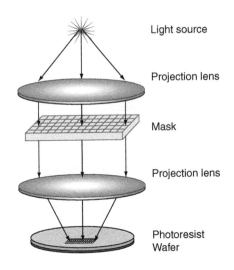

FIGURE 12.31

A typical optical lithography system.

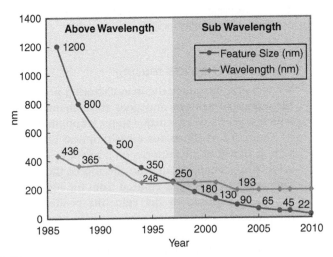

FIGURE 12.32

The sub-wavelength lithography gap: the printed feature size is smaller than the wavelength of the light shining through the mask.

the wafer. Physically, when a light with the wavelength λ passes through an aperture of the size d, the wavefronts of the light behave differently according to the relation between λ and d. When λ is much smaller than the aperture size d on the mask, the wavefronts of the light remain straight, as illustrated in Figure 12.33a. However, when λ is close to or larger than d, the light behaves

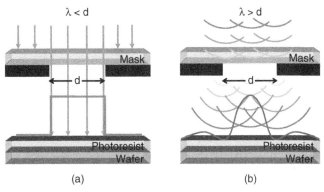

FIGURE 12.33

When a light with the wavelength λ passes through an aperture of the size d, the wavefronts of the light behave differently according to the relation between λ and d: (a) When λ is much smaller than d, the wavefronts remain straight. (b) When λ is much larger than d, diffracted wavefronts might occur.

like *waves* (instead of particles) and diffraction occurs (see Figure 12.33b), making the pattern on the wafer not exactly the same as that on the mask. As a result, intensive use of costly ***resolution enhancement techniques*** (RETs) to improve the layout accuracy becomes inevitable.

Many RETs are adopted at the post-layout stage to enhance the printability and thus the yield. The increasing design complexity, however, leaves very limited space for post-layout optimization. Therefore, it is desirable to consider the manufacturability earlier in the design flow, such as RET-aware routing.

Among the RETs, ***optical proximity correction*** (OPC) is the most popular in industry. OPC is the process of modifying the layout patterns on the mask (drawn by the designers) to compensate for the non-ideal properties of the lithography process and thus to enhance the layout printability. Figure 12.34 illustrates an example of OPC enhancement. Without OPC, the printed patterns on the wafer would be distorted from the designed pattern on the mask because of the sub-wavelength lithography. In contrast, if the patterns on the mask are enhanced by OPC, the printed patterns on the wafer could well match the original designed patterns.

However, OPC might incur a large number of extra pattern features, implying larger memory requirements to record these features and thus higher mask-making costs, such as mask synthesis, writing, and inspection verification. If a router can consider the optical effects, the number of pattern features on the final mask can greatly be reduced.

Chen and Chang proposed a **rule-based OPC-aware multilevel router** to reduce the requirements for OPC-pattern feature [Chen 2007a]. They classify the pattern distortions into three major types: **corner rounding, line-end shortening**, and **line-width shrinking**, as illustrated in Figures 12.35a–c.

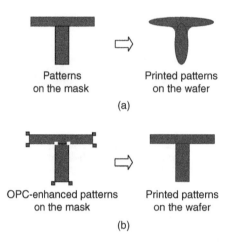

FIGURE 12.34

The effects of OPC: (a) Without OPC, the printed patterns on the wafer incur large distortions from the patterns on the mask. (b) With OPC enhancement, the printed patterns could well match the original patterns.

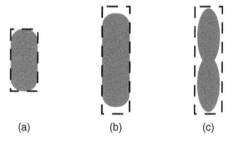

FIGURE 12.35

Three major types of pattern distortions (the dashed lines represent the ideal pattern shapes): (a) Corner rounding. (b) Line-end shortening. (c) Line-width shrinking.

For each type of distortion, the pattern features required for compensation are identified on the basis of some geometry rules, for example, the **serifs** added at corners to make the angles sharper, the **hammerheads** added at line ends to compensate for line-end shortenings, and the **line biasing** added along line sides to compensate for line-width shrinking (see Figure 12.36).

The number of pattern features required for OPC is then modeled as a cost for routing the connection. For example, as shown in Figure 12.36a, four serifs are required at the four corners to increase the fidelity of images for a line. Also, when the length of a line increases, the ends of the line become shortened, as illustrated in Figure 12.36b; therefore, two hammerheads are required at the line ends for a long line. Besides, a wider line is easier to be affected by neighboring lines than a narrower one, making the sides of a line shrink more seriously.

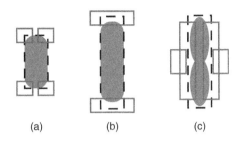

(a)   (b)   (c)

FIGURE 12.36

Three major OPC compensation pattern features: (a) Serif. (b) Hammerhead. (c) Line biasing.

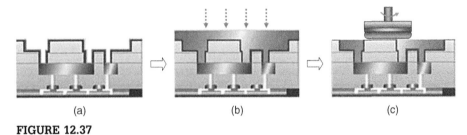

(a)   (b)   (c)

FIGURE 12.37

Damascene process: (a) Open trenches. (b) Electroplating (ECP) deposits Cu on the trenches. (c) Chemical-mechanical polishing (CMP) removes Cu that overfills the trenches.

Therefore, as shown in Figure 12.36c, some line biasing in the line sides is required for a wide line. Therefore, the total number of additional features for a line can be modeled as a function of the length and width of the line. With this function, we can incorporate the OPC cost into the original routability and wire-length costs for a router to obtain a rule-based OPC-aware routing method.

Chen, Liao, and Chang considered the OPC effects during routing to alleviate the cost of post-layout OPC operations [Chen 2008b]. They developed an analytical formula for the intensity computation from **model-based OPC** (which involves complicated simulations of various process effects) and a post-layout OPC modeling on the basis of an **inverse lithography technique**, and then incorporated the OPC costs into an OPC-friendly router. Huang *et al.* and Wu *et al.* also addressed OPC-friendly maze routing [Huang 2004b; Wu 2005b].

12.6.2.2 *CMP-aware routing*

In the modern metallization process, copper (Cu) has replaced the traditional aluminum (Al) because of its better properties, such as higher current-carrying capability, lower resistance, and lower cost. However, the process of copper is significantly different from that for traditional aluminum. The modern copper metallization process applies the **dual-Damascene process** [Luo 2005], which consists of *electroplating* (ECP) followed by the *chemical-mechanical* *polishing* (CMP). The ECP deposits the copper on the trenches, whereas the CMP removes the copper that overfills the trenches, as shown in Figures 12.37a–c.

Figure 12.38 shows a schematic diagram of the CMP process. Abrasive and corrosive chemical **slurry** that can dissolve the wafer layer is deposited on the surface of a polishing pad. Then, the polishing pad and wafer are pressed together by a dynamic, rotating polishing head. Combined with both the chemical reaction and the mechanical force, the CMP process can remove materials on the surface of the wafer and tends to make the wafer planar.

However, because of the difference in the hardness between copper and dielectric materials, the CMP planarizing process might generate topography *irregularities*, which might incur significant yield loss of copper interconnects. The studies of the CMP process have indicated that the post-CMP dielectric thickness is highly correlated to the layout pattern density, because during the polishing step, the dielectric removal rates are varied with the pattern density. A non-uniform feature density distribution on each layer might cause CMP to over polish or under polish, as illustrated in Figure 12.39.

These post-CMP thickness variations need to be carefully controlled, because the variation in one metal layer could be progressively transferred to subsequent layers during manufacturing, and finally the accumulative variation could be

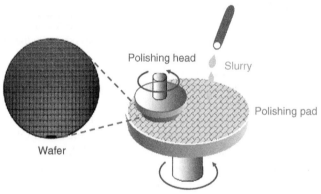

FIGURE 12.38

Schematic diagram of the CMP polisher.

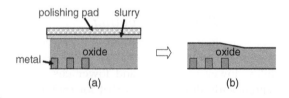

FIGURE 12.39

Layout-dependent thickness variations: (a) Pre-CMP layout. (b) Post-CMP thickness variation.

significant on the upper metal layer, which is often called the **multilayer accumulative effect** [Tian 2000].

To improve the CMP quality, modern foundries often impose recommended layout density rules (or even density gradient rules) on each layer and fill **dummy features** into layouts to reduce the variations on each layer. However, these filled dummy features might incur unwanted effects at 65nm and successive technology nodes [White 2005]. For example, they may induce high coupling capacitances to nearby interconnects and thus incur crosstalk problems. Moreover, dummy fills also significantly increase the data volume of mask, lengthening the time of the mask-making processes and thus the mask cost. Especially, these filled features would significantly increase the input data in the following time-consuming RETs, such as the OPC process.

Wire density greatly affects dummy feature filling. The layout pattern (consisting of wires and dummy features) density strongly depends on the wire density distribution, as reported in [Cho 2006]. Therefore, controlling wire density at the routing stage can alleviate the problems induced by aggressive dummy feature filling. In addition, good wire distribution can reduce the random particle short defects and also benefit the post-layout redundant-via insertion (see Section 12.6.3.2), which can translate into yield gain.

The density uniformity in different routing stages for CMP variation control has been addressed in the literature [Cho 2006; Chen 2007b; Li 2007]. Cho *et al.* considered CMP variation during global routing [Cho 2006]. They empirically showed that the number of inserted dummy features can be predicted by the wire density and observed that a path with higher pin density may not get much benefit from the wire density optimization, because there is little room for improvement (it is destined to have high wire density from the beginning). Therefore, they proposed a **minimum pin-density global-routing** algorithm to reduce the maximum wire density.

Figure 12.40 illustrates the minimum pin-density global-routing algorithm. A net from the source S to the target T to be routed is shown in

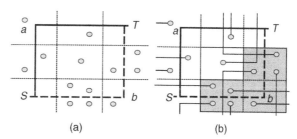

(a)   (b)

FIGURE 12.40

Minimum pin-density global routing [Cho 2006]: (a) There are two possible 1-bend paths a and b from the source S to the target T. (b) The path a with smaller pin density is better than the path b.

Figure 12.40a with a pin distribution. If only the *L*-shaped (1-bend) routing paths are allowed, there are two possible paths, *a* and *b*, with the same wire-length, but different pin densities. Because the existence of a pin implies at least one connection to other pins, a path with higher pin density like *b* would tend to have higher wire density eventually as shown in Figure 12.40b, resulting in higher final wire densities. Therefore, a path with the minimum pin density (like path *a*) leads to better wire density distribution.

Figure 12.41 shows the two-pass, top-down planarization-driven routing framework presented in [Chen 2007b], which consists of four major stages: (1) Prerouting: identify the potential density hot spots on the basis of the pin distribution and wire connection to guide the following global routing; (2) Global routing: apply prerouting-guided planarization-aware global pattern routing for nets and iteratively refine the solution; (3) Layer/track assignment: perform density-driven layer/track assignment for long segments panel by panel; and (4) Detailed routing: use segment-to-segment detailed maze routing to route short segments and reroute failed nets level by level. By handling longer nets first, the routing density for CMP can be better optimized, because the longer nets have higher density impact than the shorter ones.

In the prerouting stage, a density critical area analysis algorithm (on the basis of Voronoi diagrams [Preparata 1985]) is performed to identify the potential density hot spots. The identified density information of pins and wire connection is then used to guide the subsequent routing.

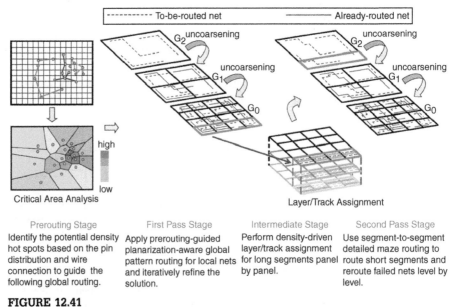

Prerouting Stage	First Pass Stage	Intermediate Stage	Second Pass Stage
Identify the potential density hot spots based on the pin distribution and wire connection to guide the following global routing.	Apply prerouting-guided planarization-aware global pattern routing for local nets and iteratively refine the solution.	Perform density-driven layer/track assignment for long segments panel by panel.	Use segment-to-segment detailed maze routing to route short segments and reroute failed nets level by level.

FIGURE 12.41

The two-pass, top-down planarization-driven routing framework.

In the first top-down (global-routing) stage, a planarization-aware global router is used to consider the density lower and upper bounds while minimizing the density gradient among global tiles. The planarization-aware cost Φ_t for each global tile t is defined as follows:

$$\Phi_t = \tilde{d}_t + \begin{cases} \kappa_p, & \text{if } d_t \geq B_u \\ \beta(2^{d_t} - 1) + (1 - \beta)(d_t - \overline{d_t})^2, & \text{if } B_l \leq d_t < B_u \\ \kappa_n, & \text{if } d_t < B_l \end{cases} \tag{12.7}$$

where d_t is the wire density of t, $\tilde{d}_t$ is the predicted hot spot cost calculated in the prerouting stage, $\overline{d}_t$ is the average wire density of tiles adjacent to t, B_l and B_u are respective density lower and upper bounds specified in foundry's density rules, and β, $0 \leq \beta \leq 1$, is a user-defined parameter. κ_p and κ_n are constants, where κ_p is a positive penalty that discourages routing through dense global tiles, and κ_n is a negative reward that encourages routing through sparse tiles. The second equation simultaneously considers the local tile density and minimizes the density gradient among adjacent regions.

The intermediate stage tries to preserve more flexibility for wire density arrangement. It consists of two phases: (1) a density-driven layer assigner evenly distributes the segments in a **panel** (row of global tiles) into layers, and (2) a density-driven track assigner balances the segment density of each track on the basis of incremental **Delaunay triangulation** (DT) [Preparata 1985]. First, the **flexibility** of a segment s_i is defined as follows:

$$\xi(s_i) = t_i + \frac{1}{\ell_i} \tag{12.8}$$

where t_i is the number of assignable tracks of s_i, and l_i is the length of s_i. If the flexibility of s_i is smaller, s_i might have a longer length or less space to insert and thus should be assigned first. Therefore, segments are inserted into tracks in the nondecreasing order of their flexibilities. Then, each segment or obstacle is represented by three points: its left-end, center, and right-end points, and then the resulting DT is analyzed. The segment is assigned to a track such that the resulting area difference among all triangles is minimized. Figure 12.42 shows a density-driven track-assignment example by inserting three segments s_1, s_2, and s_3 into tracks with obstacles O_1 (see Figure 12.42). Note that the *artificial segments* lying on the boundary are used to model the distribution of segments and obstacles in the neighborhood.

After the track assignment, the actual track position of a segment is known. Thus, classical segment-to-segment maze detailed routing is performed in the second top-down (detailed-routing) stage to connect shorter nets, and the whole routing process is finished.

12.6.3 Routing for reliability

Manufacturing reliability and yield in VLSI designs are becoming a crucial challenge as the feature sizes shrink into the nanometer scale. Both the antenna

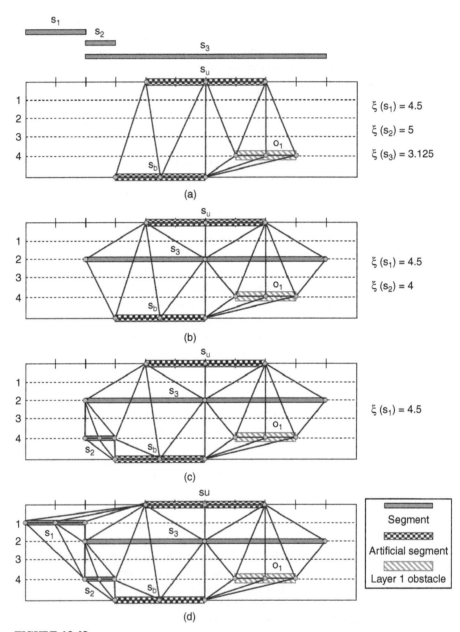

FIGURE 12.42

A density-driven track assignment example: (a) The initial Delaunay triangulation. (b) Track assignment for segment s_3. (c) Track assignment for segment s_2. (d) Track assignment for segment s_1.

effect arising in the plasma process and the via-open defect are important issues for achieving a higher reliability and yield.

12.6.3.1 *Antenna-avoidance routing*

The **antenna effect** is caused by the charges collected on the floating interconnects, which are connected to only a gate oxide. During the metallization, long floating interconnects act as temporary capacitors and store charges gained from the energy provided by fabrication steps such as plasma etching and CMP. If the collected charges exceed a threshold, the *Fowler-Nordheim* (F-N) tunneling current will discharge through the thin oxide and cause gate damage. On the other hand, if the collected charges can be released before exceeding the threshold through a low impedance path, such as diffusion, the gate damage can be avoided.

For example, considering the routing in Figure 12.43a, the interconnects are manufactured in the order of poly, metal 1, and metal 2. After manufacturing metal 1 (see Figure 12.43b), the collected charges on the right metal 1 pattern

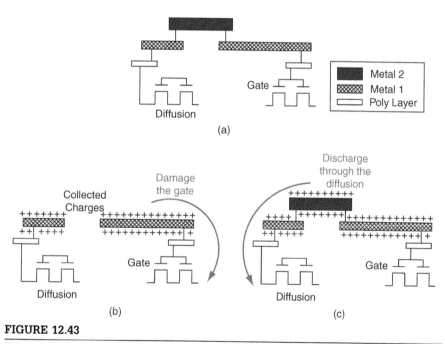

FIGURE 12.43

Illustration of the antenna effect: (a) A routing example. (b) Late stage of metal-1 pattern etching of (a), where the collected charges on the right side of the metal-1 pattern may cause damage to the connected gate oxide. (c) Late stage of metal-2 pattern etching of (a), where all the collected charges can be released through the connected diffusion on the left side.

732 CHAPTER 12 Global and detailed routing

may cause damage to the connected gate oxide. The discharging path is constructed after manufacturing metal 2 (see Figure 12.43c), and thus the charges can be released through the connected diffusion on the left side.

There are three kinds of solutions to reduce the antenna effect [Chen 2000]:

1. **Jumper insertion**: Break only signal wires with antenna violation and route to the highest level by jumper insertion. This reduces the charge amount for violated nets during manufacturing.

2. **Embedded protection diode**: Add protection diodes on every input port for every standard cell. Because these diodes are embedded and fixed, they consume unnecessary area when there is no violation at the connecting wire.

3. **Diode inserting after placement and routing**: Fix those wires with antenna violations that have enough room for "under-the-wire" diode insertion. During wafer manufacturing, all the inserted diodes are floating (or ground). One diode can be used to protect all input ports that are connected to the same output ports. However, this approach works only if there is enough room for diode insertion.

Jumper insertion is a popular way to solve the antenna problem. To avoid/fix the antenna violation, it is required that the total effective conductor connecting to a gate be less than or equal to a threshold, L_{max}. The threshold could be the wirelength limit, the wire area limit, the wire perimeter limit, the ratio of antenna strength (length, area, perimeter, etc.) to the gate size, or any model of the strength of antenna effect caused by conductors. As the example shown in Figure 12.44, we have a two-terminal net in which a is the source node and b is

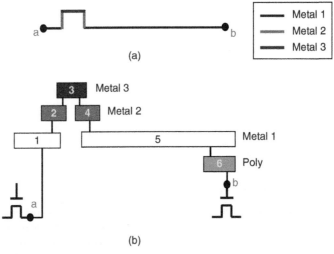

FIGURE 12.44

(a) A two-pin net. (b) The cross-sectional view.

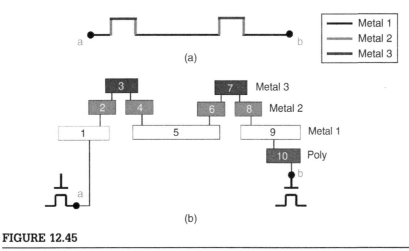

FIGURE 12.45

(a) A two-pin net with jumper insertion. (b) The cross-sectional view.

the terminal node. In this case, the antenna charge weight of b is the sum of the antenna charge weight of segments 4, 5, and 6, which may violate L_{max}. Note that once segment 3 is manufactured, a discharging path is established through segment 1 and the diffusion of the transistor a (see Figure 12.44b). If we add a jumper at the long segment 5 (see Figure 12.45), the antenna charge weight of b is just the sum of the length of segments 8, 9, and 10, which will not violate L_{max}. Thus, if we add jumpers appropriately, the antenna problem can be easily solved.

Antenna avoidance by jumper insertion has been extensively studied in the literature (*e.g.*, [Ho 2004, 2007; Wu 2005a; Su 2007]). Ho, Chang, and Chen proposed multilevel routing considering antenna effects by bottom-up jumper insertion [Ho 2004]. The work inserts jumpers only beside gate terminals, and its optimality of the use of the least jumpers to satisfy the antenna rule holds only for this special condition of inserting jumpers right beside gate terminals. Wu, Hu, and Mahapatra extended the work [Ho 2004] to handle the problem [Wu 2005a]. To fix the antenna violation of a gate terminal, the work first removes all subtrees around the node that violate the antenna rules. After all such subtrees are removed, if the sink still violates the antenna rule, the work will continually remove the heaviest branch from the sink until the antenna rules are satisfied. This approach still cannot guarantee optimal solutions under some special cases.

Su and Chang formulated the general jumper insertion for antenna avoidance (applicable at the routing stage) and/or fixing (applicable at the post-layout stage) as a **tree-cutting** problem on a routing tree and presented the first optimal algorithm for the general tree-cutting problem [Su 2007]. As usual, a net is modeled as a routing tree, where a node in the tree denotes a circuit terminal/junction (a gate, diffusion, or a junction of interconnects), and an edge denotes the interconnection between two circuit terminals or junctions. Because the interconnection connecting to a diffusion terminal will not cause any antenna violation, the algorithm focuses on those connecting to gate terminals.

Let $L(u)$ denote the sum of edge weight (could be wirelengths, wire area, wire perimeter limit, the ratio of antenna strength, etc.) between the node u and all its neighbors. The problem of jumper insertion on a routing tree for antenna avoidance/fixing can be formulated as a tree-cutting problem as follows:

Jumper Insertion on a Routing Tree for Antenna Avoidance Problem: Given a routing tree $T = (V, E)$ and an upper bound L_{max}, find the minimum set C of cutting nodes, $e \neq u$ for any $c \in C$ and $u \in V$, so that $L(u) \leq L_{\max}$, $\forall u \in V$.

As the routing-tree example shows in Figure 12.46a, u_1 and u_2 are two sink nodes, the number beside each edge denotes the antenna charge weight, and L_{max} is assumed to be 10. For this case, three jumpers suffice to solve the antenna violations; see the jumpers c_1, c_2, and c_3 shown in Figure 12.46b.

The algorithm performs in a bottom-up manner by dealing with *leaf nodes* first followed by *sub-leaf nodes* of the tree. Here, a leaf node is a node with no children, whereas a sub-leaf node is a node for which all its children are leaf nodes, and if any of its children is a gate terminal, the edges between it and its children all have weights $\leq L_{max}$. Let $p(u)$ denote the parent node of node u, and $l(e)$ (or $l(u, v)$) be the antenna charge weight of the edge $e = (u, v)$ in the routing tree.

For a leaf node u, if $l(u, p(u)) \leq L_{max}$ or u is not a gate terminal, then u satisfies the antenna rule and thus it does not need to insert any cutting nodes. However, if $l(u, p(u)) > L_{max}$ and u is a gate, then $l(u, c) = L_{max}$ gives the best position for inserting the cutting node c, as illustrated in Figure 12.47. After adding jumper c, the edge $e(u, c)$ is cut from the tree.

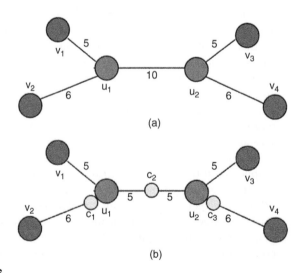

FIGURE 12.46

(a) A routing tree with two sink nodes u_1 and u_2. (b) Three jumpers c_1, c_2, c_3 are inserted to satisfy the antenna rule.

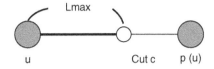

FIGURE 12.47

Optimal jumper insertion for a leaf node. The cutting node c is the optimal one among the nodes on edge $e(u, p(u))$.

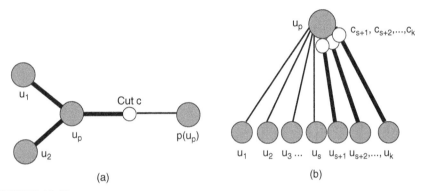

(a)   (b)

FIGURE 12.48

(a) Optimal jumper insertion for a sub-leaf node. (b) Illustration for the case of $total\_len > L_{max}$.

For a sub-leaf node u_p and its children $u_i, \forall 1 \leq i \leq k$, let $total\_len = \sum_{i=1}^{k} l(u_i, u_p)$.

There are two cases:

Case 1: $total\_len \leq L_{max}$

If $(total\_len + l(u_p, p(u_p)) > L_{max}$, the cutting node c with $l(c, u_p) + total\_len = L_{max}$ gives the best position, as shown in Figure 12.48a. After adding c, all u_p's children from the original tree are cut from the tree.

Case 2: $total\_len > L_{max}$

First sort $l(u_p, u_i) \forall 1 \leq i \leq k$ in non-decreasing order and find the maximum s such that $\sum_{i=1}^{s} l(u_p, u_i) \leq L_{max}$. Then add the cutting nodes $c_{s+1}, \ldots, c_k$ as shown in Figure 12.48b.

For the embedded protection diode, Huang *et al.* solved the diode insertion and routing problem by a minimum-cost network-flow based algorithm, called **_Diode Insertion and Routing by Min-Cost Flow_** (DIRMCF) [Huang 2004a]. As shown in Figure 12.49, the antenna-violating wires, the routing grids, and the feasible diode positions are transformed into a flow network, and then the problem is solved by the minimum-cost network-flow algorithm. Both the positions of inserted diodes and the required routing can be determined through the resulting flow.

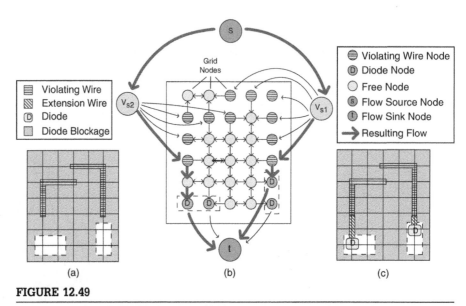

FIGURE 12.49

An example of the DIRMCF algorithm: (a) The violating wires and the routing grids. (b) The transformed flow network and the resulting flow after applying the minimum cost network-flow algorithm. (c) The inserted diodes and their corresponding routing.

Besides, because the vias of jumper insertion and the routing wires for diode insertion will both increase the driving load of the antenna violating wires (and thus the incurred *RC* delay will reduce the circuit performance), it is desirable to perform diode and jumper insertion simultaneously and consider the interaction between them to find a smaller performance degradation for the antenna fixing. Jiang and Chang [Jiang 2008] proposed a minimum-cost network-flow–based algorithm to solve the simultaneous diode/jumper insertion problem. The proposed algorithm first computes the jumper cost to fix each violating wire. Then it constructs the flow network in a similar way as the DIRMCF algorithm but integrates the jumper cost into the network. Finally, the antenna-fixed layout with the optimal fixing cost is found by applying the minimum-cost network-flow algorithm.

12.6.3.2 *Redundant-via aware routing*

In the nanometer technology, via-open defects are one of the important failures. A via may fail because of various reasons such as **random defects, electromigration, cut misalignment**, and/or **thermal stress**–induced voiding effects. The failure significantly reduces the manufacturing yield and chip performance.

To improve via reliability and yield, **redundant-via insertion** is a highly recommended technique proposed by foundries. If a via fails, a redundant via can serve as a fault-tolerant substitute for the failing one. As shown in Figure 12.50, a redundant

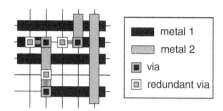

FIGURE 12.50

Double-via insertion. Each via is paired with a redundant via to form a double-via pair.

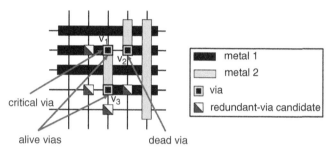

FIGURE 12.51

Illustration of redundant-via candidates, dead vias, alive vias, and critical vias. Vias v_1, v_2, and v_3 have one, zero, and three redundant-via candidates, respectively. Both v_1 and v_3 are alive vias, v_2 is a dead via, and v_1 is also called a critical via.

via can be inserted adjacent to each via to form a double-via pair. Double vias typically lead to 10 to 100 smaller failure rates than single vias.

The following gives some terminologies about vias. For a via, a **redundant-via candidate** is its adjacent position where a redundant via can be inserted. For the example shown in Figure 12.51, via v_1 has one redundant-via candidate on its left side, and via v_3 has three candidates around it. According to the number of redundant-via candidates, vias can be classified as **dead, alive**, or **critical vias**. If a via has at least one redundant-via candidate, it is an alive via; otherwise, it is called a dead via. Note that if an alive via has *exactly* one redundant-via candidate, it is also called a critical via. As shown in Figure 12.51, both vias v_1 and v_3 are alive vias, v_2 is a dead via, and v_1 is also a critical via.

Traditionally, redundant-via insertion is performed at the post-layout stage, which can be formulated as a *maximum independent set* (MIS) problem [Lee 2006], 0-1 integer linear programming (ILP) [Lee 2008], or maximum bipartite matching [Yao 2005; Chen 2008a]. However, it has been reported that if the router can minimize the number of dead and critical vias, the post-layout double-via insertion rate can be significantly improved. The reason is that the dead vias cannot be paired with redundant vias, and critical vias may not be paired because of the competition with other vias. For a routing instance from the source S to the target T shown in Figure 12.52a, an inferior routing path as

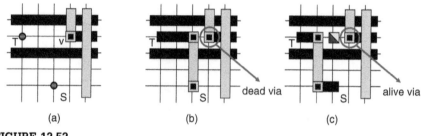

FIGURE 12.52

Redundant-via aware routing benefits the post-layout double-via insertion: (a) A detailed-routing instance for a 2-pin connection from the source S to the target T. (b) If an inferior routing path is selected, via v would become a dead via and cannot be paired. (c) For a better routing path, via v would remain alive for double-via insertion.

shown in Figure 12.52b would make via v a dead via and cannot be paired with any redundant vias. In contrast, for the better routing result as shown in Figure 12.52c, via v still remains alive for double-via insertion. Therefore, it is desirable to consider the redundant-via insertion at the routing stage to facilitate and preserve more flexibility for the post-layout double via insertion, as pointed out by [Xu 2005].

Chen *et al.* developed a redundant-via aware detailed-routing algorithm [Chen 2008a]. For each redundant-via candidate r_i of a via v, the *redundant-via cost* of r_i, $cost(r_i)$, is set as

$$cost(r_i) = \frac{1}{DoF_v} \tag{12.9}$$

where DoF_v stands for the **degree of freedom** of v and equals the number of redundant-via candidates of v. The redundant-via penalty for a connection path p is calculated as the summation of the redundant-via costs of these redundant-via candidates on p.

Figure 12.53 illustrates the routing algorithm. Figure 12.53a shows a detailed-routing instance connected from the source S to the target T. The redundant-via costs of redundant-via candidates are shown in Figure 12.53b. The router can find a better routing path by choosing one with smaller redundant-via penalty, as shown in Figure 12.53c. Finally, the routing solution would be more redundant-via friendly as shown in Figure 12.53d, which contains more alive vias and preserves more redundant-via candidates to benefit the post-layout redundant-via insertion.

12.7 CONCLUDING REMARKS

Routing is one of the most fundamental steps in the physical design flow and is typically a very complex optimization problem. Effective and efficient routing

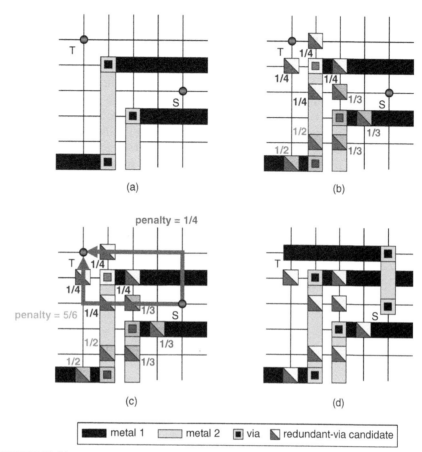

(a)

(b)

penalty = 1/4

penalty = 5/6

(c)

(d)

| ▬ metal 1 | ☐ metal 2 | ▣ via | ◨ redundant-via candidate |

FIGURE 12.53

Redundant-via aware detailed routing: (a) A detailed-routing instance connected from the source S to the target T. (b) The redundant-via costs of redundant-via candidates. (c) The router can find a better routing path with smaller redundant-via penalty. (d) The routing solution would be more redundant-via friendly by preserving more redundant-via candidates.

algorithms are essential to handle the challenges arising from the fast growing scaling of IC integration. Traditionally, the routing problem is usually solved by a two-stage approach of global routing followed by detailed routing to tackle its high complexity.

In this chapter, we have first formulated the global and detailed routing as graph-search problems and examined the general-purpose routing algorithm, which includes the maze, line-search, and A*-search routing and can be applied to both global and detailed routing. Then we have discussed the global-routing algorithms, including sequential, concurrent, and tree-based approaches. For the detailed routing, we have covered channel routing and full-chip routing and discussed the flat, hierarchical, and multilevel routing frameworks. Last, we have addressed routing for some important nanometer effects, including signal integrity, manufacturability, and reliability. As the technology nodes keep shrinking, all these effects should be considered in the earlier design stages. Considering the tradeoff between optimization flexibility and layout-information availability, routing seems to be the best stage to handle these effects.

"Old routers never die; they just fade away." With emerging design challenges (such as manufacturability, reliability, complexity, new chip architectures, and technologies), routers will keep evolving, with key techniques still remaining. It would be necessary to develop new data structures, algorithms, frameworks, and/or methods for the next-generation routers to handle the severe challenges yet to come.

12.8 **EXERCISES**

12.1. **(General-Purpose Routing)** Consider the chessboard shown in Figure 12.54. Some squares are shaded, denoting blockages. We intend to find a shortest path, if one exists, that starts at the square designated by s, after visiting the minimum number of squares, and ends at the square designated by t. The path must not pass through any shaded square. Formulate this problem as a graph-search routing problem and give an efficient algorithm to solve this problem. What is the time and space complexity of your algorithm?

12.2. **(Concurrent Global Routing)** You are asked to derive a routing algorithm for large-scale circuit designs on the basis of *integer linear programming* (ILP). ILP is typically very time-consuming for such large-scale designs. Instead of processing the whole routing region at one time, give at least two systematic approaches to divide

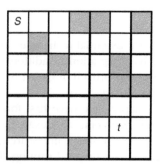

FIGURE 12.54

The graph-search problem of Exercise 12.1.

the routing region into subregions such that your algorithm can handle the routing problem subregion by subregion to reduce the problem size.

12.3. **(Routing Tree)** Given a net n with the four pins $p_1 = (6, 3)$, $p_2 = (3, 6)$, $p_3 = (1, 5)$, and $p_4 = (4, 2)$, let the estimated wirelengths by use of the ***minimum rectilinear spanning tree*** (MST) and the ***minimum rectilinear Steiner tree*** (MRST) be p and q, respectively. Find p and q.

12.4. **(Routing Tree)** Give an O(E lg V)-time algorithm to find a minimum spanning tree T of an undirected graph $G = (V, E)$ so that the maximum edge weight of T is minimum over all spanning trees of G. Analyze the time complexity of your algorithm.

12.5. **(Line-Search Routing)** For the Hightower line-search router, there is no guarantee that we can find a path if such a path exists. Give an example routing configuration for this situation.

12.6. **(Channel Routing)** Given the channel-routing instance shown in Figure 12.55,

(a) Draw the horizontal constraint graph (HCG) and the vertical constraint graph (VCG) for the given instance.

(b) Determine a tight lower bound on the channel height from the HCG.

(c) Route the instance by the dogleg channel routing algorithm. What is the final channel height?

(d) Route the instance by the constrained left-edge channel routing algorithm. What is the final channel height?

12.7. **(Channel Routing)** Design an efficient algorithm to produce optimal routing solutions for 3-layer channel routing with the VHV model. (In the VHV model, the top and the bottom layers are reserved for vertical wires, and the middle layer is reserved for horizontal wires.)

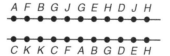

FIGURE 12.55

The channel routing instance of Exercise 12.6.

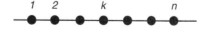

FIGURE 12.56

The routing instance of Exercise 12.8.

12.8. **(Channel Routing)** Label the terminals of the channel boundary 1, 2, ..., n in Figure 12.56, starting from the left and to the right. Let $N(i, j)$ denote the maximum number of nonintersecting connections between terminals i and j that can be routed on a single layer. Assume that there is a connection between terminals 1 and k. Give the recurrence $N(i, j)$ for the maximum number of nonintersecting connections in the layer in terms of the indices k and n. Apply dynamic programming to compute $N(1, n)$.

12.9. **(Multilevel Routing)** Given a netlist $N = \{[(1, 1), (2, 2)], [(2, 10), (2, 14)], [(6, 2), (10, 10)], [(6, 10), (10, 14)], [(10, 2), (14, 2)]\}$, where $[(p, q), (r, s)]$ denotes a route from the coordinate (p, q) to (r, s), you are asked to apply a 3-level routing (Λ-shaped multilevel routing with three levels) to route the instance N on a 16×16 chip plane. Suppose only straight and L-shaped routes are allowed during the coarsening stage, whereas maze routing is applied during uncoarsening. Also, all wire spacing (including point-to-wire spacing) must be at least 4 units. Show step by step how you obtain the routing solution.

12.10. **(Maze Routing)** Explain how you will extend the maze router for the X-architecture on which vertical, horizontal, 45°, and 135° routes are allowed for routing.

12.11. **(Programming)** This programming problem is modified from the 2007 ACM ISPD (*Int. Symp. on Physical Design*) Global Routing Contest [Nam 2007]. This programming assignment asks you to write a global router that can route 2-pin nets. To simplify the problem, we have some simplifications as follows:

1. Consider only two layers (layer 1 is for horizontal routes, and layer 2 is for vertical ones).
2. Consider only 2-pin nets.
3. Consider only tile-based coordinates. All lower left corners of the global routing regions are (0, 0). The tile width and height are ignored, because all X and Y are tile-based.
4. Consider only fixed wire width and spacing. All wire widths, wire, and via spacing are equal to 1.

(1) Input/output specification
Input format

The file format for the global routing contest is shown, with comments in italics (actual input files do not contain these comments). The example below gives an instance with two routing layers. The first line gives the problem size in terms of the number of horizontal and vertical tiles and the number of routing layers. Each global-routing tile (tile in short) has a capacity on each of its four boundaries to measure the available space.

The default capacity value of each layer is given in the second and third lines, which represents the maximum number of routing paths allowed to pass through a tile boundary. For example, the tile boundary with capacity 10 can accommodate up to 10 routing paths. The file format is as follows:

---

grid # # # *//number of horizontal tiles, vertical tiles, and layers*
vertical capacity # # *//vertical capacity by default on each layer*
horizontal capacity # # *//horizontal capacity by default on each layer*

num net # *//number of nets*
net_name net_id number_of_pins
x y layer
x y layer

. . .

[repeat for the appropriate number of nets]

Output format

All the routes in the output could only be horizontal lines, vertical lines, or via connections. For example $(18, 61, 1)-(19, 62, 1)$ is not acceptable, because it is diagonal. All the nets are written in the output file in the same order as the input file. The output file format is shown as follows:

Net net_name net_id
([x11], [y11], [z11])-([x12], [y12], [z12])
([x21], [y21], [z21])-([x22], [y22], [z22])

. . .

!

[repeat for the appropriate number of nets]

---

(2) Problem statement

Given the problem size (the number of horizontal and vertical tiles and layers), horizontal and vertical capacities on each layers, and a netlist, the global router routes all nets in the routing region. The main objective is to minimize the total number of overflows, and the second objective is to minimize the total wirelength. Here the overflow on a tile boundary is calculated as the amount of demand that exceeds the given capacity (*i.e.*, *overflow* $= \max(0, demand- capacity)$).

Following is an example of Input/Output files for Figure 12.57 with two routing layers.

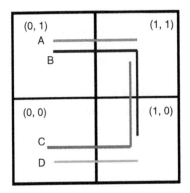

FIGURE 12.57

A routing problem and its solution.

Input file:

```
grid 2 2 2
vertical capacity 0 2
horizontal capacity 2 0

num net 4
A 0 2
0 1 1
1 1 1
B 1 2
0 1 1
1 0 1
C 2 2
0 0 1
1 1 1
D 3 2
0 0 1
1 0 1
```

Output file:

```
A 0
(0, 1, 1)-(1, 1, 1)
!
B 1
(0, 1, 1)-(1, 1, 1)
(1, 1, 1)-(1, 1, 2)
(1, 1, 2)-(1, 0, 2)
(1, 0, 2)-(1, 0, 1)
!
C 2
(0, 0, 1)-(1, 0, 1)
(1, 0, 1)-(1, 0, 2)
(1, 0, 2)-(1, 1, 2)
(1, 1, 2)-(1, 1, 1)
!
D 3
(0, 0, 1)-(1, 0, 1)
!
```

The total overflow is 0, and the total wirelength is 10.

ACKNOWLEDGMENTS

We thank Dr. Laung-Terng Wang of SynTest Technologies, Professor Cheng-Kok Koh of Purdue University, Professor Chris Chu of Iowa State University, Professor Ting-Chi Wang of National Tsing Hua University, Professor Hung-Ming Chen of National Chiao Tung University, the National Taiwan University students in the 2008 Physical Design class, and the EDA Laboratory for their very careful review of this chapter. We also thank the authors of [Nam 2007] for their help with the formulation of the programming assignment in Exercise 12.11, and Mr. Zhe-Wei Jiang of National Taiwan University for his help with Section 12.6.3.1.

REFERENCES

R12.0 Books

[Garey 1979] M. R. Garey and D. S. Johnson, *Computers and Intractability: A Guide to the Theory of NP-Completeness*, W. H. Freeman & Co., New York, 1979.
[Ho 2007] T.-Y. Ho, Y.-W. Chang, and S.-J. Chen, *Full-Chip Nanometer Routing Techniques*, Springer, Dordrecht, The Netherlands, 2007.

[Preparata 1985] F. P. Preparata and M. I. Shamos, *Computational Geometry: An Introduction,* Springer-Verlag, New York, 1985.

[Sait 1999] S. M. Sait and H. Youssef, VLSI *Physical Design Automation: Theory and Practice,* World Scientific, Singapore, 1999.

[Saxena 2007] P. Saxena, R. S. Shelar, and S. S. Sapatnekar, *Routing Congestion in VLSI Circuits: Estimation and Optimization,* Springer, New York, 2007.

R12.2 Problem Definition

[Chen 2007a] T.-C. Chen and Y.-W. Chang, Multilevel full-chip gridless routing with applications to optical proximity correction, *IEEE Trans. on Computer-Aided Design,* 26(6), pp. 1041–1053, June 2007.

[Cong 1999] J. Cong, J. Fang, and K.-Y. Khoo, An implicit connection graph maze routing algorithm for eco routing, in *Proc. IEEE/ACM Int. Conf. on Computer-Aided Design,* pp. 163–167, November 2005.

[Qusterhout 1984] J. K. Qusterhout, Corner stitching: A data structuring technique for VLSI layout tools, *IEEE Trans. on Computer-Aided Design,* 3(1), pp. 87–100, January 1984.

[Zheng 1996] S. Q. Zheng, J. S. Lim, and S. S. Iyengar, Finding obstacle avoiding shortest paths using implicit connection graphs, *IEEE Trans. on Computer-Aided Design,* 15(1), pp. 103–110, January 1996.

R12.3 General-Purpose Routing

[Akers 1967] S. B. Akers, A modification of Lee's path connection algorithm, *IEEE Trans. on Electronic Computers,* 16(1), pp. 97–98, February 1967.

[Clow 1984] G. W. Clow, A global routing algorithm for general cells, in *Proc. ACM/IEEE Design Automation Conf.,* pp. 45–51, June 1984.

[Hadlock 1977] F. O. Hadlock, A shortest path algorithm for grid graphs, *Networks,* 7(4), pp. 323–334, Winter, 1977.

[Hart 1968] P. E. Hart, N. J. Nilsson, and B. Raphael, A formal basis for the heuristic determination of minimum cost paths, *IEEE Trans. on Systems Science and Cybernetics,* 4(2), pp. 100–107, July 1968.

[Hightower 1969] D. Hightower, A solution to line routing problems on the continuous plane, in *Proc. ACM/IEEE Design Automation Conf.,* pp. 1–24, June 1969.

[Lee 1961] C. Y. Lee, An algorithm for path connection and its application, in *IRE Trans. on Electronic Computer,* 10, pp. 346–365, 1961.

[McMurchie 1995] L. McMurchie and C. Ebeling, PathFinder: A negotiation-based performance-driven router for FPGAs, *in Proc. Int. ACM Symp. on Field-Programmable Gate Arrays,* pp. 111–117, February 1995.

[Mikami 1968] K. Mikami and K. Tabuchi, A computer program for optimal routing of printed circuit connectors, in *Proc. Int. Federation for Information Processing,* pp. 1475–1478, November 1968.

[Soukup 1978] J. Soukup, Fast maze router, in *Proc. ACM/IEEE Design Automation Conf.,* pp. 100–102, June 1978.

R12.4 Global Routing

[Abel 1972] L. C. Abel, On the ordering of connections for automatic wire routing, *IEEE Trans. on Computers,* 21(11), pp. 1227–1233, November 1972.

[Chang 2008] Y.-J. Chang, Y.-T. Lee, and T.-C. Wang, NTHU-Route 2.0: A Fast and Stable Global Router, in *Proc. IEEE/ACM Int. Conf. on Computer-Aided Design,* pp. 338–343, November 2008.

[Chen 1999] H.-M. Chen, H. Zhou, F. Y. Young, D. F. Wong, H. H. Yang, and N. Sherwani, Integrated floorplanning and interconnect planning, in *Proc. IEEE/ACM Int. Conf. on computer-Aided Design*, pp. 354–357, November 1999.

[Cho 2007] M. Cho, K. Lu, K. Yuan, and D. Z. Pan, BoxRouter 2.0: Architecture and implementation of a hybrid and robust global router, in *Proc. ACM/IEEE Design Automation Conf.*, pp. 503–508, June 2007.

[Cho 2006] M. Cho, D. Z. Pan, H. Xiang, and R. Puri, Wire density driven global routing for CMP variation and timing, in *Proc. IEEE/ACM Int. Conf. on Computer-Aided Design*, pp. 487–492, November 2006.

[Chu 2004] C. Chu, FLUTE: fast lookup table based wirelength estimation technique, in *Proc. IEEE/ACM Int. Conf. on Computer-Aided Design*, pp. 696–701, November 2004.

[Garey 1977] M. R. Garey and D. S. Johnson, The rectilinear Steiner tree problem is NP-complete, *SIAM Journal Applied Mathematics*, 32(4), pp. 826–834, June 1977.

[Hanan 1966] M. Hanan, On Steiner's problem with rectilinear distance, *SIAM Journal on Applied Mathematics*, 14(2), pp. 255–265, March 1966.

[Ho 1990] J.-M. Ho, C. K. Vijayan, and C. K. Wong, New algorithms for the rectilinear Steiner tree problem, *IEEE Trans. on Computer-Aided Design*, 9(2), pp. 185–193, February 1990.

[Hsu 2008] C.-H. Hsu, H.-Y. Chen, and Y.-W. Chang, Multi-layer global routing considering via and wire capacities, in *Proc. IEEE/ACM Int. Conf. on Computer-Aided Design*, pp. 350–355, November 2008.

[Hwang 1976] F. K. Hwang, On Steiner minimal tree with rectilinear distance, *SIAM Journal on Applied Mathematics*, 30(1), pp. 104–114, January 1976.

[Kahng 1990] A. B. Kahng and G. Robins, A new class of Steiner tree heuristics with good performance: the iterated 1-Steiner approach, in *Proc. IEEE/ACM Int. Conf. on Computer-Aided Design*, pp. 428–431, November 1990.

[Kastner 2002] R. Kastner, E. Bozorgzadeh, and M. Sarrafzadeh, Pattern routing: use and theory for increasing predictability and avoiding coupling, *IEEE Trans. on Computer-Aided Design*, 21(7), pp. 777–790, November 2002.

[Kruskal 1956] J. B. Kruskal, On the shortest spanning subtree of a graph and the traveling salesman problem, in *Proc. the American Mathematical Society*, 7(1), pp. 48–50, February 1956.

[Lin 2008] C.-W. Lin, S.-Y. Chen, C.-F. Li, Y.-W. Chang, and C.-L. Yang, Obstacle-avoiding rectilinear Steiner tree construction based on spanning graphs, *IEEE Trans. Computer-Aided Design*, 27(4), pp. 643–653, April 2008.

[Lin 2007] C.-W. Lin, S.-L. Huang, K.-C. Hsu, M.-X. Lee, and Y.-W. Chang, Efficient multi-layer obstacle-avoiding rectilinear Steiner tree construction, in *Proc. IEEE/ACM Int. Conf. on Computer-Aided Design*, pp. 380–385, November 2007.

[McMurchie 1995] L. McMurchie and C. Ebeling, PathFinder: A negotiation-based performance-driven router for FPGAs, in *Proc. Int. ACM Symp. on Field-Programmable Gate Arrays*, pp. 111–117, February 1995.

[Pan 2007] M. Pan and C. N. Chu, FastRoute 2.0: A high-quality and efficient global router, in *Proc. IEEE/ACM Asian and South Pacific Design Automation Conf.*, pp. 250–255, January 2007.

[Prim 1957] R. C. Prim, Shortest connection networks and some generalizations, *Bell System Technical Journal*, 36, pp. 1389–1401, 1957.

[Raghavan 1987] P. Raghavan and C. D. Thompson, Randomized rounding: A technique for provably good algorithms and algorithmic proofs, in *Proc. Combinatorica*, pp. 365–374, December 1987.

[Roy 2007] J. A. Roy and I. L. Markov, High-performance routing at the nanometer scale, in *Proc. IEEE/ACM Int. Conf. on Computer-Aided Design*, pp. 496–502, November 2007.

[Shi 2006] Y. Shi, T. Jing, L. He, Z. Feng, and X. Hong, CDCTree: novel obstacle-avoiding routing tree construction based on current driven circuit model, in *Proc. IEEE/ACM Asia and South Pacific Design Automation Conf.*, pp. 630–635, January 2006.

[Zhou 2004] H. Zhou, Efficient Steiner tree construction based on spanning graphs, *IEEE Trans. on Computer-Aided Design*, 23(5), pp. 704–710, May 2004.

R12.5 Detailed Routing

[Burstein 1983] M. Burstein and R. Pelavin, Hierarchical wire routing, *IEEE Trans. on Computer-Aided Design*, 2(4), pp. 223–234, October 1983.

[Chang 2004] Y.-W. Chang and S.-P. Lin, MR: A new framework for multilevel full-chip routing, *IEEE Trans. on Computer-Aided Design*, 23(5), pp. 793–800, May 2004.

[Cong 2001] J. Cong, J. Fang, and Y. Zhang, Multilevel approach to full-chip gridless routing, in *Proc. IEEE/ACM Int. Conf. on Computer-Aided Design*, pp. 396–403, November 2001.

[Cong 2002] J. Cong, M. Xie, and Y. Zhang, An enhanced multilevel routing system, in *Proc. IEEE/ACM Int. Conf. on Computer-Aided Design*, pp. 51–58, November 2002.

[Deutsch 1976] D. N. Deutsch, A "dogleg" channel router, in *Proc. ACM/IEEE Design Automation Conf.*, pp. 425–433, June 1976.

[Hashimoto 1971] A. Hashimoto and J. Stevens, Wire routing by optimizing channel assignment within large apertures, in *Proc. ACM/IEEE Design Automation Conf.*, pp. 155–169, June 1971.

[Lin 2002] S.-P. Lin and Y.-W. Chang, A novel framework for multilevel routing considering routability and performance, *in Proc. IEEE/ACM Int. Conf. on Computer-Aided Design*, pp. 44–50, November 2002.

[Lin 1990] Y.-L. Lin, Y.-C. Hsu, and F.-S. Tsai, Hybrid routing, *IEEE Trans. on Computer-Aided Design*, 9(2), pp. 151–157, February 1990.

[Marek-Sadowska 1984] M. Marek-Sadowska, Global router for gate array, in *Proc. IEEE Int. Conf. on Computer Design*, pp. 332–337, October 1984.

[Szymanski 1985] T. G. Szymanski, Dogleg channel routing is NP-complete, *IEEE Trans. on Computer-Aided Design*, 4(1), pp. 31–41, January 1985.

R12.6 Modern Routing Considerations

[Chaudhary 1993] K. Chaudhary, A. Onozawa, and E. S. Kuh, A spacing algorithm for performance and crosstalk reduction, in *Proc. IEEE Int. Conf. on Computer-Aided Design*, pp. 697–702, November 1993.

[Chen 2008a] H.-Y. Chen, M.-F. Chiang, Y.-W. Chang, L. Chen, and B. Han, Full-chip routing considering double-via insertion, *IEEE Trans. on Computer-Aided Design*, 27(5), pp. 844–857, May 2008.

[Chen 2007a] T.-C. Chen and Y.-W. Chang, Multilevel full-chip gridless routing with applications to optical proximity correction, *IEEE Trans. on Computer-Aided Design*, 26(6), pp. 1041–1053, June 2007.

[Chen 2007b] H.-Y. Chen, S.-J. Chou, S.-L. Wang, and Y.-W. Chang, Novel wire density driven full-chip routing for CMP variation control, in *Proc. IEEE/ACM Int. Conf. on Computer-Aided Design*, pp. 831–838, November 2007.

[Chen 2008b] T.-C. Chen, G.-W. Liao, and Y.-W. Chang, Predictive formulae for OPC with applications to lithography-friendly routing, in *Proc. ACM/IEEE Design Automation Conf.*, pp. 510–515, June 2008.

[Chen 2000] P. H. Chen, S. Malkani, C.-M. Peng, and J. Lin, Fixing antenna problem by dynamic diode dropping and jumper insertion, in *Proc. IEEE Int. Symp. on Quality Electronic Design*, pp. 275–282, March 2000.

[Cho 2006] M. Cho, D. Z. Pan, H. Xiang, and R. Puri, Wire density driven global routing for CMP variation and timing, in *Proc. IEEE/ACM Int. Conf. on Computer-Aided Design*, pp. 487–492, November 2006.

[Cong 2001] J. Cong, D. Z. Pan, and P. V. Srinivas, Improved crosstalk modeling for noise constrained interconnect optimization, in *Proc. IEEE/ACM Asia and South Pacific Design Automation Conf.*, pp. 373–378, January 2001.

[Gao 1996] T. Gao and C.-L. Liu, Minimum crosstalk channel routing, *IEEE Trans. on Computer-Aided Design*, 15(5), pp. 465–474, May 1996.

[Ho 2004] T.-Y. Ho, Y.-W. Chang, and S.-J. Chen, Multilevel routing with antenna avoidance, in *Proc. ACM Int. Symp. on Physical Design*, pp. 34–40, April 2004.

[Ho 2005] T.-Y. Ho, Y.-W. Chang, S.-J. Chen, and D.-T. Lee, Crosstalk- and performance-driven multi-level full-chip routing, *IEEE Trans. on Computer-Aided Design*, 24(6), pp. 869–878, June 2005.

[Huang 2004a] L.-D. Huang, X. Tang, H. Xiang, D. F. Wong, and I.-M. Liu, A polynomial time-optimal diode insertion/routing algorithm for fixing antenna problem, *IEEE Trans. on Computer-Aided Design*, 23(1), pp. 141–147, January 2004.

[Huang 2004b] L.-D. Huang and D. F. Wong, Optical proximity correction (OPC)-friendly maze routing, in *Proc. ACM/IEEE Design Automation Conf.*, pp. 186–191, June 2004.

[Jiang 2008] Z.-W. Jiang and Y.-W. Chang, An optimal network-flow-based simultaneous diode and jumper insertion algorithm for antenna fixing, *IEEE Trans. on Computer-Aided Design*, 27(6), pp. 1055–1065, June 2008.

[Jiang 2000] I. H.-R. Jiang, Y.-W. Chang, and J.-Y. Jou, Crosstalk-driven interconnect optimization by simultaneous gate and wire sizing, *IEEE Trans. on Computer-Aided Design*, 19(9), pp. 999–1010, September 2000.

[Lee 2008] K.-Y. Lee, C.-K. Koh, T.-C. Wang, and K.-Y. Chao, Optimal post-routing redundant via insertion, in *Proc. ACM Int. Symp. on Physical Design*, pp. 111–117, April 2008.

[Lee 2006] K.-Y. Lee and T.-C. Wang, Post-routing redundant via insertion for yield/reliability improvement, in *Proc. ACM/IEEE Asia and South Pacific Design Automation Conf.*, pp. 303–308, January 2006.

[Li 2007] K. S.-M. Li, C.-L. Lee, Y.-W. Chang, C.-C. Su, and J. E. Chen, Multilevel full-chip routing with testability and yield enhancement, *IEEE Trans. on Computer-Aided Design*, 26(9), pp. 1625–1636, September 2007.

[Luo 2005] J. Luo, Q. Su, C. Chiang, and J. Kawa, A layout dependent full-chip copper electroplating topography model, in *Proc. IEEE/ACM Int. Conf. on Computer-Aided Design*, pp. 133–140, November 2005.

[Sakurai 1983] T. Sakurai and K. Tamaru, Simple formulas for two and three dimensional capacitance, *IEEE Trans. on Electronic Devices*, 30(2), pp. 183–185, February 1983.

[Su 2007] B.-Y. Su, Y.-W. Chang, and J. Hu, An optimal jumper insertion algorithm for antenna avoidance/fixing, *IEEE Trans. on Computer-Aided Design*, 26(10), pp. 1818–1929, October 2007.

[Tian 2000] R. Tian, D. F. Wong, and R. Boone, Model-based dummy feature placement for oxide chemical-mechanical polishing manufacturability, in *Proc. ACM/IEEE Design Automation Conf.*, pp. 667–670, June 2000.

[Vittal 1999] A. Vittal, L. H. Chen, M. Marek-Sadowska, K.-P. Wang, and S. Yang, Crosstalk in VLSI interconnections, *IEEE Trans. on Computer-Aided Design*, 18(12), pp. 1817–1824, December 1999.

[White 2005] D. White and B. Moore, An 'intelligent' approach to dummy fill, in *EE Times*, January 3, 2005.

[Wu 2005a] D. Wu, J. Hu, and R. Mahapatra, Coupling aware timing optimization and antenna avoidance in layer assignment, in *Proc. ACM. Int. Symp. on Physical Design*, pp. 20–27, April 2005.

[Wu 2005b] Y.-R. Wu, M.-C. Tsai, and T.-C. Wang, Maze routing with OPC consideration, in *Proc. ACM/IEEE Asia and South Pacific Design Automation Conf.*, pp. 198–203, January 2005.

[Xu 2005] G. Xu, L.-D. Huang, D. Z. Pan, and F. D. Wong, Redundant-via enhanced maze routing for yield improvement, in *Proc. ACM/IEEE Asia and South Pacific Design Automation Conf.*, pp. 1148–1151, January 2005.

[Yao 2005] H. Yao, Y. Cai, X. Hong, and Q. Zhou, Improved multilevel routing with redundant via placement for yield and reliability, in *Proc. ACM Great Lakes Symp. on VLSI*, pp. 143–146, April 2005.

[Zhou 1998] H. Zhou and D. F. Wong, Global routing with crosstalk constraints, in *Proc. ACM/IEEE Design Automation Conf.*, pp. 374–377, June 1998.

R12.8 Exercises

[Nam 2007] G.-J. Nam, M. Yildiz, D. Z. Pan, and P. H. Madden, ISPD placement contest updates and ISPD 2007 global routing contest, in *Proc. ACM Int. Symp. on Physical Design*, p. 167, April 2007.

Synthesis of clock and power/ground networks

13

Cheng-Kok Koh
Purdue University, West Lafayette, Indiana

Jitesh Jain
Purdue University, West Lafayette, Indiana

Stephen F. Cauley
Purdue University, West Lafayette, Indiana

ABOUT THIS CHAPTER

Clock distribution networks and power delivery systems are the two largest types of on-chip interconnect networks. They both play a crucial role in the correct operation of a circuit. A clock network delivers a synchronizing signal across the chip to coordinate the flow of data. A ***power/ground*** (P/G) supply network provides a reference voltage for determining the status of a transistor (on or off) and also the current for switching a transistor.

This chapter addresses many issues that affect the integrity of clock networks or P/G networks. We begin with a discussion of the timing, power, and robustness issues that one must consider when designing a clock network or P/G network. The chapter then examines the automated analysis, synthesis, and optimization of such large-scale interconnection networks in two sections, one for clock networks and the other one for power supply networks. For each section, we first cover the typical topologies encountered. After presenting the modeling and analysis techniques for such networks, the chapter describes algorithms to synthesize and optimize these networks.

13.1 INTRODUCTION

For a sequential circuit to operate correctly, the processing of data must occur in an orderly fashion. In a synchronous system, the order in which data are processed is coordinated by a clock signal. (To be more general, the **751** synchronization could be performed by a collection of globally distributed clock signals.) The clock signal, in the form of a periodic square wave that is globally distributed to control all sequential elements (flip-flops or latches), achieves synchronization of the circuit operation when all data are allowed to pass through the sequential elements simultaneously. A clock network is required to deliver the clock signal to all sequential elements. As the distributive nature of long interconnects becomes more pronounced because of technology scaling, the control of arrival times of the same clock edge at different sequential elements, which are scattered over the entire chip, becomes more difficult. If not properly controlled, the clock skew, defined as the difference in the clock signal delays to sequential elements, can adversely affect the performance of the systems and even cause erratic operations of the systems (*e.g.*, latching of an incorrect signal within a sequential element).

The power delivery system is in the form of two networks, one for providing the power supply voltage (V_{DD}) and one for providing the ground (*GND* or V_{SS}). We refer to them as the ***power/ground*** (P/G) network. A well-designed P/G network provides clean voltage references and adequate current for reliable and fast computation. Variations in V_{DD} and *GND*, commonly known as power supply noise, may result in logic failures or severe speed loss in the circuit. One reason for the fluctuations in V_{DD} and *GND* is that the P/G network spans the entire chip, because essentially all functional blocks (gates, memory cells, registers, etc.) on the chip have to be connected to the P/G pads. The presence of resistive (*R*) and inductive (*L*) parasitics in P/G pins and the P/G network leads to current-induced *IR* and $L \cdot di/dt$ voltage variations, respectively, when current (*I* or *i*) flows to facilitate the switching activities of transistors. Because low supply voltages are typically used in modern-day integrated circuits, the margin available to tolerate voltage variations is usually very stringent.

When considered independently, the design of a clock or P/G network already poses a formidable challenge because of stringent requirements. A well-designed clock or P/G network must also account for variations in device and interconnect parameters. What complicates matters even further is the interplay between the clock network and the P/G network. For example, in a zero-skew clock network, all sequential elements are triggered almost simultaneously. Because these elements draw current from the power network or sink current to the ground network almost simultaneously when the clock switches, the zero-skew design often leads to severe power supply noise, resulting in unacceptable degradation in performance and reliability. On the other hand, power supply noise affects the clock jitter (this refers to the time shift in the clock pulse, as well as the variations in the pulse width that arise from the time-varying operating conditions such as supply voltage), which, in turn, affects the arrival times of clock signal at different sequential elements.

We will describe in the next section many of the design considerations that one has to consider when implementing a clock network or a P/G network. These design considerations are captured either in the objective function for which a synthesis algorithm for such networks has to optimize or as constraints that the synthesis algorithm has to satisfy. Synthesis algorithms for clock networks and P/G networks are topics that we cover in Sections 13.3 and 13.4, respectively.

13.2 DESIGN CONSIDERATIONS

Consider a simple synchronous circuit that uses positive edge-triggered *flip-flops* (FF's) as the sequential elements controlled by a clock signal that has even duty cycle (*i.e.*, a single-phase clocking scheme). We use $S = \{s_1, s_2, \ldots, s_n\}$ to denote the set of clock pins of flip-flops in the circuit, with s_i being the clock pin of flip-flop FF_i.

A pair of flip-flops is *sequentially adjacent* when only combinational logic exists between the two flip-flops. Let FF_i and FF_j be two sequentially adjacent flip-flops, with the output of FF_i being fed to the data input of FF_j through a combinational logic (see Figure 13.1). We say that FF_i is the launching flip-flop and FF_j is the capturing flip-flop.

13.2.1 Timing constraints

Suppose the clock signal arrives at all the flip-flops simultaneously. In Figure 13.2, the clock signal arrives at the 12 o'clock position.[1] Here, we assume an even duty clock signal. Hence, in the first half of the clock cycle the clock signal is high ("CLK = 1"). At the 6 o'clock position, the clock switches from high to low ("CLK = 0"). After a clock period of C_P, the clock signal is again at the 12 o'clock position.

The minimum allowable clock period C_P for the synchronous system must be long enough to accommodate the time it takes to propagate the signal through the

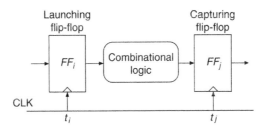

FIGURE 13.1

A simple sequential circuit.

[1]This representation was credited to Professor Mark Horowitz of Stanford University in the accompanying lecture materials of the book by [Rabaey 2003].

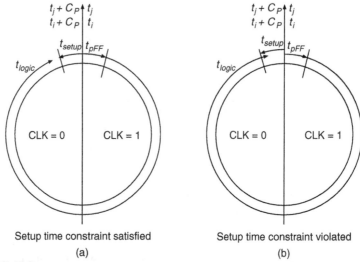

FIGURE 13.2

Setup time constraint.

launching flip-flop, denoted as t_{pFF}, and the time it takes for the signal to make its way through the combinational logic, denoted as t_{logic}. Moreover, the clock period must also account for the setup time, denoted as t_{setup}, which is the amount of time that the input to the capturing flip-flop has to stay valid before the next triggering clock edge arrives (see Figure 13.2a). This can be summarized by the following inequality, which is also known as the setup time constraint:

$$C_p \geq t_{pFF} + t_{logic} + t_{setup}$$

Of course, one should use the worst-case (or maximum) propagation delays for flip-flops and combinational logic and the maximum setup time requirement to determine the lowest operable clock period. Violation of the setup time constraint is also commonly referred to as a zero-clocking hazard, because the signal is not latched in properly by the capturing flip-flop (see Figure 13.2b).

The setup time constraint presents only one side of the story for the proper operation of a synchronous system. To ensure proper propagation of input signal through a flip-flop, the input must remain valid or hold steady for a short duration after the clock edge. That short duration is referred to as the hold time, denoted as t_{hold}. The hold time of a capturing flip-flop imposes an additional constraint on the total propagation delay of a signal through the launching flip-flop and the combinational logic as follows:

$$t_{pFF} + t_{logic} \geq t_{hold}$$

To accommodate the most extreme design corner, the preceding inequality, which is known as the hold time constraint, should be formed with the

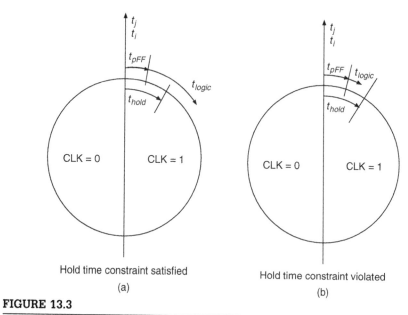

FIGURE 13.3

Hold time constraint.

maximum hold time of the capturing flip-flop and the minimum propagation delay of the launching flip-flop and the smallest propagation delay of the combinational logic (see Figure 13.3a). The violation of hold time constraint is commonly referred to as a double-clocking hazard, because two signals would have gone through the capturing flip-flop in a single clock cycle (see Figure 13.3b).

Remark: Clocking for Combinational Logic: The focus of this chapter is on clocking for sequential circuits (finite state machines or pipelined systems). Clocking is also an integral component of combinational logic: Dynamic circuits, such as Domino logic and NP CMOS logic (Zipper logic), require clock signals to synchronize the precharge phase and the evaluation phase of the circuits. It is important to realize that techniques presented in this chapter can be adapted to handle the synchronization of precharge-and-evaluate circuitry.

13.2.2 Skew and jitter

Because of the unbalanced delays in the clock distribution network, clock edges may arrive at clock pins s_i and s_j in Figure 13.1 at different times. Such *spatial variation* in the arrival times of a clock transition at two different locations on a chip is commonly known as the clock skew. Let t_i and t_j be the signal delays from the clock source to s_i and s_j, respectively. Define the clock skew between s_i and s_j, denoted by $skew_{i,j}$, to be

$$skew_{i,j} = t_i - t_j$$

By definition, $skew_{i,j}$ stays constant from cycle to cycle [Rabaey 2003]. Figure 13.4 shows two different scenarios: (a) $skew_{i,j} > 0$ and (b) $skew_{i,j} < 0$. They, respectively, correspond to the clock signal arriving at s_j earlier than and later than at s_i.

At time t_i, the clock edge arrives at flip-flop FF_i. The input of FF_i takes $t_{pFF}^{max} + t_{logic}^{max}$, in the worst case, to propagate through the flip-flop and the combinational logic. The signal must have settled down for a duration of t_{setup}^{max} before the next clock edge arrives at time $t_j + C_P$ at FF_j, the capturing flip-flop, so that it can be properly latched in. That translates into the following more general setup time constraint:

$$t_i + t_{pFF}^{max} + t_{logic}^{max} + t_{setup}^{max} \leq t_j + C_p \tag{13.1}$$

Rewriting Equation (13.1), it is clear that a *positive* clock skew, as shown in Figure 13.4a, places a lower bound on the allowable clock period (or an upper bound on the operating frequency of the circuit $1/C_P$) as follows:

$$C_p \geq skew_{i,j} + t_{pFF}^{max} + t_{logic}^{max} + t_{setup}^{max} \tag{13.2}$$

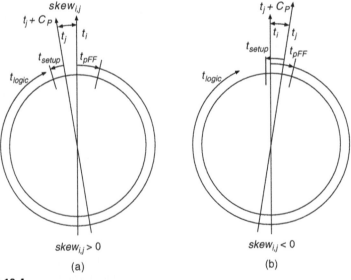

FIGURE 13.4

Setup time constraint in the presence of clock skew.

The terms in Equation (13.2) can also be rearranged into

$$skew_{i,j} \leq C_p - t_{pFF}^{\max} - t_{logic}^{\max} - t_{setup}^{\max} \qquad (13.3)$$

which shows that for a given frequency, we must bound the skew from above.

Although a positive $skew_{i,j}$ always decreases the maximum attainable clock frequency, a negative clock skew (*i.e.*, $skew_{i,j} < 0$) actually increases the effective clock period, as shown in Figure 13.4b. In other words, we may have a combinational logic block between two sequentially adjacent flip-flops that has a propagation delay longer than the given clock period.

Example 13.1 Cycle Stealing or Useful Clock Skew. Consider a 3-stage pipeline as shown in Figure 13.5a. Let $t_{logic,CLi}$ denote the delay of combinational logic block CLi. In this example, we assume that $t_{logic,CL2} = 10$ ns, $t_{logic,CL1} = t_{logic,CL3} = 8.5$ ns. If the clock signal arrives at all clock pins at the same time (Figure 13.5a), it is obvious that the fastest frequency at which the circuit can operate is 100 MHz. However, that would mean that for

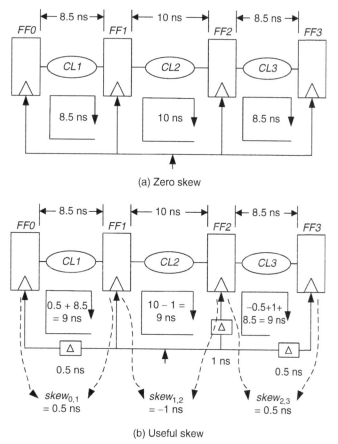

(a) Zero skew

(b) Useful skew

FIGURE 13.5

Cycle stealing.

the last 1.5 ns of the clock period, combinational logic blocks CL1 and CL3 are sitting idle. Although they maintain the correct values at FF1 and FF3 before the arrival of the next clock edge, they are not doing meaningful computation.

In the following, we will show how we can design the clock distribution network such that the system can run at 111 MHz (*i.e.*, a clock period of 9 ns). In this clock network, the clock signal arrives at FF0, FF1, FF2, and FF3 with the following skews: $skew_{0,1} = 0.5$ ns, $skew_{1,2} = -1$ ns, and $skew_{2,3} = 0.5$ ns. The effects of the skews are that CL1 and CL3 have exactly 8.5 ns for the computation, and CL2 has exactly 10 ns for the computation. In other words, CL2 steals 0.5 ns each from the clock periods for CL1 and CL3 so that the system can operate at a clock period that is smaller than the longest path delay in CL2.

However, when the skew is excessively negative, the signal may arrive so early that it races through the capturing flip-flop, resulting in double-clocking or a hold-time violation. Consider the example in Figure 13.1 again. At time t_i, the clock edge triggers flip-flop FF_i, and the input to FF_i propagates through the flip-flop and the combinational logic. Because we are considering the possibility of data racing with the clock signal, we use the shortest propagation delay through the flip-flop and the combinational logic, which is $t_{pFF}^{min} + t_{logic}^{min}$. In the worst case, the input signal at FF_j has to remain stable for t_{hold}^{max} after the clock edge of the same clock cycle arrives at t_j. Therefore, only after $t_j + t_{hold}^{max}$ may the signal from FF_i erase the input at FF_j (see Figure 13.6). This constraint is expressed as follows:

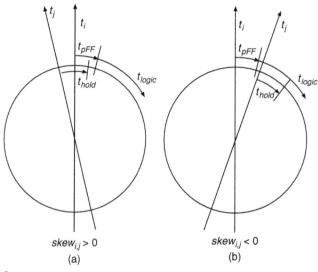

$skew_{i,j} > 0$ $skew_{i,j} < 0$

(a)   (b)

FIGURE 13.6

Hold time constraint in the presence of clock skew.

$$t_i + t_{pFF}^{min} + t_{logic}^{min} \geq t_j + t_{hold}^{max}$$

which can be rewritten as a lower bound constraint on the skew:

$$skew_{i,j} \geq t_{hold}^{max} - t_{pFF}^{min} - t_{logic}^{min} \qquad (13.4)$$

As mentioned earlier, clock skew is typically caused by unbalanced delays in the clock distribution network. Such unbalanced delays can be attributed to uneven wire lengths along the clock paths, mismatches in the numbers of buffers along the clock paths, and differences in the clock load driven by clock buffers. Although it is possible to manipulate these differences and mismatches to create useful skew for performance enhancement, skew caused by variations in the manufacturing process cannot be eliminated or exploited easily. Oxide variations, dopant variations, and width and length variations of devices affect such device parameters as the threshold voltage and parasitic capacitance of clock drivers and buffers. Variations in the interconnect thickness, widths, and spacing, and variations in interlayer dielectric thickness affect the interconnect resistance and capacitance. Because of these variations in process parameters, the clock skew between two clock pins may be different in two different dies, although the skew remains constant in a die. The challenge is to design a clock distribution network that works equally well across different dies.

Although skew caused by static variations stays constant cycle to cycle, time-varying variations in the operating condition of a circuit cause *temporal variation* of the clock period at a given point on the chip [Rabaey 2003]. The clock edge at a flip-flop may sometimes arrive earlier and sometimes later with respect to an ideal reference clock, depending on the operating condition of the circuit, as shown in Figure 13.7. Such temporal clock-period variation is referred to as *clock jitter*. In particular, we call the worst-case deviation (absolute value) of the arrival time of a clock edge at a given location with respect to an ideal reference clock edge as *absolute jitter*. Let $t_i(n)$ and $t_i(n+1)$, respectively, refer to the arrival times of the nth and $n+1$st clock edges at clock pin s_i. The cycle-to-cycle jitter (absolute value) for s_i is defined to be $t_i^{jitter}(n) = |t_i(n+1) - t_i(n) - C_P|$. Hence, the worst-case cycle-to-cycle jitter is

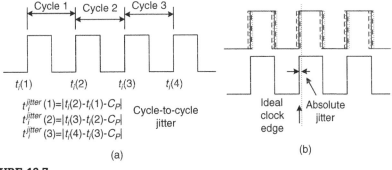

(a)

(b)

FIGURE 13.7

(a) Cycle-to-cycle jitter. (b) Absolute jitter.

twice the absolute jitter (*i.e.*, $max_n t_i^{jitter}(n) = 2T_i^{jitter}$), where T_i^{jitter} is the absolute jitter of s_i. The setup-time constraint and hold-time constraint should be updated accordingly:

$$skew_{i,j} \leq C_p - t_{pFF}^{max} - t_{logic}^{max} - t_{setup}^{max} - (T_i^{jitter} + T_j^{jitter}) \tag{13.5}$$

$$skew_{i,j} \geq t_{hold}^{max} - t_{pFF}^{min} - t_{logic}^{min} + (T_i^{jitter} + T_j^{jitter}) \tag{13.6}$$

In other words, the clock period may be shortened or lengthened by $T_i^{jitter} + T_j^{jitter}$.

A few main sources of environmental variations cause clock jitter: power supply noise, temperature gradients, substrate noise, and coupling of the clock network with adjacent signals. All these variations are caused by switching activities. The first three variations, in particular, affect the generation and propagation of clock signal, because device parameters are strong functions of these. The analog component of the clock generation circuitry and the clock buffers in the distribution network are both significant contributors to clock jitter. Switching activities, which vary from cycle to cycle, cause variations in interconnect couplings, gate capacitances, and the voltages of internal nodes of latches and flip-flops. As a consequence, the load presented to clock buffers varies from cycle to cycle, which causes jitter.

In the next section, we further elaborate on an important source of time varying variations, namely, the power supply noise.

13.2.3 *IR* **drop and** $L \cdot di/dt$ **noise**

Because of the non-zero resistance of wires connecting a transistor to a P/G pad, the transistor does not see a full V_{DD} when current flows from the P/G pad to the transistor. The voltage fluctuations, called *IR* drop, cause a degradation in drive strength. Until recently, *IR* drop has been defined mainly by considering the average current flowing through a power supply network [Lin 2001]. In a static *IR* drop analysis, we calculate the *IR* drop defined as such by performing a DC analysis of the power supply network that takes the average current as the input. Of late, designers are concerned that the voltage of a power supply network may not have sufficient time to recover because of the shrinking clock period as the clock frequency increases. There has been an increasing need for dynamic *IR* drop analysis, in which we have to consider the current waveform across multiple clock cycles [Lin 2001]. Therefore, the analysis of dynamic *IR* drop requires a transient analysis of the power supply network.

The P/G pins also have associated inductive parasitics. Moreover, wires in a P/G network typically have larger dimensions. Combined with the high operating frequencies of modern designs, the impedances of such wires are increasingly dominated by their inductive parasitics. Because of the time-varying nature of the current drawn from the power supply noise, inductance causes AC voltage fluctuations, called $L \cdot di/dt$ noise, in the supply lines. Large current spikes caused by a large number of simultaneous switchings can cause considerable $L \cdot di/dt$ noise.

In addition to causing signal integrity problems, power supply noise results in variations in performance, which may also cause hold time or setup time violations. It is important to note that this effect varies from cycle to cycle, because power supply noise is strongly related to switching activities. Moreover, the supply voltage continues to scale with device scaling. The reduction in power supply voltage is accompanied by a decrease in noise margins as well. Consequently, modern designs are more prone to failures caused by power supply noise.

Increasing wire widths is a typical approach to counter *IR* drop. Besides wire sizing, decoupling capacitors are usually inserted to reduce *IR* drop and to suppress $L \cdot di/dt$ noise. Although on-chip decoupling capacitors are indispensable for robust power supply, it is important to note that they are usually realized as MOS capacitors. Hence, they are as leaky as transistors. For energy conservation, it is crucial to restrict the use of on-chip decoupling capacitors.

13.2.4 **Power dissipation**

As the clock network switches in every cycle, its total power dissipation can be significant. The IBM Power4 processor, for example, consumes 70% power in clock networks and latches [Anderson 2001]. Because of the high switching activity, dynamic power is the most significant component of clock power:

$$P_{clk} = CV_{DD}^2 f_{clk} = CV_{DD}^2/C_p$$

The capacitive load C includes the gate capacitance of the sequential elements controlled by the clock signal, the interconnect capacitance of the clock network, and the capacitances associated with the buffers used in the clock network.

One very effective technique in reducing the switching capacitance of a clock network is through *clock gating* [Rabaey 2003]. One of many implementations of clock gating is to "AND" the clock signal with an enable signal, as shown in Figure 13.8. When the enable signal "EN" is low, all sequential elements in the

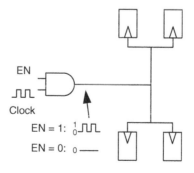

FIGURE 13.8

Clock gating for power reduction.

downstream of the AND gate will stay inactive. As the output terminals of these sequential elements remain quiet, the combinational logic between these sequential elements also stay inactive. In other words, the corresponding functional units have been disabled or clock gated. Because these sequential elements (and the ensuing combinational logic) do not participate in the switching activity, the switching capacitance of the clock network is reduced, effectively cutting down on the dynamic power dissipated by the clock network.

For clock gating to be effective, the sequential elements should be clustered such that a group of sequential elements can be gated by a single enable signal. It is not economical otherwise to have one AND gate for each sequential element. Moreover, there is an overhead of a control logic for the generation and distribution of enable signals for different clusters of clock-gated sequential elements. Propagating the clock signal through a mixture of buffers and AND gates further complicates the control of clock skew. Even for flip-flops that are not clock-gated, it may be necessary to insert an identical number of AND gates along the clock path to promote symmetry in the clock network.

Clock gating is also an active power management technique that further heightens the imbalance of switching activity at different locations of a chip. Time-varying variations such as temperature gradients may become more severe. The switching of a module between active and inactive modes also results in larger variations in the current flowing through the power supply network. All these variations are the main contributors to clock jitter.

13.2.5 Electromigration

Both the power supply network and the clock distribution network carry high currents because of the high capacitive load that they have to charge or discharge. Wires that carry high currents typically suffer from the effects of electromigration, where metal atoms migrate as a result of electrical current flowing through the metal material. Void appears in the area from where metal molecules migrate, leading to narrower line widths and eventually an open circuit. On the other hand, electromigration can also cause metal buildup in the form of hillocks in a wire, which may lead to a short with an adjacent wire. Metal lines that have existing cracks or other imperfections are particularly prone to electromigration. Circuit failure can also occur in the form of timing violation, because the resistance and capacitance of a metal wire change because of electromigration.

A good metric to characterize the reliability of a chip is the ***mean time to failure*** (MTTF). A mathematical model for the MTTF of a chip caused by electromigration is given by Black's equation [Black 1969]:

$$MTTF = (AJ^{-2})e^{E_a/kT}$$

where A is an empirically determined scaling factor, J is the current density, E_a is the activation energy that is determined by the material and its diffusion mechanism, k is Boltzmann's constant, and T is the temperature.

The typical current density at which electromigration occurs in modern on-chip interconnects is 10^6 to 10^7 A/cm^2. Because the current in a wire varies with time, the current density J in Black's equation should be replaced by the effective current density J_{eff}. For wires in a power supply network, the current flow is usually unidirectional (from a V_{DD} pad to a transistor). Thus, the effective current density is the average direct current density. For clock interconnects, the effective current in a wire is essentially zero, because the current flow is bidirectional. Treating the current as alternating, the root-mean-square current density should be used instead.

It is clear from Black's equation that the occurrence of electromigration in a design is mainly determined by the current density J and the temperature T. It is, therefore, desirable to reduce the switching activity of the circuit. We can also address the issue of electromigration by the following techniques that target the reduction of current density J. Increasing the wire width is the most straightforward way to decrease current density and increase MTTF. Multiple vias can be, and are often, used to improve reliability. The geometry of a via array should facilitate an even distribution of the current flow through all the vias. Turning corner with an oblique bend instead of a right-angled-bend also eliminates current crowding.

13.3 CLOCK NETWORK DESIGN

The two main subsystems in a clock network are: clock generation and clock delivery. Clock generation is an important topic that has been covered quite extensively in many textbooks on circuit design. This book, being an electronic design automation text, will focus on the synthesis of clock delivery networks. First, we cover some of the common clock topologies used in VLSI circuits. Second, we present the Elmore delay model, which is extensively used in the EDA community for the analysis and synthesis of clock networks. Third, we describe several basic clock synthesis algorithms, dealing with both skew scheduling and clock routing. Finally, we focus on a few fundamental clock optimization techniques, namely, buffer insertion, clock gating, wire sizing, and link insertion. For ease of presentation, the descriptions of many algorithms in this chapter may deviate from the original ones in the literature.

13.3.1 Typical clock topologies

The most economical way of distributing a clock signal is that of a tree topology. Assuming a binary tree structure, a clock tree with h levels of internal nodes can reach 2^h flip-flops at the leaf nodes.

An ideal clock tree is the *H-tree* topology [Fisher 1982; Kung 1982] as shown in Figure 13.9, where the basic building block at each level of the distribution network is a regular H-structure. Another scheme that yields equal-length interconnections is the X-tree (see also Figure 13.9), where the basic building

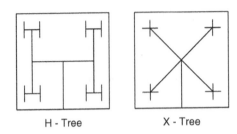

H - Tree   X - Tree

FIGURE 13.9

Symmetric H-tree and X-tree.

block is an X-structure [Bakoglu 1990]. All four corners of the H-structure and X-structure are equidistant from the center of the structure. Both the H-tree and X-tree achieve equal path length from the clock source to the leaf nodes by, respectively, repeating the H-structure and X-structure recursively top-down as shown in Figure 13.9. Such regular topologies also facilitate the addition of clock buffers in a symmetrical fashion.

H-trees (or X-trees), although effective in equalizing path lengths from a driver to a set of sinks, have serious limitations. These trees are best suited for regular layouts where the clock load is uniformly distributed over the entire chip. It is not particularly suitable for irregular placements with varying sink capacitances, which are common for cell-based designs. Moreover, a tree topology is more susceptible to the effects of variations in process parameters and operating condition because of its lack of redundancy; there exists only one unique path from the clock source to a flip-flop.

A very effective way of introducing redundant clock paths to a balanced clock tree is to add a trunk to connect all the leaf nodes of the clock tree, and branch off from the trunk to drive clock pins. In Figure 13.10, a balanced binary tree drives the center trunk at 4 positions uniformly distributed along the trunk. The binary tree is balanced, because the clock paths from the clock source to the 4 connecting points along the trunk are of equal length. The clock paths from the center trunk to the 5 clock pins are also designed to be of the same path length. The center trunk is of wider width, so that the associated resistivity is minimized. Consequently, the delays of the clock signal at various locations of the clock trunk can be minimized or considered to be uniformly equal. Hence, the clock trunk can provide a uniform clock reference for a region of the circuit.

The clock spines used in the first Intel Pentium 4 microprocessor [Kurd 2001] are variants of the clock trunk topologies. Figure 13.11 shows three clock spines that are driven at various points along the spines. The network of distributed clock buffers is almost a balanced binary tree, with the non-tree structure driven by each of the third level clock buffers being the exception. As in the case of a clock trunk topology, the spines provide a uniform clock

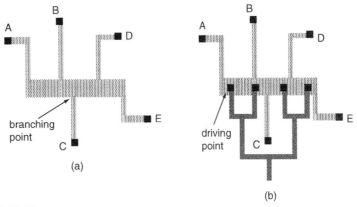

FIGURE 13.10

(a) A center trunk connecting 5 sinks. (b) A balanced binary tree driving the center trunk at 4 positions.

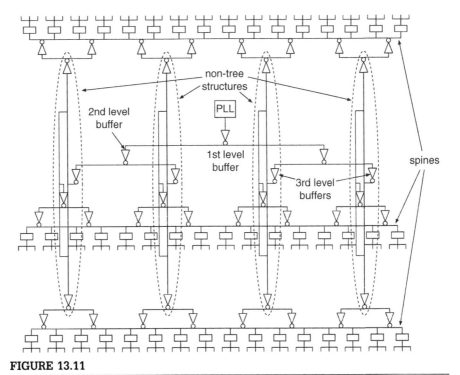

FIGURE 13.11

Clock spines in the first Intel Pentium 4 processor [Kurd 2001].

reference for all the clock trees radiating out of the spines. It is interesting to note that the clock signal goes through the same number of clock buffers before reaching the spines. Hence, the effects of device variations are minimized. However, clock skew or jitter at the clock buffers directly above the spines may result in short-circuit current between these buffers. In Figures 13.10 and 13.11, we show a trunk and a spine as a straight wire. This is an abstraction; in reality, the trunk or the spine could be a wire with multiple bends.

Compared with clock trunk topologies, a clock mesh provides significantly more alternative paths from the clock source to the flip-flops. A clock mesh is typically driven by distributed clock buffers. As shown in Figure 13.12, the clock buffers could either be located at the perimeter of the clock mesh, as in the case of the Alpha 21264 processor [Bailey 1998], or be uniformly distributed across the grid points of the clock mesh, as in the case of the IBM Power4 processor [Anderson 2001]. The distributed buffers for the mesh are typically driven by another clock network. For a mesh whose clock buffers are on the perimeter of the clock mesh, it is convenient to think of a boundary edge of the mesh as a clock trunk that connects these buffers. For a mesh whose clock buffers are uniformly distributed, it is typical to use an H-tree to distribute the clock buffers.

If a clock mesh that spans the entire chip has both high grid density and high clock buffer density, the clock signal can be considered to be available almost everywhere on the chip with minimal skew. Because of its high grid density,

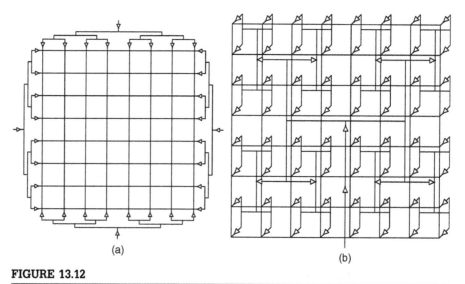

(a)   (b)

FIGURE 13.12

Two general clock mesh structures: (a) The clock drivers are at the perimeter. (b) The clock drivers are on the grid points.

most clock pins can be connected directly to the grid edges with short links. However, such a clock network incurs high power consumption. Furthermore, it is not amenable to clock gating, because most clock pins are connected to the clock mesh directly. A good compromise between power, skew, and wiring cost is to use a clock mesh with a moderate grid density and a moderate clock buffer density.

For such a clock mesh, the clock signal is no longer that freely available to flop-flops distributed across the chip. Additional clock subnetworks have to be constructed to connect them to the mesh. Figure 13.13 shows such a two-level clock distribution network: global (level) clock distribution and local (level) clock distribution. The global clock network distributes the clock signal from the clock source to various regions across the chip with a balanced H-tree driving a clock mesh. The local clock distribution subnetworks further distribute the clock signal to the flip-flops. Such a two-level design facilitates clock gating, especially when flip-flops that should be gated by the same control signal reside in close proximity.

Although this example shows a two-level clock network, it is not uncommon for a high-performance processor design to use a clock network with three or more levels of hierarchy. Different topologies can be deployed at each level. Instead of the use of a clock mesh as a global clock network, clock trunks and clock spines could also be used. Similarly, the local clock distribution subnetworks could either be a tree topology, clock trunk, clock spine, or even a clock mesh. Among these, trees are the topology of choice for the connection of the local flip-flops because of their low wiring cost. Because of the uneven

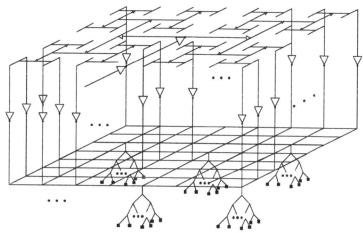

FIGURE 13.13

A hierarchical clock network with an H-tree driving a clock grid for global clock distribution, which in turn drives flip-flops through many local clock distribution networks.

distribution of flip-flops or clock load in a local region, it is common to use an asymmetric tree structure as the clock network. Figure 13.14 shows some asymmetric tree topologies constructed for the same set of clock pins under different skew requirements. As we can observe from the topologies, they become more complex and more costly as the skew constraints become more stringent (from (a) to (c)). The topologies shown here are an "abstraction" of the physical clock trees embedded in a Manhattan routing plane. Although they show the physical locations of internal nodes in the clock trees, the connections are not

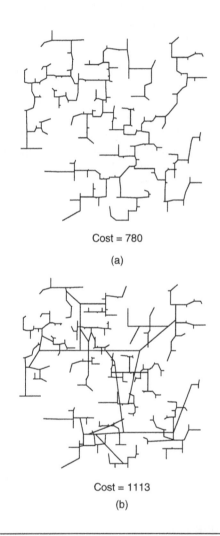

Cost = 780

(a)

Cost = 1113

(b)

FIGURE 13.14

Topologies for the same set of clock pins under various kinds of skew constraints: (a) Very relaxed skew constraints. (b) Moderate skew constraints. (c) Very tight skew constraints.

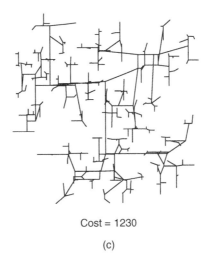

Cost = 1230

(c)

FIGURE 13.14

(Continued)

embedded with horizontal and vertical wire segments. In other words, the connections are not shown with rectilinear routing.

Most studies on clock layout are concerned with the construction of a tree topology that satisfies the skew constraints. In fact, the synthesis of such a tree will be the main focus of Section 13.3.3. Non-tree structures, such as meshes or a hybrid of meshes and trees, typically can tolerate higher degrees of parameter variations in devices and wires, as well as uncertainties in system operating conditions because of its highly interconnected nature. In other words, the behavior of non-tree structures is more predictable. However, the robustness of a non-tree structure is achieved at the expense of higher routing cost compared with a tree topology. Consequently, tree structures are preferred over non-tree structures when routing area is a premium, but non-tree structures are the preferred solutions when the predictability of clock networks is critical.

Because deep-submicron designs are more susceptible to parameter variations, more and more designs have turned to non-tree structures for clock distribution. A good tradeoff between area and process variation is the insertion of cross-links in a tree topology, as shown in Figure 13.15. Here, wires are inserted into a tree to provide alternative paths between selected pairs of nodes whose skews are deemed to be highly susceptible to process variations. We will further elaborate on the synthesis of such a non-tree structure in Section 13.3.4.

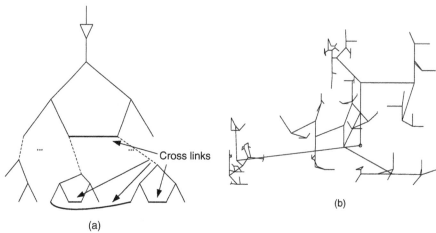

(a)

(b)

FIGURE 13.15

(a) An abstract clock tree with links inserted. (b) A clock layout for a benchmark circuit with cross-links inserted with the algorithm from [Lam 2005].

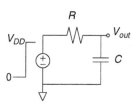

FIGURE 13.16

A simple RC circuit driven by a step input.

13.3.2 **Clock network modeling and analysis**

A key component to the synthesis of a reliable clock network is the modeling and analysis of clock signal delays. In this section, we describe a commonly used delay model for clock network synthesis, namely, the Elmore delay model [Elmore 1948]. In particular, we focus on the computation of Elmore delay for a tree.

Figure 13.16 shows a lumped RC circuit driven by a step input. The response at the capacitor is governed by the following expression:

$$v_{out}(t) = V_{DD}(1 - e^{-t/RC})$$

The time at which $v_{out}(t)$ reaches some specified critical voltage V_{crit} is given by

$$t_{crit} = RC \cdot \ln \frac{V_{DD}}{V_{DD} - V_{crit}}$$

In particular, the time at which the output voltage reaches 50% of V_{DD} is approximately $0.69 \times RC$. The rise time, which is typically defined to be the elapsed time between 10% and 90% of V_{DD}, is approximately $2.2 \times RC$.

It is, however, more difficult to calculate the 50% delay of a distributed line, or for that matter, an RC tree made up of lumped or distributed RC lines. Fortunately, a number of useful delay models have been developed to help approximate the delay of RC trees. Among these, the Elmore delay model is most commonly used.

First, we consider trees of lumped RC elements only. In [Elmore 1948], Elmore defined the delay of node i in an RC tree as

$$t_{Elmore, i} = \int_0^\infty t \cdot h_i(t)dt$$

where $h_i(t)$ is the *impulse response* to the unit impulse (applied at time 0) at time t, or equivalently, the derivative of the unit step response at time t. The 50% delay, denoted t_{50}, is the time for the monotonic step response to reach 50% of V_{DD}, and it is the median of the impulse response. In essence, the Elmore delay model uses the mean of the impulse response $h(t)$ to approximate the 50% delay of the step response.

Rubinstein, Penfield, and Horowitz [1983] derived an algorithmic approach to compute the Elmore delay of all nodes in an RC tree; the computation time is proportional to the number of lumped RC elements. Before giving the details of the approach, we first provide some definitions. Given an RC tree, there is a corresponding abstract topology. Figure 13.17 shows one such example. The tree is driven at its root (u) by a driver (or buffer), which is modeled as voltage source, labeled *src*, connected in series with an output resistance, R_u. Every internal node of the tree is connected to one or more child nodes. Because each of these connections (or tree edges) can be uniquely identified by the child node, we label the resistors between them according to the labels of the child nodes. Node u, for example, is connected to child nodes v and x by resistors R_v and R_x, respectively, in the RC tree. Each node i in the RC tree has an associated capacitor C_i.

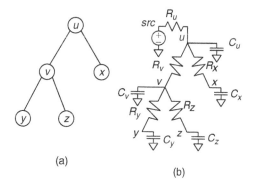

(a)

(b)

FIGURE 13.17

(a) An abstract topology. (b) The corresponding RC tree.

Let *Path(src,j)* denote the path in the RC tree T from voltage source, *src*, to node j. Moreover, let $R_{Path(src,i) \cap Path(src,j)}$ denote the total resistance along the path common to *Path(src,i)* and *Path(src,j)*. Furthermore, let T_i denote the subtree whose root node is i. The Elmore delay from *src* to i is given by the following two equivalent expressions:

$$t_{Elmore,i} = \sum_{j \in T} C_j \times R_{path(src,i) \cap path(src,j)}$$
$$= \sum_{R_k \in path(src,i)} R_k \sum_{j \in T_k} C_j \quad (13.7)$$

The second expression, *i.e.*, Equation (13.7), gives a linear time computation of the Elmore delay. Consider the example in Figure 13.17, the Elmore delays of nodes v and z can be written as:

$$t_{Elmore,v} = R_u(C_u + C_v + C_x + C_y + C_z) + R_v(C_v + C_y + C_z)$$
$$t_{Elmore,z} = R_u(C_u + C_v + C_x + C_y + C_z) + R_v(C_v + C_y + C_z) + R_z C_z$$

First, we observe that the capacitive terms in the parentheses are cumulative from right to left in the two preceding expressions or from bottom to top topologically if we traverse the RC tree. This suggests that in a bottom-up manner (or in a post-order traversal), we can compute at each node k the total downstream capacitance $C_{T_k} = \sum_{j \in T_k} C_j$. The procedure for the computation of downstream capacitance of node k in a bottom-up fashion is given in Algorithm 13.1.

Algorithm 13.1 Compute-C_T

Input: Node k

Output: Total downstream capacitance CT_k at node k

1. $C_{T_k} \leftarrow C_k$;
2. **for** each node $j \in$ Children(k) **do**
3. $C_{T_k} \leftarrow C_{T_k} +$ Compute-$C_T(j)$;
4. **end for**
5. **return** C_{T_k};

Second, $t_{Elmore,z} = t_{Elmore,v} + R_z C_z$. In other words, the Elmore delay of a node is an accumulation of RC product terms from the voltage source to the node of interest. The RC product term at node k is that of the branch resistance R_k and the total downstream capacitance C_{T_k}. Therefore, in a top-down manner from the voltage source to node i, we can sum up the RC delay of each resistor along the path. In fact, in a top-down traversal of the tree T, we can compute the Elmore delays for all nodes. A description of the recursive computation of Elmore delays of nodes in T_k in a top-down fashion is given in Algorithm 13.2.

Algorithm 13.2 Compute-Elmore-Delay

Input: Node k

Output: Elmore delay of node k

1. **if** node k is the root node

2. $t_{Elmore,k} \leftarrow 0$;

3. **else**

4. $t_{Elmore,k} \leftarrow t_{Elmore,Parent(k)} + R_k \cdot C_{T_k}$;

5. **for** each $j \in Children(k)$ **do**

6.   Compute-Elmore-Delay(j);

7. **end for**

On-chip interconnects are not lumped RC elements; they are distributed RC lines. Given an RC line whose resistance R and capacitance C are uniformly distributed over the line, we can divide the line evenly into n segments and model each segment as a lumped RC element, with resistance R/n and capacitance C/n. Assuming a step input at one end of the wire, the Elmore delay at the downstream endpoint is

$$t_{Elmore} = R/n \cdot C + R/n \cdot C(n-1)/n + \cdots + R/n \cdot C/n$$
$$= RC(n+1)/2n$$

As n approaches ∞, t_{Elmore} approaches $RC/2$. We can replace a uniformly distributed RC line as a π-type lumped RC circuit as shown in Figure 13.18, where one-half of the wire capacitance is at the downstream end. Because the uniformly distributed RC line has to present a total capacitive load of C to all upstream resistors in the tree topology, the remaining half of the wire capacitance is at the upstream end of the wire.

If one end of the wire is connected to the clock pin of a flip-flop, the capacitance associated with that end should include both the gate capacitance of the clock pin and one-half of the wire capacitance.

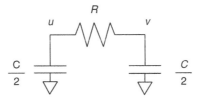

FIGURE 13.18

Modeling a uniformly distributed RC line of resistance R and capacitance C with a π-type circuit.

Remark: Is Elmore Delay Accurate Enough? The Elmore delay model has been used extensively to approximate the 50% delay point. It can easily be shown that the Elmore delay gives the 63% ($= 1 - 1/e$) delay of a simple RC circuit (with a single resistor and a single capacitor), which is an upper bound of the 50% delay. In general, the Elmore delay of a sink in an RC tree gives an upper bound on the actual 50% delay of the sink under not only the step input but also any monotonically increasing, piecewise-smooth input $u(t)$, with its derivative $\frac{d}{dt}u(t)$ being unimodal and symmetric [Gupta 1997]. The approximation of the 50% signal delay by the Elmore delay is exact only for a symmetric impulse response, where the mean is equal to the median [Gupta 1997]. Although the Elmore delay model may not be accurate, it has a high degree of *fidelity*. An optimal or near-optimal solution according to the estimator is also nearly optimal according to actual (SPICE-computed [Nagel 1975]) delay for routing constructions [Boese 1995] and wire sizing optimization [Cong 1996]. Studies in [Cong 1995b, 1998] also showed that the clock skew under the Elmore delay model has a high correlation with the actual (SPICE) skew. Figure 13.19 demonstrates the accuracy and fidelity of Elmore delay skew to actual skew for routing trees constructed under the Elmore delay [Cong 1998].

Of course, the Elmore delay model has a few disadvantages. First, the Elmore delay may not be very accurate. So, it is not suitable to be used directly for accurate circuit timing analysis. Also, it cannot handle the inductive effect, because the Elmore delay is defined for a monotonic response. More accurate delay estimation can be obtained with higher order moments.

13.3.3 Clock tree synthesis

Two problems relate to the synthesis of a clock net: (1) the determination of a *feasible clock schedule* that defines the arrival times of the clock signals at the

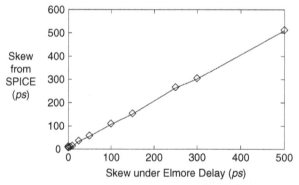

FIGURE 13.19

Elmore delay skew versus actual (SPICE simulation) delay skew for clock trees obtained by Greedy-BST/DME algorithm [Cong 1995b].

clock pins, and (2) the physical layout of the clock network that *realizes* the clock schedule. In the context of clock synthesis, a clock schedule is feasible if it meets the performance requirement without causing race hazards in the system operation. A physical clock network realizes the clock schedule if the clock signal arrives at the registers at the respective arrival times specified by the clock schedule. We refer to the first problem as that of *clock skew scheduling* and the second as that of *clock routing*.

13.3.3.1 *Clock skew scheduling*

Designers may impose additional constraints to make circuits more robust to process variations or have less power consumption. For example, we can make a circuit more robust to process variations by subtracting a *safety margin* from the upper bound constraint in Equation (13.5), or adding a safety margin to the lower bound constraint in Equation (13.6):

$$skew_{i,j} \le C_p - t_{pFF}^{max} - t_{logic}^{max} - t_{setup}^{max} - (T_i^{jitter} + T_j^{jitter}) - \delta_u \qquad (13.8)$$

$$skew_{i,j} \ge t_{hold}^{max} - t_{pFF}^{min} - t_{logic}^{min} + (T_i^{jitter} + T_j^{jitter}) + \delta_l \qquad (13.9)$$

where $\delta_u \ge 0$ and $\delta_l \ge 0$ are the safety margins. In general, safety margins may vary for different pairs of flip-flops.

For convenience, we use $l_{i,j} \le skew_{i,j} = t_i - t_j \le u_{i,j}$ to represent lower- and upper-bound skew constraints between s_i and s_j and $\mathcal{C} = \{l_{i,j} \le skew_{i,j} \le u_{i,j}\}$ to denote the set of skew constraints for all sequentially adjacent clock pins s_i and $s_j \in S$, the set of all clock pins of flip-flops. A skew schedule X is an assignment of delay values t_i to each clock pin s_i. We say that X is feasible if $skew_{i,j}$ satisfies the skew constraints in $\mathcal{C}$.

The skew bounds, each constraining the difference of a pair of variables, can be represented by a constraint graph $G_\mathcal{C} = (V, E)$ as follows [Cormen 2001]: Each clock pin in S corresponds to a vertex in the constraint graph. For the two skew constraints associated with s_i and s_j, we generate two directed edges in $G_\mathcal{C}$, $e_{i,j}$ and $e_{j,i}$. The former edge captures the lower-bound constraint and the latter edge the upper-bound constraint. The weight of $e_{i,j}$, denoted by $w_{i,j}$, is $-l_{i,j}$, and the weight of $e_{j,i}$, denoted by $w_{j,i}$, is $u_{i,j}$. We will now give the motivation for such a graph formulation.

Consider a circuit with four flip-flops with clock pins $\{s_1, s_2, s_3, s_4\}$ for example. In this circuit, FF_1 feeds its output (through some combinational logic) to FF_2 and FF_3, both of which send data to FF_4. We illustrate in Figure 13.20 the corresponding constraint graph.

First, we consider the four upper-bound skew constraints among the four clock pins: $skew_{1,2} \le u_{1,2}$, $skew_{2,4} \le u_{2,4}$, $skew_{1,3} \le u_{1,3}$, and $skew_{3,4} \le u_{3,4}$. Now, the constraint $t_i - t_j \le u_{i,j}$ can be written as $t_i \le t_j + u_{i,j}$. With this expression, we can interpret t_i as a distance label of node s_i, and say that s_i is of a distance at most $u_{i,j}$ away from s_j. That is equivalent to the existence of a

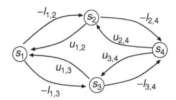

FIGURE 13.20

A constraint graph constructed from a circuit with four flip-flops represented by clock pins $\{s_1, s_2, s_3, s_4\}$. FF_1 is sequentially adjacent to FF_2 and FF_3, both of which are sequentially adjacent to FF_4.

directed edge $e_{j,i} \in E$ of weight $w_{j,i} = u_{i,j}$ from s_j to s_i for each constraint $t_i - t_j \le u_{i,j}$, as shown in Figure 13.20.

Although FF_1 and FF_4 are not sequentially adjacent, the two data paths $FF_1 \to FF_2 \to FF_4$ and $FF_1 \to FF_3 \to FF_4$ induce two transitive constraints between s_1 and s_4:

$$skew_{1,2} + skew_{2,4} = (t_1 - t_2) + (t_2 - t_4) = t_1 - t_4 \le u_{1,2} + u_{2,4}$$
$$skew_{1,3} + skew_{3,4} = (t_1 - t_3) + (t_3 - t_4) = t_1 - t_4 \le u_{1,3} + u_{3,4}$$

The tighter of the two constraints imposes an upper-bound constraint on the skew between s_1 and s_4:

$$skew_{1,4} = t_1 - t_4 \le \min\{u_{1,2} + u_{2,4}, u_{1,3} + u_{3,4}\}$$

Consequently, the upper bound on $skew_{1,4} = t_1 - t_4$ is essentially the shortest path distance from s_4 to s_1 in the constraint graph.

Next, we consider the lower bound constraints $l_{1,2} \le skew_{1,2}$, $l_{2,4} \le skew_{2,4}$, $l_{1,3} \le skew_{1,3}$, and $l_{3,4} \le skew_{3,4}$. Because $l_{i,j} \le t_i - t_j = skew_{i,j}$ is equivalent to $t_j - t_i \le -l_{i,j}$, we add a directed edge $e_{i,j} \in E$ of $w_{i,j} = -l_{i,j}$ from s_i to s_j, as shown in Figure 13.20. It is obvious that $skew_{1,4}$ is bounded from below by the negative of the shortest distance from s_1 to s_4.

It can be shown that there exists a feasible skew schedule subject to $\mathcal{C}$ if, and only if, G_C has no negative-weight cycles [Cormen 2001]. To perform a feasibility check and to compute a feasible skew schedule, we first augment the constraint graph $G_C = (V, E)$ with a source vertex S_0 and connect S_0 to each $s_i \in V$ with an outgoing edge of zero weight. Then, we apply the Bellman-Ford algorithm [Cormen 2001] to the augmented graph to compute the shortest distances from S_0 to all vertices. If the constraint graph has no negative-weight cycles, the Bellman-Ford algorithm terminates with appropriate shortest distances (or t_i's) assigned to clock pins; it returns a flag indicating that the schedule is infeasible otherwise. A careful implementation of the Bellman-Ford algorithm for the feasibility check has $O(|V||E|)$ time complexity.

The preceding formulation checks for the feasibility of a given clock frequency (or clock period C_P). We can perform clock frequency optimization (or minimization of clock period) by building on the preceding formulation with an iterative binary search procedure. Suppose we are to determine the smallest feasible C_P within the range $C_{P,\min} \leq C_P \leq C_{P,\max}$. First, we consider $C_P = (C_{P,\min} + C_{P,\max})/2$ and formulate the set of skew constraints C accordingly. If the Bellman-Ford algorithm returns a feasible schedule, we update $C_{P,\max}$ to $(C_{P,\min} + C_{P,\max})/2$; otherwise, we update $C_{P,\min}$ to $(C_{P,\min} + C_{P,\max})/2$. We repeat the binary search procedure until $C_{P,\min}$ and $C_{P,\max}$ are sufficiently close.

Another important concept related to skew scheduling in the context of clock tree synthesis is that of a *feasible skew range*. Consider a circuit with three clock pins $\{s_1, s_2, s_3\}$ of three flip-flops FF_1, FF_2, and FF_3 forming a cyclic data path $FF_1 \rightarrow FF_2 \rightarrow FF_3 \rightarrow FF_1$. The skew constraints among the three flip-flops $-10 \leq skew_{1,2} \leq 3$, $-5 \leq skew_{1,3} \leq -2$, and $1 \leq skew_{2,3} \leq 4$ are captured in the constraint graph as shown in Figure 13.21a. We refer to the range $[l_{i,j}, u_{i,j}]$ defined by the lower and upper bounds of $skew_{i,j}$ as the skew range of $skew_{i,j}$.

Interestingly, even though zero-skew is within the skew ranges of $skew_{1,2}$ and $skew_{2,3}$, a feasible skew schedule does not exist if we choose $skew_{1,2} = 0$. This is because of the transitive skew constraints $-9 \leq t_1 - t_2 \leq -3$ between s_1 and s_2 induced by the skew constraints of $skew_{1,3}$ and $skew_{2,3}$. Such stricter constraints are, in fact, captured by the shortest paths between s_1 and s_2 in the constraint graph: the upper bound of $skew_{1,2}$ is essentially the shortest path distance from s_2 to s_1 in the constraint graph, and $skew_{1,2}$ is bounded from below by the negative of the shortest distance from s_1 to s_2.

Consider a constraint graph with no negative-weight cycles. Let $d_{i,j}$ denote the shortest distance to s_j from s_i. The inequalities $-d_{i,j} \leq skew_{i,j} \leq d_{j,i}$ define the feasible skew range of $skew_{i,j}$, denoted as $FSR_{i,j}$. Given a constraint graph G_C with no negative cycles, we can build an all-pairs shortest distance matrix $D = \{d_{i,j} : s_i, s_j \in S\}$ from G_C in $O(|V|^3)$ by the Floyd-Warshall algorithm [Cormen 2001] to represent the feasible skew ranges of all skew pairs. Figure 13.21b shows the all-pairs shortest distance matrix of the original constraint graph in Figure 13.21a.

We say that a *skew commitment* is made when we narrow a nontrivial *FSR* to a single skew value. In fact, we can commit $skew_{i,j}$, the skew between s_i and s_j, to any $x \in [-d_{i,j}, d_{j,i}] = FSR_{i,j}$, because a feasible skew schedule that contains such a skew commitment always exists. However, we may not make two or more such skew commitments independently without affecting the existence of a feasible skew schedule. Algorithm 13.3 gives an $O(|V|^2)$ approach for updating matrix D after we commit $skew_{i,j}$ to a skew value of x by shrinking the feasible skew range to a single value. In general, the algorithm can be applied whenever we refine the skew range of a pair of clock pins. Figure 13.21c shows the resultant matrix D after a skew commitment of $skew_{1,2} = -3$.

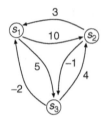

Constraint graph G_C

(a)

$$
\begin{array}{c c c c}
 & s_1 & s_2 & s_3 \\
s_1 & 0 & 9 & 5 \\
s_2 & -3 & 0 & -1 \\
s_3 & -2 & 4 & 0
\end{array}
$$

All-pairs shortest distance matrix D

(b)

$$
\begin{array}{c c c c}
 & s_1 & s_2 & s_3 \\
s_1 & 0 & 3 & 2 \\
s_2 & -3 & 0 & -1 \\
s_3 & -2 & 1 & 0
\end{array}
$$

Updated matrix D after committing $skew_{1,2} = -3$

(c)

FIGURE 13.21

An example showing the incremental skew scheduling based on an all-pairs shortest-distance matrix: (a) The constraint graph G_C. (b) The all-pairs shortest distance matrix D. (c) The all-pairs shortest distance matrix after committing the skew of s_1 and s_2 to $skew_{1,2} = -3$.

We now construct an $O(|V|^3)$ algorithm for incremental scheduling. First, the Floyd-Warshall algorithm is used to generate matrix D in $O(|V|^3)$ time complexity. As long as there exist uncommitted skews (or nontrivial feasible skew ranges), we select one of them and commit the skew to an arbitrary value in the feasible skew range. Then, we apply the incremental update procedure in

---

Algorithm 13.3 Update All-Pairs Shortest Distance Matrix

Input: A skew commitment $skew_{i,j} = x$, an all-pairs shortest distance matrix $D = \{d_{i,l}\}$
Output: An updated matrix D

1. Set $d_{i,j} = -x$ and $d_{j,i} = x$;

2. **for** each $d_{k,l}, 1 \leq k \neq l \leq n$ in D **do**

3.   Set $d_{k,1} = \min\{d_{k,l}, d_{k,i} - x + d_{j,l}, d_{k,j} + x + d_{i,l}\}$;

4. **end for**

Algorithm 13.3. In the worst case, we have to perform $|V| - 1$ skew commitments, hence the $O(|V|^3)$ time complexity.

In a sense, the incremental skew scheduler finds a spanning tree of the complete graph represented by the matrix D. Here, a skew commitment is essentially an edge of the spanning tree. (In reality, there are two edges for each skew commitment, with the edge weight being the committed skew value or the negation of it, depending on the direction of the edge.) Consequently, after at most $n - 1$ skew commitments, the spanning tree connects all sinks in the graph. Moreover, the skew between any pair of clock sinks is defined by the sum of the edge weights along the path between the two corresponding vertices in the spanning tree.

Remark: Clock Scheduling and Power Supply Noise: Very recent works on clock schedule optimization also considered current-induced noise control and minimization. If not minimized, current induced noise, which includes *IR* drop and $L \cdot di/dt$ noise, has an adverse effect on circuit performance, especially for low-power systems with reduced supply voltage, because low noise margins are a primary concern. An effective method to reduce the current-induced noise is to design the circuit such that the current drawn is fairly stable without large peaks. In many designs with a huge clock distribution network, current peaks are usually caused by the simultaneous switching of highly loaded clock lines, as well as by the switching activities in the sequential elements when they are triggered and in the ensuing combinational logic. There are several skew scheduling algorithms for spreading out these switching activities over time [Vuillod 1996; Vittal 1996; Lam 2002, 2003; Yu 2007]. The reader may refer to the relevant papers for more details.

13.3.3.2 *Clock tree routing*

We now deal with the problem of clock tree routing (*i.e.,* the construction of a tree topology that realizes the specified skew schedule). Recall that S contains the set of clock pins of flip-flops $\{s_1, \ldots, s_n\}$. For convenience, we also include the clock source, denoted as s_0 in S. In general, the clock routing problem can be formulated as follows: Given $\{l_{s_1}, \ldots, l_{s_n}\}$, a set of locations of the clock pins $\{s_1, \ldots, s_n\}$ of flip-flops and skew constraints on various pairs of flip-flops, construct with minimum wiring cost, a clock tree that satisfies the setup time and hold-time skew constraints in Equations (13.8) and (13.9). The location of the clock source, denoted as l_{s_0}, may also be given. If l_{s_0} is not given, the clock router will also determine a convenient location for s_0. There are possibly other constraints and/or objectives to the problem:

1. We want to impose a constraint on the rise/fall times of the clock signal at the sinks, because it is critical to keep the clock signal waveform clean and sharp.

2. We want the clock network to be tolerant of process variations, which cause the wire widths and device sizes on the fabricated chip to differ from the specified wire widths and device sizes, respectively, resulting in so-called *process skew* (*i.e.*, clock skew caused by process variations).

Several simplifications to the clock routing problem were made in the past to make the task easier. It is the enabling concept of (*absolute*) *global skew* that makes these simplifications possible. Indeed, these simplifications cultivate a prolific area of studies on clock tree routing.

The global skew is defined to be the maximum among all the absolute skew values between clock pins: $gskew = \max_{i,j} |t_i - t_j|$. Several simpler versions of the clock-routing problem centered around the concept of global skew are given in the following:

1. Given a set of sink locations, construct with minimum routing cost a clock routing that minimizes the global skew *gskew*.

2. Given a set of sink locations, construct with minimum routing cost a clock routing that achieves zero skew (*i.e.*, $gskew = 0$). In other words, all clock pins are required to have an identical clock delay. This is known as the zero-skew routing problem.

3. Given a set of sink locations and a skew bound $B \geq 0$, construct with minimum routing cost a clock routing that satisfies $gskew \leq B$. This is called the bounded-skew routing problem.

It is through these simplifications that we can obtain a better understanding of the general clock-routing problem, which is also known as the useful skew–routing problem, defined at the beginning of this section. We will now describe various algorithms on clock routing on the basis of the following classification: (1) zero-skew routing, (2) bounded skew routing, and (3) useful-skew routing.

In the descriptions of these algorithms, we distinguish a physical interconnect tree T from its *abstract topology* G, which is a binary tree such that all sinks are the leaf nodes of the binary tree. Every non-leaf node of G has two child nodes, with a possible exception that the root node may have only one child node. We first introduced the concept of an abstract topology in Section 13.3.2. We will now provide a more formal definition. The source driver, denoted s_0, is the root node of the tree. When l_{s_0}, the location of s_0, is given, s_0 has a singleton internal node, denoted as s_0', as its only child; otherwise, s_0

has two child nodes. Each non-root node v is connected to its parent, denoted as $p(v)$, by edge e_v. Consider any two nodes, say u and v, with a common parent node $w = p(u) = p(v)$ in the abstract topology. The signal from the source has to pass through w before reaching u and v (and their descendants).

The embedding of an abstract topology G to form an interconnect tree T involves the mapping of each internal node $v \in G$ to a location $l_v = (x_v, y_v)$ in the Manhattan plane, where (x_v, y_v) are the x- and y-coordinates, and replacing each edge $e \in G$ by a rectilinear edge or path. It is important to note that most clock-routing algorithms are concerned with only the mapping of internal nodes to physical locations and not the actual routing of wires, a task that could be accomplished by a maze router. Figure 13.22 shows an abstract topology and three of its many possible ways of embedding. Note that s_0' is the singleton child node of s_0 because the location of s_0 is given. The subtree rooted at node v in T is denoted as T_v. In T, the cost of edge e_v is its wire length, denoted by $|e_v|$. The cost of a routing tree T is the sum of its edge costs. Among the three ways of embedding the abstract topology in Figure 13.22, the embedding in (d) has the highest cost.

The objective of a clock routing algorithm is therefore twofold: to generate an abstract topology and to embed the abstract topology. For each of the clock-routing problems, namely zero-skew routing, bounded-skew routing, and useful-skew routing, we will first describe how we can embed a given abstract topology to satisfy the respective skew constraint(s). Then, we will present approaches to generate an abstract topology together with embedding.

13.3.3.3 *Zero-skew routing*

The problem of zero-skew routing was first successfully solved in [Tsay 1991] with a bottom-up approach. The *Deferred-Merge Embedding* (DME) algorithm, proposed independently in [Edahiro 1991; Chao 1992; Boese 1992], generalized the approach in [Tsay 1991] with the enabling concept of a *merging segment* to achieve zero-skew routing for a given abstract topology.

In general, given two zero-skew trees, there exists a set of locations at which two zero-skew trees can be joined with the minimum wire length such that the new tree is also of zero skew. In Figure 13.23b, for example, any point l_a on the line segment ms_a is equidistant from sinks s_1 and s_2 (*i.e.*, we obtain a zero-skew subtree rooted at l_a with sinks s_1 and s_2). For ease of illustration, the zero-skew tree in Figure 13.23 is constructed under the path length delay model. The model uses the wire length of the unique path between an ancestor node and a descendant node in the physical routing tree to estimate the delay between the two nodes. Many illustrations of different clock-routing algorithms in this section are based on the path length delay model.

Given S and an abstract topology G, the DME algorithm exploits this flexibility and embeds internal nodes of G by means of a two-phase approach: (1) a bottom-up phase that constructs for each internal node of G a set of possible placement locations of the node in a zero-skew tree T; and (2) a top-down

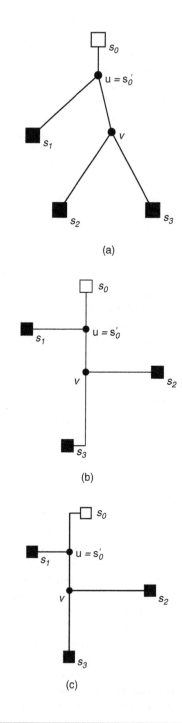

FIGURE 13.22

(a) An abstract topology. (b)–(d) Three different ways of embedding the topology.

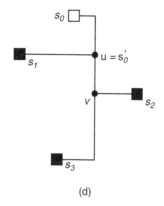

(d)

FIGURE 13.22

Continued

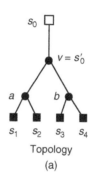

Topology

(a)

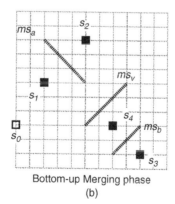

Bottom-up Merging phase

(b)

FIGURE 13.23

An illustration of the bottom-up phase of the DME algorithm: (a) Topology of a clock source s_0 and 4 sinks $s_{1..4}$. (b) Merging segments of internal nodes a, b and $v = s_0$.

embedding phase that determines for each internal node its exact location in T. Note that we do not perform actual Manhattan routing of the connections between nodes in the clock tree.

We refer to the set of possible placement locations of $v \in G$ in a minimum-cost zero-skew tree as the merging segment of v, denoted as ms_v. The segment ms_v is always a line segment (with possibly zero length when it degenerates into a point) with slope $+1$ or -1, as shown in Figure 13.23. This stems from the fact that a circle in the Manhattan routing plane is a square rotated by 45 degrees.

For each leaf node $s_i \in G$, the merging segment of s_i is simply l_{s_i}, the location of sink s_i. Suppose v is the parent node of s_i and s_j, which are at a distance of $d(l_{s_i}, l_{s_j})$ apart. Under the path length delay model, any point in ms_v should be of distance $d(l_{s_i}, l_{s_j})/2$ from l_{s_i} and l_{s_j}. The segment ms_v can be computed by taking the intersection of two Manhattan circles, both of radius $d(l_{s_i}, l_{s_j})/2$, that are centered at l_{s_i} and l_{s_j}. Because a Manhattan circle is a square that is rotated 45 degrees, the intersection must be a line segment of slope $+1$ or -1. Because it lies on the circumference of a Manhattan circle, we refer to such a line segment as a *Manhattan arc*.

Now, we consider a more general case. Let a and b be the child nodes of $v \in G$. The construction of ms_v depends on ms_a and ms_b, hence the bottom-up processing order. In particular, any point on ms_v should allow a and b to be merged with *minimum* wiring cost $|e_a| + |e_b|$, while maintaining zero skew among all sinks in T_v.

Let L denote the shortest Manhattan distance between ms_a and ms_b (i.e., $d(ms_a, ms_b) = L$). Here, the distance between two sets of points P and Q is defined as $d(P, Q) = \min\{d(p, q) | p \in P, q \in Q\}$. Let ms_v be at a distance of $x \cdot L$ from ms_a where x is between 0 and 1. Given t_a, the Elmore delay from a to its sinks in T_a, the delay from v to sinks in T_a is

$$t_a + r \cdot x \cdot L \cdot \left(C_{T_a} + \frac{c \cdot x \cdot L}{2}\right)$$

where C_{T_a} is the total capacitance of the subtree T_a, and r and c are, respectively, the unit-length resistance and capacitance of a wire of some prespecified width. When T_a happens to be just a clock pin s_i, C_{T_a} is the gate capacitance associated with clock pin and $t_a = 0$. Similarly, the Elmore delay from v to sinks in T_b is

$$t_b + r \cdot (1 - x) \cdot L \cdot \left(C_{T_b} + \frac{c \cdot (1 - x) \cdot L}{2}\right)$$

where t_b is the Elmore delay from b to its sinks in T_b, and C_{T_b} is the total capacitance of the subtree T_b. We can now solve for x as follows [Tsay 1991]:

$$x = \frac{t_b - t_a + r \cdot L \cdot (C_{T_b} + c \cdot L/2)}{r \cdot L \cdot (c \cdot L + C_{T_a} + C_{T_b})}$$

If $0 \leq x \leq 1$, we have found $|e_a| = x \cdot L$ and $|e_b| = L - |e_a|$. Clearly, the wiring cost to merge ms_a and ms_b is L, which is the minimum possible. Consequently, $C_{T_v} = C_{T_a} + C_{T_b} + c \cdot (|e_a| + |e_b|)$. Note that the preceding computation assumes both edges e_a and e_b have the same width. A simple extension can be made to achieve zero-skew merging even when e_a and e_b have different widths.

If $x < 0$ or $x > 1$, it implies that the wire delay in a wire of length L is insufficient to balance the delay difference in t_a and t_b. It is, therefore, necessary to use a wiring cost $|e_a| + |e_b| > L$ to achieve zero-skew. Without loss of generality, let $t_a > t_b$. Then, $|e_a| = 0$, and $|e_b|$ is obtained by solving the following equation [Tsay 1991]:

$$t_a = t_b + r \cdot |e_b| \cdot \left(C_{T_b} + \frac{c \cdot |e_b|}{2} \right)$$

Given $|e_a|$ and ms_a, the union of all Manhattan circles of radius $|e_a|$ that are centered at points on ms_a gives us a tilted rectangular region (a rectangle rotated by 45 degrees). All points in the tilted rectangular region, referred to as trr_a, are at a distance of at most $|e_a|$ from ms_a. Given $|e_b|$ and ms_b, trr_b can be similarly obtained. The intersection of trr_a and trr_b is the merging segment for v, which is again a Manhattan arc, as shown in Figure 13.24. Figure 13.24a shows an example for the case when $|e_a| + |e_b| = d(ms_a, ms_b)$, whereas Figure 13.24b shows an example for the case when $|e_a| = 0$ and $|e_b| > d(ms_a, ms_b)$. In the latter example, ms_v is a subset of ms_a, and it resides completely within trr_b.

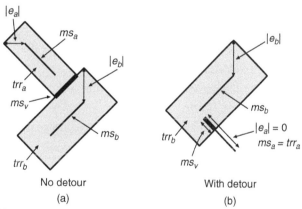

No detour
(a)

With detour
(b)

FIGURE 13.24

Intersection of trr_a and trr_b to obtain ms_v.

FIGURE 13.25

Snaking to lengthen the connection between two points.

Depending on the embedded locations of v and b, $|e_b|$ may be longer than $d(l_v, l_b)$. In that case, wire snaking or detour wiring, as shown in Figure 13.25, is necessary to implement the desirable wire length of $|e_b|$ in the final clock layout.

The bottom-up merging process computes a tree of merging segments. The process stops when we have computed ms_{s_0}, the merging segment of s_0 if l_{s_0} is not given; otherwise, the process terminates when we have computed the merging segment of s_0', the singleton child node of s_0. We refer to such a tree as a *merging tree*. Every node $v \neq s_0$ or s_0' in the merging tree also has an associated wiring cost $|e_v|$ for its connection to its parent node.

Given the merging tree, the root node s_0 is first embedded if the clock source location is not given. In such a case, any point on ms_{s_0} can be selected to be l_{s_0}; otherwise, we pick any point on $ms_{s_0'}$ that is closest to l_{s_0} to embed s_0'. Now, we proceed in a top-down fashion to embed other internal nodes of G. An internal node v is embedded at any point on ms_v that is of distance no farther than $|e_v|$ from $l_{p(v)}$, the embedded location of its parent node $p(v)$. Hence, v could be embedded at any point that lies on the intersection of the merging segment ms_v and the tilted rectangular region bounded by a Manhattan circle whose center is $l_{p(v)}$ and radius is $|e_v|$. The DME algorithm is summarized in Algorithm 13.4.

Algorithm 13.4   Deferred-Merge Embedding (DME)

Input: A set of clock pins (possibly including clock source s_0) S, an abstract topology G

Output: A zero-skew clock tree routing T

1. Initialize $ms_{s_i} = l_{s_i}$ for all $s_i \in S$;
2. **for** each merging of two subtrees T_a and T_b to form T_v based on bottom-up topological sort of abstract topology G **do**
3.   Compute $|e_a|$ and $|e_b|$;
4.   Construct trr_a from ms_a and $|e_a|$ and trr_b from ms_b and $|e_b|$;
5. $ms_v \leftarrow trr_a \cap trr_b$;
6. **end for**
7. **if** location of clock source s_0 is given
8.   Choose $l_{s_0'} \in ms_{s_0'}$ such that $d(l_{s_0'}, l_{s_0}) = d(ms_{s_0'}, l_{s_0})$;
9. **else**
10.   Choose any $l_{s_0} \in ms_{s_0}$;
11. **for** each remaining internal node $v \in G$ in top-down order **do**
12.   Let $p(v)$ be the parent node of v;
13.   Choose $l_v \in ms_v$ such that $d(l_v, l_{p(v)}) \leq |e_v|$;
14. **end for**

Figure 13.23 gives an example of a clock net with four sinks s_1–s_4. For the topology given in Figure 13.23a, the DME algorithm constructs merging segments ms_a, ms_b, and ms_v of the internal nodes a, b, and v, respectively, under the path length delay model as shown in Figure 13.23b. In this example, v is also s_0', the singleton child node of s_0, whose location is given. Each of the non-root internal nodes is embedded at a point on its merging segment that is closest to its parent node as shown in Figure 13.26.

Example 13.2 Zero-Skew Routing under Elmore Delay Model. Consider an example with 4 clock sinks [Tsay 1991], as shown in Figure 13.27b. Sink s_1 is at $(8, 0)$ with sink capacitance $c_{s_1}^g = 16$ F; sink s_2 is at $(22, 6)$ with $c_{s_2}^g = 10$ F; sink s_3 is at $(0, 10)$ with $c_{s_3}^g = 1$ F; sink

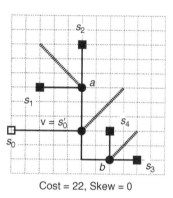

Cost = 22, Skew = 0

FIGURE 13.26

An illustration of the top-down phase of the DME algorithm: Zero-skew clock tree with a total wire length of 22 units for the example in Figure 13.23.

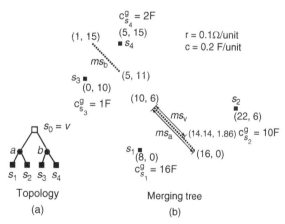

Topology
(a)

Merging tree
(b)

FIGURE 13.27

Application of DME on a 4-sink example under the Elmore delay model: (a) Abstract topology of 4 sinks. (b) Merging segments of internal nodes a, b, and $s_0 = v$.

s_4 is at $(5, 15)$ with $c_{s_4}^g = 2$ F. In this example, we assume unit-length wire resistance r and unit-length wire capacitance c to be, respectively, 0.1Ω and 0.2 F.

The abstract topology G under which a zero-skew routing tree for the 4 sinks is to be constructed is given in Figure 13.27a. Because the location of the clock source is not given, the topology is a strictly binary tree; the root $s_0 = v$ has two child nodes a and b, which, respectively, connect sink pair s_1 and s_2 and sink pair s_3 and s_4.

We now consider the merging of s_1 and s_2 (i.e., the computation of ms_a). The distance between s_1 and s_2 is $L = d(l_{s_1}, l_{s_2}) = 20$ units. Let $x \cdot L$ be the distance of ms_a from s_1. Because the child nodes of a are sinks, the delays of the sink nodes are $t_1 = t_2 = 0$ s, and the total capacitances in T_{s_1} and T_{s_2} are, respectively, $C_{T_{s_1}} = c_{s_1}^g = 16$ F and $C_{T_{s_2}} = c_{s_2}^g = 10$ F. We can solve for x as follows:

$$x = \frac{t_2 - t_1 + r \cdot L \cdot (C_{T_{s_2}} + c \cdot L/2)}{r \cdot L \cdot (c \cdot L + C_{T_{s_1}} + C_{T_{s_2}})}$$

$$= \frac{10 + 2}{4 + 16 + 10}$$

$$= 0.4$$

Therefore, $|e_{s_1}| = 0.4 \cdot 20 = 8$ units and $|e_{s_2}| = (1 - 0.4) \cdot 20 = 12$ units. We compute the merging segment ms_a by computing the intersection of the tilted rectangular regions trr_{s_1} and trr_{s_2}. The tilted rectangular region trr_{s_1} is defined by the coordinates $(0, 0)$, $(8, 8)$, $(16, 0)$, and $(8, -8)$. Essentially, it is a Manhattan circle with its center at $l_{s_1} = (8, 0)$ and a radius of 8 units. The tilted rectangular region trr_{s_2} is a Manhattan circle with its center at $l_{s_2} = (22, 6)$ and a radius of 12 units. The intersection of trr_{s_1} and trr_{s_2} gives us ms_a, which is a Manhattan arc defined by the coordinates $(10, 6)$ and $(16, 0)$, as shown in Figure 13.28.

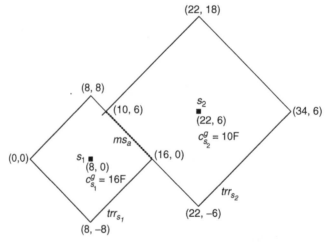

FIGURE 13.28

Computation of ms_a.

If we embed the internal node a at any point on ms_a and connect it to s_1 and s_2 with the shortest connections possible, the sink delays t_1 and t_2 from a under the Elmore delay model are both 13.44 ns, which can be computed as follows:

$$t_1 = r \cdot x \cdot L \cdot \left(C_{T_{s_1}} + \frac{c \cdot x \cdot L}{2} \right)$$
$$= 0.1 \cdot 8 \cdot (16 + 8 \cdot 0.2/2)$$
$$= 13.44 \text{ ns},$$

$$t_2 = r \cdot (1 - x) \cdot L \cdot \left(C_{T_{s_2}} + \frac{c \cdot (1 - x) \cdot L}{2} \right)$$
$$= 0.1 \cdot 12 \cdot (10 + 12 \cdot 0.2/2)$$
$$= 13.44 \text{ ns}$$

Similarly, we can compute the merging segment ms_b for the merging of s_3 and s_4. The merging segment ms_b, whose endpoints are $(1, 15)$ and $(5, 11)$, is shown in Figure 13.27b. With shortest connections to s_3 and s_4 from any point on ms_b, the sink delays t_3 and t_4 from b are both 0.96 ns.

Now, let us consider the merging of T_a and T_b. Besides r and c, the following parameters are required to determine the location of the merging segment ms_v: $t_a = t_1 = t_2 = 13.44$ ns, $t_b = t_3 = t_4 = 0.96$ ns, $L = d(ms_a, ms_b) = 10$ units, $C_{T_a} = c_{s_1}^g + c_{s_2}^g + 0.2 \cdot 20 = 16 + 10 + 4 = 30$ F, and $C_{T_b} = c_{s_3}^g + c_{s_4}^g + 0.2 \cdot 10 = 1 + 2 + 2 = 5$ F. Defining $x \cdot L$ to be the distance of ms_v from ms_a and solving for x, we obtain

$$x = \frac{t_b - t_a + r \cdot L \cdot (C_{T_b} + c \cdot L/2)}{r \cdot L \cdot (c \cdot L + C_{T_a} + C_{T_b})}$$
$$= \frac{0.96 - 13.44 + 0.1 \cdot 10 \cdot (5 + 1)}{0.1 \cdot 10 \cdot (2 + 30 + 5)}$$
$$= -0.175 < 0$$

As $x < 0$, it is necessary to use a wire length longer than $L = 10$ units to balance the delays t_a and t_b. We assign $|e_a| = 0$ and solve for $|e_b|$ as follows:

$$t_a = t_b + r \cdot |e_b| \cdot \left(C_{T_b} + \frac{c \cdot |e_b|}{2} \right)$$
$$\Rightarrow 13.44 = 0.96 + 0.1 \cdot |e_b| \cdot \left(5 + \frac{0.2 \cdot |e_b|}{2} \right)$$
$$\Rightarrow |e_b| = 18.28 \text{ units}$$

At this point, we have the total wire length of the zero-skew routing tree, which is $|e_{s_1}| + |e_{s_2}| + |e_{s_3}| + |e_{s_4}| + |e_a| + |e_b| = 8 + 12 + 6 + 4 + 0 + 18.28 = 48.28$ units.

The merging segment ms_v is obtained by taking the intersection of trr_a and trr_b, as shown in Figure 13.29. As $|e_a| = 0$, trr_a is simply ms_a. The tilted rectangular region trr_b, whose vertices are $(-17.28, 15)$, $(1, 33.28)$, $(23.28, 11)$ and $(5, -7.28)$, overlaps

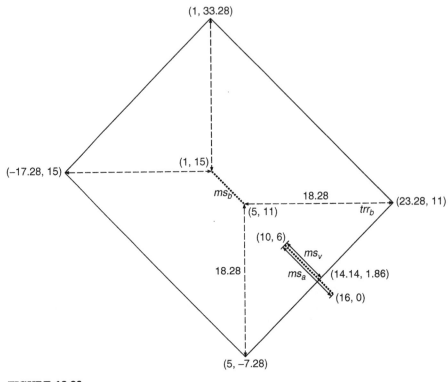

FIGURE 13.29

Computation of ms_v.

with $ms_a = trr_a$. Consequently, the two endpoints of ms_v are $(10, 6)$ and $(14.14, 1.86)$, as shown in Figure 13.29.

Now, let us turn our attention to the top-down embedding of the internal nodes of the topology G. Because the location of clock source $s_0 = v$ is not given, we have the flexibility of embedding $s_0 = v$ at any point on ms_v. Let us consider the two endpoints of ms_v for embedding:

1. We embed $s_0 = v$ at location $(10, 6)$. As $|e_a| = 0$, we also have to embed a at the same location. As $|e_b| = 18.18 > d(l_v, ms_b) = 10$, the snaking of wire may be required to balance the delay. To determine l_b, we use the intersection of ms_b and the tilted rectangular region bounded by a Manhattan circle whose center is $l_v = (10, 6)$ and radius is $|e_b| = 18.18$ units. As shown in Figure 13.30, we can embed b at any point on ms_b,

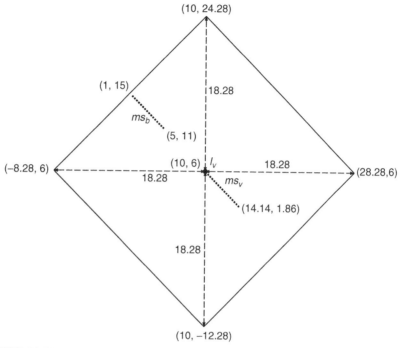

FIGURE 13.30

Embedding of internal node b.

because ms_b is completely contained within the tilted rectangular region whose vertices are $(-8.28, 6)$, $(10, 24.28)$, $(28.28, 6)$, and $(10, -12.28)$.

In Figure 13.31a, we embed b at location $(5, 11)$, which is of distance 10 units from l_v. Consequently, we require a detour of length 8.28 units in the connection from l_v to l_b. In Figure 13.31b, we embed b at location $(1, 15)$, which is of distance 18 units from l_v. As a result, the connection between l_v and l_b requires a snaking of length 0.28 units.

2. We embed $s_0 = v$ at location $(14.14, 1.86)$; that leaves the endpoint $(5, 11)$ of ms_b as the only possible location for the embedding of b. Again, we have to embed a at the same location as v. The corresponding routing tree is shown in Figure 13.31c.

It is straightforward to verify that all three zero-skew routing trees in Figure 13.31 have the same wiring cost of 48.28 units.

Because DME requires an input topology, the generation of a good abstract topology is crucial. In fact, many of the more successful approaches interleave topology construction with merging segment computation. The Greedy-DME

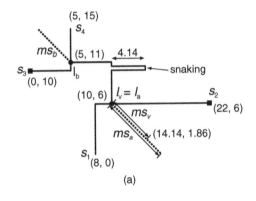

(a)

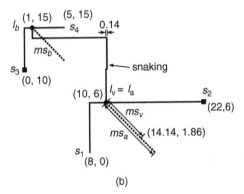

(b)

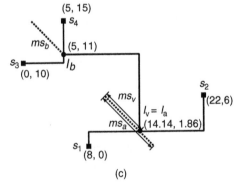

(c)

FIGURE 13.31

Three different ways of embedding the merging tree in Figure 13.27 with the same wiring cost. The embedded routing trees in (a) and (b) require detour wirings, whereas the embedding in (c) does not.

method proposed in [Edahiro 1992] is the most successful among them. Let F denote a forest of singleton merging trees, each consisting of only a single sink location. Greedy-DME iteratively finds the "nearest" pair of neighbors in F, say ms_a and ms_b, and constructs for the newly added parent node, say v, a merging segment ms_v based on a zero-skew merge of ms_a and ms_b. To account for detour wiring, the proximity of two merging segments is usually defined by the cost of merging them instead of their physical distance. The merging trees rooted at ms_a and ms_b in F are then replaced by a new merging tree rooted at ms_v. After $n-1$ merging operations, F contains a single merging tree, which also corresponds to the abstract topology of the zero-skew routing tree. The outline of Greedy-DME is similar to that of DME except for step 2, which is amended as below:

---

2. **for** each merging of two subtrees T_a and T_b that are nearest neighbors in F to form T_v **do**

In the Greedy-DME algorithm, it takes $O(n)$ iterations to construct a zero-skew clock tree, with each iteration involving an $O(n^2)$ procedure to identify the nearest-neighbor pair for merging. We can reduce the number of iterations to $O(\log n)$ by merging several nearest-neighbor pairs simultaneously [Edahiro 1993a]. In each iteration, we construct a "nearest-neighbor graph" that maintains the nearest neighbor of each merging tree in F. In nondecreasing order of distance, $|F|/k$ $(2 \le k \le 4)$ independent nearest-neighbor pairs of merging trees are chosen from the graph for zero-skew merging. The number of merging trees in F is reduced by a factor of $1/k$ after each iteration. Consequently, it takes only $O(\log_{k/(k-1)} n)$ iterations to construct a zero-skew routing tree.

The construction of a nearest-neighbor graph in each iteration has a quadratic time complexity. An approximate nearest-neighbor graph can be constructed in linear time by use of the bucket decomposition method in [Edahiro 1994]. In each iteration, the smallest square routing plane that covers all merging segments of the root nodes of the merging trees in F is uniformly partitioned into $\Theta(|F|)$ square buckets. The routing plane and buckets are all tilted by 45 degrees, as shown in Figure 13.32. Each bucket has at most eight neighboring buckets. Assuming a uniform distribution of merging segments, the number of merging segments in each bucket is $O(1)$. We restrict the nearest neighbor of a merging segment, say ms_v, to reside within the same bucket(s) as ms_v or in the adjacent buckets as shown in Figure 13.32. Consequently, an approximate nearest-neighbor graph can be constructed in linear time.

13.3.3.4 *Bounded-skew routing*

The two-phase approach taken by the DME algorithm to compute a zero-skew tree for a prescribed topology can be extended quite naturally to handle more general skew constraints. The BST/DME solutions developed in [Cong 1995a; Huang 1995; Cong 1995b] for the problem of constructing a bounded-skew

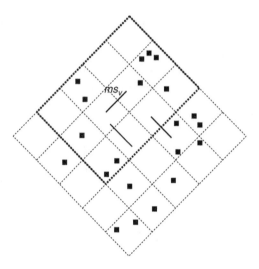

FIGURE 13.32

Bucket decomposition of a routing plane.

routing tree generalize the concept of a merging segment for zero-skew routing to that of a *merging region*. In a BST/DME algorithm, a merging region contains all candidate locations of the corresponding internal node in the bounded-skew tree. The bottom-up process constructs a tree of merging regions (in contrast to merging segments for zero-skew tree), and the top-down process then determines the exact locations of all internal nodes. Figure 13.33 shows the merging regions and routing tree constructed by BST/DME for the example in Figure 13.23 (and Figure 13.26).

The construction of the merging region for the parent node, say v, of two sinks, say s_i and s_j, is quite straightforward. First, we construct a bounded-skew merging segment ms_v^+ such that $t_i^+ = t_j^+ + B$, where t_i^+ and t_j^+ are, respectively, the delays from the merging segment ms_v^+ to s_i and s_j, and B is the skew bound. The computation of this bounded-skew merging segment is similar to that of a zero-skew merging segment. Let $L = d(l_{s_i}, l_{s_j})$ denote the shortest Manhattan distance between l_{s_i} and l_{s_j}, and $x^+ \cdot L$ the distance between ms_v^+ and l_{s_i}, where x^+ is between 0 and 1 if detour wiring is not necessary. Then, we solve for x^+ from the following expression:

$$\underbrace{r \cdot x^+ \cdot L \cdot \left(c_{s_i}^g + \frac{c \cdot x^+ \cdot L}{2} \right)}_{t_i^+} = \underbrace{r \cdot (1 - x^+) \cdot L \cdot \left(c_{s_j}^g + \frac{c \cdot (1 - x^+) \cdot L}{2} \right)}_{t_j^+} + B$$

where $c_{s_i}^g$ and $c_{s_j}^g$ are, respectively, the gate capacitances of the clock sinks s_i and s_j. For simplicity, we first consider the case that x^+ obtained from the preceding expression is, indeed, between 0 and 1. It is obvious that a tree that uses

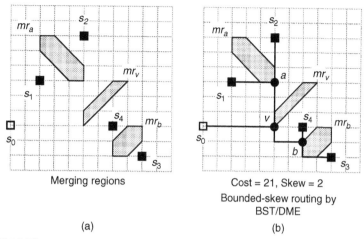

Merging regions (a)

Cost = 21, Skew = 2
Bounded-skew routing by
BST/DME
(b)

FIGURE 13.33

(a) Merging regions constructed by BST/DME for the example in Figure 13.23. (b) Embedding of internal nodes. The bounded-skew routing has a lower routing cost than the zero-skew routing in Figure 13.26.

shortest wire lengths to connect any point on ms_v^+ to s_i and s_j would have a maximum sink delay of t_i^+ and a minimum sink delay of t_j^+, satisfying the skew bound constraint B.

Similarly, we construct a bounded-skew merging segment ms_v^- such that $t_i^- = t_j^- - B$, where t_i^- and t_j^- are, respectively, the delays from the merging segment ms_v^- to s_i and s_j. The bounded-skew merging segment ms_v^- should be at a distance of $x^- \cdot L$ from l_{s_i}. Again, assume that x^- is between 0 and 1. A tree that uses shortest wire lengths to connect any point on ms_v^- to s_i and s_j would have a minimum sink delay of t_i^- and a maximum sink delay of t_j^-. The region bounded by ms_v^+ and ms_v^- within the smallest bounding box containing s_i and s_j is the merging region mr_v.

In Figure 13.33, two merging regions mr_a and mr_b are obtained by merging clock sinks under the path length delay model. One can easily verify that the bounded-skew merging segments are simply the corresponding zero-skew merging segments shifted toward or away from the corresponding clock sink.

A few important properties are associated with a merging region, say mr_v, whose two child nodes are clock sinks. By construction, $x^- \leq x^+$. Now, consider a Manhattan arc in mr_v that is of distance between x^- and x^+ from s_i, pick any point residing on this Manhattan arc, and construct a routing tree that connects this point to s_i and s_j, with the shortest wire lengths. It is fairly straightforward to verify that such a routing tree would satisfy the skew-bound constraint B. Moreover, all routing trees constructed in a similar manner with root nodes residing on this Manhattan arc would have the same maximum sink delay and the same minimum sink delay. Furthermore, $|e_{s_i}| + |e_{s_j}| = d(l_{s_i}, l_{s_j})$ regardless of where v is embedded within mr_v. Consequently, $C_{T_v} = c_{s_i}^g + c_{s_j}^g + c \cdot d(l_{s_i}, l_{s_j})$.

Now, let us consider the more complicated cases when x^+ or x^- may not be between 0 and 1:

$0 \leq x^- \leq 1 < x^+$: We force ms_v^+ to be at a distance $L = d(l_{s_i} - l_{s_j})$ from s_i. In other words, ms_v^+ coincides with l_{s_j}.

$x^- < 0$ and $1 < x^+$: We force ms_v^- and ms_v^+ to be, respectively, at distances of 0 and L from s_i. In other words, ms_v^- coincides with l_{s_i} and ms_v^+ coincides with l_{s_j}.

$x^- < 0 \leq x^+ \leq 1$: We force ms_v^- to be at a distance of 0 from s_i. In other words, ms_v^- coincides with l_{s_i}.

In other words, we always force x^- and x^+ to be between 0 and 1. Note that as we are merging two clock sinks, it is not possible for x^- and x^+ to be both less than 0 or greater than 1. Such is not the case when we construct merging regions for internal nodes higher up in the abstract topology.

Let two non-leaf nodes a and b be the child nodes of v. Given merging regions mr_a and mr_b, we follow the approach of boundary merging and embedding in [Cong 1995a; Huang 1995], where mr_v is constructed from the nearest *boundary segments* of mr_a and mr_b. A point p on the nearest boundary segment of mr_a, called a *joining segment* and denoted JS_a, can merge with a point q on joining segment JS_b if $d(p, q) = d(mr_a, mr_b)$.

Suppose mr_a and mr_b are also octilinear convex polygons as shown in Figure 13.33. Therefore, the joining segments from mr_a and mr_b are either parallel Manhattan arcs or parallel rectilinear line segments. We will now present the merging of two joining segments that are Manhattan arcs. As pointed out earlier, all bounded-skew routing trees that have their root nodes embedded on JS_a have the same maximum sink delay and the same minimum delay. Let t_a^{max} and t_a^{min} denote such delays, respectively. Similarly, t_b^{max} and t_b^{min} are defined for JS_b.

Let $L = d(JS_a, JS_b) = d(mr_a, mr_b)$ denote the distance between JS_a and JS_b. We compute a merging segment ms_v^+ such that in a routing tree obtained by making the shortest connections from a point on ms_v^+ to JS_a and JS_b, the maximum and minimum sink delays in the tree are, respectively, caused by sinks in T_a and T_b, and the skew is no greater than B. The distance of ms_v^+ from JS_a can be obtained by solving for x^+ in the following equation:

$$t_a^{max} + r \cdot x^+ \cdot L \cdot \left(C_{T_a} + \frac{c \cdot x^+ \cdot L}{2} \right)$$
$$= t_b^{min} + r \cdot (1 - x^+) \cdot L \cdot \left(C_{T_b} + \frac{c \cdot (1 - x^+) \cdot L}{2} \right) + B$$

If x^+ is between 0 and 1, ms_v^+ is at a distance of $x^+ \cdot L$ from JS_a. Similarly, we compute the location of a merging segment ms_v^- by solving for x^- in the following equation:

$$t_a^{\min} + r \cdot x^- \cdot L \cdot \left(C_{T_a} + \frac{c \cdot x^- \cdot L}{2} \right)$$

$$= t_b^{\max} + r \cdot (1 - x^-) \cdot L \cdot \left(C_{T_b} + \frac{c \cdot (1 - x^-) \cdot L}{2} \right) - B$$

If x^- is between 0 and 1, ms_v^- is at a distance of $x^- \cdot L$ from JS_a.

Suppose x^- and x^+ are both between 0 and 1, the region bounded by ms_v^+ and ms_v^- within the smallest bounding box containing JS_a and JS_b is the merging region mr_v, as shown in Figure 13.34. The merging region mr_v is constructed under the path length delay model. Each Manhattan arc is associated with a pair of numbers, the maximum and minimum sink delays from a point on that arc. The maximum (minimum) sink delay of ms_v^+ is due to some sink in T_a (T_b), whereas the maximum (minimum) sink delay of ms_v^- is due to some sink in T_b (T_a).

Similar to the merging of two clock sinks, so long as x^- and x^+ are not both less than 0 or greater than 1, we would always force x^- and x^+ to be between 0 and 1. When x^- and x^+ are both less than 0 or when they are both greater than 1, detour wiring is necessary. In the former, we set $|e_a| = 0$, make mr_v coincide with JS_a, and solve for $|e_b|$ in the following equation:

$$t_a^{\max} = t_b^{\min} + r \cdot |e_b| \cdot \left(C_{T_b} + \frac{c \cdot |e_b|}{2} \right) + B$$

In the latter, we set $|e_b| = 0$, make mr_v coincide with JS_b, and solve for $|e_a|$ in the following equation:

$$t_a^{\min} + r \cdot |e_a| \cdot \left(C_{T_a} + \frac{c \cdot |e_a|}{2} \right) = t_b^{\max} - B$$

However, detour wiring may not be necessary after all. Take the case in which x^- and x^+ are both less than 0, the tilted rectangular region defined by JS_b and

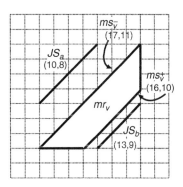

FIGURE 13.34

Merging of two Manhattan joining segments JS_a and JS_b for a skew bound of 6. The numbers associated with each Manhattan arc are the maximum and minimum sink delays from a point on that arc.

$|e_b|$ actually covers more than JS_a; it actually overlaps with mr_a. That implies that it may actually be more economical to use interior points of mr_a, instead of boundary segments, for merging. Another reason for the use of interior points is when mr_a and mr_b overlap. However, merging of interior points presents tremendous challenges, because every point in a merging region may correspond to the merging of many different pairs of interior points from its child merging regions.

To overcome the difficulty in the use of interior points for merging, we introduce the notion of *sampling segments* as in [Cong 1995b]. A merging region is represented by a set of s sampling segments, each of which is a Manhattan arc that is parallel to the 45 degrees boundary segments of the merging region. Such a sampling segment has the property that it has a constant maximum sink delay and a constant minimum sink delay on any point along the segment. Figure 13.35 illustrates the concept of sampling segments under the path length delay model. The maximum and minimum sink delays associated with each sampling segment are also shown in parentheses.

Merging of two nodes now involves two sets of sampling segments. Clearly, we can apply the approach for merging two joining segments outlined earlier to accomplish the merging of two sampling segments from two merging regions. The challenge here is that the approach would generate a set of up to s^2 merging regions for the parent node. If each of these merging regions is, in turn, sampled by s segments, the time and space complexity of this approach would grow exponentially. For an efficient and practical implementation of such an

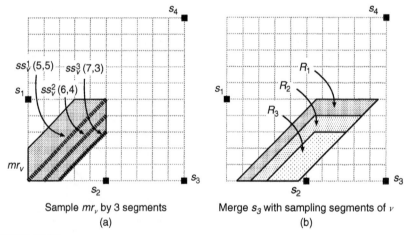

Sample mr_v by 3 segments
(a)

Merge s_3 with sampling segments of v
(b)

FIGURE 13.35

(a) Sampling of mr_v (obtained from the merging of s_1 and s_2) with $\{ss_v^1, ss_v^2, ss_v^3\}$. (b) Merging these sampling segments with sink s_3 with a skew bound of 2 units produces three merging regions where R_i is produced by merging s_3 with ss_v^i.

approach, the number of merging regions associated with a node is typically limited to a constant, say k. Each region is, in turn, sampled by s sampling segments. Therefore, the merging of two nodes will generate $k^2 s^2$ merging regions. One typical approach to pruning these $k^2 s^2$ merging regions is to keep only the best k merging regions, defined in terms of merging cost, for the parent node.

Merging with sampling segments facilitates merging of interior points. Because 45 degrees joining segments are on the boundary of a merging region, the skew associated with each joining segment is in general higher than the skew associated with other sampling segments from the same merging region. Consequently, merging of interior sampling segments may result in a larger merging region at a parent node. In Figure 13.35, R_3 is also the merging region produced by merging mr_v with s_3 by use of a boundary joining segment. In this example, R_3 is smaller than R_1 and R_2, both of which are constructed with interior sampling segments. The larger R_1 and R_2 may lead to a reduced wiring cost at the next merging step.

The generation of the $k^2 s^2$ merging region for each node in an abstract topology may not sound efficient. However, another benefit of the use of sampling segments in the merging process is that the merging of joining segments that are horizontal or vertical is obviated. Under the Elmore delay model, the merging of two parallel horizontal (or vertical) joining segments results in a merging region that is no longer octilinear. Consequently, subsequent merging operations may involve joining segments that are neither rectilinear line segments nor Manhattan arcs. Moreover, the computation of such a merging region requires $O(n)$ runtime and space complexity for a subtree that contains n sinks [Cong 1995b].

Example 13.3 Bounded-Skew Routing under Elmore Delay Model: We again consider the 4-sink example in Figure 13.27. Instead of zero-skew routing, we allow a skew bound of 2.5 ns (i.e., $B = 2.5$ ns). First, we compute x^+ and x^- for the merging of s_1 and s_2, where $x^+ \cdot d(l_{s_1}, l_{s_2})$ and $x^- \cdot d(l_{s_1}, l_{s_2})$, respectively, denote the distances of ms_a^+ and ms_a^- from l_{s_1}:

$$x^+ = \frac{B + r \cdot L \cdot (C_{T_{s_2}} + c \cdot L/2)}{r \cdot L \cdot (c \cdot L + C_{T_{s_1}} + C_{T_{s_2}})}$$

$$= \frac{2.5 + 2 \cdot (10 + 2)}{2 \cdot (4 + 16 + 10)}$$

$$= 0.44$$

$$x^- = \frac{-B + r \cdot L \cdot (C_{T_{s_2}} + c \cdot L/2)}{r \cdot L \cdot (c \cdot L + C_{T_{s_1}} + C_{T_{s_2}})}$$

$$= \frac{-2.5 + 2 \cdot (10 + 2)}{2 \cdot (4 + 16 + 10)}$$

$$= 0.36$$

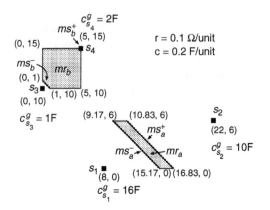

FIGURE 13.36

Merging regions of internal nodes a and b.

where $L = d(l_{s_1}, l_{s_2}) = 20$ units. We show the merging segments ms_a^+ and ms_a^- in Figure 13.36. The shaded region bounded by ms_a^+ and ms_a^- is the merging region ms_a, whose vertices are $(9.17, 6)$, $(10.83, 6)$, $(16.83, 0)$, and $(15.17, 0)$.

To compute the merging region of b, we define $x^+ \cdot d(l_{s_3}, l_{s_4})$ and $x^- \cdot d(l_{s_3}, l_{s_4})$ to be the distances of ms_b^+ and ms_b^- from l_{s_3}, respectively. We obtain $x^+ = 1.1$ and $x^- = 0.1$. Because $x^+ > 1$, we force $x^+ = 1$. In other words, ms_b^+ coincides with l_{s_4}. We also show the merging segments and merging region of b in Figure 13.36.

Now, we consider two different ways to merge a and b. First, we consider the case of merging a and b with joining segments from mr_a and mr_b that are closest (*i.e.*, $d(JS_a, JS_b) = d(mr_a, mr_b) = 8.17$ units). In this case, JS_a is simply the point $(9.17, 6)$ and JS_b is $(5, 10)$. For JS_a, we have the following parameters: $t_a^{max} = t_2 = 14.48$ ns, $t_a^{min} = t_1 = 11.98$ ns, and $C_{T_a} = 30$ F. For JS_b, we have $t_b^{max} = t_4 = 1.25$ ns, $t_b^{min} = t_3 = 0.75$ ns, and $C_{T_b} = 5$ F. Letting $L = d(JS_a, JS_b)$ and defining $x^+ \cdot L$ and $x^- \cdot L$ to be the distances of ms_v^+ and ms_v^- from JS_a, respectively, we obtain x^+ and x^- as follows:

$$
\begin{aligned}
x^+ &= \frac{t_b^{min} - t_a^{max} + B + r \cdot L \cdot (C_{T_b} + c \cdot L/2)}{r \cdot L \cdot (c \cdot L + C_{T_a} + C_{T_b})} \\
&= \frac{0.75 - 14.48 + 2.5 + 0.817 \cdot (5 + 0.817)}{0.817 \cdot (1.63 + 30 + 5)} \\
&= -0.22,
\end{aligned}
$$

$$
\begin{aligned}
x^- &= \frac{t_b^{max} - t_a^{min} - B + r \cdot L \cdot (C_{T_b} + c \cdot L/2)}{r \cdot L \cdot (c \cdot L + C_{T_a} + C_{T_b})} \\
&= \frac{1.25 - 11.98 + 2.5 + 0.817 \cdot (5 + 0.817)}{0.817 \cdot (1.63 + 30 + 5)} \\
&= -0.28
\end{aligned}
$$

Because both x^+ and x^- are negative, we force $|e_a| = 0$ (*i.e.*, ms_v is JS_a). Then, we compute $|e_b|$ with the following equation:

$$t_a^{max} = t_b^{min} + r \cdot |e_b| \cdot \left(C_{T_b} + \frac{c \cdot |e_b|}{2} \right) + B$$

$$\Rightarrow 14.48 = 0.75 + 0.1 \cdot |e_b| \cdot \left(5 + \frac{0.2 \cdot |e_b|}{2} \right) + 2.5$$

$$\Rightarrow |e_b| = 16.81 \text{ units}$$

Because $d(JS_a, JS_b) = 8.17$ units, a detour wiring of length 8.64 units is required. Figure 13.37 shows the bounded-skew routing tree constructed accordingly. The total wire length of the routing tree is 46.81 units.

Now, we consider the case of merging a and b with sampling segments from mr_a and mr_b. For any merging region mr, we always include as sampling segments the merging segments ms^+ and ms^-, which are Manhattan arcs that define the boundary segments of mr. In Figure 13.38, which shows the sampling of mr_a and mr_b, it is clear that ms_a^+,

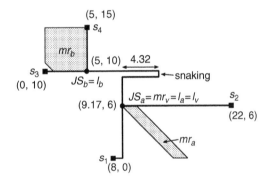

FIGURE 13.37

A bounded skew routing tree constructed based on the merging of the nearest boundary segments.

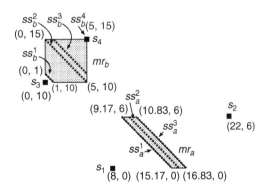

FIGURE 13.38

Sampling of the merging regions mr_a and mr_b.

ms_a^-, ms_b^+, and ms_b^- have been included as sampling segments ss_a^1, ss_a^3, ss_b^1, and ss_b^4, respectively. Each merging region has a subregion whose clock skew among its sinks is the smallest. Sampling of such a subregion is desirable, because it usually leads to a larger merging region at the parent node. In this example, both mr_a and mr_b contain the zero-skew merging segments, which are included as sampling segments ss_a^2 and ss_b^3. We also include sampling segments that contain the joining segments. Consequently, ss_b^2 is a sampling segment of mr_b. It is important to note that all sampling segments from a single merging region have the same wiring cost.

Now, we perform the pairwise merging of ss_a^i and ss_b^j, for $i \in \{1, 2, 3\}$ and $j \in \{1, 2, 3, 4\}$. It turns out that in every pairwise merging, $|e_a| = 0$ and detour wiring in e_b is required. Table 13.1 shows the different combinations of $|e_b|$, t_v^{max}, and t_v^{min} obtained by these pairwise merging operations. It should be evident from Table 13.1 that the t_v^{max} and t_v^{min} values satisfy the skew-bound constraint of 2.5 ns in all cases.

Because e_b requires detour wiring, the merging regions obtained in these pairwise mergings overlap with the sampling segments ss_a^j. Each of these merging regions (or regions that degenerate into segments) is obtained by taking the intersection of trr_b (constructed from the respective ss_b^j and $|e_a|$) and ss_a^i (see Section 13.3.3.3). The coordinates of the merging regions of v are given in Table 13.2. It is important to note that

Table 13.1 Wiring Costs and Sink Delays Obtained by Pairwise Merging of Sampling Segments from mr_a and mr_b.

	ss_a^1			ss_a^2			ss_a^3		
	$\lvert e_b \rvert$	t_v^{max} (ns)	t_v^{min} (ns)	$\lvert e_b \rvert$	t_v^{max} (ns)	t_v^{min} (ns)	$\lvert e_b \rvert$	t_v^{max} (ns)	t_v^{min} (ns)
ss_b^1	17.57	14.48	11.98	16.33	13.44	10.94	18.07	14.91	12.41
ss_b^2	16.81	14.48	11.98	15.55	13.44	10.94	17.32	14.91	12.41
ss_b^3	16.56	14.48	11.98	15.29	13.44	10.94	17.08	14.91	12.41
ss_b^4	17.70	14.48	11.98	16.46	13.44	10.94	18.20	14.91	12.41

Table 13.2 Merging Regions Obtained by Pairwise Merging of Sampling Segments from mr_a and mr_b.

	ss_a^1	ss_b^2	ss_b^3
ss_b^1	(9.17,6)–(11.87,3.30)	(10,6)–(11.66,4.34)	(10.83,6)–(12.95,3.88)
ss_b^2	(9.17,6)–(13.49,1.68)	(10,6)–(13.27,2.73)	(10.83,6)–(14.58,2.25)
ss_b^3	(9.17,6)–(12.86,2.30)	(10,6)–(12.64,3.36)	(10.83,6)–(13.95,2.88)
ss_b^4	(9.17,6)–(11.43,3.73)	(10,6)–(11.23,4.77)	(10.83,6)–(12.52,4.32)

only when detour wiring occurs does the computation of a degenerate merging region from sampling segments resemble the computation of a zero-skew merging segment. Recall that detour wiring is required for bounded-skew merging only when x^+ and x^- are both greater than 1 or both less than 0. Figure 13.39 shows the construction of 4 merging regions of v by merging ss_a^2 with all 4 sampling segments of mr_b. For ease of reference, we refer to the respective tilted rectangular region of sampling segment ss_b^j as trr_b^j and the resultant merging region as mr_v^j in the illustration.

Among these, the merging of ss_a^2 and ss_b^3 results in the lowest $|e_b|$ and, hence, a minimum-cost bounded-skew tree. Figure 13.40 shows the results of embedding v at two different locations in the merging region mr_v^3 obtained from the merging of ss_a^2 and ss_b^3.

In Figure 13.40a, v is embedded at $(12.64, 3.36)$, an endpoint of the merging region mr_v^3. Consequently, that leaves $(5, 11)$ as the only possible location for the embedding of b. When we embed v at $(10, 6)$, the possible locations for the embedding of b lie on the Manhattan arc defined by $(2.36, 13.64)$ and $(5, 11)$. Figures 13.40b,c show the results of embedding b at these two endpoints. As the embedding location of b moves from $(2.36, 13.64)$ to $(5, 11)$, the amount of snaking increases. All three embedded routing trees have the same total wiring cost of 45.29 units, which is lower than those of the bounded-skew routing tree in Figure 13.37 and the zero-skew routing trees in Figure 13.31.

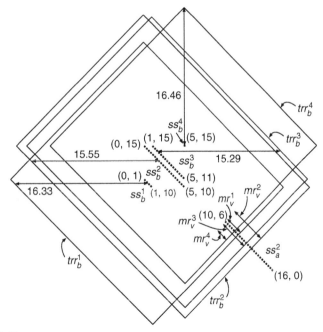

FIGURE 13.39

Construction of merging regions of v by merging ss_a^2 with all 4 sampling segments of mr_b.

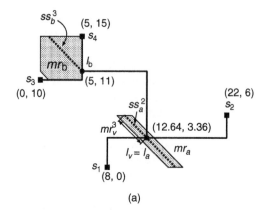

(a)

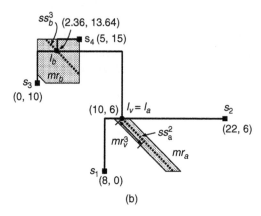

(b)

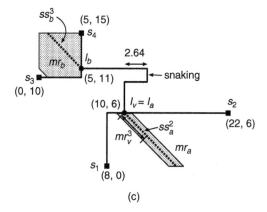

(c)

FIGURE 13.40

Bounded-skew routing trees constructed based on the merging of sampling segments.

Suppose v is not the clock source. It may be necessary to keep only a subset of the 12 merging regions for the construction of merging regions of the parent node of v. If we allow each node to be associated with only 3 merging regions, mr_v^1, mr_v^2, and mr_v^3 would be chosen on the basis of their lower merging costs. Although these merging regions overlap, it is important to note that they would now have different C_{T_v}'s, which could play an important role in determining the merging regions (and their associated merging costs) of the parent node.

For the generation of an abstract topology, we can take an approach similar to that of the Greedy-DME algorithm, with various acceleration methods incorporated. Indeed, the Greedy-BST/DME algorithm in [Huang 1995] is very similar to the Greedy-DME algorithm. However, the Greedy-BST/DME algorithm has the added flexibility that it allows merging of two subtrees at non-root nodes, whereas Greedy-DME always merges two subtrees at their root nodes. Merging with non-root nodes is an effective topology optimization method. In fact, it very closely matches the performance of the best-known heuristics for both the zero-skew and infinite-skew limiting cases (*i.e.*, Steiner routing).

Consider, for example, a clock network with eight sinks that are equally spaced on a horizontal line, as shown in Figure 13.41. Although the embedding is not shown here, it can be easily verified that the abstract topology to the left, which is obtained by the merging of T_1 and T_2 at their root nodes, can be embedded to produce a minimum-cost zero-skew tree. Although T_1 and T_2 are themselves ideal topologies for low-cost embedding for large skew bound B, the merging of them at their root nodes is quite costly (see the embedding following the abstract topology). The reason is that the root nodes (embedded at locations labeled "3" and "6," which correspond to the locations of sinks s_3 and s_6) are quite far apart. If we adjust the subtree topology such that the roots of subtrees become closer (see the abstract topology and embedding to the right), the overall tree cost would be reduced. Here, the root nodes of the adjusted topologies T_1 and T_2 are embedded at locations labeled "4" and "5" (or l_{s_4} and l_{s_5}).

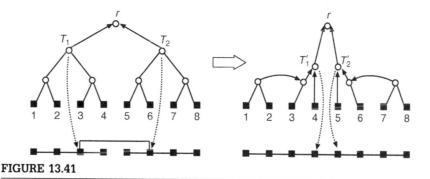

FIGURE 13.41

An example showing that given skew bound $B \gg 0$, changing the subtree topology before merging will reduce the merging cost.

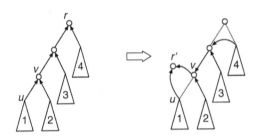

FIGURE 13.42

Repositioning the root in changing the topology.

Figure 13.42 illustrates in more detail how the tree topology is adjusted. To move the root node r to some tree edge, say $e_u = uv$, we traverse along the path from r to v. For each edge encountered along the top-down path traversal, we delete the edge and merge the two appropriate subtrees off the deleted edge. To remove the left edge of r, for example, we merge subtrees labeled "3" and "4." This newly merged tree of "3" and "4" is then combined with the subtree labeled "2" for the deletion of the parent edge of v. This tree is then merged with the subtree labeled "1" such that the new root node r' breaks the edge uv. Effectively, we have relocated the root node to be closer to the subtree labeled "1."

The re-rooting approach relocates r to uv by remerging appropriate subtrees along the path from r to v. Hence, we also know the merging region corresponding to the bounded-skew tree rooted at the parent edge of v. Indeed, a careful examination of the approach would reveal that a simple $O(n)$ top-down traversal of the abstract topology rooted at r would compute all merging regions corresponding to the relocation of the root node to all tree edges in the topology.

To incorporate such a feature in the Greedy-BST/DME algorithm, each node v in an abstract topology G (in the forest F) is associated with two merging regions, denoted as mr_v and mr_{e_v}. The former merging region mr_v is the merging region constructed for v on the basis of the sub-topology rooted at v in G. The latter merging region mr_{e_v} is the merging region if the root node of G is relocated to the parent edge of v in G (*i.e.*, e_v).

To identify in F the nearest neighbor of G, whose root node is r, we consider mr_r for the root node, as well as mr_{e_v} for all other nodes v in G. In other words, some other topologies in F, possibly rerooted as well, may be closer to G after G is rerooted. In the example given in Figure 13.41, the nearest neighbors of the sub-topologies T_1 and T_2 are $mr_{e_{s_4}}$ and $mr_{e_{s_5}}$.

Consequently, the construction of the nearest neighbor graph in Greedy-BST/DME always involves n nodes, because all nodes in all the sub-topologies in F participate in the identification of the nearest neighbors. We will examine the complexity of the Greedy-BST/DME algorithm in one of the exercises.

13.3.3.5 *Useful-skew routing*

The useful-skew routing problem refers to the problem of synthesizing a clock routing that satisfies all specified general clock skew constraints (*i.e.,* setup and hold-time constraints). There is a variant of the useful-skew routing problem that deals with a prescribed clock schedule (*i.e.,* all skew constraints are equality constraints instead of being bounded from above and below). Such a variant can be solved by a simple modification of the zero-skew routing algorithm. In the sequel, we deal with the more general problem where we maintain the flexibility of skew scheduling throughout the process of constructing a useful-skew clock routing tree [Xi 1996; Taso 2002].

In Section 13.3.3.1, we have already introduced some concepts that would be of crucial importance to useful-skew routing, namely, the feasible-skew range for a pair of clock sinks, the commitment of the skew of a pair of clock sinks, and the incremental updates of the remaining feasible skew ranges after a skew commitment. Now we will show how these concepts interact with a clock tree-embedding algorithm, called UST/DME [Tsao 2002].

Because the clock skews are also constrained to lie within a range specified by some upper and lower bounds, as in the case of bounded-skew constraint in bounded-skew routing, the underlying concept of the merging region in the UST/DME algorithm is the same as that in the BST/DME algorithm. However, the interaction of merging regions and incremental scheduling introduces a problem not encountered before. In BST/DME, when a merging region of a parent node is computed from two sampling segments of the child nodes, it implies that we have committed to some maximum and minimum sink delays associated with the sampling segments. However, the commitment does not affect the skew-bound constraint, because they are being applied to all pairs of clock sinks. That is not the case in useful-skew clock routing; a skew commitment in one subtree affects the feasible skew ranges of clock pins in another.

Therefore, the first step of the UST/DME algorithm is the construction of a constraint graph G_C from the given skew constraints C, and the second step is the computation of an all-pairs shortest distance matrix $D = \{d_{i,j} : s_i, s_j \in S\}$ from G_C to represent all feasible skew ranges $FSR_{i,j} = [-d_{i,j}, d_{j,i}]$. As before, now we consider the merging of two nodes a and b, whose parent node is v in the given abstract topology G. In other words, we construct the merging regions mr_v on the basis of mr_a and mr_b. Here, we assume that a and b are not leaf nodes. Otherwise, the merging of them is similar to the merging of two leaf nodes in BST/DME. Let the two subtrees of a be $T_{a,l}$ and $T_{a,r}$ and the two subtrees of b be $T_{b,l}$ and $T_{b,r}$.

To overcome the difficulty highlighted in the preceding paragraph (as well as the difficulties in the use of boundary segments for merging), we use a set of sampling segments to represent a merging region as in the case of BST/DME. The advantage of such a restriction is that $skew_{i,j}$ for any pair of clock pins s_i and s_j in the subtree (rooted by the sampling segment) is a constant value, say $x \in FSR_{i,j}$. We also associate each sampling segment with the leftmost clock sink in its subtree. In other words, each sampling segment is associated with the delay to its leftmost clock sink.

Let ss_a be a sampling segment in mr_a and s_i and s_j be the representative clock pins of $T_{a,l}$ and $T_{a,r}$, respectively. Because a skew commitment of $skew_{i,j} = x$ changes $FSR_{k,b}$ for the representative clock pins s_k and s_b in $T_{b,l}$ and $T_{b,r}$, respectively, we update matrix D and recompute the merging region mr_b on the basis of the updated $FSR_{k,b}$. The updated merging region mr_b is then sampled by s sampling segments, each of which is merged with ss_a to compute a merging region mr_v for v such that the skew between s_i, the representative clock pin of T_a and s_k, the representative clock pin of T_b is feasible (*i.e.*, $skew_{i,k} \in FSR_{i,k}$). The computation of the merging region mr_v is facilitated by the associated sink delays of the two sampling segments involved.

The preceding procedure will no doubt prompt the following questions:

1. Do we have to associate a sampling segment from mr_a with the delays to all sinks in its subtree instead of just the delay to the representative clock pin?
2. Does mr_v, which satisfies the constraint $skew_{i,k} \in FSR_{i,k}$, ensure that $skew_{i',k'} \in FSR_{i',k'}$ for all $s_{i'} \in T_a$ and $s_{k'} \in T_b$ in the new tree T_v?

The answers to these questions are related. As mentioned in Section 13.3.3.1, it takes at most $n - 1$ skew commitments to the incremental skew scheduler to determine a feasible skew schedule. First, we observe that for each subtree T whose root node has a nondegenerated merging region (*i.e.*, not a Manhattan arc) there have been $|S(T)| - 2$ skew commitments made, where $S(T)$ is the set of sinks in T. When we merge two sinks, for example, we have not made a skew commitment; otherwise, we would have obtained a degenerated merging region (*i.e.*, a Manhattan arc). In other words, the subtree containing the two sinks has not made any skew commitment. The selection of a sampling segment from mr_a (for the purpose of merging) is equivalent to committing the skew between the representative clock pins of $T_{a,l}$ and $T_{a,r}$. Consequently, we have now made $|S(T_a) - 1|$ skew commitments in T_a. In other words, all skews of sink pairs in T_a have been determined. Therefore, knowing the delay from ss_a to the representative clock sink in T_a would allow us to construct all the delays to other clock sinks in T_a. Note that the merging of T_a with T_b would mean that we obtain a tree T_v with $|S(T_a)| + |S(T_b)| - 2 = |S(T_v)| - 2$ skew commitments.

To answer the second question, recall that we made an analogy between skew commitments and spanning tree construction for the complete graph representing D. Under the spanning tree analogy, one should realize that the spanning tree is built in a distributive fashion similar to that in Kruskal's algorithm [Cormen 2001]. Initially, all vertices are considered as singleton components (from the perspective of the eventual spanning tree). As skew commitments are made, connected components (which are also trees) are being joined. In each connected component, the skew between any pair of clock sinks has an equality constraint. Now, consider the merging of ss_a with ss_b (*i.e.*, the construction of a merging region that satisfies $-d_{i,k} \leq skew_{i,k} \leq d_{k,i}$). Because $skew_{i,i'}$ is now committed for any $s_{i'} \in T_a$,

$$-d_{i,k} \leq skew_{i,k} = skew_{i,i'} + skew_{i',k} \leq d_{k,i}$$

or

$$-d_{i,k} - skew_{i,i'} \leq skew_{i',k} \leq d_{k,i} - skew_{i,i'}$$

The differences between the upper and lower bound constraints of $skew_{i,k}$ and $skew_{i',k}$ are exactly the same: $d_{k,i} + d_{i,k}$.

The construction of the merging region requires the computation of two Manhattan arcs as the boundary segments of the merging region. Let $L = d(ss_a, ss_b)$, t_i^a be the sink delay of s_i in T_a, and t_k^b be the sink delay of s_k in T_b. Note that for any sink $s_{i'} \in T_a$, $t_{i'}^a = t_i^a - skew_{i,i'}$. Then, assuming that $0 \leq x \leq 1$, the construction of mr_v on the basis of the delays of $s_{i'}$ and s_k, as well as the skew constraints of $skew_{i',k}$, is equivalent to finding a suitable range of x such that

$$-d_{i,k} - skew_{i,i'} \leq t_i^a - skew_{i,i'} + r \cdot x \cdot L \cdot \left(C_{T_a} + \frac{c \cdot x \cdot L}{2} \right) - \left(t_k^b + r \cdot (1 - x) \cdot L \cdot \left(C_{T_b} + \frac{c \cdot (1 - x) \cdot L}{2} \right) \right) \leq d_{k,i} - skew_{i,i'}.$$

Eliminating $skew_{i,i'}$ from the preceding expression, we obtain

$$-d_{i,k} \leq t_i^a + r \cdot x \cdot L \cdot \left(C_{T_a} + \frac{c \cdot x \cdot L}{2} \right) - \left(t_k^b + r \cdot (1 - x) \cdot L \cdot \left(C_{T_b} + \frac{c \cdot (1 - x) \cdot L}{2} \right) \right) \leq d_{k,i}$$

the inequalities necessary for the construction of merging region mr_v from s_i and s_k, the respective representative clock sinks of T_a and T_b. Consequently, merging regions constructed on the basis of different pairs of clock sinks from T_a and T_b coincide exactly.

Figure 13.43c shows the tree of merging regions and the useful skew tree that corresponds to the skew commitment $skew_{1,2} = -3$ in Figure 13.21c. First, mr_a (shaded) is constructed on the basis of the merging of s_1 and s_2. The skew from any points in mr_a to s_1 and s_2 falls in the range $[-7, -3] \subset FSR_{1,2} = [-9, -3]$ and is, therefore, feasible. Next, mr_b is constructed from the merging of $ss_a \in mr_a$ and s_3. Note that b is also the singleton child node s_0' of s_0. The skew schedule realized by the UST/DME tree under the path length delay model is $\{t_1 = 6, t_2 = 9, t_3 = 8\}$, which is feasible subject to the constraints given in Figure 13.21. In contrast, we also show a bounded-skew tree with $B = 1$ in Figure 13.43b.

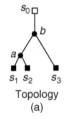

Topology
(a)

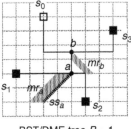

BST/DME tree $B = 1$
(b)

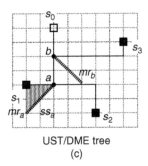

UST/DME tree
(c)

FIGURE 13.43

(a) An abstract topology G. (b) The merging tree and bounded-skew tree with $B = 1$ by BST/DME for G. (c) The merging tree and useful-skew tree by UST/DME for topology G subject to the skew constraints given in Figure 13.21.

For topology generation, we can follow the approach of Greedy-DME. One might be tempted to incorporate the flexibility that Greedy-BST/DME has in allowing merging of subtrees with non-root nodes. However, such a feature will significantly increase the computational complexity. Construction of mr_{e_v}, as defined in the Greedy-BST/DME algorithm, would require the incremental scheduler to uncommit skew commitments made in the previous merging steps. However, that would entail rebuilding the matrix D, which has an $O(n^3)$ complexity.

13.3.4 **Clock tree optimization**

In the preceding section, we focus mainly on the problems of skew scheduling and clock routing, which take into account the timing and wiring aspects of clock network synthesis. As pointed out in Section 13.2, there are other important design aspects of a clock network. This section examines other clock network optimization techniques that address some of these design considerations. Buffer insertion, for example, helps to further improve the timing characteristics of clock signals. Instead of repeating gates (*i.e.,* buffers), clocking control gates can be inserted to turn off flip-flops and their ensuing combinational logic modules when these modules are not required. Besides helping to improve the delay of a clock signal, sizing of buffers and wires can also be very effective in countering skews unintentionally introduced because of variations in manufacturing parameters. Moreover, to further enhance the tolerance of a clock tree to process variations, cross-links can be inserted to provide alternative paths for the clock signal to arrive at selected clock sinks.

13.3.4.1 *Buffer insertion in clock routing*

To drive a large load such as a clock tree, a possible solution is to use cascaded drivers that are of exponentially increasing sizes [Lin 1975]. However, the area requirement and power consumption of such drivers can be prohibitively high. A common solution to this is to break a large clock tree into multiple smaller trees, each driven by a buffer.

As shown in Figure 13.44a, a clock driver/buffer b is modeled as a switch-level RC circuit with a gate capacitance C_b, an output resistance R_b, and an intrinsic delay T_b caused by parasitics capacitance ($= T_b/R_b$) associated with the transistors. A buffer inserted in a long interconnect shields the downstream capacitance C_{down} after the output of the buffer from the upstream resistance R_{up} before the gate input of the buffer as shown in Figure 13.44b. However, it presents a load of C_b, in addition to the capacitance of the upstream interconnect, denoted C_{up}, to the upstream resistance. The new Elmore delay is

$$R_{up}(C_{up} + C_b) + T_b + (R_b + R_{down})C_{down}$$

If $R_{up}C_b + T_b + R_bC_{down} < R_{up}C_{down}$, the insertion of the buffer reduces the overall delay of the long interconnect.

Clearly, buffer insertion can play an important role in minimizing the clock phase delay, which is defined as the maximum among the sink delays. Even when delay minimization is not the main goal of clock tree synthesis, buffer insertion may help in reducing the overall wiring cost. In zero-skew routing, for example, it is always desirable to merge two subtrees with similar sink delays; when the sink delays differ greatly, wire snaking (see Figure 13.25) is commonly used to balance them by making a faster clock signal path slower. With buffer insertion, the slower clock signal path can be made faster, thereby eliminating the need for wire snaking. Moreover, compared with a clock tree

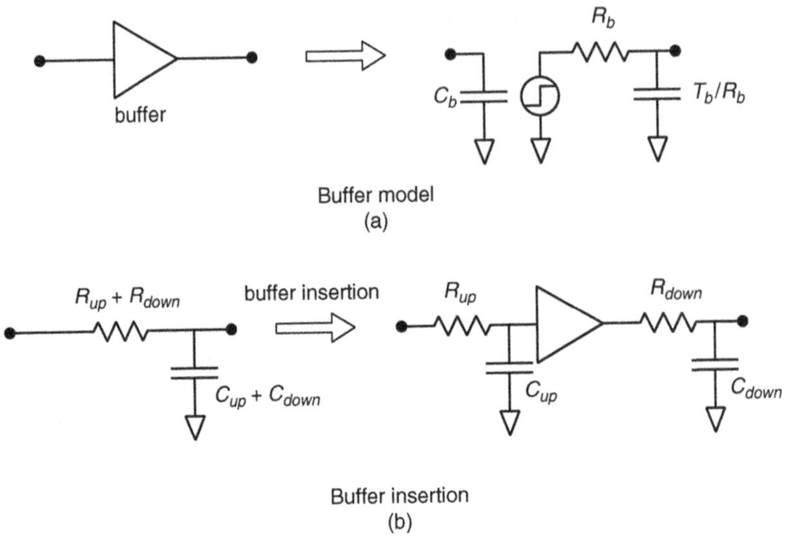

Buffer model
(a)

Buffer insertion
(b)

FIGURE 13.44

(a) A switch-level model for a clock buffer/driver. (b) The insertion of a buffer to break up a long interconnect.

with drivers only at the root node, it is easier to satisfy the rise/fall time constraints with a buffered clock tree.

With buffer insertion, clock power can also be reduced because of the reduced load presented to the clock driver and intermediate clock buffers. It is no longer necessary to have huge clock drivers, which add to the power consumption. The reduction of capacitive load also reduces the current flowing through the clock driver, improving the reliability of the interconnects near the driver output. Moreover, the current demand is now more evenly distributed across the entire chip, potentially reducing the current-induced noise in the power supply network.

The buffered clock tree construction algorithm in [Vittal 1995] is an extension of the Greedy-DME algorithm to consider the insertion of intermediate clock buffers during the construction of a zero-skew routing tree. In each merging step, three sets of possible locations for embedding an internal node of the abstract topology are computed. One of the three sets is the merging segment as in the case of DME. The other two sets correspond to the possible locations of the internal node when a buffer is inserted to drive one of the child subtrees.

Consider two subtrees T_a and T_b rooted at a and b, respectively, as shown in Figure 13.45. The merging segment ms_v of v, the parent node of a and b, can be computed as in the DME algorithm, and it corresponds to the feasible locations of v when no buffer is inserted.

(a)

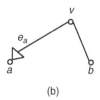

(b)

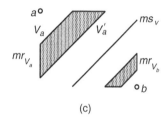

(c)

FIGURE 13.45

Insertion of a buffer at different locations along the edge e_a to drive T_a alone.

A buffer may be inserted at v to drive only T_a, but not T_b, through e_a, as shown in Figure 13.45a. The Manhattan arc V_a, as shown in Figure 13.45c contains the set of feasible locations of v when zero-skew is achieved for such a buffer location. Alternatively, the buffer may be inserted at a as shown in Figure 13.45b, and the Manhattan arc V_a', as shown in Figure 13.45c, corresponds to the set of candidate locations of v for this alternate buffer location. In this example, the insertion of a buffer at v results in a longer sink delay than inserting the buffer at a. Consequently, a is closer to V_a than to V_a'.

Because V_a and V_a' represent the locations of v when a buffer is inserted at the extreme possible locations (at the start and end of edge e_a, respectively), the "merging" region mr_{v_a} bounded by V_a and V_a' corresponds to all possible locations of v when the buffer is inserted at any point along e_a. It is important to note that the region defined by a and V_a in Figure 13.45c contains the possible locations for buffer insertion along e_a. Also note that $|e_a|$ depends on the buffer position.

Similarly, a buffer may be inserted to drive T_b alone. The merging region mr_{v_b} in Figure 13.45c captures the set of feasible locations to embed v when a buffer is inserted to drive T_b. Although not shown here, it is possible that mr_{v_a}, mr_{v_b}, and ms_v may overlap and that their relative positions may vary. Moreover, we can construct merging segments/regions on the basis of other buffering combinations; we may, for example, insert buffers into a and b simultaneously.

In all of these merging operations, it is important to distinguish between the total capacitance of a subtree and the capacitance of a DC-connected subtree. Consider the buffered interconnect in Figure 13.44b. There are two DC-connected subtrees, with one being the upstream of the buffer and the other being the downstream of the buffer. The total capacitance of the buffered interconnect is $C_{up} + C_{down} + C_b$,[2] whereas the capacitance of the DC-connected subtree corresponding to the upstream of the buffer is simply $C_{up} + C_b$. In many of the equations for computing merging segments/regions in Section 13.3.3, C_{T_a} and C_{T_b} are, in fact, the capacitances of the DC-connected subtrees rooted at a and b, respectively. For quantifying the wiring cost of the entire clock tree, we, of course, use the total capacitance of the tree. For clarity, we will use $C_{T_a}^{tot}$ to denote the total capacitance of the tree rooted at a.

The buffered clock tree construction algorithm follows the flow of the Greedy-DME algorithm with the following modifications. Instead of the use of only wire length to define merging cost, the cost of merging is a weighted combination of multiple factors such as total wire length, total buffer size, and rise time. As each internal node has multiple sets of feasible locations for its embedding (as in the case of BST/DME and UST/DME), the sink delays are not determined when these feasible locations are computed. The sink delays are determined at the next level of the merging step when the respective merging segments or merging regions (or sampling segments for ease of implementation) of the two sibling nodes that yields locally minimum zero skew merging cost are selected. Besides considering buffering in a merging operation, a buffer can also be inserted to drive the merged subtree if the sinks of the subtree do not have sharp clock edges (*i.e.*, long rise/fall time).

Most buffered clock tree construction algorithms place an upper-bound constraint on the difference in the numbers of buffers in any two source-to-sink paths in a clock tree. This is a preventive measure to minimize the effects of variations in the electrical parameters of clock buffers on the clock delays and, hence, the clock skew of the two paths. In the most restrictive case, all source-to-sink paths go through the same number of buffers. Moreover, buffers inserted at the same level are of the same size. Although these restrictions may affect the optimality in terms of signal delay and total wire length, they greatly reduce the sensitivity to process variations of clock skew.

One example of such algorithms is that in [Chen 1996], where instead of considering buffer insertion at each merging step as in [Vittal 1995], buffers

---

[2]For simplicity, we have ignored the parasitic capacitance intrinsic to the buffer.

are inserted at the roots of all subtrees simultaneously. Recall that in DME, we maintain F, a forest of subtrees. Starting with a forest of singleton subtrees, several iterations of DME-based zero-skew merging are performed until $|F|$ has been reduced by 2^k for some k, which is determined on the basis of the clock buffer strength. The stopping criterion could also be based on the rise/fall time constraint or the maximum capacitance of the subtrees in F. At this point, buffers are inserted at the roots of all subtrees in F. This is akin to clustering of nodes, followed by buffer insertion to drive each cluster.

An inserted buffer may not be connected to the root node directly. Instead, a wire may be used to connect from the buffer output to the root of the subtree such that all subtrees in S have equal sink delay. In other words, the root node of a subtree may have a tilted rectangular region to represent the set of possible locations for the inserted buffer. A future merging step that involves such a subtree will then be based on a tilted rectangular region instead of a merging segment.

Remark: Post-Silicon Tunable Buffers: Skews induced by process variations can result in significant yield loss. To address this problem, circuit designers have deployed so-called ***post-silicon tunable*** (PST) circuitry in modern commercial processor chips such as Intel P4 and Itanium [Geannopoulos 1998; Tam 2000]. The concept of post-silicon tuning is illustrated in Figure 13.46, where some of the clock buffers are delay-tunable. The delay of any of these buffers can be adjusted by activating an appropriate number of capacitors between the two inverters that form the buffer (see the schematic of a delay-tunable buffer in Figure 13.46). The post-silicon adjustment can be performed dynamically with on-chip de-skew

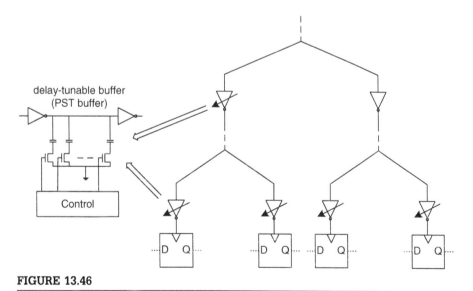

FIGURE 13.46

A clock tree with post-silicon tuning circuitry.

circuitry, whereas a fuse-based solution allows only one-time adjustment. Other post-silicon tunable techniques involve voltage biasing of the buffers.

13.3.4.2 *Clock gating*

So far, we have considered designs in which all clocked modules always receive a clock signal. However, for modules that are idle, it is not necessary to send them a clock signal. Suppose we isolate the modules from the clock source when they are in the idle mode, then the power consumption caused by the capacitance of the clock pins in these modules can be reduced. Moreover, because the flip-flops in these modules are not triggered, the output nodes of these flip-flops retain their values throughout the idle period. Consequently, the combinational logic following these flip-flops does not have any switching activity during the idle period. In other words, there is no active power consumed by idle modules. This is the clock-gating technique presented in Section 13.2.4.

In clock gating, clocking control gates (or clock gates) are inserted along with buffers in a clock tree. Consider for example a clock tree as shown in Figure 13.47, where internal node v is a candidate buffer/gate location. The activity pattern associated with each node over 6 time units is also shown, with a dark square indicating that the node should be active in this time unit and an empty square indicating an idle time unit. Note that a time unit could be multiple clock cycles. The activity pattern of an internal node is derived from the activity patterns associated with its child nodes. With a "1" to indicate an active time unit and a "0" to indicate an idle time unit, the bitwise-OR of the activity patterns of two child nodes generates the activity pattern of the parent node. Node v, for example, is active when at least one of its child nodes (*i.e.*, a and b) is active, as shown by the respective activity patterns in Figure 13.47.

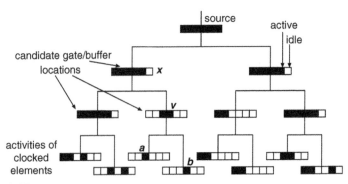

FIGURE 13.47

A clock tree topology with activity patterns for modules and candidate buffer/gate locations.

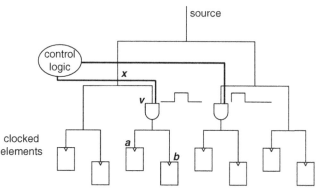

FIGURE 13.48

A gated clock tree, with two inserted clocking control gates.

The gated clock tree as shown in Figure 13.48 is the result of inserting two clock gates in the clock tree given in Figure 13.47. The clock gate at v must be active and allow the clock signal to propagate through when at least one of its sinks (*i.e.*, a or b) is active. Therefore, a gating-signal control logic module that generates control signals to turn on or off the clock gates is required.

Although clock gating reduces dynamic power, it has an overhead of the gating-signal control logic. Moreover, as long as the clock signal arrives at a clock gate, the clock gate contributes to power consumption because of its gate capacitance or the switching of its internal nodes. The other downside of clock gating is the clock skew caused by a clock gate. These are the main issues handled by gated-clock synthesis algorithms in [Téllez 1995; Oh 2001; Chen 2002; Chao 2008].

To construct a zero-skew gated-clock tree, let us again assume that we are given an abstract topology. The tasks here are to embed the internal nodes and to determine where buffers and clock gates should be inserted. We will now describe an adaptation and simplification of the approaches described in [Téllez 1995; Oh 2001; Chen 2002; Chao 2008]. The approach also closely resembles that for the construction of a buffered clock tree. Again, let v be the parent node of nodes a and b. When we merge subtrees rooted at a and b, we allow the following three choices at the root of e_a:

1. No buffer or clock gate is inserted at the root of e_a.
2. A buffer is inserted at the root of e_a.
3. A clock gate is inserted at the root of e_a. Here, we assume that similar to a buffer, a clock gate is also modeled with a switch-level RC model with gate capacitance C_{cg}, output resistance R_{cg}, and intrinsic delay T_{cg}.

We similarly afford the three choices at the root of e_b. There are, therefore, altogether 9 different ways of merging a and b, resulting in 9 zero-skew merging segments. As in all other methods that generate multiple merging segments/sampling segments when merging two subtrees, it is prudent to keep only a fixed number of sampling segments at each internal node. It is typical that the subset of sampling segments retained at each internal node is selected on the basis of the merging cost, which has to be modified significantly to account for dynamic power reduction.

Without clock gating, the power consumption of a clock tree can be obtained directly from the wiring cost (and buffering cost if buffer insertion is considered). With the insertion of a clock gate, the switching activity of the subtree after the clock gate is greatly reduced. Consider node v in Figure 13.48, the capacitance $C_{T_v}^{tot}$ *after* the clock gate switches only a third of the time. In a non-gated-clock tree, the power consumption of the subtree rooted at v has a dynamic power of $C_{T_v}^{tot} V_{DD}^2 f_{clk}$, whereas the gated subtree in Figure 13.48 has a dynamic power of only $C_{T_v}^{tot} V_{DD}^2 f_{clk}/3$. We call $C_{T_v}^{tot}/3$ the effective switching capacitance of the subtree rooted at v, with the coefficient of a third being the activity factor, denoted α_v.

In the synthesis of a gated-clock tree, the minimization of the effective switching capacitance of the entire clock tree is the main objective. To compute the effective switching capacitance of a subtree, we decompose the total capacitance of the subtree into two parts, gated capacitance and ungated capacitance. Take node x in Figure 13.48 for example—its gated capacitance, denoted GC_{T_x}, is $C_{T_v}^{tot}$. The ungated capacitance of x, denoted UGC_{T_x}, is $C_{T_x}^{tot} - GC_{T_x} = C_{T_x}^{tot} - C_{T_v}^{tot}$. Let the effective switching capacitance of the gated capacitance be $GC_{T_x}^{eff}$, which in this example is $C_{T_v}^{tot}/3$. Then, the effective switching capacitance at x, denoted $C_{T_x}^{eff}$, is $UGC_{T_x} + GC_{T_x}^{eff}$.

If we insert a buffer at x, the ungated capacitance, as well as the effective switching capacitance, at x are increased by the gate capacitance of the buffer C_b. If instead of a buffer, we insert a clock gate at x, the gated capacitance would be increased by UGC_{T_x}, whereas the ungated capacitance at x reduces to simply C_{cg}, the gate capacitance of the clock gate. Moreover, the effective switching capacitance at x is reduced by $(1 - \alpha_x)UGC_{T_x} - C_{cg}$. The following ordered operations update these capacitances correctly when a clock gate is inserted at x:

1. $C_{T_x}^{tot} \leftarrow C_{T_x}^{tot} + C_{cg}$;
2. $GC_{T_x} \leftarrow GC_{T_x} + UGC_{T_x}$;
3. $GC_{T_x}^{eff} \leftarrow GC_{T_x}^{eff} + \alpha_x UGC_{T_x}$;
4. $UGC_{T_x} \leftarrow C_{cg}$;
5. $C_{T_x}^{eff} \leftarrow UGC_{T_x} + GC_{T_x}^{eff}$.

Note that the insertion of a clock gate or a buffer creates a new DC-connected subtree, with C_{cg} or C_b, respectively, being the new C_{T_x}, which would be required for the computation of merging segments at the parent node.

The merging cost is typically defined to be some combination of the wire length $|e_a| + |e_b|$, effective switching capacitance, and possibly other metrics. An important metric that we have left out in the preceding discussion is the switching activity (and effective switching capacitance) of various clock gate control signals. Consequently, it is unwise to insert a clock gate right before a register that is enabled most of the time. To incorporate the power consumption of gating signals, we may assume a centralized gating-signal control logic and use its distance from a clock gate to estimate the wire capacitance of a gating signal.

As in the case of Greedy-DME, we may interleave the generation of abstract topology with the computation of merging segments/sampling segments. However, the efficacy of such an approach relies on the existence of neighboring subtrees that are of similar switching activity. Therefore, it is important that the high-level synthesis, logic-level synthesis, and physical-level synthesis steps all work hand-in-hand with gated-clock tree construction. All in all, the fundamental difficulty lies in the prediction of the activity pattern of various modules. Although the data activity can be quite well characterized for DSP circuits and microprocessors, it is more difficult to generate activity patterns for a general circuit/system.

13.3.4.3 *Wire sizing for clock nets*

In this section, we deal with the problem of assigning appropriate widths to wires in a clock tree to minimize the clock skew, the clock delay, and the sensitivity of the clock tree to process variations. The constraint on the maximum width of a wire is typically imposed by the available routing resources, whereas the constraint on the minimum wire width depends on the fabrication technology. Moreover, the maximum allowable current density through a wire also provides a lower bound for the wire width, so that the wire can withstand the wear-out phenomenon caused by electromigration. Note that a long wire may be divided into several segments, and each segment may have different upper and lower bounds.

Remark: Mitigating Electromigration: Wire sizing to counter electromigration can be easily handled by DME-based clock routing algorithms. When nodes a and b are merged at the parent node v, the total capacitances of subtrees T_a and T_b, which provide respective estimates of the amounts of current flow in e_a and e_b, can be used to determine for e_a and e_b appropriate wire widths that are tolerable to electromigration. Zero-skew, bounded-skew, or useful-skew merging can then be carried out with appropriate wire widths for e_a and e_b.

13.3.4.3.1 Delay sensitivity and delay minimization

From Section 13.3.2, the Elmore delay from the clock driver at the source s_0 to sink s_i in an RC tree is

$$ t_i = R_d \cdot C_{T_{s_0}} + \sum_{e_v \in Path(s_0, s_i)} |e_v| \cdot \frac{r}{w_{e_v}} \cdot \left(\frac{|e_v| \cdot w_{e_v} \cdot c_a}{2} + C_{T_v} \right) $$

where R_d is the output resistance of the clock driver. For simplicity, this formulation ignores the fringing capacitance of wires, which can be added easily. Taking the partial differential $\frac{\partial t_i}{\partial w_{e_v}}$ for any edge e_v along the s_0-s_i path,

$$ \frac{\partial t_i}{\partial w_{e_v}} = R_d \cdot c_a \cdot |e_v| - \frac{|e_v| \cdot r \cdot C_{T_v}}{w_{e_v}^2} + \sum_{e_u \in Ans(e_v)} \frac{|e_u| \cdot r \cdot c_a \cdot |e_v|}{w_{e_u}} \tag{13.10} $$

where $Ans(e_v)$ contains the path from s_0 to the parent node of v. If e_v is not along $Path(s_0, s_i)$,

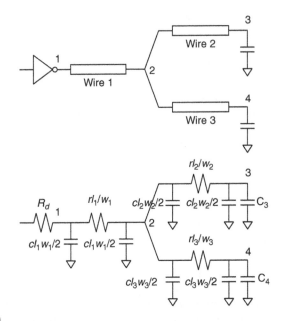

FIGURE 13.49

A clock tree with two sinks.

$$\frac{\partial t_i}{\partial w_{e_v}} = R_d \cdot c_a \cdot |e_v| + \sum_{e_u \in Ans(e_v) \cap path(s_0, s_i)} \frac{|e_u| \cdot r \cdot c_a \cdot |e_v|}{w_{e_u}} \tag{13.11}$$

Let us illustrate the computation of partial derivatives with a simple example as shown in Figure 13.49. The RC tree has three segments with wire widths of w_1, w_2, and w_3, respectively. Every wire segment of length l_i is modeled as π-type RC circuit with a resistance of rl_i/w_i and a total capacitance of cl_iw_i, where r is the resistance per square and c is the capacitance per unit area. The partial derivatives of t_3 and t_4 with respect to w_2 are:

$$\frac{\partial t_3}{\partial w_2} = R_d \cdot c \cdot l_2 - \frac{l_2 \cdot r \cdot C_3}{w_2^2} + \frac{l_1 \cdot r \cdot c \cdot l_2}{w_1}$$

$$\frac{\partial t_4}{\partial w_2} = R_d \cdot c \cdot l_2 + \frac{l_1 \cdot r \cdot c \cdot l_2}{w_1}$$

The partial differential $\frac{\partial t_i}{\partial w_{e_v}}$ evaluated at W, the currently assigned wire widths, captures the *delay sensitivity* of t_i with respect to a change in w_{e_v}. A positive sensitivity indicates that the delay increases if we widen e_v, whereas a negative sensitivity indicates that the delay decreases. When all sinks have zero delay sensitivity, the clock net is extremely tolerable to process variations, because the sink delays are not sensitive to these changes.

Having zero delay sensitivity means that we minimize all the sink delays. A sink delay can be locally minimized by setting $\frac{\partial t_l}{\partial w_{e_v}} = 0$ and solving for w_{e_v} while keeping the widths of other wires intact. Consequently, wires closer to the root should have wider wire width, because they see a larger downstream capacitance. The term $R_d \cdot c_a \cdot |e_v|$ in the partial derivative keeps the wire width from getting too large. It is also a common practice to impose an upper-bound constraint on the wire width.

It should also be obvious that the larger the downstream capacitance seen by a wire, the larger the delay sensitivity. Consequently, buffer insertion, which decouples downstream capacitance from the upstream resistance, can also greatly desensitize a clock net. Similarly, we can also define the partial derivatives of a sink delay with respect to buffer sizes and perform appropriate sizing of buffer/driver to reduce sink delay. Although the focus here is wire sizing, similar techniques can be used to perform buffer sizing. In fact, many techniques that are based on delay sensitivity perform buffer and wire sizing together [Wang 2005; Guthaus 2006].

Similar to wire sizing for mitigating electromigration, wire sizing for delay minimization can be easily incorporated in a DME-based clock routing algorithm, as in [Edahiro 1993b]. Instead of operating on an abstract topology, the heuristic approach in [Edahiro 1993b] operates on a zero-skew routing tree, and the reason for that will soon be apparent.

Consider the merging of two zero-skew subtrees T_a and T_b at the parent node v, with w_{e_a} and w_{e_b} being the wire widths of e_a and e_b, respectively. Let $L = d(ms_a, ms_b)$ be the distance between the merging segments of a and b and $x \cdot L$ be the distance of the merging segment of v from a, where $0 \le x \le 1$, assuming no detour wiring. Then, x can be computed with the following expression:

$$x = \frac{t_b - t_a + r \cdot L \cdot (C_{T_b}/w_{e_b} + c \cdot L/2)}{r \cdot L \cdot (c \cdot L + C_{T_a}/w_{e_a} + C_{T_b}/w_{e_b})}$$

Taking the derivative of x with respect to w_{e_a} yields

$$\frac{\partial x}{\partial w_{e_a}} = \frac{C_{T_a}/w_{e_a}}{c \cdot L + C_{T_a}/w_{e_a} + C_{T_b}/w_{e_b}} \left(\frac{x}{w_{e_a}}\right)$$

Recall that an optimal width assignment actually depends on both upstream resistance and downstream capacitance. Although downstream capacitances are known in the bottom-up process, we have to approximate the resistance in the upstream. We use the given zero-skew routing tree to approximate the upstream resistance from the clock source s_0 to v in the final routing tree by assuming, for example, nominal wire widths for the upstream edges. Let R_{up} denote the approximate upstream resistance. Then, the estimated delay from s_0 to a sink in T_a is

$$R_{up} \cdot c \cdot L \cdot (w_{e_a} \cdot x + w_{e_b} \cdot (1 - x)) + r \cdot x \cdot L \cdot \left(\frac{C_{T_a}}{w_{e_a}} + \frac{c \cdot x \cdot L}{2}\right) + K$$

where K is a constant independent of x and w_{e_a}. Taking the derivative of the delay with respect to w_{e_a} and setting the derivative to zero, we obtain

$$R_{up} \cdot c \cdot w_{e_a}^2 \left(\frac{w_{e_b}}{w_{e_a}} C_{T_a} \left(2 - \frac{w_{e_b}}{w_{e_a}} \right) + C_{T_b} + w_{e_b} \cdot c \cdot L \right)$$
$$= r \cdot C_{T_a} \cdot C_{T_b} + r \cdot c \cdot w_{e_b} \cdot (1 - x) \cdot L \cdot C_{T_a}$$

Assuming that wire capacitances $w_{e_b} \cdot c \cdot L$ and $w_{e_b} \cdot c \cdot (1 - x) \cdot L$ are much smaller than C_{T_b}, we discard them and obtain the following expression:

$$R_{up} \cdot c \cdot w_{e_a}^2 \left(\frac{w_{e_b}}{w_{e_a}} C_{T_a} \left(2 - \frac{w_{e_b}}{w_{e_a}} \right) + C_{T_b} \right) = r \cdot C_{T_a} \cdot C_{T_b}$$

A similar expression can be obtained by examining how the delay of a sink in T_b is affected by e_b:

$$R_{up} \cdot c \cdot w_{e_b}^2 \left(\frac{w_{e_a}}{w_{e_b}} C_{T_b} \left(2 - \frac{w_{e_a}}{w_{e_b}} \right) + C_{T_a} \right) = r \cdot C_{T_a} \cdot C_{T_b}$$

Equating the two preceding equations yields the following analytical formula for the wire width of e_a and e_b:

$$w_{e_a} = w_{e_b} = \min \left\{ \max \left\{ w_{\min}, \sqrt{\frac{r \cdot C_{T_a} \cdot C_{T_b}}{R_{up} \cdot c \cdot (C_{T_a} + C_{T_b})}} \right\}, w_{\max} \right\}$$

where $w_{\min}$ is the minimum wire width allowed by the technology or specified by electromigration constraints, and $w_{\max}$ is typically determined by the available wiring resources.

Because the newly computed wire lengths and widths may differ greatly from the original clock routing tree, it is recommended that the process be repeated for a few iterations. Because the wire widths are determined on the basis of delay sensitivity, the sensitivity of the clock tree to process variations is reduced indirectly after the iterative procedure. However, be aware that the iterations may not converge. In other words, wire lengths and widths may keep changing in successive applications of the modified DME algorithm. Restricting the procedure to just a few iterations provides a good compromise between solution quality and runtime efficiency.

Although the bottom-up wire sizing approach described in the preceding paragraphs assumes continuous wire width, it can also be modified to consider discrete wire widths. Because it is not possible to achieve arbitrary precision during fabrication, it is better to have the algorithm explicitly synthesize a clock tree with discrete wire widths to eliminate undesirable skew caused by the post-synthesis mapping of continuous widths to discrete widths.

Remark: Skew Sensitivity: What is truly of interest is *skew sensitivity*, which measures how a change in wire width can affect the clock skew. In particular, skew sensitivity caused by process variations can be used to measure how reliable a clock tree is. Skew sensitivity is simply the difference of delay sensitivity. Take the RC tree in Figure 13.49 for example, the skew sensitivity of $skew_{3,4} = t_3 - t_4$ is

$$\frac{\partial}{\partial x_2}(t_3 - t_4) = -l_2 \cdot r \cdot C_3/w_2^2 \tag{13.12}$$

It should be obvious for any two sinks, only the subtree rooted at the youngest common ancestor of the two sinks contributes to the skew sensitivity. Because the two sinks share the same path from the clock source to the youngest common ancestor, any changes along the path result in the same changes in the two sink delays, which negate each other in the computation of skew.

However, because of the definition of global clock skew as $gskew = \max_{i,j}|t_i - t_j|$, it is very costly to compute global skew sensitivity exactly. The exact approach would entail the computation of the changes in the worst-case global clock skew for each wire under consideration. Typically, an approximation method such as that in [Xi 1995] would be used.

13.3.4.3.2 Wire sizing with dynamic programming

No treatment of wire sizing is complete without the inclusion of a very popular technique to solve difficult programs in EDA, namely, dynamic programming. This solution method was applied to buffer insertion in 1990 by van Ginneken [van Ginneken 1990] for delay minimization of signal nets and has since been generalized to perform wire sizing (and buffer sizing) as well [Lillis 1995]. Although the solution we will see here is not in the same mold as those in [van Ginneken 1990; Lillis 1995], it is still based on the fundamental principle of building and enumerating solutions of larger problems from solutions of smaller problems.

Here, we assume that we are sizing an unbuffered zero-skew clock tree T. Let $\gamma = (C, t)$ denote a solution at a node v in T, where C is the total capacitance rooted at v and t is the delay from v to a sink in T_v. Consider a clock sink s_i. There is only one solution $\gamma_{s_i} = (C = c_{s_i}^g, t = 0)$ associated with it, where $c_{s_i}^g$ is the gate capacitance of s_i.

Now, consider the edge e_{s_i} connecting s_i to its parent node a. We refer to the tree with e_{s_i} as the root edge as $T_{e_{s_i}}$. Assume that a wire width of w is assigned to it. Then, $\gamma_{e_{s_i}}(w) = (C^+, t^+)$, the solution that corresponds to such a wire width assignment has the following capacitance and sink delay:

$$C^+ = C + |e_{s_i}| \cdot w \cdot c \quad \text{and} \quad t^+ = t + \frac{r \cdot c \cdot |e_{s_i}|^2}{2} + \frac{r \cdot |e_{s_i}|}{w} \cdot C$$

where $(C, t) = \gamma_{s_i}$. Because we can pick any width in the range $w_{\min} \leq w \leq w_{\max}$ to form such a solution, the solution space associated with $T_{e_{s_i}}$ can be captured by treating the capacitance and sink delay in $\gamma_{e_{s_i}}(w)$ as the coordinates of a point on a 2-D plane, of which the x-axis is capacitance and the y-axis is sink delay. The solution spaces associated with $T_{e_{s_i}}$ and $T_{e_{s_j}}$, which are connected at a, are similar to the leftmost plot in Figure 13.50.

Let $\Gamma_{e_{s_i}}$ and $\Gamma_{e_{s_j}}$ denote the solution spaces associated with $T_{e_{s_i}}$ and $T_{e_{s_j}}$, respectively. The merging at a is feasible if, and only if, there exist $\gamma_{e_{s_i}}(w) = (C_i, t_i) \in \Gamma_{e_{s_i}}$ and $\gamma_{e_{s_j}}(w') = (C_j, t_j) \in \Gamma_{e_{s_j}}$ with $t_i = t_j$. The solution associated with a obtained from the merging of $\gamma_{e_{s_i}}(w)$ and $\gamma_{e_{s_j}}(w')$ is $\gamma_a(w, w') = (C_i + C_j, t_i = t_j)$. The solution space associated with a, denoted Γ_a, is again similar to that shown in the leftmost plot of Figure 13.50. Note that $\gamma_{e_{s_i}}(w_{\min})$ has the largest sink delay and $\gamma_{e_{s_i}}(w_{\max})$ the smallest. Let $t(\gamma)$ denote the sink delay in solution γ. Then, the sink delay in Γ_a must lie in the range $[t(\gamma_{e_{s_i}}(w_{\max})), t(\gamma_{e_{s_i}}(w_{\min}))] \cap [t(\gamma_{e_{s_j}}(w_{\max})), t(\gamma_{e_{s_j}}(w_{\min}))]$.

Now, if we pick any $\gamma = (C, t) \in \Gamma_a$, we can construct a new (partial) solution space associated with T_{e_a} based on γ by assigning wire width $w_{\min} \leq w \leq w_{\max}$ to e_a. The construction of such a solution, denoted as $\Gamma_{e_a}(\gamma)$, is similar to that for $T_{e_{s_i}}$ from γ_{s_i}. Consequently, we would again obtain a solution space that is similar to the leftmost plot in Figure 13.50. However, to obtain the overall solution space associated with T_{e_a}, we would have to consider all solutions in Γ_a. In other words, $\Gamma_{e_a} = \cup_{\gamma \in \Gamma_a} \Gamma_{e_a}(\gamma)$, which is typically a 2-D region, as shown in the middle plot of Figure 13.50. It is important that we keep track of which solutions from the two edges contribute to a solution in the parent node. The arrows from the leftmost plot to the middle plot in Figure 13.50 depict this relationship.

Clearly, the computation of the wire-sizing solution spaces of internal nodes of the zero-skew routing tree quickly becomes quite unwieldy. To overcome this problem, a sampling approach [Tsai 2003] can be used. Instead of computing a continuous 2-D solution space, a sample solution that is represented by a set of horizontal line segments is calculated. Each horizontal line segment corresponds to a range of capacitances $C_{\min} \leq C \leq C_{\max}$ that achieve a particular sink delay t. There may be a few sampling segments for a particular sink delay t, because there are a few ranges of capacitances that could result in the same delay. We show two sampling segments in the middle plot of Figure 13.50.

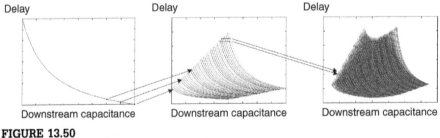

FIGURE 13.50

Wire sizing solution spaces for various nodes in a zero-skew routing tree.

The sampling of Γ_a is quite straightforward. We will now illustrate how the sampled Γ_{e_a} can be computed. It is best that we compute the sampled Γ_{e_a} together with sampled Γ_{e_b}, where a and b are siblings in the zero-skew routing tree. Doing so allows the sink delays to be sampled within the same range and facilitates the merging of the two sampled solution spaces to achieve zero-skew. In other words, the first step is to determine the minimum and maximum sink delays in Γ_{e_a} and Γ_{e_b} given the sampled solution spaces Γ_a and Γ_b. This is trivial, because for each $\gamma = (C, t) \in \Gamma_a$, the minimum (maximum) sink delay of T_{e_a} can be determined by assigning the maximum (minimum) wire width to e_a. The minimum (maximum) sink delay in sampled Γ_{e_a} is, therefore, the smallest (largest) among the minimum (maximum) sink delays obtained in this fashion.

Let $t_{e_a}^{\max}$ and $t_{e_b}^{\max}$ ($t_{e_a}^{\min}$ and $t_{e_b}^{\min}$) denote the maximum (minimum) sink delays in Γ_{e_a} and Γ_{e_b}, respectively. The sampled delays in Γ_{e_a} and Γ_{e_b} would, therefore, lie in the range $[t_{e_a}^{\min}, t_{e_a}^{\max}] \cap [t_{e_b}^{\min}, t_{e_b}^{\max}]$. Considering a sampled delay t^+ that is in this range, we now have to compute the range of capacitances that can achieve this delay. For each (C, t) from the sampled Γ_a, we can solve for w, the wire width of e_a, that achieves the sampled delay t^+ as follows:

$$w = \frac{r \cdot |e_a| \cdot C}{t^+ - t - \frac{r \cdot c \cdot |e_a|^2}{2}}$$

If $w_{\min} \leq w \leq w_{\max}$, $C^+ = C + |e_a| \cdot w \cdot c$ is in the capacitance range. The capacitance range(s) for the sampled delay t^+ is obtained by clustering the C^+'s obtained in this fashion into appropriate group(s). The solution space Γ_{e_a} is similar to that shown in the rightmost plot of Figure 13.50.

The sampled Γ_{e_b} can be obtained in a similar fashion, and together, Γ_{e_a} and Γ_{e_b} can be merged to obtain Γ_v, where v is the parent node of a and b. It is again important for each solution in Γ_v to keep track of the contributing solutions from Γ_{e_a} and Γ_{e_b}. The process continues until Γ_{s_0} is obtained.

Among the sampled delays in Γ_{s_0}, the one that meets the target delay and has the lowest capacitance is chosen. Because we have kept track of the solutions from the child nodes that contribute to a solution of the parent node, we can easily perform a top-down traversal to determine the wire width of each edge in the zero-skew routing tree.

Remark: Dynamic Programming for Delay Minimization: The biggest difference between wire sizing for delay minimization and zero-skew is that for delay minimization, the dynamic programming approach can apply pruning rules to weed out suboptimal solutions, greatly reducing the search space. Consider two solutions (C, t) and (C', t') of a tree, where the delays t and t' correspond to the maximum sink delays in these two solutions. If $C < C'$ and $t < t')$, there is no reason to consider (C', t') any further in the bottom-up process, because solutions that are built on top of it are always inferior to corresponding solutions constructed upon (C, t).

13.3.4.4 *Cross-link insertion*

Clock trees provide only one unique path from a clock source to any clock sink. Because there is a lack of redundancy in a tree structure, it is difficult to compensate for the effects of variations. The cross-link insertion technique is a promising approach that introduces redundancy while retaining the low wiring cost of a clock tree. The insertion of cross-links or interconnect wires between some nodes of a clock tree reduces the skew variability. We show examples of such clock trees in Figures 13.15 and 13.51.

Given an RC network $G = (V, E)$, we can decompose the network into a spanning tree $T = (V, E_T)$ and a set of link edges (chords) $E_L = E \backslash E_T$. In Figure 13.51 for example, the only edge in E_L is the cross-link l inserted between nodes u and w. Let R_l and C_l be the resistance and capacitance of l, it can be represented by a π-type RC element (see Figure 13.18) with two capacitors $C_l/2$ at the two ends of a resistor R_l.

Let t_i denote the Elmore delay at any node i of $T = (V, E_T)$. With the link l inserted between u and w, the delay at node i changes. We examine the changes caused by the additions of the capacitors and resistor separately. The addition of only the capacitors is fairly straightforward, because the topology remains a tree. The Elmore delay at node i after the addition of capacitors at u and w, denoted as $\tilde{t}_i$, is

$$\tilde{t}_i = t_i + \frac{C_l}{2}(R_{i,u} + R_{i,w}),$$

where $R_{i,u} = R_{Path(s_0, i) \cap Path(s_0, u)}$ $(R_{i,w} = R_{Path(s_0, i) \cap Path(s_0, w)})$ is the common resistance between the paths $Path(s_0, i)$ and $Path(s_0, u)$ $(Path(s_0, i)$ and $Path(s_0, w))$.

When the resistor R_l is inserted, the delay at node i is changed to $\hat{t}_i$ from $\tilde{t}_i$ as follows [Chan 1990]:

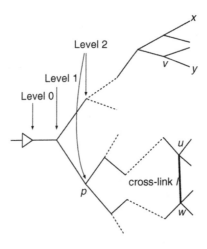

FIGURE 13.51

A clock tree with a cross-link inserted.

$$\hat{t}_i = \tilde{t}_i - \frac{\tilde{t}_u - \tilde{t}_w}{R_l + r_u - r_w} r_i,$$

where resistance r_i is the Elmore delay at node i when the capacitances at nodes u and w are $C_u = 1$ and $C_w = -1$, respectively, and all other node capacitances are zero.

Example 13.4 Adding a Cross-Link: Consider the addition of a cross-link between nodes v and x in Figure 13.17, as shown in Figure 13.52. Let the link capacitance and resistance be C_l and R_l, respectively. First, we add the link capacitance and update the Elmore delays of various nodes:

$$\tilde{t}_y = t_y + \frac{C_l}{2}(R_{y,v} + R_{y,x}), \quad \tilde{t}_z = t_z + \frac{C_l}{2}(R_{z,v} + R_{z,x}),$$

$$\tilde{t}_v = t_v + \frac{C_l}{2}(R_{v,v} + R_{v,x}), \quad \tilde{t}_x = t_x + \frac{C_l}{2}(R_{x,v} + R_{x,x}),$$

where $R_{y,v} = R_u + R_v$, $R_{y,x} = R_u$, $R_{z,v} = R_u + R_v$, $R_{z,x} = R_u$, $R_{v,v} = R_u + R_v$, $R_{v,x} = R_u$, and $R_{x,x} = R_u + R_x$.

Now, we compute the r's, the Elmore delays when $C_v = 1$, $C_x = -1$, and all other node capacitances are zero.

$$r_v = R_u(1 + (-1)) + R_v \cdot 1 = R_v, \quad r_x = R_u(1 + (-1)) + R_v \cdot (-1) = -R_x,$$
$$r_y = R_u(1 + (-1)) + R_v \cdot 1 + R_y \cdot 0 = R_v, \quad r_z = R_u(1 + (-1)) + R_v \cdot 1 + R_z \cdot 0 = R_v.$$

The computation of the new Elmore delays after link insertion is now trivial:

$$\hat{t}_y = \tilde{t}_y - \frac{\tilde{t}_v - \tilde{t}_x}{R_l + r_v - r_x} r_y, \quad \hat{t}_z = \tilde{t}_z - \frac{\tilde{t}_v - \tilde{t}_x}{R_l + r_v - r_x} r_z$$

$$\hat{t}_x = \tilde{t}_x - \frac{\tilde{t}_v - \tilde{t}_x}{R_l + r_v - r_x} r_x, \quad \hat{t}_v = \tilde{t}_v - \frac{\tilde{t}_v - \tilde{t}_x}{R_l + r_v - r_x} r_v$$

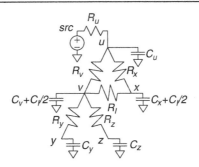

FIGURE 13.52

Insertion of a cross-link to the RC tree in Figure 13.17.

Now, we examine the effects of the additions of capacitors and resistor on the skew between u and w. The new skew between u and w, denoted as $\widehat{skew}_{u,w}$, is

$$\widehat{skew}_{u,w} = \frac{R_l}{R_l + r_u - r_w}\left(skew_{u,w} + \frac{C_l}{2}\left(R_{u,u} - R_{w,w}\right)\right)$$

It is clear that the addition of C_l almost always results in a new skew between u and w as $R_{u,u}$ is almost certainly different from $R_{w,w}$. If we neglect C_l from the preceding equation, we may make the following observations:

1. The addition of R_l reduces the magnitude of the skew between u and w (i.e., $|\widehat{skew}_{u,w}| < |skew_{u,w}|$). In other words, the skew variability of u and w is reduced. This stems from the fact that $r_u = R_{u,u} - R_{u,w}$ and $r_w = R_{u,w} - R_{w,w}$ are the Elmore delays obtained by setting $C_u = 1$ and $C_w = -1$ and zeroing all other node capacitances.
2. For the special case in which $skew_{u,w} = 0$, the skew remains zero after the addition of R_l.
3. The farther u and w are from their youngest (nearest) common ancestor, the smaller the skew variability.

These observations motivate the following procedure for link insertion [Rajaram 2004] in which links are inserted only between nodes that have zero skews (Observation 2). Moreover, only sink pairs are considered (Observation 3). To overcome the effects caused by the link capacitance, the clock tree is tuned after we divide the link capacitance and add them to the two nodes. The purpose of the tuning is to restore the original skews of all sink pairs in the tree. This can be accomplished easily by applying the DME algorithm. Consequently, the addition of link resistance will either maintain the zero skew between the two nodes or reduce the skew variability of the two nodes. Because the tuning process can be expensive, all link capacitances are inserted simultaneously, and the tuning is performed only once.

The remaining problem is that of selecting the set of sink pairs for link insertion. In the approach adopted in [Rajaram 2004], only sink pairs that satisfy the following three selection rules simultaneously are selected:

1. Observation 1 suggests that the smaller the link resistance to loop resistance ratio, $\frac{R_l}{R_l + r_u - r_w}$, the lower the skew variability. Thus, sinks u and w may be selected only when the ratio is no greater than a user-specified threshold α_{max}.
2. The effect of the link capacitance on skew is small when $\beta = |\frac{C_l}{2}(R_{u,u} - R_{w,w})|$ is small. Hence, a link may be inserted between sinks u and w only when β is no greater than a user-specified threshold β_{max}. This rule has the effect of selecting two sinks that are in close proximity and have similar path lengths from the source.
3. Observation 3 suggests that the level of the youngest common ancestor (YCA) for a sink pair should not be too high. Here, the level of a node

refers to the number of edges between the source and the node. Consequently, a sink pair may be selected only when the level of its YCA is no greater than a user-specified threshold γ_{max}.

Other link insertion algorithms can be found in [Rajaram 2004, 2005, 2006].

Example 13.5 Applying Cross-Link Selection Rules: In Figure 13.51, the level of p and the YCA of u and w is 2. Assuming that $\gamma_{max} = 3$ and that the first two selection rules are also satisfied, the cross-link connecting u and w is inserted as shown.

Although the level of the YCA of u and v is $1 \leq \gamma_{max}$, the cross-link between v and v is not inserted, because u and v are too far apart and the link resistance to loop resistance ratio is too high.

We also should not insert a cross-link between x and y, because the level of their YCA is greater than γ_{max}.

13.4 POWER/GROUND NETWORK DESIGN

Now, we shall turn our attention to the analysis and synthesis of ***power/ground*** (P/G) networks. In particular, we focus on the main design challenge highlighted in Section 13.2, namely, *IR* and $L \cdot di/dt$ power supply noise. Some techniques to suppress power supply noise, such as wire sizing, also mitigate the electromigration effects. We follow a similar organization as in the preceding section on clock network design. First, we describe some typical P/G topologies used in designs. Next, we present a random-walk method for the efficient analysis of P/G networks. Last, we focus on the automated synthesis of P/G networks.

13.4.1 Typical power/ground topologies

The design of P/G distribution networks begins with the construction of an appropriate routing topology. First, we consider a simple power distribution scheme that consists of two large concentric rings (one power and one ground) from which comblike structures can be attached (see Figure 13.53). In particular, each comblike structure is commonly used for standard-cell designs. To counter power supply noise and electromigration, the concentric rings, which are connected to V_{DD} and *GND* (or V_{SS}) pads, are typically of a larger width. Similarly, the trunk of a comb is wider than the fingers. Sometimes, the presence of larger modules such as memory blocks or bus lines would destroy the regularity shown in Figure 13.53.

Tree and mesh structures are the most common topologies for power and ground routing. Although the design of a power tree structure (see Figure 13.54)

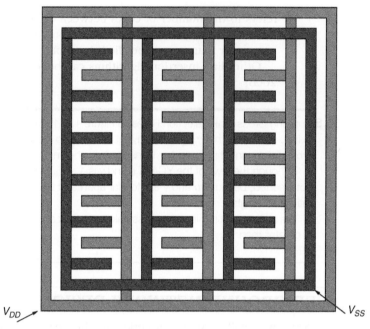

FIGURE 13.53

Interleaved power and ground routing for standard cell designs.

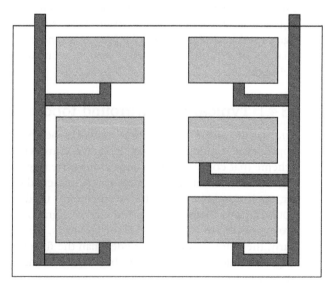

FIGURE 13.54

Local tree–based power distribution technique.

offers efficiency for resource consumption, mesh structures perform better in minimizing voltage and current variations in the supply networks. A mesh structure consists of a rectangular grid of orthogonal wires spanning the whole circuit; see Figure 13.55 for an example of a generic mesh structure. In modern-day microprocessors, the wires of the grid extend across several layers [Singh 2005]. Wires in different layers are connected through vias, solid black blocks joining two metal layers differentiated by shade, as seen in Figure 13.55.

As in the case of clock network design, it is common to see a combination of these various topologies in a single P/G network. Because of its robustness, a mesh structure typically sits at the topmost level in the hierarchy of a P/G network. Comblike structures and tree topologies, with wire width tapering, are usually used for the local distribution of power supply, because they are more frugal on routing resources.

Packaging technologies also play a significant role in enhancing the robustness of power supply. One of the most common interfaces for external power being supplied to an IC is along the periphery of the die (see Figure 13.56).

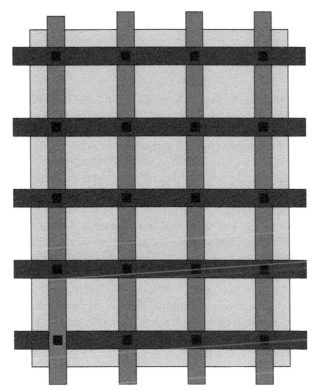

FIGURE 13.55

Typical power mesh structure.

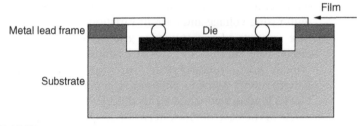

FIGURE 13.56

Use of polymer film wires to connect power to periphery of chip area.

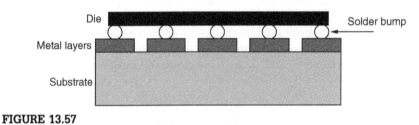

FIGURE 13.57

Flip-chip design to distribute power to any location on chip area.

Such packaging technology is typically used with comblike and tree topologies. In the comblike topology for example, the V_{DD} and GND pads on the periphery of the die can be easily connected to any point on the concentric rings.

Flip-chip packaging makes it possible to supply external power into the interior of the die area directly (see Figure 13.57). For flip-chip mounting, V_{DD} and GND pads are distributed across the topmost layer. The die is flipped upside down and connected to the substrate of the package with solder bumps. Flip-chip packaging provides two benefits: the power supply is available at any position on the chip and the parasitic inductances and capacitances of such packages are lower. Used in conjunction with a power mesh, the V_{DD} (or GND) pads usually reside on the grid points of the mesh. The pad density of a region of the chip is a function of the current demand in that region. Note that flip-chip packaging can also help address the clock distribution problems.

For power minimization, dual- or multiple-V_{DD} designs have become quite popular in recent years. For such a design, high V_{DD} is used for high-performance components, and low V_{DD} powers low-performance components to conserve energy. For a dual-V_{DD} design, for example, three P/G networks are required: V_{DD}^H (the high supply voltage), V_{DD}^L (the low supply voltage), and GND. Both V_{DD}^H and V_{DD}^L could be supplied externally. Alternately, an on-chip voltage regulator can be used to take a single externally supplied voltage $V_{DD}^H = V_{DD}$ and drop it to supply V_{DD}^L.

13.4.2 **Power/ground network analysis**

P/G network analysis can involve either the DC or the transient analysis of the network. DC analysis is required for finding the static *IR* drop or finding steady state values corresponding to the *IR* drop caused by the average current flowing through a power supply network. Transient analysis is concerned with determining the effects of switching activity, which is essentially finding voltage fluctuations on a P/G network. This is required for determining the dynamic *IR* drop and $L \cdot di/dt$ noise.

IR drop analysis of a power supply network can be done efficiently, because the resistance and capacitance matrices are both positive definite. Iterative methods, such as conjugate gradient, can be used for efficiently simulating large power distribution networks [Lin 2001]. The analysis of $L \cdot di/dt$ noise is more difficult, because including the inductance matrix in the modified nodal analysis formulation results in matrices that are not positive definite. Specifically, the use of fast iterative methods directly becomes difficult [Lin 2001]. However, by reverting to the nodal formulation, matrices that are symmetric positive definite [Chen 2001] can be formulated, and the inherent structure/conditioning can be exploited. As in the case of clock network analysis, we will focus on the modeling and analysis of P/G networks that are composed of only resistive and capacitive elements.

Traditionally, P/G networks have been modeled with large linear time-invariant models. The supply lines are modeled as distributed RC segments. For a present-day supply network, such a model can consist of millions of nodes and segments. The sources that supply power are modeled as voltage sources, whereas the drain elements that draw currents can be modeled as time-varying current sources. Figure 13.58 shows the models for power sources and drains. The Modified Nodal Analysis (MNA) of such models yields equations of the form

$$Gx + C\dot{x} = b \tag{13.13}$$

where x is the vector of node voltages, b is the vector of current sources, G is the conductance matrix, and C is the admittance matrix, which consists of capacitive elements. We assume that voltage sources have been transformed into Norton equivalent circuits. Hence, we consider only current sources in b. For an RLC network, x and C would also contain inductor currents and inductive elements, respectively.

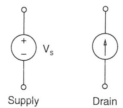

Supply   Drain

FIGURE 13.58

Models for supply and drain in a P/G network.

Applying the standard trapezoidal integration scheme in Equation (13.13) with step size b results in the following linear system of equations:

$$\underbrace{\left(G + \frac{2C}{b}\right) x^{k+1}}_{A} = \underbrace{b^{k+1} + b^k - \left(G - \frac{2C}{b}\right)x^k}_{b} \qquad (13.14)$$

where x^k is the vector of node voltages at time $k \cdot b$ and (*i.e.*, the kth time step).

The solution of Equation (13.14) involves an inversion of the coefficient matrix A. As more devices are packed on a single chip, the size of the power distribution network will increase, and consequently the matrix A tends to be of very large size. A direct solve of Equation (13.14) hence becomes impractical because of the large amount of memory and computation required. However, networks such as mesh and tree will correspond to a matrix A that is sparse and structured. In addition, A is diagonally dominant and symmetric [Kozhaya 2001]. A number of methods have been developed, that exploit these properties of A for analysis and optimization of P/G networks.

We will begin by discussing the specific sparsity structure of the matrix A that arises in the analysis of RC mesh structures. Consider a simple example of a 3×3 power mesh, as shown in Figure 13.59. The sparsity structure of matrix A for this example is as follows:

$$
\begin{pmatrix}
\times & \times & & \times & & & & & \\
\times & \times & \times & & \times & & & & \\
& \times & \times & & & \times & & & \\
\times & & & \times & \times & & \times & & \\
& \times & & \times & \times & \times & & \times & \\
& & \times & & \times & \times & & & \times \\
& & & \times & & & \times & \times & \\
& & & & \times & & \times & \times & \times \\
& & & & & \times & & \times & \times
\end{pmatrix}
$$

A multigrid-like technique was proposed in [Kozhaya 2001, 2002]. In this approach, the original large system is first mapped to a coarse grid of reduced size. The reduced system is solved, and the solution is then mapped back to the original system through interpolation. A method for hierarchical analysis

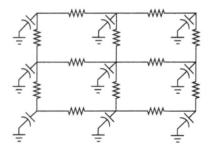

FIGURE 13.59

A 3 × 3 RC mesh structure.

of power distribution networks was provided in [Zhao 2002]. The power grid is first divided into a global grid and several local grids. Macromodels for the local grids are then generated by use of efficient numerical methods. These macromodels can then be used for simulating the global grid. In addition, several techniques based on iterative methods that use sparse linear system techniques have been proposed as well. Techniques based on preconditioned Krylov subspace methods and successive overrelaxation techniques were proposed in [Chen 2001] and [Zhong 2005], respectively.

Finally, we will examine in more detail a stochastic technique based on the random walk method [Qian 2003]. Here the problem of simulating an RC power mesh is translated into a stochastic game that will proceed iteratively until some measure of convergence is achieved. To motivate the procedure, we begin by examining a simple situation where we have a cross-shaped structure created by four resistors with a current source at the central node, as shown in Figure 13.60. The current at the central node x is described through KCL:

$$\sum_{i=1}^{4} g_i(V_i - V_x) = I_x$$

where g_i and V_i are the conductance and voltage for each bordering node, respectively. Thus, we can solve for the voltage at the central node:

$$V_x = \sum_{i=1}^{4} \frac{g_i}{\sum_{j=1}^{4} g_j} V_i - \frac{I_x}{\sum_{j=1}^{4} g_j} \qquad (13.15)$$

If we construct Equation (13.15) for each node in a resistive network and solve the linear equations simultaneously, we arrive at the exact voltage values across the network. Instead, let us form a basic stochastic game that mimics the problem of solving for these voltages.

Imagine a traveler who is currently at a position x (see Figure 13.61). With some probability $p_{x,i}$, the traveler will decide to walk down one of the available roads i. The sum of the probabilities for all of the roads that intersect at x is 1, but the individual probabilities have yet to be determined. Each node that is not labeled "HOME" is considered a hotel that must be visited (with a cost incurred, denoted as m_x), and the traveler must continue along until a final home location

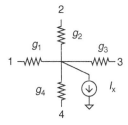

FIGURE 13.60

Portion of a resistive network.

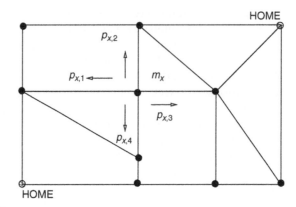

FIGURE 13.61

Random walk game.

is reached (and fixed prize amount is awarded, denoted as m_0). Thus, we can define a gain function f for the amount of money earned on the walk given that we began at a non-home node x that has n bordering locations labeled $1, 2, \ldots, n$:

$$f(x) = \sum_{i=1}^{n} p_{x,i} f(i) - m_x$$

Given that a traveler will remain at the home location once it is reached, we conclude that $f(y) = m_0$ for all home nodes y. An analogy can now be drawn between the resistive V_{DD} network and this stochastic game:

$$p_{x,i} = \frac{g_i}{\sum_{j=1}^{n} g_j}, \quad m_x = \frac{I_x}{\sum_{j=1}^{n} g_j}, \quad m_0 = V_{DD}$$

Repeated iterations of the random walk game will give the values of the voltages for each node $f(x) = V_x$ in the resistive network. To apply this technique to RC power network analysis, we have to replace each capacitor in the RC network by a resistor and a voltage-controlled current source. This simple extension is further developed in the section exercises.

13.4.3 **Power/ground network synthesis**

The synthesis of a P/G network involves the determination of its topology, the placement of power pads, the appropriate sizing of power lines, and the insertion of decoupling capacitances. When designing a power distribution network, it is important that both the maximum allowable current density for each wire and the maximum voltage drop at each node are within specifications. The maximum current density constraints must be satisfied to avoid the wearing out of metal wires, whereas the voltage drop criteria are required for maintaining correct functioning of the IC. Meanwhile, it is important that the wiring resource consumption of a P/G network is minimized.

13.4.3.1 *Topology optimization*

A variety of techniques have been proposed for topology optimization of P/G supply networks [Rothermel 1981; Syed 1982; Xiong 1986; Mitsuhashi 1992; Singh 2004, 2005]. For modern dense circuit designs, power grids can involve millions of wires, and it becomes necessary to formulate fast techniques for the design of such networks.

The synthesis technique in [Singh 2005] is based on the recursive partitioning of the chip area. The procedure starts with assigning a very coarse grid for the whole design, assuming very wide wires and large pitch. The grid is then recursively bipartitioned. The coarse grid in each of the two partitions is refined such that each wide wire is replaced by several narrower wires with smaller pitch.

The amount of computation needed to perform the analysis after each refinement step does not grow substantially as a result of the use of locality (*i.e.*, independently solving only a small local grid in each iteration). The procedure halts when the grid topology is able to satisfy the two specifications described previously.

13.4.3.2 *Power pad assignment*

We will now examine the problem of determining the optimal number of power pads and their locations when dealing with a grid topology. The objective is to minimize the number of power pads necessary to meet the two design considerations discussed earlier. A ***mixed integer linear program*** (MILP) formulation is given in [Zhao 2004]. The MILP formulation involves a set of potential power pad locations, *PC*, and a set of observation points, *OP*, on the power grid. By specifying a subset of the total number of nodes as observation points, a macromodel can be formed for the power grid system, as shown in Figure 13.62. The governing dynamics of this system can be described by the following equation:

$$I = A \cdot V + S \tag{13.16}$$

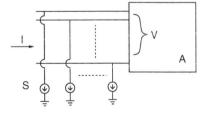

FIGURE 13.62

Macromodel schematic.

where I is a vector of currents flowing into the model through the ports, A is the conductance matrix, and V is the port voltage vector. By introducing 0-1 integer variables, we can construct constraints for the current seen at any of the potential pad locations:

$$
\begin{array}{rcl}
V_i - V_{DD} \cdot z_i & \geq & 0 \\
V_i & \leq & V_{DD} \\
V_i & \geq & V_t \\
I_t \cdot z_i - I_i & \geq & 0 \\
I_i & \geq & 0
\end{array}
\tag{13.17}
$$

Here, I_t is the maximum current allowed through a pad, V_t is the minimum voltage for any node in the power grid, and z_i is a 0-1 variable designating if location i is occupied by a pad. Thus, we can formulate the following optimization problem:

$$
\begin{array}{ll}
\min & \sum_{i \in PC} z_i \quad i \in \{0, 1\} \\
\text{subject to} & V_j \geq V_t \quad j \in OP \\
& I_i \text{ and } V_j \text{ satisfy Equation (13.16) for } j \in OP \cup PC \text{ and } i \in PC \\
& I_i \text{ and } V_j \text{ satisfy Equation (13.17) for } i \in PC
\end{array}
$$

A solution based on a branch-and-bound algorithm can be used for solving such an optimization problem. Since the MILP solution procedure can become highly expensive for a large number of variables, which in this case are the power pad locations, a heuristic can be used to aid in reducing this run time complexity. In [Zhao 2004], for example, a divide-and-conquer approach is used to divide the whole power grid into several partitions and then assign pads for each partition independently.

The heuristic presented in [Oh 1998] simultaneously performs pad assignment and P/G routing. In this approach, an attempt is made to evenly distribute the inductively induced voltage fluctuation across the pads, while minimizing routing of single-pad trees.

13.4.3.3 *Wire width optimization*

We now discuss for a given topology how the wire widths can be optimized to meet both maximum allowable voltage drop and physical breakdown constraints [Chowdhury 1985]. By varying the wire width we will be able to control the resistance of the wire and, therefore, the amount of current flowing through that path or route. This, in turn, will allow us to meet any desired maximum allowable voltage drop required by the design specifications. In this case voltage drop will be measured from the actual power pad to a power supply pin of a gate or module. In addition, for the optimization problem a minimum allowable width for each branch in the network will be imposed. This minimum width is enforced to avoid metal migration effects that will affect the physical reliability of the design. Specifically, if the cross-sectional area increases or the current density decreases, we can assume that the average lifetime of the wire will be longer. Finally, our overall goal (objective function) will be to minimize the total area required for the routing of the power network. This is equivalent to minimizing a weighted sum of the wire width needed to meet the specifications for the network.

This or similar optimization problems can be formulated as mathematical programs, which are then solved numerically. If we consider the formulation shown in [Mitsuhashi 1992] the result is a nonlinear programming approach for solving the problem. Here the possible wire widths are some discrete values based on a realistic fabrication situation. The objective function can be described as follows:

$$A = \sum_{i=1}^{n} w_i l_i = \sum_{i=1}^{n} \frac{|\rho I_i l_i^2|}{x_i}$$

where w_i and l_i are the width and length, respectively, for the n branches of the power network. The term I_i is the maximum current flow across a branch with resistivity ρ, and the voltage drop is denoted by x_i. A maximum allowable voltage drop v_j across any path p_j can be assured through the following constraints:

$$\sum_{i \in p_j} x_i \leq \Delta v_j$$

The minimum width W allowed by the fabrication process can be incorporated through these n conditions:

$$w_i = \frac{\rho I_i l_i}{x_i} \geq W$$

Finally, by putting constraints on the maximum current-wire width ratio, we can avoid the problems caused by metal migration:

$$\frac{I_i}{w_i} = \frac{x_i}{\rho l_i} \leq \sigma_i$$

where σ_i is the maximum allowable current density across branch i.

With an alternate mathematical model, a linear formulation of the optimization problem given in [Tan 2003] facilitates the use of more efficient methods of solution. Furthermore, interested readers can examine [Luenberger 2003], which provides detailed explanations of numerical schemes used to solve the optimization problems described previously.

13.4.3.4 *Decoupling capacitance*

One of the most powerful techniques for reducing power supply noise caused by *IR* drop and $L \cdot di/dt$ noise is to use decoupling capacitors across the power grid [Rao 2001]. It is commonly understood that charge is required to energize a load capacitor, where the charge in this case is supplied by the current flowing through the power supply network. However, such a mechanism introduces power supply noise caused by the intrinsic wire resistance and inductance in the network.

Alternately, charge can also be supplied by another capacitor, which can be placed in close proximity to the load capacitance. Because this additional capacitance is located near the load, it will reduce the overall size of the current loop and act to supply charge to the load. This situation is illustrated in Figure 13.63.

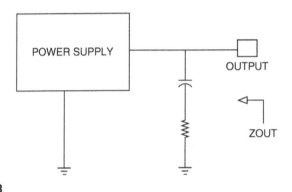

FIGURE 13.63

Decoupling network.

Consequently, current does not need to flow across most of the parasitics for the power supply network, and the current-induced noise is reduced. The needed action for a decoupling capacitor can be efficiently implemented with the gate capacitance of a transistor. With increasing noise values, it is often necessary to allocate as much as 10% of the total chip area to decoupling capacitors. Therefore, it is crucial to estimate the size and area needed for decoupling capacitor assignment.

There have been many studies into decoupling capacitor placement and optimization. First, we address the problem of estimating the amount of decoupling capacitor that each circuit module will require. The amount of decoupling capacitor required by each circuit module at the floorplanning level can be calculated as follows: suppose the maximum voltage noise that a module can tolerate is V_{noise}^{lim}. Let $Q = \int_0^\tau I(t)dt$ denote the maximum total charge that this module will draw from the power supply over τ, the duration that the current waveform $I(t)$ lasts. A greedy scheme for decoupling capacitor estimation is to allocate $C = Q/V_{noise}^{lim}$ to this module.

However, this scheme might result in overallocation of decoupling capacitors, because each decoupling capacitor in practice can be shared by several modules [Zhao 2001]. An iterative approach was proposed instead. An initial estimate of decoupling capacitor required for each module is expressed as

$$\theta = \max\left(1, \frac{V_{noise}}{V_{noise}^{lim}}\right), \quad C = \frac{(1 - \frac{1}{\theta})Q}{V_{noise}^{lim}}$$

where V_{noise} denotes the power supply noise that the module experiences. After C is added to the module, the power noise, V_{noise}, can be recalculated for all the other modules, and the remaining C are allocated accordingly. Because the decoupling capacitor at a module also helps in reducing the power supply noise at a neighboring module, the amount of decoupling capacitor allocated with this procedure can be significantly smaller than that of the greedy scheme.

Several techniques have been proposed for decoupling capacitor budgeting at the floorplanning level [Su 2003]. An iterative procedure involving circuit simulation and floorplanning was proposed in [Chen 1997]. Here, a circuit simulator is used to first analyze the switching noise and determine any hot spots (locations of excessive voltage drop). Then, the sizes of the decoupling capacitors needed to meet voltage drop specifications are determined. On-chip decoupling capacitors are usually implemented as MOS capacitors, which has a unit area capacitance of $C_{ox} = \varepsilon_{ox}/t_{ox}$, where t_{ox} is the oxide thickness and ε_{ox} is the permittivity of SiO_2. Given a decoupling capacitance C, the silicon area required for the fabrication of the decoupling capacitor is $S = C/C_{ox}$.

After this initial circuit simulation stage, the floorplanner calculates the area required to implement these decoupling capacitors, as well as possible locations for their placement. This information can then be sent back to the circuit simulator, which simulates the activity on the new floorplan with the additional added decoupling capacitors. The iterative procedure continues until the power noise is suppressed to within the required limits.

A linear programming–based technique to allocate white space for decoupling capacitors has been proposed in [Zhao 2001]. White space for decoupling capacitor placement is allocated in two stages. The first stage involves allocation of existing white space with a linear program. The objective here is to maximize the utilization of the existing white space. Suppose the white space in the existing floorplan has been partitioned into H rectilinear blocks, called white-space modules. For each white-space module w_k, we use A_k to denote its area and N_{w_k} to denote the set of circuit modules that are adjacent to it. Assume that there are M circuit modules. Let $S^{(j)}$ denote the silicon area required for the decoupling capacitor allocated to circuit module m_j for power supply noise suppression. We use N_{m_j} to denote the set of white-space modules that are adjacent to circuit module m_j. We show a floorplan of 7 rectangular circuit modules and 3 rectilinear white-space modules in Figure 13.64. In this example, $N_{w_1} = \{m_3, m_4, m_5, m_6\}$ and $N_{m_1} = \{w_3\}$.

Let $x_k^{(j)}$ be the amount of white space allocated to circuit module m_j from an adjacent white-space module w_k. The white space utilization problem can be formulated as follows:

$$\max \quad \sum_{k=1}^{H} \sum_{j \in N_{w_k}} x_k^{(j)},$$

$$\text{subject to} \quad \sum_{j \in N_{w_k}} x_k^{(j)} \leq A_k, \quad k = 1, 2, \dots, H$$

$$\sum_{k \in N_{m_j}} x_k^{(j)} \leq S^{(j)}, \quad j = 1, 2, \dots, M$$

$$x_k^{(j)} \geq 0, \quad \forall k, \forall j \in N_{w_k}$$

In the second step, the floorplan is stretched, if necessary, such that extra white space can be created to meet all the decoupling capacitor requirements of the circuit.

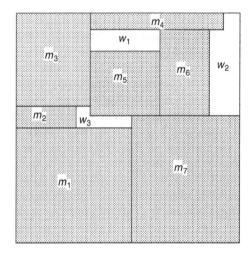

FIGURE 13.64

A floorplan of 7 circuit modules and 3 white-space modules.

All the decoupling capacitor allocation methods require the estimation of the current waveforms that result in hot spots. It is often difficult to accurately estimate the current waveforms in general situations. Hence, the allocation of decoupling capacitor is usually based on the worst-case voltage drops, resulting in an overallocation of resources for decoupling capacitors. To alleviate this problem, several sensitivity-based techniques for decoupling capacitor optimization have been proposed [Bai 2000; Su 2000; Fu 2004].

We now show how the sensitivity of a node voltage with respect to decoupling capacitances can be calculated. In [Su 2003], the authors used the adjoint network method [Director 1969] to calculate critical nodes that can benefit the most from introducing decoupling capacitance. First, an estimate of the current-induced noise at each node in the grid is constructed on the basis of both the possible tunable circuit parameters for the decoupling capacitor and the voltage drop observed at the node.[3] Then, an adjoint circuit is obtained by shorting all voltage sources and opening all current sources in the original power network. Finally, for each node with a non-zero noise measurement, appropriate current sources are applied. This allows for a measurement of the sensitivity to be computed:

$$\frac{\partial Z}{\partial C} = \int_0^T \psi_c(T-t)\dot{v}_c(t)dt \qquad (13.18)$$

---

[3]As the circuits considered in [Su 2003] are standard cell designs, the height of the decoupling capacitors are constrained to be the same as the height of standard cells. Hence, the only tunable circuit parameters are the widths of the decoupling capacitors.

where Z is the sum of the current-induced noise estimates, $\psi_c(\tau)$ is the waveform across the capacitor C in the adjoint circuit, and $v_c(t)$ is the voltage waveform from the original power-grid analysis. This information can then be used to formulate a quadratic program for determining the optimal decoupling capacitor allocation area.

13.5 CONCLUDING REMARKS

There is a rich body of work in the general areas of automated modeling and synthesis of clock distribution networks and P/G delivery systems. This chapter focuses mainly on the modeling and synthesis of clock and power/ground networks composed of RC elements. There are many existing studies that address the problem of synthesizing such networks based on more complete models that also consider inductive parasitics. These studies typically adapt many existing analysis and synthesis techniques covered in this chapter, with varying degrees of success, to address the challenge of incorporating inductive elements in the networks. Parametric variations add further challenges along a different dimension. A bigger challenge is the comodeling and cosynthesis of clock and P/G networks. As the design margin gets smaller, it is getting harder to ignore the interplay of these two large networks.

13.6 EXERCISES

13.1. (Delay Modeling) Let R_{ij} denote the resistance of the common path $Path(src, i) \cap Path(src, j)$ from the source src to nodes i and j in an RC tree T. Besides the Elmore delay, Rubinstein, Penfield, and Horowitz [Rubinstein 1983] also defined the following two delay models:

$$t_p = \sum_{k \in T} R_{kk} C_k$$
$$t_i = \left(\sum_{k \in T} R_{ki}^2 C_k \right) / R_{ii}$$

Although there is only one t_p for a given RC tree T, each node i in T has an associated t_i. It can be shown that $t_i \leq t_{Elmore,i} \leq t_p$.
(a) Write down the pseudo-code to compute t_p in linear time.
(b) Write down the pseudo-code to compute all t_i's in linear time.

13.2. (Skew Scheduling) Consider a constraint graph $G_C(V, E)$ obtained from a set of skew constraints $\mathcal{C}$. Suppose G_C has no negative cycles. Let t_i be the clock delay assigned to clock pin s_i. Suppose the given schedule is not feasible, which implies that for some edge $e_{j,i} \in E$,

$$t_i > t_j + w_{j,i}$$

When this occurs, we apply the following relaxation to change the clock delay assigned to clock pin s_i:

$$t_i \leftarrow t_j + w_{j,i}$$

(a) Show that the process of changing the clock delays assigned to clock pins with iterations of the relaxation step will converge to a feasible clock schedule.

(b) What is the worst-case time complexity for the algorithm to converge to a feasible clock schedule?

13.3. (Zero-Skew Routing) Given two Manhattan arcs, write down the pseudo-code to compute the distance between them.

13.4. (Zero-Skew Routing) Consider two merging segments ms_a and ms_b that are of distance L apart. Suppose they are merged at parent node v, with non-negative wire lengths $|e_a|$ and $|e_b|$, as computed in Section 13.3.3.3. Write down the pseudo-code to compute ms_v, the merging segment at v.

13.5. (Bounded-Skew Routing) Consider two merging regions that are polygons formed by rectilinear line segments and Manhattan arcs, as shown in Figure 13.33, for bounded-skew routing. Write down the pseudo-code to compute the distance L between them.

13.6. (Bounded-Skew Routing) In Section 13.3.3.4, we stated that for bounded-skew routing, the global skew on a boundary segment of a merging region that is a Manhattan arc is a constant.

(a) For the path length delay model, how would the global skew change along a rectilinear boundary segment?

(b) For the Elmore delay model, how would the global skew change along a rectilinear boundary segment?

13.7. (Bounded-Skew Routing) You are given a routine called Compute_merging_region, which takes in two merging regions mr_a and mr_b as input, and outputs a merging region mr_v as a result of performing a bounded-skew merge operation on mr_a and mr_b. Suppose the time complexity of this routine is $O(1)$. Write down the pseudo-code to compute mr_e for all edges $e \in G$ for a given abstract topology G. Recall that mr_e refers to the merging region if the root node of G is relocated to e (see Section 13.3.3.4). Assume that you already have the merging regions for all nodes in G. The time complexity of your algorithm should be linear.

13.8. (Bounded-Skew Routing) Use the bucket decomposition in Section 13.3.3.3 in the Greedy-BST/DME algorithm. If you also incorporate the re-rooting algorithm in Exercise 13.7, what is the time complexity of your Greedy-BST/DME algorithm?

13.9. (Useful-Skew Routing) Suppose you have been given a skew schedule. In other words, all skew constraints are equality constraints.

Modify the zero-skew routing tree algorithm in Algorithm 13.4 to perform prescribed useful-skew routing.

13.10. (Buffer Insertion) In Section 13.3.4.1, the merging region mr_{v_a} is computed by computing the merging segments V_a and V'_a, assuming that a buffer is inserted at v and a, respectively. Let C_{T_a} and C_{T_b} denote the capacitances of the DC-connected subtrees rooted at a and b, respectively. Also, let t_a and t_b denote the sink delays in the two subtrees. Let L denote the distance between a and b.

  (a) Write down the respective equations for the distances of V_a and V'_a from a.

  (b) Consider a sampling segment s in mr_{v_a}. Let d_s denote the distance of s from a. Determine $|e_a|$ and the location of the buffer along the interconnect from v to a.

13.11. (Clock Gating) In Section 13.3.4.2, we demonstrated how $C_{T_x}^{tot}$, GC_{T_x}, $GC_{T_x}^{eff}$, UGC_{T_x} and $C_{T_x}^{eff}$ can be updated. However, we ignore the parasitic capacitance T_{cg}/R_{cg} in the process.

  (a) Write down the steps to update these capacitances, taking into account T_{cg}/R_{cg}.

  (b) Suppose instead of a clock gate, we insert a buffer. How do you update these capacitances?

13.12. (Wire Sizing) In Section 13.3.4.3, we use wire sizing to reduce the delay caused by a wire. We will investigate the possibility of the use of wire sizing to slow down a wire in this problem. Consider a wire with 4 segments, with each segment modeled as a lumped RC element. For simplicity, we assume an L-type circuit for the lumped RC element (*i.e.*, the capacitance is at the downstream endpoint of the resistance). Let the width choices form the set $\{1, 2, 3, 4, 5\}$. A width choice of i has a wire resistance of value $1/i$ and a wire capacitance of value i. Determine a wire width assignment that would result in the longest wire delay. You may want to write a program to enumerate all possible assignments.

13.13. (Cross-Link Insertion) Assuming that every node in a tree has a field to record its distance (*i.e.*, number of edges) from the root node (see the definition of the level of a node in Section 13.3.4.4). Write down an algorithm to determine the youngest common ancestor (YCA) of two nodes without visiting any edges not on the paths from YCA to the two nodes.

13.14. (Random Walk) Starting from the MNA equations

$$Gx + C\dot{x} = b$$

if a backward Euler approximation (step size h) is assumed we arrive at an equation of the form:

$$\left(G+\frac{C}{b}\right)x^{k}=b^{k}+\frac{C}{b}x^{k-1}$$

(a) With the information provided in section 13.4.2 for RC networks, rewrite this equation for a node x only in terms of conductances g_i, current load I_x, and voltage values V_i.

(b) With the preceding equation, formulate the corresponding stochastic game and describe its meaning. (Hint: first reformulate to solve for V_x on the LHS and then describe what each term on the RHS would represent.)

13.15. **(Decoupling Capacitor)** Consider the floorplan shown in Figure 13.64. Suppose after solving the linear program, you realize that the decoupling capacitance requirement of m_1 cannot be fulfilled. However, there are some slacks in w_1 and w_2. How would you meet the decoupling capacitance requirement of m_1 without increasing the chip area? Suggest an iterative approach on the basis of the linear program in Section 13.4.3.4 to enhance the utilization factor of white space for decoupling capacitor deployment.

13.16. **(Decoupling Capacitor)** For a decoupling capacitor to be effective in countering the power supply noise induced by the switching activity of a circuit module, the decoupling capacitor must be placed within some distance of the circuit module. Modify the linear program in Section 13.4.3.4 to enforce the constraint on the distance between a decoupling capacitor and the circuit module that it is safeguarding.

ACKNOWLEDGMENTS

We acknowledge Ruilin Wang of Purdue University for his contributions to the sections on "Design Considerations" and "Clock Network Design." We also thank Dr. Aiqun Cao of Synopsys, Professor Jiang Hu of Texas A&M University, Professor Sheldon X.-D. Tan of the University of California at Riverside, and Dr. Chung-Wen Albert Tsao of Cadence Design Systems for reviewing the chapter.

REFERENCES
R13.0 Books

[Bakoglu 1990] H. B. Bakoglu, *Circuits, Interconnections, and Packaging for VLSI*, Addison-Wesley, Reading, MA, 1990.

[Cormen 2001] T. H. Cormen, C. E. Leiserson, R. L. Rivest, and C. Stein, *Introduction to Algorithms*, Second Edition, MIT Press, Cambridge, MA, 2001.

[Luenberger 2003] D. Luenberger, *Linear and Nonlinear Programming*, Kluwer Academic Publishers, Boston, 2003.

[Rabaey 2003] J. M. Rabaey, A. Chandrakasan, and B. Nikolič, *Digital Integrated Circuits: A Design Perspective,* Second Edition, Prentice-Hall, Upper Saddle River, NJ, 2003.
[Rao 2001] T. Rao, *Fundamentals of Microsystems Packaging,* McGraw-Hill, New York, 2001.

R13.2 Design Considerations

[Anderson 2001] C. J. Anderson, J. Petrovick, J. M. Keaty, J. Warnock, G. Nussbaum, J. M. Tendier, C. Carter, S. Chu, J. Clabes, J. DiLullo, P. Dudley, P. Harvey, B. Krauter, J. LeBlanc, P.-F. Lu, B. McCredie, G. Plum, P. J. Restle, S. Runyon, M. Scheuermann, S. Schmidt, J. Wagoner, R. Weiss, S. Weitzel, and B. Zoric, Physical design of a fourth-generation POWER GHz microprocessor, in *Proc. IEEE Int. Solid-State Circuits Conf.,* pp. 232–233, February 2001.
[Black 1969] J. R. Black, Electromigration—A brief survey and some recent results, *IEEE Trans. on Electron Devices,* 16(4), pp. 338–347, April 1969.
[Lin 2001] S. Lin and N. Chang, Challenges in power-ground integrity, in *Proc. IEEE/ACM Int. Conf. on Computer Aided Design,* pp. 651–654, November 2001.

R13.3 Clock Network Design

[Anderson 2001] C. J. Anderson, J. Petrovick, J. M. Keaty, J. Warnock, G. Nussbaum, J. M. Tendier, C. Carter, S. Chu, J. Clabes, J. DiLullo, P. Dudley, P. Harvey, B. Krauter, J. LeBlanc, P.-F. Lu, B. McCredie, G. Plum, P. J. Restle, S. Runyon, M. Scheuermann, S. Schmidt, J. Wagoner, R. Weiss, S. Weitzel, and B. Zoric, Physical design of a fourth-generation POWER GHz microprocessor, in *Proc. IEEE Int. Solid-State Circuits Conf.,* pp. 232–233, February 2001.
[Bailey 1998] D. Bailey and B. Benschneider, Clocking design and analysis for a 600-MHz Alpha microprocessor, *IEEE J. on Solid-State Circuits,* 33(11), pp. 1627–1633, November 1998.
[Boese 1992] K. D. Boese and A. B. Kahng, Zero-skew clock routing trees with minimum wire-length, in *Proc. IEEE Int. ASIC Conf.,* pp. 1.1.1–1.1.5, September 1992.
[Boese 1995] K. D. Boese, A. B. Kahng, B. A. McCoy, and G. Robins, Near optimal critical sink routing tree constructions, *IEEE Trans. on Computer- Aided Design,* 14(12), pp. 1417–1436, December 1995.
[Chan 1990] P. K. Chan and K. Karplus, Computing signal delay in general RC networks by tree/link partitioning, *IEEE Trans. on Computer-Aided Design,* 9(8), pp. 898–902, August 1990.
[Chao 1992] T.-H. Chao, Y.-C. H. Hsu, and J.-M. Ho, Zero skew clock net routing, in *Proc. ACM/IEEE Design Automation Conf.,* pp. 518–523, June 1992.
[Chao 2008] W.-C. Chao and W.-K. Mak, Low-power gated and buffered clock network construction, *ACM Trans. on Design Automation of Electronic Systems,* 13(1), Article No. 20, pp. 1–20, January 2008.
[Chen 1996] Y. P. Chen and D. F. Wong, An algorithm for zero-skew clock tree routing with buffer insertion, in *Proc. European Design and Test Conf.,* pp. 230–236, March 1996.
[Chen 2002] C. Chen, C. Kang, and M. Sarrafzadeh, Activity-sensitive clock tree construction for low power, in *Proc. Int. Symp. on Low Power Electronics and Design,* pp. 279–282, August 2002.
[Cong 1995a] J. Cong and C.-K. Koh, Minimum-cost bounded-skew clock routing, in *Proc. IEEE Int. Symp. on Circuits and Systems,* pp. 1.215–1.218, April 1995.
[Cong 1995b] J. Cong, A. B. Kahng, C.-K. Koh, and C.-W. A. Tsao, Bounded skew clock and Steiner routing under Elmore delay, in *Proc. IEEE/ACM Int. Conf. on Computer Aided Design,* pp. 66–71, November 1995.
[Cong 1996] J. Cong and L. He, Optimal wiresizing for interconnects with multiple sources, *ACM Trans. on Design Automation of Electronic Systems,* 1(4), pp. 478–511, October 1996.
[Cong 1998] J. Cong, A. B. Kahng, C.-K. Koh, and C.-W. A. Tsao, Bounded-skew clock and Steiner routing, *ACM Trans. on Design Automation of Electronic System,* 3(3), pp. 341–388, July 1998.

[Edahiro 1991] M. Edahiro, Minimum skew and minimum path length routing in VLSI layout design, *NEC Research and Development*, 32(4), pp. 569-575, October 1991.

[Edahiro 1992] M. Edahiro and T. Yoshimura, Minimum path-length equi-distant routing, in *Proc. IEEE Asia-Pacific Conf. on Circuits and Systems*, pp. 41-46, December 1992.

[Edahiro 1993a] M. Edahiro, A clustering-based optimization algorithm in zero skew routing, in *Proc. ACM/IEEE Design Automation Conf.*, pp. 612-616, June 1993.

[Edahiro 1993b] M. Edahiro, Delay minimization for zero-skew routing, in *Proc. IEEE/ACM Int. Conf. on Computer Aided Design*, pp. 563-566, November 1993.

[Edahiro 1994] M. Edahiro, An efficient zero-skew routing algorithm, in *Proc. ACM/IEEE Design Automation Conf.*, pp. 375-380, June 1994.

[Elmore 1948] W. C. Elmore, The transient response of damped linear networks with particular regard to wide-band amplifiers, *J. of Applied Physics*, 19(1), pp. 55-63, January 1948.

[Fisher 1982] A. L. Fisher and H. T. Kung, Synchronizing large systolic arrays, in *Proc. SPIE*, 341 pp. 44-52, May 1982.

[Geannopoulos 1998] G. Geannopoulos and X. Dai, An adaptive digital deskewing circuit for clock distribution networks, in *Proc. IEEE Int. Solid-State Circuits Conf.*, pp. 400-401, February 1998.

[Gupta 1997] R. Gupta, B. Tutuianu, and L. T. Pileggi, The Elmore delay as a bound for RC trees with generalized input signals, *IEEE Trans. on Computer-Aided Design*, 16(1), pp. 95-104, January 1997.

[Guthaus 2006] M. R. Guthaus, D. Sylvester, and R. B. Brown, Clock buffer and wire sizing using sequential programming, in *Proc. ACM/IEEE Design Automation Conf.*, pp. 1041-1046, July 2006.

[Huang 1995] J. H. Huang, A. B. Kahng, and C.-W. A. Tsao, On the bounded skew routing tree problem, in *Proc. ACM/IEEE Design Automation Conf.*, pp. 508-513, June 1995.

[Kung 1982] S. Y. Kung and R. J. Gal-Ezer, Synchronous versus asynchronous computation in very large scale integrated VLSI array processors, in *Proc. SPIE*, 341 pp. 53-65, May 1982.

[Kurd 2001] N. A. Kurd, J. S. Barkarullah, R. O. Dizon, T. D. Fletcher, and P. D. Madland, A multi-gigahertz clocking scheme for the Pentium(R) 4 microprocessor, *IEEE J. of Solid-State Circuits*, 36(11), pp. 1647-1653, November 2001.

[Lam 2002] W.-C. D. Lam, C.-K. Koh, and C.-W. A. Tsao, Power supply noise suppression via clock skew scheduling, in *Proc. IEEE Int. Symp. on Quality of Electronic Design*, pp. 355-360, March 2002.

[Lam 2003] W.-C. D. Lam, C.-K. Koh, and C.-W. A. Tsao, Clock scheduling for power supply noise suppression using genetic algorithm with selective gene therapy, in *Proc. IEEE Int. Symp. on Quality of Electronic Design*, pp. 327-332, March 2003.

[Lam 2005] W.-C. D. Lam, J. Jain, C.-K. Koh, V. Balakrishnan, and Y. Chen, Statistical based link insertion for robust clock network design, in *Proc. IEEE/ACM Int. Conf. on Computer Aided Design*, pp. 587-590, November 2005.

[Lillis 1995] J. Lillis, C. K. Cheng, and T. T. Y. Lin, Optimal wire sizing and buffer insertion for low power and a generalized delay model, in *Proc. IEEE/ACM Int. Conf. on Computer Aided Design*, pp. 138-143, November 1995.

[Lin 1975] H. C. Lin and L. W. Linholm, An optimized output stage for MOS integrated circuits, *IEEE J. of Solid-State Circuits*, SC-10(2), pp. 106-109, April 1975.

[Nagel 1975] L. W. Nagel, SPICE2: A computer program to simulate semiconductor circuits, *Technical Report ERL-M520*, University of California, Berkeley, CA, May 1975.

[Oh 2001] J. Oh and M. Pedram, Gated clock routing for low-power microprocessor design, *IEEE Trans. on Computer-Aided Design*, 20(6), pp. 715-722, June 2001.

[Rajaram 2004] A. Rajaram, J. Hu, and R. Mahapatra, Reducing clock skew variability via cross links, in *Proc. ACM/IEEE Design Automation Conf.*, pp. 18-23, June 2004.

[Rajaram 2005] A. Rajaram, D. Z. Pan, and J. Hu, Improved algorithms for link based non-tree clock networks for skew variability reduction, in *Proc. ACM Int. Symp. on Physical design*, pp. 55-62, April 2005.

[Rajaram 2006] A. Rajaram and D. Z. Pan, Variation tolerant buffered clock network synthesis with cross links, in *Proc. ACM Int. Symp. on Physical design,* pp. 157-164, April 2006.

[Rubinstein 1983] J. Rubinstein, P. Penfield Jr, and M. A. Horowitz, Signal delay in RC tree networks, *IEEE Trans. on Computer-Aided Design,* CAD-2(3), pp. 202-211, July 1983.

[Tam 2000] S. Tam, S. Rusu, U. N. Desai, R. Kim, J. Zhang, and I. Young, Clock generation and distribution for the first IA-64 microprocessor, *IEEE J. of Solid-State Circuits,* 35(11), pp. 1545-1552, November 2000.

[Téllez 1995] G. E. Téllez, A. Farrahi, and M. Sarrafzadeh, Activity-driven clock design for low power circuits, in *Proc. IEEE/ACM Int. Conf. on Computer Aided Design,* pp. 62-65, November 1995.

[Tsai 2003] J.-L. Tsai, T.-H. Chen, and C. C.-P. Chen, ε-Optimal minimum delay/area zero-skew clock tree wire-sizing in pseudo-polynomial time, *Proc. ACM Int. Symp. on Physical Design,* pp. 166-173, April 2003.

[Tsao 2002] C.-W. A. Tsao and C.-K. Koh, UST/DME: A clock tree router for general skew constraints, *ACM Trans. on Design Automation of Electronic Systems,* 7(3), pp. 359-379, July 2002.

[Tsay 1991] R.-S. Tsay, Exact zero skew, in *Proc. IEEE/ACM Int. Conf. on Computer Aided Design,* pp. 336-339, November 1991.

[van Ginneken 1990] L. P. P. P. van Ginneken, Buffer placement in distributed RC-tree networks for minimal Elmore delay, in *Proc. IEEE Int. Symp. On Circuits and Systems,* pp. 865-868, May 1990.

[Vittal 1995] A. Vittal and M. Marek-Sadowska, Power optimal buffered clock tree design, in *Proc. ACM/IEEE Design Automation Conf.,* pp. 497-502, June 1995.

[Vittal 1996] A. Vittal, H. Ha, F. Brewer, and M. Marek-Sadowska, Clock skew optimization for ground bounce control, in *Proc. IEEE/ACM Int. Conf. on Computer Aided Design,* pp. 395-399, November 1996.

[Vuillod 1996] P. Vuillod, L. Benini, A. Bogliolo, and G. DeMicheli, Clock-skew optimization for peak current reduction, in *Proc. Int. Symp. on Low Power Electronics and Design,* pp. 265-270, August 1996.

[Wang 2005] K. Wang, Y. Ran, H. Jiang, and M. Marek-Sadowska, General skew constrained clock network sizing based on sequential linear programming, *IEEE Trans. on Computer-Aided Design,* 24(5), pp. 773-782, May 2005.

[Xi 1995] J. G. Xi and W. W.-M. Dai, Buffer insertion and sizing under process variations for low power clock distribution, in *Proc. ACM/IEEE Design Automation Conf.,* pp. 491-496, June 1995.

[Xi 1996] J. G. Xi and W. W.-M. Dai, Useful-skew clock routing with gate sizing for low power design, in *Proc. ACM/IEEE Design Automation Conf.,* pp. 383-388, June 1996.

[Yu 2007] Z. Yu, M. C. Papaefthymiou, and X. Liu, Skew spreading for peak current reduction, in *Proc. Great Lakes Symp. on VLSI,* pp. 461-464, March 2007.

R13.4 Power/Ground Network Design

[Bai 2000] G. Bai, S. Bobba, and I. N. Hajj, Simulation and optimization of the power distribution network in VLSI circuits, in *Proc. IEEE/ACM Int. Conf. on Computer Aided Design,* pp. 481-486, November 2000.

[Chen 1997] H. H. Chen and D. D. Ling, Power supply noise analysis methodology for deep-submicron VLSI chip design, in *Proc. ACM/IEEE Design Automation Conf,* pp. 638-643, June 1997.

[Chen 2001] T.-H. Chen and C. C.-P. Chen, Efficient large-scale power grid analysis based on preconditioned Krylov-subspace iterative methods, in *Proc. ACM/IEEE Design Automation Conf.,* pp. 559-562, June 2001.

[Chowdhury 1985] S. Chowdhury and M. Breuer, The construction of minimal area power and ground nets for VLSI circuits, in *Proc. ACM/IEEE Design Automation Conf.,* pp. 794-797, June 1985.

[Director 1969] S. W. Director and R. A. Rohrer, The generalized adjoint network and network sensitivities, *IEEE Trans. on Circuit Theory*, 16(3), pp. 318-323, August 1969.

[Fu 2004] J. Fu, Z. Luo, X. Hong, Y. Cai, S. X.-D. Tan, and Z. Pan, A fast decoupling capacitor budgeting algorithm for robust on-chip power delivery, in *IEICE Trans. on Fundamentals of Electronics, Communications and Computer Science*, E87-A(12), pp. 3273-3280, December 2004.

[Kozhaya 2001] J. N. Kozhaya, S. R. Nassif, and F. N. Najm, Multigrid-like technique for power grid analysis, in *Proc. IEEE/ACM Int. Conf. on Computer Aided Design*, pp. 480-487, November 2001.

[Kozhaya 2002] J. Kozhaya, S. Nassif, and F. Najm, A multigrid-like technique for power grid analysis, *IEEE Trans. on Computer-Aided Design*, 21(10), pp. 1148-1160, October 2002.

[Lin 2001] S. Lin and N. Chang, Challenges in power-ground integrity, in *Proc. IEEE/ACM Int. Conf. on Computer Aided Design*, pp. 651-654, November 2001.

[Mitsuhashi 1992] T. Mitsuhashi and E. Kuh, Power and ground network topology optimization for cell based VLSIs, in *Proc. ACM/IEEE Design Automation Conf.*, pp. 524-529, June 1992.

[Oh 1998] J. Oh and M. Pedram, Multi-pad power/ground network design for uniform distribution of ground bounce, in *Proc. ACM/IEEE Design Automation Conf.*, pp. 287-290, June 1998.

[Qian 2003] H. Qian, S. Nassif, and S. Sapatnekar, Random walks in a supply network, in *Proc. ACM/IEEE Design Automation Conf.*, pp. 93-98, June 2003.

[Rothermel 1981] H.-J. Rothermel and D. A. Mlynski, Computation of power supply nets in VLSI layout, in *Proc. ACM/IEEE Design Automation Conf.*, pp. 37-42, June 1981.

[Singh 2004] J. Singh and S. S. Sapatnekar, Topology optimization of structured power/ground networks, in *Proc. ACM Int. Symp. on Physical design*, pp. 116-123, April 2004.

[Singh 2005] J. Singh and S. S. Sapatnekar, A fast algorithm for power grid design, in *Proc. ACM Int. Symp. on Physical design*, pp. 70-77, April 2005.

[Su 2000] H. Su, K. Gala, and S. Sapatnekar, Fast analysis and optimization of power/ground networks, in *Proc. IEEE/ACM Int. Conf. on Computer Aided Design*, pp. 447-480, November 2000.

[Su 2003] H. Su, S. Sapatnekar, and S. Nassif, Optimal decoupling capacitor sizing and placement for standard-cell layout designs, *IEEE Trans. On Computer-Aided Design*, 22(4), pp. 428-436, April 2003.

[Syed 1982] Z. A. Syed and A. E. Carnal, Single layer routing of power and ground networks in integrated circuits, *J. of Digital Systems*, VI(1), pp. 53-63, Spring 1982.

[Tan 2003] X.-D. Tan, C.-J. Shi, and F. J.-C. Lee, Reliability-constrained area optimization of VLSI power/ground networks via sequence of linear programmings, *IEEE Trans. on Computer-Aided Design*, 22(12), pp. 1678-1684, December 2003.

[Xiong 1986] X.-M. Xiong and E. S. Kuh, The scan line approach to power and ground routing, in *Proc. IEEE/ACM Int. Conf. on Computer Aided Design*, pp. 6-9, November 1986.

[Zhao 2001] S. Zhao, K. Roy, and C.-K. Koh, Decoupling capacitance allocation for power supply noise suppression, in *Proc. ACM Int. Symp. on Physical design*, pp. 66-71, April 2001.

[Zhao 2002] M. Zhao, R. Panda, S. Sapatnekar, and D. Blaauw, Hierarchical analysis of power distribution networks, *IEEE Trans. on Computer-Aided Design*, 21(2), pp. 159-168, February 2002.

[Zhao 2004] M. Zhao, Y. Fu, V. Zolotov, S. Sundareswaran, and R. Panda, Optimal placement of power supply pads and pins, in *Proc. ACM/IEEE Design Automation Conf.*, pp. 165-170, June 2004.

[Zhong 2005] Y. Zhong and M. D. F. Wong, Fast algorithms for IR drop analysis in large power grid, in *Proc. IEEE/ACM Int. Conf. on Computer Aided Design*, pp. 351-357, November 2005.

Fault simulation and test generation

James C.-M. Li
National Taiwan University, Taipei, Taiwan

Michael S. Hsiao
Virginia Tech, Blacksburg, Virginia

ABOUT THIS CHAPTER

Very large-scale integration (VLSI) circuits can be defective because of the imperfect manufacturing process. One of the most important tasks in VLSI testing is to minimize the number of defective chips shipped to customers. The quality of test patterns is critical in determining the thoroughness of testing. This requires the assessment of the quality of test patterns either developed manually or generated automatically so that a desired product quality can be achieved.

This chapter consists of two major VLSI testing topics: fault simulation and test generation. In fault simulation, we start with a discussion on fault collapsing. After an introduction of equivalent faults and dominant faults, the serial, parallel, concurrent, and differential fault simulation techniques are described, followed by qualitative comparisons between their advantages and drawbacks. These techniques trade accuracy for reduced execution time, which is crucial for managing the complexity of large designs. After fault simulation, basic *automatic test pattern generation* (ATPG) techniques, including Boolean difference, PODEM, and FAN, are described in detail. Advanced test generation techniques are also introduced to meet the demand for quality testing, including sequential ATPG, delay fault ATPG, and bridging fault ATPG. Throughout this chapter, the reader will learn about the major fault simulation and test generation techniques. This background will be valuable in selecting the test method that best meets the design needs and understands the relationship between test patterns and product quality.

14.1 INTRODUCTION

Simulation techniques have been widely used in VLSI designs for digital circuit verification, test development, design debug, and fault diagnosis. During the design stage, **logic simulation**, which has been extensively discussed in

Chapter 8, is performed to help verify whether the design meets its specifications and contains any design errors. It also helps locate design errors that may have escaped from detection during design debug.

Once the design meets its specifications and is ready for physical implementation, one must ensure that the manufactured devices will function as intended and no defective parts are shipped to customers. To achieve high product quality, typically with a defect level less than 500 *defective parts per million* (DPM), quality test patterns must be developed. At present, as we move to the nanometer design era, this has required applying **fault simulation** and *automatic test pattern generation* (ATPG) to the design that has been embedded with *design for testability* (DFT) features during test development.

In contrast to logic simulation, **fault simulation** is used to measure the effectiveness of test patterns in detecting defects that might have been introduced during the manufacturing process. This requires simulating the faulty behavior of the circuit in detecting the modeled faults of interest. (For this reason, logic simulation is generally referred to as **fault-free simulation**.) Furthermore, fault simulation is an integral component of any ATPG program.

The major difference between logic simulation and fault simulation lies in the nature of the non-idealities they deal with. Logic simulation is intended for checking whether the circuit's responses to a given set of input vectors conform to the given specifications or a known good design as the reference. Design errors may be introduced by human designers or *electronic design automation* (EDA) tools, and they should be caught before physical implementation. **Fault simulation**, on the other hand, is concerned with checking the behavior of fabricated circuits as a consequence of inevitable fabrication process imperfections. The manufacturing defects (*e.g.*, wire shorts and opens), if present, may cause the circuits to behave differently from the expected behavior. Fault simulation generally assumes that the design is functionally correct (*i.e.*, free of design errors), and it is targeted at capturing manufacturing defects. However, we note that fault simulation methods may be applied during the design verification stage as well.

The capability of fault simulation to predict the faulty circuit behavior is of great importance for test and diagnosis. First, fault simulation evaluates the effectiveness of a set of test patterns in detecting manufacturing defects. The quality of a test set is expressed in terms of **fault coverage**, which is the ratio of detected faults to the total number of faults in the circuit. In practice, the designer uses a **fault simulator** to evaluate the fault coverage of a set of input stimuli (test vectors or test patterns) with respect to the modeled faults of interest. Because fault simulation concerns the fault coverage of a test set rather than the detection of design bugs, it is also termed **fault grading**. Low fault coverage test patterns will jeopardize the manufacturing test quality and eventually lead to unacceptable field returns from customers. Second, fault simulation helps to identify undetected faults, which is especially important when the achieved fault coverage is below an acceptable level. In this case, either the designer or the ATPG has to generate additional test vectors to improve the fault coverage (*i.e.*, to detect those remaining undetected faults). Third, as part of the **test compaction** process, fault simulation identifies redundant test patterns, which may be discarded with no negative impact on the fault coverage. With the preceding capabilities and applications, fault simulation is one of the crucial components of ATPG. In fact, the implementation of an ATPG program usually starts with the fault simulator. Finally, fault simulation assists **fault diagnosis**, which determines the type and location of faults that best explain the faulty circuit behavior of the device under diagnosis. The fault simulation results are compared with the observed circuit responses to identify the most likely faults. The fault type and location information can then be used as a starting point for locating the defects that cause the circuit malfunction.

Although logic and fault simulators can provide important information about the behavior of the circuit, they require a set of test vectors with which the circuit is simulated. The objective of test generation, then, is the task of producing a set of test vectors that will uncover any defect in a chip. Figure 14.1 illustrates a high-level concept of test generation. In this figure, the circuit at the top is defect free, and for any defective chip that is functionally different from the defect-free one there must exist some input that can differentiate the two. Generating effective test patterns efficiently for a digital circuit is thus the goal of an ATPG system.

Because this problem is extremely difficult, DFT methods have been frequently used to relieve the burden on the ATPG. In this sense, a powerful ATPG can be regarded as the "Holy Grail" in testing, with which all DFT methods could potentially be eliminated. In other words, if the ATPG engine is capable of efficiently delivering high-quality test patterns that achieve high fault coverages and small test sets on large, complex chips, DFT would no longer be necessary.

Because fault simulation can help to determine those faults that could be detected by the same generated test, it becomes an essential component of ATPG. By removing those incidentally detected faults, ATPG is able to significantly

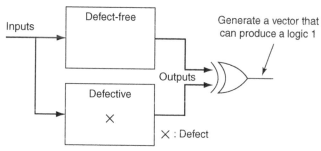

FIGURE 14.1

Conceptual view of test generation.

reduce the number of faults that it needs to consider after the generation of each new test vector, thereby improving the efficiency of the ATPG process.

For some fault models, the circuit layout information is needed. For example, wire delay values are needed to compute the longest paths, and the actual positions of gates and wires are needed to identify those likely bridges. However, because ATPG is a time-consuming process, we would like to start the ATPG process before the layout is available. In this regard, an ATPG may be performed to obtain an initial test set without the layout information. Then, after **place and route**, any faults that require circuit layout information that are undetected by the test set would be identified, and the ATPG can be invoked again to target these specific undetected faults to ensure test quality.

14.2 FAULT COLLAPSING

Fault collapsing reduces the number of faults to be considered in fault simulation and ATPG so the overall run time can be reduced. Two requirements must be met for fault collapsing to become effective. First, fault collapsing must run much faster than fault simulation or ATPG; otherwise, fault collapsing may not be worth doing. Second, the collapsed faults must be representative of all original faults modeled in the circuit. In this section, we introduce two fault-collapsing techniques: **equivalence fault collapsing** and **dominance fault collapsing**. Linear time algorithms are given to meet the first requirement. We illustrate that dominance fault collapsing produces a fewer number of faults than equivalence fault collapsing. However, from a fault coverage accuracy viewpoint, equivalence fault collapsing is more often quoted than dominance fault collapsing, because the former results in a better indication of the test quality.

14.2.1 Equivalence fault collapsing

Let two faults f and g be said to be **functionally equivalent** (or simply **equivalent**) if the faulty outputs of these two faults are identical for any input [McCluskey 1971; Abramovici 1994; Bushnell 2000]. Equivalent faults are **indistinguishable**, because there is no test pattern that can tell them apart. Consider the example of a two-input AND gate shown in Figure 14.2. The good outputs and faulty outputs of the AND gate for all four possible input combinations are listed in Table 14.1. From this table, we can see that A stuck-at zero fault (denoted as $A/0$) and C stuck-at zero fault (denoted as $C/0$) are equivalent.

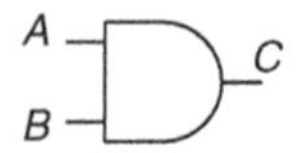

FIGURE 14.2

An example two-input AND gate.

Table 14.1 Good and faulty outputs for Figure 14.2

Input		Output						
A	B	C	A/0	C/0	B/0	A/1	C/1	B/1
0	0	0	0	0	0	0	**1**	0
0	1	0	0	0	0	**1**	1	0
1	0	0	0	0	0	0	1	**1**
1	1	1	**0**	**0**	**0**	1	1	1

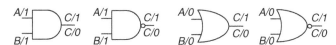

FIGURE 14.3

Equivalence collapsed fault list for four elementary gates.

This is because the faulty outputs of these two faults are always the same for all the four input combinations. On the other hand, the *A* stuck-at one fault (*A/1*) and the *C* stuck-at one fault (*C/1*) are not equivalent, because the input pattern *A* = *B* = 0 can distinguish these two faults. Another input pattern is *A* = 1 and *B* = 0. For clear illustration, the faulty outputs that are different from good outputs are underlined and highlighted in bold.

The equivalence relationship is **symmetric**. This means, if fault *f* is equivalent to fault *g*, then fault *g* is equivalent to fault *f*. The equivalence relationship is also **transitive**. That is, if fault *f* is equivalent to fault *g* and fault *g* is equivalent to fault *h*, then fault *f* is equivalent to fault *h*. For the example given in Figure 14.2, *A/0* fault is equivalent to *B/0* fault, and *B/0* fault is equivalent to *C/0* fault. These three faults {*A/0*, *B/0*, *C/0*} belong to the same **equivalence class**.

Equivalence fault collapsing reduces the set of faults that needs to be considered with the fault equivalence relation. Only one representative fault is selected from every equivalent class. Figure 14.3 shows the **equivalence collapsed fault list** for four types of elementary gates. Originally, there are six faults associated with a two-input AND gate: *A/1*, *A/0*, *B/0*, *B/1*, *C/0*, and *C/1*. After equivalence fault collapsing, the number of faults is reduced to four: *A/1*, *B/1*, *C/1*, and *C/0*. The other types of gates can be examined in the same way. Generally speaking, an *n*-input elementary gate has 2*n* and *n* + 2 stuck-at faults before and after equivalence fault collapsing, respectively. Note that the equivalence fault collapsed fault list is not unique, and there are other ways to collapse the faults than are shown in Figure 14.3. For example, {*A/0*, *A/1*, *B/1*, *C/1*} is another possible way to perform equivalence fault collapsing.

Equivalence fault collapsing can be performed by either functional analysis or structural analysis. Exhaustive functional analysis is time-consuming, because enumeration of 2^n patterns may be needed for an n-input circuit (like Table 14.1 for the AND gate shown in Figure 14.2). Therefore, in the following text, we only demonstrate a linear-time structural analysis to perform equivalence fault collapsing. The resulting equivalence collapsed fault list may not be minimal, but structural analysis is good enough for most applications.

For a fanout-free circuit consisting of elementary gates (such as buffers, inverters AND, OR, NAND, and NOR gates), equivalence fault collapsing can be performed by keeping two kinds of faults: (1) both stuck-at one and stuck-at zero faults on every primary output, and (2) one collapsed fault on each gate input whose stuck value is shown in Figure 14.3. Inverters and buffers should be treated as wires. For the example in Figure 14.4, we keep both $H/0$ and $H/1$ faults on primary output H. We also keep one fault on each gate input, such as $A/0$ and $B/0$ for OR gate G_1, etc. Note that faults on the gate outputs are removed, because they are equivalent to some other faults in the figure. For example, gate G_1 output stuck-at zero fault is equivalent to $C/0$ fault, which is again equivalent to $E/1$ fault, which is in turn equivalent to $H/0$ fault.

For circuits with fanouts, fault collapsing becomes complicated, because faults on the fanout stem are now always equivalent to the faults on the fanout branches. Figure 14.5 shows a circuit with a fanout stem E and two fanout branches L and F. According to Table 14.2, $E/0$ fault is equivalent to $F/0$ fault but not equivalent to $L/0$ fault. Also, none of the stuck-at one faults are equivalent. **Stem analysis** is required to determine equivalent faults on a fanout stem and its branches. However, stem analysis is generally not cost-effective in terms of CPU time, so the details are skipped in this chapter.

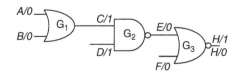

FIGURE 14.4

Equivalence fault collapsing on a fanout-free circuit.

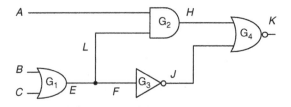

FIGURE 14.5

Equivalence fault collapsing for faults on fanouts.

Table 14.2 Good and faulty outputs for Figure 14.5

Input			Output						
A	B	C	E	E/0	F/0	L/0	E/1	F/1	L/1
0	0	0	0	0	0	0	<u>1</u>	<u>1</u>	0
0	0	1	1	<u>0</u>	<u>0</u>	1	1	1	1
0	1	0	1	<u>0</u>	<u>0</u>	1	1	1	1
0	1	1	1	<u>0</u>	<u>0</u>	1	1	<u>1</u>	1
1	0	0	0	0	0	0	0	1	0
1	0	1	0	0	0	<u>1</u>	0	0	0
1	1	0	0	0	0	<u>1</u>	0	0	0
1	1	1	0	0	0	<u>1</u>	0	0	0

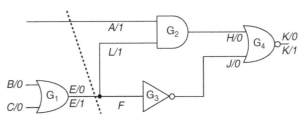

FIGURE 14.6

Equivalence collapsed fault list for Figure 14.5.

To avoid stem analysis, an approximation solution is used to partition the circuit into independent *fanout-free regions* (FFRs). Every fanout stem is treated as a primary output, so both stuck-at one and stuck-at zero faults are included in the collapsed fault list. Algorithm 14.1 introduces a simple equivalence fault-collapsing (simple_EFC) algorithm without stem analysis. Figure 14.6 shows the resulting equivalence collapsed fault list with the simple_EFC algorithm. The circuit is partitioned into two independent fanout-free regions: four faults in one region and six faults in the other. The simple_EFC algorithm reduces the number of faults from 18 to 10. Please note that inverter G_3 is ignored in this algorithm, because its input stuck-at one fault is always equivalent to its output stuck-at zero fault and *vice versa*. Also note that simple_EFC is not the only way to perform fault collapsing; other implementations of fault collapsing are possible.

Algorithm 14.1 A simple equivalence fault-collapsing algorithm

---

simple_EFC (*N*) /*N* is a netlist*/

1. fault_list = {};
2. **foreach** gate or PO or PI *g* in *N*
3.   **if** ((*g* is PO) || (*g* is PI and fanout stem)) **then**
4.   fault_list = fault_list ∪ *g* stuck-at 0 and 1;
5.   **else if** (output of gate *g* is fanout stem) **then**
6.   fault_list = fault_list ∪ *g* output stuck-at 0 and 1;
7.   **end if**
8.   **if** (gate *g* is AND) || (gate *g* is NAND) **then**
9.   fault_list = fault_list ∪ *g* input stuck-at 1;
10.   **else if** (gate *g* is OR) || (gate *g* is NOR) **then**
11.   fault_list = fault_list ∪ *g* input stuck-at 0;
12.   **end if**
13. **end foreach**
14. **return** (fault_list);

The simple_EFC algorithm can complete in linear time because it checks every gate exactly once. However, this algorithm has two drawbacks. First, the result is not optimal, because it lacks stem analysis. For example, Table 14.2 shows that $E/0$ fault is actually equivalent to $K/0$ fault, but they both appear in Figure 14.6. This small error, however, is often acceptable in most cases. Second, the relationship between the original (uncollapsed) faults and the corresponding collapsed faults is lost. For example, the link is lost between the four faults {$F/0$, $J/1$, $H/1$, $K/0$} in the same **equivalence class** and their collapsed fault $K/0$. The relation between **uncollapsed faults** and **collapsed faults** is needed when calculating the fault coverage of the circuit. Fault coverage can be calculated on the basis of either the uncollapsed faults or the collapsed faults. The **uncollapsed fault coverage** is the number of detected uncollapsed faults over the total number of uncollapsed faults, whereas the **collapsed fault coverage** is the number of detected collapsed faults over the total number of collapsed faults. Oftentimes, these two numbers are not identical but close to each other. Missing the link between the collapsed faults and the uncollapsed fault makes it difficult to convert the collapsed fault coverage to the uncollapsed fault coverage. However, modern fault simulators and ATPG programs have found an easy way to rebuild the link by performing another pass of linear-time analysis on equivalent faults.

14.2.2 Dominance fault collapsing

The equivalence collapsed fault list can be further compressed with the **fault dominance** relationship. Let the **detecting set** of fault f (denoted as T_f) be the set of all test patterns that detect fault f. Fault f dominates fault g if the detecting set of fault f contains that of fault g. That means, $T_f \supseteq T_g$. For the example in Figure 14.2, fault $C/1$ dominates fault $A/1$ because the detecting set of $C/1$ {00, 01, 10} contains the detecting set of $A/1$ {01}. The dominance relation is not symmetric but is transitive.

If a test pattern detects the **dominated fault**, then it must detect the corresponding **dominating fault**. To reduce the run time, the dominating faults can be removed from the fault list. The reduction of fault list with the fault dominance relation is called **dominance fault collapsing**. If two faults are equivalent, then they dominate each other. Therefore, the number of dominance-collapsed faults must be smaller or equal to that of equivalence-collapsed faults.

Figure 14.7 shows the **dominance collapsed fault list** of four elementary gates. Originally, there are four equivalence-collapsed faults for a two-input AND gate: $A/1$, $B/1$, $C/0$, and $C/1$. After dominance fault collapsing, the number of faults is reduced to three: $A/1$, $B/1$, and $C/0$. The other types of gates can be examined in the same way. Generally speaking, for an n-input elementary gate, there are $n + 1$ stuck-at faults after dominance fault collapsing.

For a fanout-free circuit consisting of elementary gates (such as buffers, inverters AND, OR, NAND, and NOR gates), dominance fault collapsing can be performed according to the following two rules: (1) one collapsed fault on every primary input whose value is shown in Figure 14.7, and (2) one collapsed fault on each gate output whose gate inputs are all primary inputs. Those gates whose inputs are all primary inputs are called **input gates**. Inverters and buffers should be treated as wires. Figure 14.8 shows the dominance collapsed fault list of the example fanout-free circuit. Note that no fault is needed on G_2 gate output, because G_2 is not an input gate. $E/0$ fault dominates $C/1$ fault, so the former can be removed. $E/1$ fault is equivalent to $C/0$ fault, which dominates $A/0$ fault, so both $C/0$ and $E/1$ faults are removed. The explanation of the other faults is similar so it is left as an exercise for the readers. This circuit has 14 uncollapsed faults, which are reduced to 8 equivalent faults and then to 5 dominant faults after equivalence and dominance fault collapsing, respectively.

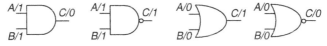

FIGURE 14.7

Dominance collapsed fault list for four elementary gates.

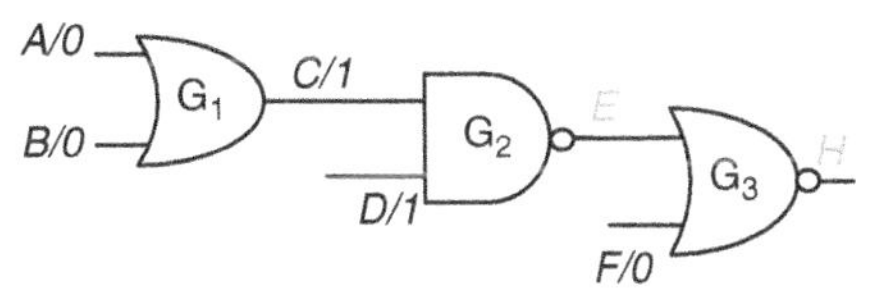

FIGURE 14.8

Dominance fault collapsing on a fanout-free circuit.

Faults on the fanout branches do not always dominate faults on the fanout stem. Consider again the example in Figure 14.5. According to Table 14.2, $F/1$ fault dominates $E/1$ fault. However, $L/1$ fault does not dominate $E/1$ fault. (Actually, $L/1$ fault has an empty detecting set so $L/1$ is a **redundant fault**. More details on redundant faults are given in the test generation section.) Again, stem analysis is needed to determine whether fanout branch faults dominate fanout stem faults.

An approximation method to avoid stem analysis is to partition the circuit into fanout-free regions and perform fault collapsing on each fanout-free region independently. A simple_DFC algorithm is shown in Algorithm 14.2. The dominance fault collapsed result is shown in Figure 14.9. The number of faults is reduced to seven. Without stem analysis, the result is not optimal because $J/0$ is equivalent to $F/1$, which dominates $E/1$.

Algorithm 14.2 A simple dominance fault-collapsing algorithm

---

simple_DFC (*N*) /*N* is a netlist*/

1. fault_list = {};
2. **foreach** gate or PI or PO *g* in *N*
3.   **if** ((*g* is PI and fanout stem) || (*g* is PO and fanout branch)) **then**
4.   fault_list = fault_list ∪ *g* output stuck-at 0 and 1;
5.   **else if** (*g* is gate) **then**
6.   **foreach** gate input *i* of gate *g*
7.   *h* = backtrace inverters starting from *i*;
8.   **if** (*h* is PI or fanout branch) **then** /* rule #1 */
9.   **if** (gate *g* is AND) || (gate *g* is NAND) **then**
10.   fault_list = fault_list ∪ *i* stuck-at 1;
11.   **else if** (gate *g* is OR) || (gate *g* is NOR)
12.   fault_list = fault_list ∪ *i* stuck-at 0;
13.   **end if**
14.   **end if**
15.   **end foreach**
16.   **if** (every input of *g* has a fault) **then** /* rule #2 */
17.   **if** (gate *g* is AND) || (gate *g* is NOR) **then**
18.   fault_list = fault_list ∪ *g* output stuck-at 0;
19.   **else if** (gate *g* is OR) || (gate *g* is NAND) **then**
20.   fault_list = fault_list ∪ *g* output stuck-at 1;
21.   **end if**
22.   **end if**
23.   **end if**
24. **end foreach**
25. **return** (fault_list);

Although the dominance collapsed fault list is smaller than the equivalence collapsed fault list, fault coverage of the former is not as representative as that of the latter. The reason is that a test pattern may detect a dominating fault without

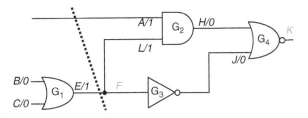

FIGURE 14.9

Dominance collapsed fault list for Figure 14.5.

detecting the dominated fault. For the example given in Figure 14.5, test pattern ABC = 100 does not detect the dominated fault $E/1$ but it detects the dominating fault $F/1$. If the dominance collapsed fault list is used during fault simulation, the dominance collapsed fault coverage may underestimate the test quality. As a result, modern fault simulators and ATPG programs favor the use of equivalence fault collapsing only.

14.3 FAULT SIMULATION

Fault simulation is a more challenging task than logic simulation because of the added dimension of complexity (*i.e.*, the behavior of the circuit containing all the modeled faults must be simulated). When simulating one fault at a time, the amount of computation is approximately proportional to the circuit size, the number of test patterns, and the number of modeled faults. Because the number of modeled faults are roughly proportional to the circuit size, the overall time complexity of fault simulation is $O(pn^2)$, for p test patterns and n logic gates, which becomes infeasible for large circuits. To improve fault simulation performance, various fault simulation techniques have been developed. In this section, we restrict our discussion to the single stuck-at fault model and illustrate the key fault simulation techniques along with qualitative comparisons between their advantages and drawbacks.

14.3.1 Serial fault simulation

Serial fault simulation is the simplest fault simulation technique. It consists of fault-free and faulty circuit simulations. Initially, fault-free logic simulation is performed on the original circuit to obtain the fault-free output responses. The fault-free responses are stored and later used to determine whether a test pattern can detect a fault or not. After fault-free simulation, a serial fault simulator simulates faults one at a time. For each fault, **fault injection** is first performed, which modifies the original circuit to mimic the circuit behavior in the presence of the fault. Then, the faulty circuit is simulated to derive the *faulty* responses of the current fault with respect to the given test patterns. This process repeats until all faults in the fault list have been simulated.

The serial fault simulation process is demonstrated with the example circuit N. In this example, the fault list comprises two faults, A stuck-at one (denoted by f) and J stuck-at zero (denoted by g), which are depicted in Figure 14.10. Note that, although both faults are drawn in the figure, only one fault is present at a time under the single stuck-at fault model. The test set consists of three test patterns (denoted by P_1, P_2, P_3, respectively, and shown in the "Input" columns of Table 14.3).

The serial fault simulator starts from fault-free simulation. The fault-free responses are $K = \{1, 1, 0\}$ for input patterns P_1, P_2, and P_3, respectively. After the fault-free responses are available, fault f is processed—fault injection is achieved by forcing A to a constant one, and the obtained faulty circuit is simulated. The circuit responses for fault f are $K_f = \{0, 0, 0\}$ with respect to the three input patterns. Compared with the fault-free responses (the "Output" column in Table 14.3), it is observed that patterns P_1 and P_2 detect fault f, but pattern P_3 does not. After fault f has been simulated, circuit N is restored by removing fault f. The next fault g is then injected by forcing J to zero. Simulation of the resulting faulty circuit is then performed to obtain the faulty outputs $K_g = \{1, 1, 1\}$ (also listed in Table 14.3). Fault g is detected by pattern P_3 but not P_1 and P_2.

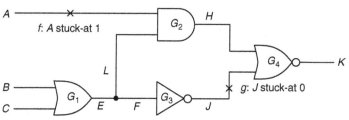

FIGURE 14.10

An example circuit with two faults.

Table 14.3 Serial Fault Simulation Results for Figure 14.10

Pat. #	Input			Internal					Output		
	A	B	C	E	F	L	J	H	K_{good}	K_f	K_g
P_1	0	1	0	1	1	1	0	0	1	0	1
P_2	0	0	1	1	1	1	0	0	1	0	1
P_3	1	0	0	0	0	0	1	0.	0	0	1

In this example, nine simulation runs are performed: three fault-free and six faulty circuit simulations. These nine simulation runs can be divided into three **simulation passes**. In each simulation pass, either the fault-free or the faulty circuit is simulated for the whole test pattern set. Thus, the first simulation pass consists of fault-free simulations for P_1, P_2, and P_3, and the second and third passes correspond to the faulty circuit simulations of faults f and g, respectively, for P_1, P_2, and P_3.

By careful inspection of the simulation results in Table 14.3, one can observe that if we are only concerned with the set of faults that are detected by the test set $\{P_1, P_2, P_3\}$, simulations of the faulty circuit with fault f for patterns P_2 and P_3 are redundant, because f is already detected by P_1. (It is assumed that the test patterns are simulated in the order P_1, P_2, and then P_3.) Halting simulation of detected faults is called **fault dropping**. For the purpose of fault grading, fault dropping dramatically improves fault simulation performance, because most faults are detected after relatively few test patterns have been applied. Fault dropping, however, should be avoided in fault diagnosis applications in which the entire fault simulation results are usually required to facilitate the identification of the fault type and location.

The simplified serial fault simulation flow is depicted in Figure 14.11. Before fault simulation, *fault collapsing* is executed to reduce the size of the fault list, denoted by F. Fault-free simulation is then performed for all test patterns to obtain the correct responses O_g. The algorithm then proceeds to fault simulation. For each fault f in F, if there exists a test pattern whose output response O_f differs from that of the corresponding good circuit O_g, f is removed from F, indicating that it is detected. When all patterns have been simulated, the remaining faults in F are the undetected faults.

The major advantage of serial fault simulation is its ease of implementation—a regular logic simulator plus fault injection and output comparison procedures will suffice. In addition, serial fault simulation can handle a wide range of fault models, as long as the fault effects can be properly injected into the circuit. The major disadvantage of serial fault simulation is its low performance. As will be discussed in the following subsections, practical fault simulation techniques exploit parallelism and/or similarities among the faulty circuits to speed up the fault simulation process.

14.3.2 **Parallel fault simulation**

Similar to parallel logic simulation, fault simulation can take advantage of the bitwise parallelism inherent in the host computer to reduce fault simulation time. For instance, in a 32-bit wide CPU, logic operations (AND, OR, or XOR) can be performed on all 32 bits at once. There are two ways to realize bitwise parallelism in fault simulation: parallelism in faults and parallelism in patterns. These two approaches are referred to as **parallel fault simulation** and **parallel pattern fault simulation**.

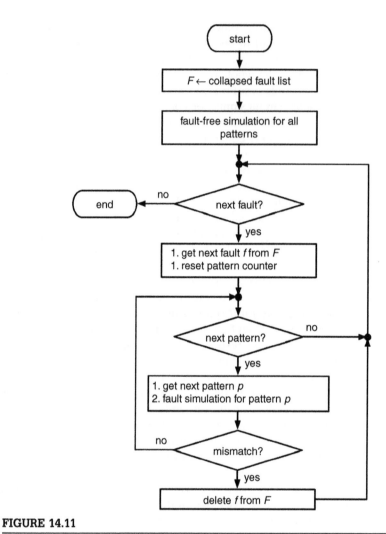

FIGURE 14.11

The serial fault simulation algorithm flow.

14.3.2.1 *Parallel fault simulation*

Parallel fault simulation was proposed in the early 1960s [Seshu 1965]. Assuming that binary logic is used, one bit is sufficient to store the logic value of a signal. Thus, in a host computer that uses w-bit wide data words, each signal is associated with a data word of which w-1 bits are allocated for w-1 faulty circuits, and the remaining bit is reserved for the fault-free circuit. This way, w-1 faulty circuits and one fault-free circuit can be processed in parallel by use of bitwise logic operations, which corresponds to a speedup factor of approximately

w-1 compared with serial fault simulation. A fault is detected if its bit value differs from that of the fault-free circuit at any of the outputs.

We will reuse the example from serial fault simulation to illustrate the parallel fault simulation process. Assuming that the width of a computer word is three bits, the first bit stores the fault-free (FF) circuit response, and the second and third bits store the faulty responses in the presence of faults f and g, respectively. The simulation results are shown in Table 14.4. Because the fault f, A stuck-at 1, uses the second bit, it is injected by forcing the second bit of the data word of signal A to 1 during fault simulation (shown in the "A_f" column with the forced value underlined—the "A" column corresponds to the fault-free case). Similarly, the "J_g" column depicts how fault g is injected by forcing the third bit to 0.

As we have mentioned, parallel fault simulation is performed by use of bitwise logic operations. For example, the logic value of signal H is obtained by a bitwise AND operation on the data words of signals A and L. (A, J, and L are circled in Table 14.4.) The faulty response of the first pattern is {1, **0**, 1}. This means that fault f is detected (the second bit), but fault g (the third bit) is not. Similarly, the outputs of P_2 and P_3 are {1 **0** 1} and {0 0 **1**}, respectively. In this example, three simulations (in one simulation pass) are performed. Compared with serial fault simulation, which requires nine simulations, parallel fault simulation saves two thirds of the simulation time.

To perform parallel fault simulation with regular parallel logic simulators, one may inject the faults by adding extra logic gates. Figure 14.12 shows how this is done for faults f and g in N. To inject f, a stuck-at one fault, an OR gate (G_f) is inserted, and to force the second bit of A_f to be one without affecting the other two bits, the side input of G_f is set to be 010. Note that the injection of fault f does not affect the fault-free circuit and the faulty circuit with fault g.

Table 14.4 Parallel fault simulation for Figure 14.10

Pat #		Input				Internal					Output	
		A	Af	B	C	E	F	L	J	Jg	H	K
P_1	FF	0	0	1	0	1	1	1	0	0	0	1
	f	**0**	**1**	1	0	1	1	1	0	0	**1**	**0**
	g	0	0	1	0	1	1	1	0	0	0	1
P_2	FF	0	0	0	1	1	1	1	0	0	0	1
	f	0	**1**	0	1	1	1	1	0	0	**1**	**0**
	g	0	0	0	1	1	1	1	0	0	0	1
P_3	FF	1	1	0	0	0	0	0	1	1	0	0
	f	1	1	0	0	0	0	0	1	1	0	0
	g	1	1	0	0	0	0	0	1	**0**	0	**1**

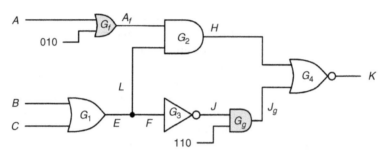

FIGURE 14.12

Fault injection for parallel fault simulation.

Similarly, injecting fault g, a stuck-at 0 fault, is achieved by adding the AND gate G_g and setting its side input to be 110.

Note that the parallel fault simulation technique is applicable to the unit or zero delay models only. More complicated delay models cannot be modeled, because several faults are evaluated at the same time. Furthermore, a simulation pass cannot terminate unless all the faults in this pass are detected. For instance, we cannot drop fault f alone after simulating pattern P_1, because fault g is not detected yet. Parallel fault simulation is best used for simulating the beginning of the test pattern sequence, when a large number of faults are detected by each pattern.

14.3.2.2 *Parallel pattern fault simulation*

Bitwise parallelism can be used to simulate test patterns in parallel. For a host computer with a w-bit data width, the signal values for a sequence of w test patterns are packed into a data word. For the fault-free or faulty circuit, w test patterns can be simulated in parallel by use of bitwise logic operations. This approach was first reported in [Waicukauski 1985], in which it is called ***parallel pattern single fault propagation*** (PPSFP), because one fault at a time is simulated. This approach is especially useful for combinational circuits or full-scan sequential circuits.

In PPSFP, logic simulations on the fault-free circuit are first performed on the first w test patterns, and the circuit outputs are recorded. Then, the faults are simulated one at a time on these w test patterns. For each fault, the simulation results are compared with the correct responses to determine whether the fault is detected. Simulation continues until the fault is detected and dropped, or all the test patterns are simulated. The faulty circuit is restored to its original state, and the next fault is processed. The same procedure repeats until all faults in the fault list are simulated.

The PPSFP results of the fault simulation example are shown in Table 14.5. The "Fault-Free" row lists the fault-free simulation results. Note that the three patterns are packed into one single word and thus are evaluated simultaneously

Table 14.5 PPSFP for Figure 14.10

		Input			Internal					Output
		A	B	C	E	F	L	J	H	K
Fault Free	P_1	0	1	0	1	1	1	0	0	1
	P_2	0	0	1	1	1	1	0	0	1
	P_3	1	0	0	0	0	0	1	0	0
f	P_1	1	1	0	1	1	1	0	1	0
	P_2	1	0	1	1	1	1	0	1	0
	P_3	1	0	0	0	0	0	1	0	0
g	P_1	0	1	0	1	1	1	0	0	1
	P_2	0	0	1	1	1	1	0	0	1
	P_3	1	0	0	0	0	0	0	0	1

by use of bitwise logic operations. The "f" row represents the simulation results with fault f injected. In PPSFP, faults are injected by activating rising or falling events, depending on the stuck-at value, at the faulty signal. Thus, fault f, A stuck-at one, is injected by activating two rising events on input A. The faulty responses are {**0**, **0**, 0}, which indicates that fault f is detected by the first and second patterns but not the third one. After fault f is simulated, fault f is removed by activating two falling events on input A at pattern P_1 and P_2. Then, fault g is injected by activating one falling event on signal J at pattern P_3. A total of three simulation runs are carried out.

Figure 14.13 illustrates the simplified PPSFP flow. Again, fault collapsing is first executed to obtain the collapsed fault list F. Then, the first w patterns are simulated on the fault-free circuit in parallel, and the good outputs (O_g) are stored. Then, each fault f in the fault list F is simulated one by one with the same w test patterns. A fault is dropped and not simulated against the remaining test patterns if its output response O_f is different from O_g. To fault simulate the next fault, the fault effect of the current fault is removed, and the next fault is injected. This process continues until all faults are either detected or simulated against all test patterns. If the number of test patterns is not an even multiple of the machine word width, only part of the machine word is used when simulating this last batch of patterns.

PPSFP is best suited for simulation of test patterns that come later in the test sequence, where the fault drop rate per pattern is lower. Parallel fault simulation does not work well in this situation, because it cannot terminate a simulation pass until all w-1 faults being processed are detected. PPSFP is not suitable for sequential circuits, because the circuit state for test pattern i in the w-bit

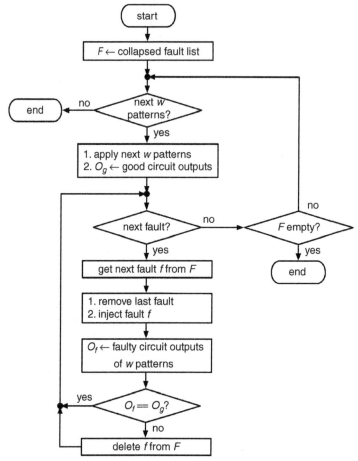

FIGURE 14.13

The PPSFP flowchart.

word depends on the previous i-1 patterns in the word, and this state is not available when the patterns are processed in parallel.

14.3.3 Concurrent fault simulation

Because a fault only affects the logic in the fanout cone from the fault site, the good circuit and faulty circuits typically only differ in a small region. **Concurrent fault simulation** exploits this fact and simulates only the differential parts of the whole circuit [Ulrich 1974]. Concurrent fault simulation is essentially an event-driven simulation with the fault-free circuit and faulty circuits simulated altogether.

In concurrent fault simulation, every gate has a **concurrent fault list**, which consists of a set of **bad gates**. A bad gate of gate x represents an imaginary copy of gate x in the presence of a fault. Every bad gate contains a fault index and the associated gate I/O values in the presence of the corresponding fault. Initially, the concurrent fault list of gate x contains **local faults** of gate x. The local faults of gate x are faults on the inputs or outputs of gate x. As the simulation proceeds, the concurrent fault list contains not only local faults but also faults propagated from previous stages (called **fault effects**). Local faults of gate x remain in the concurrent fault list of gate x until they are detected.

Figure 14.14 illustrates the concurrent simulation of the example circuit for test pattern P_1. For clear illustration, we demonstrate three faults in this example: A stuck-at one, C stuck-at zero, and J stuck-at zero faults. The concurrent fault lists with bad gates in grey are drawn beside the good gates. The fault indices are labeled in the middle of bad gates and their associated bad gate I/O values are labeled beside their I/O pins. The fault list of G_1, G_2, and G_3 initially contains their local faults: $C/0$, $A/1$, and $J/0$. When we apply the first pattern, three events occur in the primary inputs: u $\rightarrow$ 0 on A, u $\rightarrow$ 1 on B, and u $\rightarrow$ 0 on C. They are **good events**, because they happen in the good circuit. The output of good gate G_1 changes from unknown to one. In the presence of fault $C/0$, the output of faulty G_1 is the same as that of good G_1. A bad gate is **invisible** if its faulty output is the same as the good output. The bad gates C/0 and J/0 are both invisible so they are not propagated to the subsequent stages.

The output of G_2 changes from unknown to zero. In the presence of fault $A/1$, the faulty output changes from unknown to one. Because the faulty output differs from the good output, bad gate $A/1$ becomes visible. A bad gate is **visible** if its faulty output is different from the good output. The visible bad gate

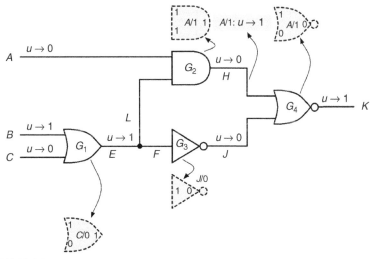

FIGURE 14.14

Concurrent fault simulation (P_1).

$A/1$ creates a bad event u $\rightarrow$ 1 on net H (in gray). A **bad event** does not occur in the good circuit; it only occurs in the faulty circuit of the corresponding fault. A new copy of bad gate $A/1$ is added to the concurrent fault list of G_4, because it has one input different from the good gate. It is said that bad gate $A/1$ **diverges from** its good gate. Finally, fault $A/1$ is detected because the faulty output K is different from the good output. At this time, we could drop detected fault $A/1$, but we keep it for illustration purposes.

Figure 14.15 illustrates the concurrent fault simulation for test pattern P_2. Two good events occur in this figure: 0 $\rightarrow$ 1 on C and 1 $\rightarrow$ 0 on B. The bad gate $C/0$, which was invisible in pattern P_1, now becomes **newly visible**. The newly visible bad gate creates a bad event, net E falls to zero, which in turn creates two divergences in G_2 and G_3. The former is invisible, but the latter creates a bad event, net J rises to one. Finally, the concurrent fault list of G_4 contains two bad gates; both faults $A/1$ and $C/0$ are detected.

For the last test pattern P_3 (Figure 14.16), two good events occur at primary inputs A and C. The bad gate $C/0$ now becomes invisible. The bad gate $C/0$ is deleted from the concurrent fault list of G_3. A bad gate **converges to** its good gate if it is not a local fault and its I/O values are identical to those of the good gate. Similarly, the other bad gates $C/0$ also converge to G_2 and G_4. Note that bad gate $C/0$ does not converge to G_1, because it is a local fault for G_1. The bad gate $A/1$ can be examined in the same way. For gate G_3, although the faulty output of bad gate $J/0$ does not change, the good event 0 $\rightarrow$ 1 on J makes bad gate $J/0$ newly visible.

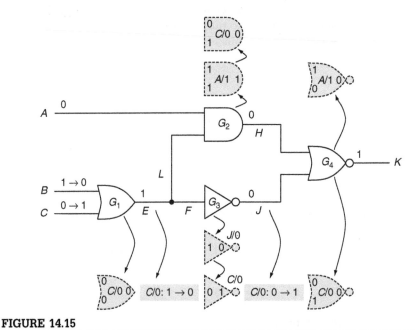

FIGURE 14.15

Concurrent fault simulation (P_2).

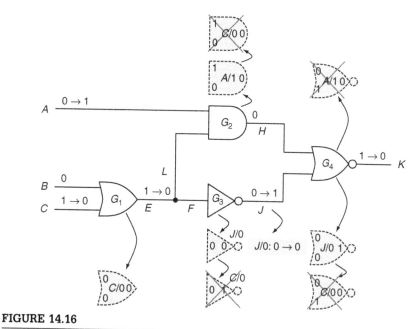

FIGURE 14.16

Concurrent fault simulation (P_3).

The newly visible event (in gray) is propagated to G_4, and a new bad gate $J/0$ diverges from G_4. Eventually, the fault $J/0$ is detected by pattern P_3.

Figure 14.17 shows a simplified concurrent fault simulation flowchart. The fault simulator applies one pattern at a time. The concurrent fault simulation is an event-driven simulation with both good events and bad events simulated at the same time. The events on the gate inputs are first analyzed. A good event affects both good and bad gates but a bad event only affects bad gates of the corresponding fault. After the analysis, events are then executed. The diverged bad gates and converged bad gates are added to or deleted from the fault list, respectively. Determining whether a bad gate diverges or converges depends on three factors: the visibility, the bad event, and the concurrent fault list (see [Abramovici 1994] for more details). After the event execution, new events are computed at the gate outputs. If an event reaches the primary outputs, detected faults can be removed from concurrent fault lists of all gates. This process repeats until there are no more test patterns, or no undetected faults.

14.3.4 Differential fault simulation

Concurrent fault simulation constructs the state of the faulty circuit from that of the same faulty circuit of the previous test pattern. Concurrent fault simulation has a potential memory problem, because the size of the concurrent fault list changes at runtime. In contrast, the single fault propagation technique

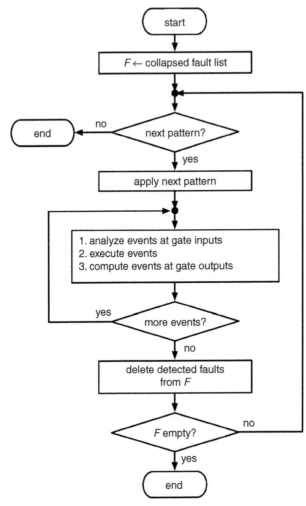

FIGURE 14.17

Concurrent fault simulation flowchart.

constructs the state of the faulty circuit from that of the good circuit. For sequential circuits, the single fault propagation technique would require a large overhead to store the states of the good circuit. Neither of the preceding two techniques are good for sequential fault simulation. Differential fault simulation combines the merits of concurrent fault simulation and single fault propagation techniques [Cheng 1989]. The idea is to simulate, in turn, every faulty circuit by tracking only the difference between a faulty circuit and the last simulated one. An event-driven simulator can easily implement differential fault simulation with the differences injected as events. This differential fault simulation technique

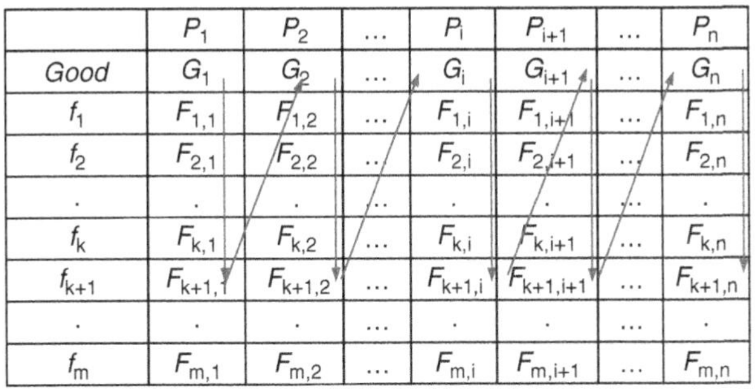

	P_1	P_2	...	P_i	P_{i+1}	...	P_n
Good	G_1	G_2	...	G_i	G_{i+1}	...	G_n
f_1	$F_{1,1}$	$F_{1,2}$	...	$F_{1,i}$	$F_{1,i+1}$	...	$F_{1,n}$
f_2	$F_{2,1}$	$F_{2,2}$	..	$F_{2,i}$	$F_{2,i+1}$	..	$F_{2,n}$
.	.	.	/..	.	.	/..	.
f_k	$F_{k,1}$	$F_{k,2}$	/...	$F_{k,i}$	$F_{k,i+1}$	/...	$F_{k,n}$
f_{k+1}	$F_{k+1,1}$	$F_{k+1,2}$	...	$F_{k+1,i}$	$F_{k+1,i+1}$	...	$F_{k+1,n}$
.	.	.	...	.	.	...	.
f_m	$F_{m,1}$	$F_{m,2}$	...	$F_{m,i}$	$F_{m,i+1}$	...	$F_{m,n}$

FIGURE 14.18

Differential fault simulation.

has been further combined with the parallel fault simulation technique, as implemented in **PROOFS** [Niermann 1992].

Figure 14.18 illustrates how differential fault simulation works. First, the first pattern P_1 is simulated on the good circuit G_1, and the good primary outputs are stored. Then the faulty circuit ($F_{1,1}$) is simulated with fault f_1 injected as an event. The first subscript indicates the fault and the second subscript indicates the pattern. The difference of states between G_1 and $F_{1,1}$ is stored. Note that only the states of storage elements, such as flip-flops, are stored, so the memory needed is small compared with concurrent fault simulation. If the primary outputs of $F_{1,1}$ and G_1 are not the same, then fault f_1 is detected. Following F_1 the second faulty circuit ($F_{2,1}$) is simulated with f_1 removed and f_2 injected. Similarly, the difference of states between F_1 and F_2 is stored. The preceding process continues until pattern P_1 has been simulated for all faults (f_1 to f_m).

Following the first pattern, the state of the good circuit G_2 is restored and the second pattern P_2 is applied. After the fault-free simulation, the primary outputs of G_2 are stored. The state of faulty circuit $F_{1,2}$ is restored by injecting the difference of G_1 and $F_{1,1}$. The fault f_1 is again injected as an event. The differential fault simulation for P_2 is the same as that of pattern P_1. Differential fault simulation goes in the direction of the arrows in Figure 14.18—G_i, $F_{1,i}$, $F_{2,i}$, ..., $F_{m,i}$, G_{i+1}, $F_{1,i+1}$, ...

Figure 14.19 shows a simplified flowchart for differential fault simulation. For every test pattern, a fault-free simulation is performed first. Then the faulty circuits are simulated one after another. The states of every circuit are restored from the last simulation. If the faulty circuit outputs are different from the good outputs, the fault is detected and dropped. The state difference of every circuit is stored. With fault dropping, the state difference of the dropped fault must be accumulated into the state differences of its next undetected fault. This process repeats until there are no test patterns or no undetected faults.

The problem with differential fault simulation is that the order of events caused by fault sites is not the same as the order of the timing of their occurrence. If the circuit behavior depends on the gate delay of the circuit, the timing

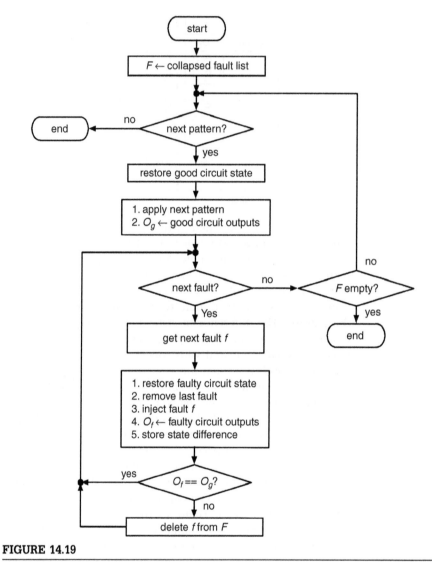

FIGURE 14.19

Differential fault simulation flowchart.

information of every event must be included. This solution, however, may potentially require high memory consumption.

14.3.5 **Comparison of fault simulation techniques**

In terms of simulation speed, it is apparent that serial fault simulation is the slowest among all the techniques. Differential fault simulation is shown to be up to twelve times faster than concurrent fault simulation and PPSFP [Cheng 1989], when the sequential circuit under test does not contain memories, such as *static random-access memories* (SRAMs) and *dynamic random-access memories* (DRAMs).

Memory use is, in general, not a problem for serial fault simulation, because it deals with one fault at a time. Similarly, parallel fault simulation and PPSFP do not require much more memory than the fault-free simulation. Concurrent fault simulation has severe memory problems, because the size of the concurrent fault list is unpredictable. Furthermore, the I/O values of every bad gate in the concurrent fault simulation must be recorded. Differential fault simulation relieves the memory management problem of concurrent fault simulation, because only the difference in storage elements is stored.

When the unknown (X) and/or high-impedance (Z) values are present in the circuit, a multiple-valued fault simulation becomes necessary. Serial fault simulation has no problem in handling multiple-valued fault simulation, because it can be realized with a regular logic simulator. In contrast, to exploit bitwise word parallelism, it is more difficult for parallel fault simulation or PPSFP to handle X or Z. In concurrent fault simulation, dealing with multiple-valued simulations is straightforward, because every bad gate is evaluated in the same way as in the fault-free simulation. Finally, differential fault simulation can simulate X or Z without a problem, because it is based on event-driven simulation.

From the aspect of delay and functional modeling capability, serial fault simulation does not encounter any difficulty. Parallel fault simulation and PPSFP cannot take delay or functional models into account, because they pack the information of multiple faults or test patterns into the same word and rely on bitwise logic operations. Being event-driven, both concurrent and differential fault simulation techniques are capable of handling functional models; however, only the former is able to process circuit delays.

When sequential circuits are of concern, serial and parallel fault simulation techniques do not have a problem. The PPSFP technique, however, is not suited for sequential circuit simulation, because a large memory space is required to store the states of the fault-free circuit. Concurrent and differential fault simulations are able to perform sequential fault simulation without difficulty.

On the basis of the previous discussions, PPSFP and parallel fault simulation techniques are currently the most popular fault simulation techniques for combinational (full-scan) circuits. On the other hand, concurrent fault simulation techniques have been widely adopted for sequential circuits embedded with memories, whereas differential fault simulation techniques are mostly suitable for sequential circuits without memories. Algorithm switching has also been used to improve performance. Parallel fault simulation can be used when the fault drop rate per test pattern is high, and then PPSFP is used when more patterns are required to drop each fault.

Even for fault simulation techniques that are efficient in time and memory, the problems of memory explosion and long simulation time still exist as *integrated circuit* (IC) complexity continues growing. To overcome the memory problem, the **multiple-pass fault simulation** approach is often adopted. The idea of multiple-pass fault simulation is to partition the faults into smaller groups, each of which is simulated independently. If the faults are well partitioned, multiple-pass fault simulation prevents the memory explosion problem. To further reduce the fault simulation time, **distributed fault simulation** approaches may be used. Distributed fault simulation divides the whole fault simulation into smaller tasks, each of which is performed independently on a separate processor.

There are several alternatives to fault simulation. The fault-sampling technique was proposed to simulate only a sampled group of faults [Butler 1974]. Critical path tracing is another alternative to fault simulation [Abramovici 1984]. Instead of performing actual fault simulation, the ***statistical fault analysis*** (STAFAN) approach proposes to use probability theory to estimate the expected value of fault coverage [Jain 1985]. These alternatives to fault simulation have also been extensively discussed in [Abramovici 1994], [Bushnell 2000], and [Wang 2006].

14.4 TEST GENERATION

First, consider the single stuck-at fault model. Figure 14.20 shows a circuit with a single stuck-at fault in which signal d is tied to logic 1 ($d/1$). A logic 0 must be applied to node d from the primary inputs of the circuit to produce a difference between the fault-free (or good) circuit and the circuit with the stuck-at fault present. Next, to observe the effect of the fault, a logic 1 must be applied to signal c. So, if the fault $d/1$ is present, it can be detected at the output e with the derived vector. Test generation attempts to generate test vectors for every possible fault in the circuit. In this example, in addition to the $d/1$ fault, faults such as $a/1$, $b/1$, and $e/1$ are also targeted by the test generator. Because some of the faults in the circuit can be logically equivalent, no test can be obtained to distinguish between them. Thus, equivalence fault collapsing as described in Section 14.2 is often used to identify equivalent faults *a priori* to reduce the number of faults that must be targeted [Abramovici 1994; Bushnell 2000; Jha 2003]. Subsequently, the ATPG is only concerned with generating test vectors for each fault in the collapsed fault list.

14.4.1 Random test generation

Random test generation (RTG) is one of the simplest methods for generating vectors. Vectors are randomly generated and fault-simulated (or fault-graded) on the ***circuit under test*** (CUT). Because no specific fault is targeted, the

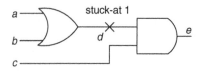

FIGURE 14.20

Example of a single stuck-at fault.

complexity of RTG is low. However, RTG often results in generating a large number of tests that achieves sub-par fault coverage because of the difficult-to-test faults.

In RTG, logic values are randomly generated at the primary inputs, with equal probability of assigning a logic 1 or logic 0 to each primary input. Thus, the random vectors are uniformly distributed in the test set. Note that the random test set is not truly random, because a pseudo-random number generator is generally used. In other words, the random test set can be repeated with the same pseudo-random number generator. Nevertheless, the vectors generated hold the necessary statistical properties of a random vector set.

The **level of confidence** one can have on a random test set T can be measured as the probability that T can detect all the stuck-at faults in the circuit. For N random vectors, the **test quality** t_N indicates the probability that all detectable stuck-at faults are detected by these N random vectors. Thus, the test quality of a random test set highly depends on the circuit under test. Consider a circuit with an eight-input AND gate (or equivalently a cone of seven two-input AND gates) illustrated in Figure 14.21. Although achieving a logic 0 at the output of the AND gate is easy, getting a logic 1 is difficult. A logic 1 would require all the inputs to be at logic 1. If the RTG assigns each primary input with an equal probability of logic 0 or logic 1, the chance of getting eight logic 1's simultaneously would only be $0.5^8 = 0.0039$. In other words, the AND gate output stuck-at-0 fault would be difficult to test by the RTG. Such faults are called **random-pattern resistant faults**.

As discussed earlier, the quality of a random test set depends on the underlying circuit. More random-pattern resistant faults will more likely reduce the quality of the random test set. To tackle the problem of targeting random-pattern resistant faults, biasing is required so the input vectors are no longer viewed as uniformly distributed. Consider the same eight-input AND gate example again. If each input of the AND gate has a much higher probability of

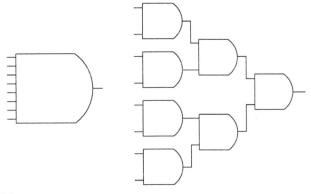

FIGURE 14.21

Two equivalent circuits.

receiving a logic 1, the probability of getting a logic 1 at the output of the AND gate significantly increases. For example, if each input has a 75% probability of receiving a logic 1, then getting a logic 1 at the output of the AND gate now becomes $0.75^8 = 0.1001$, rather than the previous 0.0039.

Determining the optimal bias values for each primary input that can achieve the highest coverage is not an easy task. Thus, rather than trying to obtain the best set of values, the objective is frequently to increase the probabilities for those difficult-to-control and difficult-to-observe nodes in the circuit. For instance, suppose a circuit has an eight-input AND gate; any fault that requires the AND gate output equal to logic 1 for detection will be considered difficult to test. It would then be beneficial to attempt to increase the probability of obtaining a logic 1 at the output of this AND gate.

Another issue regarding random test generation is the number of random vectors needed. Given a combinational circuit with n primary inputs, there are clearly 2^n possible input vectors. One can express the probability of detecting fault f by any random vector to be:

$$d_f = T_f/2^n$$

where T_f is the set of vectors that can detect fault f. Consequently, the probability that a random vector will not detect f (*i.e.*, f escapes a random vector) is: $e_f = 1 - d_f$.

Therefore, given N random vectors, the probability that none of the N vectors detect fault f is:

$$e_f^N = \left(1 - d_f\right)^N$$

In other words, the probability that at least one of N vectors will detect fault f is:

$$1 - \left(1 - d_f\right)^N$$

If the detection probability, d_f, for the hardest fault is known, N can be readily computed by solving the following inequality:

$$1 - \left(1 - d_f\right)^N \geq p$$

where p is the probability that N vectors should detect fault f.

If the detection probability is not known, it can be computed directly from the circuit. The detection probability of a fault is directly related to: (1) the controllability of the line that the fault is on and (2) the observability of the fault-effect to a primary output. The controllability and observability computations have been introduced previously in the chapter on design for testability. It is worth noting that the minimum detection probability of a detectable fault f can be determined by the output cone in which f resides. In fact, if f is detectable, it must be excited and propagated to at least one primary output, as illustrated in Figure 14.22. It is clear that all the primary inputs necessary to excite f and propagate the fault-effect must reside in the cone of the output

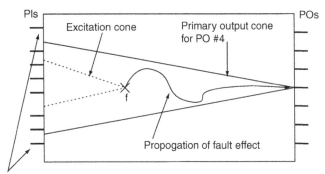

Inputs outside of PO cone are not needed for detection of fault f

FIGURE 14.22

Detection of a fault.

to which f is detected. Thus, the detection probability for f is at least $(0.5)^m$, where m is the number of primary inputs in the cone of the corresponding primary output. Taking this concept a step further, the detection probability of the most difficult fault can be obtained with the following lemma [David 1976; Shedletsky 1977].

Lemma 1: In a combinational circuit with multiple outputs, let n_{max} be the number of primary inputs that can lead to a primary output. Then, the detection probability for the most difficult detectable fault, d_{min}, is:

$$d_{\min} \geq (0.5)^{nmax}$$

Proof
The proof follows from the preceding discussion.

14.4.1.1 *Exhaustive testing*

If the combinational circuit has few primary inputs, **exhaustive testing** may be a viable option, where every possible input vector is enumerated. This may be superior to random test generation, because RTG can produce duplicated vectors and may miss certain ones.

In circuits in which the number of primary inputs is large, exhaustive testing becomes prohibitive. However, on the basis of the results of Lemma 1, it may be possible to partition the circuit and only exhaust the input vectors within each cone for each primary output. This is called **pseudo-exhaustive testing**. In doing so, the number of input vectors can be drastically reduced. When enumerating the input vectors for a given primary output cone, the values for the primary inputs that are outside the cone are simply assigned random values. Therefore, if a circuit has three primary outputs, each has a corresponding primary output cone. Note that these three primary output cones may overlap. Let $n1$, $n2$, and $n3$ be the number of primary inputs corresponding to these three cones. Then the number of pseudo-exhaustive vectors is simply at most $2^{n1} + 2^{n2} + 2^{n3}$.

14.4.2 **Theoretical Background: Boolean difference**

Consider the circuit shown in Figure 14.23. Let the target fault be the stuck-at-0 fault on primary input y. Recall the high-level concept of test generation illustrated in Figure 14.1, where the objective is to distinguish the fault-free circuit from the faulty circuit. In the example circuit shown in Figure 14.23, the faulty circuit is the circuit with y stuck at 0. Note that the circuit output can be expressed as a Boolean formula:

$$f = xy + y'z$$

Let $f2$ be the faulty circuit with the fault y/0 present. In other words,

$$f2 = f(y = 0)$$

To distinguish the faulty circuit $f2$ from the fault-free counterpart f, any input vector that can make $f \oplus f2 = 1$ would suffice. Furthermore, because the aim is test generation, the target fault must be excited. In this example, the logic value on primary input y must be logic 1 to excite the fault y/0. Putting these two conditions together, the following equation is obtained:

$$y \cdot f(y = 1) \oplus f(y = 0) = 1 \qquad (14.1)$$

Note that $f(y = 1) \oplus f(y = 0)$ indicates the exclusive-or operation on the two functions $f(y = 1)$ and $f(y = 0)$; it evaluates to logic 1 if and only if the two functions evaluate to opposing values. In terms of ATPG, this is synonymous to propagating the fault effect at node y to the primary output f. Therefore, any input vector on primary inputs x, y, and z that can satisfy Equation (14.1) is a valid test vector for fault y/0:

$$y \cdot f(y = 1) \oplus f(y = 0) = y(x \oplus z) = y(xz' + x'z) = xyz' + x'yz$$

In this running example, the two vectors $xyz = \{110, 011\}$ are candidate test vectors for fault y/0. Formally, $f(y = 1) \oplus f(y = 0)$ is called the **Boolean difference** of f with respect to y and is often written as:

$$df/dy = f(y = 1) \oplus f(y = 0)$$

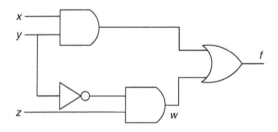

FIGURE 14.23

Example circuit to illustrate the concept of Boolean difference.

In general, if f is a function of $x_1, x_2, \ldots, x_n$, then:

$$\mathrm{d}f/\mathrm{d}x_i = f(x_1, x_2, \ldots, x_i = 1, \ldots, x_n) \oplus f(x_1, x_2, \ldots, x_i = 0, \ldots, x_n)$$

In terms of test generation, for any target fault on some fault α/v, the set of all vectors that can *propagate* the fault-effect to the primary output f is then those vectors that can satisfy:

$$\mathrm{d}f/\mathrm{d}\alpha = 1$$

(Note that this is independent of the polarity of the fault, whether it is stuck-at-0 or stuck-at-1.) Next, the constraint that the fault must be excited, α set to value v', must be added. Subsequently, the set of test vectors that can detect the fault becomes all those input values that can satisfy the following equation:

$$(\alpha = v') \cdot \mathrm{d}f/\mathrm{d}\alpha = 1 \tag{14.2}$$

Consider the same circuit shown in Figure 14.23 again. Suppose the target fault is $w/0$. The same analysis can be performed for this new fault. The set of test vectors that can detect $w/0$ is simply:

$$w \cdot \mathrm{d}f/\mathrm{d}w = 1$$
$$\Rightarrow w \cdot f(w = 1) \oplus f(w = 0) = 1$$
$$\Rightarrow w \cdot (1 \oplus xy) = 1$$
$$\Rightarrow w \cdot (xy)' = 1$$
$$\Rightarrow w \cdot (x' + y') = 1$$
$$\Rightarrow wx' + wy' = 1$$

Now, w can be expanded from the circuit shown in the figure to be $w = y' \cdot z$. Plugging this into the equation above gives us:

$$w \cdot x' + w \cdot y' = 1$$
$$\Rightarrow y' \cdot zx' + y' \cdot z \cdot y = 1$$
$$\Rightarrow x' \cdot y'z + y' \cdot z = 1$$
$$\Rightarrow y' \cdot z = 1$$

Therefore, the set of vectors that can detect $w/0$ is $\{001, 101\}$.

14.4.2.1 *Untestable faults*

If the target fault is untestable, it would be impossible to satisfy Equation (14.2). Consider the circuit shown in Figure 14.24. Suppose the target fault is $z/0$. Then the set of vectors that can detect $z/0$ are those that can satisfy:

$$z \cdot \mathrm{d}f/\mathrm{d}z = 1$$
$$\Rightarrow \quad z \cdot f(z = 1) \oplus f(z = 0) = 1$$
$$\Rightarrow \quad z \cdot (xy \oplus xy) = 1$$
$$\Rightarrow \quad z \cdot 0 = 1$$
$$\Rightarrow \quad \text{UNSATISFIABLE}$$

In other words, there exists no input vectors that can satisfy $z \cdot \mathrm{d}f/\mathrm{d}z = 1$, indicating that the fault $z/0$ is untestable.

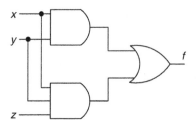

FIGURE 14.24

Example circuit for an untestable fault.

14.4.3 Designing a stuck-at ATPG for combinational circuits

In deterministic ATPG algorithms, there are two main tasks. The first is to excite the target fault, and the second is to propagate the fault-effect to a primary output. Because the logic values in both the fault-free and faulty circuits are needed, composite logic values are used. For each signal in the circuit, the values v/v_f are needed, where v denotes the value for the signal in the fault-free circuit, and v_f represents the value in the corresponding faulty circuit. Whenever $v = v_f$, v is sufficient to denote the signal value. To facilitate the manipulation of such composite values, a 5-valued algebra was proposed [Roth 1966], in which the five values are 0, 1, X, D, and $\bar{D}$; 0, 1, and X are the conventional values found in logic design for true, false, and "don't care." D represents the composite logic value 1/0 and $\bar{D}$ represents 0/1. Boolean operators such as AND, OR, NOT, and XOR can work on the 5-valued algebra as well. The simplest way to perform Boolean operations is to represent each composite value into the v/v_f form and operate on the fault-free value first, followed by the faulty value. For example, 1 AND D is 1/1 AND 1/0. AND-ing the fault-free values yields 1 AND 1 = 1, and AND-ing the faulty values yields 1 AND 0 = 0. So the result of the AND operation is 1/0 = D. As another example,

$$
\begin{aligned}
D \text{ OR } \bar{D} &= 1/0 \text{ OR } 0/1 \\
&= 1/1 \\
&= 1
\end{aligned}
$$

Tables 14.6, 14.7, and 14.8 show the AND, OR, and NOT operations for the 5-valued algebra, respectively. Operations on other Boolean conjunctives can be constructed in a similar manner.

14.4.3.1 *A naive ATPG algorithm*

A very simple and naive ATPG algorithm is shown in Algorithm 14.3, in which combinational circuits with fanout structures can be handled.

Table 14.6 AND Operation

AND	0	1	*D*	$\overline{D}$	*X*
0	0	0	0	0	0
1	0	1	D	$\overline{D}$	X
D	0	D	D	0	X
$\overline{D}$	$\overline{D}$	1	1	$\overline{D}$	X
X	X	1	X	X	X

Table 14.7 OR Operation

OR	0	1	*D*	$\overline{D}$	*X*
0	0	1	D	$\overline{D}$	X
1	1	1	1	1	1
D	D	1	D	1	X
$\overline{D}$	$\overline{D}$	1	1	$\overline{D}$	X
X	X	1	X	X	X

Table 14.8 NOT Operation

NOT	
0	1
1	0
D	$\overline{D}$
$\overline{D}$	D
X	X

Algorithm 14.3 Naive ATPG (*C*, *f*)

1. **while** a fault-effect of *f* has not propagated to a PO and all possible vector combinations have not been tried **do**
2.   pick a vector, *v*, that has not been tried;
3.   fault simulate *v* on the circuit *C* with fault *f*;
4. **end while**

Note that in an ATPG, the worst-case computational complexity is exponential, because all possible input patterns may have to be tried before a vector is found or that the fault is determined to be undetectable. One may go about line #2 of the algorithm in an intelligent fashion, so a vector is not simply selected indiscriminately. Whether or not intelligence is incorporated, some mechanism is needed to account for those attempted input vectors so no vector would be repeated. If it is possible to deduce some knowledge during the search for the input vector, the ATPG may be able to mark a set of solutions as tried and thus reduce the remaining search space. For instance, after attempting a number of input vectors, this naive ATPG realizes that any input vector with the first primary input set to logic 0 cannot possibly detect the target fault, and it can safely mark all vectors with the first primary input equal to 0 as a tried input vector. Subsequently, only those vectors with the first primary input set to 1 will be selected.

In certain cases, it may not be possible for the ATPG to deduce that all vectors with a given primary input set to some logic value would definitely not qualify to be solution vectors. However, it may be able to make an intelligent guess that input vectors with primary input #i set to some specific logic value are more likely to lead to a solution. In such a case, the ATPG would make a **decision** on primary input #i. Because the decision may actually be wrong, the ATPG may eventually have to alter its decision, trying the vectors that have the opposite Boolean value on primary input #i.

The process of making decisions and reversing decisions will result in a **decision tree**. Each node in the decision tree represents a decision variable. If only two choices are possible for each decision variable, then the decision tree is a binary tree. However, there may be cases in which multiple choices are possible in a general search tree.

Figure 14.25 shows an example decision tree. Although this figure only allows decisions to be made at the primary inputs, in general, this may not be

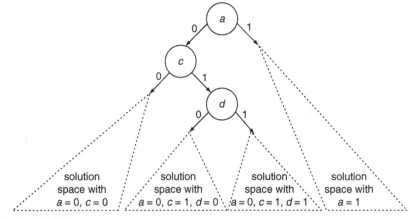

FIGURE 14.25

An example decision tree.

the case. This is used simply to allow the reader to have a clearer picture of the concept behind decision trees. At each decision, the search space is halved. For example, if the circuit has n primary inputs, then there are a total of 2^n possible vectors in the solution space. After a decision is made, the solution spaces under the two branches of a decision node are disjoint. For instance, the space under the decision $a = 1$ does not contain any vectors with $a = 0$. Note that the decision tree for a solution vector may not require the ATPG to *exhaustively* enumerate every possible vector; rather, it *implicitly* enumerates the vectors. If a solution vector exists, there must be a path along the decision tree that leads to the solution. On the other hand, if the fault is undetectable, every path in the decision tree would lead to no solution. It is important to note that a fault may be detected without having made all decisions. For example, the circuit nodes that do not play a role in exciting or propagating the fault would not have to be included in the decision process. Likewise, it may not require all decision variables before the ATPG can determine that it is on the wrong path. For example, if a certain path already sets a value on the fault site such that the fault is not excited, then no value combination on the remaining decision variables can help to excite and propagate the fault. With Figure 14.25 as an example again, suppose the path $a = 0$, $c = 1$, $d = 1$ cannot excite the target fault α. Then, the rest of the decision variables, $b, e, f, \dots$, cannot undo the effect rendered by $a = 0, c = 1, d = 1$.

14.4.3.1.1 Backtracking

Whenever a conflict is encountered (*i.e.*, a path segment in the decision tree leading to no solution), the search must not continue searching beneath that path but must go back to some earlier point and re-decide on a previous decision. If only two choices are possible for a decision variable, then some previous decision needs to be reversed if the other branch has not been explored before. This reversal of decision is called a **backtrack**. To keep track of where the search spaces have been explored and avoid repeating the search in the same

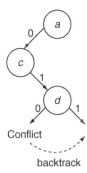

FIGURE 14.26

Backtrack on a decision.

spaces, the easiest mechanism is to reverse the most recent decision made. When reversing any decision, the signal values implied by the assignment of the previous decision variable must be undone.

Consider the decision tree illustrated in Figure 14.26 as an example. Suppose the current decisions made so far are $a = 0$, $c = 1$, $d = 0$, and this causes a conflict in detecting the target fault. Then, the search must reverse the most recently made decision, which is $d = 0$. When reversing $d = 0$ to $d = 1$, all values that resulted from $d = 0$ must be first undone. Then, the search continues with the path $a = 0$, $c = 1$, $d = 1$. If the reversal of a decision also caused a conflict (in this case, reversing $d = 0$ also caused a conflict), then it means $a = 0$, $c = 1$ actually cannot lead to any solution vector that can detect the target fault. The backtracking mechanism would then take the search to the previous decision and attempt to reverse that decision. In the running example, it would undo the decision on d, assigning d to "don't care," followed by reversing of the decision $c = 1$ and searching the portion of the search space under $a = 0$, $c = 0$. Finally, if there is no previous decision that can be reversed, the ATPG concludes that the target fault is undetectable.

Technically, whenever a decision is reversed, say $d = 0$ is reversed to $d = 1$ as shown in Figure 14.26, $d = 1$ is no longer a decision; rather, it becomes an implied value by a subset of the previous decisions made. The exact subset of decisions that implied $d = 1$ can be computed by a **conflict analysis** [Marques-Silva 1999b]. However, the details of conflict analysis are beyond the scope of this chapter and are thus omitted. The reader can refer to [Marques-Silva 1999b] for details of this mechanism. In addition, intelligent conflict analysis can also allow for **nonchronological backtracking**.

14.4.3.2 *A basic ATPG algorithm*

Given a target fault g/v in a fanout-free combinational circuit C, a simple procedure to generate a vector for the fault is shown in Algorithm 14.4, where JustifyFanoutFree() and PropagateFanoutFree() are both recursive functions.

Algorithm 14.4 Basic Fanout Free ATPG (C, g/v)

---

1. initialize circuit by setting all values to X;
2. JustifyFanoutFree(C, g, v'); /* excite the fault by justifying line g to v' */
3. PropagateFanoutFree(C, g); /* propagate fault-effect from g to a PO */

The JustifyFanoutFree(g, v) function recursively justifies the predecessor signals of g until all signals that need to be justified are, indeed, justified from the primary inputs. The simple outline of the JustifyFanoutFree routine is listed in Algorithm 14.5. In line #10 of the algorithm, controllability measures can be used to select the best input to justify. Selecting a good gate input may help to reach a primary input sooner.

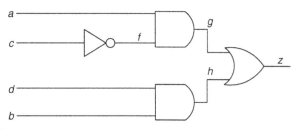

FIGURE 14.27

Example fanout-free circuit.

Consider the circuit C shown in Figure 14.27. Suppose the objective is to justify $g = 1$. According to the preceding algorithm, the following sequence of recursive calls to JustifyFanoutFree would have been made:

call #1: JustifyFanoutFree($C, g,$ 1)
call #2: JustifyFanoutFree($C, a,$ 1)
call #3: JustifyFanoutFree($C, f,$ 1)
call #5: JustifyFanoutFree($C, c,$ 0)

Algorithm 14.5 JustifyFanoutFree(C, g, v)

---

1. $g = v$;
2. **if** gate type of g == primary input **then**
3.   return;
4. **else if** gate type of g == AND gate **then**
5.   **if** v == 1 **then**
6.   **for all** inputs h of g **do**
7.   JustifyFanoutFree($C, h,$ 1);
8.   **end for**
9.   **else** {v == 0}
10. h = pick one input of g whose value == X;
11.   JustifyFanoutFree($C, h,$ 0);
12.   **end if**
13. **else if** gate type of g == OR gate **then**
14. ...
15. **end if**

After these calls to JustifyFanoutFree(), $abcd = 1X0X$ is an input vector that can justify $g = 1$.

Consider another circuit C shown in Figure 14.28. Note that the circuit is not fanout-free, but the preceding algorithm will still work for the objective of

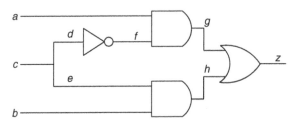

FIGURE 14.28

Example circuit with a fanout structure.

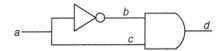

FIGURE 14.29

Circuit with a constant circuit node.

trying to justify the signal $g = 1$. According to the algorithm, the following sequence of calls to the JustifyFanoutFree function would have been made:

call #1: JustifyFanoutFree(C, g, 1)
call #2: JustifyFanoutFree(C, a, 1)
call #3: JustifyFanoutFree(C, f, 1)
call #4: JustifyFanoutFree(C, d, 0)
call #5: JustifyFanoutFree(C, c, 0)

After these five calls to JustifyFanoutFree(), $abc = 1X0$ is an input vector that can justify $g = 1$. Note that in a fanout-free circuit, the JustifyFanoutFree() routine will *always* be able to set g to the desired value v, and no conflict will ever be encountered. However, this is not always true for circuits with fanout structures, such as the circuit shown in Figure 14.29. This is because in circuits with fanout branches, two or more signals that can be traced back to the same fanout stem are **correlated**, and setting arbitrary values on these correlated signals may not always be possible. For example, in the simple circuit shown in Figure 14.29, justifying $d = 1$ is impossible, because it requires both $b = 1$ and $c = 1$, thereby causing a conflict on a.

Consider again the circuit shown in Figure 14.28. Suppose the objective is to set z = 0. On the basis of the JustifyFanoutFree() algorithm, it would first justify both $g = 0$ and $h = 0$. Now, for justifying $g = 0$, suppose it picks the signal f for justifying the objective $g = 0$; it would eventually assign $c = 1$ through the recursive JustifyFanoutFree() function. Next, for justifying $h = 0$, it no longer can choose e = 0 as a viable option, because choosing e = 0 will eventually cause a **conflict** on signal c. In other words, a different **decision** has to be made for justifying $h = 0$. In this case, $b = 0$ should be chosen. Although this example is very simple, it illustrates the possibility of making poor decisions, causing potential **backtracks** in the search. In the rest of this chapter, more discussion on avoiding conflicts will be covered.

In the preceding running example, suppose the target fault is $g/0$, and JustifyFanoutFree(C, g, 1) would have successfully excited the fault. With the fault $g/0$ excited, the next step is to propagate the fault-effect to a primary output. Similar to the JustifyFanoutFree() function, PropagateFanoutFree() is a recursive function as well, where the fault-effect is propagated one gate at a time until it reaches a primary output. Algorithm 14.6 illustrates the pseudo-code for one possible implementation of the propagate function.

Again, although the PropagateFanoutFree() routine is meant for fanout-free circuits, it is sufficient for the running example. With the PropagateFanoutFree() function on the fault-effect D at signal g, listed in Algorithm 14.5, the following calls to the JustifyFanoutFree and PropagateFanoutFree functions would have been made:

call #1: PropagateFanoutFree(C, g)
call #2: JustifyFanoutFree(C, b, 0)
call #3: JustifyFanoutFree(C, b, 0)
call #4: PropagateFanoutFree(C, z)

Algorithm 14.6 PropagateFanoutFree(C, g)

---

1. **if** g has exactly one fanout **then**
2. h = fanout gate of g;
3.   **if** none of the inputs of h has the value of X **then**
4.   backtrack;
5.   **end if**
6. **else** {g has more than one fanout}
7. h = pick one fanout gate of g that is unjustified;
8. **end if**
9. **if** gate type of h == AND gate **then**
10.   **for all** inputs, j, of h, such that $j \neq g$ **do**
11.   **if** the value on $j == X$ **then**
12.   JustifyFanoutFree(C, j, 1);
13.   **end if**
14.   **end for**
15. **else if** gate type of h == OR gate **then**
16.   **for all** inputs, j, of h, such that $j \neq g$ **do**
17.   **if** the value on $j == X$ **then**
18.   JustifyFanoutFree(C, j, 0);
19.   **end if**
20.   **end for**
21. **else if** gate type of h == ... gate **then**
22.   ...
23. **end if**
24. PropagateFanoutFree(C, h);

Because the fault-effect has successfully propagated to the primary output z, the fault $g/0$ is detected, with the vector $abc = 100$. The reader may notice that once $g/0$ has been excited, it is also propagated to z as well, because $c = 0$ also has made $b = 0$. In other words, the JustifyFanoutFree(C, b, 0) step is unnecessary. However, this is only possible if logic simulation or implication capability is embedded in the BasicFanoutFreeATPG() algorithm. For this discussion, it is not assumed that logic simulation is included.

With the same circuit shown in Figure 14.28, consider the fault $g/1$. The Basic-FanoutFreeATPG() algorithm will again be used to generate a test vector for this fault. In this case, the ATPG first attempts to justify $g = 0$, followed by propagating the fault-effect to z. During the justification of $g = 0$, the ATPG can pick either a or f as the next signal to justify. At this point, the ATPG must make a **decision**. Testability measures discussed in an earlier chapter can be used as a guide to make more intelligent decisions. In this example, choosing a is considered to be better than f, because choosing a requires no additional decisions to be made. Note that testability measures only serve as a guide to decision selection; they do not guarantee that the guidance will always lead to better decision selection.

It is important to note that in circuits with fanout structures, because the simple JustifyFanoutFree() and PropagateFanoutFree() functions described previously are meant for fanout-free circuits, will not always be applicable as illustrated in some of the earlier examples because of potential conflicts. To generate test vectors for general combinational circuits, there must be mechanisms that will allow the ATPG to avoid conflicts, as well as get out of a conflict when a conflict is encountered. To do so, the corresponding decision tree must be constructed during the search for a solution vector, and backtracks must be enforced for any conflict encountered. The following sections describe a few ATPG algorithms.

14.4.3.3 *D algorithm*

The D algorithm was proposed to tackle the generation of vectors in general combinational circuits [Roth 1966, 1967]. As indicated by the name of the algorithm, the D algorithm tries to propagate a D or $\overline{D}$ of the target fault to a primary output. Initially, every signal in the circuit has the unknown value, X. At the end of the D algorithm, some signals will be assigned 0, 1, D, or $\overline{D}$, while the rest of the signals may remain as unknown. Note that because each detectable fault can be excited, a fault-effect can always be created. In the following discussion, propagation of the fault-effect will take precedence over the justification of the signals. This allows for enhanced efficiency of the algorithm and for simpler discussion.

Before proceeding to discussing the details of the D algorithm, two important terms should be defined: the **D-frontier** and the **J-frontier**. The D-frontier consists of all the gates in the circuit whose output value is unspecified and a fault-effect (D or $\overline{D}$) is at one or more of its inputs. For this to occur, one or more inputs of the gate must currently have an unknown value, X. For example, at the start of the D algorithm, for a target fault f there is exactly one D (or $\overline{D}$) placed in the circuit

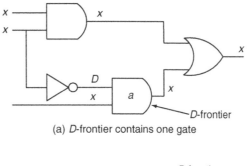

(a) *D*-frontier contains one gate

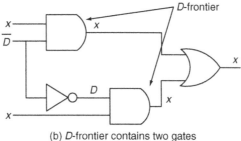

(b) *D*-frontier contains two gates

FIGURE 14.30

Illustrations of *D*-frontier.

corresponding to the stuck-at fault. All other signals currently have a "don't care" value. Thus, the *D*-frontier consists of the successor gate(s) from the line with the fault *f*. Two scenarios of a *D*-frontier are illustrated in Figure 14.30. Clearly, at any time if the *D*-frontier is empty, the fault no longer can be detected. For example, consider Figure 14.30a. If the bottom input of gate *a* is assigned a value of 0, the output of gate *a* will become 0, and the D-frontier now becomes empty. At this time, the search must backtrack and try a different search path.

The *J*-frontier consists of all the gates in the circuit whose output values are known (can be any value in the 5-valued logic except X) but is not justified by its inputs. Figure 14.31 illustrates an example of a *J*-frontier. Thus, to detect the target fault, all gates in the *J*-frontier must be justified; otherwise, some gates in the *J*-frontier must have caused a conflict, where these gates cannot be justified to the desired values.

Having discussed the two fundamental concepts of the *D*-frontier and the *J*-frontier, the explanation for the *D* algorithm can begin. The *D* algorithm begins by trying to propagate the initial *D* (or $\overline{D}$) at the fault site to a primary output. For example, in Figure 14.32, the propagation routine will set all the side inputs of the path necessary (gates $a \rightarrow b \rightarrow c$) to propagate the fault-effect to the respective noncontrolling values. These side input gates, namely x, y, and z, thus form the *J*-frontier, because they are not currently justified. Because the *D* is propagated to the primary output, the *D*-frontier eventually becomes the output gate.

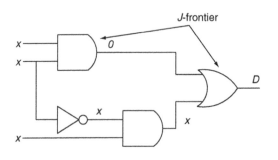

FIGURE 14.31

Illustration of *J*-frontier.

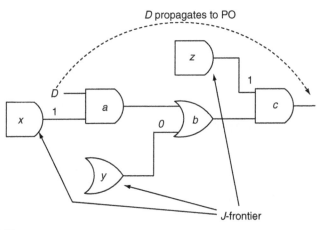

FIGURE 14.32

Propagation of *D*- and *J*-frontier.

Whenever there are paths to choose from in advancing the *D*-frontier, observability values can be used to select the corresponding gates. However, this does not guarantee that the more observable path will definitely lead to a solution. When a *D* or a $\overline{D}$ has reached a primary output, all the gates in the *J*-frontier must now be justified. This is done by advancing the *J*-frontier backward by placing predecessor gates in the *J*-frontier such that they justify the previous unjustified gates. Similar to propagation of the fault-effect, whenever a conflict occurs, a backtrack must be invoked. In addition, at each step, the *D*-frontier must be checked so the *D* (or $\overline{D}$) that has reached a primary output is still there. Otherwise, the search returns to the propagation phase and attempts to propagate the fault-effect to a primary output again. The overall procedure for the *D* algorithm is shown in Algorithms 14.7 and 14.8.

Note that the previous procedure has not incorporated any intelligence in the decision-making process. In other words, sometimes it may be possible to determine that some value assignments are not justifiable, given the current

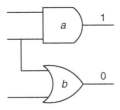

FIGURE 14.33

Conflict in the justification process.

circuit state. For instance, consider the circuit fragment shown in Figure 14.33. Justifying gate $a = 1$ and gate $b = 0$ is not possible, because $a = 1$ requires both of its inputs set to logic 1, whereas $b = 0$ requires both of its inputs set to logic 0. Noting such conflicting scenarios early can help to avoid future backtracks. Such knowledge can be incorporated into line #1 of the D-Alg-Recursion() shown in Algorithm 14.8. In particular, implications can be used to identify such potential conflicts, and they are used extensively to enhance the performance of the D algorithm (as well as other ATPG algorithms).

Algorithm 14.7 D-Algorithm(C, f)

1. initialize all gates to don't-cares;
2. set a fault-effect (D or $\overline{D}$) on line with fault f and insert it to the D-frontier;
3. J-frontier = ϕ;
4. result = D-Alg-Recursion(C);
5. **if** result == success **then**
6.   print out values at the primary inputs;
7. **else**
8.   print fault f is untestable;
9. **end if**

Consider the multiplexer circuit shown in Figure 14.28. If the target fault is f stuck-at-0, then, after initializing all gate values to X, the D algorithm places a D on line f. The algorithm then tries to propagate the fault-effect to z. First, it will place $a = 1$ in the J-frontier, followed by $b = 0$ in the J-frontier. At this time, the fault-effect has reached the primary output. Now, the ATPG tries to justify all unjustified values in the J-frontier. Because a is a primary input, it is already justified. The other signals in the J-frontier are $f = D$ and $b = 0$. For $f = D$, $d = 0$, thereby making $c = 0$. For $b = 0$, either $e = 0$ or $b = 0$ is sufficient. Whichever one it picks, the search process will terminate, as a solution has been found.

Consider the same multiplexer circuit (see Figure 14.28) again. Suppose the target fault now is f stuck-at-1. Following the similar discussion as the previous target fault $f/0$, the algorithm initializes the circuit and places a D on f. Next, to propagate the fault-effect to a primary output, it likewise inserts a = 1 and h = 0 into the J-frontier. Now, the ATPG needs to justify all the gates in the J-frontier,

---

Algorithm 14.8 D-Alg-Recursion(C)

---

1. **if** there is a conflict in any assignment or D-frontier is ϕ **then**
2.   return failure
3. **end if**
4. /* first propagate the fault-effect to a PO */
5. **if** no fault-effect has reached a PO **then**
6.   **while** not all gates in D-frontier has been tried **do**
7. g = a gate in D-frontier that has not been tried;
8.   set all unassigned inputs of g to non-controlling value and add them to the J-frontier;
9.   result = D-Alg-Recursion(C);
10.   **if** result == success **then**
11.   return (success);
12.   **end if**
13.   **end while**
14. return (failure);
15. **end if** {fault-effect has reached at least one PO}
16. **if** J-frontier is ϕ **then**
17.   return (success);
18. **end if**
19. g = a gate in J-frontier;
20. **while** g has not been justified **do**
21. j = an unassigned input of g;
22.   set j = 1 and insert j = 1 to J-frontier;
23.   result = D-Alg-Recursion(C);
24.   **if** result == success **then**
25.   return (success);
26.   **else** try the other assignment
27.   set j = 0;
28.   **end if**
29. **end while**
30. return(failure);

---

which includes $a = 1, f = D$, and $b = 0$. Because a is a primary output, it is already justified. For $f = D$, $d = 1$. For $b = 0$, suppose it selects $e = 0$. At this time, the J-frontier consists of two gate values: $d = 1$ and $e = 0$. No value assignment on c can satisfy both $d = 1$ and $e = 0$; therefore, a conflict has occurred, and backtrack on the previous decision is needed. The only decision that has been made is $e = 0$ for $b = 0$, because there were two choices possible for

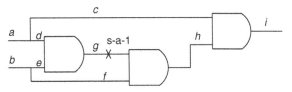

FIGURE 14.34

Example circuit.

justifying $b = 0$. At this time, the value on e is reversed, and $b = 0$ is added to the *J*-frontier. The process continues and all gate values in the *J*-frontier can be successfully justified, ending the process with the vector $abc = 101$.

Note that, in the preceding example, if some learning procedure (such as implications) is present, the decision for $b = 0$ would not result in $e = 0$, because the ATPG would have detected that $e = 0$ would conflict with $d = 1$. This knowledge could potentially improve the performance of the ATPG, which will be discussed later in this chapter.

Consider another example circuit shown in Figure 14.34. Suppose the target fault is $g/1$. After circuit initialization, the *D* algorithm places a $\overline{D}$ on g. Now, the *J*-frontier consists of $g = \overline{D}$ and the *D*-frontier consists of h. To advance the *D*-frontier, f is set to logic 1; $f = 1$ is added to the *J*-frontier, and the *D*-frontier is now i. Next, to propagate the fault-effect to the output, $c = 1$ is added to the *J*-frontier. At this time, the fault-effect has been propagated to the output, and the task is to justify the signal values in the *J*-frontier: $\{g = \overline{D}, f = 1, c = 1\}$. To justify $g = \overline{D}$, two choices are possible: $a = 0$ or $b = 0$. If $a = 0$ is selected, it is necessary to justify $f = 1$, $b = 1$. Finally, $c = 1$ remains in the *J*-frontier which is still unjustified. At this time, a contradiction has occurred ($a = 0$ and $c = 1$), and the search reverses its last decision, changing $a = 0$ to $a = 1$. The search discovers that this reversal also causes a conflict. Thus, a backtrack occurs where line b is chosen instead of a for the previous decision, so a is reset to "don't care." By assigning $b = 0$, a conflict is observed. Reversing b also cannot justify all the *J*-frontier. At this time, backtracking on b leads to no prior decisions. Thus, target fault $g/1$ is declared to be untestable.

14.4.4 PODEM

In the *D* algorithm, the decision space encompasses the entire circuit. In other words, every internal gate could be a decision point. However, noting that the end result of any ATPG algorithm is to derive a solution vector at the primary inputs and that the number of primary inputs generally is much fewer than the total number of gates, it may be possible to arrive at a very different ATPG algorithm that makes decisions only at primary inputs rather than at internal nodes of the circuit.

The **path-oriented decision-making** (PODEM) algorithm [Goel 1981] is based on this notion and makes decisions only at the primary inputs. Similar

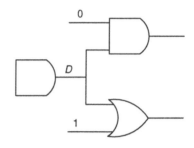

FIGURE 14.35

No X path.

to the D algorithm, a D-frontier is kept. However, because decisions are made at the primary inputs, the J-frontier is unnecessary. At each step of the ATPG search process, it checks whether the target fault is excited. If the fault is excited, it then checks whether there is an X-path from at least one fault- effect in the D-frontier to a primary output, where an X-path is a path of unspecified values from the fault-effect to a primary output. If no X-path exists, it means that all the fault-effects in the D-frontier are blocked, as illustrated in Figure 14.35, where both possible propagation paths of the D have been blocked. Otherwise, PODEM will pick the best X-path to propagate the fault-effect. Note that if the target fault has not been excited, the first steps of PODEM will be to excite the fault.

The basic flow of PODEM is illustrated in Algorithms 14.9 and 14.10. Although it is still a deterministic search algorithm, the decisions are limited to the primary inputs. All internal signals obtain their logic values by means of logic simulation (or implications) from the decision points. As a result, no conflict will ever occur at the internal signals of the circuit. The only possible conflicts in PODEM are either (1) the target fault is not excited or (2) the D-frontier becomes empty. In either of these cases, the search must backtrack.

Algorithm 14.9 PODEM(C, f)

---

1. initialize all gates to don't-cares;
2. D-frontier = ϕ;
3. result = PODEM-Recursion(C);
4. **if** result == success **then**
5.   print out values at the primary inputs;
6. **else**
7.   print fault f is untestable;
8. **end if**

Algorithm 14.10 PODEM-Recursion(C)

---

1. **if** fault-effect is observed at a PO **then**
2.   return (success);
3. **end if**
4. $(g, v) = $ getObjective(C);
5. $(pi, u) = $ backtrace (g, v);
6. logicSimulate_and_imply (pi, u);
7. result $=$ PODEM-Recursion(C);
8. **if** result $==$ success **then**
9.   return(success);
10. **end if**
11. /* backtrack */
12. logicSimulate_and_imply $(pi, \overline{u})$;
13. result $=$ PODEM-Recursion(C);
14. **if** result $==$ success **then**
15.   return(success);
16. **end if**
17. /* bad decision made at an earlier step, reset pi */
18. logicSimulate_and_imply (pi, X);
19. return(failure);

According to the algorithm in PODEM, the search starts by picking an objective, and it backtraces from the objective to a primary input by means of the best path. Controllability measures can be used here to determine which path is regarded as the best. Gradually more primary inputs will be assigned logic values. At any time the target fault becomes unexcited or the D-frontier becomes empty, a bad decision must have been made, and reversal of some previous decisions is needed. The backtracking mechanism proceeds by reversing the most recent decision. If reversing the most recent decision also causes a conflict, the recursive algorithm will continue to backtrack to earlier decisions, until no more reversals are possible, at which time the fault is determined to be undetectable.

Three important functions in PODEM-Recursion() are getObjective(), backtrace(), and logicSimulate_and_imply(). The getObjective() function returns the next objective the ATPG should try to justify. Before the target fault has been excited, the objective is simply to set the line on which the target fault resides to the value opposite to the stuck value. Once the fault is excited, the getObjective() function selects the best fault-effect from the D-frontier to propagate. The pseudo-code for getObjective() is shown in Algorithm 14.11.

Algorithm 14.11 getObjective(*C*)

---

1. **if** fault is not excited **then**
2.   return (*g*, $\bar{v}$);
3. **end if**
4. *d* = a gate in *D*-frontier;
5. *g* = an input of *d* whose value is *X*;
6. *v* = noncontrolling value of *d*;
7. return (*g*, *v*);

The backtrace() function returns a primary input assignment from which there is a path of unjustified gates to the current objective. Thus, backtrace() will never traverse through a path consisting of one or more justified gates. From the objective's point of view, the getObjective() function returns an objective, say $g = v$, which means the current value of g is unspecified and should be set to value v. If g was already specified to v, $g = v$ would have never been selected as an objective, because it is already justified. Now, if $g = x$ currently, and the objective is to set $g = v$, there must exist a path of unjustified gates from at least one primary input to g. This backtrace() function can simply be implemented as a loop from the objective to some primary inputs through a path of unspecified values. Algorithm 14.12 shows the pseudo-code for the backtrace() routine.

Finally, the logicSimulate_and_imply() function can simply be a regular logic simulation routine. The added imply is used to derive additional implications, if any, that can enhance the getObjective() routine later on.

Consider the multiplexer circuit shown in Figure 14.28 again. Consider the target fault f stuck-at-0. First, PODEM initializes all gate values to X. Then, the first objective would be to set $f = 1$. The backtrace routine selects $c = 0$ as the decision. After logic simulation, the fault is excited, together with $e = b = 0$. The D-frontier at this time is g. The next objective is to advance the D-frontier, thus getObjective() returns $a = 1$. Because a is already a primary input, backtrace() will simply return $a = 1$. After simulating $a = 1$, the fault-effect is successfully propagated to the primary output z, and PODEM is finished with this target fault with the computed vector $abc = 1X0$. Table 14.9 shows the series of objectives and backtraces for this example.

Table 14.9 PODEM Objectives and Decisions for *f* Stuck-At-0

getObjective()	backtrace()	logicSim()	D-frontier
f = 1	*c* = 0	*d* = 0, *f* = D, *e* = 0, *h* = 0	*g*
a = 1	*a* = 1	*g* = D, *z* = D	*f*/0 detected

Algorithm 14.12 backtrace(*C*)

---

    1. *i* = *g*;
    2. num_inversion = 0;
    3. **while** *i* ≠ primary input **do**
    4.   *i* = an input of *i* whose value is *X*;
    5.   **if** *i* is an inverted gate type **then**
    6.   num_inversion ++;
    7.   **end if**
    8. **end while**
    9. **if** num_inversion == odd **then**
   10.   *v* = $\bar{v}$;
   11. **end if**
   12. return (i, *v*);

Consider the circuit shown in Figure 14.29. Suppose the target fault is *b* stuck-at-0. After circuit initialization, the first objective is $b = 1$ to excite the fault. The backtrace() returns $a = 0$. After logic simulation, although the target fault is excited, there is no *D*-frontier, because $c = d = 0$. At this time, PODEM reverses its last decision $a = 0$ to $a = 1$. After logic simulating $a = 1$, the target fault is not excited and the *D*-frontier is still empty. PODEM backtracks but there is no prior decision point. Thus, it concludes that fault *b*/0 is undetectable. Table 14.10 shows the steps made for this example, and Figure 14.36 shows the corresponding decision tree.

Consider again the circuit shown in Figure 14.34 with the target fault *g*/1. After circuit initialization, the first objective is to excite the fault; in other words, the objective is $g = 0$. The backtrace() function backtraces from the objective backward to a primary input via a path of "don't cares." Suppose the

---

Table 14.10 PODEM Objectives and Decisions for *b* Stuck-At-0

getObjective()	backtrace()	logicSim()	D-frontier
b = 1	a = 0	b = 1, c = 0, d = 0	{}
a = 1 (reversal)	–	b = 0, c = 1, d = 0	{}

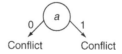

FIGURE 14.36

Decision tree for fault *b*/0.

backtrace reaches $a = 0$. After logic simulation, $g = 0$, $c = d = 0$, and $i = 0$. The D-frontier is h. However, note that there is no path of "don't cares" from any fault-effect in the D-frontier to a primary output! If the PODEM algorithm is modified to check that any objective has at least a path of "don't cares" to one or more primary outputs, some needless searches can be avoided. For instance, in this example, if the next objective was $f = 1$, even after the decision of $b = 1$ is made, the target fault still would not have been detected, because there was no path to propagate the fault-effect to a primary output even before the decision $b = 1$ was made. In other words, the search could immediately backtrack on the first decision $a = 0$. In this case, $a = 1$, and the objective is still $g = 0$. Backtrace() will now return $b = 0$. After logic simulation, $g = 0$, $c = 1$, $f = 0$, $b = 0$, $i = 0$. Again, there is no propagation path possible. As there is no earlier decision to backtrack to, the ATPG concludes that fault $g/1$ is untestable. Table 14.11 shows the steps for this example.

14.4.5 FAN

Although PODEM reduces the number of decision points from the number of gates in the circuit to the number of primary inputs, it still can make an excessive number of decisions. Furthermore, because PODEM targets one objective at a time, the decision process may sometimes be too localized and miss the global picture. The **fanout-oriented test generation** (FAN) algorithm [Fujiwara 1983] extends the PODEM-based algorithm to remedy these shortcomings.

To reduce the number of decision points, FAN first identifies the **headlines** in the circuit, which are the output signals of fanout-free regions. Because of the fanout-free nature of each cone, all signals outside the cone that do not conflict with the headline assignment would never require a conflicting value assignment on the primary inputs of the corresponding fanin cone. In other words, any value assignment on the headline can always be justified by its fanin cone. This allows the backtrace() function to backtrace to either headlines or primary inputs. Because each headline has a corresponding fanin cone with several primary inputs, this allows the number of decision points to be reduced.

Consider the circuit shown in Figure 14.37. If the current objective is to set $z = 1$, the corresponding decision tree based on the PODEM algorithm will involve many decisions at the primary inputs, such as $a = 1$, $c = 1$, $d = 1$, $e = 1$, $f = 1$. On the other hand, the decision based on the FAN algorithm is

Table 14.11 PODEM Objectives and Decisions for g Stuck-At-1

getObjective()	backtrace()	logicSim()	D-frontier
$g = 1$	$a = 0$	$g = D$, $c = 0$, $d = 0$, $i = 0$	$\{h\}$ (but no X-path to PO)
$a = 1$ (reversal)	–	$c = 1$, $d = 1$	$\{\}$

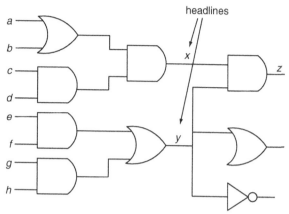

FIGURE 14.37

Circuit with identified headlines.

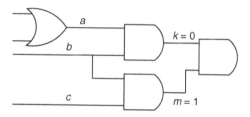

FIGURE 14.38

Multiple backtrace to avoid potential conflicts.

significantly smaller, involving only two decisions: $x = 1$ and $y = 1$. If $z = 1$ was not the first objective, there would have been other decisions made earlier. In other words, if there were a poor decision made in an earlier step, PODEM would need to reverse and backtrack many more decisions compared with FAN.

The next improvement that FAN makes over PODEM is the simultaneous satisfaction of multiple objectives, as opposed to only one target objective at each step. Consider the circuit fragment shown in Figure 14.38. Without taking into account multiple objectives, the backtrace() routine may choose the easier path in trying to justify $k = 0$. The easier path may be through the fanout stem b. However, this would cause a conflict later on with the other objective $m = 1$. In FAN, multiple objectives are taken into account, and the backtrace routine scores the nodes visited from each objective in the current set of objectives. The nodes along the path with the best scores are chosen. In this example, $a = 0$ will be chosen rather than $b = 0$, even if $a = 0$ is less controllable.

A powerful implication engine can have a significant impact on the performance of ATPG algorithms. Thus, much effort has been invested over the years in the efficient computation of implications. The quality of implications was improved with the computation of indirect implications in **SOCRATES** [Schulz 1988]. **Static learning** was extended to **dynamic learning** in [Schulz 1989 and Kunz 1993], where some nodes in the circuit already had value assignments during the learning process. A 16-valued logic was introduced in [Rajski 1990 and Cox 1994]. Reduction lists were used to dynamically determine the gate values. In [Chakradhar 1993], the authors proposed a transitive closure procedure based on the implication graph. **Recursive learning** was later proposed in [Kunz 1994] in which a complete set of pairwise implications could be computed. To keep the computational costs low, a small recursion depth can be enforced in the recursive learning procedure. Finally, implications to capture time frame information in sequential circuits in a graphical representation were proposed in [Zhao 2001] to compactly store the implications in sequential circuits.

The implications can be used to quickly identify untestable faults [Iyer 1996a,b; Zhao 2001; Hsiao, 2002; Syal 2003]. This will allow the ATPG not to specifically target these faults that can often consume much of the ATPG computational resources. For more information on implication and untestable fault identification, refer to [Bushnell 2000, Jha 2003, and Wang 2006].

14.5 ADVANCED TEST GENERATION

Thus far, the discussions have focused primarily on the basic ATPG algorithms. As circuits have become increasingly larger and more complex, more powerful ATPG algorithms are needed. In particular, the handling of sequential circuits is a must, because not all circuits may have the luxury of having a full-scan inserted. Next, deterministic ATPGs may face tremendous hurdles when dealing with the need to generate a sequence of many vectors. In this regard, simulation-based ATPGs may be better suited. Finally, the stuck-at fault model may be insufficient in capturing defects that occur at the deep-submicron or nanoscale designs. Such defects include delay faults and bridging faults. This section addresses how the basic ATPG can be extended to deal with these issues.

14.5.1 Sequential ATPG: Time frame expansion

Test generation for sequential circuits bears much similarity with that for combinational circuits. However, one vector may be insufficient to detect the target fault, because the excitation and propagation conditions may necessitate some of the flip-flop values to be specified at certain values. The general model for a sequential circuit is shown in Figure 14.39, where flip-flops constitute the memory/state elements of the design. All the flip-flops receive the same clock signal, so no multiple clocks are assumed in the circuit model.

Figure 14.40 illustrates an example of a sequential circuit that is unrolled into several time frames, also called an *iterative logic array* (ILA) of the

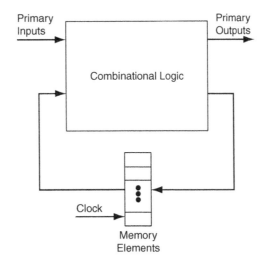

FIGURE 14.39

Model of a sequential circuit.

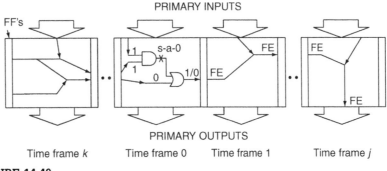

FIGURE 14.40

An iterative logic array (ILA) model.

circuit. For each time frame, the flip-flop inputs from the previous time frame are often referred to as **pseudo primary inputs** with respect to that time frame, and the output signals to feed the flip-flops to the next time frame are referred to as **pseudo primary outputs**. Note that in any unrolled circuit, a target fault is present in every time frame.

When the test generation begins, the first time frame is referred to as time frame 0. An ATPG search similar to a combinational circuit is carried out. At the end of the search, a combinational vector is derived, where the input vector consists of primary inputs and pseudo primary inputs. The fault-effect for the target fault may be sensitized to either a primary output of the time frame or a pseudo primary output. If at least one pseudo primary input has been specified, then the search must attempt to justify the needed flip-flop values in time frame -1. Similarly, if fault-effects only propagate to pseudo primary outputs, the ATPG must try to propagate the fault-effects across time frame $+1$. Note that this results in a **test sequence** of vectors. As opposed to combinational circuits, in which a single vector is sufficient to detect a detectable fault, in sequential circuits a test sequence is often needed.

One question naturally arises: Should the ATPG first attempt the fault excitation via several time frames -1, -2, etc., or should the ATPG attempt to propagate the fault-effect through time frames 1, 2, etc.? It can be observed that in propagating the fault-effect in time frame 1, the search may place additional values on the flip-flops between the boundary of time frames 0 and 1. These added constraints propagate backward and may add additional values needed at the pseudo primary inputs at time frame 0. In other words, if the ATPG first justifies the pseudo primary inputs at time frame 0, it would have missed the additional constraints placed by the propagation. Therefore, the ATPG first tries to propagate the fault-effect to a primary output via several time frames, with all the intermediate flip-flop values propagated back to time frame 0. Then, the ATPG proceeds to justify all the pseudo primary input values at time frame 0.

Although easy to understand, the process can be very complex, for example, if the fault-effect has propagated forward for three time frames: time frames 1, 2, and 3. Now in time frame 4, suppose the ATPG successfully propagates the fault-effect to a primary output (*i.e.*, it has derived a vector at time frame 4), it must go back to time frame 3 to make sure the values assigned to the flip-flops at the boundary between time frames 3 and 4 are, indeed, possible. It must perform this check for time frames 2, 1, and 0. If at any time frame a conflict occurs, the vector derived at time frame 4 is actually invalid, because it is not justifiable from the previous vectors. At this time, a backtrack occurs in time frame 4, and the ATPG must try to find a different solution vector #4. This process is repeated.

One way to reduce the complexity discussed is to try to propagate the fault-effect in an unrolled circuit instead of propagating the fault-effect time frame by time frame. In doing so, a k-frame combinational circuit is obtained, say $k = 256$, and the ATPG views the entire 256-frame circuit as one large combinational circuit. However, the ATPG must keep in mind that the target fault is present in all 256 time frames. This eliminates the need to check for state boundary justifiability and allows the ATPG to propagate the fault-effect across multiple time frames at a time.

When the fault-effect has been propagated to at least one primary output, the pseudo primary inputs at time frame 0 must be justified. Again, the justification can be performed in a similar process of viewing an unrolled 256-frame circuit. As before, the ATPG must ensure that the fault is present in every time frame of the unrolled circuit.

HITEC [Niermann 1991] is a popular sequential test generator that performs the search similar to the discussed methods with a 9-valued algebra. In addition, it uses the concept of **dominators** to help reduce the search complexity. A dominator for a target fault is a gate in the circuit through which the fault-effect must traverse [Kirkland 1987]. Therefore, for a given target fault, all inputs of any dominator gate that are not in the fanout cone of the fault must be assigned to noncontrolling values to detect the fault.

The concept of controllability and observability metrics can be extended to sequential circuits such that the backtrace routine would prefer to backtrace toward primary inputs and those easy-to-justify flip-flops. The use of sequential testability metrics allows the ATPG to narrow the search space by favoring the easy-to-reach states and avoiding getting into difficult-to-justify states.

The computational complexity of a sequential ATPG is intuitively higher than that of the combinational ATPG. Therefore, aggressive learning can help to reduce the computational cost. For instance, if a known subset of unreachable states is available, this information can be used to allow the ATPG to backtrack much sooner when an intermediate state is unreachable. This can avoid successive justification of an unreachable state. Likewise, if a justification sequence has been successfully computed for state S before, and a different target fault requires the same state S, the previous justification sequence can be used to guide the search. Note that, because the target faults are different, the justification sequence may not simply be copied from the solution for one fault to another.

For large circuits, deterministic ATPGs may suffer from a potentially large number of backtracks. Thus, in the past two decades, effort on **simulation-based ATPGS** has yielded much success, presenting themselves as a viable alternative to deterministic ATPGs. One class of nondeterministic ATPGs is the **genetic algorithm–based** (GA-based) ATPG. There have been numerous GA-based ATPGs proposed over the years. For example, **CONTEST** [Agrawal 1989] targets test generation in three phases, each having its own distinct fitness measure. **GATEST** [Rudnick 1994] distinguishes fault detection from those that only propagate to flip-flop boundaries. **DIGATE** [Hsiao 1996] targets individual faults and uses distinguishing sequences to help propagate the faults from flip-flops to a primary output.

STRATEGATE [Hsiao 1997; 2000] addresses fault excitation by justifying the needed state as well. Although GA-based ATPGs have achieved success, the underlying fault simulation engine may incur excessive computational cost. In recent years, approaches that use logic simulation rather than fault simulation have been proposed [Pomeranz 1995; Guo 1999; Giani 2001; Sheng 2002; Wu 2004]. Logic-simulation–based test generators usually target some inherent "property" in the fault-free circuit and try to derive test vectors that exercise these properties. In general, the property used often relates to the states reached by the test sequence.

14.5.2 Delay fault ATPG

Today's integrated circuits are seeing an escalating clock rate, shrinking dimensions, increasing chip density, etc. Consequently, there arises a class of defects that would affect the functionality of the design if the chip were run at a high speed.

In other words, the design is functionally correct when it is operated at a slow clock. This type of defect is referred to as a **delay defect**. Although the conventional stuck-at testing can catch some delay defects, the stuck-at fault model is insufficient to model delay defects satisfactorily. This has prompted engineers and researchers to propose a variety of methods and fault models for detecting speed failures. Among the fault models are the *transition fault* [Levendel 1986; Waicukauski 1987; Cheng 1993], the *path-delay fault* [Smith 1985], and the *segment delay fault* [Heragu 1996]. The path-delay fault model considers the cumulative effect of the delays along a specific combinational path in the circuit. If the cumulative delay in a faulty circuit exceeds the clock period for the path, then the test pattern that can exercise this path will fail the chip. The segment delay fault model targets path segments instead of complete paths.

Because a transition has to be launched to propagate across a given path, two vectors are needed. The first vector initializes the circuit nodes, and the second vector launches a transition at the start of a path and ensures that the transition is propagated along the given path. Given a path P, a signal is an **on-input** of P if it is on P. Conversely, a signal is an **off-input** of P if it is an input to a gate in P but is not an on-input of P. A path-delay fault can be a rising fault, where a rising transition is at the start of the path, or a falling fault, where a falling transition is at the start of the path. The rising and falling path-delay faults are denoted with the up-arrow $\uparrow$ and the down-arrow $\downarrow$ before path P, respectively. For example, $\uparrow g_1 g_4 g_7$ is a rising path that traverses through gates g_1, g_4, and g_7.

Delay tests can be applied three different ways: **launch-on-capture** (also called broad-side [Savir 1994] or double-capture [Wang 2006]), **launch-on-shift** (also called skewed-load [Savir 1993]), and **enhanced-scan** [Dervisoglu 1991]. In launch-on–capture-based testing, the first n-bit vector is scanned into the circuit with n scan flip-flops at a slow speed, followed by another clock that creates the transition. Finally, an at-speed functional clock is applied that captures the response. Thus, only one vector has to be stored per test, and the second vector is directly derived from the initial vector by pulsing the clock. In launch-on–shift-based testing, the first $n-1$ bits of an n-bit vector are shifted in at a slow speed. The final nth shift is performed, and it is also used to launch the transition. This is followed by an at-speed quick capture. Similar to launch-on-capture, only one vector has to be stored per test, because the second vector is simply the shifted version of the first vector. Finally, in enhanced-scan testing, both vectors in the vector pair (V_1, V_2) have to be stored in the tester memory. The first vector V_1 is loaded into the scan chain, followed by its immediate application to initialize the circuit under test. Next, the second vector is scanned in, followed by an immediate application and capture of the response. Note that the node values in the circuit are preserved during the shifting-in of the second vector V_2. To achieve this, a **hold-scan design** [Dervisoglu 1991] is required.

Because both launch-on-capture and launch-on-shift place constraints on what the second vector can be, they will achieve lower fault coverage compared with enhanced-scan. However, enhanced-scan comes at a price of

hold-scan cells (enhanced-scan cells [Wang 2006]), which consume more chip area. This may not be viewed as a huge negative in microprocessors and some custom-designed circuits, because hold-scan cells are used to prevent the combinational logic from seeing the values being shifted. This is done because the intermediate state of the scan cells may cause contention in some of the signals in the logic, as well as reducing the power consumption in the combinational logic during the shifting of the data in scan cells. In addition, hold-scan cells also help increase the diagnostic capability on failing chips in which the data captured in the scan chain can be retrieved.

In terms of test data volume, enhanced-scan tests may actually require less storage to achieve the same delay fault coverage. In other words, for launch-on-capture or launch-on-shift to achieve the same level of fault coverage, many more patterns may have to be applied.

Unlike stuck-at faults, where a fault is either detected or not detected by a given test vector, a path-delay fault may be detected by different test patterns (consisting of two vectors) with differing levels of quality. In other words, some test patterns can detect a path-delay fault only with certain restrictions in place. Higher quality test patterns place more restrictions on sensitization of the path. On the other hand, similar to stuck-at faults, some paths may be untestable if the sensitization requirement for a given path is not satisfiable.

For designs with two interactive clock domains, modifications can be made to allow for tests. For example, the following at-speed delay test approaches can be used for both launch-on-capture and launch-on-shift architectures: **one-hot double-capture, aligned double-capture**, and **staggered double-capture** [Bhawmik 1997; Wang 2006, 2007b].

If tests were possible for all the paths in a circuit, we would not need any additional test vectors for capturing the delay defects. However, because very few paths are robustly testable, and the number of path-delay faults is exponential to the number of circuit lines, other delay fault models have been proposed. For example, transition tests have been generated to improve the detection of speed failures in microprocessors [Tendulkar 2002], as well as *application-specific integrated circuits* (ASICs) [Hsu 2001]. These reasons make transition faults popular in industry.

Similar to the stuck-at fault model, two transition faults are possible at each node of the circuit: *slow-to-rise* and *slow-to-fall*. A test pattern for a transition fault consists of a pair of vectors (V_1, V_2), where V_1 (called the *initial vector*) is required to set the target node to an initial value and V_2 (called the *test vector*) is required to launch the corresponding transition at the target node and also propagate the fault effect to a primary output [Waicukauski 1987; Savir 1993].

Transition tests can also be applied in three different ways as for the other delay fault models discussed earlier: launch-on-capture, launch-on-shift, and enhanced scan. As with path-delay tests, because both launch-on-capture and launch-on-shift place constraints on what the second vector can be, they will achieve lower transition fault coverage compared with enhanced-scan.

14.5.3 **Bridging fault ATPG**

Recall that bridging faults are those faults that involve a short between two signals in the circuit. Given a circuit with n signals, there are potentially $n \times (n - 1)$ possible bridging faults. However, practically, only those signals that are locally close on the die are more likely to be bridged. Therefore, the total number of bridging faults can be reduced to be linear in the number of signals in the circuit.

Consider two signals x and y in the circuit that are bridged. This bridging fault will not be excited unless different values are placed on x and y. Note that the actual voltage at x and y may be different because of the resistance value of the bridge. Subsequently, the logic that takes x as its input may interpret the logic value differently from the logic that takes y as its input. To reduce the complexity, five common bridging fault models are often used:

1. AND bridge—The faulty value of the bridge for x' and y' is taken to be the logical AND of x and y in the original fault-free circuit.
2. OR bridge—The faulty value of the bridge for x' and y' is taken to be the logical OR of x and y in the original fault-free circuit.
3. x DOM y bridge—x dominates y; in other words, the faulty value of the bridge for both x' and y' is taken to be the logic value of x in the fault-free circuit.
4. x DOM1 y bridge—x dominates y if $x = 1$; in other words, the faulty value of x' is unaffected, but the faulty value for y' is taken to be the logical OR of x and y in the fault-free circuit.
5. x DOM0 y bridge—x dominates y if $x = 0$; in other words, the faulty value of x' is unaffected, but the faulty value for y' is taken to be the logical AND of x and y in the fault-free circuit.

Figure 14.41 illustrates the faulty circuit models corresponding to each of these five bridge types. If a path exists between x and y, then the bridging fault is said to be a **feedback-bridging fault**. Otherwise, it is a **non-feedback-bridging fault**. Figure 14.42 illustrates a feedback-bridging fault. In this figure, if $abc = 110$, then in the fault-free circuit $z = 0$. If the bridge is an AND-bridge, then a cycle would result. In other words, a becomes 0 and in turn makes z = 1. Because $a = 1$ initially, it will again try to drive $z = 0$, resulting in an infinite loop around the bridge. For the following discussion, only non-feedback bridging faults will be considered.

Testing for bridging faults is similar to a constrained stuck-at ATPG. In other words, when testing for the AND-bridge(x, y), either (1) x/0 has to be detected with $y = 0$ or (2) y/0 has to be detected with $x = 0$ [Williams 1973]. A conventional stuck-at ATPG can be modified to handle the added constraint. Likewise, the ATPG can be modified for other bridging fault types.

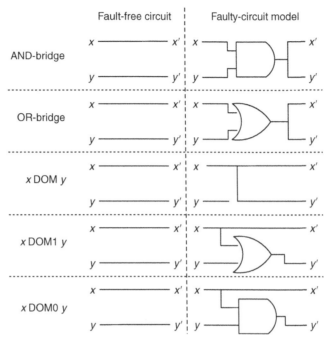

FIGURE 14.41

Bridging fault models.

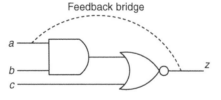

FIGURE 14.42

A feedback bridging fault.

14.6 CONCLUDING REMARKS

For fault simulation, both event-driven simulation and compiled-code simulation techniques can be found in commercially available ***electronic design automation*** (EDA) applications. The fault simulators can be stand-alone tools or used as an integrated feature in the ATPG programs. As a stand-alone tool, concurrent fault simulation with the event-driven simulation technique is used in Verifault-XL (from Cadence Design Systems [Cadence 2008]) and TurboFault (from SynTest Technologies [SynTest 2008]). As an integrated feature in ATPG, bitwise parallel simulation with the compiled-code simulation technique is widely used in modern commercial ATPG programs, including Encounter Test (from Cadence Design Systems), FastScan (from Mentor Graphics [Mentor 2008]), TetraMAX (from Synopsys [Synopsys 2008]), and TurboScan (from SynTest Technologies).

As we move to the nanometer age, we have started to see nanometer designs that contain hundreds of millions of transistors. We anticipate the semiconductor industry will completely adopt the scan method for quality considerations. As a result, it is becoming imperative that advanced techniques for both logic simulation and fault simulation be developed to address the high-performance and high-capacity issues, in particular, for addressing new fault models, such as transition faults [Waicukauski 1986], path-delay faults [Schulz 1989], bridging faults [Li 2003], and small delay defects [Sato 2005; Hamada 2006]. At the same time, more innovations are needed in developing advanced concurrent fault simulation techniques, because at present designs based on the scan method are still not 100% scan testable. Fault simulation with functional patterns is important for at-speed test applications to detect small delay faults and achieve the ***parts-per-million*** (PPM) defect level goals.

The theory and implementation of an ATPG engine have also been described in detail in the second half of this chapter. Several algorithms were laid out with specific examples given. Advanced ATPG algorithms were discussed where sequential ATPG and ATPG for non-stuck-at faults were covered. Test generation remains to be an important research area as circuit sizes and complexities continue to increase. New and powerful algorithms are needed to cope with the increased complexity. In addition, with nanoscale feature sizes, new defect types and hence new fault models will be needed in future ATPGs.

Should there be defective chips that were uncovered by the test set, fault diagnosis and failure analysis are often subsequently performed to identify the causes and further reduce the defect level in the future. To ease the burden of fault diagnosis and failure analysis, adding ***design-for-debug-and-diagnosis*** (DFD), ***design-for-reliability*** (DFR), ***design-for-manufacturability*** (DFM), and ***design-for-yield*** (DFY) features can be implemented in the design. These features and techniques are extensively discussed in [Wang 2006, 2007a]. Finally, successful ATPG algorithms not only can help in the area of manufacturing tests, but they also provide much insight to other EDA problems, such as synthesis and verification.

14.7 EXERCISES

14.1. (Equivalence Fault Collapsing) How many uncollapsed single stuck-at faults are there in circuit *M* shown in Figure 14.43 Please perform equivalence fault collapsing with the simple_EFC algorithm. How many equivalence collapsed faults do you have?

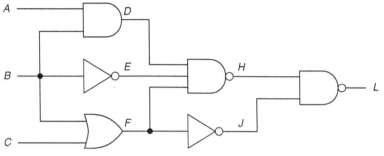

FIGURE 14.43

Circuit *M*.

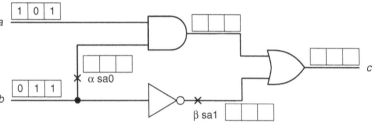

FIGURE 14.44

An example circuit *K*.

14.2. (Dominance Fault Collapsing) Continued from Exercise 14.1. Please perform dominance fault collapsing with the simple_DFC algorithm. How many dominance collapsed faults do you have?

14.3. (Dominance Fault Collapsing) For the circuit in Figure 14.9, please explain why $K/0$ and $K/1$ faults can be removed from the dominance collapsed fault list. Also explain why $F/1$ and $F/0$ can be removed.

14.4. (Parallel-Pattern Single-Fault Propagation) For circuit K shown in Figure 14.44 and two given stuck-at faults shown in Figure 14.44, use the parallel-pattern single-fault propagation fault simulation technique to identify which faults can be detected by the given test patterns.

14.5. (Parallel Fault Simulation) Repeat Exercise 14.4 by use of parallel fault simulation.

14.6. (Concurrent Fault Simulation) Repeat Exercise 14.5 with concurrent fault simulation.

14.7. (Random Test Generation) Given a circuit with three primary outputs, $x, y,$ and z, the fanin cone of x is $\{a, b, c\}$, the fanin cone of y is $\{c, d, e, f\}$, and the fanin cone of z is $\{e, f, g\}$. Devise a pseudo-exhaustive test set for this circuit. Is this test set the minimal pseudo-exhaustive test set?

14.8. (Random Test Generation) With the circuit shown in Figure 14.28, compute the detection probabilities for each of the following faults:

a. $e/0$

b. $e/1$

c. $c/0$

14.9. (Boolean Difference) With the circuit shown in Figure 14.28, compute the set of all vectors that can detect each of the following faults using Boolean difference:

a. $e/0$

b. $e/1$

c. $c/0$

14.10. (Boolean Difference) Assume a single-output combinational circuit, where the output is denoted as f. If two faults, a and b, are indistinguishable, it means that there does not exist a vector that can detect only one and not the other. Show that $f_a \oplus f_b = 0$ if they are indistinguishable.

14.11. (D Algorithm) Construct the table for the XNOR operation for the 5-valued logic similar to Tables 14.6, 14.7, and 14.8.

14.12. (D Algorithm) Consider a three-input AND gate g. Suppose g is a D-frontier. What are all the possible value combinations the three inputs of g can take such that g is a valid D-frontier?

14.13. (PODEM) With the circuit shown in Figure 14.28, compute a test vector that can detect each of the following faults by use of PODEM:

a. $e/0$

b. $e/1$

c. $c/0$

14.14. (FAN) Consider the circuit shown in Figure 14.37. Suppose the constraint that $y = 1 \rightarrow x = 0$ is given. How could one use this knowledge to reduce the search space when trying to generate vectors in the circuit? For example, suppose the target fault is $y/0$.

14.15. (Sequential ATPG) Consider the circuit shown in Figure 14.45. The target fault is $a/0$.

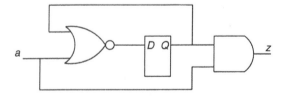

FIGURE 14.45

Example sequential circuit.

a. Generate a test sequence for the target fault by use of only 5-valued logic.

b. Generate a test sequence for the target fault by use of 9-valued logic.

14.16. (Sequential ATPG) Given a sequential circuit, is it possible that two stuck-at faults, $a/0$ and $a/1$, are both detected by the same vector v_t in a test sequence $v_0, v_1, \ldots, v_k$?

14.17. (Sequential ATPG) Consider an **iterative logic array** (ILA) expansion of a sequential circuit, where the initial pseudo primary inputs are fully controllable. Show that the states reachable in successive time frames of the ILA shrink monotonically.

14.18. (Bridging Faults) Consider a bridging fault between the outputs of an AND gate $x = ab$ and an OR gate $y = c + d$. What values to $abcd$ would induce the largest current in the bridge?

ACKNOWLEDGMENTS

We thank Professor Hank Walker of the University of A&M for contributing a portion of the Fault Simulation section; and Professor Xiaoqing Wen of Kyushu Institute of Technology and Professor Charles E. Stroud of Auburn University for reviewing the text and providing helpful comments.

REFERENCES

R14.0 Books

[Abramovici 1994] M. Abramovici, M. A. Breuer, and A. D. Friedman, *Digital Systems Testing and Testable Design, Revised Printing,* IEEE Press, Piscataway, NJ, 1994.

[Bushnell 2000] M. L. Bushnell and V. D. Agrawal, *Essentials of Electronic Testing for Digital, Memory, and Mixed-Signal VLSI circuits,* Springer Science, New York, 2000.

[Holland 1975] J. H. Holland, *Adaptation in Natural and Artificial Systems,* University of Michigan Press, Ann Arbor, MI, 1975.

[Jha 2003] N. Jha and S. Gupta, *Testing of Digital Systems,* Cambridge University Press, London, 2003.

[Wang 2006] L.-T. Wang, C.-W. Wu, and X. Wen, editors, *VLSI Test Principles and Architectures: Design for Testability,* Morgan Kaufmann, San Francisco, 2006.

[Wang 2007a] L.-T. Wang, C. E. Stroud, and N. A. Touba, editors, *System-on-Chip Test Architectures: Nanometer Design for Testability,* Morgan Kaufmann, San Francisco, November 2007.

R14.1 Fault Collapsing

[McCluskey 1971] E. J. McCluskey and F. W. Clegg, Fault equivalence in combinational logic networks, *IEEE Trans. on Computers,* C-20(11), pp. 1286–1293, November 1971.

R14.2 Fault Simulation

[Abramovici 1984] M. Abramovici, P. R. Menon, and D. T. Miller, Critical path tracing: An alternative to fault simulation, *IEEE Design & Test of Computers*, 1(1), pp. 83-93, February 1984.

[Butler 1974] T. T. Butler, T. G. Hallin, J. J. Kulzer, and K. W. Johnson, LAMP: Application to switching system development, *Bell System Technical J.*, 53, pp. 1535-1555, October 1974.

[Cheng 1989] W. T. Cheng and M. L. Yu, Differential fault simulation: A fast method using minimal memory, in *Proc. ACM/IEEE Design Automation Conf.*, pp. 424-428, June 1989.

[Goel 1980] P. Goel, Test generation cost analysis and projections, in *Proc. ACM/IEEE Design Automation Conf.*, pp. 77-84, June 1980.

[Jain 1985] S. K. Jain and V. D. Agrawal, Statistical fault analysis, *IEEE Design & Test of Computers*, 2(1), pp. 38-44, February 1985.

[Niermann 1992] T. M. Niermann, W.-T. Cheng, and J. H. Patel, PROOFS: A fast, memory-efficient sequential circuit fault simulator, *IEEE Trans. on Computer-Aided Design*, 11(2), pp. 198-207, February 1992.

[Schulz 1989] M. Schulz, F. Fink, and K. Fuchs, Parallel pattern fault simulation of path delay faults, in *Proc. ACM/IEEE Design Automation Conf.*, pp. 357-363, June 1989.

[Seshu 1965] S. Sesuh and D. N. Freeman, On improved diagnosis program, *IEEE Trans. on Electronic Computers*, Vol. EC-14(1), pp. 76-79, February 1965.

[Ulrich 1974] E. G. Ulrich and T. Baker, Concurrent simulation of nearly identical digital networks, *IEEE Trans. on Computers*, 7(4), pp. 39-44, April 1974.

[Waicukauski 1985] J. A. Waicukauski, E. B. Eichelberger, D. O. Forlenza, E. Lindbloom, and T. McCarthy, Fault Simulation for Structured VLSI, in *Proc. VLSI System Design*, 6(12), pp. 20-32, December 1985.

[Waicukauski 1986] J. A. Waicukauski, E. Lindbloom, B. K. Rosen, and V. S. Iyengar, Transition fault simulation by parallel pattern single fault propagation, in *Proc. IEEE Int. Test Conf.*, pp. 542-549, September 1986.

R14.3 Test Generation

[Breuer 1971] M. A. Breuer, A random and an algorithmic technique for fault detection test generation for sequential circuits, *IEEE Trans. on Computers*, 20(11), pp. 1364-1370, November 1971.

[Chakradhar 1993] S. T. Chakradhar, V. D. Agrawal, and S. G. Rothweiler, A transitive closure algorithm for test generation, *IEEE Trans. on Computer-Aided Design*, 12(7), pp. 1015-1028, July 1993.

[Cox 1994] H. Cox and J. Rajski, On necessary and non-conflicting assignments in algorithmic test pattern generation, *IEEE Trans. on Computer-Aided Design*, 13(4), pp. 515-530, April 1994.

[David 1976] R. David and G. Blanchet, About random fault detection of combinational networks, *IEEE Trans. on Computers*, C-25(6), pp. 659-664, June 1976.

[Fujiwara 1983] H. Fujiwara and T. Shimono, On the acceleration of test generation algorithms, *IEEE Trans. on Computers*, C-32(12), pp. 1137-1144, December 1983.

[Goel 1981] P. Goel, An implicit enumeration algorithm to generate tests for combinational logic circuits, *IEEE Trans. on Computers*, C-30(3), pp. 215-222, March 1981.

[Hsiao 2002] M. S. Hsiao, Maximizing impossibilities for untestable fault identification, in *Proc. Design, Automation, and Test in Europe Conf.*, pp. 949-953, March 2002.

[Iyer 1996a] M. A. Iyer and M. Abramovici, FIRE: A fault independent combinational redundancy algorithm, *IEEE Trans. VLSI Syst.*, 4(2), pp. 295-301, June 1996.

[Iyer 1996b] M. A. Iyer, D. E. Long, and M. Abramovici, Identifying sequential redundancies without search, in *Proc. ACM/IEEE Design Automation Conf.*, pp. 457-462, June 1996.

[Kunz 1993] W. Kunz and D. K. Pradhan, Accelerated dynamic learning for test pattern generation in combinational circuits, *IEEE Trans. on Computer-Aided Design*, 12(5), pp. 684-694, May 1993.

[Kunz 1994] W. Kunz and D. K. Pradhan, Recursive learning: A new implication technique for efficient solutions to CAD problems—test, verification, and optimization, *IEEE Trans. on Computer-Aided Design*, 13(9), pp. 1149-1158, September 1994.

[Lisanke 1987] R. Lisanke, F. Brglez, A. J. Degeus, and D. Gregory, Testability-driven random test-pattern generation, *IEEE Trans. on Computer-Aided Design*, 6(6), pp. 1082-1087, November 1987.

[Marques-Silva 1999] J. P. Marques-Silva and K. A. Sakallah, GRASP: A search algorithm for propositional satisfiability, *IEEE Trans. on Computers*, 48(5), pp. 506-521, May 1999.

[Muth 1976] P. Muth, A nine-valued circuit model for test generation, *IEEE Trans. on Computers*, C-25(6), pp. 630-636, June 1976.

[Rajski 1990] J. Rajski and H. Cox, A method to calculate necessary assignments in ATPG, in *Proc. IEEE Int. Test Conf.*, pp. 25-34, October 1990.

[Roth 1966] J. P. Roth, Diagnosis of automata failures: A calculus and a method, in *IBM J. Research and Development*, 10(4), pp. 278-291, July 1966.

[Roth 1967] J. P. Roth, W. G. Bouricius, and P. R. Schneider, Programmed algorithms to compute tests to detect and distinguish between failures in logic circuits, *IEEE Trans. on Electron. Comput.*, EC-16(10), pp. 567-579, October 1967.

[Schnurmann 1975] H. D. Schnurmann, E. Lindbloom, and R. G. Carpenter, The weighted random test-pattern generator, *IEEE Trans. on Computers*, 24(7), pp. 695-700, July 1975.

[Schulz 1988] M. H. Schulz, E. Trischler, and T. M. Sarfert, SOCRATES: A highly efficient automatic test pattern generation system, *IEEE Trans. on Computer-Aided Design*, 7(1), pp. 126-137, January 1988.

[Schulz 1989] M. H. Schulz and E. Auth, Improved deterministic test pattern generation with applications to redundancy identification, *IEEE Trans. on Computer-Aided Design*, 8(7), pp. 811-816, July 1989.

[Seshu 1965] S. Seshu and D. N. Freeman, The diagnosis of synchronous sequential switching systems, *IEEE Trans. on Electron. Comput.*, 11, pp. 459-465, August 1962.

[Shedletsky 1977] J. J. Shedletsky, Random testing: Practicality vs. verified effectiveness, in *Proc. IEEE Int. Symp. on Fault-Tolerant Computing*, pp. 175-179, June 1977.

[Syal 2003] M. Syal and M. S. Hsiao, A novel, low-cost algorithm for sequentially untestable fault identification, in *Proc. ACM/IEEE Design, Automation, and Test in Europe Conf.*, pp. 316-321, March 2003.

[Zhao 2001] J. Zhao, J. A. Newquist, and J. H. Patel, A graph traversal based framework for sequential logic implication with an application to C-cycle redundancy identification, in *Proc. IEEE Int. Conf. on VLSI Design*, pp. 163-169, January 2001.

R14.4 Advanced Test Generation

[Agrawal 1989] V. D. Agrawal, K.-T. Cheng, and P. Agrawal, A directed search method for test generation using a concurrent simulator, *IEEE Trans. on Computer-Aided Design*, 8(2), pp. 131-138, February 1989.

[Bhawmik 1997] S. Bhawmik, Method and Apparatus for Built-In Self-Test with Multiple Clock Circuits, U.S. Patent No. 5,680,543, October 21, 1997.

[Cheng 1993] K.-T. Cheng, S. Devadas, and K. Keutzer, Delay-fault test generation and synthesis for testability under a standard scan design methodology, *IEEE Trans. on Computer-Aided Design*, 12(8), pp. 1217-1231, August 1993.

[Dervisoglu 1991] B. Dervisoglu and G. Stong, Design for testability: Using scanpath techniques for path-delay test and measurement, in *Proc. IEEE Int. Test Conf.*, pp. 365-374, October 1991.

[Giani 2001] A. Giani, S. Sheng, M. S. Hsiao, and V. Agrawal, Efficient spectral techniques for sequential ATPG, in *Proc. IEEE Design, Automation, and Test in Europe Conf.*, pp. 204-208, March 2001.

[Guo 1999] R. Guo, S. M. Reddy, and I. Pomeranz, Proptest: A property based test pattern generator for sequential circuits using test compaction, in *Proc. ACM/IEEE Design Automation Conf.*, pp. 653–659, June 1999.

[Heragu 1996] K. Heragu, J. H. Patel, and V. D. Agrawal, Segment delay faults: A new fault model, in *Proc. IEEE VLSI Test Symp.*, pp. 32–39, April 1996.

[Hsiao 1996] M. S. Hsiao, E. M. Rudnick, and J. H. Patel, Automatic test generation using genetically engineered distinguishing sequences, in *Proc. IEEE VLSI Test Symp.*, pp. 216–223, April 1996.

[Hsiao 1997] M. S. Hsiao, E. M. Rudnick, and J. H. Patel, Sequential circuit test generation using dynamic state traversal, in *Proc. European Design and Test Conf.*, pp. 22–28, February 1997.

[Hsiao 2000] M. S. Hsiao, E. M. Rudnick, and J. H. Patel, Dynamic state traversal for sequential circuit test generation, *ACM Trans. on Design Automation of Electronic Systems*, 5(3), pp. 548–565, July 2000.

[Hsu 2001] F. F. Hsu, K. M. Butler, and J. H. Patel, A case study of the Illinois scan architecture, in *Proc. IEEE Int. Test Conf.*, pp. 538–547, October 2001.

[Kirkland 1987] T. Kirkland and M. R. Mercer, A topological search algorithm for ATPG, in *Proc. ACM/IEEE Design Automation Conf.*, pp. 502–508, June 1987.

[Levendel 1986] Y. Levendel and P. Menon, Transition faults in combinational circuits: Input transition test generation and fault simulation, in *Proc. Fault-Tolerant Computing Symp.*, pp. 278–283, July 1986.

[Niermann 1991] T. M. Niermann and J. H. Patel, HITEC: A test generation package for sequential circuits, in *Proc. European Design Automation Conf.*, pp. 214–218, February 1991.

[Pomeranz 1995] 1Pomeranz and S. M. Reddy, LOCSTEP: A logic simulation based test generation procedure, in *Proc. Fault-Tolerant Computing Symp.*, pp. 110–119, June 1995.

[Rudnick 1994] E. M. Rudnick, J. H. Patel, G. S. Greenstein, and T. M. Niermann, Sequential circuit test generation in a genetic algorithm framework, in *Proc. ACM/IEEE Design Automation Conf.*, pp. 698–704, June 1994.

[Savir 1993] J. Savir and S. Patil, Scan-based transition test, *IEEE Trans. on Computer-Aided Design*, 12(8), pp. 1232–1241, August 1993.

[Savir 1994] J. Savir and S. Patil, On broad-side delay test, in *Proc. IEEE VLSI Test Symp.*, pp. 284–290, April 1994.

[Sheng 2002] S. Sheng, K. Takayama, and M. S. Hsiao, Effective safety property checking based on simulation-based ATPG, in *Proc. ACM/IEEE Design Automation Conf.*, pp. 813–818, June 2002.

[Smith 1985] G. L. Smith, Model for delay faults based upon paths, in *Proc. IEEE Int. Test Conf.*, pp. 342–349, October 1985.

[Tendulkar 2002] N. Tendulkar, R. Raina, R. Woltenburg, X. Lin, B. Swanson, and G. Aldrich, Novel techniques for achieving high at-speed transition fault coverage for Motorola's microprocessors based on PowerPC instruction set architecture, in *Proc. IEEE VLSI Test Symp.*, pp. 3–8, April 2002.

[Waicukauski 1987] J. A. Waicukauski, E. Lindbloom, B. K. Rosen, and V. S. Iyengar, Transition fault simulation, *IEEE Design & Test of Computers*, 4(2), pp. 32–38, April 1987.

[Wang 2007b] L.-T. Wang, P.-C. Hsu, and X. Wen, Multiple-Capture DFT System for Detecting or Locating Crossing Clock-Domain Faults During Scan-Test, U.S. Patent No. 7,260,756, August 21, 2007.

[Williams 1973] M. J. Y. Williams and J. B. Angel, Enhancing testability of large-scale integrated circuits via test points and additional logic, *IEEE Trans. on Computers*, C-22(1), pp. 46–60, January 1973.

[Wu 2004] Q. Wu and M. S. Hsiao, Efficient ATPG for design validation based on partitioned state exploration histories, in *Proc. IEEE VLSI Test Symp.*, pp. 389–394, April 2004.

R14.5 Concluding Remarks

[Cadence 2008] Cadence Design Systems, http://www.cadence.com, April 2004.

[Hamada 2006] S. Hamada, T. Maeda, A. Takatori, Y. Noduyama, and Y. Sato, Recognition of sensitized longest paths in transition-delay test, in *Proc. IEEE Int. Test Conf.*, Paper 11.1, October 2006.

[Li 2003] Z. Li, X. Lu, W. Qiu, W. Shi, and D. M. H. Walker, A circuit level fault model for resistive bridges, *ACM Trans. on Design Automation of Electronic Systems*, 8(4), pp. 546-559, October 2003.

[Mentor 2008] Mentor Graphics, http://www.mentor.com, 2008.

[Sato 2005] Y. Sato, S. Hamada, T. Maeda, A. Takatori, Y. Nozuyama, and S. Kajihara, Invisible delay quality-SDQM model lights up what could not be seen, in *Proc. IEEE Int. Test Conf.* Paper 47.1, November 2005.

[Synopsys 2008] Synopsys, http://www.synopsys.com, October 2003.

[SynTest 2008] SynTest Technologies, http://www.syntest.com, 2008.

[Waicukauski 1986] J. A. Waicukauski, E. Lindbloom, B. K. Rosen, and V. S. Iyengar, Transition fault simulation by parallel pattern single fault propagation, in *Proc. IEEE Int. Test Conf.*, pp. 542-549, September 1986.

Index

Printed and bound by CPI Group (UK) Ltd, Croydon, CR0 4YY

03/10/2024

01040317-0020